Birkhäuser

Trends in Mathematics

Trends in Mathematics is a series devoted to the publication of volumes arising from conferences and lecture series focusing on a particular topic from any area of mathematics. Its aim is to make current developments available to the community as rapidly as possible without compromise to quality and to archive these for reference.

Proposals for volumes can be submitted using the Online Book Project Submission Form at our website www.birkhauser-science.com.

Material submitted for publication must be screened and prepared as follows:

All contributions should undergo a reviewing process similar to that carried out by journals and be checked for correct use of language which, as a rule, is English. Articles without proofs, or which do not contain any significantly new results, should be rejected. High quality survey papers, however, are welcome.

We expect the organizers to deliver manuscripts in a form that is essentially ready for direct reproduction. Any version of TEX is acceptable, but the entire collection of files must be in one particular dialect of TEX and unified according to simple instructions available from Birkhäuser.

Furthermore, in order to guarantee the timely appearance of the proceedings it is essential that the final version of the entire material be submitted no later than one year after the conference.

Vladimir Baranovsky • Nicolas Guay
Travis Schedler
Editors

Representation Theory and Algebraic Geometry

A Conference Celebrating the Birthdays of Sasha Beilinson and Victor Ginzburg

Birkhäuser

Editors
Vladimir Baranovsky
Department of Mathematics
University of California, Irvine
Irvine, CA, USA

Nicolas Guay
Department of Math. & Stat. Sciences
University of Alberta
Edmonton, AB, Canada

Travis Schedler
Department of Mathematics
Imperial College London
London, UK

ISSN 2297-0215 ISSN 2297-024X (electronic)
Trends in Mathematics
ISBN 978-3-030-82009-1 ISBN 978-3-030-82007-7 (eBook)
https://doi.org/10.1007/978-3-030-82007-7

Mathematics Subject Classification: 14L30, 14M15, 20C08, 14F10, 14F07, 18M20, 17B37, 32G24

This book is published under the imprint Birkhäuser, www.birkhauser-science.com by the registered company Springer Nature Switzerland AG
The registered company address is: Gewerbestrasse 11, 6330 Cham, Switzerland

Preface

The conference “Interactions between Representation Theory and Algebraic Geometry” was held at the University of Chicago on August 21–25, 2017. It brought together about 150 participants from several major universities in the USA and abroad, more than half of whom were junior mathematicians. It featured 21 talks by eminent mathematicians from the USA, Europe, and Asia on topics at the cutting edge of mathematical research. Some junior participants gave introductory talks in the evening to try to make the presentations by the main speakers more accessible to the general audience, in addition to giving poster presentations of their research.

The research articles in these proceedings are dedicated, as was the conference, to Alexander Beilinson and Victor Ginzburg, two visionaries in the fields of Representation Theory and Algebraic Geometry, in honor of their 60th birthdays. Their work and mentoring has influenced a great number of researchers and will continue to shape the development of those branches of mathematics for many years to come. Their influence can be perceived throughout this volume, for instance where important roles are played by D-modules and perverse sheaves, Grassmannians and their affine and loop group analogues, categorical approaches to representation theory, versions of Hecke algebras, symplectic algebraic geometry, and many other topics.

The chapters have been organized thematically as follows: the first three deal with the subjects of groups, algebras, and categories and their representation theory; the next four deal with D-modules and perverse sheaves, particularly on flag varieties and their generalizations; the final two deal with analogous varieties defined by quivers and their relationships to representation theory, cohomology theories, and symplectic geometry.

Irvine, CA, USA — Vladimir Baranovsky

Edmonton, AB, Canada — Nicolas Guay

London, UK — Travis Schedler

Contents

Part I
Groups, Algebras, Categories, and Their Representation Theory

On Semisimplification of Tensor Categories

Pavel Etingof and Victor Ostrik

Contents

To Sasha Beilinson and Vitya Ginzburg on their 60th birthdays with admiration

---

P. Etingof (✉)
Department of Mathematics, Massachusetts Institute of Technology, Cambridge, MA, USA
e-mail: etingof@math.mit.edu

V. Ostrik
Department of Mathematics, University of Oregon, Eugene, OR, USA

Laboratory of Algebraic Geometry, National Research University Higher School of Economics, Moscow, Russia
e-mail: vostrik@uoregon.edu

V. Baranovsky et al. (eds.), *Representation Theory and Algebraic Geometry*, Trends in Mathematics, https://doi.org/10.1007/978-3-030-82007-7_1

1 Introduction

The notion of the *semisimplification* of a spherical tensor category was introduced in [BW], although in the context of algebraic geometry, it can be traced back to the notion of numerical equivalence of cycles in the theory of motives; see, e.g., [Ja]. More generally, various *adequate equivalence relations* in the same theory can be considered as examples of *tensor ideals* in the symmetric tensor category of Chow motives.

Recall that a morphism $f : X \to Y$ in a spherical tensor category $\mathcal{C}$ over a field $\mathbf{k}$ is called *negligible* if for any morphism $g : Y \to X$, one has $\mathrm{Tr}(f \circ g) = 0$. One can show that the collection $\mathcal{N}$ of negligible morphisms is a tensor ideal; thus, one can define an additive monoidal category $\overline{\mathcal{C}} := \mathcal{C}/\mathcal{N}$. One can show that $\overline{\mathcal{C}}$ is, in fact, semisimple abelian, with simple objects being the indecomposable objects of $\mathcal{C}$ of nonzero dimension, and it is called the *semisimplification* of $\mathcal{C}$. Moreover, this definition can be generalized to pivotal categories in which the left and right dimension of indecomposables vanish simultaneously and even to Karoubian (not necessarily abelian) monoidal categories in which the trace of a nilpotent endomorphism is zero.

The semisimplification construction is a rich source of semisimple tensor categories. In the simplest cases, when the classification of indecomposables in $\mathcal{C}$ is tame, the semisimplification can be described explicitly. Admittedly, this happens rather rarely: most of the time, the classification of indecomposables is wild, and the corresponding semisimplified category $\overline{\mathcal{C}}$ is somewhat unmanageable, i.e., may have uncountably many simple objects even if $\mathcal{C}$ is finite (e.g., this happens already for $\mathcal{C} = \mathrm{Rep}_{\mathbf{k}}((\mathbb{Z}/p)^2)$, where $\mathbf{k}$ is an uncountable field with $\mathrm{char}(\mathbf{k}) = p > 2$). However, in this case, we may consider the tensor subcategory of $\mathcal{C}$ generated by a given object X, which is much more manageable (in particular, always has a finite or countable set of isomorphism classes of simple objects); in particular, it is an interesting question when this subcategory is fusion (i.e., has finitely many simple objects) and what it looks like in this case.

The goal of this paper is to develop a number of tools for studying semisimplifications of tensor categories and to apply them to compute the semisimplifications and their tensor subcategories generated by particular objects in a number of specific examples.

Specifically, in Sect. 2, we review the basic theory of tensor ideals and semisimplifications.

In Sect. 3, we give some general results about semisimplifications. In particular, we discuss semisimplifications of Tannakian categories in characteristic zero, reductive envelopes of algebraic groups, and the generalized Jacobson-Morozov lemma (following André and Kahn), compatibility of semisimplification with equivariantization and with surjective tensor functors.

In Sect. 4, we use classical results of modular representation theory (the Green correspondence) to show that the semisimplification of the category $\mathrm{Rep}\, G$ of representations of a finite group G in characteristic $p > 0$ is naturally equivalent to that of the normalizer of its p-Sylow subgroup and compute the semisimplification of $\mathrm{Rep}\, G$ when the Sylow subgroup is cyclic of order p (in particular for $G = S_{n+p}$

with $0 \leq n < p$). We then use this result and the work of Harman to compute the semisimplification of the abelian envelope of the Deligne category $\underline{\mathrm{Rep}}^{\mathrm{ab}}(S_n)$.

In Sect. 5, we compute the semisimplifications of some non-symmetric categories in characteristic zero, namely, the category of representations of the Kac-De Concini quantum group $U_q(\mathfrak{b})$, where $\mathfrak{b}$ is the Borel subalgebra of $\mathfrak{sl}_2$ when q is generic and when q is a root of unity.

In Sect. 6, we study surjective tensor functors between Verlinde categories attached to simple algebraic groups in characteristic p; interesting examples of such functors, which are attached to pairs of simple algebraic groups $G \supset K$ where K contains a regular unipotent element of G, are obtained from the semisimplification construction.

In Sect. 7, we study objects of finite type in semisimplifications of categories of group representations in characteristic p, i.e., objects generating fusion subcategories. We give a number of nontrivial examples of objects of finite type and study the fusion categories they generate.

In Sect. 8, we determine the semisimplifications of the tilting categories of $GL(n)$, $SL(n)$, and $PGL(n)$ in characteristic 2.[1]

Finally, in the appendix, we classify categorifications of the representation ring and Verlinde ring for $SO(3)$. This is used in Sect. 5.

2 Preliminaries

2.1 *Tensor Ideals*

Let $\mathbf{k}$ be a field and let $\mathcal{C}$ be a $\mathbf{k}-$linear monoidal category. Recall that a *tensor ideal* I in $\mathcal{C}$ is a collection of subspaces $I(X, Y) \subset \mathrm{Hom}(X, Y)$ for all $X, Y \in \mathcal{C}$ such that for all $X, Y, Z, T \in \mathcal{C}$

(1) for $\alpha \in I(X, Y)$ and $\beta \in \mathrm{Hom}(Y, Z), \gamma \in \mathrm{Hom}(Z, X)$, we have $\alpha \circ \gamma \in I(Z, Y)$ and $\beta \circ \alpha \in I(X, Z)$;
(2) for $\alpha \in I(X, Y), \beta \in \mathrm{Hom}(Z, T)$, we have $\alpha \otimes \beta \in I(X \otimes Z, Y \otimes T)$ and $\beta \otimes \alpha \in I(Z \otimes X, T \otimes Y)$.

If I is a tensor ideal in $\mathcal{C}$, then one can define a new $\mathbf{k}-$linear monoidal category $\mathcal{C}'$ (the *quotient* of $\mathcal{C}$ by I) as follows: the objects of $\mathcal{C}'$ are the objects of $\mathcal{C}$; $\mathrm{Hom}_{\mathcal{C}'}(X, Y) := \mathrm{Hom}_{\mathcal{C}}(X, Y)/I(X, Y)$; the composition of morphisms is the same as in $\mathcal{C}$ (note that condition (1) ensures that the composition is well defined); the tensor product is the same as in $\mathcal{C}$ (well defined, thanks to condition (2)).

Moreover, the identity map on the objects and morphisms induces a canonical quotient monoidal functor $\mathcal{C} \to \mathcal{C}'$.

It is clear that if $\mathcal{C}$ is rigid, pivotal, spherical, braided, and symmetric, then so is $\mathcal{C}'$.

[1]We refer the reader to [BEEO] where the results of Sect. 8 are generalized to arbitrary characteristic.

2.2 Semisimplification of a Spherical Tensor Category

We recall the theory of semisimplifications of spherical tensor categories, due to Barrett and Westbury [BW]. We give proofs for reader's convenience.

Let $\mathbf{k}$ be an algebraically closed field and $\mathcal{C}$ be a spherical tensor category over $\mathbf{k}$ (see [EGNO, Subsection 4.7]).

Definition 2.1 A morphism $f : X \to Y$ in $\mathcal{C}$ is called *negligible* if for any morphism $g : Y \to X$, one has $\mathrm{Tr}(f \circ g) = 0$.

Lemma 2.2 ([B2], Exercise 3(ii), Subsection 2.18) *Let $X = \oplus_i X_i$ and $Y = \oplus_j Y_j$ be decompositions of X, Y into indecomposable objects and $f = \oplus_{i,j} f_{ij}$ be a morphism $X \to Y$, where $f_{ij} : X_i \to Y_j$. Then f is negligible if and only if for each i, j either $\dim Y_j = 0$ or f_{ij} is not an isomorphism (equivalently, either $\dim X_i = 0$ or f_{ij} is not an isomorphism).*

Proof First let us prove the lemma when X, Y are indecomposable. If $f : X \to Y$ is not an isomorphism, then for any $g : Y \to X$, the morphism $f \circ g : Y \to Y$ is not an isomorphism, either; otherwise, f is injective (hence not surjective) and $Y \cong \mathrm{Im} f \oplus \mathrm{Ker} g$, with both summands nonzero, giving a contradiction. Hence, $f \circ g$ is nilpotent and $\mathrm{Tr}(f \circ g) = 0$. Also, if f is an isomorphism (so $\dim X = \dim Y$), then for any $g : Y \to X$, one has $f \circ g = \lambda \mathrm{Id} + h$, where $\lambda \in \mathbf{k}$ and $h : Y \to Y$ is nilpotent. Hence, $\mathrm{Tr}(f \circ g) = \lambda \dim Y = \lambda \dim X$. If $\dim X = \dim Y = 0$, this is always zero, while if $\dim X = \dim Y \neq 0$, then we can take $g = f^{-1}$ (so that $\lambda = 1$), and $\mathrm{Tr}(f \circ g) = \dim Y \neq 0$, as desired.

Now consider the general case. Suppose the condition of the lemma is satisfied and $g : Y \to X$ is a morphism, $g = (g_{ji})$. Then $\mathrm{Tr}(f \circ g) = \sum_{i,j} \mathrm{Tr}(f_{ij} \circ g_{ji})$. If either $\dim Y_j = 0$ or f_{ij} is not an isomorphism (equivalently, either $\dim X_i = 0$ or f_{ij} is not an isomorphism) for all i, j, then by the indecomposable case, $\mathrm{Tr}(f_{ij} \circ g_{ji}) = 0$ for all i, j; hence, $\mathrm{Tr}(f \circ g) = 0$. However, if for some i, j this condition is violated, then we can take $g_{ji} = f_{ij}^{-1}$ and $g_{pq} = 0$ for $(p, q) \neq (i, j)$, so that $\mathrm{Tr}(f \circ g) = \dim X_i = \dim Y_j$. This implies the lemma. □

Let $\mathcal{N}(\mathcal{C})$ be the collection of negligible morphisms of $\mathcal{C}$.

Lemma 2.3 *$\mathcal{N}(\mathcal{C})$ is a tensor ideal in $\mathcal{C}$.*

Proof It is clear that a linear combination of negligible morphisms is negligible. Also, it is easy to see that $f \circ a, b \circ f$ are negligible for any a, b (when these compositions make sense). It remains to show that the tensor products $a \otimes f$ and $f \otimes b$ are negligible. Let us prove this for $a \otimes f$, where $a : Z \to T$; the case of $f \otimes b$ is similar. Let $g : T \otimes Y \to Z \otimes X$. Then $\mathrm{Tr}((a \otimes f) \circ g) = \mathrm{Tr}(f \circ g')$, where $g' := \mathrm{Tr}_T((a \otimes \mathrm{Id}) \circ g)$. Hence, $\mathrm{Tr}((a \otimes f) \circ g) = 0$ and $a \otimes f$ is negligible, as desired. □

Thus, we can define a spherical tensor category $\overline{\mathcal{C}} := \mathcal{C}/\mathcal{N}(\mathcal{C})$.

Proposition 2.4 *The category $\overline{\mathcal{C}}$ is a semisimple tensor category. The simple objects of $\overline{\mathcal{C}}$ are the indecomposable objects of $\mathcal{C}$ of nonzero dimension.*

Proof It is clear that indecomposable objects of $\overline{\mathcal{C}}$ are images of indecomposable objects of $\mathcal{C}$. More precisely, if $X, Y \in \mathcal{C}$ are indecomposable, then by Lemma 2.2, $\mathrm{Hom}_{\overline{\mathcal{C}}}(X, Y) = 0$ if $X \ncong Y$ or $\dim X = 0$ or $\dim Y = 0$ (i.e., if $\dim X = 0$, then $X = 0$ in $\overline{\mathcal{C}}$), and $\dim \mathrm{Hom}_{\overline{\mathcal{C}}}(X, Y) = 1$ if $X \cong Y$ and $\dim X \neq 0$. This implies the proposition. □

Definition 2.5 The category $\overline{\mathcal{C}}$ is called the *semisimplification* of $\mathcal{C}$.

Note that the category $\overline{\mathcal{C}}$ comes equipped with a natural monoidal functor $\mathbf{S} : \mathcal{C} \to \overline{\mathcal{C}}$, which we call the *semisimplification functor*. This functor, however, is not a tensor functor, since it is not left or right exact, in general. We will denote the image $\mathbf{S}(X)$ of an object X under this functor by $\overline{X}$.

2.3 Generalization to Pivotal Karoubian Categories

The above results generalize to pivotal tensor categories [EGNO, Subsection 4.7] such that $\dim^L X = 0$ if and only if $\dim^R X = 0$ for any indecomposable object $X \in \mathcal{C}$ (an example of such a category which is not spherical is the category of representations of the Taft Hopf algebra). Namely, in such a category, for any endomorphism $h : X \to X$ of an indecomposable object X, one has $\mathrm{Tr}^L(h) = 0$ if and only if $\mathrm{Tr}^R(h) = 0$. Thus, if $f : X \to Y$ is a morphism between arbitrary objects of $\mathcal{C}$, then the condition that for any $g : Y \to X$, one has $\mathrm{Tr}^L(f \circ g) = 0$ is equivalent to the condition that for any $g : Y \to X$, one has $\mathrm{Tr}^R(f \circ g) = 0$. One then defines f to be negligible if any of these two equivalent conditions is satisfied. Then Lemmas 2.2, 2.3, and Proposition 2.4 generalize verbatim, with analogous proofs.

Moreover, the above results also extend to the case when $\mathcal{C}$ is a Karoubian rigid monoidal category in which the trace of a nilpotent endomorphism is zero, a necessary condition for $\mathcal{C}$ to be embeddable into an abelian tensor category.[2] For instance, the well-known construction of the fusion categories attached to a simple Lie algebra $\mathfrak{g}$ (in characteristic zero or p bigger than the Coxeter number), [EGNO, Subsection 8.18.2], starts with the category of tilting modules for the corresponding (quantum) group (which is Karoubian) and takes a quotient by the tensor ideal of negligible morphisms. Note that in this special case, negligible morphisms happen to be those that factor through negligible objects (i.e., direct sums of simple objects of dimension 0); this is not the case in general (e.g., for $\mathrm{Rep}_{\mathbf{k}}(\mathbb{Z}/p)$).

[2]Note that this condition is not necessarily satisfied: e.g., if $\mathrm{char}(\mathbf{k}) = p$, $t \in \mathbf{k}$, and $\underline{\mathrm{Rep}}_{\mathbf{k}}(S_t)$ is the Karoubian Deligne category of representations of S_t [EGNO, Subsection 9.12], then this property holds only if $t \in \mathbb{F}_p \subset \mathbf{k}$; namely, if σ is the cyclic permutation on $X^{\otimes p}$, where X is the tautological object, then $(1 - \sigma)^p = 0$ but $\mathrm{Tr}(1 - \sigma) = t^p - t$.

To summarize, we have the following result. Let $\mathcal{C}$ be a pivotal category, let $\dim^L(X) := \mathrm{Tr}^L(\mathrm{Id}_X)$, $\dim^R(X) := \mathrm{Tr}^R(\mathrm{Id}_X)$ for $X \in \mathcal{C}$, and call a morphism $f : X \to Y$ negligible if for any $g : Y \to X$, one has $\mathrm{Tr}^L(f \circ g) = 0$.

Theorem 2.6 *Let $\mathcal{C}$ be a $\mathbf{k}$–linear Karoubian rigid monoidal category such that all morphism spaces are finite dimensional.*[3] *Assume that $\mathcal{C}$ is equipped with a pivotal structure such that*

(1) the left trace Tr^L of any nilpotent endomorphism is zero;
(2) $\dim^L X = 0$ if and only if $\dim^R X = 0$ for an indecomposable $X \in \mathcal{C}$.

Then negligible morphisms are characterized as in Lemma 2.2 and form a tensor ideal $\mathcal{N}(\mathcal{C})$. Moreover, $\mathcal{C}/\mathcal{N}(\mathcal{C})$ is a semisimple tensor category, whose simple objects are the indecomposable objects of $\mathcal{C}$ of nonzero dimension.

Proof First of all, (1) implies that the right trace of any nilpotent endomorphism in $\mathcal{C}$ is zero, since $\mathrm{Tr}^L(f) = \mathrm{Tr}^R(f^*)$; see [EGNO, Proposition 4.7.3].

Hence, for an endomorphism $h : X \to X$, $\mathrm{Tr}^L(h) = 0$ if and only if $\mathrm{Tr}^R(h) = 0$. Indeed, by decomposing X into generalized eigenobjects of h, we may assume that $h = \lambda \mathrm{Id} + h_0$, where h_0 is nilpotent. Then $\mathrm{Tr}^L(h) = \lambda \dim^L X$ and $\mathrm{Tr}^R(h) = \lambda \dim^R X$ (as $\mathrm{Tr}^L(h_0) = \mathrm{Tr}^R(h_0) = 0$), so our claim follows from (2).

The rest of the proof is parallel to the spherical abelian case. □

Example 2.7

1. If $\mathcal{C}$ is semisimple, then $\overline{\mathcal{C}} \cong \mathcal{C}$. Moreover, in this case, for any tensor category $\mathcal{D}$, one has $\overline{\mathcal{C} \boxtimes \mathcal{D}} \cong \mathcal{C} \boxtimes \overline{\mathcal{D}}$.
2. If $\mathrm{char}(\mathbf{k}) = p > 0$ and $\mathcal{C} = \mathrm{Rep}_{\mathbf{k}}(\mathbb{Z}/p)$, then $\overline{\mathcal{C}}$ is the Verlinde category Ver_p introduced by Gelfand-Kazhdan and Georgiev-Mathieu; see [O] and references therein.
3. Let $\mathrm{char}(\mathbf{k}) = 0$ and $\mathcal{C} = \mathrm{Rep}\, GL(n|1)$, $n \geq 1$. Then
$$\overline{\mathcal{C}} = \mathrm{Rep}(GL(n-1) \times GL(1) \times GL(1)) \boxtimes \mathrm{Supervec},$$
where Supervec is the category of supervector spaces; see [H], Theorem 4.13.
4. Let $G = (\mathbb{Z}/2\mathbb{Z})^2$ and $\mathrm{char}(\mathbf{k}) = 2$. Then it is well known that indecomposable representations of G over $\mathbf{k}$ of nonzero mod 2 (i.e., odd) dimension are precisely $\Omega^n(\mathbf{1})$, $n \in \mathbb{Z}$, where Ω is the Heller shift operator; see, e.g., [B1, Theorem 4.3.3]. Also one deduces from [B1, Corollary 3.1.6] that
$$\Omega^n(\mathbf{1}) \otimes \Omega^m(\mathbf{1}) \simeq \Omega^{n+m}(\mathbf{1}) \oplus \text{a projective module}.$$
Thus, $\overline{\mathrm{Rep}_{\mathbf{k}}(G)} = \mathrm{Vec}_{\mathbb{Z}} = \mathrm{Rep}\, GL(1)$.

[3] Note that any Karoubian linear category with finite dimensional morphism spaces satisfies the Krull-Schmidt theorem, which says that any object has a unique decomposition into a direct sum of indecomposables (up to a non-unique isomorphism); for this reason, such categories are sometimes called Krull-Schmidt categories.

Remark 2.8

1. It is clear that if $\mathcal{C}$ is symmetric or braided, then so is $\overline{\mathcal{C}}$ and the functor **S**.
2. If $\mathcal{C}$ is finite, then $\overline{\mathcal{C}}$ may be infinite (see Example 2.7(4)) and can, in fact, be unmanageably large, since the problem of classifying indecomposable objects in finite abelian categories is often wild (in fact, this is already so for $\mathrm{Rep}_{\mathbf{k}}(\mathbb{Z}/p)^2$, where $\mathrm{char}(\mathbf{k}) = p > 2$).

Remark 2.9

1. Let $\mathcal{C} = \mathrm{Rep}\, H$, where H is a finite dimensional Hopf algebra over a field **k** of characteristic zero. Then condition (2) of Theorem 2.6 (that $\dim^L X = 0$ if and only if $\dim^R X = 0$) holds for any pivotal structure. Indeed, we may assume that $\mathbf{k} = \mathbb{C}$. A pivotal structure on $\mathrm{Rep}\, H$ is given by a grouplike element $g \in H$ such that $gxg^{-1} = S^2(x)$ for $x \in H$ and $\dim^L X = \mathrm{Tr}_X(g)$, $\dim^R X = \mathrm{Tr}_X(g^{-1})$. But g has finite order, so the eigenvalues of g are roots of unity; hence, $\dim^R X = \overline{\dim^L X}$, as desired. We expect that the same holds for any finite tensor category over a field of characteristic zero.

   However, the above condition can be violated for categories of finite dimensional modules or comodules over an infinite dimensional Hopf algebra. For example, let $\mathcal{C}$ be the category of finite dimensional representations of $U_q(\mathfrak{b})$, $q \in \mathbb{C}^\times$, where $\mathfrak{b} \subset \mathfrak{sl}_3$ is a Borel subalgebra. Recall that a pivotal structure on $\mathcal{C}$ is defined by the element $K = q^{2\rho}$. Let X be the $U_q(\mathfrak{b})$ subrepresentation of the adjoint representation of $U_q(\mathfrak{sl}_3)$ (with highest weight $\alpha_1 + \alpha_2$) spanned by the vectors whose weights are positive roots. Then $\dim^L X = 2q^2 + q^4$ and $\dim^R X = 2q^{-2} + q^{-4}$. So if $q^2 = -2$, then $\dim^L X = 0$ but $\dim^R X = -3/4 \neq 0$.

   The same happens in characteristic p, even for a finite dimensional Hopf algebra. Namely, we can take the same example. Note that $q^2 = -2$ is then a root of unity (or some order dividing $p-1$), so one may replace $U_q(\mathfrak{b})$ with the corresponding small quantum group $\mathfrak{u}_q(\mathfrak{b})$.
2. Condition (1) of Theorem 2.6 holds true if $\mathcal{C}$ is an abelian tensor category, since the quantum trace is additive on exact sequences; see, e.g., [EGNO, Proposition 4.7.5]. Moreover, assume that there exists a pivotal tensor functor $\mathcal{C} \to \mathcal{D}$, where the category $\mathcal{D}$ satisfies condition (1) of Theorem 2.6 (e.g., $\mathcal{D}$ is abelian). Then obviously the category $\mathcal{C}$ also satisfies condition (1) of Theorem 2.6. This observation was used by U. Jannsen to prove that the category of numerical motives is semisimple; see [Ja]. Moreover, the assumption on finite dimensionality of morphism spaces in $\mathcal{C}$ in Theorem 2.6 can be dropped if there exists a pivotal monoidal functor $F : \mathcal{C} \to \mathcal{D}'$, where all morphism spaces in $\mathcal{D}'$ are finite dimensional, since the tensor ideal of morphisms sent by F to zero consists of negligible morphisms, which implies finite dimensionality of morphism spaces in $\mathcal{C}/\mathcal{N}(\mathcal{C})$.

   Here is an example of such a situation. Take any collection of morphisms in a symmetric tensor category $\mathcal{D}$, compute some of relations between them, and define $\mathcal{C}$ to be the Karoubian envelope of the universal symmetric monoidal category generated by morphisms satisfying these relations. Then we have an obvious symmetric monoidal functor $\mathcal{C} \to \mathcal{D}$; hence, the semisimplification of $\mathcal{C}$ is a semisimple symmetric tensor category.

3 General Results on Semisimplification of Tensor Categories

3.1 Splitting of the Semisimplification Functor for Tannakian Categories in Characteristic Zero, Reductive Envelopes, and the Jacobson-Morozov Lemma

For Tannakian categories in characteristic zero, André and Kahn showed that the semisimplification functor $\mathbf{S}$ admits a splitting $\mathbf{S}^*$ and used it to show the existence and uniqueness (up to conjugation) of the reductive envelope of any affine proalgebraic group in characteristic zero. In this subsection, we review this theory (cf. [AK, S]).

Theorem 3.1 ([AK], Theorem 1, Theorem 2) *If* $\mathrm{char}(\mathbf{k}) = 0$ *and* $\mathcal{C} = \mathrm{Rep}\, G$ *is a Tannakian category over* $\mathbf{k}$ *(where* G *is an affine proalgebraic group over* $\mathbf{k}$*), then the functor* $\mathbf{S} : \mathcal{C} \to \overline{\mathcal{C}}$ *admits a splitting* $\mathbf{S}^* : \overline{\mathcal{C}} \to \mathcal{C}$*, a surjective tensor functor such that* $\mathbf{S}^*(\overline{X}) \cong X$ *for each indecomposable* $X \in \mathcal{C}$*, and* $\mathbf{S} \circ \mathbf{S}^* \cong \mathrm{Id}$ *as a symmetric tensor functor.*

Now let $\mathcal{C}$ be as above and F be the forgetful functor $\mathcal{C} \to \mathrm{Vec}$. Then $F \circ \mathbf{S}^* : \overline{\mathcal{C}} \to \mathrm{Vec}$ is a fiber functor, so by the Tannakian formalism [DM], we have $\overline{\mathcal{C}} = \mathrm{Rep}\, \overline{G}$, where $\overline{G} := \mathrm{Aut}(F \circ \mathbf{S}^*)$ is a reductive affine proalgebraic group, equipped with a homomorphism $\psi_G : G \to \overline{G}$ (defined up to conjugation in $\overline{G}$) giving rise to the functor $\mathbf{S}^*$. Moreover, since $\mathbf{S}^*$ is surjective, ψ_G is an inclusion.

Definition 3.2 ([AK]) The group $\overline{G}$ equipped with the homomorphism ψ_G (defined up to conjugation) is called the *reductive envelope* of G.

Theorem 3.3 ([AK], Theorem 3, Theorem 4) *The reductive envelope* $\overline{G}$ *enjoys the following universal property: If* $\phi : G \to L$ *is a homomorphism from* G *to a reductive proalgebraic group* L*, then there exists a homomorphism* $\overline{\phi} : \overline{G} \to L$ *such that* $\phi = \overline{\phi} \circ \psi_G$*. Moreover,* $\overline{\phi}$ *is unique up to conjugation in* L *by elements commuting with* $\phi(G)$*.*

Proof The morphism ϕ gives rise to a symmetric tensor functor $\Phi : \mathrm{Rep}\, L \to \mathrm{Rep}\, G$. Consider the functor $\Phi' := \mathbf{S} \circ \Phi$. Even though $\mathbf{S}$ may not be exact on any side, the functor Φ' is exact since the category $\mathrm{Rep}\, L$ is semisimple (as L is reductive). Thus, $\Phi' : \mathrm{Rep}\, L \to \mathrm{Rep}\, \overline{G}$ is a symmetric tensor functor. Hence, by Tannakian formalism [DM], it comes from a homomorphism $\phi' : \overline{G} \to L$ defined uniquely up to conjugation in L. Moreover, consider the functor $\mathbf{S}^* \circ \Phi' = \mathbf{S}^* \circ \mathbf{S} \circ \Phi$. This functor is exact since its source is a semisimple category, so it is a symmetric tensor functor, and it follows from Proposition 13.7.1 of [AK] that it is naturally isomorphic to Φ as a tensor functor. This means that the homomorphisms ϕ and $\phi' \circ \psi_G$ are conjugate under L: $\phi(g) = \ell\phi'(\psi_G(g))\ell^{-1}$ for some $\ell \in L$ and all $g \in G$. Hence, $\phi(g) = \widetilde{\phi}(\psi_G(g))$ for all $g \in G$, where $\widetilde{\phi}(a) := \ell\phi'(a)\ell^{-1}$.

Finally, let us show that the homomorphism $\widetilde{\phi}$ in the theorem is determined uniquely up to conjugation in L (automatically by elements commuting with $\phi(G)$). To this end, let $\widetilde{\Phi} : \operatorname{Rep} L \to \operatorname{Rep} G$ be the functor defined by $\widetilde{\phi}$. Then $\mathbf{S}^* \circ \widetilde{\Phi} = \Phi$; hence, postcomposing with $\mathbf{S}$, we get $\widetilde{\Phi} = \mathbf{S} \circ \Phi$. Thus, $\widetilde{\Phi}$ is uniquely determined, and hence $\widetilde{\phi}$ is determined up to conjugation, as desired. □

Remark 3.4 A geometric proof of the existence and properties of the reductive envelope is given in [S].

Example 3.5 Consider the special case $G = \mathbb{G}_a$. In this case, the indecomposable representations of G are unipotent Jordan blocks J_n of sizes $n = 1, 2, 3\ldots$, so it is easy to see that $\overline{\operatorname{Rep} G} \cong \operatorname{Rep} SL(2)$ (as the Grothendieck ring of $\overline{\operatorname{Rep} G}$ coincides with that of $\operatorname{Rep} SL(2)$ and the dimensions of nonzero objects of $\overline{\operatorname{Rep} G}$ are positive). So in this case, the existence of $\overline{G}$ is easy (namely, $\overline{G} = SL(2)$), and the existence of the splitting $\mathbf{S}^*$ is also straightforward (namely, $\mathbf{S}^*$ is induced by the standard inclusion $\psi_G : \mathbb{G}_a \hookrightarrow SL(2)$ as upper triangular matrices with ones on the diagonal). Thus, Theorem 3.3 in this case tells us that any homomorphism $\phi : \mathbb{G}_a \to L$ for a reductive proalgebraic group L uniquely (up to conjugacy) factors through a homomorphism $\widetilde{\phi} : SL(2) \to L$. As pointed out in [AK, S], this implies the celebrated *Jacobson-Morozov lemma*:

Proposition 3.6 *Let L be a reductive algebraic group over $\mathbf{k}$ and $u \in L$ a unipotent element. Then there exists a homomorphism $\theta : SL(2) \to L$ such that $\theta\begin{pmatrix} 1 & 1 \\ 0 & 1 \end{pmatrix} = u$. Moreover, θ is unique up to conjugation by the centralizer Z_u of u.*

Proof Let G be the 1-parameter unipotent subgroup of L generated by u and $\phi : G \to L$ be the corresponding embedding. Identify G with $\mathbb{G}_a$ by sending u to 1. Then it remains to apply Theorem 3.3 and set $\theta = \widetilde{\phi}$. □

Note that when G is an algebraic group, then $\overline{G}$ is typically only a proalgebraic group (of infinite type), which can be very large. In fact, this is already so when $G = \mathbb{G}_a^2$, since the problem of classifying pairs of commuting matrices is well known to be wild; i.e., the case $G = \mathbb{G}_a$ (leading to the Jacobson-Morozov lemma) is a rare exception. In other words, the whole category $\operatorname{Rep} \overline{G}$ is typically unmanageable. However, it makes sense to consider tensor subcategories of this category generated by a single object, which are more manageable. Namely, we have the following corollary.

Corollary 3.7 *Let G be an affine algebraic group over $\mathbf{k}$ and $V \in \operatorname{Rep} G$ a faithful representation of G (so that $G \hookrightarrow GL(V)$). Then there exists a reductive algebraic group $G_V \subset GL(V)$ containing G (a quotient of $\overline{G}$) such that the subcategory $\mathcal{C}_V$ of $\overline{\operatorname{Rep} G} \cong \operatorname{Rep} \overline{G}$ tensor generated by $\overline{V}$ is naturally equivalent to $\operatorname{Rep} G_V$.*

Proof Let $F_V : \mathcal{C}_{\overline{V}} \to \mathrm{Vec}$ be the restriction of the fiber functor of $\mathrm{Rep}\,\overline{G}$ to $\mathcal{C}_{\overline{V}}$. Let $G_V := \mathrm{Aut}(F_V)$. Then $G_V \subset GL(V)$ is a reductive subgroup such that $\mathrm{Rep}\,G_V = \mathcal{C}_{\overline{V}}$. Moreover, G_V is a quotient of $\overline{G}$; hence, we have a natural homomorphism $G \to G_V$, which is obviously injective, as desired. □

Definition 3.8 We will call G_V the *reductive envelope of G inside $GL(V)$*.

Remark 3.9

1. Let $\mathcal{C} = \mathrm{Rep}_{\mathbf{k}}\,\mathbb{Z}/p$, where $\mathrm{char}(\mathbf{k}) = p \geq 5$. Then a tensor functor $\mathbf{S}^* : \overline{\mathcal{C}} \to \mathcal{C}$ does not exist, since $\overline{\mathcal{C}} = \mathrm{Ver}_p$ contains objects of non-integer Frobenius-Perron dimension. Also, if $\mathcal{C} = \mathrm{Rep}\,GL(n|1)$ over $\mathbf{k}$ of characteristic zero, then a symmetric functor $\mathbf{S}^*$ as is Theorem 3.1 does not exist, either. Indeed, if V is the vector representation of $GL(n|1)$, then $\wedge^{n-1}\mathbf{S}(V) = 0$ (cf. Example 2.7(3)), while $\wedge^{n-1}V \neq 0$ (it is a negligible but nonzero object in $\mathcal{C}$). In fact, it is clear that a splitting functor $\mathbf{S}^*$ with the properties stated in Theorem 3.1 cannot exist if $\mathcal{C}$ has indecomposable objects of dimension 0.
2. Note that the existence of the group $\overline{G}$ such that $\mathrm{Rep}\,\overline{G} \cong \overline{\mathrm{Rep}\,G}$ follows from Deligne's theorem [D1, Theorem 7.1], since $\overline{\mathrm{Rep}\,G}$ is a symmetric tensor category over $\mathbf{k}$ in which nonzero objects have positive integer dimensions. This is, in fact, used in the proof of Theorem 3.1 in [AK].

Moreover, using a more general version of Deligne's theorem for supergroups, [D2], one can see that if G is an affine proalgebraic supergroup over $\mathbf{k}$ of characteristic zero and $z \in G$ an element of order ≤ 2 acting on $O(G)$ by parity, and $\mathrm{Rep}(G, z)$ is the category of representations of G on superspaces on which z acts by parity, then $\overline{\mathrm{Rep}(G, z)} = \mathrm{Rep}(\overline{G}, \overline{z})$ for some reductive proalgebraic supergroup $\overline{G}$, i.e., one whose representation category is semisimple; see [H, Theorem 2.2]. In particular, for each $V \in \mathrm{Rep}(G, z)$, $\overline{V}$ generates a category $\mathrm{Rep}(G_V, \overline{z})$, where G_V is a reductive algebraic supergroup (a quotient of $\overline{G}$). This means that the connected component of the identity G_V^0 of G_V is of the form G'_V/C, where C is a finite central subgroup, and $G'_V = G_V^+ \times G_V^-$, where G_V^+ is a usual reductive group, and $\mathrm{Lie}G_V^-$ is a direct sum of Lie superalgebras of type $\mathfrak{osp}(1|2n)$; see [W].

In fact, as was explained to us by Thorsten Heidersdorf, the symmetric structure of the category is not essential in the André-Kahn theorem on the existence of splitting of the semisimplification functor. Namely, Theorem 12.1.1 and 13.2.1 of [AK] immediately imply the following theorem:

Theorem 3.10 *Let $\mathcal{C}$ be a Karoubian pivotal category as in Theorem 2.6, such that the ideal $\mathcal{N}(\mathcal{C})$ of negligible morphisms in $\mathcal{C}$ coincides with the nilpotent radical* $\mathrm{rad}(\mathcal{C})$ *of $\mathcal{C}$ (where* $\mathrm{rad}(\mathcal{C})(X, Y)$ *is the intersection of the radical of the algebra* $\mathrm{End}(X \oplus Y)$ *with* $\mathrm{Hom}_{\mathcal{C}}(X, Y)$*); in other words, $\mathcal{C}$ has no nonzero indecomposable objects of zero dimension. Then the semisimplification functor $\mathbf{S} : \mathcal{C} \to \overline{\mathcal{C}}$ admits a monoidal splitting $\mathbf{S}^* : \overline{\mathcal{C}} \to \mathcal{C}$.*

Corollary 3.11 *Let $\mathcal{C}$ be a Karoubian pivotal category as in Theorem 2.6. Then following conditions are equivalent:*

(i) Any indecomposable direct summand in a tensor product of two indecomposable objects of nonzero dimension also has nonzero dimension; in other words, indecomposables of nonzero dimension span a full monoidal subcategory of $\mathcal{C}$.
(ii) The semisimplification functor $\mathbf{S} : \mathcal{C} \to \overline{\mathcal{C}}$ admits a monoidal splitting $\mathbf{S}^* : \overline{\mathcal{C}} \to \mathcal{C}$.

Proof It follows from Theorem 3.10 that (i) implies (ii). To prove that (ii) implies (i), assume that (i) fails, and let X, Y be simple objects of nonzero dimension such that $X \otimes Y \cong Z \oplus T$ where $Z \neq 0$ is indecomposable and has dimension zero. Then $\mathbf{S}(X) \otimes \mathbf{S}(Y) \cong \mathbf{S}(T)$. So if there is a monoidal splitting $\mathbf{S}^*$, then applying $\mathbf{S}^*$ to the last equality, we get $X \otimes Y \cong T$. Thus, $T \cong Z \oplus T$, which contradicts the Krull-Schmidt theorem. □

In particular, this implies existence of reductive envelopes for quantum groups. Namely, let H be a Hopf algebra over a field $\mathbf{k}$ in which the squared antipode is given by conjugation by a character $\chi \in H^*$ (thereby defining a pivotal structure on $\mathrm{Corep}(H)$), with no indecomposable finite dimensional comodules $M \neq 0$ of zero dimension. Then the category $\mathrm{Corep}(H)$ satisfies the assumptions of Theorem 3.10. Therefore, we obtain

Proposition 3.12 *There exists a unique universal cosemisimple Hopf algebra $\overline{H}$ with a surjective homomorphism $s : \overline{H} \to H$ (in other words, any Hopf algebra homomorphism $H' \to H$ from a cosemisimple Hopf algebra H' uniquely factors through s). Namely,* $\mathrm{Corep}(\overline{H}) = \overline{\mathrm{Corep}(H)}$.

The proof is analogous to the case when H is commutative over $\mathbb{C}$ (i.e., the case of proalgebraic groups). Heuristically writing $G = \mathrm{Spec}(H)$ and $\overline{G} = \mathrm{Spec}(\overline{H})$, we may say that the reductive quantum group $\overline{G}$ is the reductive envelope of the quantum group G.

Example 3.13 Let q be a transcendental number or, more generally, a complex number which is not a root of any polynomial with positive integer coefficients, say a positive real number or its image under an automorphism of $\mathbb{C}$. Let B_q be the quantum Borel subgroup of the quantum group G_q attached to a simple complex algebraic group G. The category $\mathrm{Rep}(B_q) = \mathrm{Corep}(O(B_q))$ has a pivotal structure given by the element $q^{2\rho}$ of the corresponding quantum enveloping algebra. Therefore, the dimension (both left and right) of any nonzero representation of B_q is a Laurent polynomial in q with positive integer coefficients. Thus, this dimension is nonzero. Hence, Proposition 3.12 applies, and $H := O(B_q)$ is the quotient of some cosemisimple Hopf algebra $\overline{H} := O(\overline{B_q})$ for some reductive quantum group $\overline{B}_q$, such that every representation of B_q factors canonically through $\overline{B_q}$.

3.2 Compatibility of Semisimplification with Equivariantization

Now let $\mathcal{C}$ be a tensor category and L be a finite group acting on $\mathcal{C}$. Let $\mathcal{C}^L$ be the L-equivariantization of $\mathcal{C}$ [EGNO, Subsection 4.15]. The following lemma is easy [EGNO, Exercise 4.15.3].

Lemma 3.14 *If*

$$1 \to N \to G \to L \to 1$$

is an exact sequence of groups, then L acts naturally on $\mathrm{Rep}_{\mathbf{k}} N$, *and* $(\mathrm{Rep}_{\mathbf{k}} N)^L \cong \mathrm{Rep}_{\mathbf{k}} G$.

Clearly, any action of L on $\mathcal{C}$ descends to its action on the semisimplification $\overline{\mathcal{C}}$.

Proposition 3.15 *If $|L| \neq 0$ in* $\mathbf{k}$ *and L preserves the spherical structure of $\mathcal{C}$, then L-equivariantization commutes with semisimplification. In other words, we have a natural equivalence of tensor categories $\overline{\mathcal{C}^L} \cong \overline{\mathcal{C}}^L$.*

Proof We have a natural forgetful functor $F : \mathcal{C}^L \to \mathcal{C}$. We claim that if $X, Y \in \mathcal{C}^L$ and $f : X \to Y$ is negligible, then $F(f)$ is negligible. Indeed, recall that $\mathrm{Hom}(F(X), F(Y))$ carries a natural action of L and that F defines an isomorphism $\mathrm{Hom}(X, Y) \cong \mathrm{Hom}(F(X), F(Y))^L$. Now let $h \in \mathrm{Hom}(F(Y), F(X))$, and let us show that $\mathrm{Tr}(F(f) \circ h) = 0$. Let $\overline{h} = |L|^{-1} \sum_{\gamma \in L} \gamma(h)$. Then $\overline{h} = F(g)$ for a unique $g \in \mathrm{Hom}(X, Y)$. Thus, since $F(f)$ commutes with L and the action of L preserves traces, we have

$$\mathrm{Tr}(F(f) \circ h) = \mathrm{Tr}(F(f) \circ \overline{h}) = \mathrm{Tr}(F(f) \circ F(g)) = 0,$$

as desired. Thus, the functor F descends to a tensor functor $\overline{F} : \overline{\mathcal{C}^L} \to \overline{\mathcal{C}}$. Moreover, for any $T \in \overline{\mathcal{C}^L}$, the object $\overline{F}(T)$ has a natural structure of an L-equivariant object (coming from that of T), so the functor $\overline{F}$ factors naturally through a tensor functor $E : \overline{\mathcal{C}^L} \to \overline{\mathcal{C}}^L$.

Suppose $T \in \overline{\mathcal{C}^L}$ is simple. Then $T = \overline{X}$, where $X \in \mathcal{C}^L$ is an indecomposable object of nonzero dimension. Thus, $X = \mathrm{Ind}_{L_Z}^L(\rho \otimes Z)$, where Z is an indecomposable object of $\mathcal{C}$ of nonzero dimension, L_Z is the stabilizer of Z in L, and ρ is an irreducible representation of L_Z over $\mathbf{k}$. Then $E(T) = \mathrm{Ind}_{L_Z}^L(\rho \otimes \overline{Z})$. Thus, $E(T)$ is simple (since so is $\overline{Z}$, and $L_{\overline{Z}} = L_Z$).

It remains to show that E is essentially surjective, i.e., every simple object of $\overline{\mathcal{C}}^L$ is of the form $E(T)$. To this end, note that every simple object of $\overline{\mathcal{C}}^L$ has the form $W = \mathrm{Ind}_{L_V}^L(\rho \otimes V)$, where $V = \overline{X}$ is a simple object of $\overline{\mathcal{C}}$ and ρ is an irreducible representation of L_V. Since $|L_V| \neq 0$ in $\mathbf{k}$, we have $\dim \rho \neq 0$ in $\mathbf{k}$. Hence, $W = E(T)$, where $T = \mathrm{Ind}_{L_V}^L(\rho \otimes X)$ is a simple object of $\overline{\mathcal{C}^L}$ (as $L_V = L_X$). The proposition is proved. □

Corollary 3.16 *In the setup of Lemma 3.14, assume that $|L| \neq 0$ in $\mathbf{k}$. Then* $\overline{\mathrm{Rep}_{\mathbf{k}}\, G} \cong \overline{\mathrm{Rep}_{\mathbf{k}}\, N}^{L}$.

Proof This follows from Proposition 3.15 and Lemma 3.14. □

Remark 3.17 Similarly to Theorem 2.6, Proposition 3.15 and its proof generalize to Karoubian pivotal categories satisfying the assumptions of Theorem 2.6.

3.3 Compatibility of Negligible Morphisms with Surjective Tensor Functors

Let $\mathcal{C}, \mathcal{D}$ be finite spherical tensor categories [EGNO, Section 6] and $F : \mathcal{C} \to \mathcal{D}$ a surjective tensor functor [EGNO, Subsection 6.3]. Let $I : \mathcal{D} \to \mathcal{C}$ be the right adjoint of F. Note that I is exact since F maps projectives to projectives [EGNO, Theorem 6.1.16].

Definition 3.18 The *index* of F is $d := \dim I(\mathbf{1})$.

Definition 3.19 Let us say that I is *dimension-scaling* if $\dim I(V) = d \dim V$ for all $V \in \mathcal{D}$.

Proposition 3.20 *If F has a nonzero index and I is dimension-scaling, then*

(i) $\dim F(Y) = \dim Y$ *for all* $Y \in \mathcal{C}$;
(ii) for any negligible morphism f in $\mathcal{C}$, the morphism $F(f)$ is negligible in $\mathcal{D}$.

Proof We have a functorial isomorphism $\varepsilon_Y : I(F(Y)) \to I(\mathbf{1}) \otimes Y$. Indeed,

$$\mathrm{Hom}(X, I(F(Y))) = \mathrm{Hom}(F(X), F(Y)) = \mathrm{Hom}(F(X) \otimes F(Y)^*, \mathbf{1}) =$$

$$\mathrm{Hom}(F(X \otimes Y^*), \mathbf{1}) = \mathrm{Hom}(X \otimes Y^*, I(\mathbf{1})) = \mathrm{Hom}(X, I(\mathbf{1}) \otimes Y).$$

Since I is dimension-scaling, we have

$$d \dim F(Y) = \dim I(F(Y)) = \dim(I(\mathbf{1}) \otimes Y) = d \dim Y.$$

Since $d \neq 0$, this implies (i).

Now let us prove (ii). For this, note that if $f : X \to Y$ is a morphism in $\mathcal{C}$, then $\varepsilon_Y \circ I(F(f)) \circ \varepsilon_X^{-1} = \mathrm{Id}_{I(\mathbf{1})} \otimes f$. Hence, the morphism $I(F(f))$ is negligible. □

Lemma 3.21 *If $h : V \to V$ is a morphism in $\mathcal{D}$, then one has $Tr(I(h)) = dTr(h)$.*

Proof By decomposing V into generalized eigenobjects of h, we may assume that h has a single eigenvalue λ. Then $h = \lambda \mathrm{Id} + h_0$, where h_0 is nilpotent, so $I(h) = \lambda \mathrm{Id} + I(h_0)$. Since $I(h_0)$ is nilpotent, the desired statement reduces to the identity $\dim I(V) = d \dim V$ for all $V \in \mathcal{D}$, which holds since I is dimension-scaling. □

Now let $g : F(Y) \to F(X)$ be a morphism. Then by Lemma 3.21,

$$d\mathrm{Tr}(F(f) \circ g) = \mathrm{Tr}(I(F(f) \circ g)) = \mathrm{Tr}(I(F(f)) \circ I(g)).$$

But this is zero, since $I(F(f))$ is negligible. Since $d \neq 0$, this implies that $\mathrm{Tr}(F(f) \circ g) = 0$, i.e., $F(f)$ is negligible, yielding (ii). □

Proposition 3.20 immediately implies

Corollary 3.22 *If F has a nonzero index and I is dimension-scaling, then F descends to a tensor functor $\overline{F} : \overline{\mathcal{C}} \to \overline{\mathcal{D}}$.*

Now let H be an involutive finite dimensional Hopf algebra over any algebraically closed field $\mathbf{k}$ (i.e., $S^2 = \mathrm{Id}$, where S is the antipode of H) and K be a Hopf subalgebra in H; for example, H is cocommutative (e.g., a group algebra). Then $\mathcal{C} = \mathrm{Rep}\, H$ and $\mathcal{D} = \mathrm{Rep}\, K$ are finite spherical tensor categories, where dimensions are the usual dimensions (projected to $\mathbf{k}$). Restriction from H to K defines a surjective tensor functor $F : \mathcal{C} \to \mathcal{D}$. Let $I : \mathrm{Rep}(K) \to \mathrm{Rep}(H)$ be the right adjoint to this functor, i.e., the induction functor, $I(V) = \mathrm{Hom}_K(H, V)$.

Recall that by the Nichols-Zoeller theorem [NZ], H is a free K-module, of some rank d.

Corollary 3.23 *Assume that $d \neq 0$ in $\mathbf{k}$. Then any negligible morphism $f : X \to Y$ of H-modules is also negligible as a morphism of K-modules. Thus, F defines a tensor functor: $\overline{\mathrm{Rep}\, H} \to \overline{\mathrm{Rep}\, K}$.*

Proof We have $\dim I(V) = d \dim V$, i.e., I is dimension-scaling. Thus, the result follows from Proposition 3.20. □

4 Semisimplification of Representation Categories of Finite Groups in Characteristic p

4.1 The Result

Let $\mathrm{char}(\mathbf{k}) = p > 0$. Let G be a finite group and P be a Sylow p-subgroup of G. Let $N_G(P)$ be the normalizer of P in G. Since $[G : N_G(P)] \neq 0$ in $\mathbf{k}$, Corollary 3.23 implies

Proposition 4.1 *Let $f : X \to Y$ be a negligible morphism of G-modules. Then f is negligible as a morphism of $N_G(P)$-modules. Thus, the restriction functor F : $\mathrm{Rep}_{\mathbf{k}}\, G \to \mathrm{Rep}_{\mathbf{k}}\, N_G(P)$ descends to a tensor functor between semisimplifications $\overline{F} : \overline{\mathrm{Rep}_{\mathbf{k}}\, G} \to \overline{\mathrm{Rep}_{\mathbf{k}}\, N_G(P)}$.*

Our main result in this section is the following theorem.

Theorem 4.2 *The functor $\overline{F}$ in Proposition 4.1 is an equivalence of tensor categories.*

Theorem 4.2 is proved in the next subsection.

Let $L = N_G(P)/P$.

Corollary 4.3 *One has* $\overline{\mathrm{Rep}_{\mathbf{k}}\, G} \cong (\overline{\mathrm{Rep}_{\mathbf{k}}\, P})^L$.

Proof This follows from Theorem 4.2 and Corollary 3.16, since $|L| \neq 0$ in $\mathbf{k}$ (as $P \subset G$ is a p-Sylow subgroup). □

4.2 Proof of Theorem 4.2

To prove Theorem 4.2, we will use the theory of vertices of modular representations and the Green correspondence (see, e.g., [A, Chapter III]), which we will now recall. Let M be a finite dimensional representation of a finite group G over a field $\mathbf{k}$ of characteristic p. Let H be a subgroup of G.

Definition 4.4 We say that M is relatively H-projective if M is a direct summand in $\mathrm{Ind}_H^G V$ for some finite dimensional H-module V.

Proposition 4.5 (see, e.g., [A], Section 9) *For each indecomposable G-module M, the minimal subgroups $H \subset G$ such that M is relatively H-projective are conjugate, and they are p-groups.*

Definition 4.6 The minimal subgroup $H \subset G$ such that M is relatively H-projective (well defined up to conjugation, thanks to Proposition 4.5) is called the *vertex* of M.

Proposition 4.7 ([G], Theorem 9) *If* $\dim M \neq 0$ *in* $\mathbf{k}$, *then the vertex of M is the Sylow p-subgroup $P \subset G$.*

Proof The result is well known, but we give a proof for reader's convenience. Let H be the vertex of M, so M is a direct summand of $\mathrm{Ind}_H^G V$ for some $H-$module V. For the sake of contradiction, assume that H is not conjugate to P. We will prove:

(a) any direct summand of $\mathrm{Ind}_H^G V$ has dimension zero.
This is a contradiction with our assumption on H, since M is one of such direct summands. We deduce (a) from the following stronger statement:
(b) any direct summand of $\mathrm{Res}_P^G \mathrm{Ind}_H^G V$ has dimension zero.

To prove (b), recall that by the Mackey formula (see [A, III.8, Lemma 7])

$$\mathrm{Res}_P^G \mathrm{Ind}_H^G V = \bigoplus_{s \in P\backslash G/H} \mathrm{Ind}_{P \cap sHs^{-1}}^P \mathrm{Res}_{P \cap sHs^{-1}}^{sHs^{-1}} s(V).$$

By the assumption $P \cap sHs^{-1}$ is strictly contained in P for any s. Since P is $p-$group, $P \cap sHs^{-1}$ is a subnormal subgroup of P. Thus, by Green's indecomposability theorem (see [A, III.8, Theorem 8]), the functor $\mathrm{Ind}_{P \cap sHs^{-1}}^P$ sends indecomposable modules to indecomposable ones. In particular, any direct summand of $\mathrm{Ind}_{P\cap sHs^{-1}}^{P}\mathrm{Res}_{P\cap sHs^{-1}}^{sHs^{-1}}s(V)$ is induced from $P \cap sHs^{-1}$ and has dimension divisible by the index $[P : P \cap sHs^{-1}]$, hence vanishes in $\mathbf{k}$; see [A, III.8, Lemma 4]. The result follows. □

Theorem 4.8 *[A] (Green's correspondence) For each p-subgroup $H \subset N_G(P)$, there is a bijection between indecomposable representations of G with vertex H and indecomposable representations of $N_G(P)$ with vertex H, given by $X \mapsto X^\circ$ for $X \in \mathrm{Rep}_\mathbf{k} G$, such that $X|_{N_G(P)} = X^\circ \oplus N$, where N is a direct sum of indecomposable $N_G(P)$-modules with vertices other than H.*

We can now prove Theorem 4.2. Let $T \in \overline{\mathrm{Rep}\, G}$ be a simple object. Then $T = \overline{X}$, where $X \in \mathrm{Rep}\, G$ is an indecomposable module of nonzero dimension. Hence, by Proposition 4.7, the vertex of X is P. Hence, by Theorem 4.8, $X|_{N_G(P)} = X^\circ \oplus N$, where N is a direct sum of indecomposable $N_G(P)$-modules whose vertices are different from P. Then by Proposition 4.7, the dimension of each of these indecomposable modules is zero; hence, N is negligible. This means that $\overline{F}(T) = \overline{X^\circ}$, which is a simple object of $\overline{\mathrm{Rep}\, N_G(P)}$. This shows that the functor $\overline{F}$ is injective.

Now let $Z \in \overline{\mathrm{Rep}\, N_G(P)}$ be a simple object. Then $Z = \overline{Y}$ for some $Y \in \mathrm{Rep}\, N_G(P)$. But Y is a direct summand in $(\mathrm{Ind}_{N_G(P)}^G Y)|_{N_G(P)}$. Hence, Z is a direct summand in $\overline{F}(\overline{\mathrm{Ind}_{N_G(P)}^G Y})$, proving that $\overline{F}$ is surjective.

Thus, $\overline{F}$ is an equivalence, as claimed.

4.3 The Case of Sylow Subgroup of Prime Order

Let us now consider the simplest nontrivial special case of Theorem 4.2, when the p-Sylow subgroup of G has order p.

Corollary 4.9 *If $P = \mathbb{Z}/p$, then $\overline{\mathrm{Rep}_\mathbf{k}\, G} = (\mathrm{Ver}_p)^L$, where Ver_p is the Verlinde category (see Example 2.7(2)).*

Note that if $p = 2$, then $\mathrm{Ver}_p = \mathrm{Vec}$. Thus, if $P = \mathbb{Z}/2$, then Corollary 4.9 says that $\overline{\mathrm{Rep}_\mathbf{k}\, G} = \mathrm{Rep}_\mathbf{k}\, L$.

So let us consider the case $p > 2$ and compute the category $(\mathrm{Ver}_p)^L$ more explicitly. Note that $\mathrm{Ver}_p = \mathrm{Ver}_p^+ \boxtimes \mathrm{Supervec}$ and Ver_p has no nontrivial symmetric tensor autoequivalences [O], while Ver_p^+ has no nontrivial tensor automorphisms of the identity functor. Thus, from Corollary 4.9, we get

$$\overline{\mathrm{Rep}_\mathbf{k}\, G} = \mathrm{Ver}_p^+ \boxtimes \mathrm{Supervec}^L.$$

The group of tensor automorphisms of the identity functor of Supervec is $\mathbb{Z}/2$. Hence, actions of L on Supervec correspond to elements $H^2(L, \mathbb{Z}/2)$. Let $c \in H^2(L, \mathbb{Z}/2)$ be the element corresponding to the action as above, and let us compute c. Since the action of L on $\mathbb{Z}/p$ factors through an action of $\mathbb{Z}/(p-1)$, the element c is pulled back from a canonical element $\bar{c} \in H^2(\mathbb{Z}/(p-1), \mathbb{Z}/2) = \mathbb{Z}/2$.

Proposition 4.10 *The element $\bar{c}$ is nontrivial.*

Proof It suffices to show that the pullback of $\bar{c}$ to $\mathbb{Z}/2 \subset \mathbb{Z}/(p-1)$ is nontrivial. For this purpose, it suffices to consider the semisimplification of $\mathrm{Rep}_{\mathbf{k}}\, D_p$, where $D_p := \mathbb{Z}/2 \ltimes \mathbb{Z}/p$ is the dihedral group. In $\overline{\mathrm{Rep}_{\mathbf{k}}\, D_p}$, we have an invertible object X of vector space dimension $p-1$, which has composition series $\mathbf{k}_+, \mathbf{k}_-, \dots, \mathbf{k}_+, \mathbf{k}_-$, where $\mathbf{k}_+$ is the trivial representation of $\mathbb{Z}/2$ and $\mathbf{k}_-$ is the sign representation, and it suffices to show that X has order > 2. But we have $X = X^* \otimes \mathbf{k}_-$. Thus, X cannot have order 2, as desired. □

Let $\widetilde{L}$ be the central extension of L by $\mathbb{Z}/2$ defined by the cocycle c, and let z be the generator of the central subgroup $\mathbb{Z}/2 \subset \widetilde{L}$.

Corollary 4.11 *If p is odd and $P = \mathbb{Z}/p$, then*

$$\overline{\mathrm{Rep}_{\mathbf{k}}\, G} \cong \mathrm{Ver}_p^+ \boxtimes \mathrm{Rep}_{\mathbf{k}}(\widetilde{L}, z),$$

where $\mathrm{Rep}_{\mathbf{k}}(\widetilde{L}, z)$ is the category of representations of $\widetilde{L}$ on supervector spaces, so that z acts by the parity operator.

4.4 The Case of the Symmetric Group S_{p+n}, where $n < p$

If $p = 2$, then we have $\overline{\mathrm{Rep}_{\mathbf{k}}\, S_2} = \overline{\mathrm{Rep}_{\mathbf{k}}\, S_3} = \mathrm{Vec}_{\mathbf{k}}$. So consider the case $p > 2$. Let $G = S_{p+n}$, where $0 \leq n < p$. Then $P = \mathbb{Z}/p$, and $N_G(P) = S_n \times \mathbb{Z}/(p-1) \ltimes \mathbb{Z}/p$. Thus, by Corollary 4.11,

$$\overline{\mathrm{Rep}_{\mathbf{k}}\, S_{p+n}} \cong \mathrm{Rep}_{\mathbf{k}}\, S_n \boxtimes \mathrm{Ver}_p^+ \boxtimes \mathrm{Rep}_{\mathbf{k}}(\mathbb{Z}/2(p-1), z),$$

where z is the element of order 2 in $\mathbb{Z}/2(p-1)$. In particular, for $n \geq 2$, the group of invertible objects of this category is $\mathbb{Z}/2 \times \mathbb{Z}/2(p-1)$.

In particular, we obtain the following proposition.

Proposition 4.12 *If $n < p$, then the restriction functor*

$$\mathrm{Res} : \mathrm{Rep}_{\mathbf{k}}\, S_{n+p} \to \mathrm{Rep}_{\mathbf{k}}(S_n \times S_p)$$

induces an equivalence $\overline{\mathrm{Rep}_{\mathbf{k}}\, S_{n+p}} \to \overline{\mathrm{Rep}_{\mathbf{k}}(S_n \times S_p)}$.

Proof The functor Res descends to a tensor functor $\overline{\mathrm{Rep}_{\mathbf{k}}\, S_{n+p}} \to \overline{\mathrm{Rep}_{\mathbf{k}}(S_n \times S_p)}$ by Corollary 3.22, and this tensor functor is an equivalence since the inclusion $S_n \times S_p \hookrightarrow S_{n+p}$ induces an isomorphism of the normalizers of the Sylow p-subgroups. □

Let us now describe the functor $\mathbf{S}$ more explicitly, in the special case $n = 0$, i.e., $\mathcal{C} = \mathrm{Rep}_{\mathbf{k}}\, S_p$, where $p > 2$. It is well known that in this case, we have a unique non-semisimple block $\mathcal{B}$ of defect 1, namely, the block of the trivial representation. The blocks of defect zero consist of objects of dimension 0, so they are killed by $\mathbf{S}$. So let us first consider the images under $\mathbf{S}$ of the simple objects of $\mathcal{B}$. These objects have the form $\wedge^i V_{p-2}$, $i = 0, \dots, p-2$, where V_{p-2} is the $p-2$-dimensional irreducible representation of S_p which is the middle composition factor in the permutation representation. To compute the image $\mathbf{S}(V_{p-2})$ of V_{p-2}, denote by L_i $i = 1, 3, 5, \dots, p-2$ the simple objects of Ver_p^+ (so that $L_1 = \mathbf{1}$) and by χ the generator of $\mathrm{Rep}_{\mathbf{k}}(\mathbb{Z}/2(p-1))$. The object $\mathbf{S}(V_{p-2})$ has to be simple and has dimension -2, so it has the form $L_{p-2} \otimes \chi^m$, where m is even as $\dim \chi = -1$. Moreover, $S^2 V_{p-2}$ contains $\mathbf{1}$ as a direct summand, which implies that $m = 0$ or $m = p-1$. Finally, $\wedge^{p-2} V_{p-2} = \mathrm{sign}$ is the sign representation of S_p, so $\wedge^{p-2}(L_{p-2} \otimes \chi^m) = \chi^{m(p-2)}$ is nontrivial, which implies that $m = p-1$. Thus,

$$\mathbf{S}(V_{p-2}) = L_{p-2} \otimes \chi^{p-1}.$$

This means that

$$\mathbf{S}(\wedge^i V_{p-2}) = L_{p-1-i} \otimes \chi^{p-1}$$

for odd $i \leq p-2$ and

$$\mathbf{S}(\wedge^i V_{p-2}) = L_{i+1}$$

for even $i \leq p-2$.

Now consider the representation V_{p-1} of S_p on the space of functions on $[1, p]$ modulo constants. Then $\mathbf{S}(V_{p-1})$ has dimension -1, so it is of the form χ^m for some odd m. Moreover, it is well known that $\wedge^i V_{p-1}$ is indecomposable for $i \leq p-1$. Since it is not invertible for $0 < i < p-1$, we see that $\chi^{mi} \neq \mathbf{1}, \chi^{p-1}$ for any $0 < i < p-1$. Also $\chi^{m(p-1)} = \chi^{p-1}$. This implies that the order of χ^m is $2(p-1)$, so we may assume that $m = 1$ by making a suitable choice of χ. Thus, for a suitable choice of χ, we have

$$\mathbf{S}(V_{p-1}) = \chi.$$

The suitable choice of χ is well defined only up to the change $\chi \to \chi^p$, since the group $\mathbb{Z}/2(p-1)$ has an automorphism of order 2 (sending 1 to p) which acts trivially on $\mathbb{Z}/(p-1) = \mathrm{Aut}(\mathbb{Z}/p)$. Thus, the well-defined question is to determine χ^2, which is a character of $\mathrm{Aut}(\mathbb{Z}/p)$, naturally identified with $\mathbb{F}_p^\times$. Then it is easy to show by a direct calculation that χ^2 is the natural inclusion $\mathbb{F}_p^\times \hookrightarrow \mathbf{k}^\times$ coming from the inclusion of fields $\mathbb{F}_p \hookrightarrow \mathbf{k}$.

Thus, we obtain

Proposition 4.13 *The category $\overline{\mathrm{Rep}_{\mathbf{k}} S_p}$ is generated by $\overline{V_{p-2}}$ and $\overline{V_{p-1}}$. In other words, the simple objects of $\overline{\mathrm{Rep}_{\mathbf{k}} S_p}$ have the form $\overline{V_{p-1}}^{\otimes m} \otimes \wedge^i \overline{V}_{p-2}$, where*

$0 \le m \le p-2$ and $0 \le i \le p-2$ (so the total number of simple objects is $(p-1)^2$).

4.5 Application: The Semisimplification of the Deligne Category $\underline{\mathrm{Rep}}^{\mathrm{ab}} S_n$

Let n be a nonnegative integer and $\mathbf{k}$ be a field of characteristic zero. Let $\underline{\mathrm{Rep}} S_n$ denote the Karoubian Deligne category over $\mathbf{k}$ defined in [D3] (its main property is that it can be interpolated to non-integer values of n in $\mathbf{k}$). This category has a tensor ideal I such that $\underline{\mathrm{Rep}} S_n/I = \mathrm{Rep}_{\mathbf{k}} S_n$. Moreover, it is known (see [D3, CO]) that $\underline{\mathrm{Rep}} S_n$ has an abelian envelope $\underline{\mathrm{Rep}}^{\mathrm{ab}} S_n$; in particular, the trace of any nilpotent endomorphism in $\underline{\mathrm{Rep}} S_n$ vanishes. Since I consists of morphisms factoring through negligible objects (i.e., direct sums of indecomposable objects of dimension zero), and $\mathrm{Rep}_{\mathbf{k}} S_n$ is semisimple, we see that $I = \mathcal{N}(\mathcal{C})$ is the full ideal of negligible morphisms (i.e., every negligible morphism factors through a negligible object), and the semisimplification $\overline{\underline{\mathrm{Rep}} S_n}$ coincides with $\mathrm{Rep}_{\mathbf{k}} S_n$.

The question of describing the semisimplification of the abelian envelope $\underline{\mathrm{Rep}}^{\mathrm{ab}} S_n$ is more interesting. The answer is given by the following theorem.

Theorem 4.14

(i) The restriction functor

$$\mathrm{Res} : \underline{\mathrm{Rep}}^{\mathrm{ab}} S_n \to \mathrm{Rep}_{\mathbf{k}} S_n \boxtimes \underline{\mathrm{Rep}}^{\mathrm{ab}} S_0$$

induces an equivalence between the semisimplifications of these categories.

(ii) We have an equivalence of symmetric tensor categories

$$\overline{\underline{\mathrm{Rep}}^{\mathrm{ab}} S_n} \cong \mathrm{Rep}_{\mathbf{k}} S_n \boxtimes \mathrm{Rep}_{\mathbf{k}}(GL(1) \times SL(2), (-1,-1)).$$

Proof We will use the approach of [Ha] to Deligne categories. Namely, let us take $\mathbf{k} = \mathbb{C}$. Then, according to [Ha, Theorem 1.1(b)], we have

$$\underline{\mathrm{Rep}}^{\mathrm{ab}} S_n = \lim_{p\to\infty} \mathrm{Rep}_{\overline{\mathbb{F}}_p} S_{n+p},$$

where lim denotes an appropriate ultrafilter limit (i.e., ultraproduct). More precisely, this means that $\underline{\mathrm{Rep}}^{\mathrm{ab}} S_n$ is the tensor subcategory in the appropriate ultrafilter limit tensor generated by the "permutation" object P (the analog of the permutation representation). It is easy to see that the ultrafilter limit commutes with the semisimplification, so (i) follows from Proposition 4.12.

By virtue of (i), it suffices to check (ii) for $n = 0$. In this case, according to Sect. 4.4, $\overline{\mathrm{Rep}_{\overline{\mathbb{F}}_p} S_{n+p}} = \overline{\mathrm{Rep}_{\overline{\mathbb{F}}_p} S_p}$ is generated by $\overline{V_{p-2}}$ and $\overline{V_{p-1}}$. In the ultrafilter limit, the sequences of representations V_{p-2} and V_{p-1} converge to the objects V_{-2} and V_{-1} of $\underline{\mathrm{Rep}}^{\mathrm{ab}} S_0$ (of dimensions -2 and -1, respectively), defined by the (non-split) exact sequences

$$0 \to \mathbf{1} \to P \to V_{-1} \to 0, \ 0 \to V_{-2} \to V_{-1} \to \mathbf{1} \to 0$$

(in particular, V_{-2} is simple). Thus, by Proposition 4.13, the category $\overline{\underline{\mathrm{Rep}}^{\mathrm{ab}} S_n}$ is generated by $\overline{V_{-2}}$ and $\overline{V_{-1}}$. Moreover, since

$$\overline{V_{p-2}} = L_{p-2} \otimes \chi^{p-1},$$

we find that $\overline{V_{-2}}$ generates a subcategory with Grothendieck ring of $\mathrm{Rep}_{\mathbf{k}}\, SL(2)$. Since $\dim V_{-2} = -2$, this is the category $\mathrm{Rep}_{\mathbf{k}}(SL(2), -1)$. Similarly, since $\overline{V_{p-1}}$ is invertible of order $2(p-1)$, we see that $\overline{V_{-1}}$ is invertible of infinite order, so since its dimension is -1, it generates $\mathrm{Rep}_{\mathbf{k}}(GL(1), -1)$. Thus, together these two objects generate the category $\mathrm{Rep}_{\mathbf{k}}\, S_n \boxtimes \mathrm{Rep}_{\mathbf{k}}(GL(1) \times SL(2), (-1, -1))$, as claimed. □

5 Semisimplification of Some Non-Symmetric Categories

Let $\mathrm{char}(\mathbf{k}) = 0$, $q \in \mathbf{k}^\times$, and H_q be the Hopf algebra generated by the grouplike element g and element E with defining relation $gEg^{-1} = qE$ and coproduct defined by $\Delta(E) = E \otimes g + 1 \otimes E$. Then $S(E) = -Eg^{-1}$, so $S^2(E) = gEg^{-1} = qE$. Let $\mathcal{C}_q \subset \mathrm{Rep}\, H_q$ be the category of finite dimensional representations of H_q on which g acts semisimply with eigenvalues being powers of q. This category has a pivotal structure defined by the element g.

5.1 *Generic q*

First assume that q is not a root of unity. Then for any $V \in \mathcal{C}_q$, $E|_V$ is nilpotent, since E maps eigenvectors of g with eigenvalue λ to those with eigenvalue λq. Thus, the indecomposable objects of $\mathcal{C}_q$ are V_{m_1,m_2}, where $m_1 \geq m_2$ are integers, namely, Jordan blocks for E of size $m_1 - m_2 + 1$ containing a nonzero vector v with $gv = q^{m_1}v$, $Ev = 0$. Then $\dim V_{m_1,m_2} = q^{m_2} + \ldots + q^{m_1}$, which is never zero, so there is no nonzero negligible objects. It is easy to see that the tensor product of V_{m_1,m_2} obeys the same fusion rules as representations of $GL_{\mathbf{q}}(2)$ with highest weights (m_1, m_2), where $\mathbf{q}^2 = q$. From this, we obtain

Proposition 5.1 *One has* $\overline{\mathcal{C}_q} \cong \mathrm{Rep}\, GL_{\mathbf{q}}(2)$.

Proof Let us construct a tensor functor $T : \mathrm{Rep}\, GL_{\mathbf{q}}(2) \to \mathcal{C}_q$ such that $\mathbf{S} \circ T$ is an equivalence $\mathrm{Rep}\, GL_{\mathbf{q}}(2) \to \overline{\mathcal{C}_q}$. For this purpose, consider the Hopf algebra

$U_{\mathbf{q}}(\mathfrak{gl}_2)$ with generators g_1, g_2, e, f such that g_1, g_2 are commuting grouplike elements and

$$g_1 e g_1^{-1} = \mathbf{q} e, \ g_1 f g_1^{-1} = \mathbf{q}^{-1} f, \ g_2 e g_2^{-1} = \mathbf{q}^{-1} e, \ g_2 f g_2^{-1} = \mathbf{q} f,$$

$$[e, f] = \frac{g_1 g_2^{-1} - g_2 g_1^{-1}}{\mathbf{q} - \mathbf{q}^{-1}},$$

$$\Delta(e) = e \otimes g_1 g_2^{-1} + 1 \otimes e, \ \Delta(f) = f \otimes 1 + g_2 g_1^{-1} \otimes f.$$

Let us realize $\operatorname{Rep} GL_{\mathbf{q}}(2)$ as the category of finite dimensional representations of $U_{\mathbf{q}}(\mathfrak{gl}_2)$ on which g_1, g_2 act semisimply with eigenvalues being powers of $\mathbf{q}$. Let J be the twist for $U_{\mathbf{q}}(\mathfrak{gl}_2)$ which acts on $v \otimes w$ by $\mathbf{q}^{-rs}$ when $g_1 v = \mathbf{q}^r v$ and $g_2 w = \mathbf{q}^s w$. Then the conjugated coproduct $\Delta_J(a) := J^{-1}\Delta(a)J$ of the element e has the form

$$\Delta_J(e) = e \otimes g_1 + g_1^{-1} \otimes e.$$

Thus, setting $\bar{e} := g_1 e$, we have

$$\Delta_J(\bar{e}) = \bar{e} \otimes g_1^2 + 1 \otimes \bar{e}.$$

We therefore have an inclusion of Hopf algebras $\psi : H_q \hookrightarrow U_{\mathbf{q}}(\mathfrak{gl}_2)^J$ given by $\psi(g) = g_1^2$, $\psi(E) = \bar{e}$, which defines the desired tensor functor T. □

5.2 *Roots of Unity*

Now consider the case when q is a root of unity of some order n, which is more interesting. For simplicity, assume that $n \geq 3$ is odd, and let $\mathbf{q}$ be a root of unity of order $2n$ such that $\mathbf{q}^2 = q$. In this case, by definition, $\mathcal{C}_q = \operatorname{Rep} H_q/(g^n - 1)$ is the category of finite dimensional representations of the quotient Hopf algebra $H_q/(g^n - 1)$. Note that the action of E on objects of $\mathcal{C}_q$ no longer needs to be nilpotent. Namely, E^n is a central element which can act on a simple module by an arbitrary scalar. However, if $E^n = \lambda \neq 0$ on some simple module V, then given an eigenvector $v \in V$ of g with eigenvalue γ, the elements $v, Ev, \ldots, E^{n-1}v$ are a basis of V, so V has dimension $\gamma(1 + q + q^2 + \ldots + q^{n-1}) = 0$. Thus, the action of E on any non-negligible indecomposable module must be nilpotent. This shows that the non-negligible indecomposable modules are still V_{m_1,m_2}, but now $d := m_1 - m_2 + 1$ is not divisible by n, and also $m := m_1$ is defined only up to a shift by n. We will denote this module by $V(m, d)$. Thus, the simple objects of $\overline{\mathcal{C}_q}$ are $V(m, d)$, where $0 \leq m \leq n-1$ and $d \geq 1$, not divisible by n. Note that

$V(m,1) \otimes V(r,d) = V(r,d) \otimes V(m,1) = V(r+m,d)$ (with addition mod n); thus, $\overline{V(m,1)} \otimes \overline{V(r,d)} = \overline{V(r,d)} \otimes \overline{V(m,1)} = \overline{V(m+r,d)}$.

To compute the fusion rules in $\overline{\mathcal{C}_q}$, consider the Hopf subalgebra $K_q \subset H_q$ generated by g and E^n (this Hopf algebra is commutative and cocommutative, as E^n is a primitive element). Let χ be the generating character of the cyclic group generated by g such that $\chi(g) = q$. Then the Green ring of the category of finite dimensional representations of $K_q/(g^n - 1)$ with nilpotent action of E^n is $R[\mathbb{Z}/n] = R[\chi]/(\chi^n - 1)$, where R is the representation ring of $SL(2)$. Moreover, if $X \in \mathcal{C}_q$ is a negligible indecomposable module over $H_q/(g^n - 1)$, then its restriction to $K_q/(g^n - 1)$ lies in the ideal of $R[\mathbb{Z}/n]$ generated by $1 + \chi + \ldots + \chi^{n-1}$. Thus, we have a natural homomorphism

$$\theta : \mathrm{Gr}(\overline{\mathcal{C}_q}) \to R[\chi]/(1 + \chi + \ldots + \chi^{n-1}).$$

Let us now compute $\theta(\overline{V(m,d)})$. First, it is clear that $\theta(\overline{V(m,1)}) = \chi^m$. Also, for a simple object $X \in \overline{\mathcal{C}_q}$, let $\nu(X) \in \mathbb{Z}/2n$ be defined by $\nu(\overline{V(m,d)}) = 2m - d + 1$. Then for any direct summand Z in $X \otimes Y$, we have $\nu(Z) = \nu(X) + \nu(Y)$ (since the representations $V(m,d)$ extend to $GL_{\mathbf{q}}(2)$, where the order of $\mathbf{q}$ is $2n$, and $\mathbf{q}^{2m-d+1} = \mathbf{q}^{m_1+m_2}$ is determined by the action of the central element $g_1 g_2$). Thus, the subcategory $\mathcal{C}_q^0$ spanned by $V(m,d)$ with $2m - d + 1 = 0$ modulo $2n$ is a tensor subcategory of $\mathcal{C}_q$. Moreover, it is easy to check that the restriction

$$\theta : \mathrm{Gr}(\overline{\mathcal{C}_q^0}) \to R[\chi]/(1 + \chi + \ldots + \chi^{n-1})$$

is injective.

Now, the basis of $\mathrm{Gr}(\overline{\mathcal{C}_q^0})$ is formed by $\overline{V(m, 2rn + 2m + 1)}$, $r \geq 0$. Consider first the case $r = 0$, $0 \leq m \leq \frac{n-3}{2}$. In this case, we get

$$\theta(\overline{V(m, 2m+1)}) = \chi^m + \chi^{m-1} + \ldots. + \chi^{-m}.$$

This means that the collection of $(n-1)/2$ objects

$$\overline{V(m, 2m+1)}, 0 \leq m \leq (n-3)/2$$

span a tensor subcategory, whose Grothendieck ring is that of $\mathrm{Ver}_{\mathbf{q}}^+$, the even part of the category $\mathrm{Ver}_{\mathbf{q}}$ (the fusion category attached to $U_{\mathbf{q}}(\mathfrak{sl}_2)$).

Now, let $W_i \in R$ be a unique irreducible representation of $SL(2)$ with $\dim(W_i) = i + 1$. Then it is easy to see (by looking at bases of representations) that

$$\theta(\overline{V(0, 2rn+1)}) = W_{2r+1} - W_{2r},\ \theta(\overline{V(-1, 2rn-1)}) = W_{2r-1} - W_{2r},\ r \geq 1.$$

This means that the collection of objects $\overline{V(0, 2rn+1)}$, $\overline{V(-1, 2rn-1)}$, $r \geq 1$ spans a tensor subcategory with Grothendieck ring of $\mathrm{Rep}\, PGL(2)$, with $\overline{V(0,2rn+1)} \mapsto U_{4r+1}$, $\overline{V(-1,2rn-1)} \mapsto U_{4r-1}$, with U_s denoting the irreducible representation of $PGL(2)$ of dimension s. Indeed, let us evaluate the characters of W_i at the point $-x$. Then we have

$$\overline{V(0,2rn+1)} \mapsto x^{2r+1}+x^{2r}\ldots+x^{-2r-1}, \overline{V(-1,2rn-1)} \mapsto x^{2r}+x^{2r-1}\ldots.+x^{-2r},$$

which implies the statement.

We also note that the object $V(n-1,n-1)$ is invertible and has order 2.

The analysis of the case when n is even is similar, using Theorem A.3.

Thus, we obtain

Theorem 5.2 *The Grothendieck ring of* $\overline{\mathcal{C}_q}$ *is isomorphic to the Grothendieck ring of the category*

$$\mathrm{Vec}_{\mathbb{Z}/n} \boxtimes \mathrm{Ver}_{\mathbf{q}} \boxtimes \mathrm{Rep}\, PGL(2).$$

Corollary 5.3

(i) *The category spanned by* $\overline{V(0,2rn+1)}$, $\overline{V(-1,2rn-1)}$ *is a tensor category equivalent to* Rep $OSp(1|2)$.

(ii) *The category spanned by* $\overline{V(m,2m+1)}$, $\overline{V(m,2m+1)}\otimes\overline{V(n-1,n-1)}$, $0 \le m \le (n-3)/2$ *is a tensor category equivalent to* $\mathrm{Ver}_{\mathbf{q}}$.

Proof Part (i) follows from Theorems 5.2 and A.1(ii) (since the generating object $V(-1,2n-1)$ corresponding to U_3 has dimension -1).

Part (ii) follows from Theorems 5.2, A.3, and Remark A.4(iii). □

Thus, we expect that there is an equivalence of tensor categories

$$\overline{\mathcal{C}_q} \cong \mathrm{Vec}_{\mathbb{Z}/n} \boxtimes \mathrm{Ver}_{\mathbf{q}} \boxtimes \mathrm{Rep}\, OSp(1|2).$$

Note that this does not immediately follow from Theorem 5.2 since the external tensor product $\mathcal{C} \boxtimes \mathcal{D}$ might have nontrivial associators (for instance, this is the case when both categories $\mathcal{C}$ and $\mathcal{D}$ are pointed).

6 Surjective Symmetric Tensor Functors Between Verlinde Categories $\mathrm{Ver}_p(G)$

Let G be a simple algebraic group over $\mathbb{Z}$, $h = h(G)$ the Coxeter number of G, and $p \ge h$ a prime. Let $\mathbf{k}$ be an algebraically closed field of characteristic p. Let $\mathrm{Ver}_p(G) = \mathrm{Ver}_p(G,\mathbf{k})$ be the associated Verlinde category of G, i.e., the semisimplification of the category Tilt(G) of tilting modules for G over $\mathbf{k}$. For example, $\mathrm{Ver}_p(SL(2)) = \mathrm{Ver}_p$.

Similarly, one defines $\mathrm{Ver}_p(G)$ when G is connected reductive. In this case, we should require that $p \ge h_i$ for all i, where h_i are the Coxeter numbers of all simple constituents of G. Note that $\mathrm{Ver}_p(G)$ is a fusion category (i.e., finite) if and only if G is semisimple.

We would like to construct surjective symmetric tensor functors $\mathrm{Ver}_p(G) \to \mathrm{Ver}_p(K)$ for simple G. To this end, suppose that $\phi : K \hookrightarrow G$ is an embedding of reductive algebraic groups. In this case, we have the following proposition.

Proposition 6.1 *Let p be sufficiently large, and let T be a tilting module for G. Then $T|_K$ is also a tilting module.*

Proof The module T occurs as a direct summand in $V^{\otimes m}$, where V is the direct sum of the irreducible G-modules whose highest weights generate the cone of dominant weights for G. Hence, $T|_K$ is a direct summand in $V^{\otimes m}|_K$. But $V|_K$ is a direct sum of simple K-modules with small highest weights (compared to p), which are therefore tilting. Thus, $T|_K$ is tilting. □

Proposition 6.2 *Let p be sufficiently large, and let K contain a regular unipotent element of G (equivalently, a principal $SL(2)$-subgroup of G). Then for any negligible tilting module T over G, the restriction $T|_K$ is negligible.*

Proof Let $u \in K(\mathbf{k})$ be a regular unipotent element of G and $U \cong \mathbb{Z}/p$ be the subgroup generated by u. Then by Jantzen [J], E13, $T|_U$ is projective, hence negligible. This implies that $T|_K$ is negligible. □

Corollary 6.3 *If K contains a regular unipotent element of G, then for large enough p, we have a surjective tensor functor $F : \mathrm{Ver}_p(G) \to \mathrm{Ver}_p(K)$.*

Proof By Proposition 6.1, we have a monoidal functor

$$\mathrm{Res} : \mathrm{Tilt}(G) \to \mathrm{Tilt}(K),$$

and by Proposition 6.2, it maps negligible objects to negligible ones. Hence, this functor descends to a tensor functor between the semisimplifications $\overline{\mathrm{Res}} : \overline{\mathrm{Tilt}(G)} \to \overline{\mathrm{Tilt}(K)}$. This implies the required statement, since $\overline{\mathrm{Tilt}(G)} \cong \mathrm{Ver}_p(G)$ (and similarly for K), so we can take $F = \overline{\mathrm{Res}}$, and it is clear that this functor is surjective. □

Corollary 6.3 raises a question of classification of pairs $K \subset G$, where G is simple, K is connected reductive, and K contains a regular unipotent element of G. Let us call such a pair a *principal pair*. It is clear that it suffices to classify the corresponding pairs of Lie algebras (which we also call principal); namely, a principal pair of groups $K \subset G$ is determined by a principal pair of Lie algebras $\mathfrak{k} \subset \mathfrak{g}$ and a central subgroup in G. The question of classification of principal pairs of Lie algebras is solved by the following theorem.

Theorem 6.4 *[SS] The principal pairs of Lie algebras $\mathfrak{k} \subset \mathfrak{g}$ (with a proper inclusion) are given by the following list:*

(1) $\mathfrak{sp}(2n) \subset \mathfrak{sl}(2n)$, $n \geq 2$;
(2) $\mathfrak{so}(2n+1) \subset \mathfrak{sl}(2n+1)$, $n \geq 2$;

(3) $\mathfrak{so}(2n+1) \subset \mathfrak{so}(2n+2)$, $n \geq 3$;
(4) $G_2 \subset \mathfrak{so}(7)$;
(5) $G_2 \subset \mathfrak{so}(8)$;
(6) $G_2 \subset \mathfrak{sl}(7)$;
(7) $F_4 \subset E_6$.
(8) $\mathfrak{sl}_2 \subset \mathfrak{g}$ *for any simple* $\mathfrak{g}$.

Namely, the subalgebras (1), (2), (3), (5), and (7) are obtained as fixed points of a Dynkin diagram automorphism, (4) is obtained by composing (5) and (3), and (6) is obtained by composing (5) and (2).

Note that Theorem 6.4 holds not only in characteristic zero but also in sufficiently large characteristic (for each fixed $\mathfrak{g}$).

Question 6.5 Suppose that the groups $K \subsetneq G$ are fixed. Is it true that for large enough p, all surjective tensor functors $F : \mathrm{Ver}_p(G) \to \mathrm{Ver}_p(K)$ are given by Corollary 6.3 (up to autoequivalences of $\mathrm{Ver}_p(G)$ and $\mathrm{Ver}_p(K)$)?

7 Objects of Finite Type in Semisimplifications

Let $\mathcal{D}$ be a semisimple tensor category and $X \in \mathcal{D}$. Let us say that X is of finite type if the number of isomorphism classes of simple objects occurring in tensor products of X and X^* is finite; i.e., X generates a fusion subcategory $\mathcal{D}_X \subset \mathcal{D}$. If $\overline{\mathcal{C}}$ is the semisimplification of a category $\mathcal{C}$, and $X \in \mathcal{C}$, we will say that X is of finite type if so is $\overline{X}$. It is an interesting question which objects of $\mathcal{C}$ are of finite type. Note that according to Example 2.7(4), X does not have to be of finite type even if $\mathcal{C}$ is the representation category of a finite group (e.g., $\mathcal{C} = \mathrm{Rep}_{\mathbf{k}}(\mathbb{Z}/2)^2$ for $\mathrm{char}(\mathbf{k}) = 2$).

Yet, a lot of interesting representations of finite groups do turn out to be of finite type and generate interesting fusion categories. The goal of this subsection is to give some examples of such representations.

Let H be an affine algebraic group over an algebraically closed field $\mathbf{k}$ of characteristic zero. Let V be a rational representation of H. Let H_V be the reductive envelope of H inside $GL(V)$ defined in Definition 3.8. Assume that H contains a regular unipotent element of H_V (e.g., $H = U_n$, the maximal unipotent subgroup of $SL(n)$ and $V = \mathbf{k}^n$; then $H_V = SL(V)$). Note that all this data is defined over some finitely generated subring $R \subset \mathbf{k}$, hence can be reduced modulo p for sufficiently large p; namely, given a homomorphism $\psi : R \to \overline{\mathbb{F}_p}$, we have $\psi(R) = \mathbb{F}_q$, where $q = p^r$ for some r, and we have a chain of finite groups $H(\mathbb{F}_q) \subset H_V(\mathbb{F}_q) \subset GL(V(\mathbb{F}_q))$. Let $V_\psi = V(\overline{\mathbb{F}}_p)$; it is a representation of these finite groups over $\overline{\mathbb{F}_p}$. Let $\mathcal{C} := \mathrm{Rep}_{\overline{\mathbb{F}_p}} H(\mathbb{F}_q)$.

Theorem 7.1 *For large enough p, the category* $\overline{\mathcal{C}}_{\overline{V_\psi}}$ *generated by* $\overline{V_\psi}$ *is a quotient of* $\mathrm{Ver}_p(H_V) = \mathrm{Ver}_p(H_V, \overline{\mathbb{F}_p})$. *In particular, the object* V_ψ *is of finite type in* $\mathcal{C}$.

Proof We have an additive monoidal restriction functor

$$\mathrm{Res} : \mathrm{Tilt}(H_V(\overline{\mathbb{F}_p})) \to \mathrm{Rep}_{\overline{\mathbb{F}}_p} H(\mathbb{F}_q),$$

hence an additive monoidal functor

$$\mathbf{S} \circ \mathrm{Res} : \mathrm{Tilt}(H_V(\overline{\mathbb{F}_p})) \to \overline{\mathrm{Rep}_{\overline{\mathbb{F}}_p} H(\mathbb{F}_q)}.$$

Moreover, the image of a negligible module under the functor Res is negligible, as it is already so after restricting to the group $\mathbb{Z}/p$ generated by a regular unipotent element of H_V contained in $H(\mathbb{F}_q)$ [J, E13]. Hence, the functor $\mathbf{S} \circ \mathrm{Res}$ descends to a tensor functor $\widetilde{F} : \mathrm{Ver}_p(H_V) \to \overline{\mathrm{Rep}_{\overline{\mathbb{F}}_p} H(\mathbb{F}_q)}$ (this functor is automatically exact since the source category is semisimple). Moreover, the functor $\widetilde{F}$ lands in $\overline{\mathcal{C}}_{\overline{V_\psi}}$, so we get a surjective tensor functor $F : \mathrm{Ver}_p(H_V) \to \overline{\mathcal{C}}_{\overline{V_\psi}}$. In particular, in this case, $\overline{\mathcal{C}}_{\overline{V_\psi}}$ is a quotient of $\mathrm{Ver}_p(H_V)$, thus a fusion category if H_V is semisimple.

Moreover, even if H_V is not semisimple but only reductive, $\overline{\mathcal{C}}_{\overline{V_\psi}}$ is still a fusion category, since one-dimensional representations of H_V obviously have finite order when restricted to the finite group $H(\mathbb{F}_q)$. □

Conjecture 7.2 For sufficiently large p, the surjective tensor functor F : $\mathrm{Ver}_p(H_V) \to \overline{\mathcal{C}}_{\overline{V_\psi}}$ is an equivalence.

Remark 7.3 Let $\mathcal{C}$ be a symmetric tensor category over a field $\mathbf{k}$ of characteristic $p > 0$, $\overline{\mathcal{C}}$ be its semisimplification, and $X \in \mathcal{C}$. According to Conjecture 1.3 of [O], there should be a Verlinde fiber functor $F : \overline{\mathcal{C}}_{\overline{X}} \to \mathrm{Ver}_p$ (this is actually a theorem if X is of finite type; see [O]). So, in particular, assuming this conjecture, we can define the number $d(X) := \mathrm{FPdim}(F(\overline{X}))$, the Frobenius-Perron dimension of $F(\overline{X})$. A more refined invariant is the full decomposition of $F(\overline{X})$ into the simple objects $L_1, \ldots, L_{p-1}$ of Ver_p: $F(\overline{X}) = \sum_i a_i(X)L_i$. It is an interesting question how to compute these invariants for a given X (actually, this question can also be asked in characteristic zero, with Ver_p replaced by Supervec). Also, one can define the affine group scheme $G_X = \mathrm{Aut}(F)$ in Ver_p (or Supervec), and its dimension $\delta(X)$ is another interesting invariant of X. Note that X is of finite type if and only if $\delta(X) = 0$. Also note that if $X = V_\psi$ in the setting of Theorem 7.1, then the above invariants can be easily computed using the results of [EOV].

We also have the following proposition.

Proposition 7.4 *Let G be a finite group and V a representation of G over an algebraically closed field $\mathbf{k}$ of characteristic p of dimension $d < p$. Suppose that there exists an element $g \in G$ such that the restriction of V to the cyclic group generated by g is indecomposable. Then V is of finite type.*

Proof We may assume that V is faithful, i.e., $G \subset GL(d)$. Let u be the unipotent part of g. Then u is a power of g and a regular unipotent element of $GL(d)$ (as it acts indecomposably on V). Hence, the restriction functor $\mathrm{Tilt}_p(GL(d)) \to \mathrm{Rep}_{\mathbf{k}}(G)$ descends to a tensor functor $\mathrm{Ver}_p(GL(d)) \to \overline{\mathrm{Rep}_{\mathbf{k}}(G)}$. In particular, the tensor category generated by V in the semisimplification of $\mathrm{Rep}_{\mathbf{k}}(G)$ is finite, as desired.

□

This proposition can be generalized as follows, with a similar proof:

Proposition 7.5 *Let G be a finite group and V a faithful representation of G over an algebraically closed field $\mathbf{k}$ of characteristic p of dimension $d < p$. Suppose that $K \subset GL(d)$ is a reductive subgroup containing G, such that G contains a regular unipotent element of K. Then V is of finite type.*

8 Semisimplification of Tilt$(GL(n))$ when char(k) = 2

In this section, we describe the category $\overline{\mathcal{C}}$ when $\mathcal{C} = \mathrm{Tilt}(GL(n))$ and $\mathrm{char}(\mathbf{k}) = 2$.

First recall Lucas' theorem in elementary number theory:

Let $a \in \mathbb{Z}$, $b \in \mathbb{Z}_+$ with p-adic expansions

$$a = \sum_i a_i p^i, \ b = \sum_i b_i p^i,$$

with $0 \le a_i, b_i \le p - 1$.

Proposition 8.1 (Lucas' Theorem) *One has*

$$\binom{a}{b} = \prod_i \binom{a_i}{b_i} \ \mathit{mod}\ p.$$

In particular, $\binom{a}{b}$ is not divisible by p if and only if $b_i \le a_i$ for all i.

Let $V = \mathbf{k}^n$ be the tautological representation of $GL(n)$. Recall that the indecomposable objects of the category $\mathcal{C}$ are the indecomposable direct summands in tensor products of the fundamental modules $\wedge^\ell V$, $1 \le \ell \le n$. Moreover, it is well known that we can take ℓ to be only powers of 2. Indeed, if $\ell = 2^{k_1} + \cdots + 2^{k_r}$ with $0 \le k_1 < \cdots < k_r$ is the binary expansion of ℓ, then by Lucas' theorem, the multinomial coefficient

$$N := \frac{\ell!}{2^{k_1}! \ldots 2^{k_r}!} = \binom{\ell}{2^{k_1}}\binom{\ell - 2^{k_1}}{2^{k_2}} \cdots$$

is odd. Now pick a subset of coset representatives $C \subset S_\ell$ mapping bijectively onto the quotient $S_\ell/(S_{2^{k_1}} \times \cdots \times S_{2^{k_r}})$, and define the operator $P := \sum_{g \in C} g$ on the space $\wedge^{2^{k_1}} V \otimes \cdots \otimes \wedge^{2^{k_s}} V$. Since $|C| = N$, it is easy to see (e.g., by picking a basis of V) that $P^2 = P$ and $\mathrm{Im}(P) = \wedge^\ell V$, which shows that $\wedge^\ell V$ is naturally a direct summand in $\wedge^{2^{k_1}} V \otimes \cdots \otimes \wedge^{2^{k_s}} V$, as desired.

This shows that the semisimplification $\overline{\mathcal{C}}$ is generated by the objects $X_m := \overline{\wedge^{2^m} V}$, with $0 \le m \le \log_2 n$. Note that $\dim_{\mathbf{k}} X_m = \binom{n}{2^r}$, which by Lucas' theorem is odd if and only if the m-th digit (from the right) in the binary expansion of n is 1. Thus, we can keep only X_m with such values of m. In other words, $\overline{\mathcal{C}}$ is generated by $X_{m_1}, \ldots, X_{m_s}$, where $n = 2^{m_1} + \cdots + 2^{m_s}$, $0 \le m_1 < \cdots < m_s$, is the binary expansion of n.

Proposition 8.2 *Let $n_j = 2^{m_1} + \cdots + 2^{m_j}$, $1 \le j \le s$, and let $Y_j := \overline{\wedge^{n_j} V}$. Then Y_j is invertible. Moreover, $Y_j = X_{m_j} \otimes Y_{j-1}$ (where $Y_0 := \mathbf{1}$), so X_{m_j} are invertible as well. Hence, the category $\overline{\mathcal{C}}$ is pointed.*

Proof To prove that Y_j is invertible, it suffices to show that the module $\wedge^{n_j} V \otimes (\wedge^{n_j} V)^*$ has a unique indecomposable direct summand of odd dimension, namely, $\mathbf{1}$ (which is a direct summand using the evaluation and coevaluation maps). To this end, it suffices to show that this is so after restriction of this representation to any subgroup $G \subset GL(n)$. Take $G = GL(n_j) \times GL(n - n_j)$. Then $V = V' \oplus V''$, where $\dim_{\mathbf{k}} V' = n_j$ and $\dim_{\mathbf{k}} V'' = n - n_j$, and

$$\wedge^{n_j} V|_G = \oplus_{i=0}^{n_j} \wedge^{n_j - i} V' \otimes \wedge^i V''.$$

Note that $n - n_j$ is divisible by $2^{m_j+1} > n_j$; hence, by Lucas' theorem, $\wedge^i V''$ is even dimensional for any $0 < i \le n_j$. This means that any odd-dimensional indecomposable direct summand in the G-module $\wedge^{n_j} V \otimes (\wedge^{n_j} V)^*$ is $\wedge^{n_j} V' \otimes (\wedge^{n_j} V')^* = \mathbf{1}$.

Now, Y_j is a direct summand in $X_{m_j} \otimes Y_{j-1}$. Since X_{m_j} is simple and Y_{j-1} is invertible, we get $Y_j = X_{m_j} \otimes Y_{j-1}$, i.e., $X_{m_j} = Y_j \otimes Y_{j-1}^*$ is invertible. □

Proposition 8.3 *The objects X_{m_j}, $j = 1, \ldots, s$ are multiplicatively independent. In other words, we have $\overline{\mathcal{C}} = \mathrm{Vec}_{\mathbf{k}}(\mathbb{Z}^s)$, where the group $\mathbb{Z}^s$ is generated by the isomorphism classes of the objects X_{m_j} (or Y_j).*

Proof Assume the contrary, i.e., that we have a nontrivial relation

$$X_{m_1}^{\otimes p_1} \otimes \cdots \otimes X_{m_\ell}^{\otimes p_\ell} \cong X_{m_1}^{\otimes q_1} \otimes \cdots \otimes X_{m_\ell}^{\otimes q_\ell}, \tag{1}$$

where $\ell \le s$ and $p_i, q_i \in \mathbb{Z}_{\ge 0}$ with $p_i q_i = 0$ for $1 \le i \le \ell$ and $p_\ell \ne 0$ (so $q_\ell = 0$). Let $r = 2^{m_\ell} + \cdots + 2^{m_s}$, so that $n - r = 2^{m_1} + \cdots + 2^{m_{\ell-1}}$. Consider the subgroup $G = GL(r) \times GL(n - r) \subset GL(n)$. Then $V = V' \oplus V''$, where $\dim_{\mathbf{k}} V' = r$ and $\dim_{\mathbf{k}} V'' = n - r$. For each $1 \le j \le \ell$, we have

$$\wedge^{2^{m_j}} V|_G = \oplus_{i=0}^{2^{m_j}} \wedge^i V' \otimes \wedge^{2^{m_j} - i} V'',$$

Since r is divisible by 2^{m_ℓ} and $n - r < 2^{m_\ell}$, all the indecomposable summands in this direct sum have even dimension except $i = 0$ for $j < l$ and $i = 2^{m_\ell}$ for $j = \ell$.

Thus, the only odd-dimensional indecomposable direct summand of $\wedge^{2^{m_j}} V|_G$ is a trivial representation of $GL(r)$ except for $j = \ell$, in which case $GL(r)$ acts on this summand by the determinant character. Thus, $GL(r)$ acts trivially on the unique odd-dimensional indecomposable direct summand on the right-hand side of (1) but by $\det^{p\ell}$ on such summand on the left-hand side, which is a contradiction. □

Corollary 8.4 *We have* $\overline{\mathrm{Tilt}(SL_n)} = \mathrm{Vec}_{\mathbf{k}}(\mathbb{Z}^{s-1})$ *where* $\mathbb{Z}^{s-1}$ *is generated by the isomorphism classes of* $X_{m_1}, \dots, X_{m_s}$ *modulo the relation* $X_{m_1} \dots X_{m_s} = 1$, *and* $\overline{\mathrm{Tilt}(PGL_n)} = \mathrm{Vec}_{\mathbf{k}}(\mathbb{Z}^{s-1})$ *where* $\mathbb{Z}^{s-1}$ *is the group of* $X_{m_1}^{\otimes n_1} \dots X_{m_s}^{\otimes n_s}$, $n_1 \dots, n_s \in \mathbb{Z}$, *where* $\sum_i 2^{m_i} n_i = 0$.

Proof Straightforward from Proposition 8.3. □

Appendix A: Categorifications of Based Rings Attached to $SO(3)$

The goal of this Appendix is to deduce some classification results on categorifications of certain based rings from the results of [MPS]. We assume that the base field $\mathbf{k}$ is algebraically closed of characteristic zero.

A.1 We consider the based ring K_∞ (see [EGNO, Chapter 3]) with basis $X_i, i \in \mathbb{Z}_{\geq 0}$ and with multiplication determined by

$$X_0 = 1, \ X_1X_i = X_iX_1 = X_{i-1} + X_i + X_{i+1}, \ i \geq 1.$$

It is a classical fact that K_∞ is isomorphic to the representation ring of the group $SO(3)$ via the map sending X_i to a unique irreducible representation of dimension $2i + 1$.

We will consider *pivotal categorifications* of K_∞, that is, semisimple pivotal tensor categories $\mathcal{C}$ equipped with an isomorphism of based rings $K(\mathcal{C}) \simeq K_\infty$ (cf. [EGNO, 4.10]). Any such category $\mathcal{C}$ is automatically spherical since every object of $\mathcal{C}$ is self-dual. Let $X \in \mathcal{C}$ be an object such that its class $[X]$ corresponds to $X_1 \in K_\infty$. Let $d \in \mathbf{k}$ be the dimension of X. There exists $\mathbf{q} \in \mathbf{k}$ such that $d = [3]_{\mathbf{q}} = \mathbf{q}^2 + 1 + \mathbf{q}^{-2}$.

Theorem A.1

(i) Assume that $\mathbf{q}^2 = 1$ *or that* $\mathbf{q}^2$ *is not a root of 1. Then* $\mathcal{C}$ *is equivalent to the category* $\mathrm{Rep}(SO(3)_{\mathbf{q}})$ *(see [MPS, Section 4]).*

(ii) Assume $\mathbf{q}^2 = -1$. *Then* $\mathcal{C}$ *is equivalent to the category* $\mathrm{Rep}(OSp(1|2))$ *(see [MPS, Section 4]).*

Proof Let $\mathcal{C}_0$ be the monoidal subcategory of $\mathcal{C}$ generated by X and by (nonzero) morphisms $\mathbf{1} \to X \otimes X$, $X \otimes X \to \mathbf{1}$, $X \to X \otimes X$, $X \otimes X \to X$. Thus:

$$\text{objects of } \mathcal{C}_0 = X^{\otimes n}, \ n \in \mathbb{Z}_{\geq 0},$$

$$\text{morphisms of } \mathcal{C}_0 = \text{morphisms in } \mathcal{C} \text{ which are linear combinations}$$

$$\text{of tensor products and compositions of the four morphisms above.}$$

Let $\mathcal{N}$ be the ideal of negligible morphisms in $\mathcal{C}_0$, and let $\tilde{\mathcal{C}} = \mathcal{C}_0/\mathcal{N}$ be the quotient. Clearly,

$$\begin{aligned} \dim \mathrm{Hom}_{\tilde{\mathcal{C}}}(X^{\otimes m}, X^{\otimes n}) &\leq \dim \mathrm{Hom}_{\mathcal{C}_0}(X^{\otimes m}, X^{\otimes n}) \\ &\leq \dim \mathrm{Hom}_{\mathcal{C}}(X^{\otimes m}, X^{\otimes n}). \end{aligned} \tag{2}$$

The category $\tilde{\mathcal{C}}$ is an example of a (possibly twisted) *trivalent category*, as defined in [MPS, Section 7] (thus, $\tilde{\mathcal{C}}$ satisfies the assumptions of [MPS, Definition 2.1] except, possibly, the rotational invariance of the morphism $X \to X \otimes X$). Moreover, the numbers $\dim \mathrm{Hom}_{\tilde{\mathcal{C}}}(\mathbf{1}, X^{\otimes k})$ are bounded by the numbers $d_k = \dim \mathrm{Hom}_{\mathcal{C}}(\mathbf{1}, X^{\otimes k})$, which are easily computable using the isomorphism $K(\mathcal{C}) \simeq K_\infty$. In particular, $d_k = 1, 0, 1, 1, 3$ for $k = 0, 1, 2, 3, 4$. Since $d \neq 2$, [MPS, Proposition 7.1] implies that $\tilde{\mathcal{C}}$ is not twisted, that is, $\tilde{\mathcal{C}}$ is a trivalent category in the sense of [MPS, Definition 2.1]. Thus, by [MPS, Theorem A], $\tilde{\mathcal{C}}$ is equivalent to $\mathrm{Rep}(SO(3)_{\mathbf{q}})$ or $\mathrm{Rep}(OSp(1|2))$; in particular, the Grothendieck ring $K(\tilde{\mathcal{C}})$ of (the Karoubian envelope of) $\tilde{\mathcal{C}}$ is isomorphic to $K_\infty = K(\mathcal{C})$. Thus, the inequalities in (2) are, in fact, equalities, and the category $\mathcal{C}$ is equivalent to the Karoubian envelope of $\tilde{\mathcal{C}}$. The result follows. □

Remark A.2

(i) We expect that the assumption on $\mathbf{q}$ in Theorem A.1 is automatically satisfied, i.e., there is no categorification of K_∞ where $\mathbf{q}^2 \neq \pm 1$ is a root of 1. Moreover, it seems likely that the assumption on pivotality of $\mathcal{C}$ can also be dropped.
(ii) D. Copeland and H. Wenzl recently obtained a classification of *ribbon categorifications* of the based rings $K(\mathrm{Rep}(SO(n)_{\mathbf{q}}))$ for any n. In particular, this implies Theorem A.1 (and Theorem A.3 below) under an additional assumption that the category $\mathcal{C}$ is braided.

A.2. Fusion Categories For an integer $l \geq 2$, we consider the based ring K_l with basis X_i, $i = 0, \ldots, l$ and with multiplication determined by

$$X_0 = 1, \ X_1X_i = X_{i-1} + X_i + X_{i+1}, \ i = 1, \ldots l-1, \ X_1X_l = X_{l-1}.$$

The ring K_l can be considered as a truncated version of the ring K_∞. It is well known that the ring K_l has categorifications of the form $\mathrm{Rep}(SO(3)_{\mathbf{q}}) = \mathrm{Ver}_{\mathbf{q}}^+$, where $\mathbf{q}$ is a suitable root of 1.

Theorem A.3 *Let $\mathcal{C}$ be a pivotal fusion category which is a categorification of K_l where $l > 2$. Then there is a tensor equivalence $\mathcal{C} \simeq \mathrm{Rep}(SO(3)_{\mathbf{q}})$ where $\mathbf{q}$ is a primitive root of 1 of degree $4(l+1)$.*

Proof We start by classifying homomorphisms $\phi : K_l \to \mathbf{k}$. Any such homomorphism is uniquely determined by $\phi(X_1)$; if $\phi_{\mathbf{q}}(X_1) = [3]_{\mathbf{q}} = \mathbf{q}^2 + 1 + \mathbf{q}^{-2}$, then $\phi_{\mathbf{q}}(X_i) = [2i+1]_{\mathbf{q}} = \frac{\mathbf{q}^{2i+1} - \mathbf{q}^{-2i-1}}{\mathbf{q} - \mathbf{q}^{-1}}$; in particular, the existence of $\phi_{\mathbf{q}}$ is equivalent to the equation

$$[3]_{\mathbf{q}}[2l+1]_{\mathbf{q}} = [2l-1]_{\mathbf{q}} \iff \mathbf{q}^{2(l+1)} = \pm 1,\ \mathbf{q}^2 \neq 1.$$

Clearly, $\phi_{\mathbf{q}} = \phi_{\mathbf{q}'}$ if and only if $\mathbf{q}^2 = \mathbf{q}'^{\pm 2}$. One computes easily the *formal codegree* $f_{\phi_{\mathbf{q}}}$ (see, e.g., [O1, Section 2.3]) of $\phi_{\mathbf{q}}$:

$$f_{\phi_{\mathbf{q}}} = \begin{cases} l+1 & \text{if } \mathbf{q}^2 = -1, \\ -\frac{2(l+1)}{(\mathbf{q}-\mathbf{q}^{-1})^2} & \text{if } \mathbf{q}^2 \neq -1. \end{cases}$$

The category $\mathcal{C}$ is spherical, as all its objects are self-dual. Hence, by [O1, Corollary 2.15], the *dimension field* (i.e., the subfield of $\mathbf{k}$ generated by the dimensions of the objects) of $\mathcal{C}$ contains all $f_{\phi_{\mathbf{q}}}$; thus, the degree of the dimension field over the rationals is $\geq \frac{1}{2}\varphi(2(l+1))$, where φ is the Euler function (this is the degree of the field generated by $\mathbf{q}^2 + \mathbf{q}^{-2}$, where $\mathbf{q}^2$ is a primitive root of 1 of degree $2(l+1)$). It follows that the dimension homomorphism $K_l = K(\mathcal{C}) \to \mathbf{k}$ is $\phi_{\mathbf{q}}$, with $\mathbf{q}^2$ being a primitive root of 1 of degree r, where r divides $2(l+1)$ and $\varphi(r) = \varphi(2(l+1))$. Thus, either $r = 2(l+1)$ or $r = l+1$. The latter case is possible only if $l+1$ is odd, and in this case, $\phi_{\mathbf{q}}(X_{l/2}) = 0$, so $\phi_{\mathbf{q}}$ cannot be the dimension homomorphism.

Thus, we have proved that the dimension homomorphism

$$K_l = K(\mathcal{C}) \to \mathbf{k}$$

coincides with the dimension homomorphism

$$K_l = K(\mathrm{Rep}(SO(3)_{\mathbf{q}}) \to \mathbf{k},$$

where $\mathbf{q}$ is a primitive root of 1 of degree $4(l+1)$.

The rest of the proof is parallel to the proof of Theorem A.1. We consider the subcategory $\mathcal{C}_0$ of $\mathcal{C}$ generated by the morphisms

$$\mathbf{1} \to X \otimes X,\ X \otimes X \to \mathbf{1},\ X \to X \otimes X,\ X \otimes X \to X$$

and its quotient $\tilde{\mathcal{C}}$ by negligible morphisms. Then one deduces from [MPS, Theorem A] that the (Karoubian envelope of) the category $\tilde{\mathcal{C}}$ is equivalent to $\mathrm{Rep}(SO(3)_{\mathbf{q}})$, which has the same Grothendieck ring as $\mathcal{C}$. This implies that $\tilde{\mathcal{C}} \cong \mathcal{C}$, and the result follows. □

Remark A.4

(i) The categorifications of K_l with $l = 2$ are completely classified in [EGO]. This case is somewhat different from the case $l > 2$; see [MPS, Section 7].

(ii) It is conjectured that any fusion category has a pivotal structure. Thus, we expect that the pivotality assumption in Theorem A.3 is superfluous.

(iii) Another family of truncations of the ring K_∞ is given by rings $\tilde{K}_l,\ l \geq 1$ with basis $X_0, \ldots, X_l$ and with multiplication

$$X_0 = 1,\ X_1X_i = X_{i-1} + X_i + X_{i+1},\ i = 1, \ldots l-1,\ X_1X_l = X_{l-1} + X_l.$$

Such rings are also categorified by $\mathrm{Rep}(SO(3)_{\mathbf{q}})$ where $\mathbf{q}$ is a suitable root of 1. It is easy to see that there are no other categorifications $\mathcal{C}$ of $\tilde{K}_l$, since $\mathcal{C} \boxtimes \mathrm{Vec}_{\mathbb{Z}/2\mathbb{Z}}$ would have been an example of a Temperley-Lieb category generated by the object $X_l \boxtimes \mathbf{1}$.

Acknowledgments The authors are grateful to D. Benson, I. Entova-Aizenbud, T. Heidersdorf, A. Kleshchev, D. Nakano, V. Serganova, and N. Snyder for useful discussions. The work of P.E. and V.O. was partially supported by the NSF grants DMS-1502244 and DMS-1702251. The work of V.O. has also been funded by the Russian Academic Excellence Project '5-100'.

References

[A] J. L. Alperin, Local representation theory, Modular Representations as an Introduction to the Local Representation Theory of Finite Groups, Cambridge Studies in Advanced Mathematics, 11, Cambridge University Press, 1993.

[AK] Y. Andre, B. Kahn, with an appendix by P. O'Sullivan, Nilpotence, radicaux et structures monoídales, arXiv:math/0203273, Rendiconti del Seminario Matematico dell'Universita di Padova 108 (2002), 107–291.

[BEEO] J. Brundan, I. Entova-Aizenbud, P. Etingof, V. Ostrik, Semisimplification of the category of tilting modules for GL(n), Adv. Math. 375 (2020), 107331.

[BW] J. Barrett and B. Westbury, Spherical categories, Adv. Math. 143 (1999), 357–375.

[B1] D. J. Benson, Representations and Cohomology, I: Basic representation theory of finite groups and associative algebras, Cambridge University Press, 1995.

[B2] D. J. Benson, Modular Representation theory, New trends and methods. SLNM 1081 (1984).

[CO] J. Comes, V. Ostrik, On the Deligne category $\underline{\mathrm{Rep}}^{\mathrm{ab}} S_d$, arXiv:1304.3491, Algebra Number Theory 8 (2014), no. 2, 473–496.

[D1] P. Deligne, Catégories tannakiennes, in : The Grothendieck Festschrift, vol. 2, Birkhäuser P.M. 87 (1990), 111–198.

[D2] P. Deligne. Catégories tensorielles, Mosc. Math. J., 2(2):227–248, 2002.

[D3] P. Deligne, La catégorie des représentations du groupe symétrique S_t, lorsque t nest pas un entier naturel, in: Algebraic Groups and Homogeneous Spaces, in: Tata Inst. Fund. Res. Stud. Math., Tata Inst. Fund. Res., Mumbai, 2007, 209–273.

[DM] P. Deligne, J. Milne, Tannakian categories, Lecture notes in Math. 900, 1981.

[EGO] P. Etingof, S. Gelaki, V. Ostrik, Classification of fusion categories of dimension pq, Int. Math. Res. Not. **2004**, no. 57, p. 3041–3056.

[EOV] P. Etingof, V. Ostrik, S. Venkatesh, Computations in symmetric fusion categories in characteristic p, arXiv:1512.02309, Int. Math. Res. Not. IMRN 2017, no. 2, p.468–489.

[EGNO] P. Etingof, S. Gelaki, D. Nikshych, V. Ostrik, Tensor categories, AMS, 2015.

[G] J. A. Green, On indecomposable representations of a finite group, Math. Zeitschrift, v. 70, p. 430–445, 1959.

[H] T. Heidersdorf, On supergroups and their semisimplified representation categories, arXiv:1512.03420.

[Ha] N. Harman, Deligne categories as limits in rank and characteristic, arXiv:1601.03426.

[Ja] U. Jannsen, Motives, numerical equivalence and semi-simplicity, Inv. Math. 107 (1992), 447–452.

[J] J. C. Jantzen, Representations of algebraic groups, 2nd edition, American Mathematical Society, Providence, RI, 2003.

[MPS] S. Morrison, E. Peters and N. Snyder, Categories generated by a trivalent vertex, Selecta Math. (N.S.) 23 (2017) no. 2, 817–868.

[NZ] W. Nichols and M. B. Zoeller, A Hopf algebra freeness theorem. Amer. J. Math. 111 (1989), 381–385.

[O1] V. Ostrik, Pivotal fusion categories of rank 3, Mosc. Math. J. **15** (2015), no. 2, p. 373–396.

[O] V. Ostrik, On symmetric fusion categories in positive characteristic, arXiv:1503.01492.

[S] P. O'Sullivan, The generalized Jacobson-Morozov theorem, Memoirs of the AMS, v.973, 2010.

[SS] J. Saxl, G. Seitz, Subgroups of algebraic groups containing regular unipotent elements, Journal of the London Mathematical Society / Volume 55 / Issue 02 / April 1997, pp 370–386.

[W] R. Weissauer, Semisimple algebraic tensor categories, arXiv:0909.1793.

Totally Aspherical Parameters for Cherednik Algebras

Ivan Losev

Dedicated to Vitya Ginzburg, on his 60th birthday, with admiration.

Contents

MSC 2010: 16G99

1 Introduction

Let W be a complex reflection group and $\mathfrak{h}$ be its reflection representation. The rational Cherednik algebra $H_c(W)$ is a filtered deformation of the skew-group ring $S(\mathfrak{h} \oplus \mathfrak{h}^*)\#W$ depending on a parameter c that is a collection of complex numbers.

I. Losev (✉)
Department of Mathematics, Yale University, New Haven, CT, USA

V. Baranovsky et al. (eds.), *Representation Theory and Algebraic Geometry*, Trends in Mathematics, https://doi.org/10.1007/978-3-030-82007-7_2

Inside $H_c(W)$, there is a so-called spherical subalgebra $eH_c(W)e$, where $e \in \mathbb{C}W$ is the averaging idempotent, that deforms $S(\mathfrak{h} \oplus \mathfrak{h}^*)^W$. The parameter c is called *spherical* if $H_c(W)$ and $eH_c(W)e$ are Morita equivalent (via the bimodule $H_c(W)e$). The parameter c is called *aspherical* if it is not spherical. The goal of this paper is to investigate parameters that are "as aspherical as possible."

The algebra $H_c(W)$ has a triangular decomposition, $H_c(W) = S(\mathfrak{h}^*) \otimes \mathbb{C}W \otimes S(\mathfrak{h})$, that is analogous to the triangular decomposition of the universal enveloping algebra of a semisimple Lie algebra. Thanks to this decomposition, it makes sense to consider the category $\mathcal{O}$ for $H_c(W)$: by definition, the category $\mathcal{O}_c(W)$ consists of all $H_c(W)$-modules M that are finitely generated over $S(\mathfrak{h}^*)$ and have locally nilpotent action of $\mathfrak{h}$.

Definition 1.1 We say that c is *totally aspherical* if, for $M \in \mathcal{O}_c(W)$, the following two conditions are equivalent:

(i) M is torsion as an $S(\mathfrak{h}^*)$-module.
(ii) $eM = 0$.

We will see below, Proposition 2.7, that c is totally aspherical if and only if the algebra $eH_c(W)e$ is simple.

Let us point out that (ii) always implies (i). So, informally, c is totally aspherical if the multiplication by e kills as many modules in $\mathcal{O}_c(W)$ as it theoretically can. On the other hand, there are totally aspherical parameters that are spherical. Those are the parameters where there are no $S(\mathfrak{h}^*)$-torsion modules in $\mathcal{O}_c(W)$. But this case is not interesting: the category $\mathcal{O}_c(W)$ is semisimple.

For example, we can take $W = \mathfrak{S}_n$ (the symmetric group). Here, c is a single complex number, and we will show that all parameters in the interval $(-1, 0)$ are totally aspherical. More generally, we will establish the existence of sufficiently many (in a suitable sense) of totally aspherical parameters for $W = G(\ell, 1, n) := \mathfrak{S}_n \ltimes (\mathbb{Z}/\ell\mathbb{Z})^n$.

We apply the notion of the total asphericity to relate two remarkable modules in $\mathcal{O}_c(W)$: the projective object P_{KZ} and what we call the Harish-Chandra module.

One can construct objects in $\mathcal{O}_c(W)$ as follows. We have the induction functor $\Delta_c : S(\mathfrak{h})\#W\text{-mod}_{fd,ln} \to \mathcal{O}_c(W)$ from the category of the finite dimensional $S(\mathfrak{h})\#W$-modules with locally nilpotent $\mathfrak{h}$-action. It is defined by $\Delta_c(M) := H_c(W) \otimes_{S(\mathfrak{h})\#W} M$. For example, take an irreducible W-module λ; it can be viewed as an $S(\mathfrak{h})\#W$-module with zero action of $\mathfrak{h}$. The object $\Delta_c(\lambda) \in \mathcal{O}_c(W)$ is called a Verma module. It has a unique simple quotient to be denoted by $L_c(\lambda)$. The objects $L_c(\lambda)$ form a complete collection of simples in $\mathcal{O}_c(W)$. Another interesting example is for $M = S(\mathfrak{h})/(S(\mathfrak{h})^W_+)$, where $S(\mathfrak{h})^W_+$ is the ideal in $S(\mathfrak{h})^W$ of all polynomials without constant term and $(S(\mathfrak{h})^W_+)$ stands for the ideal in $S(\mathfrak{h})$ generated by $S(\mathfrak{h})^W_+$. Recall that M is a graded module isomorphic to the regular representation of W. The module $\mathfrak{H}_c := \Delta_c(M)$ will be called the *Harish-Chandra module*, by analogy with a D-module on a semisimple Lie algebra; see [HK]. A question one can ask is whether this natural module has any nice categorical properties, for example, whether it is projective.

On the other hand, in [GGOR], the authors introduced a crucial tool to study the category $\mathcal{O}_c(W)$, the KZ (Knizhnik-Zamolodchikov) functor. This is a functor $\mathrm{KZ} : \mathcal{O}_c(W) \to \mathcal{H}_q(W)$-mod, where we write $\mathcal{H}_q(W)$ for the Hecke algebra of the group W corresponding to a parameter q computed from c (we do not need a precise formula at this point). The functor KZ is exact. On the level of vector spaces, to a module $M \in \mathcal{O}_c(W)$, this functor assigns the fiber M_x of M at a generic point $x \in \mathfrak{h}$. So KZ is given by $\mathrm{Hom}_{\mathcal{O}_c(W)}(P_{\mathrm{KZ},c}, \bullet)$, where $P_{\mathrm{KZ},c}$ is a projective object equipped with an epimorphism $\mathcal{H}_q(W) \twoheadrightarrow \mathrm{End}_{\mathcal{O}_c(W)}(P_{\mathrm{KZ},c})^{opp}$ (which is an isomorphism if and only if $\dim \mathcal{H}_q(W) = |W|$[1]). The object $P_{\mathrm{KZ},c}$ is decomposed as

$$\bigoplus_{\lambda \in \mathrm{Irr}(W)} P_c(\lambda)^{\mathsf{rk} L_c(\lambda)}. \tag{1.1}$$

Here, we write $P_c(\lambda)$ for the projective cover of $L_c(\lambda)$, and rk stands for the generic rank over $S(\mathfrak{h}^*)$ (=the dimension of a general fiber). The object $P_{KZ,c}$ has very nice categorical properties, for example, its indecomposable summands are precisely the indecomposable projectives that are also injectives. On the other hand, it is difficult to construct $P_{KZ,c}$ explicitly.

Around 2005, Ginzburg and, independently, Rouquier asked the question when $P_{KZ,c}$ is isomorphic to $\mathfrak{H}_c$ (unpublished). In this paper, we establish the following criterion for $P_{KZ,c} \cong \mathfrak{H}_c$.

Theorem 1.2 *The following two conditions are equivalent:*

(a) $P_{KZ,c} \cong \mathfrak{H}_c$.
(b) *c is totally aspherical.*

Corollary 1.3 *For $W = \mathfrak{S}_n$, we have $P_{KZ,c} \cong \mathfrak{H}_c$ for $c \in (-1, 0)$.*

This paper is organized as follows. We start Sect. 2 by recalling some generalities on the rational Cherednik algebras and their categories $\mathcal{O}$. Then we investigate some properties of $\mathcal{O}_c(W)$ for a totally aspherical parameter c and use those properties to prove Theorem 1.2. We finish the section by recalling a general conjecture on the locus, where the equivalent conditions of Theorem 1.2 hold.

In Sect. 3, we study the case of the groups $W = G(\ell, 1, n)$. We first prove Corollary 1.3. Then we extend it to the groups $G(\ell, 1, n)$ for $\ell > 1$, Theorem 3.1. The proof of this theorem is based on results from [BL].

[1] After this paper was written, Etingof, [E], checked that this is the case for all complex reflection groups W.

2 P_{KZ} vs Harish-Chandra Module

2.1 *Reminder on Cherednik Algebras*

Let W be a complex reflection group and $\mathfrak{h}$ its reflection representation. We write S for the set of reflections in W. For $s \in S$, we choose elements $\alpha_s \in \mathfrak{h}^*, \alpha_s^\vee \in \mathfrak{h}$ such that $s\alpha_s = \lambda_s s, s\alpha_s^\vee = \lambda_s^{-1}\alpha_s^\vee$ with $\lambda_s \in \mathbb{C} \setminus \{1\}$ and $\langle \alpha_s, \alpha_s^\vee \rangle = 2$.

Let $c : S \to \mathbb{C}$ be a function constant on the conjugacy classes. By definition, [EG, Section 1.4], [GGOR, Section 3.1], the rational Cherednik algebra $H_c(= H_c(W) = H_c(W, \mathfrak{h}))$ is the quotient of the skew-group algebra $T(\mathfrak{h} \oplus \mathfrak{h}^*)\#W$ by the following relations:

$$[x, x'] = [y, y'] = 0, [y, x] = \langle y, x \rangle - \sum_{s \in S} c(s)\langle x, \alpha_s^\vee \rangle \langle y, \alpha_s \rangle s, \quad x, x' \in \mathfrak{h}^*, y, y' \in \mathfrak{h}.$$

Let us recall some structural results about H_c. The algebra H_c is filtered (say, with $\deg \mathfrak{h} = \deg \mathfrak{h}^* = 1, \deg W = 0$), and its associated graded is $S(\mathfrak{h} \oplus \mathfrak{h}^*)\#W$, [EG, Section 1.2]. This yields the triangular decomposition $H_c = S(\mathfrak{h}^*) \otimes \mathbb{C}W \otimes S(\mathfrak{h})$, [GGOR, Section 3.2].

Consider the element $\delta := \prod_s \alpha_s^{\ell_s} \in S(\mathfrak{h}^*)^W$, where ℓ_s stands for the order of s so that the element δ is W-invariant. Since $\operatorname{ad}\delta$ is locally nilpotent, the quotient $H_c[\delta^{-1}]$ is well-defined. There is a natural isomorphism $H_c[\delta^{-1}] \cong D(\mathfrak{h}^{reg})\#W$, [EG, Section 1.4], [GGOR, Section 5.1].

Consider the averaging idempotent $e := |W|^{-1} \sum_{w \in W} w \in \mathbb{C}W \subset H_c$. The *spherical subalgebra* of H_c, by definition, is eH_ce. When the algebras eH_ce and H_c are Morita equivalent (automatically, via the bimodule H_ce), we say that the parameter c is *spherical*. Otherwise, we say that c is *aspherical*.

There is an *Euler element* $h \in H_c$ satisfying $[h, x] = x, [h, y] = -y, [h, w] = 0$. So the operator $[h, \cdot]$ defines a grading on H_c to be called the Euler grading. The Euler element is constructed as follows. Pick a basis $y_1, \ldots, y_n \in \mathfrak{h}$, and let $x_1, \ldots, x_n \in \mathfrak{h}^*$ be the dual basis. Then

$$h = \sum_{i=1}^n x_i y_i + \frac{n}{2} - \sum_{s \in S} \frac{2c(s)}{1 - \lambda_s} s. \tag{2.1}$$

2.2 *Reminder on Categories $\mathcal{O}$*

Following [GGOR, Section 3.2], we consider the full subcategory $\mathcal{O}_c(W)$ of H_c-mod consisting of all modules M that are finitely generated over $S(\mathfrak{h}^*)$ and with locally nilpotent action of $\mathfrak{h}$. Equivalently, we can require that the modules in $\mathcal{O}_c(W)$ are finitely generated over H_c (and $\mathfrak{h}$ still acts locally nilpotently). In the category $\mathcal{O}_c(W)$, we have analogs of Verma modules that are indexed by the irreducible representations λ of W. By definition, the Verma module indexed by λ is $\Delta_c(\lambda) := H_c \otimes_{S(\mathfrak{h})\#W} \lambda$, where $\mathfrak{h}$ acts by 0 on λ. It is easy to see that this module is in $\mathcal{O}_c(W)$. Using the Euler element, one shows that $\Delta_c(\lambda)$ has a unique simple quotient to be denoted by $L_c(\lambda)$. The objects $L_c(\lambda)$ form a complete collection of simples in $\mathcal{O}_c(W)$.

Often we drop W from the notation and just write $\mathcal{O}_c$. Let us note that all finite dimensional modules lie in $\mathcal{O}_c$, thanks to the presence of the Euler element.

Definition 2.1

- To a module $M \in \mathcal{O}_c$, we can assign its *associated variety* $\mathrm{V}(M)$ that, by definition, is the support of M (as a coherent sheaf) in $\mathfrak{h}$. Clearly, $\mathrm{V}(M)$ is a closed W-stable subvariety.
- The dimension of the variety $\mathrm{V}(M)$ is called the *Gelfand-Kirillov* (shortly, GK) dimension of M.
- Finally to a module M in $\mathcal{O}_c$, one can assign its *generic rank*, $\mathsf{rk} M$, the generic rank of M viewed as an $S(\mathfrak{h}^*)$-module.

The category $\mathcal{O}_c$ has enough projective objects; see, for example, [GGOR, Theorem 2.19]. For $\lambda \in \operatorname{Irr} W$, let $P_c(\lambda)$ denote the projective cover of $L_c(\lambda)$. Following [GGOR], we consider the projective object P_{KZ} defined by

$$P_{\mathrm{KZ}} = \bigoplus_{\lambda \in \operatorname{Irr} W} P_c(\lambda)^{\mathsf{rk} L_c(\lambda)}.$$

We write $P_{\mathrm{KZ},c}$ when we want to indicate the parameter. The importance of this projective is that it defines a quotient functor from $\mathcal{O}_c$ to the category of modules over a suitable Hecke algebra.

We will also need the category $\mathcal{O}$ over eH_ce (to be denoted by $\mathcal{O}_c^{sph}$); see [GL, Section 3]. By definition, this category consists of all finitely generated eH_ce-modules N such that

(1) $ehe(= he = eh)$ acts on N locally finitely.
(2) the sum of the positive graded components of eH_ce (with respect to the Euler grading) acts locally nilpotently on N.

The following lemma is a consequence of [GL, Proposition 3.2.1].

Lemma 2.2 *The functor $M \mapsto eM$ is a quotient functor $\mathcal{O}_c \to \mathcal{O}_c^{sph}$. The left adjoint and right inverse is given by $N \mapsto H_ce \otimes_{eH_ce} N$.*

For $N \in \mathcal{O}_c^{sph}$, we define the generic rank, $\mathsf{rk} N$, as the generic rank of N viewed as an $S(\mathfrak{h}^*)^W$-module. Note that $\mathsf{rk} N = \mathsf{rk}\, eN$.

We have naive duality functors $\mathcal{O}_c(W, \mathfrak{h}) \xrightarrow{\sim} \mathcal{O}_{c^*}(W, \mathfrak{h}^*)$, $\mathcal{O}_c^{sph}(W, \mathfrak{h}) \xrightarrow{\sim} \mathcal{O}_{c^*}^{sph}(W, \mathfrak{h}^*)$, where $c^*(s) := c(s^{-1})$; see [GGOR, 4.2] for the former. Namely, consider the isomorphism $\varphi : H_c(W, \mathfrak{h}) \to H_{c^*}(W, \mathfrak{h}^*)^{opp}$ that is the identity on $\mathfrak{h}, \mathfrak{h}^*$ and maps w to w^{-1}, $w \in W$. Let us note that φ restricts to an isomorphism

$eH_c(W,\mathfrak{h})e \xrightarrow{\sim} (eH_{c^*}(W,\mathfrak{h}^*)e)^{opp}$. For $M \in \mathcal{O}_c(W,\mathfrak{h})$, the restricted dual $M^\vee$ (the sum of the duals of the generalized eigenspaces for h) is a left $H_{c^*}(W,\mathfrak{h}^*)$-module that lies in the category $\mathcal{O}$. The functor $M \mapsto M^\vee$ is the duality functor that we need. A duality between the spherical categories is defined in a similar fashion.

Lemma 2.3 *The following claims are true.*

(1) *The duality $\mathcal{O}_c(W,\mathfrak{h}) \xrightarrow{\sim} \mathcal{O}_{c^*}(W,\mathfrak{h}^*)$ preserves the GK dimensions and generic ranks.*
(2) *A similar claim holds for the spherical categories $\mathcal{O}$.*
(3) *Moreover, the duality functors $\mathcal{O}_c(W,\mathfrak{h}) \xrightarrow{\sim} \mathcal{O}_{c^*}(W,\mathfrak{h}^*)$, $\mathcal{O}_c^{sph}(W,\mathfrak{h}) \xrightarrow{\sim} \mathcal{O}_{c^*}^{sph}(W,\mathfrak{h}^*)$ intertwine the functors of multiplication by e.*

Proof We can read the GK dimension and the GK multiplicity of a graded $S(\mathfrak{h}^*)$-module from the dimensions of the graded components (by the definitions of those invariants). Note that the GK multiplicity is the same as the generic rank.

Choose a section $\sigma : \mathbb{C}/\mathbb{Z} \to \mathbb{C}$. Introduce a grading on $M \in \mathcal{O}_c$ by declaring that the degree of the generalized eigenspace for h with eigenvalue α is $\alpha - \sigma(\alpha)$. Then we get the grading on $M^\vee$ compatible with the Euler grading on $H_{c^*}(W,\mathfrak{h}^*)$ and such that $\dim M_n^\vee = \dim M_n$. This completes the proof of (1). The proof of (2) is similar.

The claim that $(eM)^\vee = e(M^\vee)$ follows directly from the construction. □

We will also need induction and restriction functors for categories $\mathcal{O}_c$ introduced in [BE]. Let W' be a parabolic subgroup of W. Then we have an exact functor $\mathrm{Res}_W^{W'} : \mathcal{O}_c(W) \to \mathcal{O}_c(W')$. The target category is the category $\mathcal{O}$ for $H_c(W')$, where the parameter is obtained by restricting c to $W' \cap S$. We will need two facts about the restriction functor:

(a) If W' is the stabilizer of a generic point in an irreducible component of $\mathrm{V}(M)$, then $\mathrm{Res}_W^{W'}(M)$ is a nonzero finite dimensional module.
(b) The functor $\mathrm{Res}_W^{W'}$ induces a functor between the quotients, $\mathcal{O}_c^{sph} \to \mathcal{O}_c^{sph}(W')$.

(a) was established in [BE, Section 3.8], while (b) is in [GL, Section 3.5].

2.3 Totally Aspherical Parameters

Recall that we call a parameter c *totally aspherical* if $eM = 0$ for all $S(\mathfrak{h}^*)$-torsion modules $M \in \mathcal{O}_c(W)$. Equivalently, c is totally aspherical if and only if $\mathrm{V}(N) = \mathfrak{h}/W$ for all $N \in \mathcal{O}_c^{sph}(W)$. Here and below, we write $\mathrm{V}(N)$ for the associated variety of $N \in \mathcal{O}_c^{sph}(W)$ in $\mathfrak{h}/W$.

Lemma 2.4 *If c is totally aspherical, then all modules in $\mathcal{O}_c^{sph}(W)$ are free over $S(\mathfrak{h}^*)^W$.*

Proof Repeating the argument from [EGL, Section 3.2], we see that all modules in $\mathcal{O}_c^{sph}(W)$ are Cohen-Macaulay over $S(\mathfrak{h}^*)^W$. Since all modules are torsion-free over $S(\mathfrak{h}^*)^W$, this implies that they are projective over $S(\mathfrak{h}^*)^W$. But they are also graded and hence are free. □

Corollary 2.5 *Let c be a totally aspherical parameter, and let $M \in \mathcal{O}_c^{sph}(W)$. Then*

$$\dim M^{S(\mathfrak{h})_+^W} = \mathsf{rk} M,$$

where we write $M^{S(\mathfrak{h})_+^W}$ for the subspace $\{m \in M | S(\mathfrak{h})_+^W m = 0\}$.

Proof Recall the duality $\bullet^\vee : \mathcal{O}_c^{sph}(W, \mathfrak{h}) \xrightarrow{\sim} \mathcal{O}_{c^*}^{sph}(W, \mathfrak{h}^*)$ explained in Sect. 2.2. The parameter c^* is totally aspherical for $H_{c^*}(W, \mathfrak{h}^*)$; this follows from Lemma 2.3. So $\dim M^\vee / S(\mathfrak{h})_+^W M^\vee = \mathsf{rk} M^\vee = \mathsf{rk} M$, the first equality follows from Lemma 2.4 and the second one from Lemma 2.3. But $\dim M^{S(\mathfrak{h})_+^W} = \dim M^\vee / S(\mathfrak{h})_+^W M^\vee$ by the construction of $\bullet^\vee$. □

2.4 *Harish-Chandra Module vs P_{KZ}*

In this section, we prove Theorem 1.2.

Proof of Theorem 1.2 For $M \in \mathcal{O}_c(W)$, we have

$$\mathrm{Hom}_{\mathcal{O}_c(W)}(\mathfrak{H}_c, M) \cong \mathrm{Hom}_{S(\mathfrak{h})\#W}(S(\mathfrak{h})/(S(\mathfrak{h})_+^W), M) \cong e(M^{S(\mathfrak{h})_+^W}) = (eM)^{S(\mathfrak{h})_+^W}. \tag{2.2}$$

Let us prove the implication $(b) \Rightarrow (a)$. It follows from Corollary 2.5 that

$$\dim \mathrm{Hom}_{\mathcal{O}_c(W)}(\mathfrak{H}_c, M) = \mathsf{rk}(eM).$$

The right-hand side coincides with $\mathsf{rk}(M)$ which, in turn, equals $\dim \mathrm{Hom}_{\mathcal{O}_c(W)}(P_{KZ,c}, M)$. So we conclude that

$$\dim \mathrm{Hom}_{\mathcal{O}_c(W)}(\mathfrak{H}_c, M) = \dim \mathrm{Hom}_{\mathcal{O}_c(W)}(P_{KZ,c}, M). \tag{2.3}$$

Since $P_{KZ,c}$ is projective, equivalently, the functor $\mathrm{Hom}_{\mathcal{O}_c(W)}(P_{KZ,c}, \bullet)$ is exact; we see that the functor $\mathrm{Hom}_{\mathcal{O}_c(W)}(\mathfrak{H}_c, \bullet)$ is exact; equivalently, $\mathfrak{H}_c$ is projective. Equation (2.3) now implies $\mathfrak{H}_c \cong P_{KZ,c}$.

Let us prove (a)⇒(b). Assume there is a simple object $M \in \mathcal{O}_c(W)$ that is $S(\mathfrak{h}^*)$-torsion (equivalently, $\mathrm{Hom}_{\mathcal{O}_c(W)}(P_{KZ,c}, M) = 0$) such that $eM \neq 0$. Since $S(\mathfrak{h})_+^W$ acts locally nilpotently, we see that $(eM)^{S(\mathfrak{h})_+^W} \neq 0$. Thanks to (2.2), we see that $\mathrm{Hom}_{\mathcal{O}_c(W)}(\mathfrak{H}_c, M) \neq 0$. So $\mathfrak{H}_c \ncong P_{KZ,c}$. □

Remark 2.6 Note that if c is not totally aspherical, then $\mathfrak{H}_c$ is not projective. Assume the contrary. We have

$$\dim \operatorname{Hom}_{\mathcal{O}_c(W)}(\mathfrak{H}_c, M) = \dim(eM)^{S(\mathfrak{h})^W_+} = \dim(eM)^\vee / S(\mathfrak{h})^W_+(eM)^\vee$$
$$\geqslant \mathsf{rk}(eM)^\vee = \mathsf{rk}(eM) = \mathsf{rk}(M).$$

So, for a torsion-free $L_c(\lambda)$, we see that the multiplicity of $P_c(\lambda)$ in $\mathfrak{H}_c$ is not less than the multiplicity of $P_c(\lambda)$ in $P_{\mathrm{KZ},c}$. But both $P_{\mathrm{KZ},c}$ and $\mathfrak{H}_c$ have a filtration of length $\sum_{\lambda\in \operatorname{Irr} W} \dim \lambda$ with Verma quotients. This length is independent of the choice of the filtration and hence $\mathfrak{H}_c \cong P_{\mathrm{KZ},c}$. Contradiction.

2.5 Total Asphericity vs Simplicity

Here, we are going to provide one more equivalent formulation of total asphericity.

Proposition 2.7 *A parameter c is totally aspherical if and only if the algebra eH_ce is simple.*

Let us introduce the following condition on c.

(*) $e_{W'}M' = 0$ for any parabolic subgroup $W' \subset W$ different from $\{1\}$ and any finite dimensional $H_c(W')$-module M'. Here, we write $e_{W'}$ for the averaging idempotent in $\mathbb{C}W'$.

The proposition follows from the next two lemmas.

Lemma 2.8 *The parameter c is totally aspherical if and only if (*) holds.*

This lemma will also be important in Sect. 3 to determine aspherical values for the groups $W = G(\ell, 1, n)$.

Lemma 2.9 *The algebra eH_ce is simple if and only if (*) holds.*

Proof of Lemma 2.8 Recall the restriction functors $\operatorname{Res}^{W'}_W : \mathcal{O}_c(W) \to \mathcal{O}_c(W'), \mathcal{O}^{sph}_c(W) \to \mathcal{O}^{sph}_c(W')$ defined in [BE, Section 3.5] and [GL, Section 3.5]. We have $\operatorname{Res}^{W'}_W(e\bullet) \cong e_{W'}\operatorname{Res}^{W'}_W(\bullet)$; see [GL, Section 3.5.2]. Also for any nonzero $M \in \mathcal{O}_c(W)$, there is W' such that $\operatorname{Res}^{W'}_W(M)$ is nonzero finite dimensional; see [BE, Section 3.8]. So if $e_{W'}M' = 0$ for any parabolic subgroup W' and any finite dimensional module $M' \in \mathcal{O}_c(W')$, then c is totally aspherical for $H_c(W)$.

Now suppose that c is totally aspherical. Recall the functor $\operatorname{Ind}^W_{W'}$ from [BE, Section 3.5], a right adjoint of $\operatorname{Res}^{W'}_W$. It was proved in [S] (under some additional assumptions on W) and later in [L3] in the full generality that $\operatorname{Ind}^W_{W'}$ is also a left adjoint of $\operatorname{Res}^{W'}_W$. Let N be a finite dimensional $e_{W'}H_c(W')e_{W'}$-module. The $H_c(W')$ module $\tilde{N} := H_c(W')e_{W'} \otimes_{e_{W'}H_c(W')e_{W'}} N$ is also finite dimensional because $H_c(W')e_{W'}$ is finitely generated over $e_{W'}H_c(W')e_{W'}$. The module

$\mathrm{Ind}_{W'}^{W}(\tilde{N})$ is nonzero and has proper associated variety by Shan and Vasserot [SV, Proposition 2.7]. So $e\,\mathrm{Ind}_{W'}^{W}(\tilde{N}) = 0$. The module $\mathrm{Res}_{W}^{W'} \circ \mathrm{Ind}_{W'}^{W}(\tilde{N})$ is not killed by $e_{W'}$ because it admits a nonzero homomorphism to $\tilde{N}$. On the other hand,

$$e_{W'}\,\mathrm{Res}_{W}^{W'} \circ \mathrm{Ind}_{W'}^{W}(\tilde{N}) = \mathrm{Res}_{W}^{W'}(e\,\mathrm{Ind}_{W'}^{W}(\tilde{N})) = 0.$$

This contradiction completes the proof. □

Proof of Lemma 2.9 Let $\mathcal{J} \subset eH_ce$ be a proper two-sided ideal. Its associated variety in $(\mathfrak{h} \oplus \mathfrak{h}^*)/W$, i.e., the subvariety defined by $\mathrm{gr}\,\mathcal{J}$, is proper. Pick a generic point (with stabilizer, say, W') in an irreducible component of the associated variety of $\mathcal{J}$, and consider the functor $\bullet_{\dagger}$ from [L1] at that point. Then $\mathcal{J}_{\dagger}$ is a proper ideal of finite codimension in $e_{W'}H_c(W')e_{W'}$, by a direct analog of [L1, Theorem 3.4.6] for spherical subalgebras (that holds with the same proof). So (*) does not hold.

Now assume that $e_{W'}H_c(W')e_{W'}$ has a finite dimensional representation. Let $\mathcal{I}$ be the annihilator, and set $\mathcal{B} := e_{W'}H_c(W')e_{W'}/\mathcal{I}$. Then we apply (the spherical version of) the functor $\bullet^{\dagger}$ associated to W' from [L1, Section 3.7] to $\mathcal{B}$ and get a Harish-Chandra H_c-bimodule $\mathcal{B}^{\dagger}$ with a homomorphism $eH_c(W)e \to \mathcal{B}^{\dagger}$. The kernel is a proper two-sided ideal; its associated variety is the image of $(\mathfrak{h} \oplus \mathfrak{h}^*)^{W'}$ in $(\mathfrak{h} \oplus \mathfrak{h}^*)/W$. So eH_ce is not simple. □

2.6 Shifts

There is a conjecture of Rouquier saying that there are "sufficiently many" totally aspherical parameters. Namely, let us write $\mathfrak{p}$ for the space of the W-invariant functions $c : S \to \mathbb{C}$. Let us write c_s for a function that maps $s' \in S$ to 1 if s' is conjugate to s and to 0 else. For a reflection hyperplane H and $i \in \{0, \ldots, \ell_H - 1\}$ (where ℓ_H is the cardinality of the pointwise stabilizer W_H) define an element $h_{H,i} \in \mathfrak{p}^*$ by

$$h_{H,i}(c) = \frac{1}{\ell_H} \sum_{s \in W_H \setminus \{1\}} \frac{2c_s}{\lambda_s - 1} \lambda_s^{-i} \tag{2.4}$$

The elements $h_{H,i}$ with $i \neq 0$ form a basis in $\mathfrak{p}^*$ and $\sum_{i=0}^{\ell_H - 1} h_{H,i} = 0$. Let $\mathfrak{p}^*_{\mathbb{Z}}$ denote the integral lattice in $\mathfrak{p}^*$ spanned by the elements $h_{H,i} - h_{H,0}$, and let $\mathfrak{p}_{\mathbb{Z}}$ be dual to $\mathfrak{p}^*_{\mathbb{Z}}$. For example, when $W = \mathfrak{S}_n$, we get $h_{H,1} - h_{H,0} = c$, and the lattice $\mathfrak{p}_{\mathbb{Z}}$ consists of integers.

In general, the meaning of the lattice $\mathfrak{p}_{\mathbb{Z}}$ is as follows. Recall that from a Cherednik parameter c, one can produce a parameter q for the Hecke algebra of W. Two Cherednik parameters c_1, c_2 produce the same Hecke parameter if $c_1 - c_2 \in \mathfrak{p}_{\mathbb{Z}}$.

It was conjectured by Rouquier (unpublished) that every coset $c + \mathfrak{p}_{\mathbb{Z}}$ contains a parameter c with $P_{KZ,c} = \mathfrak{H}_c$, i.e., (see Theorem 1.2) a totally aspherical parameter. In the next section, we will prove a theorem that implies this conjecture for the groups $W = G(\ell, 1, n)$.

3 Cyclotomic Case

3.1 *Type A*

Here, we are going to prove Corollary 1.3. Our proof is based on Lemma 2.8.

Proof of Corollary 1.3 Recall that the algebra $H_c(\mathfrak{S}_m)$ has a finite dimensional representation if and only if c is rational with denominator m, and in this case, there is a unique finite dimensional irreducible representation, and every finite dimensional representation is completely reducible; see [BEG]. Further, by *loc.cit.*, the finite dimensional irreducible representation is annihilated by e if and only if $c \in (-1, 0)$. All parabolic subgroups of $\mathfrak{S}_n$ are conjugate to ones of the form $\mathfrak{S}_{n_1} \times \ldots \times \mathfrak{S}_{n_k}$ with $\sum_{i=1}^k n_i \leqslant n$. Now Corollary 1.3 follows from Lemma 2.8. $\square$

Let us point out that the totally aspherical parameters outside $(-1, 0)$ are either irrational or have denominator bigger than n (and so the category $\mathcal{O}_c(\mathfrak{S}_n)$ is semisimple).

3.2 *Quantized Quiver Varieties*

We proceed to the case of the groups $G(\ell, 1, n)$. The spherical subalgebras in this case can be realized as quantized quiver varieties (i.e., as Hamiltonian reductions of the algebras of differential operators on spaces of quiver representations). In this subsection, we recall how this is done.

Let $Q = (Q_0, Q_1, t, h)$ be the affine Dynkin quiver of type A with ℓ vertices (and some orientation). We label the vertices cyclically by numbers $0, \ldots, \ell - 1$ with 0 corresponding to the extending vertex. Let $\epsilon_i \in \mathbb{C}^{Q_0}$ denote the coordinate vector (=the simple root) at the vertex i and $\delta = \sum_{i \in Q_0} \epsilon_i$ be the indecomposable imaginary root. Consider the representation space $R = R(Q, v, \epsilon_0)$ with dimension vector $v = (v_i)_{i \in Q_0}$ and one-dimensional co-framing $w = \epsilon_0$ at the extending vertex, explicitly, $R = \bigoplus_{a \in Q_1} \mathrm{Hom}(V_{t(a)}, V_{h(a)}) \oplus V_0^*$, where $\dim V_i = v_i$. On the space R, we have a natural action of the group $G = \prod_{i \in Q_0} \mathrm{GL}(V_i)$.

To the action of G on R, we can associate several varieties/algebras obtained by Hamiltonian reduction. Let us recall the construction. We have a moment map $\mu : T^*R \to \mathfrak{g}^*$ defined as follows. Its dual map $\mu^* : \mathfrak{g} \to \mathbb{C}[T^*R]$ sends $x \in \mathfrak{g}$

to the velocity vector field $x_R \in \mathrm{Vect}(R)$ viewed as an element of $\mathbb{C}[T^*R]$ via a natural inclusion $\mathrm{Vect}(R) \hookrightarrow \mathbb{C}[T^*R]$. Consider the variety $X^0(v) := T^*R///_0 G(= \mu^{-1}(0)//G)$. When $v_i = n$ for all i (i.e., $v = n\delta$), this variety is identified with $(\mathfrak{h} \oplus \mathfrak{h}^*)/W$, where $W = G(\ell, 1, n)$; see, e.g., [Go2, 7.7]. The identification can be made equivariant with respect to the $\mathbb{C}^\times$-actions by dilations: the action on $\mu^{-1}(0)//G$ is induced from the action on T^*R by dilations.

Next, take a generic enough character θ of the group G. We can form the GIT reduction $X^\theta(v) := T^*R///_0^\theta G(= \mu^{-1}(0)^{\theta-ss}//G)$. This is a smooth symplectic variety equipped with a natural morphism $X^\theta(v) \to X^0(v)$. When $v = n\delta$, this morphism is a resolution of singularities; see, e.g., [Go2, 7.8].

One can also quantize the variety $X^0(n\delta)$ using the quantum Hamiltonian reduction. Define the symmetrized quantum comoment map $\Phi(x) := \frac{1}{2}(x_R + x_{R^*}) : \mathfrak{g} \to D(R)$, where $\mathfrak{g}$ stands for the Lie algebra of G and $D(R)$ for the algebra of linear differential operators on R. Note that $\Phi(x)$ does not depend on the orientation of Q. Pick a collection $\lambda = (\lambda_0, \ldots, \lambda_{\ell-1})$ of complex numbers, and view it as a character of $\mathfrak{g}$ via $\langle \lambda, x\rangle := \sum_{i=0}^{\ell-1} \lambda_i \operatorname{tr}(x_i)$. Then set

$$\mathcal{A}_\lambda(n\delta) := D(R)///_\lambda G(= [D(R)/D(R)\{\Phi(x) - \langle \lambda, x\rangle\}]^G).$$

This is a filtered algebra with $\operatorname{gr} \mathcal{A}_\lambda(n\delta) = \mathbb{C}[X^0(n\delta)]$. Moreover, we have a filtered algebra isomorphism $\mathcal{A}_\lambda(n\delta) \cong eH_c(W)e$, where one recovers λ from c as follows. We can encode the parameter c as $(\kappa, c_1, \ldots, c_{\ell-1})$, where $\kappa \in \mathbb{C}$ is the parameter corresponding to a reflection in $\mathfrak{S}_n$ and c_i corresponds to $i \in \mathbb{Z}/\ell\mathbb{Z}$.

According to [Go1] (see also [L2, Section 6.2]), we get

$$\begin{aligned}\lambda_k &= \frac{1}{\ell}(1 + 2\sum_{j=1}^{\ell-1} c_j \exp(2\pi\sqrt{-1}j/\ell)), k \neq 0,\\ \lambda_0 &= \frac{1}{\ell}(1 + 2\sum_{j=1}^{\ell-1} c_j) + \kappa - \frac{1}{2}.\end{aligned} \tag{3.1}$$

We can also produce quantizations of $X^\theta(v)$ using quantum Hamiltonian reduction. Namely, we microlocalize the algebra $D(R)$ to a sheaf (in the conical topology) D_R of filtered algebras on T^*R. Then we set

$$\mathcal{A}_\lambda^\theta(v) := D_R///_\lambda^\theta G(= [D_R/D_R\{\Phi(x) - \langle \lambda, x\rangle\}|_{(T^*R)^{\theta-ss}}]^G)$$

(note that this definition differs from [BL] by a shift of λ because here we use the symmetrized quantum comoment map). This is a sheaf of filtered algebras on $X^\theta(v)$ with $\operatorname{gr} \mathcal{A}_\lambda^\theta(v) = \mathcal{O}_{X^\theta(v)}$ and $\Gamma(\mathcal{A}_\lambda^\theta(n\delta)) = \mathcal{A}_\lambda(n\delta)$. In general, we set $\mathcal{A}_\lambda(v) := \Gamma(\mathcal{A}_\lambda^\theta(v))$; the right-hand side does not depend on the choice of θ by Braden et al. [BPW, Section 3.3]. We also set $\mathcal{A}_\lambda^0(v) = [D(R)/D(R)\{\Phi(x) - \langle \lambda, x\rangle\}]^G$, $X(v) := \operatorname{Spec}(\mathbb{C}[X^\theta(v)])$. Note that there is a natural homomorphism $\mathcal{A}_\lambda^0(v) \to \mathcal{A}_\lambda(v)$. Let

ρ_v denote the natural morphism $X^\theta(v) \to X(v)$; it is a resolution of singularities for any v. We often write ρ instead of ρ_v.

We can consider the derived global section functor

$$R\Gamma : D^b(\mathcal{A}^\theta_\lambda(v)\text{-mod}) \to D^b(\mathcal{A}_\lambda(v)\text{-mod}).$$

Here, we write $\mathcal{A}_\lambda(v)$-mod for the category of $\mathcal{A}_\lambda(v)$-modules and $\mathcal{A}^\theta_\lambda(v)$-mod for the category of all quasi-coherent sheaves of $\mathcal{A}^\theta_\lambda(v)$-modules (as usual, by a quasi-coherent sheaf, one means a sheaf that is the union of its coherent subsheaves, where "coherent" is the same as "good" in [MN, Section 4.5]). Note that $\mathcal{A}^\theta_\lambda(v)$-mod is the quotient of the category of the (G, λ)-equivariant $D(R)$-modules by the Serre category of all D-modules with θ-unstable support, [BL, Proposition 2.13].

The functor $R\Gamma$ restricts to $R\Gamma : D^b_{\rho^{-1}(0)}(\mathcal{A}^\theta_\lambda(v)\text{-mod}) \to D^b_{fin}(\mathcal{A}_\lambda(v)\text{-mod})$. Here, the superscript "$\rho^{-1}(0)$" indicates the category of all complexes with coherent homology supported on $\rho^{-1}(0)$. Similarly, the subscript "fin" means the category of all complexes with finite dimensional homology.

3.3 Main Result in the Cyclotomic Case

Let us state our main result. Define the subalgebra $\mathfrak{a}^\lambda \subset \mathfrak{g}(Q) = \hat{\mathfrak{sl}}_\ell$ as follows. It is generated by the Cartan subalgebra $\mathfrak{t}$ of $\mathfrak{g}(Q)$ and the root subspaces $\mathfrak{g}(Q)_\beta$ for real roots $\beta = \sum_{i \in Q_0} b_i \epsilon_i$ such that

$$\sum_{i \in Q_0} b_i \lambda_i + \frac{1}{2} b_0 \in \mathbb{Z}.$$

Theorem 3.1 *Suppose that $v = n\delta$ and*

(1) $|\langle \lambda, \beta \rangle| \leqslant \frac{1}{2}|b_0|$ *for all roots $\beta = (b_i)_{i \in Q_0}$ of $\mathfrak{a}^\lambda$,*
(2) *and $\langle \lambda, \delta \rangle - \frac{1}{2}$ is either in $(-1, 0)$ or an integer or has denominator bigger than n (irrational numbers have denominator $+\infty$, by convention).*

Then the parameter c computed from λ using (3.1) is totally aspherical.

Corollary 3.2 *Rouquier's conjecture mentioned in Sect. 2.6 is true for the groups $G(\ell, 1, n)$: every coset $c + \mathfrak{p}_\mathbb{Z}$ contains a totally aspherical parameter.*

Proof of Corollary 3.2 Note that, under the isomorphism $\mathfrak{p} \cong \mathbb{C}^{Q_0}$ given by (3.1), the lattice $\mathfrak{p}_\mathbb{Z}$ coincides with $\mathbb{Z}^{Q_0}$; a computation is somewhat implicitly contained, e.g., in [GL, 2.3.1]. So it is enough to show that every $\lambda \in \mathbb{C}^{Q_0}$ admits an integral shift satisfying the conditions of Theorem 3.1.

If $\mathfrak{a}^\lambda = \mathfrak{t}$, then the proof is easy; we just need to achieve condition (2). Assume $\mathfrak{a}^\lambda \neq \mathfrak{t}$. Choose a system of simple roots, $\beta_1, \ldots, \beta_k$, for $\mathfrak{a}^\lambda$ consisting of positive roots for $\mathfrak{g}(Q)$. We claim that it is enough to show that there is an element $\lambda' \in \lambda + \mathbb{Z}^{Q_0}$ satisfying

$$0 \leqslant \langle \lambda' + \omega_0^\vee/2, \beta_i \rangle \leqslant (\omega_0, \beta_i), \quad \forall i = 1, \ldots, k. \tag{3.2}$$

Here, $\omega_0^\vee$ denotes the fundamental coweight corresponding to 0, and $(\cdot, \cdot)$ stands for the usual invariant scalar product (so that $\langle \omega_0^\vee, \bullet \rangle = (\omega_0, \bullet)$). Since the β_i's are roots of $\mathfrak{a}^\lambda$, we see that $\langle \lambda' + \omega_0^\vee/2, \beta_i \rangle$ is an integer. Clearly, (3.2) implies (1). Now suppose the denominator d of $\langle \lambda + \omega_0^\vee/2, \delta \rangle$ satisfies $d \leqslant n$. Then $d\delta$ is integral on $\lambda + \omega_0^\vee/2$, and hence $(\omega_0, \beta_i) \in \{0, \ldots, d-1\}$. Indeed, $(\omega_0, \beta_i) \geqslant 0$ because β_i is positive, and if $(\omega_0, \beta_i) \geqslant d$, then $\beta_i - d\delta$ is a positive root of $\mathfrak{a}^\lambda$, and hence β_i is not a simple root for $\mathfrak{a}^\lambda$. It follows that $d\delta$ is the sum of some of β_i's and hence (3.2) also implies (2).

Now let us prove that (3.2) holds for some $\lambda' \in \lambda + \mathbb{Z}^{Q_0}$. For simplicity, consider the case when $k = \ell$, the general case ($k < \ell$) is analogous. We can apply an element from $\mathfrak{S}_\ell$ to λ; this does not change the inequalities (3.2). Thanks to this, we can assume that $\beta_i = \epsilon_i + d_i\delta, i = 1, \ldots, \ell$. Set $x_i := \langle \lambda' + \omega_0^\vee/2, \epsilon_i \rangle$. Then (3.2) can be rewritten as

$$0 \leqslant x_i + d_i(x_1 + \ldots + x_\ell) \leqslant d_i + \delta_{i,\ell}. \tag{3.3}$$

Set $s := x_1 + \ldots + x_\ell$ so that $s = \langle \lambda' + \frac{\omega_0^\vee}{2}, \delta \rangle$ and $x_i + d_i s$ is an integer for all i. From $\beta_i = \epsilon_i + d_i\delta, i = 1, \ldots, \ell$, and $d\delta = \sum_{i=1}^{\ell} \beta_i$, we deduce that $1 + \sum_{i=1}^{\ell} d_i = d$. The coset $s + \mathbb{Z}$ is determined by λ because changing λ by adding an element of $\mathbb{Z}^{Q_0}$ results in changing s by an integer. The number ds equals $\langle \lambda' + \omega_0^\vee/2, d\delta \rangle$ hence is an integer. The numbers d, ds are coprime because d is precisely the denominator of $\langle \lambda + \omega_0^\vee/2, \delta \rangle$ and hence of s.

Summing the inequalities (3.3), we get $0 \leqslant ds \leqslant d$. This specifies $s \in [0, 1]$ uniquely (with the exception of the non-interesting case $d = 1$). One can then rewrite (3.3) as

$$\begin{aligned} &0 \leqslant x_i + d_i s \leqslant d_i, \quad i = 1, \ldots, \ell - 1, \\ &0 \leqslant ds - \sum_{i=1}^{\ell-1}(x_i + d_i s) \leqslant d - \sum_{i=1}^{\ell-1} d_i \end{aligned} \tag{3.4}$$

It is now clear that these inequalities have a solution $(x_1, \ldots, x_\ell)$ in a given coset in $\mathbb{Q}^\ell/\mathbb{Z}^\ell$. □

Let us describe key ideas of the proof of Theorem 3.1. Lemma 2.8 reduces the problem to checking the absence of the finite dimensional representations. To show that $\mathcal{A}_\lambda(n\delta)$ has no finite dimensional representations, it is enough to prove that $R\Gamma(M) = 0$ for any $\mathcal{A}_\lambda^\theta(n\delta)$-module M supported on $\rho_{n\delta}^{-1}(0)$. The structure of the irreducible $\mathcal{A}_\lambda^\theta(v)$-modules supported on $\rho_v^{-1}(0)$ was studied in [BL]. We will see that under condition (1), the algebra $\mathcal{A}_\lambda(n\delta)$ has no finite dimensional representations; this is a key step. From here and results of Sect. 3.1, we will see that (1) and (2) guarantee the absence of the finite dimensional representations of the slice algebras (which in this case means the spherical subalgebras for the parabolic subgroups).

Let us discuss our key step in more detail. In [BL, Section 5.2], we have shown that the category $\bigoplus_v D^b(\mathcal{A}^\theta_{\lambda_v}(v)\text{-mod}_{\rho_v^{-1}(0)})$ carries a categorical action (in a weak sense) of $\mathfrak{a}^\lambda$ (here, the parameters λ_v depend on v in a suitable way; note that parameters λ we use in this paper differ from those in [BL] by a shift). This means that there are endo-functors E_α, F_α indexed by simple roots α of $\mathfrak{a}^\lambda$ that define a representation of $\mathfrak{a}^\lambda$ (in the usual sense) on K_0. Each pair E_α, F_α defines a categorical $\mathfrak{sl}_2$-action in the strong sense (of Chuang-Rouquier, [CR]; see also Rouquier, [R]). Moreover, E_α, F_α define an $\mathfrak{sl}_2$-crystal structure on the set of simples with crystal operators $\tilde{e}_\alpha, \tilde{f}_\alpha$. We will see that the condition $|\langle\lambda, \alpha\rangle| \leqslant \frac{1}{2}(\omega_0, \alpha)$ guarantees that we have $R\Gamma(M) = 0$ for a simple M lying in the image of $\tilde{f}_\alpha$. We will first do this when α is a simple root for $\mathfrak{g}(Q)$, Sect. 3.4, and then extend this result to the general case.

Remark 3.3 It does not seem that Theorem 3.1 describes all totally aspherical parameters: it should not be true that condition (1) specifies precisely the parameters where $\mathcal{A}_\lambda(n\delta)$ has no finite dimensional representations. [BL, Conjecture 9.8] reduces the question of when $\mathcal{A}_\lambda(n\delta)$ has no finite dimensional representation to a problem in the integrable representation theory of $\mathfrak{a}^\lambda$.

Remark 3.4 In the case when $n = 1$, all parameters c where $P_{KZ,c} \cong \mathfrak{H}_c$ were described in [T]: by [T, Theorem C], this isomorphism holds if and only if c is in the set $\mathcal{F}$ introduced in [T, Definition 4.4].

3.4 Functors E_i, F_i

An important role in the proof of the main theorem of [BL] is played by Webster's functors E_i, F_i between the derived categories of modules over the sheaves $\mathcal{A}^\theta_\lambda(v)$.

Fix $i \in Q_0$ such that $\lambda_i \in \mathbb{Z} + \frac{1}{2}(w_i + \sum_{a,t(a)=i} v_{h(a)} + \sum_{a,h(a)=i} v_{t(a)})$, and assume that $\theta_k > 0$ for all $k \in Q_0$. We are going to recall the functors

$$F_i : D^b(\mathcal{A}^\theta_\lambda(v)\text{-mod}) \rightleftarrows D^b(\mathcal{A}^\theta_{\lambda'}(v + \epsilon_i)\text{-mod}) : E_i$$

introduced in [W]. Here, λ' is a parameter recovered from λ (below we will explain how).

Assume i is a source (otherwise, we can invert some arrows; this does not change λ and the sheaf $\mathcal{A}^\theta_\lambda(v)$). Set $\tilde{w}_i := w_i + \sum_{a,t(a)=i} v_{h(a)}$. Consider the reduction $\mathcal{A}^{\theta_i}_{\lambda_i}(v, i) := D_R ///^{\theta_i}_{\lambda_i} \mathrm{GL}(V_i)$. Since i is a source and $\theta_i > 0$, this reduction is

$$D^{\tilde{\lambda}_i}_{\mathrm{Gr}(v_i,\tilde{w}_i)} \otimes D_{\underline{R}},$$

where the notation is as follows. We write $\underline{R}$ for

$$\bigoplus_{a,t(a)\neq i} \mathrm{Hom}(V_{t(a)}, V_{h(a)}) \oplus \bigoplus_{j\neq i} \mathrm{Hom}(V_j, \mathbb{C}^{w_j}),$$

where $w_j = \delta_{j0}$. So $R = \mathrm{Hom}(V_i, \mathbb{C}^{\tilde{w}_i}) \oplus \underline{R}$. The superscript $\tilde{\lambda}_i$ above indicates the differential operators with coefficients in the line bundle $\mathcal{O}(\tilde{\lambda}_i)$ on $\mathrm{Gr}(v_i, \tilde{w}_i)$, where $\tilde{\lambda}_i := \lambda_i - \tilde{w}_i/2$. The category $\mathcal{A}^{\theta}_{\lambda}(v)$-mod is the quotient of $D_R///^{\theta_i}_{\lambda_i} \mathrm{GL}(V_i)\text{-mod}^{\underline{G},\underline{\lambda}}$ by a Serre subcategory; the superscript $(\underline{G}, \underline{\lambda})$ means "$\underline{\lambda}$-twisted $\underline{G}$-equivariant modules." Here, we write $\underline{G}$ for $\prod_{j\neq i} \mathrm{GL}(V_i)$, $\underline{\lambda} := (\tilde{\lambda}_j)_{j\neq i}$. The formula for $\tilde{\lambda}_j$ is

$$\tilde{\lambda}_j := \lambda_j - \frac{1}{2}(w_j + \sum_{t(a)=j} v_{h(a)} - \sum_{h(a)=j} v_{t(a)}).$$

Note that $\mathcal{A}^{\theta}_{\lambda}(v)$ is the quantum Hamiltonian reduction of $D(R)$ for the quantum comoment map $x \mapsto x_R$ and level $\tilde{\lambda} = (\tilde{\lambda}_j)_{j\in Q_0}$. The Serre subcategory in $\mathcal{A}^{\theta_i}_{\lambda_i}(v, i)\text{-mod}^{\underline{G},\underline{\lambda}}$ we have to mod out consists of all modules whose singular support is contained in the image of $\mu_i^{-1}(0)^{\theta_i-ss} \setminus \mu^{-1}(0)^{\theta-ss}$ in $T^*R///^{\theta_i} \mathrm{GL}(V_i)$.

One can define functors, [W],

$$F_i : D^b_{\underline{G},\underline{\lambda}}(\mathcal{A}^{\theta_i}_{\lambda_i}(v, i)\text{-mod}) \rightleftarrows D^b_{\underline{G},\underline{\lambda}}(\mathcal{A}^{\theta_i}_{\lambda_i}(v + \epsilon_i, i)\text{-mod}) : E_i \tag{3.5}$$

in a standard way (pull-push), using the correspondence $\mathrm{Fl}(v_i, v_i+1, \tilde{w}_i)$ consisting of 2 step flags (or, more precisely, $\mathrm{Fl}(v_i, v_i + 1, \tilde{w}_i) \times \delta_{\underline{R}}$, where $\delta_{\underline{R}}$ stands for the diagonal in $\underline{R}^2$). Here, we write $D^b_{\underline{G},\underline{\lambda}}(\bullet)$ for the $(\underline{G}, \underline{\lambda})$-equivariant derived category in the sense of Bernstein and Lunts. Of course, the functors E_i, F_i make sense for the non-equivariant derived categories as well.

The category $D^b(\mathcal{A}^{\theta}_{\lambda}(v)\text{-mod})$ is the quotient of $D^b_{\underline{G},\underline{\lambda}}(\mathcal{A}^{\theta_i}_{\lambda_i}(v, i)\text{-mod})$ by the full subcategory of all complexes whose homology satisfies the support condition mentioned above; see [BL, Section 7.1].

As Webster checked in [W], the functors E_i, F_i descend to functors between the quotient categories $D^b(\mathcal{A}^{\theta}_{\lambda}(v)\text{-mod})$, $D^b(\mathcal{A}^{\theta}_{\lambda'}(v + \epsilon_i)\text{-mod})$. Here, we write λ' for the parameter producing the vector $\tilde{\lambda}$, but for the dimension $v + \epsilon_i$, we have $\lambda'_k = \lambda_k + n_{ik}/2$, where n_{ik} is the number of arrows connecting i to k.

Let us write $R\Gamma_i$ for the derived global section functor

$$D^b_{\underline{G},\underline{\lambda}}(\mathcal{A}^{\theta_i}_{\lambda_i}(v, i)\text{-mod}) \to D^b_{\underline{G},\underline{\lambda}}(\mathcal{A}_{\lambda_i}(v, i)\text{-mod}),$$

where $\mathcal{A}_{\lambda_i}(v,i) := \Gamma(\mathcal{A}^{\theta_i}_{\lambda_i}(v,i))$. We have

$$\mathcal{A}_{\lambda_i}(v,i) = D^{\tilde{\lambda}_i}(\mathrm{Gr}(v_i,\tilde{w}_i)) \otimes D(\underline{R}).$$

The following lemma gives a sufficient condition for a $\mathcal{A}^{\theta_i}_{\lambda_i}(v,i)$-module to be annihilated by $R\Gamma_i$. This is our first result toward the proof of Theorem 3.1.

Lemma 3.5 *Suppose $2v_i \leqslant \tilde{w}_i$ (and still i is a source). If $|\lambda_i| \leqslant \tilde{w}_i/2 - v_i$, then $R\Gamma_i$ annihilates every $\mathcal{A}^{\theta_i}_{\lambda_i}(v,i)$-module M that appears as a subquotient in $H_\bullet(F_i N)$ for $N \in \mathcal{A}^{\theta_i}_{\lambda_i}(v-\epsilon_i,i)\text{-mod}^{\underline{G},\lambda}$.*

Proof It is enough to prove this claim for the usual (non-equivariant) derived categories.

Let $M' \in \mathcal{A}^{\theta_i}_{\tilde{w}_i/2}(v,i)\text{-mod}$ be the twist of M by the line bundle $\mathcal{O}(-\tilde{\lambda}_i)$ on $\mathrm{Gr}(v_i,\tilde{w}_i)$. It follows from the Beilinson-Bernstein theorem that $\Gamma_i : \mathcal{A}^{\theta_i}_{\tilde{w}_i/2}(v,i)\text{-mod} \to \mathcal{A}_{\tilde{w}_i/2}(v,i)\text{-mod}$ is a category equivalence.

The functor $F_i : D^b(\mathcal{A}_{\tilde{w}_i/2}(v-\epsilon_i,i)\text{-mod}) \to D^b(\mathcal{A}_{\tilde{w}_i/2}(v,i)\text{-mod})$ can be realized as a tensor product $\mathcal{B} \otimes^L_{D(\mathrm{Gr}(v_i-1,\tilde{w}_i))} \bullet$, where $\mathcal{B}$ is a complex of $D(\mathrm{Gr}(v_i,\tilde{w}_i))$-$D(\mathrm{Gr}(v_i-1,\tilde{w}_i))$- bimodules with Harish-Chandra homology (both algebras are quotients of $U(\mathfrak{gl}(\tilde{w}_i))$ and so the notion of a HC bimodule does make sense). Since $2v_i \leqslant w_i$, the Gelfand-Kirillov dimension of $D(\mathrm{Gr}(v_i-1,\tilde{w}_i))$ is less than that of $D(\mathrm{Gr}(v_i,\tilde{w}_i))$. It follows that there is a proper two-sided ideal I in $D(\mathrm{Gr}(v_i,\tilde{w}_i))$ annihilating all Harish-Chandra $D(\mathrm{Gr}(v_i,\tilde{w}_i))$-$D(\mathrm{Gr}(v_i-1,\tilde{w}_i))$-bimodules. Namely, the regular $D(\mathrm{Gr}(v_i,\tilde{w}_i))$-bimodule has finite length, and for I, we take the minimal two-sided ideal such that the GK dimension of $D(\mathrm{Gr}(v_i,\tilde{w}_i))/I$ does not exceed that of $D(\mathrm{Gr}(v_i,\tilde{w}_i))$.

Note that $R\Gamma_i(M) = \mathrm{RHom}_A(B,\Gamma_i(M'))$, where we write A for $D(\mathrm{Gr}(v_i,\tilde{w}_i))$ and B for the translation $D(\mathrm{Gr}(v_i,\tilde{w}_i))$-$D^{\tilde{\lambda}_i}(\mathrm{Gr}(v_i,\tilde{w}_i))$-bimodule. Then

$$\mathrm{RHom}_A(B,\Gamma_i(M')) = \mathrm{RHom}_{A/I}(B/IB,\Gamma_i(M')).$$

So to show that $R\Gamma_i(M) = 0$, it remains to check that $B/IB = 0$.

Assume the contrary. The bimodule B and the ideal I are HC bimodules. Applying a suitable restriction functor (see [BL, Section 5.4]) to B/IB, we get a nonzero finite dimensional bimodule over the slice algebras to $D^{\tilde{\lambda}_i}(\mathrm{Gr}(v_i,\tilde{w}_i))$, $D(\mathrm{Gr}(v_i,\tilde{w}_i))$. Recall that a slice algebra to $D^{\lambda'_i}(v_i,\tilde{w}_i)$ for $\lambda'_i \in \mathbb{C}$ has the form $D^{\lambda'_i+k}(v_i-k,\tilde{w}_i-2k)$; this follows, for example, from [BL, Section 5.4]. From $|\lambda_i| \leqslant \tilde{w}_i/2 - v_i$, we see that the slice algebras for $D^{\lambda_i}(\mathrm{Gr}(v_i,\tilde{w}_i))$ have no finite dimensional representations. This proves $B = IB$ and completes the proof of the lemma. □

The lemma has the following important corollary, which gives a sufficient condition for a $\mathcal{A}^{\theta}_{\lambda}(v)$-module to be annihilated by $R\Gamma$.

Corollary 3.6 *Suppose $2v_i \leqslant \tilde{w}_i$ (and still i is a source). If $|\lambda_i| \leqslant \tilde{w}_i/2 - v_i$, then $R\Gamma$ annihilates every simple $\mathcal{A}^\theta_\lambda(v)$-module M_0 that appears in the homology of $F_i N_0$, where N_0 is a simple $\mathcal{A}^\theta_{\lambda'}(v-\epsilon_i)$-module supported on $\rho^{-1}_{v-\epsilon_i}(0)$ (and λ' is a suitable parameter so that the functor F_i makes sense).*

Proof Let $\pi^\theta(v) : D(R)\text{-mod}^{G,\tilde{\lambda}} \twoheadrightarrow \mathcal{A}^\theta_\lambda(v)\text{-mod}$ denote the quotient functor; it factors as $\pi^{\underline{\theta}}(v) \circ \pi^{\theta_i}(v,i)$, where $\pi^{\theta_i}(v,i) : D(R)\text{-mod}^{G,\tilde{\lambda}} \twoheadrightarrow \mathcal{A}^{\theta_i}_{\lambda_i}(v,i)\text{-mod}^{\underline{G},\underline{\lambda}}\text{-mod}$ and $\pi^{\underline{\theta}}(v) : \mathcal{A}^{\theta_i}_{\lambda_i}(v,i)\text{-mod}^{\underline{G},\underline{\lambda}}\text{-mod} \to \mathcal{A}^\theta_\lambda(v)\text{-mod}$ are the quotient functors; see [BL, Section 7.1]. We also have derived analogs of these functors (between the corresponding equivariant derived categories) to be denoted by the same letters. The functors $\pi^\theta(v), \pi^{\underline{\theta}}(v), \pi^{\theta_i}(v,i)$ have derived right adjoints, for example, $R\pi^\theta(v)^*$ lifts a complex of $\mathcal{A}^\theta_\lambda(v)$-modules to that of $(G,\tilde{\lambda})$-equivariant $D_R|_{T^*R^{\theta-ss}}$-modules and then takes derived global sections; compare with [MN, Lemma 4.6]. So we see that $R^j\Gamma(\bullet) = R^j\pi^\theta(v)^*(\bullet)^G$ for all j (we pass to the cohomology because a priori the two derived functors map to slightly different categories). So it is enough to prove that $R^j\pi^\theta(v)^*(M_0)^{\mathrm{GL}(v_i)} = 0$.

From the description of $R\pi^\theta(v)^*$ (and similar descriptions of the other two derived functors), we deduce that $R\pi^\theta(v)^*(\bullet)^{\mathrm{GL}(v_i)} = R\Gamma_i \circ R\pi^{\underline{\theta}}(v)^*(\bullet)$. So it suffices to show that $R\Gamma_i \circ R\pi^{\underline{\theta}}(v)^*(M_0) = 0$. By Bezrukavnikov and Losev [BL, Corollary 6.2] and the main result of [BL], N_0 is regular holonomic. By the proof of [BL, Corollary 6.2], $F_i N_0$ is the direct sum of simple $\mathcal{A}^{\theta_i}_{\lambda_i}(v,i)$-modules with homological shifts. So it is sufficient to show that $R\Gamma_i \circ R\pi^{\underline{\theta}}(v)^* \circ F_i(N_0) = 0$.

Recall that $\pi^{\underline{\theta}}(v-\epsilon_i) \circ E_i \cong E_i \circ \pi^{\underline{\theta}}(v)$. Also recall that (up to a homological shift) F_i is right adjoint to E_i. So we have an isomorphism $F_i \circ R\pi^{\underline{\theta}}(v-\epsilon_i)^* \cong R\pi^{\underline{\theta}}(v)^* \circ F_i$ (up to a homological shift). So it suffices to show that $R\Gamma_i$ annihilates $F_i \circ R\pi^{\underline{\theta}}(v-\epsilon_i)^*(N_0)$. But this follows from Lemma 3.5. □

3.5 Absence of Finite Dimensional Representations

Lemma 3.7 *Let θ be a generic stability condition. Then the following conditions are equivalent.*

(a) *The algebra $\mathcal{A}_\lambda(v)$ has no finite dimensional representations.*
(b) *$R\Gamma(M) = 0$ for any simple $\mathcal{A}^\theta_\lambda(v)$-module M supported on $\rho_v^{-1}(0)$.*

Proof We can view $R\Gamma$ as a functor $D^-(\mathcal{A}^\theta_\lambda(v)\text{-mod}) \to D^-(\mathcal{A}_\lambda(v)\text{-mod})$. This functor has a left adjoint and right inverse functor, namely, the derived localization functor

$$L\,\mathrm{Loc} : D^-(\mathcal{A}_\lambda(v)\text{-mod}) \to D^-(\mathcal{A}^\theta_\lambda(v)\text{-mod}).$$

The functors $R\Gamma, L\,\mathrm{Loc}$ restrict to the subcategories $D^-_{\rho^{-1}(0)}(\mathcal{A}^\theta_\lambda(v)\text{-mod})$, $D^-_{fin}(\mathcal{A}_\lambda(v)\text{-mod})$. So $R\Gamma$ is a quotient functor

$$D^{-}_{\rho^{-1}(0)}(\mathcal{A}^{\theta}_{\lambda}(v)\text{-mod}) \twoheadrightarrow D^{-}_{fin}(\mathcal{A}_{\lambda}(v)\text{-mod}).$$

Our claim follows. □

Now take θ with all positive coordinates; it is generic. Consider the system of simple roots for $\mathfrak{a}^{\lambda}$ consisting of minimal roots whose pairing with θ is positive. All these roots are positive. Moreover, the condition in (1) of Theorem 3.1 is equivalent to $|\langle\lambda,\alpha\rangle| \leqslant (\omega_0,\alpha)/2$ for all simple roots α of $\mathfrak{a}^{\lambda}$.

In [BL, Section 5.2], we have constructed endo-functors E_α, F_α of $\bigoplus_v D^b(\mathcal{A}^{\theta}_{\lambda_v}(v)\text{-mod})$ that preserve $\bigoplus_v D^b_{\rho_v^{-1}(0)}(\mathcal{A}^{\theta}_{\lambda_v}(v)\text{-mod})$ (we write λ_v instead of λ, because the parameter depends on v; see Sect. 3.4). The construction goes as follows. Pick an element σ in the Weyl group $W(Q)$ of Q. Then we have a Lusztig-Maffei-Nakajima (LMN) isomorphism $\sigma_* : X^{\theta}(v) \xrightarrow{\sim} X^{\sigma\theta}(\sigma\cdot v)$, where we write $\sigma\cdot v$ for the dimension vector corresponding to the weight $\sigma\nu$. This isomorphism lifts to that of quantizations: $\sigma_* : \mathcal{A}^{\theta}_{\lambda}(v) \xrightarrow{\sim} \mathcal{A}^{\sigma\theta}_{\sigma\lambda}(\sigma\cdot v)$; see [BL, Section 2.2] (note that here we use the symmetrized quantum comoment map so the parameter transforms linearly). The categories $\mathcal{A}^{\theta}_{\lambda}(v)\text{-mod}, \mathcal{A}^{\theta'}_{\lambda}(v)\text{-mod}$ are naturally equivalent as long as θ,θ' lie in the same Weyl chamber for $\mathfrak{a}^{\lambda}$; see [BL, Section 5.1.4]. Now pick a simple root α, and choose θ' inside the Weyl chamber for $\mathfrak{a}^{\lambda}$ containing θ and such that $\ker\alpha$ is a wall for the $\mathfrak{g}(Q)$-chamber of θ'. So there is $\sigma\in W(Q)$ such that $\sigma\alpha$ is a simple root for $\mathfrak{g}(Q)$ and all entries of $\sigma\theta'$ are positive. Then we set $F_\alpha = \sigma_*^{-1}\circ F_i\circ\sigma_*, E_\alpha = \sigma_*^{-1}\circ E_i\circ\sigma_*$.

Proposition 3.8 *Suppose (1) of Theorem 3.1 holds. Then the algebra $\mathcal{A}_{\lambda}(n\delta)$ has no finite dimensional irreducible representations.*

Proof We will need the following facts from [BL]:

(i) For fixed α, the functors E_α, F_α define an $\mathfrak{sl}_2$-action on $\bigoplus_v K_0(\mathcal{A}^{\theta}_{\lambda_v}(v)\text{-mod}_{\rho_v^{-1}(0)})$. The classes of simples form a dual perfect basis.
(ii) The simples annihilated by all E_α live in dimensions v corresponding to extremal weights for $\mathfrak{g}(Q)$.

(i) is in [BL, Sections 5.2, 9.1.2], while (ii) is a consequence of the main theorem of [BL], Theorem 1.1 there (i) allows to define crystal operators on the set of simples in $\bigoplus_v \mathcal{A}^{\theta}_{\lambda_v}(v)\text{-mod}$. It also shows that if a simple in $\mathcal{A}^{\theta}_{\lambda_v}(v)\text{-mod}_{fin}$ does not lie in the image of $\tilde{f}_\alpha$, then it is annihilated by $\tilde{e}_\alpha$ and hence by E_α.

Now let σ be as in the construction of E_α, F_α that was recalled above. Let $v' := \sigma\cdot v$. The condition $|\langle\lambda,\alpha\rangle| \leqslant \frac{(\omega_0,\alpha)}{2}$ is easily seen to be equivalent to $|(\sigma\lambda)_i| = |\langle\sigma\lambda,\alpha_i\rangle| \leqslant \frac{1}{2}(\sigma\nu,\alpha_i)$. The latter is nothing else but $\frac{1}{2}\tilde{w}'_i - v'_i$, where $\tilde{w}'_i$ is constructed from v'. By Corollary 3.6, the functor $R\Gamma$ annihilates all simple $\mathcal{A}^{\theta}_{\lambda}(n\delta)$-modules lying in the image of $\tilde{f}_\alpha$.

So, if $R\Gamma(M)\neq 0$, then $E_\alpha M = 0$ for all simple roots α of $\mathfrak{a}^{\lambda}$. However, this is impossible by (ii): the weight $\omega_0 - n\delta$ cannot be extremal for $\mathfrak{g}(Q)$. □

3.6 Proof of the Main Theorem

We need to show that, under conditions (1) and (2) of Theorem 3.1, there are no finite dimensional $e_{W'}H_c(W')e_{W'}$-modules for any parabolic subgroup $W' \subset W = G(\ell, 1, n)$. The parabolic subgroups of W have the form $G(\ell, 1, n_0) \times \mathfrak{S}_{n_1} \times \ldots \times \mathfrak{S}_{n_k}$, where $n_0 + n_1 + \ldots + n_k \leqslant n$. The slice algebras are of the form $\mathcal{A}_\lambda(n_0\delta) \otimes \mathcal{A}_\kappa(n_1) \ldots \otimes \mathcal{A}_\kappa(n_k)$, where we write $\mathcal{A}_\kappa(n_i)$ for the spherical subalgebra in $H_\kappa(\mathfrak{S}_{n_i})$. This follows from [BL, Section 5.4]. If (1) holds, then $\mathcal{A}_\lambda(n_0\delta)$ has no finite dimensional irreducible representations (when $n_0 > 0$). If (2) holds, then the algebras $\mathcal{A}_\kappa(n_i)$ do not. This completes the proof of Theorem 3.1.

Remark 3.9 One can state a sufficient condition for the algebra $\mathcal{A}_\lambda(v)$ to be simple generalizing Theorem 3.1 for an arbitrary quiver of finite or affine type, thanks to results of [BL, L4]. We are not going to do this explicitly.

Acknowledgments I would like to thank Raphael Rouquier who explained me a conjectural isomorphism in Theorem 1.2. I am also grateful to Gwyn Bellamy, Roman Bezrukavnikov, Pavel Etingof, Victor Ginzburg, and Iain Gordon for stimulating discussions. And I would like to thank the referee for many comments that helped me to improve the exposition. This work was supported by the NSF under Grant DMS-1161584.

References

[BEG] Yu. Berest, P. Etingof, V. Ginzburg, *Finite-dimensional representations of rational Cherednik algebras*. Int. Math. Res. Not. 2003, no. 19, 1053–1088.

[BE] R. Bezrukavnikov, P. Etingof, *Parabolic induction and restriction functors for rational Cherednik algebras*. Selecta Math., 14(2009), 397–425.

[BL] R. Bezrukavnikov, I. Losev, *Etingof conjecture for quantized quiver varieites*. Available at https://web.northeastern.edu/iloseu/bezpaper.pdf

[BPW] T. Braden, N. Proudfoot, B. Webster, *Quantizations of conical symplectic resolutions I: local and global structure*. arXiv:1208.3863.

[E] P. Etingof, *Proof of the Broué-Malle-Rouquier conjecture in characteristic zero (after I. Losev and I. Marin - G. Pfeiffer)*, arXiv:1606.08456.

[CR] J. Chuang and R. Rouquier, *Derived equivalences for symmetric groups and* $\mathfrak{sl}_2$*-categorifications*. Ann. Math. (2) 167(2008), n.1, 245–298.

[EG] P. Etingof and V. Ginzburg. *Symplectic reflection algebras, Calogero-Moser space, and deformed Harish-Chandra homomorphism*, Invent. Math. **147** (2002), 243–348.

[EGL] P. Etingof, E. Gorsky, I. Losev, *Representations of Rational Cherednik algebras with minimal support and torus knots*. arXiv:1304.3412.

[GGOR] V. Ginzburg, N. Guay, E. Opdam and R. Rouquier, *On the category $\mathcal{O}$ for rational Cherednik algebras*, Invent. Math., **154** (2003), 617–651.

[Go1] I. Gordon. *A remark on rational Cherednik algebras and differential operators on the cyclic quiver*. Glasg. Math. J. 48(2006), 145–160.

[Go2] I. Gordon. *Symplectic reflection alegebras*. Trends in representation theory of algebras and related topics, 285–347, EMS Ser. Congr. Rep., Eur. Math. Soc., Zürich, 2008.

[GL] I. Gordon, I. Losev, *On category $\mathcal{O}$ for cyclotomic Rational Cherednik algebras*. J. Eur. Math. Soc. 16 (2014), 1017–1079.

[HK] R. Hotta, M. Kashiwara, The invariant holonomic system on a semisimple Lie algebra. Invent. Math. 75 (1984), 327–358.

[L1] I. Losev, *Completions of symplectic reflection algebras*. Selecta Math., 18(2012), N1, 179–251.

[L2] I. Losev, *Isomorphisms of quantizations via quantization of resolutions*. Adv. Math. 231(2012), 1216–1270.

[L3] I. Losev, *On isomorphisms of certain functors for Cherednik algebras*. Repres. Theory, 17 (2013), 247–262.

[L4] I. Losev. *Wall-crossing functors for quantized symplectic resolutions: perversity and partial Ringel dualities*. Pure Appl. Math. Q. 13 (2017), no. 2, 247–289.

[MN] K. McGerty and T. Nevins, *Derived equivalence for quantum symplectic resolutions*. Selecta Math. 20(2014), 675–717.

[R] R. Rouquier, *2-Kac-Moody algebras*. arXiv:0812.5023.

[S] P. Shan. *Crystals of Fock spaces and cyclotomic rational double affine Hecke algebras*. Ann. Sci. Ecole Norm. Sup. **44** (2011), 147–182.

[SV] P. Shan and E. Vasserot, *Heisenberg algebras and rational double affine Hecke algebras*. J. Amer. Math. Soc. 25(2012), 959–1031.

[T] S. Thelin. *An algebraic approach to the KZ-functor for rational Cherednik algebras associated with cyclic groups*. J. Algebra, 471 (2017), 113–148.

[W] B. Webster. *A categorical action on quantized quiver varieties*. arXiv:1208.5957.

Microlocal Approach to Lusztig's Symmetries

Michael Finkelberg and Vadim Schechtman

To Victor Ginzburg on his 60th birthday

Contents

M. Finkelberg (✉)
Department of Mathematics, National Research University Higher School of Economics, Russian Federation, Moscow, Russia

Skolkovo Institute of Science and Technology, Institute for Information Transmission Problems of RAS, Moscow, Russia

V. Schechtman
Institut de Mathématiques de Toulouse, Université Paul Sabatier, Toulouse, France
e-mail: schechtman@math.ups-tlse.fr

V. Baranovsky et al. (eds.), *Representation Theory and Algebraic Geometry*, Trends in Mathematics, https://doi.org/10.1007/978-3-030-82007-7_3

1 Introduction

1.1 Coxeter Categories

Let $\mathfrak{g}$ be a semisimple Lie algebra and $\mathfrak{h}$ a Cartan subalgebra. Let $\mathfrak{h}^{\text{reg}} \subset \mathfrak{h}$ be the complement to the root hyperplane arrangement. For an integrable $\mathfrak{g}$-module V, C. De Concini and C. Procesi [DCP] have introduced an integrable *Casimir* connection with coefficients in the trivial vector bundle $V \otimes \mathcal{O}_{\mathfrak{h}^{\text{reg}}}$ (it was later rediscovered by J. Millson and V. Toledano Laredo [MT] and J. Felder, Y. Markov, V. Tarasov, and A. Varchenko [FMTV]) and conjectured that its monodromy can be expressed in terms of the action of the quantum Weyl group [LU2, SO] on the corresponding Weyl module W_V over the corresponding quantum group $\mathbf{U}_v(\mathfrak{g})$. This conjecture was later independently formulated and proved by V. Toledano Laredo for v in the formal neighborhood of 1. The key notion introduced in his proof was the notion of a (quasi-)*Coxeter category*. The original definition of this notion is of combinatorial nature. We suggest a more topological version of this definition in Sect. 3. It is a collection of local systems of restriction functors on the open strata of hyperplane arrangements arising from the root hyperplanes of a root system, compatible under Verdier specialization. This approach makes it clear, for example, that the category of representations of a rational Cherednik algebra carries a Coxeter structure; see [BE].

One of the main examples of a Coxeter category is a category of integrable representations of a quantum group. We consider the category $\mathcal{C}$ of representations of Lusztig's small quantum group. It has a geometric realization as the category $\mathcal{FS}$ of factorizable sheaves [BFS]. This is the category of certain compatible collections of perverse sheaves on the configuration spaces of a Riemann surface. One of our key observations is that the category $\mathcal{FS}$ carries a natural Coxeter structure (in our topological definition).

We conjecture that the equivalence $\Phi \colon \mathcal{FS} \xrightarrow{\sim} \mathcal{C}$ of [BFS] takes the Coxeter structure on $\mathcal{FS}$ to the Coxeter structure on $\mathcal{C}$. This is essentially a reformulation of De Concini-Toledano Laredo conjecture (hence, it follows from the results of

V. Toledano Laredo for v in the formal neighborhood of 1). Roughly, it says that the monodromy in the vanishing cycles of factorizable sheaves acts by Lusztig's symmetries.

1.2 Vanishing Cycles and Lusztig's Symmetries

Let us formulate the last point of Sect. 1.1 more precisely. We choose a Borel subalgebra $\mathfrak{h} \subset \mathfrak{b} \subset \mathfrak{g}$. The corresponding set of simple roots is denoted by I; for $i \in I$, the corresponding simple root is denoted by α_i. We fix a Weyl group invariant symmetric bilinear form $?\cdot?$ on $\mathfrak{h}^*$ such that the square length of a *short* root is $\alpha_i \cdot \alpha_i = 2$.

We fix a primitive root of unity ζ of degree ℓ; for simplicity in this introduction, we assume that ℓ is not divisible by 2 and 3. We consider an integral dominant weight $\lambda \in \mathfrak{h}^*$ such that $0 \leq \langle\lambda, \check{\alpha}_i\rangle < \ell$ for any $i \in I$ (pairings with simple coroots).

For $\beta = \sum_{i\in I} b_i\alpha_i \in \mathbb{N}[I]$, we consider the configuration space $\mathbb{A}^\beta$ of colored divisors on the complex affine line $\mathbb{A}^1$. The open subspace $\overset{\diamond}{\mathbb{A}}{}^\beta \subset \mathbb{A}^\beta$ of multiplicity free divisors on $\mathbb{A}^1\backslash\{0\}$ carries a one-dimensional local system $\mathcal{J}^\beta_\lambda$ with the following monodromies: $\zeta^{-2\alpha_i\cdot\alpha_j}$ when a point of color i goes counterclockwise around a point of color $j \neq i$; $-\zeta^{-\alpha_i\cdot\alpha_i}$ when two points of color i trade their positions going around a half circle counterclockwise; and $\zeta^{2\lambda\cdot\alpha_i}$ when a point of color i goes around 0 counterclockwise. We denote by $\mathcal{J}^\beta_\lambda$ the Goresky-MacPherson extension of $\mathcal{J}^\beta_\lambda$ to $\mathbb{A}^\beta$ (a perverse sheaf).

We have a pairing $\langle\cdot,\cdot\rangle\colon \mathfrak{h}\times\mathbb{A}^\beta \to \mathbb{A}^1$ given in the coordinates $(t_{i,r})^{1\leq r\leq b_i}_{i\in I}$ on $\mathbb{A}^\beta$, and $(z_j)_{j\in I}$ in the basis of fundamental coweights on $\mathfrak{h}$, as follows: $\langle(z_j),(t_{i,r})\rangle = \sum_{i\in I} z_i \sum_{r=1}^{b_i} t_{i,r}$. The vanishing cycle $\Phi^\beta_\lambda := \Phi_{\langle\cdot,\cdot\rangle}\, \mathrm{p}^\circ \mathcal{J}^\beta_\lambda$ of the pullback of $\mathcal{J}^\beta_\lambda$ to $\mathfrak{h}\times\mathbb{A}^\beta$ is a perverse sheaf supported on $\mathfrak{h} \simeq \mathfrak{h}\times\{\beta\cdot 0\}$.

We conjecture that Φ^β_λ is smooth along $\mathfrak{h}^{\mathrm{reg}} \subset \mathfrak{h}$. In order to describe its monodromy on $\mathfrak{h}^{\mathrm{reg}}$, recall that one of the main results of [BFS] is a canonical identification of the stalk $(\Phi^\beta_\lambda)_{C_0}$ at the fundamental Weyl chamber in $\mathfrak{h}^{\mathrm{reg}}_{\mathbb{R}}$ with the weight space $L^\lambda_{\lambda-\beta}$ of the irreducible module with highest weight λ over Lusztig's big quantum group $\dot{\mathbf{U}}_\zeta$ (note that the restriction of L^λ to Lusztig's small quantum group $\dot{\mathbf{u}}_\zeta$ remains irreducible since λ is an ℓ-restricted weight). We conjecture that the local system $\Phi^\beta_\lambda|_{\mathfrak{h}^{\mathrm{reg}}}$ is given by the following representation of the fundamental groupoid of $\mathfrak{h}^{\mathrm{reg}}$: the stalk at a Weyl chamber wC_0 in $\mathfrak{h}^{\mathrm{reg}}_{\mathbb{R}}$ is $L^\lambda_{w(\lambda-\beta)}$ (w runs through the Weyl group W), and the half monodromies around the walls are given by Lusztig's symmetries $T'_{i,\pm1}$ and $T''_{i,\pm1}$ (see Sect. 4 for details).

1.3 Organization of the Paper

Here is the outline of the paper.

In Sect. 2, we consider an elementary example of type A_2. We compare the action of Lusztig's symmetries in the "almost extremal" weight spaces of integrable modules over the corresponding quantum group (i.e., the weights obtained from the extremal ones by subtracting a root) with the monodromy action in the vanishing cycles of related perverse sheaves on the plane.

In Sect. 3, we propose a topological reformulation of Toledano Laredo's notion of Coxeter category. It is very similar to Deligne's topological reformulation [D] of the notion of braided tensor category.

In Sect. 4, we recall the (algebraic) construction of the Coxeter structure on the category of integrable modules over Lusztig's big quantum group, in terms of Lusztig's symmetries. It gives rise to the Coxeter structure on the category $\mathcal{C}$ of representations of Lusztig's small quantum group.

In Sect. 5, we recall very concisely the category of factorizable sheaves $\mathcal{FS}$ introduced in [BFS] and the equivalence $\Phi\colon \mathcal{FS} \xrightarrow{\sim} \mathcal{C}$ (see Sects. 5.2 and 5.3). We also take an opportunity to correct a confusion in [BFS] between Langlands dual types (see Sect. 5.1). Then we go on to construct a Coxeter structure on $\mathcal{FS}$. The construction goes through the De Rham realization of $\mathcal{FS}$ and works only for v sufficiently close to 1, but we expect it to work for arbitrary v. The construction also uses some results on iterated specialization and microlocalization over hyperplane arrangements presented in Sect. 6 which might be of independent interest. A more systematic approach to these questions is developed in [FKS].

Finally, in Sect. 7, we formulate the main conjecture that the equivalence $\Phi\colon \mathcal{FS} \xrightarrow{\sim} \mathcal{C}$ intertwines the Coxeter structures on $\mathcal{FS}$ and on $\mathcal{C}$.

2 An Example

2.1 Algebra

We follow the notations of [LU2]. Let $\mathbf{U}$ be the quantum universal enveloping algebra of type A_2, over the ring $\mathcal{A} = \mathbb{Z}[v^{\pm 1}]$. The positive (resp. negative) subalgebra $\mathbf{U}^+$ (resp. $\mathbf{U}^-$) is generated by the divided powers $E_i^{(r)}$ (resp. $F_i^{(r)}$), $i = 1, 2$, $r \in \mathbb{N}$. Let $\Lambda = (\mu_1, \mu_2) \in \mathbb{N}^2$ be a dominant highest weight such that $\mu_1 \geq 1 \leq \mu_2$ and $L(\Lambda)$ the corresponding integrable $\mathbf{U}$-module with the highest vector v. We will be interested in the weight spaces $L(\Lambda)_{(\mu_1-1,\mu_2-1)}$, $L(\Lambda)_{(-\mu_1+1,\mu_1+\mu_2-2)}$, $L(\Lambda)_{(\mu_1+\mu_2-2,-\mu_2+1)}$, $L(\Lambda)_{(\mu_2-1,-\mu_1-\mu_2+2)}$, $L(\Lambda)_{(-\mu_1-\mu_2+2,\mu_1-1)}$, *and* $L(\Lambda)_{(-\mu_2+1,-\mu_1+1)}$ (these weights form a single Weyl group orbit). They have canonical bases $(F_1F_2\mathsf{v}, F_2F_1\mathsf{v})$, $(F_1^{(\mu_1)}F_2\mathsf{v}, F_2F_1^{(\mu_1)}\mathsf{v})$, $(F_2^{(\mu_2)}F_1\mathsf{v}, F_1F_2^{(\mu_2)}\mathsf{v})$, $(F_2^{(\mu_1+\mu_2-1)}F_1^{(\mu_1)}\mathsf{v}, F_1F_2^{(\mu_1+\mu_2-1)}$

$F_1^{(\mu_1-1)}$v), $(F_1^{(\mu_1+\mu_2-1)}F_2^{(\mu_2)}$v, $F_2F_1^{(\mu_1+\mu_2-1)}F_2^{(\mu_2-1)}$v), *and*$(F_1^{(\mu_2-1)}F_2^{(\mu_1+\mu_2-1)}$ $F_1^{(\mu_1)}$v $= F_2^{(\mu_1)}F_1^{(\mu_1+\mu_2-1)}F_2^{(\mu_2-1)}$v, $F_1^{(\mu_2)}F_2^{(\mu_1+\mu_2-1)}F_1^{(\mu_1-1)}$v $= F_2^{(\mu_1-1)}$ $F_1^{(\mu_1+\mu_2-1)}F_2^{(\mu_2)}$v), respectively.

We are interested in the action of Lusztig's symmetries $T'_{1,2,\pm}$ on the above weight spaces.

Lemma 2.2 $T'_{1\pm}(F_1F_2\mathsf{v}) = -v^{\pm(\mu_1+1)}F_1^{(\mu_1)}F_2\mathsf{v}$, $T'_{1\pm}(F_2F_1\mathsf{v}) = -v^{\pm\mu_1}F_1^{(\mu_1)}F_2\mathsf{v} + F_2F_1^{(\mu_1)}\mathsf{v}$;
$T'_{2\pm}(F_2F_1\mathsf{v}) = -v^{\pm(\mu_2+1)}F_2^{(\mu_2)}F_1\mathsf{v}$, $T'_{2\pm}(F_1F_2\mathsf{v}) = -v^{\pm\mu_2}F_2^{(\mu_2)}F_1\mathsf{v} + F_1F_2^{(\mu_2)}\mathsf{v}$;
$T'_{2\pm}(F_2F_1^{(\mu_1)}\mathsf{v}) = -v^{\pm(\mu_1+\mu_2)}F_2^{(\mu_1+\mu_2-1)}F_1^{(\mu_1)}\mathsf{v}$,
$T'_{2\pm}(F_1^{(\mu_1)}F_2\mathsf{v}) = -v^{\pm\mu_2}F_2^{(\mu_1+\mu_2-1)}F_1^{(\mu_1)}\mathsf{v} + F_1F_2^{(\mu_1+\mu_2-1)}F_1^{(\mu_1-1)}\mathsf{v}$;
$T'_{1\pm}(F_1F_2^{(\mu_2)}\mathsf{v}) = -v^{\pm(\mu_1+\mu_2)}F_1^{(\mu_1+\mu_2-1)}F_2^{(\mu_2)}\mathsf{v}$,
$T'_{1\pm}(F_2^{(\mu_2)}F_1\mathsf{v}) = -v^{\pm\mu_1}F_1^{(\mu_1+\mu_2-1)}F_2^{(\mu_2)}\mathsf{v} + F_2F_1^{(\mu_1+\mu_2-1)}F_2^{(\mu_2-1)}\mathsf{v}$;
$T'_{1\pm}(F_1F_2^{(\mu_1+\mu_2-1)}F_1^{(\mu_1-1)}\mathsf{v}) = -v^{\pm(\mu_2+1)}F_1^{(\mu_2)}F_2^{(\mu_1+\mu_2-1)}F_1^{(\mu_1-1)}\mathsf{v}$,
$T'_{1\pm}(F_2^{(\mu_1+\mu_2-1)}F_1^{(\mu_1)}\mathsf{v}) = -v^{\pm1}F_1^{(\mu_2)}F_2^{(\mu_1+\mu_2-1)}F_1^{(\mu_1-1)}\mathsf{v} + F_1^{(\mu_2-1)}F_2^{(\mu_1+\mu_2-1)}F_1^{(\mu_1)}\mathsf{v}$;
$T'_{2\pm}(F_2F_1^{(\mu_1+\mu_2-1)}F_2^{(\mu_2-1)}\mathsf{v}) = -v^{\pm(\mu_1+1)}F_2^{(\mu_1)}F_1^{(\mu_1+\mu_2-1)}F_2^{(\mu_2-1)}\mathsf{v}$,
$T'_{2\pm}(F_1^{(\mu_1+\mu_2-1)}F_2^{(\mu_2)}\mathsf{v}) = -v^{\pm1}F_2^{(\mu_1)}F_1^{(\mu_1+\mu_2-1)}F_2^{(\mu_2-1)}\mathsf{v} + F_2^{(\mu_1-1)}F_1^{(\mu_1+\mu_2-1)}F_2^{(\mu_2)}\mathsf{v}$.

Proof We consider two subalgebras $\mathbf{U}_1, \mathbf{U}_2 \subset \mathbf{U}$ of type A_1: the first one is generated by $E_1^{(r)}, F_1^{(r)}$, $r \in \mathbb{N}$, and the second one is generated by $E_2^{(r)}, F_2^{(r)}$, $r \in \mathbb{N}$. To prove the first formula, we consider the $\mathbf{U}_1$-submodule M_1 of $L(\Lambda)$ with the highest vector F_2v and canonical base F_2v, F_1F_2v, …, $F_1^{(\mu_1)}F_2$v, $F_1^{(\mu_1+1)}F_2$v. We also consider another $\mathbf{U}_1$-submodule M'_1 of $L(\Lambda)$ with the highest vector $\mathsf{w}^+ := (v^{\mu_1} - v^{-\mu_1})F_1F_2\mathsf{v} - (v^{\mu_1+1} - v^{-\mu_1-1})F_2F_1\mathsf{v}$ and the lowest vector (in the same canonical base) $\mathsf{w}^- = (v - v^{-1})F_1^{(\mu_1)}F_2\mathsf{v} - (v^{\mu_1+1} - v^{-\mu_1-1})F_2F_1^{(\mu_1)}\mathsf{v}$. In effect, it is straightforward that $E_1\mathsf{w}^+ = 0$, and it follows from [LU2, Lemma 42.1.2.(d)] that $F_1\mathsf{w}^- = 0$; hence, $F_1^{(\mu_1-1)}\mathsf{w}^+ = a\mathsf{w}^-$ for some $a \in \mathcal{A}$. The fact that $a = 1$ follows by comparing the coefficients of F_1F_2v in w^+ and of $F_1^{(\mu_1)}F_2$v in w^-. Now according to [LU2, Proposition 5.2.2.(a)], $T'_{1\pm}\mathsf{w}^+ = \mathsf{w}^-$, $T'_{1\pm}F_1F_2\mathsf{v} = -v^{\pm(\mu_1+1)}F_1^{(\mu_1)}F_2\mathsf{v}$. From this, we deduce the first two formulas. The other formulas are proved similarly. Say, to prove the fifth and sixth formulas, we consider the $\mathbf{U}_2$-submodule M_2 of $L(\Lambda)$ with the highest vector $F_1^{(\mu_1)}$v and canonical base $F_1^{(\mu_1)}$v, $F_2F_1^{(\mu_1)}$v, …, $F_2^{(\mu_1+\mu_2-1)}F_1^{(\mu_1)}$v, $F_2^{(\mu_1+\mu_2)}F_1^{(\mu_1)}$v. We also consider another $\mathbf{U}_2$-submodule M'_2 of $L(\Lambda)$ with the highest vector $\mathsf{x}^+ := (v^{\mu_1+\mu_2} - v^{-\mu_1-\mu_2})F_1^{(\mu_1)}F_2\mathsf{v} - (v^{\mu_2} - v^{-\mu_2})F_2F_1^{(\mu_1)}\mathsf{v}$ and the lowest vector (in the same canonical base) $\mathsf{x}^- = (v^{\mu_1+\mu_2} - v^{-\mu_1-\mu_2})F_1F_2^{(\mu_1+\mu_2-1)}F_1^{(\mu_1-1)}\mathsf{v} - (v^{\mu_1} - v^{-\mu_1})F_2^{(\mu_1+\mu_2-1)}F_1^{(\mu_1)}\mathsf{v}$. Then $T'_{2\pm}\mathsf{x}^+ = \mathsf{x}^-$, $T'_{2\pm}F_2F_1^{(\mu_1)}\mathsf{v} = -v^{\pm(\mu_1+\mu_2)}F_2^{(\mu_1+\mu_2-1)}F_1^{(\mu_1)}\mathsf{v}$. From this, we deduce the fifth and sixth formulas. And so on. □

2.3 Topology

Let $\mathbb{A}_{\mathbb{R}}$ be a two-dimensional real vector space, and let $\mathbb{A}$ be its complexification with coordinates (t_1, t_2) stratified by three lines: $t_1 = 0,\ t_2 = 0,\ and t_1 - t_2 = 0$. Let $\mathcal{L}$ be the shriek extension of the one-dimensional local system on the complement of the three lines with monodromies $v^{2\mu_1}, v^{2\mu_2}, and v^{2\mu_3}$. In applications to algebra, $2\mu_3 = 2$. The dual vector space $\mathbb{A}^*$ has coordinates (z_1, z_2), and the dual stratification consists of the lines $z_1 = 0,\ z_2 = 0, and\ z_1 + z_2 = 0$. This is the root hyperplane arrangement of type A_2. There are six real chambers of this arrangement: C_0 is the dominant chamber containing an interior point $z^{(e)} = (1, 1)$; the other chambers have interior points $z^{(1)} = (-1, 2),\ z^{(21)} = (-2, 1),\ z^{(121)} = z^{(212)} = (-1, -1),\ z^{(12)} = (1, -2), and\ z^{(2)} = (2, -1)$. The chambers are naturally numbered by the Weyl group W of type A_2 generated by simple reflections $s_1 and s_2$. For $w \in W$, we have $C_w \ni z^{(w)}$, say $C_{s_1s_2s_1}$ (C_{121} for short) contains $z^{(121)}$. We have *six real* affine lines $\ell_w,\ w \in W$, in $\mathbb{A}_{\mathbb{R}}$ given by equations $z^{(w)} = 1$. For example, ℓ_e and ℓ_1 are given by the equations $t_1 + t_2 = 1$ and $-t_1 + 2t_2 = 1$, respectively. More generally, for $\varepsilon \in \mathbb{R},\ \varepsilon > 0$, let us denote by $\ell_{w,\varepsilon}$ the real straight line given by the equation $z^{(w)} = \varepsilon$.

The microlocalization (Fourier transform) $\boldsymbol{\mu}\mathcal{L}$ is a certain constructible complex on $\mathbb{A}^*$. We will be interested only in its restriction to the complement of the three lines in $\mathbb{A}^*$, which is a two-dimensional local system. Let us describe this local system explicitly.

The stalk $\boldsymbol{\mu}^{(w)}\mathcal{L}$ at $z^{(w)}$ equals the vanishing cycles $\Phi_{z^{(w)}}\mathcal{L}$. Let i_w denote the inclusion $\ell_w \hookrightarrow \mathbb{A}$. Then $\Phi_{z^{(w)}}\mathcal{L}$ may be identified with $H^1(\ell_w, i_w^*\mathcal{L})$. It is a two-dimensional vector space with the base *dual* to the basis ℓ'_w, ℓ''_w of 1-cycles with coefficients in $i_w^*\mathcal{L}^*$.

The 1-cycles are defined as follows: ℓ'_e is the interval between the points $(1, 0)$ and $(1/2, 1/2)$; ℓ''_e is the interval between the points $(1/2, 1/2)$ and $(0, 1)$; ℓ'_1 is the interval between the points $(1, 1)$ and $(0, 1/2)$; ℓ''_1 is the interval between the points $(0, 1/2)$ and $(-1, 0)$; ℓ'_{21} is the interval between the points $(0, 1)$ and $(-1/2, 0)$; ℓ''_{21} is the interval between the points $(-1/2, 0)$ and $(-1, -1)$; ℓ'_{121} is the interval between the points $(-1, 0)$ and $(-1/2, -1/2)$; ℓ''_{121} is the interval between the points $(-1/2, -1/2)$ and $(0, -1)$; ℓ'_2 is the interval between the points $(0, -1)$ and $(1/2, 0)$; ℓ''_2 is the interval between the points $(1/2, 0)$ and $(1, 1)$; ℓ'_{12} is the interval between the points $(-1, -1)$ and $(0, -1/2)$; ℓ''_{12} is the interval between the points $(0, -1/2)$ and $(1, 0)$; ℓ'_{212} is the interval between the points $(-1, 0)$ and $(-1/2, -1/2)$; ℓ''_{212} is the interval between the points $(-1/2, -1/2)$ and $(0, -1)$. Note that $\ell''_{212} = \ell''_{121},\ \ell'_{212} = \ell'_{121}$. The dual basis to ℓ'_w, ℓ''_w will be denoted by ϕ'_w, ϕ''_w (in particular, $\phi''_{212} = \phi''_{121},\ \phi'_{212} = \phi'_{121}$).

More generally, for any $\varepsilon > 0$, we have canonical isomorphisms

$$\Phi_{z^{(w)}}\mathcal{L} = H^1(\ell_w, i_w^*\mathcal{L}) \xrightarrow{\sim} H^1(\ell_{w,\varepsilon}, i_{w,\varepsilon}^*\mathcal{L})$$

where $i_{w,\varepsilon} : \ell_{w,\varepsilon} \hookrightarrow \mathbb{A}$, and we can define similar parallelly transported bases in $H_1(\ell_{w,\varepsilon}, i^*_{w,\varepsilon}\mathcal{L}^*)$.

For two neighboring chambers $C_y, C_w,\ y, w \in W$, let $\gamma^{\pm}_{y,w}$ be a path from C_y to C_w obtained from a straight line interval modified near the wall between these two chambers by going around it in the positive (resp. negative) imaginary halfspace. We will keep the same notation for the induced operator (half monodromy along $\gamma^{\pm}_{y,w}$) from $\Phi_{z(y)}\mathcal{L}$ to $\Phi_{z(w)}\mathcal{L}$.

Lemma 2.4 $\gamma^{\pm}_{e,1}\phi'_e = -v^{\pm(\mu_1+\mu_3)}\phi''_1,\ \gamma^{\pm}_{e,1}\phi''_e = -v^{\pm\mu_1}\phi''_1 + \phi'_1;$
$\gamma^{\pm}_{e,2}\phi''_e = -v^{\pm(\mu_2+\mu_3)}\phi'_2,\ \gamma^{\pm}_{e,2}\phi'_e = -v^{\pm\mu_2}\phi'_2 + \phi''_2;$
$\gamma^{\pm}_{1,21}\phi'_1 = -v^{\pm(\mu_2+\mu_1)}\phi''_{21},\ \gamma^{\pm}_{1,21}\phi''_1 = -v^{\pm\mu_2}\phi''_{21} + \phi'_{21};$
$\gamma^{\pm}_{2,12}\phi''_2 = -v^{\pm(\mu_1+\mu_2)}\phi'_{12},\ \gamma^{\pm}_{2,12}\phi'_2 = -v^{\pm\mu_1}\phi'_{12} + \phi''_{12};$
$\gamma^{\pm}_{21,121}\phi'_{21} = -v^{\pm(\mu_3+\mu_2)}\phi''_{121},\ \gamma^{\pm}_{21,121}\phi''_{21} = -v^{\pm\mu_3}\phi''_{121} + \phi'_{121};$
$\gamma^{\pm}_{12,212}\phi''_{12} = -v^{\pm(\mu_3+\mu_1)}\phi'_{212},\ \gamma^{\pm}_{12,212}\phi'_{12} = -v^{\pm\mu_3}\phi'_{212} + \phi''_{212}.$

Proof All the formulas being similar, we prove the first two. For the transposed map between dual spaces, we must check that

$$\gamma^{\pm}_{1,e}\ell'_1 = \ell''_e,\ \gamma^{\pm}_{1,e}\ell''_1 = -v^{\pm\mu_1}\ell''_e - v^{\pm(\mu_1+\mu_3)}\ell'_e.$$

(Note that the second equality is equivalent to $\gamma^{\pm}_{1,e}(\ell'_1 + v^{\mp\mu_1}\ell''_1) = -v^{\pm\mu_3}\ell'_e$.) To prove it, we rotate the line ℓ_1 clockwise in $\mathbb{A}_{\mathbb{R}}$ with the point (0, 1/2) fixed and observe what happens with the real cycles ℓ'_1 and ℓ''_1. At some critical moment, the rotated line becomes parallel to the t_1-axis; at this moment, we must pass for a short time into the complex upper (or lower) halfspace, and at the end, we get the line parallel to ℓ_e. We see that at the end of this rotation, ℓ'_1 turns into ℓ''_e, whereas ℓ''_1 stretches and after the critical moment turns into the necessary linear combination of ℓ'_e and ℓ''_e. □

Remark 2.5 Writing down the composition $\gamma^{\pm}_{21,121} \circ \gamma^{\pm}_{1,21} \circ \gamma^{\pm}_{e,1}$ in our bases as the product of matrices, we find

$$\begin{pmatrix} 0 & 1 \\ -v^{\pm(\mu_3+\mu_2)} & -v^{\pm\mu_3} \end{pmatrix}\begin{pmatrix} 0 & 1 \\ -v^{\pm(\mu_2+\mu_1)} & -v^{\pm\mu_2} \end{pmatrix}\begin{pmatrix} 0 & 1 \\ -v^{\pm(\mu_1+\mu_3)} & -v^{\pm\mu_1} \end{pmatrix} =$$
$$\begin{pmatrix} v^{\pm(\mu_1+\mu_2+\mu_3)} & 0 \\ 0 & v^{\pm(\mu_1+\mu_2+\mu_3)} \end{pmatrix},$$

cf. [LU1, Corollary 5.9].

Remark 2.6 In case $\mu_1 = \mu_2 = \mu_3 = 1$, all the six weight spaces considered in Sect. 2.1 coincide with $L(1, 1)_{(0,0)}$ with the base $F_1F_2\mathsf{v},\ F_2F_1\mathsf{v}$. In this base, the operator $T'_{1\pm}$ of the first line of Lemma 2.2 corresponding to the operator $\gamma^{\pm}_{e,1}$ of Lemma 2.4 has the matrix $\begin{pmatrix} -v^{\pm 2} & -v^{\pm 1} \\ 0 & 1 \end{pmatrix}$, while the operator $T'_{2\pm}$ of the second line of Lemma 2.2 corresponding to the operator $\gamma^{\pm}_{e,2}$ of Lemma 2.4 has the matrix $\begin{pmatrix} 1 & 0 \\ -v^{\pm 1} & -v^{\pm 2} \end{pmatrix}$. Note that $(\gamma^{-}_{e,1})^{-1} = \gamma^{+}_{e,1}$ and $(\gamma^{-}_{e,2})^{-1} = \gamma^{+}_{e,2}$.

2.7 Discussion

We set $\mu_3 = 1$. The theory of factorizable sheaves [BFS] provides a canonical isomorphism $\Phi_{z(e)}\mathcal{L} \simeq L(\Lambda)_{(\mu_1-1,\mu_2-1)}$. The stalks of microlocalization at the other chambers $\Phi_{z(w)}\mathcal{L}$ *do not* have an algebraic interpretation in the framework of this theory.[1] However, the comparison of Lemmas 2.2 and 2.4 shows that the monodromy of the local system $\boldsymbol{\mu}\mathcal{L}$ (as the automorphism group of $\Phi_{z(e)}\mathcal{L} \simeq L(\Lambda)_{(\mu_1-1,\mu_2-1)}$) can be expressed in terms of Lusztig's symmetries $T'_{1,2\pm}, T''_{1,2\pm}$. In fact, the comparison of Lemmas 2.2 and 2.4 suggests a much more precise relation, in particular, between a natural topological basis in $\Phi_{z(e)}\mathcal{L}$ and the canonical basis on the algebraic side. Unfortunately, we have no clue how to define such a topological basis in general. However, the relation between the monodromy and Lusztig's symmetries seems to generalize. This is the subject of the main body of the note.

3 Coxeter Categories

3.1 Notations

Let us set up a few notations related to a simple Lie algebra $\mathfrak{g}$ with Cartan subalgebra $\mathfrak{h}$ and Borel subalgebra $\mathfrak{h} \subset \mathfrak{b} \subset \mathfrak{g}$. The set of simple coroots is denoted by I; for $i \in I$, the corresponding simple coroot is denoted $\check{\alpha}_i$ or sometimes simply i. The corresponding simple root is denoted α_i or sometimes i'. We fix a Weyl group invariant symmetric bilinear form ?·? on $\mathfrak{h}^*$ such that the square length of a *short* root is $\alpha_i \cdot \alpha_i = 2$. This bilinear form gives rise to an isomorphism $\mathfrak{h} \xrightarrow{\sim} \mathfrak{h}^*$ so that the coroot lattice Y generated by $\{\check{\alpha}_i\}_{i\in I}$ embeds into $\mathfrak{h}^*$. We then have $\check{\alpha}_i \cdot \check{\alpha}_i \in \{2, 1, \frac{2}{3}\}$ and $\alpha_i \cdot \alpha_i \in \{2, 4, 6\}$. We set $d_i = \alpha_i \cdot \alpha_i/2$. Let d be the ratio of the square lengths of the long and short roots, so that $d \in \{1, 2, 3\}$. We set $\check{d}_i = d/d_i$. Then

$$\langle \alpha_i, \check{\alpha}_j \rangle = \frac{\alpha_i \cdot \alpha_j}{d_j} = d_i \check{\alpha}_i \cdot \check{\alpha}_j = d \frac{\check{\alpha}_i \cdot \check{\alpha}_j}{\check{d}_i}.$$

[1] see Conjecture 7.3 however.

3.2 The Fundamental Groupoid of $\mathfrak{h}_{D'}^{\mathrm{reg}}$

We follow the notations of [AT]. Let D be the Dynkin diagram of the simple Lie algebra $\mathfrak{g}$ with Cartan $\mathfrak{h}$ (so that I is the set of vertices of D). The root system of $\mathfrak{h} \subset \mathfrak{g}$ is $R_D \subset \mathfrak{h}^*$. The complement in $\mathfrak{h}$ to the root hyperplanes is the open subset $\mathfrak{h}^{\mathrm{reg}}$.

Given a subset D' of the set of vertices of D, we denote by $\mathfrak{h}_{D'}$ the quotient of $\mathfrak{h}$ by the center of the corresponding Levi subalgebra $\mathfrak{l}_{D'} \subset \mathfrak{g}$. In other words, $\mathfrak{h}_{D'}^* \subset \mathfrak{h}_D^*$ is spanned by the simple roots corresponding to the vertices from D'. We denote by $\mathfrak{h}_{D'}^{\mathrm{reg}}$ the complement in $\mathfrak{h}_{D'}$ to the root hyperplanes of the root subsystem $R_{D'}$ corresponding to $D' \subset D$.

We recall the Salvetti presentation of the fundamental groupoid of $\mathfrak{h}_{D'}^{\mathrm{reg}}$, cf. [SA]. Let $\mathfrak{h}_{D',\mathbb{R}}^{\mathrm{reg}}$ denote the set of real points of $\mathfrak{h}_{D'}^{\mathrm{reg}}$. It is a union of the connected components called *chambers*. We fix a chamber $C_0^{D'}$ formed by the points with *positive* coordinates (in the basis of fundamental coweights). The Weyl group $W_{D'}$ acts on the set $\mathbf{C}^{D'}$ of chambers simply transitively on the left. The choice of $C_0^{D'}$ identifies $\mathbf{C}^{D'}$ with $W_{D'}$ and defines the *right* action of $W_{D'}$ on $\mathbf{C}^{D'}$ (transferred from the right action of $W_{D'}$ on itself). The set of walls of $C_0^{D'}$ is canonically identified with the set of vertices of D'. The left action of $W_{D'}$ on $\mathbf{C}^{D'}$ extends this identification to any chamber. For $i \in D'$, the right action of a simple reflection works as follows: $C \cdot s_i$ is a unique neighboring chamber C' having the s_i-wall in common with C.

The set of objects of the fundamental groupoid $\Pi(\mathfrak{h}_{D'}^{\mathrm{reg}})$ is $\mathbf{C}^{D'}$. Given a straight line interval γ connecting the endpoints $\gamma_1 \in C_1$ and $\gamma_2 \in C_2$ and intersecting only one wall at a time, we define the morphisms $\gamma^{\pm} \in \mathrm{Mor}_{\Pi(\mathfrak{h}_{D'}^{\mathrm{reg}})}(C_1, C_2)$ as follows. The path γ^+ (resp. γ^-) coincides with γ away from the small neighborhoods of its intersection with walls, where γ^+ (resp. γ^-) goes around the intersection in the positive (resp. negative) imaginary direction in $\mathfrak{h}_{D'}^{\mathrm{reg}}$. According to Salvetti, $\Pi(\mathfrak{h}_{D'}^{\mathrm{reg}})$ is generated by the set of morphisms $\gamma^{\pm}$ with relations $\beta^{\pm} = \gamma^{\pm}$ provided γ_1, β_1 lie in the same chamber C_1 and γ_2, β_2 lie in the same chamber C_2.

3.3 The Fundamental Groupoid of $N_{\mathfrak{h}_{D'/D''}/\mathfrak{h}_{D'}}^{\mathrm{reg}}$

Given a third subdiagram $D'' \subset D' \subset D$, we consider the linear subspace $\mathfrak{h}_{D'/D''} \subset \mathfrak{h}_{D'}$ spanned by the *fundamental coweights* in $D' - D'' \subset D'$. For example, $\mathfrak{h}_{D/D''} \subset \mathfrak{h}_D$ is the center of the Levi $\mathfrak{l}_{D''} \subset \mathfrak{g}$. We have an exact sequence

$$0 \longrightarrow \mathfrak{h}_{D'/D''} \longrightarrow \mathfrak{h}_{D'} \longrightarrow \mathfrak{h}_{D''} \longrightarrow 0$$

which may serve as another definition of $\mathfrak{h}_{D'/D''}$.

We denote by $\mathfrak{h}^{\mathrm{reg}}_{D'/D''}$ the complement in $\mathfrak{h}_{D'/D''}$ to the root hyperplanes (roots in $R_{D'}$) not containing $\mathfrak{h}_{D'/D''}$. The connected components of the real part $\mathfrak{h}^{\mathrm{reg}}_{D'/D'',\mathbb{R}}$ are called *chambers*; the set of chambers is denoted $\mathbf{C}^{D'/D''}$. It is naturally isomorphic to the set of parabolics in $\mathfrak{g}_{D'}$ containing the standard Levi $\mathfrak{l}_{D''}$; see, e.g., [MW, I.1.10]. We say that a chamber $C \in \mathbf{C}^{D'}$ is *adjacent* to $\mathfrak{h}_{D'/D''} \subset \mathfrak{h}_{D'}$ if the intersection of the closure $\overline{C}$ with $\mathfrak{h}_{D'/D''}$ has the maximal (real) dimension $\dim \mathfrak{h}_{D'/D''}$; then this intersection is the closure of a chamber in $\mathbf{C}^{D'/D''}$ to be denoted $\pi(C)$. The set of chambers adjacent to $\mathfrak{h}_{D'/D''}$ is denoted $\mathbf{A}^{D'/D''}$. Thus, we have a projection $\pi: \mathbf{A}^{D'/D''} \to \mathbf{C}^{D'/D''}$.

The natural projection $\mathrm{pr}: \mathfrak{h}_{D'} \to \mathfrak{h}_{D''}$ (see Sect. 3.2) works in the bases of fundamental coweights as follows: $\mathrm{pr}\, \check{\omega}_i = 0$ if $i \in D' - D'' \subset D'$; and if $i \in D'' \subset D'$, then $\check{\omega}_i$ goes to the corresponding fundamental coweight in $\mathfrak{h}_{D''}$. Given a chamber $C \in \mathbf{A}^{D'/D''}$, its projection $\mathrm{pr}\, C$ is a chamber in $\mathbf{C}^{D''}$. Thus, we have a projection $\mathrm{pr}: \mathbf{A}^{D'/D''} \to \mathbf{C}^{D''}$.

Lemma 3.4 *The product* $\mathrm{pr} \times \pi : \mathbf{A}^{D'/D''} \to \mathbf{C}^{D''} \times \mathbf{C}^{D'/D''}$ *establishes a one-to-one correspondence.* □

Definition 3.5

(a) For a chamber $C \in \mathbf{C}^{D''}$, we define a subgroupoid $\Pi_{\mathrm{pr}^{-1}(C)}(\mathfrak{h}^{\mathrm{reg}}_{D'}) \subset \Pi(\mathfrak{h}^{\mathrm{reg}}_{D'})$ as follows: the objects are $\mathrm{pr}^{-1}(C) \subset \mathbf{A}^{D'/D''} \subset \mathbf{C}^{D'}$, and the morphisms are generated by $\gamma^{\pm}$ where γ is a straight line interval *parallel* to $\mathfrak{h}_{D'/D''}$ (i.e., such that $\mathrm{pr}\,\gamma$ is a point).
(b) For a chamber $C \in \mathbf{C}^{D'/D''}$, we define a subgroupoid $\Pi_{\pi^{-1}(C)}(\mathfrak{h}^{\mathrm{reg}}_{D'}) \subset \Pi(\mathfrak{h}^{\mathrm{reg}}_{D'})$ as follows: the objects are $\pi^{-1}(C) \subset \mathbf{A}^{D'/D''} \subset \mathbf{C}^{D'}$, and the morphisms are generated by $\delta^{\pm}$ where δ is a straight line interval inside the union of closures of chambers adjacent to C.
(c) A subgroupoid $\Pi_{\mathbf{A}^{D'/D''}}(\mathfrak{h}^{\mathrm{reg}}_{D'}) \subset \Pi(\mathfrak{h}^{\mathrm{reg}}_{D'})$ is generated by all the groupoids in (a,b) above. That is, its objects are $\mathbf{A}^{D'/D''}$, and the morphisms are all the possible products of morphisms in (a,b) above (see Fig. 1).

Lemma 3.6

(a) For any $C \in \mathbf{C}^{D''}$, π *induces an equivalence* $\Pi_{\mathrm{pr}^{-1}(C)}(\mathfrak{h}^{\mathrm{reg}}_{D'}) \xrightarrow{\sim} \Pi(\mathfrak{h}^{\mathrm{reg}}_{D'/D''})$.
(b) For any $C \in \mathbf{C}^{D'/D''}$, pr *induces an equivalence* $\Pi_{\pi^{-1}(C)}(\mathfrak{h}^{\mathrm{reg}}_{D'}) \xrightarrow{\sim} \Pi(\mathfrak{h}^{\mathrm{reg}}_{D''})$.
(c) The natural projection morphisms $\pi: \Pi_{\mathbf{A}^{D'/D''}}(\mathfrak{h}^{\mathrm{reg}}_{D'}) \to \Pi(\mathfrak{h}^{\mathrm{reg}}_{D'/D''}$ *and* $\mathrm{pr}: \Pi_{\mathbf{A}^{D'/D''}}(\mathfrak{h}^{\mathrm{reg}}_{D'}) \to \Pi(\mathfrak{h}^{\mathrm{reg}}_{D''})$ *give rise to an equivalence* $\Pi_{\mathbf{A}^{D'/D''}}(\mathfrak{h}^{\mathrm{reg}}_{D'}) \xrightarrow{\sim} \Pi(\mathfrak{h}^{\mathrm{reg}}_{D'/D''}) \times \Pi(\mathfrak{h}^{\mathrm{reg}}_{D''})$.

Proof (M. Kapranov) The relations in the Salvetti complex [SA] follow from a cell decomposition of the complement which is glued out of intervals, $2n$-gons (for any codim 2 cell where n hyperplanes meet) and so on, and the relations in the fundamental groupoid are obtained from the 2-skeleton, i.e., from these $2n$-gons.

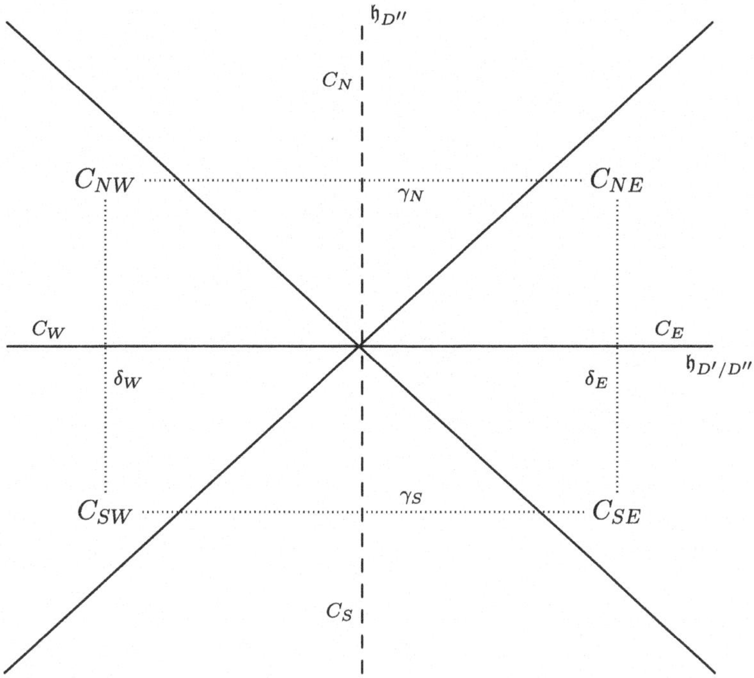

Fig. 1 Topographical example

So the two-dimensional case implies the general one. See example with $n = 3$ in Fig. 1. We keep "vertical" and "horizontal" $2k$-gons (contributing to $\Pi_{\pi^{-1}(C)}(\mathfrak{h}^{\mathrm{reg}}_{D'})$ or $\Pi_{\mathrm{pr}^{-1}(C)}(\mathfrak{h}^{\mathrm{reg}}_{D'})$) intact and replace the remaining $2n$-gons with rectangles like the dotted one in Fig. 1. It follows that the "horizontal" and "vertical" morphisms commute. This produces a two-dimensional CW-subcomplex of the complement which is (the 2-skeleton of) the product of two separate two-dimensional subcomplexes. □

3.7 Specialization

We consider the normal bundle $N_{\mathfrak{h}_{D'/D''}/\mathfrak{h}_{D'}} \xrightarrow{\sim} \mathfrak{h}_{D'/D''} \times \mathfrak{h}_{D''}$ and its open subspace $N^{\mathrm{reg}}_{\mathfrak{h}_{D'/D''}/\mathfrak{h}_{D'}} = \mathfrak{h}^{\mathrm{reg}}_{D'/D''} \times \mathfrak{h}^{\mathrm{reg}}_{D''} \subset N_{\mathfrak{h}_{D'/D''}/\mathfrak{h}_{D'}}$ with Poincaré groupoid $\Pi_{\mathbf{A}_{D'/D''}}(\mathfrak{h}^{\mathrm{reg}}_{D'})$. Then the Verdier specialization of a local system on $\mathfrak{h}^{\mathrm{reg}}_{D'}$ along $\mathfrak{h}_{D'/D''}$ will be a well-defined local system on $N^{\mathrm{reg}}_{\mathfrak{h}_{D'/D''}/\mathfrak{h}_{D'}}$. At the level of representations of Poincaré groupoids, the specialization is nothing but restriction to $\Pi_{\mathbf{A}_{D'/D''}}(\mathfrak{h}^{\mathrm{reg}}_{D'})$.

Definition 3.8

(A) A *pure Coxeter category of type D* is the collection of the following data:

(a) A category $\mathcal{C}_{D'}$ for any subset $D' \subset D$;
(b) For $D'' \subset D'$, a local system of restriction functors $F_{D'D''}: \mathcal{C}_{D'} \to \mathcal{C}_{D''}$ on $\mathfrak{h}^{\text{reg}}_{D'/D''}$;
(c) For $D''' \subset D'' \subset D'$, an isomorphism of local systems of functors:

$$\phi_{D'D''D'''}: \operatorname{Sp}_{\mathfrak{h}_{D'/D''}} F_{D'D'''} \xrightarrow{\sim} F_{D''D'''} \circ F_{D'D''}$$

on $\mathfrak{h}^{\text{reg}}_{D'/D''} \times \mathfrak{h}^{\text{reg}}_{D''/D'''}$ which satisfy the natural "cocycle" or "pentagon" identity associated with $D^{iv} \subset D''' \subset D'' \subset D'$.
(d) In case D''' is disjoint from D' (i.e., no vertex of D''' is connected by an edge to a vertex of D', and $D''' \cap D' = \emptyset$), we have a canonical isomorphism $\mathfrak{h}^{\text{reg}}_{D'/D''} = \mathfrak{h}^{\text{reg}}_{(D' \cup D''')/(D'' \cup D''')}$, and we are given a homomorphism of local systems of endomorphism algebras $\eta: \operatorname{End}(F_{D'D''}) \to \operatorname{End}(F_{(D' \cup D''')(D'' \cup D''')})$.

(B) A *tensor Coxeter category of type D* is the additional datum of braided balanced tensor structures on $\mathcal{C}_{D'}$ such that

(i) The pullback of $F_{D'D''}$ to the universal cover $\widetilde{\mathfrak{h}}^{\text{reg}}_{D'/D''}$ is a (trivial) local system of functors equipped with tensor structures $\widetilde{F}_{D'D''}: \mathcal{C}_{D'} \to \mathcal{C}_{D''}$. (But we *do not* require them to respect the balance and braiding. Also, the monodromy isomorphisms of stalks $\gamma_*: (F_{D'D''})_x \xrightarrow{\sim} (F_{D'D''})_y$ where $\gamma: x \longrightarrow y$ is a path in $\Pi(\mathfrak{h}^{\text{reg}}_{D'/D''})$ *are not* required to be morphisms of tensor functors. Neither are the isomorphisms (Ad) above required to respect the tensor structure.);
(ii) The isomorphisms of (c) above pulled back to $\widetilde{\mathfrak{h}}^{\text{reg}}_{D'/D''} \times \widetilde{\mathfrak{h}}^{\text{reg}}_{D''/D'''}$ are isomorphisms of tensor functors;
(iii) Let $\gamma_0 \in \pi_1(\mathfrak{h}^{\text{reg}}_{D'/D''})$ be the generator of the center; geometrically, it is a loop $\exp(2\pi i\theta) \cdot x,\ 0 \le \theta \le 1,\ x \in \mathfrak{h}^{\text{reg}}_{D'/D''}$. The automorphism $\gamma_{0*}: F_{D'D''} \xrightarrow{\sim} F_{D'D''}$ (it is the automorphism induced by the $\mathbb{C}^*$-monodromic structure on the sheaf $F_{D'D''}$) is inverse to the ratio of the balance automorphisms of the identity functors of $\mathcal{C}_{D''}$ and $\mathcal{C}_{D'}$.[2]

[2]We thank A. Appel and V. Toledano Laredo for correcting mistakes in the original version of the definition.

3.9 Comparison with the Appel-Toledano Laredo Coxeter Braided Tensor Categories

If we impose an additional assumption that the local systems in Definition 3.8(Ab) are lifted from $\mathfrak{h}^{\mathrm{reg}}_{D'/D''}/W_{D'/D''}$ (quotient with respect to the free action of the finite group $W_{D'/D''} := \mathrm{Norm}_{L_{D'}}(L_{D''})/L_{D''}$ (normalizer of the Levi subgroup $L_{D''}$ in $L_{D'} \subset G$, modulo $L_{D''}$)), then we get an equivalent version of [AT, Definitions 3.10, 4.1].

In the example of factorizable sheaves $\mathcal{FS}_D$ (Sect. 5.4), the balance on an irreducible sheaf $\mathcal{L}(\lambda)$ is multiplication by $\zeta^{\lambda\cdot(\lambda+2\rho)}$. Factorizable sheaves $\mathcal{FS}_\emptyset$ for Levi=Cartan also have a nontrivial braiding and balance; namely, on an irreducible sheaf $\mathcal{L}_\emptyset(\mu)$, the balance is multiplication by $\zeta^{\mu\cdot(\mu+2\rho)}$. The ratio of these two balances on a weight component $\mathcal{L}(\lambda)^\alpha$ of $\mathcal{L}(\lambda)$ is $\zeta^{\lambda\cdot(\lambda+2\rho)-(\lambda-\alpha)\cdot(\lambda-\alpha+2\rho)}$ and coincides with the monodromy automorphism of the monodromic sheaf $\mathcal{L}(\lambda)^\alpha$.

The identity $\Delta_i(T_i) = R_i^{21}\cdot(T_i\otimes T_i)$, and, more generally, for any $D'\subset D$, $\Delta_{D'}(T_{w_0^{D'}}) = R_{D'}^{21}\cdot(T_{w_0^{D'}}\otimes T_{w_0^{D'}})$, implies $\Delta_{D'}(T_{w_0^{D'}})^2(T^2_{w_0^{D'}}\otimes T^2_{w_0^{D'}})^{-1} = R_{D'}^{12}\circ R_{D'}^{21}$, which in view of Definition 3.8(Biii) is nothing but the usual relation between the braiding and the balance.

4 Algebra

4.1 Lusztig's Symmetries

Given $\zeta\in\mathbb{C}$, $\zeta^6\neq 1$, we consider Lusztig's small quantum group $\mathbf{u}_{D'}$ (see, e.g., [BFS, 0.2.12]). We extend it by the projectors to the weight spaces 1_λ, $\lambda\in X$, to obtain the algebra $\dot{\mathbf{u}}_{D'}$ such that $\mathrm{Rep}(\dot{\mathbf{u}}_{D'}) = \mathcal{C}_{D'}$ (notations of [BFS, 0.2.11-0.2.13]). The algebra $\dot{\mathbf{u}}_{D'}$ is a subalgebra of Lusztig's big quantum group ${}_R\dot{\mathbf{U}}_{D'}$ [LU2, Chapter 31] (where $R\colon \mathbb{Z}[v^{\pm 1}]\to\mathbb{C},\ v\mapsto\zeta$), generated by $E_i = E_i^{(1)}$, $F_i = F_i^{(1)}$, $i\in D'$, and 1_λ, $\lambda\in X$. According to [LU2, Chapters 33,35], if ζ is a root of unity (primitive of order ℓ), there is a reductive algebraic group $\check{G}_{D',\zeta}$ with Cartan torus $\check{T}_{D',\zeta}\subset\check{G}_{D',\zeta}$ and a tensor functor $\mathrm{Fr}^*\colon \mathrm{Rep}(\check{G}_{D',\zeta})\to\mathrm{Rep}({}_R\dot{\mathbf{U}}_{D'})$ (pullback with respect to the quantum Frobenius homomorphism). Note that the character lattice $X^*(\check{T}_{D',\zeta})$ is naturally a sublattice of the weight lattice X.

Lusztig's symmetries $T'_{i,e}, T''_{i,e}$, $i\in D'$, $e=\pm 1$, of ${}_R\dot{\mathbf{U}}_{D'}$ [LU2, 41.1.8] clearly preserve the subalgebra $\dot{\mathbf{u}}_{D'}$ and restrict to the same named symmetries of this subalgebra. We define a functor $T_{\mathbf{u}}$ from $\Pi(\mathfrak{h}^{\mathrm{reg}}_{D'})$ to the category of $\mathbb{C}$-algebras on generators as follows: $T_{\mathbf{u}}(C) = \dot{\mathbf{u}}_{D'}$ for any $C\in\mathbf{C}$; for γ a straight line interval connecting the endpoints in two neighboring chambers C_1, C_2 with the common wall of type s_i, $i\in D'$, we set $T_{\mathbf{u}}(\gamma^+) = T'_{i,1}$ (resp. $T''_{i,-1}$) and $T_{\mathbf{u}}(\gamma^-) = T'_{i,-1}$ (resp. $T''_{i,1}$), if γ goes from a Bruhat smaller chamber to the bigger one (resp. from a Bruhat bigger chamber to the smaller one). According to [LU2, Theorem 39.4.3], $T_{\mathbf{u}}$ is well defined.

Given an integrable ${}_R\dot{\mathbf{U}}_{D'}$-module M with Lusztig's symmetries $T'_{i,e}, T''_{i,e}\colon M \to M$ [LU2, 41.2.3], we define a functor T_M from $\Pi(\mathfrak{h}^{\mathrm{reg}}_{D'})$ to the category of $\mathbb{C}$-vector spaces on generators as follows: $T_M(C) = M$ for any $C \in \mathbf{C}$; for γ a straight line interval connecting the endpoints in two neighboring chambers C_1, C_2 with the common wall of type s_i, $i \in D'$, we set $T_M(\gamma^+) = T'_{i,1}$ (resp. $T''_{i,-1}$) and $T_M(\gamma^-) = T'_{i,-1}$ (resp. $T''_{i,1}$), if γ goes from a Bruhat smaller chamber to the bigger one (resp. from a Bruhat bigger chamber to the smaller one). According to [LU2, Proposition 41.2.4], T_M is well defined.

Let ${}_R\mathcal{C}_{D'}$ (resp. $\mathcal{C}_{D'}$) denote the category of integrable ${}_R\dot{\mathbf{U}}_{D'}$-modules (resp. $\dot{\mathbf{u}}_{D'}$-modules), and let $\Upsilon\colon {}_R\mathcal{C}_{D'} \to \mathcal{C}_{D'}$ stand for the restriction functor. In the previous paragraph, we have defined the local system ${}_RF_{D',\emptyset}$ on $\mathfrak{h}^{\mathrm{reg}}_{D'}$ of restriction functors ${}_R\mathcal{C}_{D'} \to \mathrm{Vect}_X = {}_R\mathcal{C}_{\emptyset}$ to the category of X-graded $\mathbb{C}$-vector spaces.

Proposition 4.2 *There exists a unique local system $F^{\mathcal{C}}_{D',\emptyset}$ on $\mathfrak{h}^{\mathrm{reg}}_{D'}$ of restriction functors $\mathcal{C}_{D'} \to \mathrm{Vect}_X = \mathcal{C}_{\emptyset}$ such that ${}_RF_{D',\emptyset} = F^{\mathcal{C}}_{D',\emptyset} \circ \Upsilon$.*

Proof According to [AG, Theorem 4.7], we view $\mathcal{C}_{D'}$ as the category of Hecke eigen-objects in ${}_R\mathcal{C}_{D'}$. That is, an object of $\mathcal{C}_{D'}$ is an object M of ${}_R\mathcal{C}_{D'}$ endowed with a collection of isomorphisms $\alpha_V\colon \mathrm{Fr}^*(V) \otimes M \xrightarrow{\sim} \mathrm{Res}^{\check{T}_{D',\zeta}}_{\check{G}_{D',\zeta}}(V) \otimes M$, $V \in \mathrm{Rep}(\check{G}_{D',\zeta})$. Since Lusztig's symmetries act on $\mathrm{Fr}^*(V)$ and on $\mathrm{Res}^{\check{T}_{D',\zeta}}_{\check{G}_{D',\zeta}}(V)$, a Hecke eigen-object (M, α) gives rise to a representation $T_{(M,\alpha)}$ of $\Pi(\mathfrak{h}^{\mathrm{reg}}_{D'})$. Hence, the action of $\Pi(\mathfrak{h}^{\mathrm{reg}}_{D'})$ on ${}_RF_{D',\emptyset}$ canonically extends to the action of $\Pi(\mathfrak{h}^{\mathrm{reg}}_{D'})$ on $F_{D',\emptyset}$. □

Remark 4.3 For example, if $\lambda \in X$ is a dominant ℓ-restricted weight (recall that ℓ is the order of ζ), then the irreducible $\dot{\mathbf{u}}_{D'}$-module $L^{\lambda}_{\mathbf{u}}$ with highest weight λ is the restriction of the irreducible ${}_R\dot{\mathbf{U}}_{D'}$-module $L^{\lambda}_{\mathbf{U}}$ with highest weight λ, and $T_{L^{\lambda}_{\mathbf{u}}} = T_{L^{\lambda}_{\mathbf{U}}}$.

4.4 A Coxeter Structure on ${}_R\mathcal{C}$, $\mathcal{C}$

We need to define the local systems of restriction functors $F^{\mathcal{C}}_{D'D''}$, not just $F^{\mathcal{C}}_{D',\emptyset}$ as in the previous subsection. To this end, we restrict the action of the fundamental groupoid $\Pi(\mathfrak{h}^{\mathrm{reg}}_{D'})$ defined in Proposition 4.2 to the subgroupoid $\Pi_{\mathrm{pr}^{-1}(C_0^{D''})}(\mathfrak{h}^{\mathrm{reg}}_{D'}) \subset \Pi(\mathfrak{h}^{\mathrm{reg}}_{D'})$; see Definition 3.5(a). More precisely, we consider a wall between two neighboring chambers c_1, c_2 of $\mathfrak{h}^{\mathrm{reg}}_{D'/D'',\mathbb{R}}$. Let C_1 (resp. C_2) be a (unique) chamber of $\mathfrak{h}^{\mathrm{reg}}_{D',\mathbb{R}}$ adjacent to c_1 (resp. c_2) such that $\mathrm{pr}(C_1) = \mathrm{pr}(C_2) = C_0^{D''}$ (the fundamental chamber of $\mathfrak{h}^{\mathrm{reg}}_{D'',\mathbb{R}}$). Let γ be a straight line interval going from c_1 to c_2, and let Γ

be its lift going from C_1 to C_2 parallel to $\mathfrak{h}_{D'/D''}$. In the notations of Sect. 4.1, we set $T_M^{D'/D''}(\gamma^+) = \mathrm{Id}_M$, $T_M^{D'/D''}(\gamma^-) = T_M^{-1}(\Gamma^-) \circ T_M(\Gamma^+)$.

Lemma 4.5 *The action of $\Pi(\mathfrak{h}_{D'/D''}^{\mathrm{reg}})$ on a ${}_R\dot{\mathbf{U}}_{D'}$-module M commutes with the action of ${}_R\dot{\mathbf{U}}_{D''} \subset {}_R\dot{\mathbf{U}}_{D'}$.*

Proof Note that (since c_1 and c_2 are neighbors) the types of walls intersected by Γ are all in D'' except for exactly one $d \in D' - D''$. Let C_1' be the chamber adjacent to c_1 such that $\mathrm{pr}(C_1') = w_0^{D''} C_0^{D''}$. Then $C_2 = w_0^{D'' \sqcup d} C_1'$. Let Δ be a straight line interval going from C_1' to C_1 and ending at the starting point of Γ, and let $\Gamma\Delta$ be the concatenation of Γ and Δ. Then $T_M^{-1}(\Gamma^-) \circ T_M(\Gamma^+) = T_M^{-1}(\Delta^+) \circ T_M(\Delta^-) \circ T_M^{-1}(\Gamma\Delta^-) \circ T_M(\Gamma\Delta^+)$. It suffices to prove that $T_{\mathbf{U}}^{-1}(\Gamma^-) \circ T_{\mathbf{U}}(\Gamma^+) = \mathrm{Id}_{{}_R\dot{\mathbf{U}}_{D''}}$. According to [LU1, Corollary 5.9] or [KT], we have $T_{\mathbf{U}}^{-1}(\Gamma\Delta^-) \circ T_{\mathbf{U}}(\Gamma\Delta^+)(E_i) = \tilde{K}_i^{-2} E_i$, $T_{\mathbf{U}}^{-1}(\Gamma\Delta^-) \circ T_{\mathbf{U}}(\Gamma\Delta^+)(F_i) = F_i \tilde{K}_i^2$ for any $i \in D'' \sqcup d$ and $T_{\mathbf{U}}^{-1}(\Delta^-) \circ T_{\mathbf{U}}(\Delta^+)(E_i) = \tilde{K}_i^{-2} E_i$, $T_{\mathbf{U}}^{-1}(\Delta^-) \circ T_{\mathbf{U}}(\Delta^+)(F_i) = F_i \tilde{K}_i^2$ for any $i \in D''$. □

Now by Lemma 3.6(a) (and Proposition 4.2), we obtain the desired local system of restriction functors ${}_R F_{D'D''} \colon {}_R\mathcal{C}_{D'} \to {}_R\mathcal{C}_{D''}$ (resp. $F^{\mathcal{C}}_{D'D''} \colon \mathcal{C}_{D'} \to \mathcal{C}_{D''}$) on $\mathfrak{h}_{D'/D''}^{\mathrm{reg}}$. The isomorphisms of Lemma 3.6(c) give rise to the isomorphisms of Definition 3.8(Ac). The conditions of Definition 3.8(B) are satisfied trivially.

Remark 4.6 The Coxeter structure on ${}_R\mathcal{C}$ studied in [TL] differs from ours by the twist by an invertible local system. More precisely, for a weight component $M_\lambda \subset M$, in the setup of Sect. 4.1, the Coxeter structure of [TL, 4.1.3] $T_{M_\lambda}^{TL}(\gamma^+) = \zeta^{d_i \langle \check{\alpha}_i, \lambda\rangle^2/4} T''_{i,+1} = (-1)^{\langle \check{\alpha}_i, \lambda\rangle} \zeta^{d_i \langle \check{\alpha}_i, \lambda\rangle} \zeta^{d_i \langle \check{\alpha}_i, \lambda\rangle^2/4} T'_{i,+1} = (-1)^{\langle \check{\alpha}_i, \lambda\rangle} \zeta^{d_i \langle \check{\alpha}_i, \lambda\rangle + d_i \langle \check{\alpha}_i, \lambda\rangle^2/4} T_{M_\lambda}(\gamma^+)$ (the second equality is [LU2, 5.2.3.b)]) for γ going through an s_i-wall from a Bruhat smaller chamber to a Bruhat bigger one; $T_{M_\lambda}^{TL}(\gamma^-) = \zeta^{d_i \langle \check{\alpha}_i, \lambda\rangle^2/4} T''_{i,+1} = \zeta^{d_i \langle \check{\alpha}_i, \lambda\rangle^2/4} T_{M_\lambda}(\gamma^-)$ for γ going through an s_i-wall from a Bruhat bigger chamber to a Bruhat smaller one; the remaining two half monodromies are the inverses of the above two.

Note that if $s_i\lambda = \lambda$, then the scalar factors above are identically equal to one. We define $M_{W\lambda} := \bigoplus_{\mu \in W\lambda} M_\mu$, the direct sum over the Weyl group orbit of λ. Since $T_{M_{W\lambda}}^{TL}(\gamma^\pm)$ arise from a local system on $\mathfrak{h}^{\mathrm{reg}}/W$ [TL] (i.e., ${}_R F_{D\emptyset}^{TL}$ possesses a W-equivariant structure), it follows that ${}_R F_{D\emptyset}$ also possesses a W-equivariant structure.

5 Topology

5.1 Erratum to [BFS]

We take this opportunity to correct a blunder pertaining to the non-simply laced case of [BFS]. Let us define the quantities $i' \cdot j'$ as $\alpha_i \cdot \alpha_j$ in the sense of Sect. 3.1. Then throughout [BFS] in all formulas, the occurrences of $i \cdot j$ should be replaced by $i' \cdot j'$.

For example:

- in [BFS, Part 0, 2.1]: $\langle i, j' \rangle = 2i' \cdot j'/i' \cdot i'$, and $d_i = i' \cdot i'/2$;
- in the relation [BFS, Part 0, 2.7(d)], one should replace $\zeta^{i \cdot j}$ by $\zeta^{i' \cdot j'}$,
  Thus, in [BFS, Part 0, 2.7] $\tilde{K}_i = K_i^{d_i}$ as before, and $\zeta_i = \zeta^{d_i}$, but the meaning of d_i should be changed: d_i is defined *not* as half the square length of the coroot $\check{\alpha}_i$ but as half the square length of the root α_i.
- On the geometric side, the monodromy of the cohesive local system corresponding to a full counterclockwise turn of a point i around j should be $\zeta^{-2i' \cdot j'}$, cf. [BFS, Part 0, 3.10] and Sect. 5.2

To summarize, the main assertion of [BFS] (reviewed below in more details) consists of two parts: first, an equivalence of the geometric category $\mathcal{FS}$ with a category of graded modules over the algebra $\mathfrak{u}$ defined in [BFS, Part 0, 2.7]. This assertion is true, and our correction just replaces the root system by the dual one on both sides. The second assertion is an identification of $\mathfrak{u}$ with Lusztig's small quantum group. This identification is described in [BFS, Part 0, 2.12, or Part II, 12.5] and should be corrected: the "geometric" algebra $\mathfrak{u}$ is isomorphic to the "Langlands dual" Lusztig's algebra connected with the *dual* root system.

This replacement of the root system by its dual is a rather subtle point. Its origin lies in the definition of the braiding in [LU2], cf. the proof of [LU2, Lemma 32.2.3].

Also, there is a misprint in the definition of a balance in [BFS, IV.6.6]: $n(\lambda)$ must be replaced by $2n(\lambda) = \lambda \cdot (\lambda + 2\rho)$.

5.2 A Review of [BFS]: Cohesive System and Algebra $\mathfrak{u}^-$

For $\beta \in \mathbb{N}[I]$, we consider the configuration space $\mathbb{A}^\beta$ of colored divisors on the complex affine line $\mathbb{A}^1$. The open subspace $\overset{\circ}{\mathbb{A}}{}^\beta \subset \mathbb{A}^\beta$ of multiplicity free divisors carries a one-dimensional *cohesive local system* $\mathcal{J}^\beta$ with the following monodromies: $\zeta^{-2\alpha_i \cdot \alpha_j}$ when a point of color i goes counterclockwise around a point of color $j \neq i$ and $-\zeta^{-\alpha_i \cdot \alpha_i}$ when two points of color i trade their positions going around a half circle counterclockwise. We denote by $\mathcal{I}^\beta$ the Goresky-MacPherson extension of $\mathcal{J}^\beta$ to $\mathbb{A}^\beta$ (a perverse sheaf). Given two disjoint open discs $\mathbb{A}^1 \supset D(p_i, \varepsilon_i)$, $i = 1, 2$, with centers in p_i of radii ε_i, and $\beta_{1,2} \in \mathbb{N}[I]$, we have an open embedding $m\colon D(p_1, \varepsilon_1)^{\beta_1} \times D(p_2, \varepsilon_2)^{\beta_2} \hookrightarrow \mathbb{A}^{\beta_1+\beta_2}$ and a canonical isomorphism $\psi\colon m^*\mathcal{I}^{\beta_1+\beta_2} \xrightarrow{\sim} \mathcal{I}^{\beta_1} \boxtimes \mathcal{I}^{\beta_2}$. We denote by r the closed embedding $\mathbb{A}^\beta_\mathbb{R} \hookrightarrow \mathbb{A}^\beta$; we keep the same notation for $D(p, \varepsilon)^\beta_\mathbb{R} \hookrightarrow D(p, \varepsilon)^\beta$ in case $p \in \mathbb{R}$. We consider the *real hyperbolic stalk* $\Phi_\mathbb{R}(\mathcal{I}^\beta) := H^\bullet_c(\mathbb{A}^\beta_\mathbb{R}, r^*\mathcal{I}^\beta)$. According to [BFS, Theorem I.3.9], $\Phi_\mathbb{R}(\mathcal{I}^\beta)$ lives in cohomological degree 0. According to [BFS, Theorem I.3.5], we have a canonical isomorphism $\Phi_\mathbb{R}(\mathcal{I}^\beta)^* \simeq$

$\Phi_{\mathbb{R}}(\mathcal{D}\mathcal{I}^\beta)$ where $\mathcal{D}$ stands for the Verdier duality. We have canonical isomorphisms $\Phi_{\mathbb{R}}(\mathcal{I}^\beta) \simeq H_c^\bullet(D(p,\varepsilon)_{\mathbb{R}}^\beta, r^*\mathcal{I}^\beta)$ for arbitrary $p \in \mathbb{R}$, $\varepsilon \in \mathbb{R}_{>0}$.

The isomorphism ψ^{-1} above gives rise to the multiplication map $\Phi_{\mathbb{R}}(\mathcal{I}^{\beta_1}) \otimes \Phi_{\mathbb{R}}(\mathcal{I}^{\beta_2}) \simeq H_c^\bullet(D(1,\varepsilon_1)_{\mathbb{R}}^{\beta_1}, r^*\mathcal{I}^{\beta_1}) \otimes H_c^\bullet(D(0,\varepsilon_2)_{\mathbb{R}}^{\beta_2}, r^*\mathcal{I}^{\beta_2}) \to H_c^\bullet(\mathbb{A}_{\mathbb{R}}^{\beta_1+\beta_2}, r^*\mathcal{I}^{\beta_1+\beta_2}) = \Phi_{\mathbb{R}}(\mathcal{I}^{\beta_1+\beta_2})$. The above self-duality gives rise to the comultiplication map $\Phi_{\mathbb{R}}(\mathcal{I}^{\beta_1+\beta_2}) \to \Phi_{\mathbb{R}}(\mathcal{I}^{\beta_1}) \otimes \Phi_{\mathbb{R}}(\mathcal{I}^{\beta_2})$. According to [BFS, I,II], the twisted graded Hopf algebra $\Phi_{\mathbb{R}}(\mathcal{I}) := \bigoplus_{\beta\in\mathbb{N}[I]} \Phi_{\mathbb{R}}(\mathcal{I}^\beta)$ is naturally isomorphic to $\mathfrak{u}^-$, the negative part of the small quantum group at $v = \zeta$.

5.3 A Review of [BFS]: Factorizable Sheaves

We have an open subset $\overset{\circ}{\mathbb{A}}{}^\beta \supset \overset{\diamond}{\mathbb{A}}{}^\beta$ of configurations of distinct colored points in $\mathbb{A}^1 \setminus \{0\}$. It carries a one-dimensional cohesive local system $\mathcal{J}_\lambda^\beta$ with the monodromies around diagonals same as the ones of $\mathcal{J}^\beta$ and also the monodromy $\zeta^{2\lambda\cdot\alpha_i}$ when a point of color i goes around 0 counterclockwise (here λ is a weight). We denote by $\mathcal{I}_\lambda^\beta$ the Goresky-MacPherson extension of $\mathcal{J}_\lambda^\beta$ to $\mathbb{A}^\beta$ (a perverse sheaf). Denoting by $A(p,\varepsilon)$ the complement in $\mathbb{A}^1$ to the closure of $D(p,\varepsilon)$ (an open annulus), we have an open embedding $m\colon A(0,\varepsilon)^{\beta_1} \times D(0,\varepsilon)^{\beta_2} \hookrightarrow \mathbb{A}^{\beta_1+\beta_2}$ and a canonical isomorphism $\psi\colon m^*\mathcal{I}_\lambda^{\beta_1+\beta_2} \xrightarrow{\sim} \mathcal{I}_{\lambda-\beta_2}^{\beta_1} \boxtimes \mathcal{I}_\lambda^{\beta_2}$. A *factorizable sheaf* of highest weight λ is a collection of perverse sheaves $\mathcal{M}^\beta$ on $\mathbb{A}^\beta$ equipped with factorization isomorphisms $m^*\mathcal{M}^{\beta_1+\beta_2} \xrightarrow{\sim} \mathcal{I}_{\lambda-\beta_2}^{\beta_1} \boxtimes \mathcal{M}^{\beta_2}$. In particular, since for $p_1 \in \mathbb{R}$ big enough, and ε_1 small enough, $D(p_1,\varepsilon_1) \subset A(0,\varepsilon)$, and the restriction of $\mathcal{I}_{\lambda-\beta_2}^{\beta_1}$ from $A(0,\varepsilon)^{\beta_1}$ to $D(p_1,\varepsilon_1)^{\beta_1}$ is canonically isomorphic to $\mathcal{I}^{\beta_1}$, we obtain isomorphisms

$$\mathcal{M}^{\beta_1+\beta_2}|_{D(p_1,\varepsilon_1)^{\beta_1}\times D(0,\varepsilon)^{\beta_2}} \xrightarrow{\sim} \mathcal{I}^{\beta_1} \boxtimes \mathcal{M}^{\beta_2} \tag{5.1}$$

Let $a\colon \mathbb{A}^\beta \to \mathbb{A}^1$ be the addition and $\Phi_a(\mathcal{M}^\beta)$ the corresponding vanishing cycles. It is a perverse sheaf on the hypersurface $a = 0$, but since $\mathcal{M}^\beta$ is smooth along coordinate-diagonal stratification, $\Phi_a(\mathcal{M}^\beta)$ is supported at the origin $\{\beta\cdot 0\} \subset \mathbb{A}^\beta$, so we will view $\Phi_a(\mathcal{M}^\beta)$ just as a vector space. It is canonically isomorphic to $H_c^\bullet(\mathbb{A}_{\mathbb{R}+}^\beta, r_+^*\mathcal{M}^\beta)$ where $r_+\colon \mathbb{A}_{\mathbb{R}+}^\beta \hookrightarrow \mathbb{A}^\beta$ is the closed embedding of the "real half-space" formed by the real configurations in the preimage $a^{-1}(\mathbb{R}_{\geq 0})$. Since the vanishing cycles commute with duality, we have a canonical isomorphism $\Phi_a(\mathcal{M}^\beta)^* \simeq \Phi_a(\mathcal{D}\mathcal{M}^\beta)$ (see, e.g., [BFS, Theorem 0.6.3]). The isomorphism (5.1) gives rise to the map $\Phi_{\mathbb{R}}(\mathcal{I}^{\beta_1}) \otimes \Phi_a(\mathcal{M}^{\beta_2}) \simeq H_c^\bullet(D(p_1,\varepsilon_1)_{\mathbb{R}}^{\beta_1}, r^*\mathcal{I}^{\beta_1}) \otimes H_c^\bullet(D(0,\varepsilon)_{\mathbb{R}+}^{\beta_2}, r_+^*\mathcal{M}^{\beta_2}) \to H_c^\bullet(\mathbb{A}_{\mathbb{R}+}^{\beta_1+\beta_2}, r_+^*\mathcal{M}^{\beta_1+\beta_2}) = \Phi_a(\mathcal{M}^{\beta_1+\beta_2})$, i.e., to the action of $\mathfrak{u}^- \simeq \Phi_{\mathbb{R}}(\mathcal{I})$ on $\Phi_a(\mathcal{M}) := \bigoplus_{\beta\in\mathbb{N}[I]} \Phi_a(\mathcal{M}^\beta)$. The above self-duality gives rise to the coaction $\Phi_a(\mathcal{M}^{\beta_1+\beta_2}) \to \Phi_{\mathbb{R}}(\mathcal{I}^{\beta_1}) \otimes \Phi_a(\mathcal{M}^{\beta_2})$; equivalently, $\Phi_{\mathbb{R}}(\mathcal{I}^{\beta_1})^* \otimes \Phi_a(\mathcal{M}^{\beta_1+\beta_2}) \to$

$\Phi_a(\mathcal{M}^{\beta_2})$. Taking into account the isomorphism $\Phi_{\mathbb{R}}(\mathcal{I}^{\beta_1})^* = (\mathfrak{u}^-_{-\beta_1})^* \simeq \mathfrak{u}^+_{\beta_1}$, we obtain an action of $\mathfrak{u}^+$ on $\Phi_a(\mathcal{M})$. We assign to $\Phi_a(\mathcal{M}^\beta)$ the weight $\lambda - \beta$. This, together with the action of $\mathfrak{u}^\pm$, defines the action of $\dot{\mathfrak{u}}$ on $\Phi_a(\mathcal{M})$ (an isomorphism of $\dot{\mathfrak{u}}$ and Lusztig's small quantum group $\dot{\mathbf{u}} = \dot{\mathbf{u}}_D$ is established in [BFS, Theorem 2.13], cf. Sect. 5.1). The resulting functor from the category $\mathcal{FS}$ of factorizable sheaves to the category $\mathcal{C}$ of $\dot{\mathbf{u}}$-modules (to be denoted Φ) is an equivalence of categories.

5.4 A Coxeter Structure on $\mathcal{FS}$

The diagram D is the Dynkin diagram of an irreducible Cartan datum $(I, \cdot)$ of finite type. The category of factorizable sheaves introduced in [BFS, 0.4.6] will be denoted by $\mathcal{FS}_D$. For a subdiagram $D' \subset D$, we denote by $\mathcal{FS}_{D'}$ a similarly defined category with grading by the weight lattice X. That is, compared to the definition of $\mathcal{FS}_D$, the lattice of weights is always X, while the set of colors is $D' \subset D = I$. The braided balanced tensor structure on $\mathcal{FS}_{D'}$ is introduced in [BFS, 0.5.9, 0.5.10, IV.6.6].

In order to construct the local systems $F^{\mathcal{FS}}_{D'D''}$ of restriction functors $\mathcal{FS}_{D'} \to \mathcal{FS}_{D''}$, we vary the definition [BFS, 0.6.7, 0.6.8] of the vanishing cycles functor Φ in the following way. Let $\mathbb{N}[D'] \ni \beta = \sum_{j\in D'} b_j\alpha_j$, and let $\mathbb{A}^\beta = \prod_{j\in D'}(\mathbb{A}^1)^{(b_j)}$ be the configuration space of D'-colored effective divisors on the affine line $\mathbb{A}^1$ with coordinate t, of degree ν. For $D'' \subset D'$, we have a pairing $\langle\cdot,\cdot\rangle\colon \mathfrak{h}_{D'/D''} \times \mathbb{A}^\beta \to \mathbb{A}^1$ given in the coordinates $(t_{j,s})_{j\in D',\ 1\le s\le b_j}$ on $\mathbb{A}^\beta$ and $(z_j)_{j\in D'-D''}$ in the basis of fundamental coweights on $\mathfrak{h}_{D'/D''}$ as follows: $\langle(z_j),(t_{j,s})\rangle := \sum_{j\in D'-D''} z_j \sum_{s=1}^{b_j} t_{j,s}$. The decomposition $\beta = \beta'' + {}'\beta := \sum_{j\in D''} b_j\alpha_j + \sum_{k\in D'-D''} b_k\alpha_k$ gives rise to the direct product decomposition $\mathbb{A}^\beta = \mathbb{A}^{\beta''} \times \mathbb{A}^{'\beta}$. Clearly, given a perverse sheaf $\mathcal{M}^\beta$ on $\mathbb{A}^\beta$, the vanishing cycle $\Phi_{\langle\cdot,\cdot\rangle}\mathcal{M}^\beta$ is a perverse sheaf supported on $\mathfrak{h}_{D'/D''} \times \mathbb{A}^{\beta''} \times 0^{'\beta} \simeq \mathfrak{h}_{D'/D''} \times \mathbb{A}^{\beta''}$.

Let us write ζ in the form $\zeta = \exp(\pi i \varkappa)$.

Theorem 5.5 *If $\mathcal{M}^\beta$ is a part of data of a factorizable sheaf $\mathcal{M}$ and $\varkappa$ is sufficiently close to 0, then $\Phi_{\langle\cdot,\cdot\rangle}\mathcal{M}^\beta|_{\mathfrak{h}^{\mathrm{reg}}_{D'/D''}\times\mathbb{A}^{\beta''}}$ is smooth along $\mathfrak{h}^{\mathrm{reg}}_{D'/D''}$.*

Proof It suffices to consider an irreducible $\mathcal{M}$ and hence an irreducible $\mathcal{M}^\beta$. We may and will assume $D' = D$. Then $\mathcal{M}^\beta$ is isomorphic to the Goresky-MacPherson sheaf $\mathcal{I}^\beta_\lambda$ of Sect. 5.3 for a certain weight λ. For $\beta = \sum_{i\in I} b_i\alpha_i$, we consider an *unfolding* $\pi\colon J \twoheadrightarrow I$ such that for any $i \in I$, $\sharp\pi^{-1}(i) = b_i$. Then the product of symmetric groups $\Sigma_\pi := \prod_{i\in I}\mathfrak{S}_{b_i}$ acts on the affine space $\mathbb{A}^J$, and $\mathbb{A}^\beta = \mathbb{A}^J/\Sigma_\pi$. We denote the natural projection $\mathbb{A}^J \to \mathbb{A}^\beta$ by π as well. We denote by $\overset{\diamond}{\mathbb{A}}{}^J \subset \mathbb{A}^J$ the complement to the diagonals in $(\mathbb{A}^1 \setminus \{0\})^J$. We consider the one-dimensional local system $\mathcal{J}^J_\lambda$ on $\overset{\diamond}{\mathbb{A}}{}^J$ with the following monodromies: $\zeta^{-2\alpha_i\cdot\alpha_j}$ when a point of color i goes counterclockwise around a point of color $j \ne i$ and

$\zeta^{2\lambda\cdot\alpha_i}$ when a point of color i goes around 0 counterclockwise. We denote by $\mathcal{I}^J_\lambda$ the Goresky-MacPherson extension of $\mathcal{J}^J_\lambda$ to $\mathbb{A}^J$. Then $\mathcal{I}^J_\lambda$ carries an evident Σ_π-equivariant structure, so Σ_π acts on the perverse sheaf $\pi_*\mathcal{I}^J_\lambda$, and $\mathcal{I}^\beta_\lambda$ is nothing but the subsheaf $(\pi_*\mathcal{I}^J_\lambda)^{\Sigma_\pi,-}$ of Σ_π-antiinvariants in $\pi_*\mathcal{I}^J_\lambda$ (see [BFS, II.6.13]).

For $D'' \subset D = I$, let $J'' := \pi^{-1}(D'') \subset J$ and $'J = J \setminus J''$, so that $\mathbb{A}^{J''} \hookrightarrow \mathbb{A}^J$ (we set the remaining coordinates to be all zeros). The construction of Sect. 5.4 gives rise to a linear map $m\colon \mathfrak{h}_{D/D''} \times \mathbb{A}^{J''} \to T^*_{\mathbb{A}^{J''}}\mathbb{A}^J = (\mathbb{A}^{'J})^* \times \mathbb{A}^{J''}$, and we have $\Phi_{\langle\cdot,\cdot\rangle}\mathcal{M}^\beta = (\pi_* m^\circ \boldsymbol{\mu}_{\mathbb{A}^{J''}/\mathbb{A}^J}\mathcal{I}^J_\lambda)^{\Sigma_\pi,-}$ where $\boldsymbol{\mu}$ is the microlocalization functor [KS1].

Now by Kashiwara-Schapira theorem identifying the microlocalization and the Fourier transform (see [KS1, Proposition 8.6.3], [BRYL, KS2]), we have $\boldsymbol{\mu}_{\mathbb{A}^{J''}/\mathbb{A}^J}\mathcal{I}^J_\lambda \simeq \mathrm{FT}_{\mathbb{A}^{J''}}\,\mathrm{Sp}_{\mathbb{A}^{J''}}\mathcal{I}^J_\lambda$. The specialization $\mathrm{Sp}_{\mathbb{A}^{J''}}\mathcal{I}^J_\lambda$ is a Σ_π-equivariant perverse sheaf all of whose simple constituents are of the form $\mathcal{I}^{'J_1}_{\lambda_1} \boxtimes \mathcal{I}^{J''_2}_{\lambda_2}$ for certain subsets $'J_1 \subset {}'J,\ J''_2 \subset J''$. It suffices to consider these simple constituents. We have $\mathrm{FT}_{\mathbb{A}^{J''}}(\mathcal{I}^{'J_1}_{\lambda_1} \boxtimes \mathcal{I}^{J''_2}_{\lambda_2}) \simeq (\mathrm{FT}_0\,\mathcal{I}^{'J_1}_{\lambda_1}) \boxtimes \mathcal{I}^{J''_2}_{\lambda_2}$.

The Fourier transform $({}'\pi_*(m^\circ\,\mathrm{FT}_0\,\mathcal{I}^{'J_1}_{\lambda_1})|_{\mathfrak{h}^{\mathrm{reg}}_{D/D''}})^{\Sigma'_\pi,-}$ in De Rham setting (i.e., the corresponding D-module) is calculated in [FMTV, Theorem 3.2]. It is a version of Casimir connection

$$\nabla = d - \varkappa \sum_{\alpha\in{}'R^+} \frac{\alpha\cdot\alpha}{2}\frac{d\alpha}{\alpha} f_\alpha e_\alpha \tag{5.2}$$

along $\mathfrak{h}^{\mathrm{reg}}_{D/D''}$. In particular, the required smoothness assertion follows.

In fact, the authors of [FMTV] work with the De Rham complex of "flag logarithmic forms," but [KV, Corollary 6.10] implies that if $\varkappa$ is sufficiently close to 0, this subcomplex is quasi-isomorphic to the De Rham complex $DR(\mathcal{I}^{'J}_{\lambda_1})$. □

Conjecture 5.6 The conclusion of Theorem 5.5 holds true for an arbitrary $\zeta \in \mathbb{C}^*$.

5.7 *Restriction Functors*

The collection of perverse sheaves $\Phi_{\langle\cdot,\cdot\rangle}\mathcal{M}^\nu|_{\mathfrak{h}^{\mathrm{reg}}_{D'/D''}\times\mathbb{A}^{\nu''}}$, $\nu \in Y^+$, enjoys the D''-factorization property, i.e., may be viewed as a local system over $\mathfrak{h}^{\mathrm{reg}}_{D'/D''}$ of objects of $\mathcal{FS}_{D''}$. This provides the desired construction of the local systems $F^{\mathcal{FS}}_{D'D''}$ of restriction functors $\mathcal{FS}_{D'} \to \mathcal{FS}_{D''}$ required by Definition 3.8(Ab), in the formal neighborhood of 0 with respect to $\varkappa$.

If Conjecture 5.6 is true, we get these local systems for every ζ.

5.8 Iterated Vanishing Cycles

The isomorphisms of Definition 3.8(Ac) are a particular case of the following construction. Let $\langle\cdot,\cdot\rangle_W : W \times V \to \mathbb{A}^1$ be a bilinear pairing between two complex vector spaces. Let $U \subset W$ be a linear subspace. We denote the restriction of $\langle\cdot,\cdot\rangle$ to $U \times V$ by $\langle\cdot,\cdot\rangle_U$. Let $\mathcal{M}$ be a perverse sheaf on V smooth along a central hyperplane arrangement. We will view $\Phi_{\langle\cdot,\cdot\rangle_W}\mathcal{M}$ as a perverse sheaf on $W \times W^\perp$ (where $W^\perp \subset V$ is the annihilator of W). Note that the pairing $\langle\cdot,\cdot\rangle_W$ descends to the well-defined pairing $\langle\cdot,\cdot\rangle_{W/U}$ between W/U and $U^\perp$.

Theorem 5.9 *There is a canonical isomorphism*

$$\mathrm{Sp}_{U\times W^\perp}\Phi_{\langle\cdot,\cdot\rangle_W}\mathcal{M} \xrightarrow{\sim} \Phi_{\langle\cdot,\cdot\rangle_{W/U}}\Phi_{\langle\cdot,\cdot\rangle_U}\mathcal{M}$$

of perverse sheaves on $U \times W/U \times W^\perp$.

Proof of Theorem 5.9. Let $V^\perp \subset W$ be the kernel of the bilinear pairing $\langle,\rangle$: $W \times V \to \mathbb{A}^1$. The smooth base change for the projection $W \to W/V^\perp$ reduces the claim to a construction of a canonical isomorphism

$$\mathrm{Sp}_{Y\times X^\perp/Y\times Y^\perp}\mu_{Y/V}\mathcal{M} \xrightarrow{\sim} \mu_{Y\times X^\perp/X\times X^\perp}\mu_{X/V}\mathcal{M} \tag{5.3}$$

of perverse sheaves on $Y \times (X/Y)^* \times (V/X)^*$, where $Y := W^\perp \subset X := U^\perp \subset V$, and μ is the microlocalization functor [KS1]. This isomorphism will be proved in the next section; see Theorem 6.4. □

6 Iterated Specialization and Microlocalization

6.1 Iterated Specialization

Fix a complex or real vector space V equipped with a finite central hyperplane arrangement $\mathcal{H} = \{H_i\}$. A linear subspace $V' \subset V$ is called a *flat* if it is an intersection of some hyperplanes from $\mathcal{H}$. A filtration

$$\ldots \subset V_{i+1} \subset V_i \subset \ldots \subset V_0 = V$$

is called *admissible* if all V_i are flats or 0.

Let $V'' \subset V' \subset V$ be an admissible filtration. $\mathcal{H}$ induces a central arrangement on V'/V''. Let $Sh_{\mathcal{H}}(V'/V'')$ denote the category of constructible sheaves smooth along the corresponding stratification, and let $D^b_{\mathcal{H}}(V'/V'')$ denote the bounded derived category of complexes whose cohomology belongs to $Sh_{\mathcal{H}}(V'/V'')$.

If V is complex and $\mathcal{H}$ is real, which means that all H_i are given by real equations, then the abelian subcategory $Perv_{\mathcal{H}}(V) \subset D^b_{\mathcal{H}}(V)$ admits a description in terms of linear algebra (quiver) data, cf. [KSCH].

Given a flat $W \subset V$, we have the specialization functor [V, KS1]

$$\mathrm{Sp}_W : D^b_{\mathcal{H}}(V) \longrightarrow D^b_{\mathcal{H}}(W \oplus V/W)$$

whose value on $\mathcal{M} \in D^b_{\mathcal{H}}(V)$ can be described as follows [V, Section 9]. We fix Hermitian metrics on V and on V/W. Let p denote the natural projection $V \to V/W$. Let $\xi = (w, u) \in W \oplus V/W$. We fix a sufficiently small $\varepsilon > 0$ and $0 < \rho \ll \varepsilon$. We set $U_{\varepsilon,\rho} = \{v \in V : \rho\|w - v\| + \|\rho u - p(v)\| < \varepsilon\rho\}$: an open subset of V. Then the stalk $\mathrm{Sp}_W(\mathcal{M})_\xi = R\Gamma(U_{\varepsilon,\rho}, \mathcal{M})$.

The following lemma is a consequence of this description of specialization, cf. [FKS, Theorem 3.17].

Lemma 6.2

(a) Let $V_2 \subset V_1 \subset V$ be an admissible filtration. We have a natural isomorphism of functors $D^b_{\mathcal{H}}(V) \longrightarrow D^b_{\mathcal{H}}(V_2 \oplus V_1/V_2 \oplus V/V_1)$,

$$\phi_{12} : \mathrm{Sp}_{V_1/V_2}\mathrm{Sp}_{V_2} \xrightarrow{\sim} \mathrm{Sp}_{V_2}\mathrm{Sp}_{V_1}.$$

Let us abbreviate the notation as $\phi_{12} : \mathrm{Sp}_1\mathrm{Sp}_2 \xrightarrow{\sim} \mathrm{Sp}_2\mathrm{Sp}_1$.

(b) Let $V_3 \subset V_2 \subset V_1 \subset V$ be an admissible filtration. The various isomorphisms ϕ from (a) satisfy the pentagon relation $\phi_{23} \circ \phi_{13} \circ \phi_{12} = \phi_{12} \circ \phi_{23} : \mathrm{Sp}_1\mathrm{Sp}_2\mathrm{Sp}_3 \longrightarrow \mathrm{Sp}_3\mathrm{Sp}_2\mathrm{Sp}_1$. □

6.2.1 Cube

Let us explain the relation (b). We have eight categories related to the subquotients of V:

$$D^b_{\mathcal{H}}(V)$$

$$D^b_{\mathcal{H}}(V_3 \oplus V/V_3) \quad D^b_{\mathcal{H}}(V_2 \oplus V/V_2) \quad D^b_{\mathcal{H}}(V_1 \oplus V/V_1)$$

$$D^b_{\mathcal{H}}(V_3 \oplus V_2/V_3 \oplus V/V_2) \quad D^b_{\mathcal{H}}(V_3 \oplus V_1/V_3 \oplus V/V_1) \quad D^b_{\mathcal{H}}(V_2 \oplus V_1/V_2 \oplus V/V_1)$$

$$D^b_{\mathcal{H}}(V_3 \oplus V_2/V_3 \oplus V_1/V_2 \oplus V/V_1),$$

they are in bijection with the vertices of a cube. The functors Sp act from a category to all the categories one level below it. There are six longest paths from the category $D^b_{\mathcal{H}}(V)$ to the category $D^b_{\mathcal{H}}((V_3 \oplus V_2/V_3 \oplus V_1/V_2 \oplus V/V_1)$; these paths are in bijection with the symmetric group S_3 (this is an instance of a well-known geometric fact: the longest paths on an n-cube are in bijection with the symmetric group S_n).

The paths are connected by homotopies arising from the natural transformations ϕ; this is a weak Bruhat order on S_3. Among these six paths, there are two which are equal. The pentagon (b) above is a hexagon where one of the natural transformations on the right is the identity.

6.3 Iterated Microlocalization

We also have the microlocalization functor which may be defined as the composition

$$\boldsymbol{\mu}_{W/V} : D^b_{\mathcal{H}}(V) \xrightarrow{\mathrm{Sp}_W} D^b_{\mathcal{H}}(W \oplus V/W) \xrightarrow{\mathrm{FT}_W} D^b_{\mathcal{H}}(W \oplus (V/W)^*)$$

where FT_W is the Fourier-Sato transformation [KS1]. In order to keep track of the ambient space, from now on, we will use another notation for the specialization: $\mathrm{Sp}_W = \mathrm{Sp}_{W/V}$.

Theorem 6.4 *Let V be a complex vector space, and let $Y \subset X \subset V$ be an admissible filtration. For $\mathcal{M} \in Perv_{\mathcal{H}}(V)$, there exists a canonical isomorphism*

$$\psi_{XY} : \mathrm{Sp}_{Y\times X^\perp/Y\times Y^\perp}\, \boldsymbol{\mu}_{Y/V}\mathcal{M} \xrightarrow{\sim} \boldsymbol{\mu}_{Y\times X^\perp/X\times X^\perp}\boldsymbol{\mu}_{X/V}\mathcal{M} \tag{6.1}$$

of perverse sheaves on $Y \times (X/Y)^* \times (V/X)^*$.

These isomorphisms satisfy the pentagon relation connected with an admissible filtration $Y \subset X \subset Z \subset V$.

6.5 Proof of Theorem 6.4

The rest of this section is devoted to the proof of this theorem.

The following properties are consequences of the definition.

6.5.1 Subspace

For a linear subspace $Y \subset V$ and a perverse sheaf $\mathcal{M}$ on V, we have a canonical isomorphism $\boldsymbol{\mu}_{Y/V}\mathcal{M} \xrightarrow{\sim} \boldsymbol{\mu}_{Y/Y\times(V/Y)}\, \mathrm{Sp}_{Y/V}\, \mathcal{M}$.

6.5.2 Product

For a perverse sheaf $\mathcal{M}$ on the product of two vector spaces $Y \times Z$, monodromic along the projection $Y \times Z \to Y$, we have a canonical isomorphism $\boldsymbol{\mu}_{Y/Y\times Z}\mathcal{M} \xrightarrow{\sim} FS_{Y/Y\times Z}\mathcal{M}$ between the microlocalization and the Fourier-Sato transform on the vector bundle $Y \times Z \to Y$.

6.5.3 *D*-modules

We will have to work on the D-module side of the Riemann-Hilbert correspondence, so we recall the definition of the specialization functor in this context. Given a linear subspace $Y \subset V$, we choose a complementary subspace $Z \subset V$, $V \simeq Y \oplus Z$, with linear coordinates $z_1, \ldots, z_d$. Then the ring of differential operators D_V has a grading such that $\deg z_i = 1$, $\deg \partial_{z_i} = -1$, $\deg y = \deg \partial_y = 0$ for any $i = 1, \ldots, d$, and any linear coordinates y on Y. Let $F^\bullet D_V$ be the corresponding descending filtration. Note that $\mathrm{gr}^F D_V$ is canonically isomorphic to $D_{Y\times(V/Y)}$. Let M be a regular holonomic D-module on V. It possesses a unique (descending) Malgrange-Kashiwara filtration $\ldots F^{-1}M \supset F^0M \supset F^1M \supset \ldots$ compatible with the filtration on D_V such that (a) for $j > 0$ and $k \gg 0$, we have $F^{\pm k \pm j}M = F^{\pm j}D_V F^{\pm k}M$; (b) the generalized eigenvalues of the Euler vector field $\sum_{1\le i\le d} z_i\partial_{z_i}$ on $F^kM/F^{k+1}M$ have real parts in $[k, k+1)$ for any $k \in \mathbb{Z}$.

The specialization of M is defined as a $D_{Y\times(V/Y)}$-module $\mathrm{Sp}_{Y/V} M := \mathrm{gr}^F M$. We say that M is *potentially monodromic* along Y if the Malgrange-Kashiwara filtration $F^\bullet M$ is compatible with some grading $G^\bullet M$ compatible with the grading on D_V. Equivalently, we can choose a complementary subspace $Z \subset V$ such that M is monodromic along the corresponding projection $V \to Y$. Any such choice defines the isomorphisms $Y \times Z \xrightarrow{\sim} V$ and $\mathrm{Sp}_{Y/V} M \xrightarrow{\sim} M$.

6.5.4 *D*-modules on a Product Space

For a regular holonomic D-module on the product of two vector spaces $Y \times Z$, monodromic along the projection $Y \times Z \to Y$, its microlocalization $\boldsymbol{\mu}_{Y/Y\times Z}M \xrightarrow{\sim} FT_{Y/Y\times Z}$ is the following regular holonomic $D_{Y\times Z^*}$-module. We choose some linear coordinates $y_1, \ldots, y_e$ on Y and $z_1, \ldots, z_d$ on Z. Let $\xi_1, \ldots, \xi_d$ be the dual coordinates on Z^*. Then we have an isomorphism $D_{Y\times Z^*} \xrightarrow{\sim} D_{Y\times Z}$: $y_i \mapsto y_i$, $\partial_{y_i} \mapsto \partial_{y_i}$, $\xi_j \mapsto \partial_{z_j}$, $\partial_{\xi_j} \mapsto -z_j$ (independent of the choice of coordinates), and $\boldsymbol{\mu}_{Y/Y\times Z}M$ is nothing but M viewed as a $D_{Y\times Z^*}$-module via this isomorphism.

Now we proceed to the construction of isomorphism (6.1) in the D-module setting. All the D_V-modules below are assumed to be regular holonomic and smooth along some central hyperplane arrangement in V.

Lemma 6.6 *If a regular holonomic D_V-module M is potentially monodromic along both Y and X, then there is a canonical isomorphism* (6.1).

Proof The microlocalization $\mu_{Y/V}M$ is monodromic along the projection $Y \times Y^\perp \to Y$ and potentially monodromic along $Y \times X^\perp$. Any choice of the direct complement $Z \subset X\colon X = Y \oplus Z$ as in the definition of potential monodromicity gives rise to the isomorphisms $Y \times Z^* \times X^\perp \xrightarrow{\sim} Y \times Y^\perp$ and $\mathrm{Sp}_{Y\times X^\perp/Y\times Y^\perp}\,\mu_{Y/V}M \xrightarrow{\sim} \mu_{Y/V}M$. So it remains to construct an isomorphism $\mu_{Y/V}M \xrightarrow{\sim} \mu_{Y\times X^\perp/X\times X^\perp}\mu_{X/V}M$. We choose a direct complement $S \subset V\colon V = X \oplus S$ as in the definition of potential monodromicity. Then the required isomorphism follows from the explicit formulas in Sect. 6.5.4. One can check that it does not depend on the choices of the complements Z and S. □

Lemma 6.7 *If a regular holonomic D_V-module M is potentially monodromic along Y, then there is a canonical isomorphism* (6.1).

Proof By Sect. 6.5.1, we can replace the RHS of (6.1) by $\mu_{Y\times X^\perp/X\times X^\perp}\mu_{X/V}$ $\mathrm{Sp}_{X/V}\,M$. But $\mathrm{Sp}_{X/V}\,M$ is potentially monodromic along both Y and X. So it remains to use Lemma 6.6 and construct an isomorphism

$$\mathrm{Sp}_{Y\times X^\perp/Y\times Y^\perp}\,\mu_{Y/V}M \xrightarrow{\sim} \mu_{Y/X\times(V/X)}\,\mathrm{Sp}_{X/V}\,M \tag{6.2}$$

We choose a direct complement $Z \subset X\colon X = Y \oplus Z$ and a direct complement $S \subset V\colon V = X \oplus S$ such that $Z \oplus S$ is as in the definition of potential monodromicity. We choose the linear coordinates $z_1, \ldots, z_d$ in Z, and $s_1, \ldots, s_e$ in S, and the dual coordinates $\xi_1, \ldots, \xi_d$ in Z^*, and $\eta_1, \ldots, \eta_e$ in S^*. The D_V-module M has a grading $G^\bullet M$ compatible with the grading on D_V such that $\deg z_i = \deg s_j = 1$, $\deg \partial_{z_i} = \deg \partial_{s_j} = -1$, $\deg y_k = \deg \partial_{y_k} = 0$. According to Sect. 6.5.4, the microlocalization $\mu_{Y/V}M$ amounts to the substitution $\xi_i \mapsto \partial_{z_i}$, $\eta_j \mapsto \partial_{s_j}$, $\partial_{\xi_i} \mapsto -z_i$, $\partial_{\eta_j} \mapsto -s_j$, $y_k \mapsto y_k$, $\partial_{y_k} \mapsto \partial_{y_k}$. To compute the specialization $\mathrm{Sp}_{X/V}\,M$, we use the Malgrange-Kashiwara filtration $F^\bullet M$ as in Sect. 6.5.3. To compute the LHS of (6.2), we use the unique filtration $'F^\bullet M$ compatible as in Sect. 6.5.3 with the grading on D_V such that $\deg z_i = -1$, $\deg \partial_{z_i} = 1$, $\deg s_j = \deg \partial_{s_j} = \deg y_k = \deg \partial_{y_k} = 0$ (note that it is *not* the Malgrange-Kashiwara grading of D_V, but rather the Fourier image of one).

The construction of the isomorphism (6.2) amounts to the construction of the isomorphism

$$\mathrm{gr}^F M \xrightarrow{\sim} \mathrm{gr}^{'F} M \tag{6.3}$$

Note that both $F^\bullet M$ and $'F^\bullet M$ are compatible with the grading $G^\bullet M$, that is, $F^i M = \oplus_j(F^i M \cap G^j M)$ and $'F^i M = \oplus_j('F^i M \cap G^j M)$ for any $i \in \mathbb{Z}$. From the uniqueness of $'F^\bullet M$ and $F^\bullet M$, it follows that $'F^k M \cap G^j M = F^{k+j} M \cap G^j M$. Thus, the desired isomorphism (6.3) is the direct sum of natural isomorphisms $('F^k M \cap G^j M)/('F^{k+1} M \cap G^j M) = (F^{k+j} M \cap G^j M)/(F^{k+j+1} M \cap G^j M)$. □

6.8 The End of the Proof

Now we can finish the construction of isomorphism (6.1) for an arbitrary D_V-module M. It suffices to construct the isomorphism (6.2) for arbitrary M (not necessarily potentially monodromic along Y). By Sect. 6.5.1, we can replace the RHS of (6.2) by $\mu_{Y/X\times(V/X)}\operatorname{Sp}_{Y/X\times(V/X)}\operatorname{Sp}_{X/V}M$. By Lemma 6.2, we can replace this by $\mu_{Y/X\times(V/X)}\operatorname{Sp}_{Y\times(X/Y)/Y\times(V/Y)}\operatorname{Sp}_{Y/V}M$. However, $\operatorname{Sp}_{Y/V}M$ is already potentially monodromic along Y, so by Lemma 6.7, we have $\operatorname{Sp}_{Y\times X^\perp/Y\times Y^\perp}\mu_{Y/V}\operatorname{Sp}_{Y/V}M \xrightarrow{\sim} \mu_{Y/X\times(V/X)}\operatorname{Sp}_{Y\times(X/Y)/Y\times(V/Y)}\operatorname{Sp}_{Y/V}M$. One last application of Sect. 6.5.1 allows to replace the LHS by $\operatorname{Sp}_{Y\times X^\perp/Y\times Y^\perp}\mu_{Y/V}M$.

This completes the proof of Theorem 6.4 and hence that of Theorem 5.9.

Remark 6.9 The statements of Lemma 6.2 and Theorem 6.4 look similar, but the proofs are very different: one is topological; another is De Rham (via D-modules). Let us comment on this discrepancy. On the one hand, Lemma 6.2 has an easy proof in De Rham setting as well. On the other hand, let $Perv_{\mathcal{H}}(V)\subset D^b_{\mathcal{H}}(V)$ denote the subcategory of perverse sheaves. If all $H_i\in\mathcal{H}$ are given by real equations, then $Perv_{\mathcal{H}}(V)$ admits an explicit description in terms of linear algebra (quiver) data, cf. [KSCH]. The specialization and microlocalization functors can be described in this language, and a topological proof of Lemma 6.2 (resp. Theorem 6.4) is given in [FKS, Theorem 3.17] (resp. [FKS, Theorem 5.6]).

7 Discussion

7.1 Desiderata

We have gone to all this trouble just to conjecture that the functor Φ of [BFS] takes the Coxeter structure of Sect. 5.4 to the one of Sect. 4.4. By [FMTV, Theorem 3.2] and Fourier=microlocalization, this would imply that the monodromy of the Casimir connection is given by the Lusztig symmetries for any ζ. Note that the functor Φ of [BFS] is nothing but the stalk at the fundamental chamber $C_0^{D'}$ of the local system of the restriction functors $F^{\mathcal{FS}}_{D'\emptyset}\colon \mathcal{FS}_{D'}\to\mathrm{Vect}_X$ of Sect. 5.4.

7.2 Tilted Functors Φ

Recall the setup of Sect. 5.3. Let us choose a point $z^{(w)}$ in a chamber $C_w\subset\mathfrak{h}_{\mathbb{R}}^{\mathrm{reg}}$, $w\in W$. Instead of $\Phi_a(\mathcal{M}^\beta)$, let us consider the spaces of "tilted" vanishing cycles $\Phi_w(\mathcal{M}^\beta):=\Phi_{\langle z^{(w)},?\rangle}(\mathcal{M}^\beta)\simeq H_c^\bullet(\mathbb{A}^\beta_{\mathbb{R}w},r_w^*\mathcal{M}^\beta)$ where $r_w\colon\mathbb{A}^\beta_{\mathbb{R}w}\hookrightarrow\mathbb{A}^\beta$ is the closed embedding of the "tilted halfspace" formed by the real configurations in the preimage $\langle z^{(w)}, ?\rangle^{-1}(\mathbb{R}_{\geq 0})$. In particular, $\Phi_a \simeq \Phi_e$.

For a real p_1 such that $\langle z^{(w)}, \beta \cdot p_1\rangle$ is positive and big enough, similar to Sect. 5.3, we obtain the map

$$\Phi_{\mathbb{R}}(\mathcal{J}^{\beta_1}) \otimes \Phi_w(\mathcal{M}^{\beta_2}) \simeq H_c^\bullet(D(p_1, \varepsilon_1)_{\mathbb{R}}^{\beta_1}, r^*\mathcal{J}^{\beta_1}) \otimes H_c^\bullet(D(0,\varepsilon)_{\mathbb{R}w}^{\beta_2}, r_w^*\mathcal{M}^{\beta_2}) \to$$

$$H_c^\bullet(\mathbb{A}_{\mathbb{R}w}^{\beta_1+\beta_2}, r_w^*\mathcal{M}^{\beta_1+\beta_2}) = \Phi_w(\mathcal{M}^{\beta_1+\beta_2}),$$

i.e., the action of $\mathfrak{u}^- \simeq \Phi_{\mathbb{R}}(\mathcal{J})$ on $\Phi_w(\mathcal{M}) := \bigoplus_{\beta\in\mathbb{N}[I]} \Phi_w(\mathcal{M}^\beta)$.

The self-duality of Φ_w gives rise to the action of $\mathfrak{u}^+$ on $\Phi_w(\mathcal{M})$ similar to Sect. 5.3. We assign to $\Phi_w(\mathcal{M}^\beta)$ the weight $w(\lambda-\beta)$, and using the isomorphisms $T'_{w\pm}: \mathfrak{u}^+ \xrightarrow{\sim} T'_{w\pm}(\mathfrak{u}^+) \subset \mathfrak{u}$, $\mathfrak{u}^- \xrightarrow{\sim} T'_{w\pm}(\mathfrak{u}^-) \subset \mathfrak{u}$, we obtain the action of $T'_{w\pm}(\mathfrak{u}^+), T'_{w\pm}(\mathfrak{u}^-)$ on $\Phi_w(\mathcal{M})$. This, together with the above grading, defines an action of $\dot{\mathfrak{u}}$ on $\Phi_w(\mathcal{M})$, i.e., gives rise to two functors $\Phi_{w\pm}: \mathcal{FS} \to \mathcal{C}$.

Given a straight line interval γ_w from $z^{(e)}$ to $z^{(w)}$, we obtain the corresponding "half monodromy" transformations $\gamma^\pm_{w,\beta}(\mathcal{M})\colon \Phi_e(\mathcal{M}^\beta) \xrightarrow{\sim} \Phi_{w\pm}(\mathcal{M}^\beta)$ (independent of the choice of γ) for any factorizable sheaf $\mathcal{M}$ and $\beta \in \mathbb{N}[I]$.

The following conjecture is a reformulation of Sect. 7.1.

Conjecture 7.3 The maps $\{\gamma^\pm_{w,\beta}(\mathcal{M}),\ \mathcal{M} \in \mathcal{FS}\}$ define two natural transformations of functors $\gamma_w^+ : \Phi \xrightarrow{\sim} \Phi_{w+}$ and $\gamma_w^- : \Phi \xrightarrow{\sim} \Phi_{w-}$.

As we have already mentioned, the main theorem of [TL] implies this conjecture for $\zeta = \exp(\hbar)$, $\hbar$ being a formal parameter.

Acknowledgments We are grateful to R. Fedorov, S. Khoroshkin, B. Feigin, A. Postnikov, G. Rybnikov, L. Rybnikov, V. Toledano Laredo, D. Gaitsgory, D. Kazhdan, M. Kapranov, A. Braverman, R. Bezrukavnikov, and A.Varchenko for the inspiring discussions and M. Kashiwara for an important reference.

In fact, this note arose from a question asked by R. Fedorov in the summer 2012.

The research of M.F. has been funded within the framework of the HSE University Basic Research Program and the Russian Academic Excellence Project '5-100'.

References

[AT] A. Appel, V. Toledano Laredo, *Coxeter categories and quantum groups*, Selecta Math. (N.S.) **25** (2019), no. 3, Art. 44, 97pp.

[AG] S. Arkhipov, D. Gaitsgory, *Another realization of the category of modules over the small quantum group*, Adv. Math. **173** (2003), 114–143.

[BFS] R. Berzukavnikov, M. Finkelberg, V. Schechtman, *Factorizable sheaves and quantum groups*, Lecture Notes in Math. **1691** (1998).

[BE] R. Bezrukavnikov, P. Etingof, *Parabolic induction and restriction functors for rational Cherednik algebras*, Selecta Math. (N.S.) **14** (2009), no. 3–4, 397–425.

[BRYL] J.-L. Brylinski, *Transformations canoniques, dualité projective, théorie de Lefschetz, transformations de Fourier et sommes trigonométriques*, Astérisque **340–341** (1986), 3–134.

[D] P. Deligne, *Une description de catégorie tressée (inspiré par Drinfeld)*, Letter to V. Drinfeld (1990).

[DCP] C. De Concini, C. Procesi, *Hyperplane arrangements and holonomy equations*, Selecta Math. (N.S.) **1** (1995), no. 3, 495–535.

[FMTV] G. Felder, Y. Markov, V. Tarasov, A. Varchenko, *Differential equations compatible with KZ equations*, Math. Phys. Anal. Geom. **3** (2000), 139–177.

[FKS] M. Finkelberg, M. Kapranov, V. Schechtman, *Fourier-Sato transform on hyperplane arrangements*, arxiv:1712.07432.

[KSCH] M. Kapranov, V. Schechtman, *Perverse sheaves over real hyperplane arrangements*, Ann. of Math. (2) **183**, no. 2 (2016), 619–679.

[KS1] M. Kashiwara, P. Schapira, *Sheaves on manifolds*, Grundlehren der Mathematischen Wissenschaften **292** Springer-Verlag, Berlin (1990), x+512pp.

[KS2] M. Kashiwara, P. Schapira, *Integral transforms with exponential kernels and Laplace transform*, J. Amer. Math. Soc. **10** (1997), 939–972.

[KT] S. Khoroshkin, V. Tolstoy, *Twisting of quantum (super-) algebras*, Generalized symmetries in physics (Clausthal, 1993), World Sci. Publ., River Edge, NJ (1994), 42–54.

[KV] S. Khoroshkin, A. Varchenko, *Quiver D-modules and homology of local systems over an arrangement of hyperplanes*, Int. Math. Res. Pap. **2006**, Art. ID 69590, 116pp.

[LU1] G. Lusztig, *Canonical bases arising from quantized enveloping algebras. II*, Progress of Theoretical Physics **102** (1990), 175–201.

[LU2] G. Lusztig, *Introduction to quantum groups*, Progress in Math. **110** (1993).

[MT] J. Millson, V. Toledano Laredo, *Casimir operators and monodromy representations of generalised braid groups*, Transform. Groups **10** (2005), no. 2, 217–254.

[MW] C. Moeglin, J.-L. Waldspurger, *Spectral decomposition and Eisenstein series. Une paraphrase de l'Écriture [A paraphrase of Scripture]*, Cambridge Tracts in Mathematics **113**, Cambridge University Press, Cambridge (1995), xxviii+338 pp.

[SA] M. Salvetti, *Topology of the complement of real hyperplanes in* $\mathbb{C}^N$, Invent. Math. **88** (1987), no. 3, 603–618.

[SO] Ya. Soibelman, *Algebra of functions on a compact quantum group and its representations* (Russian), Algebra i Analiz **2** (1990), no. 1, 190–212; translation in Leningrad Math. J. **2** (1991), no. 1, 161–178.

[TL] V. Toledano Laredo, *Quasi-Coxeter algebras, Dynkin diagram cohomology, and quantum Weyl groups*, Int. Math. Res. Pap. IMRP 2008, Art. ID rpn009, 167 pp.

[V] J.-L. Verdier, *Spécialisation de faisceaux et monodromie modérée*, Astérisque **101–102** (1983), 332–364.

Part II
D-Modules and Perverse Sheaves, Particularly on Flag Varieties and Their Generalizations

Fourier-Sato Transform on Hyperplane Arrangements

Michael Finkelberg, Mikhail Kapranov, and Vadim Schechtman

To our friends and teachers Sasha Beilinson and Vitya Ginzburg

Contents

M. Finkelberg (✉)
Department of Mathematics, National Research University Higher School of Economics, Russian Federation, Moscow, Russia

Skolkovo Institute of Science and Technology, Institute for Information Transmission Problems of RAS, Moscow, Russia

M. Kapranov
Kavli IPMU, Chiba, Japan
e-mail: mikhail.kapranov@ipmu.jp

V. Schechtman
Institut de Mathématiques de Toulouse, Université Paul Sabatier, Toulouse, France
e-mail: schechtman@math.ups-tlse.fr

V. Baranovsky et al. (eds.), *Representation Theory and Algebraic Geometry*, Trends in Mathematics, https://doi.org/10.1007/978-3-030-82007-7_4

1 Introduction

1.1 Setup and Goals

The theory of perverse sheaves can be said to provide an interpolation between homology and cohomology (or to mix them in a self-dual way). Since homology, sheaf-theoretically, can be understood as cohomology with compact support, interesting operations on perverse sheaves usually combine the functors of the types $f_!$ and f_* or, dually, the functors of the types $f^!$ and f^* in the classical formalism of Grothendieck.

An important context when this point of view can be pushed quite far is that of perverse sheaves $\mathcal{F}$ on a complex affine space $\mathbb{C}^n$ smooth with respect to the stratification given by an arrangement $\mathcal{H}$ of hyperplanes with real equations [KS1]. Denoting by $i_{\mathbb{R}} : \mathbb{R}^n \hookrightarrow \mathbb{C}^n$ the embedding, we associate to such an $\mathcal{F}$ its *hyperbolic stalks*

$$E_A(\mathcal{F}) = R\Gamma(A, i_A^* i_{\mathbb{R}}^! \mathcal{F}).$$

Here $i_A : A \hookrightarrow \mathbb{R}^n$ is the embedding of a face (stratum) of the real arrangement. It is remarkable that the $E_A(\mathcal{F})$ reduce to single vector spaces, not complexes (while the ordinary stalks of $\mathcal{F}$ are of course complexes, $\mathcal{F}$ being a complex of sheaves). This type of phenomena was originally observed by T. Braden in the context of varieties with a $\mathbb{C}^*$-action [Br, DG].

It was shown in [KS1] that the vector spaces $E_A(\mathcal{F})$ together with natural linear maps γ_{AB}, δ_{BA} ("generalization and specialization") connecting them determine the perverse sheaf $\mathcal{F}$ uniquely. Moreover, the category $\mathrm{Perv}(\mathbb{C}^n, \mathcal{H})$ of perverse sheaves of the above type is equivalent to the category $\mathrm{Hyp}(\mathcal{H})$ formed by linear algebra data

$(E_A, \gamma_{AB}, \delta_{BA})$ satisfying an explicit set of conditions. We call such linear algebra data *hyperbolic sheaves*; see Sect. 2D.

The goal of this paper is to develop the beginnings of a "hyperbolic calculus," describing the effect of several standard operations on perverse sheaves directly in terms of hyperbolic sheaves. These operations include forming vanishing cycles, specialization, and Fourier-Sato transform. To illustrate the importance of such questions, recall [BFS] that the weight components of the highest weight modules (e.g., Verma, or their irreducible quotients) over quantized Kac-Moody algebras have interpretation as the spaces of vanishing cycles $\Phi_f(\mathcal{F})$ for appropriate $\mathcal{F} \in \mathrm{Perv}(\mathbb{C}^n, \mathcal{H})$ and f. In this case, $\mathcal{H}$ is a so-called discriminantal arrangement, $\mathcal{F}$ is an extension of a one-dimensional local system on the generic stratum, and f is a linear function. The monodromy of Fourier-Sato transforms of these sheaves is related to the action of Lusztig's symmetries on the corresponding representations [FS].

1.2 Pattern of the Results

To identify the effect of each operation on perverse sheaves above, we produce a new hyperbolic sheaf out of a given one. Our constructions and results fall into the following pattern.

(1) Each vector space of the new hyperbolic sheaf is identified with the 0th cohomology space of an otherwise acyclic complex formed by some of the vector spaces $E_A \otimes \mathrm{or}_A$ (here or_A is the orientation space), with the differential formed out of either the γ_{AB} or the δ_{BA}. So there are two versions of the answer, the γ-answer and the δ-answer, in each case.
(2) The complexes in (1) are subquotients of the two fundamental complexes (Proposition 2.12) calculating $R\Gamma_c(\mathbb{C}^n, \mathcal{F})$ and $R\Gamma(\mathbb{C}^n, \mathcal{F})$. These complexes are sums over all the faces A of the spaces $E_A \otimes \mathrm{or}_A$, and their differentials are formed out of the γ_{AB} and δ_{AB}, respectively. The $R\Gamma_c(\mathbb{C}^n, \mathcal{F})$ and $R\Gamma(\mathbb{C}^n, \mathcal{F})$ typically have more than one nonzero cohomology, but the subquotients we take turn out to be acyclic outside degree 0.
(3) The choice of subquotient is obtained by taking not all but some summands $E_A \otimes \mathrm{or}_A$. The selection rule, depending on the problem, reflects the geometry of the problem in some rough ("tropical") way.
(4) In each case, there is also a companion real statement, about complexes of sheaves on $\mathbb{R}^n$ constructible w.r.t. the stratification by the faces. This real statement is proved first, and the statement for perverse sheaves is deduced from it.

1.3 Structure of the Paper

In § 1, we recall the basics of the description of $\mathrm{Perv}(\mathbb{C}^n, \mathcal{H})$ by hyperbolic sheaves.

§ 2 is devoted to the calculation of the space of vanishing cycles $\Phi_f(\mathcal{F})$ in terms of hyperbolic sheaves. Here $f : \mathbb{C}^n \to \mathbb{C}$ is a linear function with real coefficients. The selection rule for subquotients of $R\Gamma_c(\mathbb{C}^n, \mathcal{F})$ and $R\Gamma(\mathbb{C}^n, \mathcal{F})$ consists in taking all faces $B \subset \mathbb{R}^n$ on which $f \geq 0$.

§ 3 describes the specialization of $\mathcal{F} \in \mathrm{Perv}(\mathbb{C}^n, \mathcal{H})$ along a $\mathbb{C}$-vector subspace $L_{\mathbb{C}} \subset \mathbb{C}^n$ with real equations. This is a perverse sheaf $\nu_L(\mathcal{F})$ on the normal bundle $T_L\mathbb{C}^n$ which is itself a vector space. In this case, we have the real subspace $L_{\mathbb{R}}$ and the product arrangement $\nu_L(\mathcal{H})$ in $T_{L_{\mathbb{R}}}\mathbb{R}^n$. We further have the specialization at the level of faces which is a monotone map of posets

$$\nu : \{\text{faces of } \mathcal{H}\} \to \{\text{faces of } \nu_L(\mathcal{H})\}.$$

The selection rule for subquotients of $R\Gamma_c(\mathbb{C}^n, \mathcal{F})$ and $R\Gamma(\mathbb{C}^n, \mathcal{F})$ consists in taking all faces A with $\nu(A) = B$ being a fixed face B of $\nu_L(\mathcal{H})$. This produces complexes calculating the hyperbolic stalk of $\nu_L(\mathcal{F})$ at B.

We also give a description of the specialization for constructible sheaves of $\mathbb{R}^n$ as the direct image under an appropriate cellular map $q : \mathbb{R}^n \to T_{L_{\mathbb{R}}}\mathbb{R}^n$. This allows us to identify (in our particular case) different possible (and, in general, non-equivalent) definitions of the bispecialization functor [ST, T] for a flag of subspaces $N \subset M \subset V$.

In § 4, we give a similar description of the Fourier-Sato transform $\mathrm{FS}(\mathcal{F})$ which is a perverse sheaf on the dual space $(\mathbb{C}^n)^*$. It is smooth with respect to an appropriate arrangement $\mathcal{H}^\vee$. Each face $A^\vee$ on $\mathcal{H}^\vee$ gives a natural strictly convex cone $V(A^\vee) \subset \mathbb{R}^n$. The selection rule for subquotients of $R\Gamma_c(\mathbb{C}^n, \mathcal{F})$ and $R\Gamma(\mathbb{C}^n, \mathcal{F})$ consists in taking all faces $B \subset V(A^\vee)$ for a fixed $A^\vee$. This produces complexes calculating the hyperbolic stalk of $\mathrm{FS}(\mathcal{F})$ at $A^\vee$.

Combining the descriptions of the specialization and of the Fourier-Sato transform at the level of hyperbolic sheaves, one obtains a description of the microlocalization $\mu_L(\mathcal{F})$ along a linear subspace with real equations. The final §5 is dedicated to comparison, in our linear case, of several possible definitions of the second microlocalization of Kashiwara and Laurent; see [L, ST, T].

2 Real and Complex Data Associated with Perverse Sheaves

2.1 The Real Setup

Let $V_{\mathbb{R}} = \mathbb{R}^n$ be a finite-dimensional vector space over $\mathbb{R}$ and $\mathcal{H}$ be a finite central arrangement of hyperplanes in $V_{\mathbb{R}}$. We denote by $\mathcal{S}_{\mathbb{R}} = \mathcal{S}_{\mathbb{R},\mathcal{H}}$ the poset of *faces* of $\mathcal{H}$; see, e.g., [KS1], §2A. Faces form a real stratification of $V_{\mathbb{R}}$ into (a disjoint union of) locally closed polyhedral cones. The order $\leq$ on $\mathcal{S}_{\mathbb{R}}$ is by inclusion of closures: $A \leq B$ means $A \subset \overline{B}$. For an integer $p \geq 0$, we use the notation $A <_p B$ to signify that $A \leq B$ and $\dim(B) = \dim(A) + p$, in particular, $A <_0 B$ means $A = B$. We denote by $i_A : A \to V_{\mathbb{R}}$ the embedding of a face A.

Let $\mathbf{k}$ be a field and $\mathrm{Vect}_{\mathbf{k}}$ be the category of finite-dimensional $\mathbf{k}$-vector spaces. For any poset S, we denote by $\mathrm{Rep}(S)$ the abelian category of *representations of* S *over* $\mathbf{k}$, i.e., of covariant functors from S (considered as a category) to $\mathrm{Vect}_{\mathbf{k}}$. By $D^b(\mathrm{Rep}(S))$, we denote the bounded derived category of $\mathrm{Rep}(S)$.

For a topological space X, we denote by Sh_X the category of sheaves of $\mathbf{k}$-vector spaces on X and by $D^b(X)$ the derived category of Sh_X.

We denote by $\mathrm{Sh}(V_{\mathbb{R}}, \mathcal{S}_{\mathbb{R}})$ the abelian category formed by sheaves of $\mathbf{k}$-vector spaces on $V_{\mathbb{R}}$ which are constructible with respect to the stratification $\mathcal{S}_{\mathbb{R}}$. Let also $D^b(V_{\mathbb{R}}, \mathcal{S}_{\mathbb{R}})$ be the full subcategory in the bounded derived category of sheaves of $\mathbf{k}$-vector spaces on $V_{\mathbb{R}}$ formed by complexes with all cohomology sheaves lying in $\mathrm{Sh}(V_{\mathbb{R}}, \mathcal{S}_{\mathbb{R}})$. For $\mathcal{G} \in D^b(V_{\mathbb{R}}, \mathcal{S}_{\mathbb{R}})$ and a face A, we denote

$$\mathcal{G}_A = R\Gamma(A, \mathcal{G}) := R\Gamma(A, i_A^* \mathcal{G}) \in D^b(\mathrm{Vect}_{\mathbf{k}}) \tag{2.1}$$

the stalk of $\mathcal{G}$ at A. Thus, $\mathcal{G}_A$ is a complex which is a single vector space, if $\mathcal{G}$ is a single sheaf. The following is well known.

Proposition 2.2

(a) *We have an equivalence of categories*

$$\mathrm{Sh}(V_{\mathbb{R}}, \mathcal{S}_{\mathbb{R}}) \longrightarrow \mathrm{Rep}(\mathcal{S}_{\mathbb{R}}), \quad \mathcal{G} \mapsto \big(\mathcal{G}_A, \gamma_{AB} : \mathcal{G}_A \to \mathcal{G}_B,\ A \leq B\big).$$

Here γ_{AB} is the generalization map.

(b) *The natural functor $D^b(\mathrm{Sh}(V_{\mathbb{R}}, \mathcal{S}_{\mathbb{R}})) \to D^b(V_{\mathbb{R}}, \mathcal{S}_{\mathbb{R}})$ is an equivalence. In particular:*

(c) *We have an equivalence of categories $D^b(\mathrm{Sh}(V_{\mathbb{R}}, \mathcal{S}_{\mathbb{R}})) \to D^b(\mathrm{Rep}(\mathcal{S}_{\mathbb{R}}))$.*

In view of (b), we can interpret the equivalence in (c) as sending a complex of sheaves $\mathcal{G}$ to the collection of complexes of vector spaces $\mathcal{G}_A$ defined by (2.1) and generalization maps (morphisms of complexes) γ_{AB} connecting them.

By a *cell*, we mean a topological space B homeomorphic to $\mathbb{R}^d$ for some d. For a cell B, we denote by $\mathrm{or}_B = H_c^{\dim(B)}(B, \mathbf{k})$ the one-dimensional *orientation vector* $\mathbf{k}$*-space* of B. For two cells B, C, we set $\mathrm{or}_{B/C} = \mathrm{or}_C \otimes \mathrm{or}_B^*$ and call it the *relative orientation space* of C and B.

In particular, any face $B \in \mathcal{S}_{\mathbb{R}}$ is a cell and so we have the space or_B. When B, C are two faces such that $B <_1 C$, we have a canonical "incidence isomorphism"

$$\varepsilon_{BC} : \mathrm{or}_B \to \mathrm{or}_C .$$

It can be seen as a canonical trivialization of $\mathrm{or}_{C/B}$. If $B <_1 C_1, C_2 <_1 D$ is a square of codimension 1 inclusion of faces, then the diagram

$$\begin{array}{ccc} \mathrm{or}_B & \xrightarrow{\varepsilon_{B,C_1}} & \mathrm{or}_{C_1} \\ {\scriptstyle \varepsilon_{B,C_2}}\downarrow & & \downarrow{\scriptstyle \varepsilon_{C_1,D}} \\ \mathrm{or}_{C_2} & \xrightarrow[\varepsilon_{C_2,D}]{} & \mathrm{or}_D \end{array} \tag{2.3}$$

is anti-commutative.

Let $j_A : A \to V_{\mathbb{R}}$ be the embedding of a face A. If $A <_1 A'$ are two faces of $\mathcal{H}$, we have a canonical morphism $\xi_{AA'} : j_{A!}\underline{\mathbf{k}}_A \longrightarrow j_{A'!}\underline{\mathbf{k}}_{A'}[1]$ in $D^b(V_{\mathbb{R}}, \mathcal{S}_{\mathbb{R}})$. Viewed as an element of $\mathrm{Ext}^1(j_{A!}\underline{\mathbf{k}}_A, j_{A'!}\underline{\mathbf{k}}_{A'})$, it represents the extension given by the subsheaf in $(j_{A'})_* \underline{\mathbf{k}}_{A'}$ formed by sections which vanish on all codimension 1 faces of A' except A. The morphisms $\xi_{AA'}$ anticommute in squares of codimension 1 embeddings, just like the morphisms $\varepsilon_{AA'}$ in (2.3).

Proposition 2.4 *For $\mathcal{G} \in D^b(V_{\mathbb{R}}, \mathcal{S}_{\mathbb{R}})$, the following are equivalent:*

(i) $\mathcal{G}$ corresponds to the data $(\mathcal{G}_A, \gamma_{AB})$.
(ii) We have a resolution of $\mathcal{G}$ (a complex over $D^b(Sh_{V_{\mathbb{R}}})$ with total object $\mathcal{G}$) of the form

$$\bigoplus_{\dim(A)=0} j_{A!}(\underline{\mathcal{G}_A}_A) \xrightarrow{\gamma\otimes\xi} \bigoplus_{\dim(A)=1} j_{A!}(\underline{\mathcal{G}_A}_A)[1] \xrightarrow{\gamma\otimes\xi} \bigoplus_{\dim(A)=2} j_{A!}(\underline{\mathcal{G}_A}_A)[2] \xrightarrow{\gamma\otimes\xi} \cdots,$$

the direct sums ranging over all faces of $\mathcal{H}$ of given dimension.

Proof See, e.g., [KS1] Eq. (1.12). □

Corollary 2.5 *If $\mathcal{G} \in D^b(V_{\mathbb{R}}, \mathcal{S}_{\mathbb{R}})$ corresponds to $(\mathcal{G}_A, \gamma_{AB})$, then*

$$R\Gamma_c(V_{\mathbb{R}}, \mathcal{G}) \simeq \mathrm{Tot}\left\{ \bigoplus_{\dim(A)=0} \mathcal{G}_A \otimes \mathrm{or}_A \xrightarrow{\gamma\otimes\varepsilon} \bigoplus_{\dim(A)=1} \mathcal{G}_A \otimes \mathrm{or}_A \xrightarrow{\gamma\otimes\varepsilon} \cdots \right\}$$

(the cohomology with compact supports is calculated by the cellular cochain complex).

Proof This follows because $R\Gamma_c(V_{\mathbb{R}}, j_{A!}\underline{\mathbf{k}}_A) = \mathrm{or}(A)[-\dim(A)]$ (cohomology of a cell with compact support). □

2.2 The Complex Setup

Let $V_{\mathbb{C}} = \mathbb{C}^n$ be the complexification of $V_{\mathbb{R}}$ and $\mathcal{H}_{\mathbb{C}}$ the arrangement of hyperplanes in $V_{\mathbb{C}}$ formed by the $H_{\mathbb{C}}$, the complexifications of the hyperplanes $H \in \mathcal{H}$. By a *flat* of $\mathcal{H}_{\mathbb{C}}$, we will mean a subspace of the form $L = \bigcap_{H\in J} H_{\mathbb{C}}$ for a subset $J \subset \mathcal{H}$

(with $J = \emptyset$ or $J = \mathcal{H}$ allowed). Flats form a poset $\mathrm{Fl}(\mathcal{H}_{\mathbb{C}})$ ordered by inclusion. Because $\mathcal{H}$ is assumed central, $\mathrm{Fl}(\mathcal{H}_{\mathbb{C}})$ has 0 as the minimal element and $V_{\mathbb{C}}$ as the maximal element.

For a flat L, we denote its *generic part* by

$$L^{\circ} = L \setminus \bigcup_{H \in \mathcal{H},\ H_{\mathbb{C}} \not\supset L} L \cap H_{\mathbb{C}}. \tag{2.6}$$

The subsets L° form a stratification of $V_{\mathbb{C}}$ which we denote by $\mathcal{S}_{\mathbb{C}} = \mathcal{S}_{\mathbb{C},\mathcal{H}}$. We view it as a poset, isomorphic to the poset of flats.

Note that faces can be defined as connected components of $L^{\circ}_{\mathbb{R}} = L^{\circ} \cap V_{\mathbb{R}}$ for strata L° of $\mathcal{S}_{\mathbb{C}}$. We therefore have the morphism of posets ("complexification")

$$c : \mathcal{S}_{\mathbb{R}} \longrightarrow \mathcal{S}_{\mathbb{C}}.$$

We denote by $D^b(V_{\mathbb{C}}, \mathcal{S}_{\mathbb{C}})$ the full subcategory in the bounded derived category of sheaves of $\mathbf{k}$-vector spaces on $V_{\mathbb{C}}$ formed by complexes whose cohomology sheaves are constructible with respect to $\mathcal{S}_{\mathbb{C}}$. This category has a perfect duality given by passing from $\mathcal{F}$ to $\mathcal{F}^*$, the Verdier dual complex. Inside it, we have $\mathrm{Perv}(V_{\mathbb{C}}, \mathcal{S}_{\mathbb{C}})$ the abelian subcategory of perverse sheaves. We normalize the conditions of (middle) perversity so that $\underline{\mathbf{k}}_{V_{\mathbb{C}}}[n]$, the constant sheaf put in degree $(-n)$, is perverse. This normalization agrees with that of [BBD] and differs by shift from that of [KS1]. The abelian category $\mathrm{Perv}(V_{\mathbb{C}}, \mathcal{S}_{\mathbb{C}})$ is closed under Verdier duality.

2.3 *Real Data: Stalks and Hyperbolic Stalks*

Let $i_{\mathbb{R}}; V_{\mathbb{R}} \to V_{\mathbb{C}}$ be the embedding. It induces exact functors of triangulated categories

$$i_{\mathbb{R}}^*, i_{\mathbb{R}}^! : D^b(V_{\mathbb{C}}, \mathcal{S}_{\mathbb{C}}) \longrightarrow D^b(V_{\mathbb{R}}, \mathcal{S}_{\mathbb{R}}).$$

To every complex $\mathcal{F} \in D^b(V_{\mathbb{C}}, \mathcal{S}_{\mathbb{C}})$ and every face $A \in \mathcal{S}_{\mathbb{R}}$, we can associate therefore two complexes of vector spaces, which we call the *stalk* and the *hyperbolic stalk* of $\mathcal{F}$ at A:

$$\mathcal{F}_A = (i_{\mathbb{R}}^* \mathcal{F})_A = R\Gamma(A, i_A^* i_{\mathbb{R}}^* \mathcal{F}), \quad E_A(\mathcal{F}) = (i_{\mathbb{R}}^! \mathcal{F})_A = R\Gamma(A, i_A^* i_{\mathbb{R}}^! \mathcal{F}).$$

For any pair of faces $A \leq B$, we have the generalization maps (morphisms of complexes) for $i_{\mathbb{R}}^* \mathcal{F}$ and $i_{\mathbb{R}}^! \mathcal{F}$:

$$r_{AB} : \mathcal{F}_A \longrightarrow \mathcal{F}_B, \quad \gamma_{AB} : E_A(\mathcal{F}) \longrightarrow E_B(\mathcal{F}). \tag{2.7}$$

By the Duality Theorem (see [KS1] Prop. 4.6 or [BFS] Pt. I, Thm. 3.9), we have natural isomorphisms

$$E_A(\mathcal{F}^*) \simeq E_A(\mathcal{F})^*. \tag{2.8}$$

which imply the following.

Proposition 2.9

(a) We have a canonical identification $E_A(\mathcal{F}) \simeq R\Gamma(A, i_A^! i_{\mathbb{R}}^* \mathcal{F})$.
(b) The hyperbolic stalk $E_A(\mathcal{F})$ *is identified with the complex*

$$\mathcal{F}_{\geq A} := \mathrm{Tot}\left\{ \mathcal{F}_A \xrightarrow{\digamma \otimes \varepsilon} \bigoplus_{B >_1 A} \mathcal{F}_B \otimes \mathrm{or}_{B/A} \xrightarrow{\digamma \otimes \varepsilon} \bigoplus_{B >_2 A} \mathcal{F}_B \otimes \mathrm{or}_{B/A} \xrightarrow{\digamma \otimes \varepsilon} \cdots \right\}$$

with the differential $\digamma \otimes \varepsilon$ *having matrix elements* $\digamma_{BC} \otimes \varepsilon_{BC}$, $B <_1 C$. □

For a dual statement, expressing ordinary stalks through hyperbolic stalks, see Corollary 2.14.

Proof Part (a) follows from (2.8) and the fact that Verdier duality interchanges i^* and $i^!$. Part (b) follows by interpreting $i_A^! i_{\mathbb{R}}^* \mathcal{F}$ as $\underline{R\Gamma}_A(i_{\mathbb{R}}^* \mathcal{F})$, the complex of sheaves formed by (derived) sections with support in A. The stalk of this complex at any $a \in A$ can be seen as

$$R\Gamma_{\{a\}}(D, i_{\mathbb{R}}^* \mathcal{F}) = R\Gamma_c(D, i_{\mathbb{R}}^* \mathcal{F}),$$

where $D \subset V_{\mathbb{R}}$ is a small transverse open ball (of complementary dimension) to A centered at a. The situation is similar to that of Corollary 2.5 (with a ball instead of a vector space), and the same argument gives the result. □

It was proved in [KS1] Prop. 4.9(a) that for $\mathcal{F} \in \mathrm{Perv}(V_{\mathbb{C}}, \mathcal{S}_{\mathbb{C}})$, the complex $i_{\mathbb{R}}^!(\mathcal{F})$ is exact in degrees $\neq 0$, and so the functor

$$\mathrm{Perv}(V_{\mathbb{C}}, \mathcal{S}_{\mathbb{C}}) \to \mathrm{Sh}(V_{\mathbb{R}}, \mathcal{S}_{\mathbb{R}}), \quad \mathcal{F} \mapsto \mathcal{E}(\mathcal{F}) := \underline{H}^0(i_{\mathbb{R}}^! \mathcal{F}) = \underline{\mathbb{H}}^0_{V_{\mathbb{R}}}(\mathcal{F}) \tag{2.10}$$

is an exact functor of abelian categories. In particular, each $E_A(\mathcal{F})$ reduces to a single vector space. Further, (2.8) allows us to define maps of vector spaces

$$\delta_{BA} = \delta_{BA}^{\mathcal{F}} : E_B(\mathcal{F}) \longrightarrow E_A(\mathcal{F}), \ A \leq B, \quad \delta_{BA}^{\mathcal{F}} := (\gamma_{AB}^{\mathcal{F}^*})^*.$$

which form an anti-representation of $\mathcal{S}_{\mathbb{R}}$, i.e., a contravariant functor $(\mathcal{S}_{\mathbb{R}}, \leq) \to \mathrm{Vect}_{\mathbf{k}}$. This leads to the following concept.

2.4 Hyperbolic Sheaves

By a *hyperbolic sheaf* on $\mathcal{H}$, we will mean a datum

$$\mathcal{Q} = \big(E_A, \gamma_{AB} : E_A \to E_B,\ \delta_{BA} : E_B \to E_A,\ A \leq B\big)$$

where E_A, $A \in \mathcal{S}_\mathbb{R}$, are finite-dimensional $\mathbf{k}$-vector spaces, (γ_{AB}) form a representation of $\mathcal{S}_\mathbb{R}$, and (δ_{BA}) form an anti-representation so that the following additional conditions hold:

(i) For each $B \leq A$, $\gamma_{BA}\delta_{AB} = \mathrm{Id}_{E_A}$. This allows us to define for arbitrary $A, B \in \mathcal{S}_\mathbb{R}$, the "flopping operator"

$$\phi_{AB} := \gamma_{CB}\delta_{AC} :\ E_A \longrightarrow E_B.$$

Here $C \in \mathcal{S}_\mathbb{R}$ is any face such that $C \leq A, B$, and the definition does not depend on the choice of C.

(ii) Let us call a triple of faces (A, B, C) *collinear* if there exist points $x \in A$, $y \in B$, $z \in C$ lying on the same straight line, with $y \in [x, z]$. Then for any such collinear triple, we must have

$$\phi_{AC} = \phi_{BC}\,\phi_{AB}.$$

(iii) Let A, B be two faces. Let us say that they are *neighbors* if they have the same dimension d, and there exists a face $C \leq A, C \leq B$, with $\dim C = d - 1$ (a *wall* separating A and B). Such a wall is unique if it exists. For any such pair of neighbors, we require that ϕ_{AB} is an isomorphism.

We denote by $\mathrm{Hyp}(\mathcal{H})$ the abelian category formed by hyperbolic sheaves on $\mathcal{H}$. This category has a perfect duality

$$\mathcal{Q} = (E_A, \gamma_{AB}, \delta_{BA}) \ \mapsto\ \mathcal{Q}^* = (E_A^*, \delta_{BA}^*, \gamma_{AB}^*).$$

The main result of [KS1] can be formulated as follows.

Theorem 2.11 *The functor*

$$\mathcal{F} \mapsto \mathcal{Q}(\mathcal{F}) = \big(E_A(\mathcal{F}), \gamma_{AB} : E_A(\mathcal{F}) \to E_B(\mathcal{F}),\ \delta_{BA} : E_B(\mathcal{F}) \to E_A(\mathcal{F}), A \leq B\big)$$

defines an equivalence $\mathrm{Perv}(V_\mathbb{C}, \mathcal{S}_\mathbb{C}) \to \mathrm{Hyp}(\mathcal{H})$. *This equivalence commutes with duality:* $\mathcal{Q}(\mathcal{F}^*) \simeq \mathcal{Q}(\mathcal{F})^*$. □

The goal of this paper is to describe various features of perverse sheaves explicitly, in terms of the linear algebra data given by the associated hyperbolic sheaves.

Let us first note the following.

Proposition 2.12 *If* $\mathcal{F} \in \operatorname{Perv}(V_{\mathbb{C}}, \mathcal{S}_{\mathbb{C}})$ *corresponds to a hyperbolic sheaf* $\mathcal{Q}(E_A, \gamma_{AB}, \delta_{BA})$*, then*

$$R\Gamma_c(V_{\mathbb{C}}, \mathcal{F}) \simeq \left\{ \bigoplus_{\dim(A)=0} E_A \otimes \operatorname{or}_A \xrightarrow{\gamma\otimes\varepsilon} \bigoplus_{\dim(A)=1} E_A \otimes \operatorname{or}_A \xrightarrow{\gamma\otimes\varepsilon} \cdots \right\},$$

$$R\Gamma(V_{\mathbb{C}}, \mathcal{F}) \simeq \left\{ \bigoplus_{\operatorname{codim}(A)=0} E_A \otimes \operatorname{or}_A \xrightarrow{\delta\otimes\varepsilon} \bigoplus_{\operatorname{codim}(A)=1} E_A \otimes \operatorname{or}_A \xrightarrow{\delta\otimes\varepsilon} \cdots \right\}.$$

Proof The first quasi-isomorphism follows from Corollary 2.5 and the lemma below. The second quasi-isomorphism follows from the first one by applying the Verdier duality. □

Lemma 2.13 *For any* $\mathcal{F} \in D^b(V_{\mathbb{C}}, \mathcal{S}_{\mathbb{C}})$*, we have*

$$R\Gamma_c(V_{\mathbb{C}}, \mathcal{F}) \simeq R\Gamma_c(V_{\mathbb{R}}, i_{\mathbb{R}}^! \mathcal{F}).$$

Proof of the Lemma: Let $i_{0,\mathbb{C}} : \{0\} \to V_{\mathbb{C}}$ and $i_{0,\mathbb{R}} : \{0\} \to V_{\mathbb{R}}$ be the embeddings of the origin. Any $\mathcal{F} \in D^b(V_{\mathbb{C}}, \mathcal{S}_{\mathbb{C}})$ is $\mathbb{R}_+$*-conic*, i.e., each cohomology sheaf of $\mathcal{F}$ is locally constant on each orbit of the scaling action of $\mathbb{R}_{>0}$ on $V_{\mathbb{C}}$. This implies that

$$R\Gamma_c(V_{\mathbb{C}}, \mathcal{F}) \simeq R\Gamma_{\{0\}}(V_{\mathbb{C}}, \mathcal{F}) = R\Gamma(V_{\mathbb{C}}, i_{0,\mathbb{C}}^! \mathcal{F}).$$

Similarly, $i_{\mathbb{R}}^! \mathcal{F}$ is $\mathbb{R}_+$-conic on $V_{\mathbb{R}}$ and

$$R\Gamma_c(V_{\mathbb{R}}, i_{\mathbb{R}}^! \mathcal{F}) \simeq R\Gamma_{\{0\}}(V_{\mathbb{R}}, i_{\mathbb{R}}^! \mathcal{F}) = R\Gamma(V_{\mathbb{C}}, i_{0,\mathbb{R}}^! i_{\mathbb{R}}^! \mathcal{F}),$$

which is the same as the above because $i_{\mathbb{R}} i_{0,\mathbb{R}} = i_{0,\mathbb{C}}$. □

We can now complement Proposition 2.9 by a "Koszul dual" statement.

Corollary 2.14 *For* $\mathcal{F} \in \operatorname{Perv}(V_{\mathbb{C}}, \mathcal{S}_{\mathbb{C}})$ *the ordinary stalk* $\mathcal{F}_A$*,* $A \in \mathcal{S}_{\mathbb{R}}$ *is expressed through hyperbolic stalks as follows:*

$$\mathcal{F}_A \simeq \left\{ \bigoplus_{\substack{B\geq A \\ \operatorname{codim}(B)=0}} E_B \otimes \operatorname{or}_{B/A} \xrightarrow{\delta\otimes\varepsilon} \bigoplus_{\substack{B\geq A \\ \operatorname{codim}(B)=1}} E_B \otimes \operatorname{or}_{B/A} \xrightarrow{\delta\otimes\varepsilon} \cdots \right\}.$$

Proof For $A = 0$, this is the second identification of Proposition 2.12, since $\mathcal{F}_0 = R\Gamma(U, \mathcal{F})$ for a small convex open $U \ni 0$, and this complex is independent of U, so is the same for $U = V_{\mathbb{C}}$.

For an arbitrary A, the statement reduces to the above by considering the quotient arrangement $\mathcal{H}/L_{\mathbb{R}}$ in $V_{\mathbb{R}}/L_{\mathbb{R}}$, where $L_{\mathbb{R}}$ is the $\mathbb{R}$-linear span of A. Faces of $\mathcal{H}/L_{\mathbb{R}}$ are in bijection with faces B of $\mathcal{H}$ such that $B \geq A$.

The arrangement $\mathcal{H}/L_{\mathbb{R}}$ represents the transversal slice M to A; the restriction $\mathcal{F}|_{M_{\mathbb{C}}}$ to the complexified transversal slice is, by Kapranov and Schechtman [KS1] Prop. 5.3, represented by the hyperbolic sheaf $\mathcal{Q}^{\geq A}$ formed by E_B, $B \geq A$, so the calculation of

$$\mathcal{F}_A = R\Gamma(M_{\mathbb{C}}, \mathcal{F}|_{M_{\mathbb{C}}}) = (\mathcal{F}|_{M_{\mathbb{C}}})_0$$

reduces to the above case. □

3 Vanishing Cycles in Terms of Hyperbolic Sheaves

The standard microlocal approach to the study of perverse sheaves on any stratification is in terms of the local systems of vanishing cycles on the generic parts of conormal bundles to the strata; see [MV, KS2]. Our first result provides an explicit description of the fibers of these local systems for perverse sheaves from $\mathrm{Perv}(V_{\mathbb{C}}, \mathcal{S}_{\mathbb{C}})$.

3.1 Background on Vanishing Cycles

We recall that for any (polynomial) function $f : V_{\mathbb{C}} \to \mathbb{C}$ and any perverse sheaf $\mathcal{F}$ on $V_{\mathbb{C}}$, we have a perverse sheaf $\Phi_f(\mathcal{F})$ on $V_{\mathbb{C}}$ supported on the hypersurface $\{f = 0\}$ and known as the *perverse sheaf of vanishing cycles*; see [Be, De]. We will use the following real analytic interpretation of this perverse sheaf [KS2]. This interpretation reflects the intuitive meaning of the term "vanishing cycles."

Proposition 3.1 *We have an isomorphism in the derived category of sheaves on $V_{\mathbb{C}}$:*

$$\Phi_f(\mathcal{F}) \simeq i^* \underline{R\Gamma}_{\{\Re(f)\geq 0\}}(\mathcal{F}),$$

where i is the closed embedding of the subset $\{f = 0\}$ into $\{\Re(f) \geq 0\}$. □

We will be interested in the case when f is linear. More precisely, let $L^\circ \in \mathcal{S}_{\mathbb{C}}$ be a stratum, i.e., the generic part of a flat L, as in (2.6). The conormal bundle to L° is

$$T^*_{L^\circ} V_{\mathbb{C}} = L^\circ \times (V_{\mathbb{C}}/L)^* \subset V_{\mathbb{C}} \times V_{\mathbb{C}}^* = T^* V_{\mathbb{C}}.$$

A hyperplane $\Pi \subset V_{\mathbb{C}}$ is said to be *transversal to $\mathcal{S}_{\mathbb{C}}$ at L* if $L \subset \Pi$, and $L' \in \mathrm{Fl}(\mathcal{H}_{\mathbb{C}})$ with $L' \subset \Pi$ implies $L' \subset L$. Let us call a *polarization at L* a linear function $f : V_{\mathbb{C}} \to \mathbb{C}$ such that $\Pi := \mathrm{Ker}\, f$ is transversal to $\mathcal{S}_{\mathbb{C}}$ at L. Polarizations of L form an open subset $\mathrm{Pol}(L) \subset (V_{\mathbb{C}}/L)^*$, and we define the generic part of the conormal bundle to L° as

$$(T^*_{L^\circ} V_{\mathbb{C}})^\circ \;=\; L^\circ \times \mathrm{Pol}(L).$$

Proposition 3.2 *Let $\mathcal{F} \in \mathrm{Perv}(V_{\mathbb{C}}, \mathcal{S}_{\mathbb{C}})$. If $L \in \mathrm{Fl}(\mathcal{H}_{\mathbb{C}})$ and $f \in \mathrm{Pol}(L)$, then $\Phi_f(\mathcal{F})$ is supported on L. In particular, being perverse, it reduces to a local system in degree $(-\dim(L))$ on L°.*

Proof Let $x \in \{f = 0\} \subset V_{\mathbb{C}}$, and suppose $x \notin L$. Since $f \in \mathrm{Pol}(L)$, the hyperplane $\Pi = \{f = 0\}$ cannot contain any flats L' which are not contained in L. So x is not contained in any flat other than $V_{\mathbb{C}}$ itself, which means that near x the perverse sheaf $\mathcal{F}$ is reduced to a local system in degree $(-n)$, and so $\Phi_f(\mathcal{F})_x = 0$. □

We now describe the stalks of the local system $\Phi_f(\mathcal{F})$ at the maximal faces of $L_{\mathbb{R}}$.

3.2 The Complex Result

Theorem 3.3 *Let $\mathcal{F} \in \mathrm{Perv}(V_{\mathbb{C}}, \mathcal{S}_{\mathbb{C}})$ and $\mathcal{Q} = (E_A, \gamma_{AB}, \delta_{BA})$ be the corresponding hyperbolic sheaf as in Theorem 2.11. Suppose further that $f \in \mathrm{Pol}(L)$ is real, i.e., takes $V_{\mathbb{R}}$ to $\mathbb{R}$. Let A be a connected component of $L^\circ_{\mathbb{R}}$, so A is a face of $\mathcal{H}$. Consider the complex*

$$E^\bullet_{f,A} \;=\; \left\{ E_A \overset{\gamma\otimes\varepsilon}{\to} \bigoplus_{B>_1 A,\ f|_B\geq 0} E_B \otimes \mathrm{or}_{B/A} \overset{\gamma\otimes\varepsilon}{\to} \bigoplus_{B>_2 A,\ f|_B\geq 0} E_B \otimes \mathrm{or}_{B/A} \overset{\gamma\otimes\varepsilon}{\to} \cdots \right\}$$

with the differential $\gamma\otimes\varepsilon$ having matrix elements $\gamma_{BC}\otimes\varepsilon_{BC}$, $B >_1 C$. Then $E^\bullet_{f,A}$ is exact outside of the leftmost term, and its leftmost cohomology is identified with the vector space $\Phi_f(\mathcal{F})_a[-\dim(L)]$ for any $a \in A$.

The theorem implies that the shifted space of vanishing cycles is identified with the subspace

$$E_{f,A} \;=\; H^0(E^\bullet_{f,A}) \;=\; \bigcap_{B>_1 A,\ f|_B\geq 0} \mathrm{Ker}(\gamma_{AB}) \;\subset\; E_A.$$

It also implies the following.

Corollary 3.4 *Consider the complex*

$$\check{E}^\bullet_{f,A} \;=\; \left\{ \cdots \overset{\delta\otimes\varepsilon}{\to} \bigoplus_{B>_2 A,\ f|_B\geq 0} E_B \otimes \mathrm{or}_{B/A} \overset{\delta\otimes\varepsilon}{\to} \bigoplus_{B>_1 A,\ f|_B\geq 0} E_B \otimes \mathrm{or}_{B/A} \overset{\delta\otimes\varepsilon}{\to} E_A \right\}$$

with the differential $\delta\otimes\varepsilon$ having matrix elements $\delta_{CB}\otimes\varepsilon_{CB}$, $B >_1 C$. Then $E^\bullet_{f,A}$ is exact outside of the rightmost term, and its rightmost cohomology is identified with *the vector space* $\Phi_f(\mathcal{F})_a[-\dim(L)]$ *for any* $a \in A$. *In other words,*

$$E_{f,A} \simeq \operatorname{Coker}\Big(\sum \delta_{BA} : \bigoplus_{B>_1 A,\ f|_B \geq 0} E_B \longrightarrow E_A\Big).$$

Proof of the Corollary: The vanishing cycle functor commutes with Verdier duality. Therefore, the vector spaces $\Phi_f(\mathcal{F})_a[-\dim(L)]$ and $\Phi_f(\mathcal{F}^*)_a[-\dim(L)]$ are canonically dual to each other. On the other hand, the hyperbolic sheaf corresponding to $\mathcal{F}^*$ is, by Theorem 2.11, identified with $\mathcal{Q}^* = (E_A^*, \delta_{BA}^*, \gamma_{AB}^*)$. Our statement follows by combining this with Theorem 3.3 for $\mathcal{F}$ and $\mathcal{F}^*$. □

Remark 3.5 Theorem 3.3 and Corollary 3.4 can be interpreted as follows. The same graded space $E_{f,A}^\bullet$ possesses two differentials going in the opposite directions: one induced by the maps γ and the other one induced by the maps δ. It is natural therefore to form the "Laplacian" $\Delta = \delta\gamma + \gamma\delta$ out of them.

In the examples we have calculated, $\Delta : E_{f,A}^i \to E_{f,A}^i$ is an isomorphism for $i > 0$. This of course implies the acyclicity statements above. One may wonder if this stronger property (Laplacian being an isomorphism for $i > 0$) holds more generally.

3.3 *The Real Analog*

Before proving Theorem 3.3, we establish its real counterpart.

Let $\mathcal{G} \in D^b(V_\mathbb{R}, \mathcal{S}_\mathbb{R})$, and let $(\mathcal{G}_A, \gamma_{AB})$ be the complex of representations of $\mathcal{S}_\mathbb{R}$ corresponding to $\mathcal{G}$ by Proposition 2.2. That is, $\mathcal{G}_A$ is the ordinary stalk of $\mathcal{G}$ at A, and γ_{AB} is the generalization map.

Given a nonzero $f \in V_\mathbb{R}^*$, we have the real hyperplane $\Pi = \{f = 0\} \subset V_\mathbb{R}$. The arrangement $\mathcal{H}$ cuts out an arrangement $\mathcal{H} \cap \Pi$ in Π. We denote by $\mathcal{S}_{\mathbb{R},\Pi}$ the stratification of Π into cells of $\mathcal{H} \cap \Pi$. We then have the real version of the vanishing cycle sheaf. It is the complex of sheaves

$$i_\Pi^* \underline{R\Gamma}_{f \geq 0}(\mathcal{G}) \in D^b(\Pi, \mathcal{S}_{\mathbb{R},\Pi}).$$

Here $i_\Pi : \Pi \to V_\mathbb{R}$ is the embedding.

Proposition 3.6

(a) *Let* C' *be a cell of* $\mathcal{H} \cap \Pi$ *and* C *be the unique cell of* $\mathcal{H}$ *such that* $C' = C \cap \Pi$. *The stalk of* $\underline{R\Gamma}_{f \geq 0}(\mathcal{G})$ *at* C' *is quasi-isomorphic to the total complex of the double complex*

$$\Big\{\mathcal{G}_C \xrightarrow{\gamma\otimes\varepsilon} \bigoplus_{D>_1 C,\ f|_D \geq 0} \mathcal{G}_D \otimes \operatorname{or}_{D/C} \xrightarrow{\gamma\otimes\varepsilon} \bigoplus_{D>_2 C,\ f|_D \geq 0} \mathcal{G}_D \otimes \operatorname{or}_{D/C} \xrightarrow{\gamma\otimes\varepsilon} \cdots\Big\}$$

(b) Let $C'_1 \leq C'_2$ be an inclusion of cells of $\mathcal{H} \cap \Pi$. The generalization map

$$\gamma_{C'_1, C'_2} : \underline{R\Gamma}_{f\geq 0}(\mathcal{G})_{C_1} \longrightarrow \underline{R\Gamma}_{f\geq 0}(\mathcal{G})_{C_2}$$

is given by the maps $\gamma_{DD'}$ for $\mathcal{G}$ which induce a morphism of complexes in (a).

Proof Let $x \in C'$ and U be a small open ball centered at x. By definition,

$$\underline{R\Gamma}_{f\geq 0}(\mathcal{G})_{C'} = R\Gamma(U, U \cap \{f < 0\}; \mathcal{G})$$

The relative cellular cochain complex representing this is precisely the complex in (a). Part (b) also follows immediately. □

3.4 Proof of Theorem 3.3

Let f be as in the theorem. Considering f as a complex functional on $V_{\mathbb{C}}$, we have the complex hyperplane $\Pi_{\mathbb{C}} = \{f = 0\} \subset V^*_{\mathbb{C}}$ and the perverse sheaf $\Phi_f(\mathcal{F})$ on $\Pi_{\mathbb{C}}$. By Proposition 3.1, we can express the hyperbolic stalk of $\Phi_f(\mathcal{F})$ at a cell $C' \in \mathcal{S}_{\mathbb{R},\Pi}$ as

$$E_{C'}(\Phi_f(\mathcal{F})) = (\underline{R\Gamma}_{\Pi_{\mathbb{R}}} \underline{R\Gamma}_{\Re(f)\geq 0}(\mathcal{F}))_{C'} = (\underline{R\Gamma}_{f\geq 0} \underline{R\Gamma}_{V_{\mathbb{R}}}(\mathcal{F}))_{C'}.$$

Now, the complex (actually a sheaf) $\mathcal{G} = \underline{R\Gamma}_{V_{\mathbb{R}}}(\mathcal{F})$ on $V_{\mathbb{R}}$ is given by the stalks E_B and generalization maps γ_{BC} from the hyperbolic sheaf $\mathcal{Q}$. So applying Proposition 3.6 to this $\mathcal{G}$ and to the cell $C' = A$ as in the formulation of theorem, we get the statement. □

Remark 3.7 It is worth noticing the following contrast between Proposition 3.6 and Theorem 3.3. If $\mathcal{G}$ is an arbitrary sheaf (not a complex) on $V_{\mathbb{R}}$, then Proposition 3.6 gives, in general, a complex with several nontrivial cohomology spaces, because $\underline{R\Gamma}_{f\geq 0}(\mathcal{G})$ need not reduce to a single sheaf. However, in the case when $\mathcal{G}$ has the form $\mathcal{G} = \underline{R\Gamma}_{V_{\mathbb{R}}}(\mathcal{F})$ for a perverse sheaf $\mathcal{F} \in \mathrm{Perv}(V_{\mathbb{C}}, \mathcal{S}_{\mathbb{C}})$, this complex is, by Theorem 3.3, quasi-isomorphic to a single vector space in degree 0.

A more immediate instance of such special behavior of the sheaves $\underline{R\Gamma}_{V_{\mathbb{R}}}(\mathcal{F})$ can be seen from the property (i) of hyperbolic sheaves in Sect. 2D: the condition $\delta_{AB}\gamma_{BA} = \mathrm{Id}$ implies that each γ_{BA} is surjective.

4 Specialization and Hyperbolic Sheaves

4.1 Generalities on Specialization

We recall the necessary material from [KS2] §4.1-4.2. Let X be a C^∞-manifold, $M \subset X$ a locally closed submanifold, and $T_M X$ the normal bundle to M in X. Any subset $S \subset X$ gives rise to its *normal cone* with center M, which is a closed subset $C_M S \subset T_M X$ depending only on the closure $\overline{S}$. We will need the following example.

Example 4.1 Let X be a finite-dimensional $\mathbb{R}$-vector space and $M \subset X$ be an $\mathbb{R}$-vector subspace. Then $T_M X = M \times (X/M)$. If S is also an $\mathbb{R}$-vector subspace, then, with respect to the above identification,

$$C_M(S) = (M \cap S) \times \big(S/(M \cap S)\big).$$

For any complex of sheaves $\mathcal{G} \in D^b(\mathrm{Sh}_X)$, we have its *specialization* at M which is an $\mathbb{R}_{>0}$-conic complex of sheaves $\nu_M(\mathcal{G}) \in D^b(T_M X)$. We will later recall its definition in the case we need.

4.2 The Case of Sheaves on Arrangements

We will study this construction in two related cases, related to the data of a real arrangement $(V_\mathbb{R}, \mathcal{H})$.

Complex case: $X = V_\mathbb{C}$, $M = L_\mathbb{C}$ a complex flat of $\mathcal{H}$ and $\mathcal{G} = \mathcal{F} \in \mathrm{Perv}(V_\mathbb{C}, \mathcal{S}_\mathbb{C})$ a perverse sheaf smooth with respect to $\mathcal{S}_\mathbb{C}$.

Real case: $X = V_\mathbb{R}$, $M = L_\mathbb{R}$ is a real flat and $\mathcal{G} \in D^b(V_\mathbb{R}, \mathcal{S}_\mathbb{R})$ is any complex smooth with respect to the cell decomposition $\mathcal{S}_\mathbb{R}$.

In each of these cases, the normal bundle is itself a vector space:

$$T_{L_\mathbb{C}} V_\mathbb{C} = L_\mathbb{C} \times (V_\mathbb{C}/L_\mathbb{C}), \quad T_{L_\mathbb{R}} V_\mathbb{R} = L_\mathbb{R} \times (V_\mathbb{R}/L_\mathbb{R}). \tag{4.2}$$

The subspace $L_\mathbb{R}$ carries the *induced arrangement* $\mathcal{H} \cap L_\mathbb{R}$ formed by the hyperplanes $H \cap L_\mathbb{R}$ for $H \in \mathcal{H}$, $H \not\supset L_\mathbb{R}$. The quotient space $V_\mathbb{R}/L_\mathbb{R}$ carries the *quotient arrangement* $\mathcal{H}/L_\mathbb{R}$ formed by the hyperplanes $H/L_\mathbb{R}$ for $H \in \mathcal{H}$, $H \supset L_\mathbb{R}$. We equip $T_{L_\mathbb{R}} V_\mathbb{R}$ with the *product arrangement*

$$\nu_L \mathcal{H} := (\mathcal{H} \cap L_\mathbb{R}) \oplus (\mathcal{H}/L_\mathbb{R}) =$$

$$\big\{(H \cap L_\mathbb{R}) \times V_\mathbb{R}/L_\mathbb{R},\ H \not\supset L_\mathbb{R}\big\} \cup \big\{L_\mathbb{R} \times (H_\mathbb{R}/L_\mathbb{R}),\ H \supset L_\mathbb{R}\big\}.$$

We have a surjective map $\mathcal{H} \to \nu_L(\mathcal{H})$ between (the sets of hyperplanes of) the two arrangements. Two hyperplanes H, H' of $\mathcal{H}$ can give the same hyperplane of $\nu_L(\mathcal{H})$, if $H \cap L_{\mathbb{R}} = H' \cap L_{\mathbb{R}}$ is the same hyperplane in $L_{\mathbb{R}}$.

We denote by

$$\mathcal{S}^{\nu}_{\mathbb{R}} = \mathcal{S}_{1,\mathbb{R}} \times \mathcal{S}_{2,\mathbb{R}}, \quad \mathcal{S}^{\nu}_{\mathbb{C}} = \mathcal{S}_{1,\mathbb{C}} \times \mathcal{S}_{2,\mathbb{C}} \tag{4.3}$$

the stratification of $T_{L_{\mathbb{R}}} V_{\mathbb{R}}$ by the faces of $\nu_L(\mathcal{H})$ and the stratification of $T_{L_{\mathbb{C}}} V_{\mathbb{C}}$ by the generic parts of the complex flats of $\nu_L(\mathcal{H})$. Here $\mathcal{S}_{1,\mathbb{R}}$ is the stratification of $L_{\mathbb{R}}$ by the faces of $\mathcal{H} \cap L$, while $\mathcal{S}_{2,\mathbb{R}}$ is the stratification of $V_{\mathbb{R}}/L_{\mathbb{R}}$ by the faces of $\mathcal{H}/L$ and similarly for $\mathcal{S}_{i,\mathbb{C}}$.

Proposition 4.4

(a) If $\mathcal{F} \in D^b(V_{\mathbb{C}}, \mathcal{S}_{\mathbb{C}})$, then $\nu_{L_{\mathbb{C}}}\mathcal{F} \in D^b(T_{L_{\mathbb{C}}} V_{\mathbb{C}}, \mathcal{S}^{\nu}_{\mathbb{C}})$.
(b) If $\mathcal{G} \in D^b(V_{\mathbb{R}}, \mathcal{S}_{\mathbb{R}})$, then $\nu_{L_{\mathbb{R}}}\mathcal{G} \in D^b(T_{L_{\mathbb{R}}} V_{\mathbb{R}}, \mathcal{S}^{\nu}_{\mathbb{R}})$.

Proof We treat only the real case (b), the complex case (a) being identical. In the proof, we simply write V for the ambient vector space $V_{\mathbb{R}}$, as well as L for a real flat and so on. We denote by $\mathrm{SS}(\mathcal{G}) \subset T^*V$ the microsupport of the complex $\mathcal{G}$ and similarly for complexes of sheaves on other spaces; see [KS2] Ch. VI. The statement that $\mathcal{G} \in D^b(V, \mathcal{S})$, resp. that $\nu_L(\mathcal{G}) \in D^b(T_L V, \mathcal{S}^{\nu})$, is equivalent to

$$\mathrm{SS}(\mathcal{G}) \subset \bigcup_{P \in \mathrm{Fl}(\mathcal{H})} T^*_P V, \quad \text{resp.} \quad \mathrm{SS}(\nu_L(\mathcal{G})) \subset \bigcup_{Q \in \mathrm{Fl}(\nu_L(\mathcal{H}))} T^*_Q(L \times (V/L)).$$

So we deduce the second inclusion from the first. By Theorem 6.4.1 of [KS2], for any manifold X, a submanifold M, and a complex of sheaves $\mathcal{G}$ on X, we have

$$\mathrm{SS}(\nu_M(\mathcal{G})) \subset C_{T^*_M X}(\mathrm{SS}(\mathcal{G})) \subset T_{T^*_M X} T^*X \overset{(!)}{\simeq} T^*(T_M X).$$

Here $C_{T^*_M X}(\mathrm{SS}(\mathcal{G}))$ is the normal cone to $\mathrm{SS}(\mathcal{G}) \subset T^*X$, and the identification (!) looks, in our concrete case, as follows.

We have $T^*V = V \times V^*$, and $T^*_L V = L \times L^{\perp}$. Therefore,

$$\begin{aligned} T_{T^*_L V} T^*V &= T_{L \times L^{\perp}}(V \times V^*) = (L \times L^{\perp}) \times \big((V/L) \times L^*\big), \\ T^*(T_L V) &= T^*\big(L \times (V/L)\big) = \big(L \times (V/L)\big) \times (L^* \times L^{\perp}), \end{aligned}$$

and (!) identifies factors number 1, 2, 3, 4 of the first product with factors number 1, 4, 2, 3 of the second one.

With this understanding, we need to prove that for any flat P of $\mathcal{H}$, the normal cone $C_{T^*_L V}(T^*_P V)$ is contained in the union of $T^*_Q\big(L \times (V/L)\big)$ over flats Q of the product arrangement in $L \times (V/L)$. In fact, it is contained in a single $T^*_Q\big(L \times (V/L)\big)$, where Q is the product flat $(P \cap L) \times \big(P/(P \cap L)\big)$, as follows from Example 4.1. This finishes the proof of Proposition 4.4. □

4.3 Specialization of Faces as a Continuous Map

Given a face A of $\mathcal{H}$, the intersection $\overline{A} \cap L_{\mathbb{R}}$ is the closure of a unique face of the arrangement $\mathcal{H} \cap L_{\mathbb{R}}$ which we denote by $\nu'_L(A)$. Further, the image of A in $V_{\mathbb{R}}/L_{\mathbb{R}}$ is a face of the quotient arrangement $\mathcal{H}/L_{\mathbb{R}}$ which we denote by $\nu''_L(A)$. The pair $\nu_L(A) = (\nu'_L(A), \nu''_L(A))$ is then a face of the product arrangement $\nu_L(\mathcal{H})$ which we call the *specialization* of A.

Proposition 4.5 *The closure of $\nu_L(A)$ is identified with the normal cone $C_{L_{\mathbb{R}}}(A)$. Thus, $\nu_L(A)$ is the interior (complement of the boundary) of $C_{L_{\mathbb{R}}}(A)$.*

Proof This is similar to Example 4.1. □

Example 4.6 The concept of specialization is illustrated in Fig. 1, where $\mathcal{H}$ consists of *five* lines in the plane, $L_{\mathbb{R}}$ is the horizontal line, and $\mathcal{H}/L_{\mathbb{R}}$ is the coordinate arrangement of two lines in $\mathbb{R}^2$. The three open sectors (colored red) on top, together with the open half-lines bounding them, specialize to the upward half-line (also colored red) in $\mathbb{R}^2$. The open sector (colored blue) with one side being the positive part of $L_{\mathbb{R}}$ specializes to the first quadrant in $\mathbb{R}^2$ (also colored blue).

The following is obvious.

Proposition 4.7 *The correspondence $A \mapsto \nu_L(A)$ defines a surjective monotone map $\nu_L : \mathcal{S}_{\mathbb{R}} \to \mathcal{S}^{\nu}_{\mathbb{R}}$ between the posets of faces of $\mathcal{H}$ and $\nu_L(\mathcal{H})$ such that* $\dim \nu_L(A) \leq \dim A$. □

We now form the "geometric realization" of the morphism of posets ν_L to construct a continuous map $q : V_{\mathbb{R}} \to L_{\mathbb{R}} \times (V_{\mathbb{R}}/L_{\mathbb{R}})$ from $V_{\mathbb{R}}$ to the normal bundle. That is, choose a point x_A in each face A of $\mathcal{H}$. Then we have the barycentric subdivision of V into based simplicial convex cones

$$C(A_1, \cdots, A_p) = \mathbb{R}_{>0} \cdot x_{A_1} + \cdots + \mathbb{R}_{>0} \cdot x_{A_p}$$

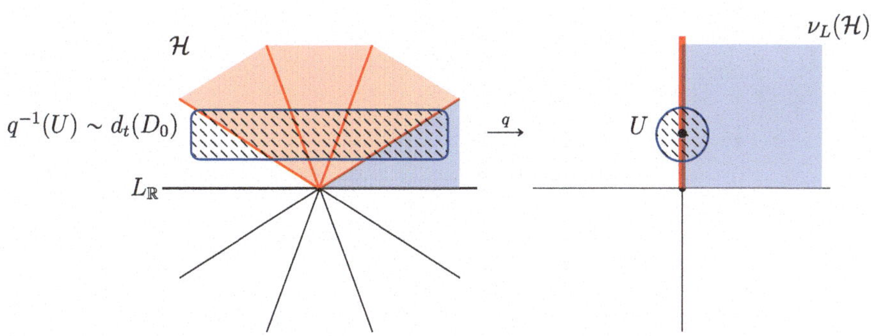

Fig. 1 Specialization of faces

corresponding to all increasing chains $A_1 < \cdots < A_p$ in $\mathcal{S}_{\mathbb{R}}$. In particular, each A is the union of the $C(A_1, \cdots, A_p)$ with $A_p = A$. Similarly, choose a point y_B in each face B of $\nu_L(\mathcal{H})$. Then we have the barycentric subdivision of $L \times (V/L)$ into similarly defined based simplicial convex cones $C(B_1, \cdots, B_p)$ for all chains $B_1 < \cdots < B_p$ in $\mathcal{S}^{\nu}_{\mathbb{R}}$. For each chain $A_1 < \cdots < A_p$, we define

$$p_{A_1,\cdots,A_p} : C(A_1, \cdots, A_p) \longrightarrow C(\nu_L(A_1), \cdots, \nu_L(A_p))$$

to be the unique $\mathbb{R}$-linear map taking x_{A_i} to $y_{\nu_L(A_i)}$.

Proposition 4.8 *q is a continuous, proper, piecewise linear surjective map. Further, each face A of $\mathcal{H}$ is mapped by q to $\nu_L(A)$ in a surjective, piecewise-linear way.*

Proof Clear from construction. □

4.4 The Real Result

In this subsection, we deal only with the real situation so we write V for $V_{\mathbb{R}}$, etc. Let $\mathcal{G} \in D^b(V, \mathcal{S})$ be a constructible complex.

Theorem 4.9 *The specialization $\nu_L(\mathcal{G})$ is identified with the topological direct image $Rq_*\mathcal{G}$ where q is the map from Proposition 4.8.*

Proof We first recall the definition [KS2, §4.1-2] of $\nu_L(\mathcal{G})$ in terms of the *normal deformation* $\widetilde{V}_L$ which, in our linear case, reduces to a single chart.

Choose a linear complement L' to L in V so $V = L \oplus L'$. Then L' is identified with V/L, and $T_L V$ is also identified with $L \oplus L'$, i.e., with V. We write a general vector of V as $v = (l, l')$ with $l \in L$ and $l' \in L'$. Then we define the commutative diagram with Cartesian squares:

$$\begin{array}{ccccccc} & & & \multicolumn{4}{c}{\overset{p}{\frown\!\!\!\longrightarrow}} \\ T_L V = V \times \{0\} & \xrightarrow{\;s\;} & \widetilde{V}_L := V \times \mathbb{R} & \xleftarrow{\;j\;} & \Omega & \xrightarrow{\;\widetilde{p}\;} & V \\ \Big\downarrow & & {\scriptstyle\tau}\Big\downarrow & & \Big\downarrow{\scriptstyle\widetilde{\tau}} & & \\ 0 & \longrightarrow & \mathbb{R} & \longleftarrow & \mathbb{R}_{>0}. & & \end{array} \tag{4.10}$$

where

$$p(l, l', t) = (l, t \cdot l'), \quad \tau(l, l', t) = t, \quad l \in L,\ l' \in L',\ t \in \mathbb{R}.$$

The space Ω is defined as $\tau^{-1}(\mathbb{R}_{>0}) = V \times \mathbb{R}_{>0}$, and $\widetilde{p}$ is the restriction of p to Ω.

After that, the specialization is defined by

$$\nu_L(\mathcal{G}) \;=\; s^* Rj_* \widetilde{p}^*(\mathcal{G}) \;\in\; D^b(\mathrm{Sh}_{T_M X}).$$

Let now $\xi = (l, l')$ be a point of $L \oplus L' = T_L V = \tau^{-1}(0)$. By definition, the stalk of $\nu_L(\mathcal{G})$ at ξ is

$$\nu_L(\mathcal{G})_\xi \;=\; R\Gamma(D \cap \Omega, p^*\mathcal{G})$$

where $D \subset V \times \mathbb{R}$ is a small $(n+1)$-dimensional open ball around $(\xi, 0) = (l, l', 0)$. Now, $\Omega = V \times \mathbb{R}_{>0}$. For each $t > 0$, consider the slice $D_t = D \cap (V \times \{t\})$. The restriction of p to D_t is the dilation $d_t : (l, l') \mapsto (l, t \cdot l')$ in the direction of L'.

Since D is a ball, the intersections $D_t \cap \Omega$ are nonempty for t lying in an open interval of the form $(0, \varepsilon)$ for some $\varepsilon > 0$ (the radius of D). For such t, we have that $D_t \cap \Omega = D_t$ is the slice over t. Since D is a small ball, these nonempty slices together with the complexes $d_t^*\mathcal{G}$ form a topologically trivial family over $(0, \varepsilon)$. This means that we can replace the cohomology of $D \cap \Omega$ (the union of all slices D_t, $t \in (0, \varepsilon)$) by the cohomology of any single slice, i.e.,

$$\nu_L(\mathcal{G})_\xi \;\simeq\; R\Gamma(D_t, d_t^*\mathcal{G})$$

for any sufficiently small $t > 0$. We can further replace D_t for such t with 0th slice $D_0 = D \cap (V \times \{0\})$. This slice is just a small n-dimensional open ball in $L \oplus L' = V$ around (l, l'). This gives

$$\nu_L(\mathcal{G})_\xi \;\simeq\; R\Gamma(D_0, d_t^*\mathcal{G}) \;=\; R\Gamma(d_t(D_0), \mathcal{G}), \quad 0 < t \ll 1.$$

When $t \to 0$, the open sets $d_t(D_0)$ become more and more flattened. We compare them with open sets of the form $q^{-1}(U)$ where U is a small ball in $T_L V = L \oplus L'$ around $d_t(\xi) = (l, t \cdot l')$. More precisely, we notice that $d_t(D_0)$ and $q^{-1}(U)$ become homotopy equivalent relatively to the stratification by the faces; see Fig. 1. This means that we have identifications (the last one expressing the conic nature of $Rq_*(\mathcal{G})$):

$$\nu_L(\mathcal{G})_\xi \;\simeq\; R\Gamma(q^{-1}(U), \mathcal{G}) \;=\; Rq_*(\mathcal{G})_{d_t(\xi)} \;\simeq\; Rq_*(\mathcal{G})_\xi .$$

This identifies the stalks. The same considerations show that the generalization maps between the stalks match as well. The theorem is proved. □

Assume now that $\mathcal{G}$ is given by a complex of representations $G = (\mathcal{G}_A, \gamma_{AA'})$ of $\mathcal{S}_\mathbb{R}$. So the complexes $\mathcal{G}_A$ are the stalks of $\mathcal{G}$, and the $\gamma_{AA'}$ are the generalization maps. For any face $B \in \mathcal{S}_\mathbb{R}^\nu$ of $\nu_L(\mathcal{H})$, define a complex

$$\mathcal{G}_{L,B} \;=\; \mathrm{Tot}\left\{ \bigoplus_{\substack{\nu_L(A)=B \\ \dim(A)=\dim(B)}} \mathcal{G}_A \otimes \mathrm{or}_{A/B} \xrightarrow{\gamma\otimes\varepsilon} \bigoplus_{\substack{\nu_L(A)=B \\ \dim(A)=\dim(B)+1}} \mathcal{G}_A \otimes \mathrm{or}_{A/B} \xrightarrow{\gamma\otimes\varepsilon} \cdots \right\}. \tag{4.11}$$

Let $B <_d B'$ be two faces of $\nu_L(\mathcal{H})$. We define a morphism of complexes

$$\gamma^L_{B,B'} : \mathcal{G}_{L,B} \to \mathcal{G}_{L,B'}$$

as follows. Let $A \in \mathcal{S}_{\mathbb{R}}$ be such that $\nu_L(A) = B$ and $\dim(A) = \dim(B) + p$, so that $\mathcal{G}_A \otimes \mathrm{or}_{A/B}$ is a summand in the pth term of $\mathcal{G}_{L,B}$. Similarly, let $A' \in \mathcal{S}_{\mathbb{R}}$ be such that $\nu_L(A') = B'$ and $\dim(A') = \dim(B') + p$, so that $\mathcal{G}'_A \otimes \mathrm{or}_{A'/B'}$ is a summand in the pth term of $\mathcal{G}_{L,B'}$. If $A \leq A'$, then $A <_d A'$, and the identification of the quotient spaces

$$\mathrm{Lin}_{\mathbb{R}}(A')/\mathrm{Lin}_{\mathbb{R}}(A) \xrightarrow{\simeq} \mathrm{Lin}_{\mathbb{R}}(B')/\mathrm{Lin}_{\mathbb{R}}(B)$$

gives, passing to the determinants and transposing, an isomorphism

$$\sigma^*_{AA'} : \mathrm{or}_{A/B} \longrightarrow \mathrm{or}_{A'/B'} .$$

We define the matrix element

$$(\gamma^L_{B,B'})^{A'}_A : \mathcal{G}_A \otimes \mathrm{or}_{A/B} \longrightarrow \mathcal{G}_{A'} \otimes \mathrm{or}_{A'/B}$$

to be equal to $\gamma_{AA'} \otimes \sigma^*_{AA'}$ if $A < A'$ and to 0 otherwise.

Corollary 4.12 *Each $\gamma^L_{BB'}$ is indeed a morphism of complexes, and the data $(\mathcal{G}_{L,B}, \gamma^L_{BB'})$ is a complex of representations of $\mathcal{S}_{v,\mathbb{R}}$, the poset of faces of the arrangement $\nu_L(\mathcal{H})$. This complex of representations describes the constructible complex $\nu_L(\mathcal{G})$.*

Proof Choose any point $b \in B$. Since q is a proper map, the stalk of $Rq_*(\mathcal{G})$ at b is identified with $R\Gamma(q^{-1}(b), \mathcal{G})$. Now $\mathcal{G}_{L,B}$ is nothing but the cellular cochain complex calculating $R\Gamma(q^{-1}(b), \mathcal{G})$. We similarly identify the generalization maps. □

Remark 4.13 At the formal algebraic level, the property that $\gamma^L_{BB'}$ is indeed a morphism of complexes simply reflects the fact that the differential in $R\Gamma(V, \mathcal{G})$, the cellular cochain complex, satisfies $d^2 = 0$. More precisely, we have an identification (isomorphism, not just a quasi-isomorphism) of cellular cochain complexes

$$R\Gamma(V, \mathcal{G}) \simeq R\Gamma(L \times (V/L), Rq_*(\mathcal{G})) \simeq R\Gamma(L \times (V/L), \nu_L(\mathcal{G})).$$

The RHS of this identification represents the same complex in a "block" form, with blocks (stalks of $\nu_L(\mathcal{G})$) parameterized by faces B of $\nu_L(\mathcal{H})$. The fact that the maps $\gamma^L_{BB'}$ between the blocks are morphisms of complexes is implied by the fact that the total differential squares to 0.

4.5 *Bispecialization*

We first consider the general situation studied in [ST, T]. Let $N \subset M \subset X$ be a flag of C^∞ submanifolds in a C^∞ manifold X. In the normal bundle $T_N X$, we have the submanifold (subbundle) $T_N M$. In the normal bundle $T_M X$, we have the submanifold N, embedded into M (the zero section of $T_M X$). It turns out that the normal bundles of these new submanifolds are identified.

Proposition 4.14 *We have identifications*[1]

$$T_{T_N M}(T_N X) \overset{(1)}{\simeq} T_N M \oplus (T_M X)|_N \overset{(2)}{\simeq} T_N(T_M X).$$

Proof The statement is a part of Prop. 2.1 of [T]. For convenience of the reader, we give a sketch of the proof. The identification (1) is a particular case of the well-known fact which generalizes, to vector bundles, the identification (4.2) for vector spaces: If $L \subset V$ is a C^∞ vector subbundle in a C^∞ vector bundle over a C^∞-manifold B, then $T_L V \simeq L \oplus (V/L)$. To see (2), we recognize, inside $T_N(T_M X)$, two subbundles: first, $T_N M$ (the normal bundle to N inside the zero section of $T_M X$) and, second, $(T_M X)|_N$ (the restriction to N of the normal bundle). Inspection in local coordinates shows that these two subbundles form a direct sum decomposition. □

In this context, Schapira and Takeuchi [ST, T] defined a functor

$$\nu_{NM} : D^b(X) \longrightarrow D^b(T_N \oplus (T_X M)|_N)$$

called *bispecialization*. It is defined, similar to the usual specialization, through the *binormal deformation* $\widetilde{X}_{NM}$, recalled below. On the other hand, we can iterate the specialization functors, getting a diagram of functors between derived categories of sheaves on the manifolds in question:

$$\begin{array}{ccc} D^b(X) & \xrightarrow{\ \nu_N\ } & D^b(T_N X) \\ {\scriptstyle \nu_M}\downarrow & \searrow^{\nu_{NM}} & \downarrow{\scriptstyle \nu_{T_N M}} \\ D^b(T_M X) & \xrightarrow[\nu_N]{} & D^b(T_N M \oplus (T_M X)|_N). \end{array} \tag{4.15}$$

This diagram is not (2-)commutative, i.e., the two composite functors (iterated specializations) are not isomorphic.

[1]The notation $\oplus$ here and below means direct sum of vector bundles, i.e., fiber product over N.

Example 4.16 Let $X = \mathbb{R}^2$ with coordinates x, y, let M be the line $y = 0$, and let N be the origin $(0, 0)$. Let $P \subset X$ be the parabola $y = x^2$ and $\mathcal{G} = \underline{\mathbf{k}}_P$ be the constant sheaf on P. We identify all three manifolds $T_N X$, $T_M X$, and $T_N M \oplus (T_M X)|_N$ back with $\mathbb{R}^2$ with the same coordinates. Then $\nu_N(\mathcal{G})$ is the constant sheaf on the horizontal line $y = 0$ (the tangent line to P), and $\nu_{T_N M}(\nu_N(\mathcal{G}))$ is again the constant sheaf on the line $y = 0$. On the other hand, $\nu_M(\mathcal{G})$ is supported on the vertical half-line $x = 0$, $y \geq 0$ (since P is contained in the upper half plane $y \geq 0$ and does not meet M except for $x = 0$). So $\nu_N(\nu_M(\mathcal{G}))$ will be again supported on this half-line.

Nevertheless, in the linear case, all three possible functors are identified.

Theorem 4.17 *Let $X = V$ be an $\mathbb{R}$-vector space and $N \subset M \subset V$ be a flag of $\mathbb{R}$-linear subspaces. Let $\mathcal{H}$ be an arrangement of hyperplanes in V and $\mathcal{S}_\mathbb{R}$ the corresponding stratification by faces. Then for $\mathcal{G} \in D^b(V, \mathcal{S}_\mathbb{R})$, we have canonical quasi-isomorphisms*

$$\nu_N(\nu_M(\mathcal{G})) \simeq \nu_{T_N M}(\nu_N(\mathcal{G})) \simeq \nu_{NM}(\mathcal{G}).$$

In other words, the diagram (4.15) *becomes 2-commutative if the top left corner is replaced by $D^b(V, \mathcal{S}_\mathbb{R})$.*

Proof Enlarging $\mathcal{H}$ if necessary, we can assume that N and M are flats of $\mathcal{H}$. The space $T_N M \oplus (T_M X)|_N$ is identified with vector space $V'' = N \oplus (M/N) \oplus (V/M)$ which carries the *triple product arrangement*

$$\nu_{NM}(\mathcal{H}) := (\mathcal{H} \cap N) \oplus ((\mathcal{H} \cap M)/N) \oplus (\mathcal{H}/M).$$

Denote by $\mathcal{S}_\mathbb{R}^{\nu_{NM}}$ the stratification given by the faces of this arrangement. Also denote $\mathcal{S}_\mathbb{R}^{\nu_N}$ and $\mathcal{S}_\mathbb{R}^{\nu_M}$ the stratifications given by the faces of the arrangements $\nu_N(\mathcal{H})$ and $\nu_M(\mathcal{H})$. Now notice that specialization of faces gives a *commutative* diagram of morphisms of posets which we then use to construct a commutative diagram of proper piecewise linear maps:

$$\begin{array}{ccc} \mathcal{S}_\mathbb{R} & \xrightarrow{\nu_N} & \mathcal{S}_\mathbb{R}^{\nu_N} \\ {\scriptstyle \nu_M}\downarrow & & \downarrow{\scriptstyle \nu_{T_N M}} \\ \mathcal{S}_\mathbb{R}^{\nu_M} & \xrightarrow[\nu'_N]{} & \mathcal{S}_\mathbb{R}^{\nu_{NM}} \end{array} \qquad \begin{array}{ccc} V & \xrightarrow{q_N} & T_N V \\ {\scriptstyle q_M}\downarrow & & \downarrow{\scriptstyle q_{T_N M}} \\ T_M V & \xrightarrow{q'_N} & T_N M \oplus (T_M V)|_N. \end{array}$$

The direct images in this second diagram correspond, by Theorem 4.9, to the specialization functors on the outer edges of the diagram (4.15). This shows that the outer rim of (4.15) is 2-commutative.

We now show that the composite functor given by the outer rim of (4.15) is isomorphic to ν_{NM}. (This will also give another proof of the commutativity of the outer rim.) For this, we recall the explicit form of the binormal deformation diagram; see [T] Eq. (2.20). We choose a complement L' to N in M and a complement L'' to M in V, thus identifying V, as well as $T_N M \oplus (T_M V)|_N$, with $N \oplus L' \oplus L''$. So we write elements of either of this spaces as (n, l', l''). Then the "bi"-analog of the diagram (4.10) has the form

$$\begin{array}{ccccccc} T_N M \oplus (T_M V)|_N = V \times \{(0,0)\} & \xrightarrow{\ s\ } & \widetilde{V}_{NM} = V \times \mathbb{R}^2 & \xleftarrow{\ j\ } & \Omega & \xrightarrow{\ \widetilde{p}\ } & V \\ \downarrow & & \downarrow{\scriptstyle \tau} & & \downarrow{\scriptstyle \widetilde{\tau}} & & \\ 0 & \longrightarrow & \mathbb{R}^2 & \longleftarrow & \mathbb{R}^2_{>0}, & & \end{array}$$

(with $p = \widetilde{p} \circ j^{-1}$ drawn as a curved arrow from $\widetilde{V}_{NM}$ to V)

with

$$p((n, l', l''), (t', t'')) = (n,\ t'l',\ t't''l''), \quad \tau((n, l', l''), (t', t'')) = (t', t''),$$

so the restriction of p to $V \times \{(t', t'')\}$ is the map

$$p_{(t',t'')} : (n, l', l'') \mapsto (n,\ t'l',\ t't''l'').$$

The bispecialization is defined as $\nu_{NM}(\mathcal{G}) = s^* Rj_* \widetilde{p}^* \mathcal{G}$ with respect to this diagram, so its stalk at (n, l', l'') is $R\Gamma(D \cap \Omega, p^*\mathcal{G})$ where D is a small open $(n+2)$-dimensional ball around $((n, l', l''), (0, 0))$. We slice D into n-dimensional balls $D_{(t',t'')} = D \cap \tau^{-1}(t', t'')$. □

Lemma 4.18 *For sufficiently small* $\varepsilon > 0$*, the slices* $D_{(t',t'')}$*, together with the restrictions* $p^*\mathcal{G}|_{D_{(t',t'')}} = p^*_{(t',t'')}\mathcal{G}$*, form a topologically trivial family over the product of open intervals* $(0, \varepsilon) \times (0, \varepsilon)$.

Proof of the Lemma: For $u', u'' > 0$, let

$$d_{(u',u'')} : V \to V, \quad (n, l', l'') \mapsto (n,\ u'l',\ u''l'')$$

be the bi-dilation in the last two variables. Then $p_{(t',t'')} = d_{c(t',t'')}$, where $c : \mathbb{R}^2 \to \mathbb{R}^2$ is the map

$$(t', t'') \mapsto (u', u'') = (t',\ t't'').$$

Now, c maps the open square $(0, \varepsilon)^2$ homeomorphically onto the open triangular wedge ∇_ε of slope ε; see Fig. 2.

For small $t', t'' > 0$, we can identify the slices $D_{(t',t'')}$ with $D_{(0,0)}$ (alternatively, we could have taken D to be the product of balls in V and in $\mathbb{R}^2$ so that the slices would not change at all).

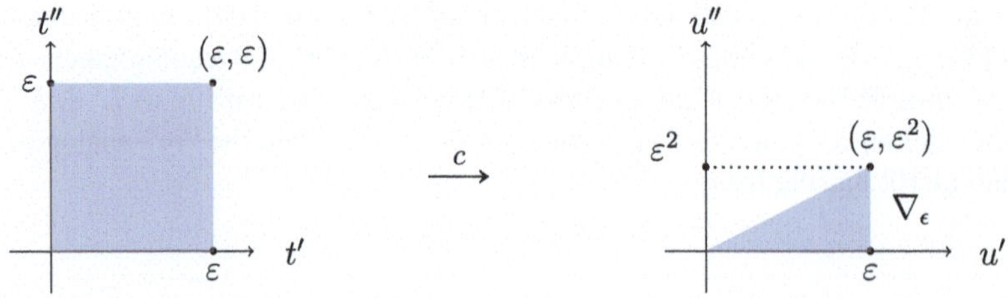

Fig. 2 The wedge ∇_ϵ

We recall that $\mathcal{G}$ is smooth with respect to a hyperplane arrangement $\mathcal{H}$ (so the slopes of the hyperplanes are fixed). On the other hand, the slope of the wedge ∇_ϵ is shrinking as $\varepsilon \to 0$. Therefore, for sufficiently small ε, we will have that for all $(u', u'') \in \nabla_\epsilon$, the topological structure of $d^*_{(u',u'')}\mathcal{G}$ on $D_{(0,0)}$ will stabilize. This proves the lemma.

The lemma implies that the stalk of $\nu_{NM}(\mathcal{G})$ at (n, l', l'') can be written as

$$R\Gamma(D_{(0,0)}, p^*_{(t',t'')}\mathcal{G}) = R\Gamma(p_{(t',t'')}(D_{(0,0)}), \mathcal{G})$$

for any sufficiently small positive t', t''.

It remains to similarly analyze the two outer composite functors (iterated specializations) in (4.15) and to find that they correspond to the choice of $0 < t' \ll t'' \ll 1$, resp. $0 < t'' \ll t' \ll 1$. Because of the topological triviality of the family over all $(t', t'') \in (0, \varepsilon) \times (0, \varepsilon)$, all three results are the same. □

4.6 The Complex Result

We now consider the complex situation: that of a perverse sheaf $\mathcal{F} \in \mathrm{Perv}(V_\mathbb{C}, \mathcal{S}_\mathbb{C})$ and the corresponding hyperbolic sheaf $\mathcal{Q} = (E_A, \gamma_{AA'}, \delta_{A'A})$. Let $\mathcal{Q}^\nu = (E^\nu_B, \gamma^\nu_{BB'}, \delta^\nu_{B'B})$ be the hyperbolic sheaf corresponding to $\nu_{L_\mathbb{C}}(\mathcal{F}) \in \mathrm{Perv}(T_{L_\mathbb{C}}V_\mathbb{C}, \mathcal{S}_{\nu,\mathbb{C}})$. Here B, B' are faces of the product arrangement $\nu_L(\mathcal{H})$.

Theorem 4.19

(a) The hyperbolic stalk E^ν_B is identified as

$$E^\nu_B \simeq \left\{ \bigoplus_{\substack{\nu_L(A)=B \\ \dim(A)=\dim(B)}} E_A \otimes \mathrm{or}_{A/B} \xrightarrow{\gamma\otimes\varepsilon} \bigoplus_{\substack{\nu_L(A)=B \\ \dim(A)=\dim(B)+1}} E_A \otimes \mathrm{or}_{A/B} \xrightarrow{\gamma\otimes\varepsilon} \cdots \right\}$$

That is, the complex in the RHS is exact everywhere except the leftmost term, where the kernel is identified with E^ν_B.

(a') We also have an identification

$$E_B^\nu \simeq \left\{ \cdots \xrightarrow{\delta\otimes\varepsilon} \bigoplus_{\substack{\nu_L(A)=B \\ \dim(A)=\dim(B)+1}} E_A \otimes \mathrm{or}_{A/B} \xrightarrow{\delta\otimes\varepsilon} \bigoplus_{\substack{\nu_L(A)=B \\ \dim(A)=\dim(B)}} E_A \otimes \mathrm{or}_{A/B} \right\}.$$

That is, the complex in the RHS is exact everywhere except the rightmost term, where the cokernel is identified with E_B^ν.

(b) The maps $\gamma_{BB'}^\nu$ are induced by the maps $\gamma_{AA'}$ which induce morphisms of complexes in (a), similar to Corollary 4.12.

(b') The maps $\delta_{B'B}^\nu$ are induced by the map $\delta_{A'A}$ which induce morphisms of complexes in (a').

Proof We first prove parts (a) and (b). Let $i_{\mathbb{R}} : V_{\mathbb{R}} \to V_{\mathbb{C}}$ and $i_{\mathbb{R},\nu} : T_{L_{\mathbb{R}}} V_{\mathbb{R}} \to T_{L_{\mathbb{C}}} V_{\mathbb{C}}$ be the embeddings of the real parts. Put

$$\mathcal{G} = i_{\mathbb{R}}^! \mathcal{F}, \quad \mathcal{G}_\nu = i_{\mathbb{R},\nu}^! \nu_{L_{\mathbb{C}}}(\mathcal{F}).$$

These are ordinary sheaves (not just complexes) on $V_{\mathbb{R}}$ and $T_{L_{\mathbb{R}}} V_{\mathbb{R}}$, smooth with respect to $\mathcal{S}_{\mathbb{R}}$ and $\mathcal{S}_{\mathbb{R}}^\nu$, respectively. Their stalks are given by the E_A and E_B^ν, and their generalization maps are given by the $\gamma_{AA'}$ and $\gamma_{BB'}^\nu$, respectively. Note that we have a canonical morphism

$$\nu_{L_{\mathbb{R}}}(\mathcal{G}) = \nu_{L_{\mathbb{R}}}(i_{\mathbb{R}}^! \mathcal{F}) \xrightarrow{\beta} i_{\mathbb{R},\nu}^! \nu_{L_{\mathbb{C}}}(\mathcal{F}) = \mathcal{G}_\nu;$$

see [KS2] Prop. 4.2.5. So our statements will follow from Corollary 4.12 if we establish the following. □

Proposition 4.20 *For any $\mathcal{F} \in D^b(V_{\mathbb{C}}, \mathcal{S}_{\mathbb{C}})$, the morphism $\beta : \nu_{L_{\mathbb{R}}}(i_{\mathbb{R}}^! \mathcal{F}) \to i_{\mathbb{R},\nu}^! \nu_{L_{\mathbb{C}}}(\mathcal{F})$ is a quasi-isomorphism.*

Proof of Proposition 4.20 Since $\nu_{L_{\mathbb{R}}}$ and $\nu_{L_{\mathbb{C}}}$ commute with Verdier duality, it is enough to show that for any $\mathcal{F} \in D^b(V_{\mathbb{C}}, \mathcal{S}_{\mathbb{C}})$, the dual morphism $\alpha : i_{\mathbb{R},\nu}^* \nu_{L_{\mathbb{C}}}(\mathcal{F}) \to \nu_{L_{\mathbb{R}}}(i_{\mathbb{R}}^* \mathcal{F})$ is a quasi-isomorphism. Such a morphism is defined for any $\mathcal{F} \in D^b(\mathrm{Sh}_{V_{\mathbb{C}}})$ whatsoever; see [KS2] Prop. 4.2.5. So we show that it is a quasi-isomorphism for a more general class of complexes. Namely, $V_{\mathbb{C}}$ has the *product stratification* $\mathcal{S}_{\mathbb{R}} \times \mathcal{S}_{\mathbb{R}}$ formed by the cells of the form $A' + iA'' \subset V_{\mathbb{C}} = V_{\mathbb{R}} + iV_{\mathbb{R}}$, where A' and A'' are arbitrary faces of $\mathcal{H}$ and $i = \sqrt{-1}$. This stratification refines $\mathcal{S}_{\mathbb{C}}$, so $D^b(V_{\mathbb{C}}, \mathcal{S}_{\mathbb{C}}) \subset D^b(V_{\mathbb{C}}, \mathcal{S}_{\mathbb{R}} \times \mathcal{S}_{\mathbb{R}})$. Therefore, it suffices to prove: □

Lemma 4.21 *For any $\mathcal{F} \in D^b(V_{\mathbb{C}}, \mathcal{S}_{\mathbb{R}} \times \mathcal{S}_{\mathbb{R}})$, the morphism $\alpha : i_{\mathbb{R},\nu}^* \nu_{L_{\mathbb{C}}}(\mathcal{F}) \to \nu_{L_{\mathbb{R}}}(i_{\mathbb{R}}^* \mathcal{F})$ is a quasi-isomorphism.*

Proof of Lemma 4.21 The stratification on $V_{\mathbb{R}}$ induced by $i_{\mathbb{R}}$ from $\mathcal{S}_{\mathbb{R}} \times \mathcal{S}_{\mathbb{R}}$ is $\mathcal{S}_{\mathbb{R}}$. This means that the specializations maps of the posets of faces are compatible, and therefore we have a commutative diagram

$$\begin{array}{ccc} V_{\mathbb{R}} & \xrightarrow{q_{\mathbb{R}}} & L_{\mathbb{R}} \times (V_{\mathbb{R}}/L_{\mathbb{R}}) \\ {\scriptstyle i_{\mathbb{R}}}\downarrow & & \downarrow{\scriptstyle i_{\mathbb{R},\nu}} \\ V_{\mathbb{C}} & \xrightarrow[q_{\mathbb{C}}]{} & L_{\mathbb{C}} \times (V_{\mathbb{C}}/L_{\mathbb{C}}), \end{array}$$

where $q_{\mathbb{R}}$ and $q_{\mathbb{C}}$ are the proper maps constructed in Proposition 4.8. So our statement follows from Theorem 4.9 by proper base change. □

This finishes the proof of Proposition 4.20 and of parts (a) and (b) of Theorem 4.19.

Now, parts (a') and (b') of Theorem 4.19 follow from (a) and (b) because $\nu_{L_{\mathbb{C}}}$ commutes with Verdier duality whose effect on hyperbolic sheaves exchanges γ and δ; see Theorem 2.11. Theorem 4.19 is proved.

5 Fourier Transform and Hyperbolic Sheaves

5.1 Generalities on the Fourier-Sato Transform

Let W be a finite-dimensional $\mathbb{R}$-vector space and W^* the dual space. We denote by $D^b_{\text{con}}(E) \subset D^b(E)$ the full subcategory formed by complexes $\mathcal{G}$ which are *conic*, i.e., such that each sheaf $\underline{H}^j(\mathcal{G})$ is locally constant on any orbit of the scaling action of $\mathbb{R}_{>0}$ on W.

Set

$$P = \{(x, f) \in W \times W^* \mid f(x) \geq 0\} \overset{i_P}{\hookrightarrow} W \times W^*$$

and denote by p_1, p_2 the projections of P to W and W^*, respectively. The *Fourier-Sato transform* is an equivalence of categories

$$\text{FS} : D^b_{\text{con}}(W) \longrightarrow D^b_{\text{con}}(W^*), \quad \text{FS}(\mathcal{G}) = Rp_{2!}(p_1^*\mathcal{G});$$

see [KS2] Def. 3.7.8. The base change theorem implies at once the following.

Proposition 5.1 *Let $f \in W^*$. The stalk of* FS$(\mathcal{G})$ *at f is found as*

$$\text{FS}(\mathcal{G})_f \simeq R\Gamma_c(P_f, \mathcal{G}),$$

where $P_f = p_2^{-1}(f) = \{x \in W | f(x) \geq 0\}$. (Thus, P_f is a closed half-space for $f \neq 0$ and $P_f = W$ for $f = 0$.) □

5.2 The Dual Arrangement

We specialize the above to the two situations related to an arrangement of hyperplanes $\mathcal{H}$ in $V_{\mathbb{R}}$. We denote $n = \dim_{\mathbb{R}} V_{\mathbb{R}}$.

(1) $W = V_{\mathbb{R}}$ and $\mathcal{G} \in D^b(V_{\mathbb{R}}, \mathcal{S}_{\mathbb{R}})$. In this case, we would like to find the stalks of $\mathrm{FS}(\mathcal{G})$.

(2) $W = V_{\mathbb{C}}$ and $\mathcal{G} \in \mathrm{Perv}(V_{\mathbb{C}}, \mathcal{S}_{\mathbb{C}})$. We identify $W^* = \mathrm{Hom}_{\mathbb{C}}(V, \mathbb{C})$ with the real dual $\mathrm{Hom}_{\mathbb{R}}(V_{\mathbb{C}}, \mathbb{R})$ by means of the form

$$(x, f) \mapsto \Re(f(x)), \quad x \in V_{\mathbb{C}}, f \in V_{\mathbb{C}}^*.$$

In this case, it is known (see [KS2] Ch. X) that $\mathrm{FS}(\mathcal{G})[-n]$ is a perverse sheaf on $V_{\mathbb{C}}^*$ with respect to some stratification. We would like to relate this stratification to an arrangement of hyperplanes and to find the hyperbolic stalks of $\mathrm{FS}(\mathcal{G})$.

This leads to the following definition.

Definition 5.2 The *dual arrangement* $\mathcal{H}^\vee$ of hyperplanes in $V_{\mathbb{R}}^*$ consists of orthogonals $l^\perp$ where l is a one-dimensional flat of $\mathcal{H}$. We denote by $\mathcal{S}_{\mathbb{R}}^\vee$ the stratification of $V_{\mathbb{R}}^*$ into faces of $\mathcal{H}^\vee$ and by $\mathcal{S}_{\mathbb{C}}^\vee$ the stratification of $V_{\mathbb{C}}^*$ into generic parts of the complex flats of $\mathcal{H}^\vee$.

Proposition 5.3 *We have an inclusion $\mathcal{H} \subset \mathcal{H}^{\vee\vee}$ (as sets of hyperplanes in $V_{\mathbb{R}}$).*

Proof One-dimensional flats of $\mathcal{H}^\vee$ are the orthogonals $M^\perp$, where M runs over hyperplanes in $V_{\mathbb{R}}$ which are sums of one-dimensional flats of $\mathcal{H}$. Such M are therefore precisely the hyperplanes of $\mathcal{H}^{\vee\vee}$. Now the statement means that each hyperplane $H \in \mathcal{H}$ can be obtained as a sum of one-dimensional flats of $\mathcal{H}$. This is indeed the case, since we have assumed from the outset that $\mathcal{H}$ is central, i.e., the intersection of all $H \in \mathcal{H}$ is 0. □

Example 5.4

(a) Call an arrangement $\mathcal{H}$ *reflexive*, if $\mathcal{H}^{\vee\vee} = \mathcal{H}$. A sufficient condition for this is that the set of flats of $\mathcal{H}$ is closed not only under intersections but also under sums, i.e., it forms a lattice. This follows from the proof of Proposition 5.3. Examples of reflexive arrangements include any arrangement with $\dim(V_{\mathbb{R}}) \leq 2$, as well as any direct sum of such arrangements.

(b) In general, forming the union of the arrangements

$$\mathcal{H} \subset \mathcal{H}^{\vee\vee} \subset \mathcal{H}^{\vee\vee\vee\vee} \subset \cdots$$

amounts to closing $\mathcal{H}$ under the operations of sum and intersection, i.e., to forming the lattice of subspaces generated by $\mathcal{H}$ and taking all $(n-1)$-dimensional elements of it. Such a lattice (and therefore the above union) is typically infinite. For instance, for $n = 3$, we start with a finite set of lines in

$\mathbb{R}P^2$, form all their intersection points, and then draw new lines through these points and so on.

(c) Let $V_{\mathbb{R}} = \mathbb{R}^n$ with coordinates $x_1, \cdots, x_n$. Take $\mathcal{H}$ to be the arrangement of the following hyperplanes:

$$\{x_i = 0\},\ i = 1, \cdots, n, \quad \{x_i = x_{i+1}\},\ i = 1, \cdots, n-1.$$

There are $\binom{n+1}{2}$ one-dimensional flats of $\mathcal{H}$; they have the form

$$L_{[i,j]} = \{x \mid x_i = x_{i+1} = \cdots = x_j;\ x_k = 0, k \notin [i,j]\}, \quad 1 \leq i \leq j \leq n.$$

On the other hand, consider $\mathbb{R}^{n+1}$ with coordinates $y_0, \cdots, y_n$, and let $W_{\mathbb{R}} = \mathbb{R}^{n+1}/\mathbb{R}\cdot(1, \cdots, 1)$. Thus, $W_{\mathbb{R}} = \mathfrak{h}^*$ is the space of weights for the Lie algebra $\mathfrak{sl}_{n+1}(\mathbb{R})$. We have an isomorphism $V_{\mathbb{R}} \to W_{\mathbb{R}}^*$ which takes the ith basis vector $e_i \in V_{\mathbb{R}}$, $i = 1, \cdots, n$, to the functional $y \mapsto y_{i-1} - y_i$ (simple co-root). This isomorphism takes $L_{[i,j]}$ to the co-root hyperplane $\{y_{i-1} = y_j\}$. Therefore, the dual arrangement $\mathcal{H}^\vee$ is the co-root arrangement in $\mathfrak{h}^*$.

Next, flats of $\mathcal{H}^\vee$ are in bijection with equivalence relations R on the set $\{0, 1, \cdots, n\}$. The flat corresponding to R has the form

$$M_R = \{y \mid y_i = y_j \text{ whenever } i \equiv_R j.\}.$$

It is one-dimensional if and only if R has only 2 equivalence classes, both non-empty. Thus, there are $2^{n-1} - 1$ one-dimensional flats of $\mathcal{H}^\vee$, and so the double dual arrangement $\mathcal{H}^{\vee\vee}$ consists of $2^{n-1} - 1$ hyperplanes and is much bigger than $\mathcal{H}$.

Proposition 5.5

(a) If $\mathcal{G} \in D^b(V_{\mathbb{R}}, \mathcal{S}_{\mathbb{R}})$, then $\mathrm{FS}(\mathcal{G}) \in D^b(V_{\mathbb{R}}^*, \mathcal{S}_{\mathbb{R}}^\vee)$.
(b) If $\mathcal{F} \in \mathrm{Perv}(V_{\mathbb{C}}, \mathcal{S}_{\mathbb{C}})$, then $\mathrm{FS}(\mathcal{F})[-n] \in \mathrm{Perv}(V_{\mathbb{C}}^*, \mathcal{S}_{\mathbb{C}}^\vee)$.

Proof As in the proof of Proposition 4.4, the real and complex case are completely parallel, so we treat the real case, dropping the subscript $\mathbb{R}$. The microsupport of $\mathcal{G}$ is contained in the union of the $T_L^*V = L \times L^\perp$ over $L \in \mathrm{Fl}(\mathcal{H})$. Now, the effect of FS on microsupports is via the identification ("Legendre transform")

$$T^*V = V \times V^* \longrightarrow V^* \times V = T^*V^*.$$

This identification takes T_L^*V to $T_{L^\perp}^*V^*$. This means that $\mathrm{FS}(\mathcal{G})$ is smooth with respect to the stratification $\mathcal{S}^*$ formed by the generic parts

$$L_\circ^\perp = L^\perp \setminus \bigcup_{L_1^\perp \not\supset L^\perp} L_1^\perp, \quad L \in \mathrm{Fl}(\mathcal{H}).$$

Now, $\mathcal{S}^\vee$ refines $\mathcal{S}^*$, so $\mathrm{FS}(\mathcal{G})$ is smooth with respect to $\mathcal{S}^\vee$. $\square$

5.3 *Big and Small Dual Cones*

Let $A^\vee \in \mathcal{S}^\vee_{\mathbb{R}}$ be a face. Its *big dual cone* is defined as

$$U(A^\vee) \ = \ \{x \in V \,|\, f(x) \geq 0, \ \forall f \in A^\vee\} \ \subset \ V_{\mathbb{R}}. \tag{5.6}$$

It is a closed polyhedral cone in V with nonempty interior, the union of the closures of (in general, several) chambers of $\mathcal{H}$.

The *small dual cone* of $A^\vee$ is defined as

$$V(A^\vee) \ = \ \bigcap_{B^\vee \geq A^\vee \text{ chamber}} U(B^\vee). \tag{5.7}$$

It is a strictly convex (not containing $\mathbb{R}$-linear subspaces) closed polyhedral cone in $V_{\mathbb{R}}$. Note that $U(A^\vee) = V(A^\vee)$ if $A^\vee$ is a chamber but $U(A^\vee)$ can be strictly larger than $V(A^\vee)$ in general. For example, if $A^\vee$ is a half-line (*one*-dimensional face) of $\mathcal{H}^\vee$, then $U(A^\vee)$ is a closed half-space in $V_{\mathbb{R}}$, while V_A is strictly convex, cf. Fig. 3.

The next statement is clear from the definitions.

Proposition 5.8 *If $A_1^\vee \leq A_2^\vee$, then $U(A_1^\vee) \supset U(A_2^\vee)$ and $V(A_1^\vee) \subset V(A_2^\vee)$.* □

Proposition 5.9 *Let $f \in A^\vee$ be arbitrary. Then:*

(a) $U(A^\vee)$ is the union of all faces B of $\mathcal{H}$ such that $f|_B \geq 0$ (non-strict inequality).
(b) $V(A^\vee)$ is the union of 0 and all the faces B of $\mathcal{H}$ such that $f|_B > 0$ (strict inequality everywhere).

Proof

(a) Since $A^\vee$ is a face of $\mathcal{H}^\vee$, for each $f \in A^\vee$, the pattern of signs (positive, negative or zero) of f on faces of $\mathcal{H}$ is the same. So the requirement that $f|_B \geq$

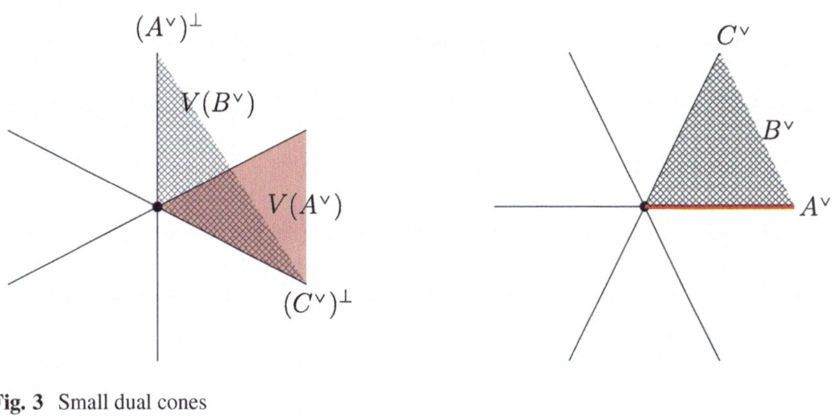

Fig. 3 Small dual cones

0 for each $f \in A^\vee$ (appearing in the definition of $U(A^\vee)$) is equivalent to the requirement that $f|_B \geq 0$ for any particular choice of $f \in A^\vee$ (appearing in the statement of the proposition).

(b) Let V' be the union of the faces in question. If $B \neq 0$ is a face of $\mathcal{H}$ such that $B \subset V'$, i.e., that $f|_B > 0$, then $g|_B > 0$ for any $g \in A^\vee$, by definition of the dual arrangement. This means that for any $B^\vee \geq A^\vee$ and any $g \in B^\vee$ sufficiently close to $A^\vee$, we still have $g|_B > 0$. This further implies (again, by the definition of the dual arrangement) that for any $B^\vee \geq A^\vee$ and any $g \in B^\vee$ whatsoever, we still have $g|_B > 0$. This means that $B \subset V(B^\vee)$ for any $B^\vee \geq A^\vee$, in other words, that $B \subset V(A^\vee)$. We proved that $V' \subset V(A^\vee)$.

Conversely, suppose $B \subset V(A^\vee)$. For any chamber $B^\vee \geq A^\vee$ and any $g \in B^\vee$, the restriction $g|_B$ cannot vanish, since that would mean that g is not inside a chamber of a dual arrangement. Therefore, $g|_B > 0$ everywhere. Now, if $f \in A^\vee$ and $A^\vee$ is not a chamber, then looking at g varying in a small transverse ball to $A^\vee$ near f in $V_\mathbb{R}^*$, we see that all such $g|_B$ must be positive and therefore $f|_B$ must be positive. In other words, we proved that $V(A^\vee) \subset V'$. □

Corollary 5.10 *We have*

$$U(A^\vee) = \bigcup_{B^\vee \geq A^\vee} U(B^\vee) = \bigcup_{B^\vee \geq A^\vee} V(B^\vee).$$

□

We now analyze the nature of the covering of $U(A^\vee)$ by the $U(B^\vee)$, $B^\vee \geq A^\vee$. All $B^\vee \geq A^\vee$ are in bijection with faces of the *quotient arrangement* $\mathcal{H}^\vee/A^\vee$ in the quotient space $V_\mathbb{R}^*/\mathrm{Lin}_\mathbb{R}(A^\vee)$, cf. [KS1] §2B. We denote by $B^\vee/A^\vee$ the face of $\mathcal{H}^\vee/A^\vee$ corresponding to $B^\vee \geq A^\vee$.

Proposition 5.11 *Let $A^\vee \in \mathcal{S}_\mathbb{R}^\vee$ and $B \in \mathcal{S}_\mathbb{R}$. Then:*

(a) There is a closed convex polyhedral cone $K(A^\vee, B) \subset V_\mathbb{R}^*/\mathrm{Lin}_\mathbb{R}(A^\vee)$, a union of faces of $\mathcal{H}^\vee/A^\vee$, which has the following property:

For $B^\vee \geq A^\vee$ we have $B \subset U(B^\vee)$ if and only $B^\vee/A^\vee \subset K(A^\vee, B)$.

(b) The cone $K(A^\vee, B)$ coincides with the whole $V_\mathbb{R}^*/\mathrm{Lin}_\mathbb{R}(A^\vee)$ if and only if $B \subset V(A^\vee)$.

Proof

(a) Let $U(B) \subset V_\mathbb{R}^*$ be the dual cone to B, i.e., the set of $f \in V_\mathbb{R}^*$ such that $f|_B \geq 0$. It is a convex, closed polyhedral cone in $V_\mathbb{R}^*$ which is a union of faces of $\mathcal{H}^\vee$. In fact, the condition $B^\vee \subset U(B)$ is equivalent to $B \subset U(B^\vee)$, both meaning that $(b^\vee, b) \geq 0$ for each $b^\vee \in B^\vee$ and $b \in B$.
Let also $(V_\mathbb{R}^*)^{\geq A^\vee} \subset V_\mathbb{R}^*$ be the union of all faces $B^\vee$ of $\mathcal{H}^\vee$ such that $B^\vee \geq A^\vee$. It is a convex, open polyhedral cone in $V_\mathbb{R}^*$. The intersection $U(B) \cap (V_\mathbb{R}^*)^{\geq A^\vee}$

is then a convex polyhedral cone which is closed in $(V_{\mathbb{R}}^*)^{\geq A^\vee}$. Since this cone is a union of faces $B^\vee \geq A^\vee$, it projects to a convex closed polyhedral cone in $V_{\mathbb{R}}^*/\mathrm{Lin}_{\mathbb{R}}(A^\vee)$ which we denote $K(A^\vee, B)$. By construction, $K(A^\vee, B)$ satisfies the required property.

(b) This is a reformulation of the formula (5.7) defining $V(A^\vee)$. □

Note an appealing numerical corollary of Proposition 5.11. For any subset $Z \subset V$, we denote by $\mathbf{1}_Z : V \to \mathbb{R}$ its characteristic function, equal to 1 on Z and to 0 elsewhere. □

Corollary 5.12 *(Inclusion—Exclusion Formulas) We have the identities*

$$\mathbf{1}_{U(A^\vee)} = \sum_{B^\vee \supset A^\vee} (-1)^{\dim(B^\vee)-\dim(A^\vee)} \mathbf{1}_{V(A^\vee)}, \qquad (a)$$

$$\mathbf{1}_{V(A^\vee)} = \sum_{B^\vee \supset A^\vee} (-1)^{\dim(B^\vee)-\dim(A^\vee)} \mathbf{1}_{U(A^\vee)}. \qquad (b)$$

Identities of this general nature (representing the characteristic function of a convex polytope as an alternating sum of characteristic functions of simplices or cones) are familiar in the theory of convex polytopes [V, FL] and the theory of automorphic forms; see, e.g., [Ar], §11.

We note the similarity of the identities (a) and (b) with Proposition 2.9(b) and Corollary 2.14 relating the usual stalks and hyperbolic stalks of a perverse sheaf. In fact, we will use a "categorified" version of these identities to relate the usual and hyperbolic stalks of the Fourier-Sato transform.

Proof of Corollary 5.12 (b) Write the RHS of the proposed identity as $\sum_B c_B \mathbf{1}_B$ with B running over faces of $\mathcal{H}$. Part (a) of Proposition 5.11 implies that

$$c_B = \sum_{\substack{B^\vee \geq A^\vee \\ B^\vee/A^\vee \subset K(A^\vee, B)}} (-1)^{\dim(B^\vee)-\dim(A^\vee)} = (-1)^{n-\dim(A^\vee)} \chi\big(H_c^\bullet(K(A^\vee, B), \mathbf{k}\big)$$

is the signed (calculated from the top) Euler characteristic of the cohomology with compact support of the cone $K(A^\vee, B)$. This signed Euler characteristic is equal to 0 unless $K(A^\vee, B)$ is the entire vector space, in which case it is 1. By Part (b) of Proposition 5.11, this happens precisely when $B \subset V(A^\vee)$, so the identity is proved.

(a) is a formal consequence of (b) in virtue of the identity

$$\sum_{B^\vee \geq A^\vee} (-1)^{\dim(B^\vee)-\dim(A^\vee)} = 1$$

(the Euler characteristic of the link of $A^\vee$). □

5.4 The Real Result

Theorem 5.13 *Let $\mathcal{G} \in D^b(V_{\mathbb{R}}, \mathcal{S}_{\mathbb{R}})$ be represented by a complex $(\mathcal{G}_A, \gamma_{AB})$ of representations of $(\mathcal{S}_{\mathbb{R}}, \leq)$.*

(a) The stalk $\mathrm{FS}(\mathcal{G})_{A^\vee}$ of $\mathrm{FS}(\mathcal{G})$ at a face $A^\vee \in \mathcal{S}_{\mathbb{R}}^\vee$ is identified with the complex

$$\mathcal{U}_{A^\vee} := \mathrm{Tot}\left\{ \bigoplus_{\substack{\dim(B)=0,\\ B\subset U(A^\vee)}} \mathcal{G}_B \otimes \mathrm{or}(B) \xrightarrow{\gamma\otimes\varepsilon} \bigoplus_{\substack{\dim(B)=1,\\ B\subset U(A^\vee)}} \mathcal{G}_B \otimes \mathrm{or}(B) \xrightarrow{\gamma\otimes\varepsilon} \cdots \right\}.$$

(b) Let $A_1^\vee \leq A_2^\vee$ be two faces of $\mathcal{H}^\vee$. Then the inclusion $U(A_1^\vee) \supset U(A_2^\vee)$ (Proposition 5.8) exhibits $\mathcal{U}(A_2^\vee)$ as a quotient complex of $\mathcal{U}(A_1^\vee)$, and the the generalization map $\gamma_{A_1^\vee, A_2^\vee} : \mathrm{FS}(\mathcal{G})_{A_1^\vee} \to \mathrm{FS}(\mathcal{G})_{A_2^\vee}$ of $\mathrm{FS}(\mathcal{G})$ is identified with the quotient map $\mathcal{U}_{A_1^\vee} \to \mathcal{U}_{A_2^\vee}$.

Proof

(a) Let $f \in A^\vee$. By Proposition 5.1, we have

$$\mathrm{FS}(\mathcal{G})_{A^\vee} \simeq \mathrm{FS}(\mathcal{G})_f \simeq R\Gamma_c(P_f, \mathcal{G}).$$

We now use the resolution of $\mathcal{G}$ given by Proposition 2.4(ii). The pth term of this resolution is the direct sum of $j_{B!}\, \underline{\mathcal{G}_B}_B[p]$ where B runs over p-dimensional faces of $\mathcal{H}$.

□

Lemma 5.14 *Let B be a face of $\mathcal{H}$ and E be any* **k***-vector space. We have natural quasi-isomorphisms*

$$R\Gamma_c(P_f, j_{B!}\underline{E}_B)[\dim B] \simeq \begin{cases} E \otimes \mathrm{or}(B), & \text{if } B \subset P_f, \\ 0, & \text{if } B \not\subset P_f. \end{cases}$$

Proof of the Lemma: The case $B \subset P_f$ follows from the canonical identification $R\Gamma_c(B, \mathbf{k}) \simeq \mathrm{or}(B)[-\dim(B)]$ (compactly supported cohomology of a cell with constant coefficients). Suppose $B \not\subset P_f$. If B does not meet P_f at all, then the statement is obvious. If B does meet P_f, then the intersection $B \cap P_f$ is homeomorphic to a closed half-space in a Euclidean space, i.e., to a Cartesian product of several open intervals (0, 1) and one half-open interval [0, 1). So our statement follows from the fact that $H_c^\bullet([0,1), \mathbf{k}) = 0$. □

Applying this lemma to the resolution of $\mathcal{G}$ given by Proposition 2.4(ii), we obtain a complex representing $R\Gamma_c(P_f, \mathcal{G})$ whose pth term is the sum of the $\mathcal{G}_B \otimes \mathrm{or}_B$ for B running over p-dimensional faces $B \subset P_f$ and the differential is formed by the maps $\gamma \otimes \varepsilon$. By Proposition 5.11 (a), the condition $B \subset P_f$ is equivalent to $B \subset U(A^\vee)$. This proves part (a) of Theorem 5.13.

We now prove part (b). Let $f_1 \in A_1^\vee$ and $f_2 \in A_2^\vee$ be a small deformation of f_1. As in the proof of (a), we can write our generalization map as

$$\gamma_{A_1^\vee, A_2^\vee} : R\Gamma_c(P_{f_1}, \mathcal{G}) \longrightarrow R\Gamma_c(P_{f_2}, \mathcal{G}).$$

As before, consider first the case $\mathcal{G} = j_{B!}\underline{E}_B$ for some face B and some $\mathbf{k}$-vector space E. In this case, we find that $\gamma_{A_1^\vee, A_2^\vee}$ is equal to the identity map, if B is contained in P_{f_2} (and therefore in P_{f_1}), and it is equal to 0 otherwise (since the target is the zero vector space). That is, claim (b) obviously holds in this case. The case of general $\mathcal{G}$ is now obtained from this by considering the resolution of $\mathcal{G}$ given by Proposition 2.4(ii). Theorem 5.13 is proved.

5.5 The Complex Result

Let $\mathcal{F} \in \mathrm{Perv}(V_\mathbb{C}, \mathcal{S}_\mathbb{C})$ correspond to a hyperbolic sheaf $\mathcal{Q} = (E_A, \gamma_{AB}, \delta_{BA})$. By Proposition 5.5, $\mathrm{FS}(\mathcal{F})[-n]$ lies in $\mathrm{Perv}(V_\mathbb{C}^*, \mathcal{S}_\mathbb{C}^\vee)$ and so is described by a hyperbolic sheaf which we denote $\mathcal{Q}^\vee = (E_{A^\vee}^\vee, \gamma_{A^\vee, 'A^\vee}, \delta_{'A^\vee, A^\vee})$. Here $A^\vee \leq {}'A^\vee$ are faces of the arrangement $\mathcal{H}^\vee$.

It turns out that the hyperbolic stalks $E_{A^\vee}^\vee$ are governed by the small dual cones $V(A^\vee)$.

Theorem 5.15

(a) The space $E_{A^\vee}^\vee$ is quasi-isomorphic to the complex

$$\mathcal{V}_{A^\vee} = \left\{ \bigoplus_{\substack{\dim(B)=0,\\ B\subset V(A^\vee)}} E_B \otimes \mathrm{or}_{V/B} \xrightarrow{\gamma\otimes\varepsilon} \bigoplus_{\substack{\dim(B)=1,\\ B\subset V(A^\vee)}} E_B \otimes \mathrm{or}_{V/B} \xrightarrow{\gamma\otimes\varepsilon} \cdots \right\}.$$

In other words, $\mathcal{V}_{A^\vee}$ is exact everywhere except the leftmost term, where the cohomology (kernel) is identified with $E_{A^\vee}^\vee$.

(a') The space $E_{A^\vee}^\vee$ is also quasi-isomorphic to the complex

$$\mathcal{V}_{A^\vee}^\dagger = \left\{ \cdots \xrightarrow{\delta\otimes\varepsilon} \bigoplus_{\substack{\dim(B)=1,\\ B\subset V(A^\vee)}} E_B \otimes \mathrm{or}_{V/B} \xrightarrow{\delta\otimes\varepsilon} \bigoplus_{\substack{\dim(B)=0,\\ B\subset V(A^\vee)}} E_B \otimes \mathrm{or}_{V/B} \right\}.$$

In other words, $\mathcal{V}_{A^\vee}^\dagger$ is exact everywhere except the rightmost term, where the cohomology (cokernel) is identified with $E_{A^\vee}^\vee$.

(b) Let $A_1^\vee \leq A_2^\vee$ be two faces of $\mathcal{H}^\vee$. Then the embedding $V(A_1^\vee) \subset V(A_2^\vee)$ realizes $\mathcal{V}_{A_1^\vee}$ as a quotient complex of $\mathcal{V}_{A_2^\vee}$, and the map $\delta_{A_2^\vee, A_1^\vee} : E_{A_2^\vee}^\vee \to E_{A_1^\vee}^\vee$ is identified with the quotient map $\mathcal{V}_{A_2^\vee} \to \mathcal{V}_{A_1^\vee}$.

(b') In the situation of (b), the embedding $V(A_1^\vee) \subset V(A_2^\vee)$ realizes $\mathcal{V}^\dagger_{A_1^\vee}$ as a subcomplex of $\mathcal{V}^\dagger_{A_2^\vee}$, and the map $\gamma_{A_1^\vee, A_2^\vee}$ is identified with the embedding $\mathcal{V}^\dagger_{A_1^\vee} \to \mathcal{V}^\dagger_{A_2^\vee}$.

Remark 5.16

(a) Note that for $A^\vee = 0$, the cone $V(A^\vee)$ is equal to $\{0\}$; therefore, $E_0^\vee$ is identified with E_0.
(b) Let $A^\vee \neq 0$. Then, by Proposition 5.9(b), one can re-write the complex $\mathcal{V}_{A^\vee}$ as

$$E_0 \otimes \mathrm{or}_V \xrightarrow{\gamma\otimes\varepsilon} \bigoplus_{\substack{\dim(B)=1\\ f|_B>0}} E_B \otimes \mathrm{or}_{V/B} \xrightarrow{\gamma\otimes\varepsilon} \bigoplus_{\substack{\dim(B)=2\\ f|_B>0}} E_B \otimes \mathrm{or}_{V/B} \xrightarrow{\gamma\otimes\varepsilon} \cdots,$$

where $f \in A^\vee$ is an arbitrary element. Similarly for $\mathcal{V}^\dagger_{A^\vee}$.

The proof of Theorem 5.15 is based on the following preliminary result which shows that the big dual cones $U(A^\vee)$ govern the ordinary stalks, not hyperbolic stalks of $\mathrm{FS}(\mathcal{F})$,

Proposition 5.17

(a) If $A^\vee \in \mathcal{S}_\mathbb{R}$ is any face, then the ordinary stalk $\mathrm{FS}(\mathcal{F})_{A^\vee}$ is quasi-isomorphic to the complex

$$\mathrm{FS}(\mathcal{F})_{A^\vee} \simeq \left\{ \bigoplus_{\substack{\dim(B)=0,\\ B\subset U(A^\vee)}} E_B \otimes \mathrm{or}_{V/B} \xrightarrow{\gamma\otimes\varepsilon} \bigoplus_{\substack{\dim(B)=1,\\ B\subset U(A^\vee)}} E_B \otimes \mathrm{or}_{V/B} \xrightarrow{\gamma\otimes\varepsilon} \cdots \right\}.$$

(b) The generalization maps for the $\mathrm{FS}(\mathcal{F})_{A^\vee}$ are induced by the projections of the complexes in (a), similar to Theorem 5.13(b).

Proof of Proposition 5.17: Our statement will follow from Theorem 5.13, if we establish the following. □

Proposition 5.18 *For any $\mathcal{F} \in \mathrm{Perv}(V_\mathbb{C}, \mathcal{S}_\mathbb{C})$, we have an identification*

$$\mathrm{FS}(i_\mathbb{R}^! \mathcal{F}) \simeq i_\mathbb{R}^* \mathrm{FS}(\mathcal{F}),$$

where $i_\mathbb{R}$ on the right means the embedding $V_\mathbb{R}^* \to V_\mathbb{C}^*$.

Proof of Proposition 5.18: We first recall the behavior of the Fourier-Sato transform with respect to an arbitrary $\mathbb{R}$-linear map $\phi : W_1 \to W_2$ of $\mathbb{R}$-vector spaces. Denoting ${}^t\phi : W_2^* \to W_1^*$ the transposed map, we have, for any conic complex $\mathcal{G}$ on W_2:

$$\mathrm{FS}(\phi^! \mathcal{G}) \simeq R({}^t\phi)_* \mathrm{FS}(\mathcal{G})$$

see [KS2] Prop. 3.7.14.

We specialize this to $\phi = i_{\mathbb{R}} : V_{\mathbb{R}} \to V_{\mathbb{C}}$ and $\mathcal{G} = \mathcal{F}$. In this case,

$$ {}^t\phi \ = \ \Re : V_{\mathbb{C}}^* \longrightarrow V_{\mathbb{R}}^* $$

is the real part map. So after replacing V^* by V and $\mathrm{FS}(\mathcal{F})$ by $\mathcal{F}$, Proposition 5.18 reduces to the following. □

Lemma 5.19 *For any $\mathcal{F} \in D^b(V_{\mathbb{C}}, \mathcal{S}_{\mathbb{C}})$, we have an identification $i_{\mathbb{R}}^*\mathcal{F} \simeq R\,\Re_*(\mathcal{F})$, where $\Re : V_{\mathbb{C}} \to V_{\mathbb{R}}$ is the real part map for V.*

Proof of Lemma 5.19: We consider $\Re : V_{\mathbb{C}} \to V_{\mathbb{R}}$ as a real vector bundle over $V_{\mathbb{R}}$. The complex $\mathcal{F}$, being constructible with respect to the complexification of a real hyperplane arrangement, is conic with respect to this vector bundle structure. Therefore, the stalk at $x \in V_{\mathbb{R}}$ of $i_{\mathbb{R}}^*\mathcal{F}$ which is $R\Gamma(U, \mathcal{F})$ for a small open $U \subset V_{\mathbb{C}}$ containing x is equal to $R\Gamma(\Re^{-1}(U \cap V_{\mathbb{R}}), \mathcal{F})$ which is the stalk of $R\,\Re_*(\mathcal{F})$ at x. □

This finishes the proof of Propositions 5.18 and 5.17.

Proof of Theorem 5.15: We prove (a') and (b'). Parts (a) and (b) follow by Verdier duality.

We denote $\mathcal{K} = \mathrm{FS}(\mathcal{F})$, and let $\mathcal{L} = \mathcal{K}^* = \mathrm{FS}(\mathcal{F}^*)$ be the Verdier dual perverse sheaf. By definition, $E^{\vee}_{A^{\vee}}$ is the stalk at $A^{\vee}$ of

$$ i_{\mathbb{R}}^!\mathcal{K} \simeq (i_{\mathbb{R}}^*\mathcal{L})^* \simeq \big(i_{\mathbb{R}}^*\mathrm{FS}(\mathcal{F}^*)\big)^*. $$

First, we recall that $\mathcal{F}^*$ is represented by the hyperbolic sheaf $(E_A^*, \delta_{BA}^*, \gamma_{AB}^*)$. Applying Proposition 5.17 to $\mathcal{F}^*$, we write the stalk of $i_{\mathbb{R}}^*\mathcal{L}$ at $A^{\vee}$ as

$$ \mathcal{L}_{A^{\vee}} \simeq \Bigg\{ \bigoplus_{\substack{\dim(B)=0\\ B\subset U(A^{\vee})}} E_B^* \otimes \mathrm{or}_{V/B} \overset{\delta^*\otimes\varepsilon}{\longrightarrow} \bigoplus_{\substack{\dim(B)=1\\ B\subset U(A^{\vee})}} E_B^* \otimes \mathrm{or}_{V/B} \overset{\delta^*\otimes\varepsilon}{\longrightarrow} \cdots \Bigg\}. \tag{5.20} $$

Further, for $A_1^{\vee} \leq A_2^{\vee}$, we have $U(A_1^{\vee}) \supset U(A_2^{\vee})$, and Proposition 5.17 implies that the generalization map $F_{A_1^{\vee}, A_2^{\vee}} : \mathcal{L}_{A_1^{\vee}} \to \mathcal{L}_{A_2^{\vee}}$ is given by the projection of the corresponding complexes in (5.20).

We now recall the following general procedure on finding the stalks and generalization maps of the Verdier dual complex. See, e.g., [KS1] Prop. 1.11. We formulate it here for complexes on $V_{\mathbb{R}}^*$ constructible with respect to $\mathcal{S}_{\mathbb{R}}^{\vee}$. □

Lemma 5.21 *Let $\mathcal{M} \in D^b(V_{\mathbb{R}}^*, \mathcal{S}_{\mathbb{R}}^{\vee})$ correspond to a complex $(\mathcal{M}_{A^{\vee}}, F_{A_1^{\vee}, A_2^{\vee}})$ of representations of $\mathcal{S}_{\mathbb{R}}^{\vee}$. Then:*

(a) The stalk of $\mathcal{M}^*$ at $A^\vee$ is identified with the complex

$$\mathcal{D}_{A^\vee} = \mathrm{Tot}\left\{\cdots \stackrel{F^*\otimes\varepsilon^*}{\longrightarrow} \bigoplus_{C^\vee >_1 A^\vee} (\mathcal{M}_{C^\vee})^* \otimes \mathrm{or}_{C^\vee} \stackrel{F^*\otimes\varepsilon^*}{\longrightarrow} (\mathcal{M}_{A^\vee})^* \otimes \mathrm{or}_{A^\vee}\right\},$$

with the horizontal grading associating to the summand $(\mathcal{M}_{A^\vee})^* \otimes \mathrm{or}_{A^\vee}$ degree $-\dim(A^\vee)$. The horizontal differential $F^* \otimes \varepsilon^*$ has, as the matrix element corresponding to $C_2^\vee >_1 C_1^\vee \geq A^\vee$, the tensor product of the dual maps to $F_{C_1^\vee, C_2^\vee}$ and to $\varepsilon_{C_1^\vee, C_2^\vee}$.

(b) For two faces $A_1^\vee \leq A_2^\vee$, the generalization map $(\mathcal{M}^*)_{A_1^\vee} \to (\mathcal{M}^*)_{A_2^\vee}$ of $\mathcal{M}^*$ is identified with the projection of the complexes $\mathcal{D}_{A_1^\vee} \to \mathcal{D}_{A_2^\vee}$. □

Applying part (a) of the lemma to $\mathcal{M} = \mathcal{L}$ and substituting, instead of each $\mathcal{L}_{C^\vee}$, its expansion (5.20), we identify (quasi-isomorphically) $E^\vee_{A^\vee}$ with the total complex of the following double complex. We denote this total complex $\mathcal{E}_{A^\vee}$.

$$\begin{array}{ccccc}
\cdots & \xrightarrow{\mathrm{emb}\otimes\varepsilon} & \bigoplus_{C^\vee >_1 A^\vee} \bigoplus_{\substack{\dim(B)=0\\ B\subset U(C^\vee)}} E_B \otimes \mathrm{or}_{V/B} \otimes \mathrm{or}_{C^\vee} & \xrightarrow{\mathrm{emb}\otimes\varepsilon} & \bigoplus_{\substack{\dim(B)=0\\ B\subset U(C^\vee)}} E_B \otimes \mathrm{or}_{V/B} \otimes \mathrm{or}_{A^\vee} \\
 & & \uparrow {\scriptstyle \delta\otimes\varepsilon\otimes\mathrm{Id}} & & \uparrow {\scriptstyle \delta\otimes\varepsilon\otimes\mathrm{Id}} \\
\cdots & \xrightarrow{\mathrm{emb}\otimes\varepsilon} & \bigoplus_{C^\vee >_1 A^\vee} \bigoplus_{\substack{\dim(B)=1\\ B\subset U(C^\vee)}} E_B \otimes \mathrm{or}_{V/B} \otimes \mathrm{or}_{C^\vee} & \xrightarrow{\mathrm{emb}\otimes\varepsilon} & \bigoplus_{\substack{\dim(B)=1\\ B\subset U(C^\vee)}} E_B \otimes \mathrm{or}_{V/B} \otimes \mathrm{or}_{A^\vee} \\
 & & \uparrow {\scriptstyle \delta\otimes\varepsilon\otimes\mathrm{Id}} & & \uparrow {\scriptstyle \delta\otimes\varepsilon\otimes\mathrm{Id}} \\
\cdots & \xrightarrow{\mathrm{emb}\otimes\varepsilon} & \bigoplus_{C^\vee >_1 A^\vee} \bigoplus_{\substack{\dim(B)=2\\ B\subset U(C^\vee)}} E_B \otimes \mathrm{or}_{V/B} \otimes \mathrm{or}_{C^\vee} & \xrightarrow{\mathrm{emb}\otimes\varepsilon} & \bigoplus_{\substack{\dim(B)=2\\ B\subset U(C^\vee)}} E_B \otimes \mathrm{or}_{V/B} \otimes \mathrm{or}_{A^\vee} \\
 & & \uparrow {\scriptstyle \delta\otimes\varepsilon\otimes\mathrm{Id}} & & \uparrow {\scriptstyle \delta\otimes\varepsilon\otimes\mathrm{Id}} \\
 & & \vdots & & \vdots
\end{array}$$

Here the vertical differentials are dual to those in $\mathcal{L}_{C^\vee}$, i.e., given by the δ maps. Matrix elements of the horizontal differential are dual to the F maps for $\mathcal{L}$, and those F maps are given by the projections. So each matrix element in question is in fact the product of an *embedding* of δ-complexes and the ε map of orientation torsors.

For two faces $A_1^\vee \leq A_2^\vee$, the generalization map $\gamma_{A_1^\vee, A_2^\vee} : E^\vee_{A_1^\vee} \to E^\vee_{A_2^\vee}$ is identified, by part (b) of Lemma 5.21, with the projection $\mathcal{E}_{A_1^\vee} \to \mathcal{E}_{A_2^\vee}$.

We now compare $\mathcal{E}_{A^\vee}$ with the complex $\mathcal{V}^\dagger_{A^\vee}$ from the formulation of Theorem 5.15(a'). Let B be a face of $\mathcal{H}$. The summand corresponding to B in $\mathcal{V}^\dagger_{A^\vee}$ is either $E_B \otimes \mathrm{or}_{V/B}$ or 0 depending on whether $B \subset V(A^\vee)$ or not. On the other hand, $\mathcal{E}_{A^\vee}$ has many summands associated with B; they are labeled by $C^\vee > A^\vee$ such that $B \subset U(C^\vee)$. By Proposition 5.11, such $C^\vee$ are in bijection with faces of the closed polyhedral cone $K(A^\vee, B)$. So in the double complex above, the summand $E_B \otimes \mathrm{or}_{V/B}$ is multiplied by a combinatorial complex which is easily found to calculate the cohomology with compact support $H_c^\bullet(K(A^\vee, B), \mathbf{k})$. If $B \not\subset V(A^\vee)$, then, by the same Proposition 5.11, $K(A^\vee, B)$ is a proper closed cone with nonempty interior in $V_\mathbb{R}^*/\mathrm{Lin}_\mathbb{R}(A^\vee)$, and so its cohomology with compact support vanishes entirely. If $B \subset V(A^\vee)$, then $K(A^\vee, B) = V_\mathbb{R}^*/\mathrm{Lin}_\mathbb{R}(A^\vee)$ so it has the top cohomology with compact support identified with $\mathrm{or}_{V/A^\vee}$, so the part of $\mathcal{E}_{A^\vee}$ corresponding to B is quasi-isomorphic to $E_B \otimes \mathrm{or}_{V/B}$. Moreover, we see that these quasi-isomorphisms combine into a quasi-isomorphism between $\mathcal{E}_{A^\vee}$ and $\mathcal{V}^\dagger_{A^\vee}$. This shows part (a') of Theorem 5.17. Part (b') follows by noticing that the projection $\mathcal{E}_{A_1^\vee} \to \mathcal{E}_{A_2^\vee}$ corresponds, under our quasi-isomorphism, to the embedding $\mathcal{V}^\dagger_{A_1^\vee} \to \mathcal{V}^\dagger_{A_2^\vee}$. Theorem 5.17 is proved.

6 Applications to Second Microlocalization

6.1 *Microlocalization*

If $M \subset X$ is a C^∞ submanifold of a C^∞ manifold, as in §Sect. 4A, then for any $\mathcal{G} \in D^b(X)$, the *microlocalization* of $\mathcal{G}$ along M is defined as

$$\mu_M(\mathcal{G}) = \mathrm{FS}_M(\nu_M(\mathcal{G})) \in D^b(T_M^* X);$$

see [KS2] Ch. 4. Here FS_M is the relative Fourier-Sato transform on the vector bundle $T_M X \to M$.

If $X = V$ is a real vector space with an arrangement $\mathcal{H}$, if $\mathcal{G} \in D^b(V, \mathcal{S}_\mathbb{R})$ and M is a vector subspace, then our descriptions of the Fourier-Sato transform and the specialization functors can be combined to obtain a combinatorial description of $\mu_M(\mathcal{G})$. We leave this to the reader, establishing instead some compatibility properties of various approaches to "second microlocalization" of Kashiwara and Laurent; see [L] and references therein. For convenience, we give a brief general introduction.

6.2 *Iterated Microlocalization*

Lemma 6.1 *Let (W, ω) be a symplectic $\mathbb{R}$-vector space and $L_1, L_2 \subset W$ be Lagrangian vector subspaces. Then the restriction of ω gives an identification*

$$\left(\frac{L_1}{L_1 \cap L_2}\right)^* \simeq \left(\frac{L_2}{L_1 \cap L_2}\right).$$

Proof Consider the restriction of ω to the subspace $L_1 + L_2$. Its kernel on this subspace is

$$(L_1 + L_2)^\perp = L_1^\perp \cap L_2^\perp = L_1 \cap L_2.$$

Therefore, the restriction of ω makes

$$\frac{L_1 + L_2}{L_1 \cap L_2} = \frac{L_1}{L_1 \cap L_2} \oplus \frac{L_2}{L_1 \cap L_2}$$

into a symplectic vector space decomposed into the direct sum of two Lagrangian subspaces. So these Lagrangian subspaces become dual to each other. □

Let now (S, ω) be a C^∞ symplectic manifold and $\Lambda_1, \Lambda_2 \subset S$ be two (smooth) Lagrangian submanifolds. We say that Λ_1 and Λ_2 *intersect cleanly* (in the symplectic sense), if, locally, near each $x \in \Lambda_1 \cap \Lambda_2$, there is a symplectomorphism of a neighborhood of x in S to a neighborhood of 0 in a symplectic vector space W, sending Λ_i to linear Lagrangian subspaces L_i as above. This implies that $\Lambda_1 \cap \Lambda_2$ is smooth.

Corollary 6.2 *If Λ_1, Λ_2 intersect cleanly, then the restriction of ω gives an identification*

$$T^*_{\Lambda_1 \cap \Lambda_2} \Lambda_1 \simeq T_{\Lambda_1 \cap \Lambda_2} \Lambda_2.$$

□

Now let X be a C^∞ manifolds and $M, N \subset X$ be two smooth submanifolds. We assume that they intersect cleanly in the sense that they can locally be brought by a diffeomorphism to two vector subspaces in a vector space. Then $S = T^*X$ has two Lagrangian submanifolds $\Lambda_1 = T^*_M X$, $\Lambda_2 = T^*_N X$ which intersect cleanly in the symplectic sense. Given a complex of sheaves $\mathcal{G} \in D^b(X)$, we have microlocalizations

$$\mu_M(\mathcal{G}) \in D^b(\Lambda_1), \quad \mu_N(\mathcal{G}) \in D^b(\Lambda_2)$$

and we can specialize and microlocalize further, getting two complexes of sheaves

$$\mu_{\Lambda_1 \cap \Lambda_2} \mu_M(\mathcal{G}) \in D^b(T^*_{\Lambda_1 \cap \Lambda_2} \Lambda_1), \quad \nu_{\Lambda_1 \cap \Lambda_2} \mu_N(\mathcal{G}) \in D^b(T_{\Lambda_1 \cap \Lambda_2} \Lambda_2)$$

on two spaces which are identified by Corollary 6.2, so we can consider them as living on the same space. One can then formulate

Second Microlocalization Problem 6.3 Under which conditions on M, N and $\mathcal{G}$ can we guarantee that

$$\mu_{\Lambda_1 \cap \Lambda_2} \mu_M(\mathcal{G}) \simeq \nu_{\Lambda_1 \cap \Lambda_2} \mu_N(\mathcal{G}) \quad ?$$

6.3 Bi-microlocalization

Let us restrict to the case $N \subset M$. In this case, we have

Proposition 6.4 *We have identifications*

$$T^*_{\Lambda_1 \cap \Lambda_2} \Lambda_1 \;=\; T_{\Lambda_1 \cap \Lambda_2} \Lambda_2 \;\simeq\; T^*_N M \oplus (T^*_M X)|_N.$$

Proof Obviously, $\Lambda_1 \cap \Lambda_2$ projects, under $T^*X \to X$, to N. Looking at the fibers of this projection, we find that $\Lambda_1 \cap \Lambda_2 = (T^*_M X)|_N$. Looking at the Cartesian square

$$\begin{array}{ccc} \Lambda_1 \cap \Lambda_2 & \longrightarrow & T^*_M X = \Lambda_1 \\ {\scriptstyle \rho}\big\downarrow & & \big\downarrow{\scriptstyle \pi} \\ N & \longrightarrow & M \end{array}$$

with π being a smooth fibration (projection of a vector bundle), we find that

$$T^*_{\Lambda_1 \cap \Lambda_2} \Lambda_1 \;\simeq\; \rho^* T^*_N M \;\simeq\; T^*_N M \oplus (T^*_M X)|_N. \qquad \square$$

We already considered the situation of a flag $N \subset M \subset X$ in discussing bispecialization $\nu_{NM}(\mathcal{G})$ in Sect. 4E. Further, in this context, Schapira and Takeuchi [ST, T] have defined the *bimicrolocalization*

$$\mu_{NM}(\mathcal{G}) \;=\; \mathrm{FS}_N(\nu_{NM}(\mathcal{G})) \;\in\; D^b(T^*_N M \oplus (T^*_M X)|_N).$$

Here FS_N is the relative Fourier-Sato transform on the vector bundle $T_N M \oplus (T_M X)|_N \to N$. So we have the following specialization-microlocalization diagram:

$$\begin{array}{ccc} D^b(X) & \xrightarrow{\quad \mu_M \quad} & D^b(T^*_M X = \Lambda_1) \\ {\scriptstyle \mu_N}\big\downarrow & \searrow^{\mu_{NM}} & \big\downarrow{\scriptstyle \mu_{\Lambda_1 \cap \Lambda_2}} \\ D^b(T^*_N X = \Lambda_2) & \xrightarrow[\nu_{\Lambda_1 \cap \Lambda_2}]{} & D^b\big((T^*_N X \oplus (T^*_M X)|_N)) = T^*_{\Lambda_1 \cap \Lambda_2} \Lambda_1 = T_{\Lambda_1 \cap \Lambda_2} \Lambda_2\big) \end{array} \tag{6.5}$$

which gives three possible "second microlocalizations."

6.4 Comparisons in the Linear Case

Theorem 6.6 *Let $X = V$ be an $\mathbb{R}$-vector space, $N \subset M \subset V$ be vector subspaces, and $\mathcal{H}$ an arrangement of hyperplanes in V with the corresponding face stratification $\mathcal{S}_{\mathbb{R}}$. Then the diagram* (6.5) *is canonically 2-commutative if we replace $D^b(V)$ with $D^b(V, \mathcal{S}_{\mathbb{R}})$.*

In the complex situation, when $V = V_{\mathbb{C}}$ is a $\mathbb{C}$-vector space, $N \subset M \subset V_{\mathbb{C}}$ are $\mathbb{C}$-subspaces, and $D^b(V, \mathcal{S}_{\mathbb{R}})$ is replaced by $\mathrm{Perv}(V_{\mathbb{C}}, \mathcal{S}_{\mathbb{C}})$, the commutativity of the outer square of (6.5) was proved in [FS] using the $\mathcal{D}$-module techniques.

We will deduce Theorem 6.6 from the following result.

Theorem 6.7 (P. Schapira) *Let B be a C^∞-manifold and V be a smooth $\mathbb{R}$-vector bundle on B. Let $M \subset V$ be a vector subbundle. Then, the Fourier-Sato transforms on V and $T_M(V) = M \oplus (V/M)$ are compatible with specializations. In other words, the following diagram of functors is canonically 2-commutative:*

$$\begin{array}{ccc} D^b_{\mathrm{con}}(V) & \xrightarrow{\mathrm{FS}_V} & D^b(V^*) \\ \Big\downarrow{\scriptstyle \nu_M} & & \Big\downarrow{\scriptstyle \nu_{M^\perp}} \\ D^b(M \oplus (V/M)) & \xrightarrow[P_{12} \circ \mathrm{FS}_{M \oplus (V/M)}]{} & D^b(M^\perp \oplus M^*). \end{array}$$

Here P_{12} is the permutation of the two direct summands in $M^* \oplus M^\perp$.

The notation $\oplus$ here and below means direct sum of vector bundles, i.e., fiber product over B.

We note that the diagram in Theorem 6.7 can be seen as a particular case of the outer rim of the diagram (6.5) for the case when $X = V$, when $M \subset V$ is our subbundle and $N = B$ is the zero section of V. In other words, Theorem 6.7 can be seen as a parameterized version (with arbitrary base B instead of $B = \mathrm{pt}$) of a particular case of Theorem 6.6 corresponding to $N = 0$.

6.5 Proof of Theorem 6.7

The following proof is an adaptation of the argument communicated to us by P. Schapira.

We consider three pairs

$$M^\perp \subset V^*, \quad M \oplus M^\perp \subset V \oplus V^*, \quad M \subset V,$$

and the corresponding normal deformations which are related by the natural projections:

$$\widetilde{V}^*_{M^\perp} \longleftarrow \widetilde{V \oplus V^*}_{M\oplus M^\perp} \longrightarrow \widetilde{V}_M. \tag{6.8}$$

Each of the three normal deformations fits into its own diagram of the form (4.10) whose spaces and maps will be decorated by the subscripts $M^\perp$, $M \oplus M^\perp$, and M. In particular, the projections of the three spaces in (6.8) to the line $\mathbb{R}$ will be denoted $\tau_{M^\perp}, \tau_{M\times M^\perp}$, and τ_M. These projections commute with the maps in (6.8). The coordinate in $\mathbb{R}$ will be denoted t.

Now, the Fourier-Sato transform on any vector bundle W is defined using the region

$$P = P_W = \{(x, f) \in W \oplus W^* \,|\, f(x) \geq 0\},$$

cf. Sect. 5A. We apply this to $W = V$ and $W = M \oplus (V/M)$ and denote the corresponding regions

$$P_V \subset V \oplus V^*, \quad P_{M\oplus(V/M)} \subset M \oplus (V/M) \oplus M^* \oplus (V/M)^*.$$

We want to lift P_V into a region $\overline{\mathcal{P}} \subset \widetilde{V \oplus V^*}_{M\oplus M^\perp}$ which specializes, for $t > 0$, to P_V and, for $t = 0$, to $P_{M\oplus(V/M)}$.

For this, we consider the region $\Omega_{M\oplus M^\perp} \subset \widetilde{V \oplus V^*}_{M\oplus M^\perp}$, defined as the preimage $\tau^{-1}_{M\oplus M^\perp}(\mathbb{R}_{>0})$, cf. (4.10). It is identified with $V \oplus V^* \times \mathbb{R}_{>0}$. Let $\mathcal{P} \subset \Omega$ be the image of $P_V \times \mathbb{R}_{>0}$.

Proposition 6.9 *The closure $\overline{\mathcal{P}}$ of $\mathcal{P}$ in $\widetilde{V \oplus V^*}_{M\oplus M^\perp}$ is the union of $\mathcal{P}$ and $P_{M\oplus(V/M)} \subset \tau^{-1}_{M\oplus M^\perp}(0)$.*

Proof The statement is local in B. So we can assume that there exists a complement M' to M and to write $V = M \oplus M'$. We then identify, as in (4.10),

$$\widetilde{V \oplus V^*}_{M\oplus M^\perp} = M \oplus M' \oplus M^* \oplus M'^* \times \mathbb{R}$$

and the projection $p_{M\oplus M^\perp} : \widetilde{V \oplus V^*}_{M\oplus M^\perp} \to V \oplus V^*$ can be written as

$$\begin{aligned} p_{M\oplus M^\perp} : M \oplus M' \oplus M^* \oplus M'^* \times \mathbb{R} &\longrightarrow M \oplus M' \oplus M^* \oplus M'^*, \\ (m, m', \phi, \phi', t) &\mapsto (m, tm', t\phi, \phi'). \end{aligned} \tag{6.10}$$

Recall that the identification $\Omega_{M\oplus M^\perp} \to V \oplus V^* \times \mathbb{R}_{>0}$ is given by the map $(p_{M\oplus M^\perp}, \tau_{M\oplus M^\perp})$, the second component being projection to t. It follows from (6.10) that for any $t > 0$, the image, under $p_{M\oplus M^\perp}$, of $P_V \times \{t\}$ is P_V. Therefore, the inverse of $p_{M\oplus M^\perp}, \tau_{M\oplus M^\perp}$) identifies $P_V \times \mathbb{R}_{\geq 0}$ with $P_V \times \mathbb{R}_{\geq 0}$, where for $t = 0$ our choice of complement has identified P_V with $P_{M\oplus M'} = P_{M\oplus(V/M)}$. □

We now consider the following diagram:

$$
\begin{array}{ccccc}
V^* & \xleftarrow{\;p_{2,V}\;} & P_V & \xrightarrow{\;p_{1,V}\;} & V \\
\uparrow{\scriptstyle \widetilde{p}_{M^\perp}} & & \uparrow & & \uparrow{\scriptstyle \widetilde{p}_{M}} \\
\Omega_{M^\perp} & \xleftarrow{\;\rho_2\;} & \mathcal{P} & \xrightarrow{\;\rho_1\;} & \Omega_M \\
\downarrow{\scriptstyle j_{M^\perp}} & & \downarrow & & \downarrow{\scriptstyle j_M} \\
\widetilde{V}^*_{M^\perp} & \xleftarrow{\;\pi_2\;} & \widetilde{\mathcal{P}} & \xrightarrow{\;\pi_1\;} & \widetilde{V}_M \\
\uparrow{\scriptstyle s_{M^\perp}} & & \uparrow & & \uparrow{\scriptstyle s_M} \\
M^\perp \oplus M^* & \xleftarrow[\;p_{2,M\oplus(V/M)}\;]{} & P_{M\oplus(V/M)} & \xrightarrow[\;p_{1,M\oplus(V/M)}\;]{} & M \oplus (V/M)
\end{array}
$$

Given $\mathcal{G} \in D^b_{\mathrm{con}}(V)$, we have that

$$
\begin{aligned}
\nu_{M^\perp}\mathrm{FS}_V(\mathcal{G}) &= s^*_{M^\perp} R(j_{M^\perp})_* \widetilde{p}^*_{M^\perp} (p_{2,V})_! \, p^{-1}_{1,V}(\mathcal{G}), \\
\mathrm{FS}_{M\oplus(V/M)}\nu_M(\mathcal{G}) &= (p_{2,M\oplus(V/M)})_! \, p^*_{1,M\oplus(V/M)} \, s^*_M \, R(j_M)_* \, \widetilde{p}^*_M(\mathcal{G})
\end{aligned}
$$

are given by moving along the two boundary paths of this diagram from the northeast to the southwest corner. We identify these functors using the base change theorem for the Cartesian squares forming this diagram.

6.6 Proof of Theorem 6.6

We write the diagram (6.5) in our case as follows:

$$
\begin{array}{ccc}
D^b(V, \mathcal{S}_{\mathbb{R}}) & \xrightarrow{\;\mu_M\;} & D^b_{\mathrm{con}}(M \times (V/M)^*) \\
\downarrow{\scriptstyle \mu_N} & \overset{\mu_{NM}}{\searrow} & \downarrow{\scriptstyle \mu_{N\times(V/M)^*}} \\
D^b_{\mathrm{con}}(N \times (V/N)^*) & \xrightarrow[\;\nu_{N\times(V/M)^*}\;]{} & D^b_{\mathrm{bicon}}(N \times (M/N)^* \times (V/M)^*).
\end{array}
\tag{6.11}
$$

Here and below, the subscript "con" means complexes which are $\mathbb{R}_{>0}$-conic with to the second argument, and "bico" means complexes which are $(\mathbb{R}_{>0})^2$-biconic with respect to the second and third arguments.

We recall that μ_{NM} is the composition

$$D^b(V, \mathcal{S}_{\mathbb{R}}) \xrightarrow{\nu_{NM}} D^b_{\text{bicon}}(N \times (M/N) \times (V/M)) \xrightarrow{\text{FS}_{(M/N)\times(V/M)}} D^b_{\text{bicon}}(N \times (M/N)^* \times (V/M)^*).$$

We now prove the 2-commutativity of each of the two triangles in (6.11).

Upper Triangle We write each μ as the composition of the corresponding FS and ν and apply Theorem 4.17 to decompose ν_{NM} as the composition of two specializations. After this, we represent the two paths in the triangle as the two boundary paths in the following diagram:

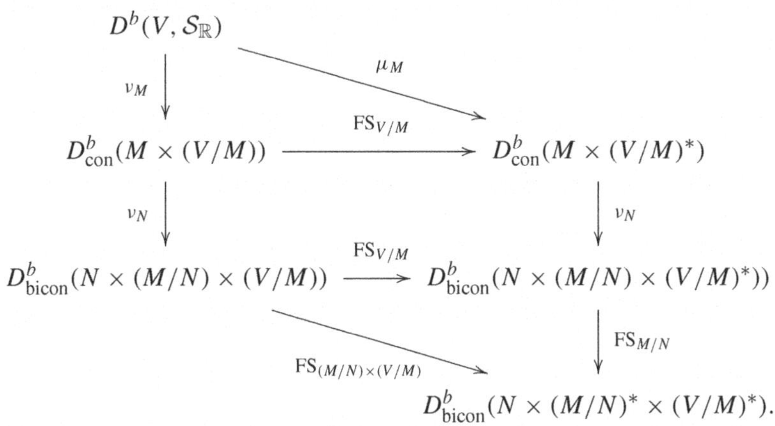

In this diagram, the top triangle commutes by definition of μ_M, and the commutativity of the bottom triangle expresses the fact that the Fourier-Sato transform of biconic sheaves on the direct sum of vector bundles can be done in stages, cf. [KS2] Prop. 3.7.15. The commutativity of the middle square follows because specialization along N and the Fourier-Sato transform along V/M operate in different factors so they are independent of each other and can be permuted.

Lower Triangle As before, by unraveling the definitions of various μ and applying Theorem 4.17, we represent the two paths in the triangle as the two boundary paths in the following diagram:

$$
\begin{array}{ccc}
 & & D^b(V, \mathcal{S}_{\mathbb{R}}) \\
 & \swarrow^{\mu_N} & \downarrow \nu_N \\
D^b_{\text{con}}(N \times (V/N)^*) & \xleftarrow{\ \mathrm{FS}_{V/N}\ } & D^b_{\text{con}}(N \times (V/N)) \\
\downarrow \nu_{N\times(V/M)^*} & & \downarrow \nu_{N\times(M/N)} \\
D^b_{\text{bicon}}(N \times (M/N) \times (V/M)) & \xleftarrow[\mathrm{FS}_{(M/N)\times(V/M)}]{} & D^b_{\text{bicon}}(N \times (M/N) \times (V/M))
\end{array}
$$

The commutativity of the top triangle in this diagram is the definition of μ_N. The commutativity of the lower square is an instance of Theorem 6.7 for the trivial vector bundle over $B = N$ with fiber V/N and the trivial subbundle with fiber M/N. Theorem 6.6 is proved.

Acknowledgments We would like to thank P. Schapira for remarks on a preliminary draft of the paper and for communicating to us a proof of Theorem 6.7. We are also grateful to Peng Zhou for several corrections.

The research of M.F. was supported by the grant RSF 19-11-00056.

The research of M.K. was supported by the World Premier International Research Center Initiative (WPI Initiative), MEXT, Japan.

V. S. thanks Kavli IPMU for support of a visit during the preparation of this paper.

References

[Ar] J. Arthur. An introduction to the trace formula. In: "Harmonic Analysis, Trace Formula and Schimura Varieties" (J. Arthur, D. Ellwood, R. Kottwitz Eds.) *Clay Math. Proceedings* **4**, Amer. Math. Soc. (2005) 3–263.

[Be] A. Beilinson. How to glue perverse sheaves. In: K-theory, arithmetic and geometry (Moscow, 1984), *Lecture Notes in Math.* **1289**, Springer-Verlag, (1987) 42–51.

[BBD] A. Beilinson, J. Bernstein, P. Deligne. Faisceaux pervers. *Astérisque* **100** (1983).

[BFS] R. Bezrukavnikov, M. Finkelberg, V. Schechtman. Factorizable sheaves and quantum groups. *Lecture Notes in Math.* **1691**, Springer-Verlag, (1998).

[Br] T. Braden, Hyperbolic localization of intersection cohomology, *Transform. Groups*, **8** (2003), 209–216.

[De] P. Deligne, Le formalisme des cycles évanescents. SGA 7, Exp. 13, 14 *Lecture Notes in Math.* **340**, Springer-Verlag (1973).

[DG] V. Drinfeld, D. Gaitsgory, On a theorem of Braden. *Transform. Groups* **19** (2014) 313–358.

[FL] T. Finis, E. Lapid. On the spectral side of the Arthur's trace formula—combinatorial setup. *Ann. Math.* **174** (2011) 197–223.

[FS] M. Finkelberg, V. Schechtman. Microlocal approach to Lusztig's symmetries. *arXiv:*1401.5885.

[KS1] M. Kapranov, V. Schechtman. Perverse sheaves over real hyperplane arrangements. *Ann. Math.* **183** (2016) 619–679.

[KS2] M. Kashiwara, P. Schapira, Sheaves on Manifolds. *Grundlehren der Mathematischen Wissenschaften* **292**, Springer-Verlag, (1990).

[L] Y. Laurent. Théorie de la Deuxième Microlocalization dans le Domaine Complexe. *Progress in Math.* **53**, Birkhäuser, Boston, (1985).

[MV] R. D. McPherson, K. Vilonen. Elementary construction of perverse sheaves. *Invent. Math.* **84** (1986) 403–435.

[ST] P. Schapira, K. Takeuchi. Déformation binormale et bispecialization. *C. R. Acad. Sci.* **319** (1994) 707–712.

[T] K. Takeuchi. Binormal deformation and bimicrolocalization. *Publ.RIMS Kyoto Univ.* **32** (1996) 277–322.

[V] A. N. Varchenko, Combinatorics and topology of the arrangement of affine hyperplanes in the real space. *Funct. Anal. Appl.* **21** (1987) 11–22.

A Quasi-Coherent Description of the Category D-mod$(\mathrm{Gr}_{\mathrm{GL}(n)})$

Alexander Braverman and Michael Finkelberg

To Sasha Beilinson and Vitya Ginzburg

Contents

A. Braverman (✉)
Department of Mathematics, University of Toronto and Perimeter Institute of Theoretical Physics, Waterloo, ON, Canada

Skolkovo Institute of Science and Technology, Moscow, Russia
e-mail: braval@math.toronto.edu

M. Finkelberg
Department of Mathematics, National Research University Higher School of Economics, Russian Federation, Moscow, Russia

Skolkovo Institute of Science and Technology, Institute for Information Transmission Problems of RAS, Moscow, Russia

V. Baranovsky et al. (eds.), *Representation Theory and Algebraic Geometry*, Trends in Mathematics, https://doi.org/10.1007/978-3-030-82007-7_5

1 Introduction and Statement of the Results

1.1 General Notation

In general, we work over $\mathbb{C}$.

For a (derived) stack $\mathcal{Y}$, we denote by $\mathrm{QCoh}(\mathcal{Y})$ the derived category of quasi-coherent sheaves on $\mathcal{Y}$ and by $D\text{-mod}(\mathcal{Y})$ the derived category of D-modules on $\mathcal{Y}$. In addition, we are going to denote by $\mathrm{IndCoh}(\mathcal{Y})$ the derived category of ind-coherent sheaves on $\mathcal{Y}$; this category coincides with $\mathrm{QCoh}(\mathcal{Y})$ when $\mathcal{Y}$ is a classical (non-derived) smooth stack, but in general, the two are different (we are going to use [AG15] as our main reference for the notion and properties of ind-coherent sheaves).

Let $\mathcal{O} = \mathbb{C}[[z]], \mathcal{K} = \mathbb{C}((z))$. Set $\mathcal{D} = \mathrm{Spec}(\mathcal{O}), \mathcal{D}^* = \mathrm{Spec}(\mathcal{K})$. By a local system of rank n on $\mathcal{D}^*$, we shall mean a vector bundle $\mathcal{E}$ on $\mathcal{D}^*$ of rank n endowed with a connection $\nabla\colon \mathcal{E} \to \mathcal{E} \otimes \Omega^1_{\mathcal{D}^*}$. We denote by $\mathrm{LocSys}_n(\mathcal{D}^*)$ the stack of local systems of rank n on $\mathcal{D}^*$.

For an algebraic group G over $\mathbb{C}$, we denote by $\mathrm{Gr}_G = G(\mathcal{K})/G(\mathcal{O})$ the affine Grassmannian of G (viewed as an ind-scheme).

1.2 The Main Conjecture: GL(n)-case

Let $\mathcal{W}_n$ denote the stack which classifies the following data:

(1) A local system $\mathcal{E}_i$ on $\mathcal{D}^*$ of rank i for any $i = 1, \ldots, n$.
(2) A morphism $\kappa_i\colon \mathcal{E}_i \to \mathcal{E}_{i+1}$ for any $i < n$.

This stack maps naturally to the stack $\mathrm{LocSys}_n(\mathcal{D}^*)$) (this map sends the above data to $\mathcal{E}_n$). The trivial local system defines a map $\mathrm{pt}/\mathrm{GL}(n) \to \mathcal{W}_n$, and we let $\mathcal{W}_n^{\mathrm{triv}}$ product

$$\begin{array}{ccc} \mathcal{W}_n^{\mathrm{triv}} & \longrightarrow & \mathcal{W}_n \\ \downarrow & & \downarrow \\ \mathrm{pt}/\mathrm{GL}(n) & \longrightarrow & \mathrm{LocSys}_n(\mathcal{D}^*). \end{array}$$

It is worthwhile to note that $\mathcal{W}_n^{\mathrm{triv}}$ is a dg-stack.

The following is a slightly corrected version of a conjecture formulated in [BF19]:

Conjecture 1.3 *The category* $\mathrm{IndCoh}(\mathcal{W}_n^{\mathrm{triv}})$ *is equivalent to the category* $D\text{-mod}(\mathrm{Gr}_{\mathrm{GL}(n)})$. *This equivalence respects the natural action of the tensor category* $\mathrm{Rep}(\mathrm{GL}(n))$ *on both sides.*

It is explained in [BF19] how to "deduce" Conjecture 1.3 from quantum field theory considerations. In this paper, we are not going to discuss this physical motivation at all: instead, we are going to present some mathematical evidence for it (mostly in the case $n = 2$).

1.4 *The Main Conjecture:* GL(2)*-case*

In this subsection, we would like to strengthen Conjecture 1.3 in the case of GL(2). First, let us ask a natural question for arbitrary n. Namely, it is clear that the category $\mathrm{IndCoh}(\mathcal{W}_n^{\mathrm{triv}})$ lives over $\prod_{i=1}^{n-1} \mathrm{LocSys}_i(\mathcal{D}^*)$. How to see this structure on $D\text{-mod}(\mathrm{Gr}_{\mathrm{GL}(n)})$?

We don't know the answer to this question except for the case $n = 2$. To explain the answer, we need to recall the statement of geometric local class field theory (due to G. Laumon, cf. [Lau]):

Theorem 1.5 *There is a natural equivalence of monoidal categories* $D\text{-mod}(\mathcal{K}^\times) \simeq \mathrm{QCoh}(\mathrm{LocSys}_1(\mathcal{D}^*))$.[1]

Theorem 1.5 implies that the structure of "living over $\mathrm{LocSys}_1(\mathcal{D}^*)$" on a category $\mathcal{C}$ is the same as a strong action of $\mathcal{K}^\times$ on $\mathcal{C}$ (see, e.g., [Gai17, 4.1.2]). Thus, to answer our question for $n = 2$, it is enough to describe a strong action of $\mathcal{K}^\times$ on the category $D\text{-mod}(\mathrm{Gr}_{\mathrm{GL}(n)})$. Since the group $\mathrm{GL}(2, \mathcal{K})$ acts strongly on $D\text{-mod}(\mathrm{Gr}_{\mathrm{GL}(2)})$, it is enough to describe a map $\mathcal{K}^\times \to \mathrm{GL}(2, \mathcal{K})$. The relevant map is given by

$$x \overset{\eta}{\mapsto} \begin{pmatrix} x & 0 \\ 0 & 1 \end{pmatrix}$$

So, we get the following conjecture:

Conjecture 1.6 *The category* $\mathrm{IndCoh}(\mathcal{W}_2^{\mathrm{triv}})$ *is equivalent to the category* $D\text{-mod}(\mathrm{Gr}_{\mathrm{GL}(2)})$. *This equivalence respects the natural action of the tensor category* $\mathrm{Rep}(\mathrm{GL}(2))$ *on both sides. In addition, the action of the tensor category* $\mathrm{QCoh}(\mathrm{LocSys}_1(\mathcal{D}^*)) \simeq D\text{-mod}(\mathcal{K}^\times)$ *on* $D\text{-mod}(\mathrm{Gr}_{\mathrm{GL}(2)})$ *coming from the natural*

[1] In this case, the equivalence actually holds on the level of abelian categories, but the equivalence of Conjecture 1.3 only has a chance to hold on the derived level. Also in this case, there is no difference between QCoh and IndCoh.

projection $\mathcal{W}_2^{\mathrm{triv}}/\mathrm{GL}(2) \to \mathrm{LocSys}_1(\mathcal{D}^*)$ *under the above equivalence corresponds to the action of* $D\text{-mod}(\mathcal{K}^\times)$ *coming from the embedding* $\eta\colon \mathcal{K}^\times \to \mathrm{GL}(2,\mathcal{K})$ *defined above.*

1.7 Fiberwise Version

We don't know how to prove Conjecture 1.6 either. The purpose of this paper is to prove a weaker statement: namely, we are going to show that the fibers both of $\mathrm{IndCoh}(\mathcal{W}_2^{\mathrm{triv}})$ and of $D\text{-mod}(\mathrm{Gr}_{\mathrm{GL}(2)})$ over any $\mathcal{E} \in \mathrm{LocSys}_1(\mathcal{D}^*)$ are equivalent. Let us look at these fibers in more detail.

Denote by π the natural projection $\mathcal{W}_2^{\mathrm{triv}} \to \mathrm{LocSys}_1(\mathcal{D}^*)$. Let $\mathcal{E} \in \mathrm{LocSys}_1(\mathcal{D}^*)$. Let us first work with QCoh instead of IndCoh. Then the fiber of $\mathrm{QCoh}(\mathcal{W}_2^{\mathrm{triv}})$ over $\mathcal{E}$ (which we shall denote by $\mathrm{QCoh}(\mathcal{W}_2^{\mathrm{triv}})_{\mathcal{E}}$) is equivalent to $\mathrm{QCoh}(\pi^{-1}(\mathcal{E}))$.[2] Assume that $\mathcal{E}$ is non-trivial. Then any morphism from $\mathcal{E}$ to the trivial local system of rank 2 is 0; in other words, away from the trivial local system (of rank 1), the natural map $\mathcal{W}_2^{\mathrm{triv}} \to \mathrm{LocSys}_1(\mathcal{D}^*) \times \mathrm{pt}/\mathrm{GL}(2)$ is an isomorphism. Hence, $\pi^{-1}(\mathcal{E}) = \mathrm{pt}/\mathrm{GL}(2)$, and in this case, $\mathrm{QCoh}(\mathcal{W}_2^{\mathrm{triv}})_{\mathcal{E}}$ is equivalent to $\mathrm{Rep}(\mathrm{GL}(2))$.

On the other hand, assume that $\mathcal{E}$ is trivial. Then $\pi^{-1}(\mathcal{E})$ is a dg-stack equivalent to $(\mathbb{V} \times \mathbb{V}[-1])/\mathrm{GL}(\mathbb{V})$ where $\mathbb{V}$ is a two-dimensional vector space over $\mathbb{C}$ (this follows from the fact the dg-scheme classifying $f \in \mathbf{O}_{\mathcal{D}^*}$ such that $df = 0$ is $\mathbb{A}^1 \times \mathbb{A}^1[-1]$).

Let us go back to the IndCoh story. Assume that we have a morphism $\pi : \mathcal{Y} \to \mathcal{X}$ of (dg) stacks; assume moreover that $\mathcal{X}$ is a smooth classical stack. In this case, the fiber of $\mathrm{IndCoh}(\mathcal{Y})$ over a point $x \in \mathcal{X}$ is described in Section 2 of [AG15]. We are not going to reproduce that general answer here as it will require introducing more cumbersome notation; let us just explain what this answer amounts to in the case when $\mathcal{Y} = \mathcal{W}_2^{\mathrm{triv}}$ and $\mathcal{X} = \mathrm{LocSys}_1(\mathcal{D}^*)$.

Let $\mathcal{E}$ be a rank 1 local system on $\mathcal{D}^*$ as above. First, if $\mathcal{E}$ is non-trivial, then it is easy to see that the fiber $\mathrm{IndCoh}(\mathcal{W}_2^{\mathrm{triv}})_{\mathcal{E}}$ of $\mathrm{IndCoh}(\mathcal{W}_2^{\mathrm{triv}})$ over $\mathcal{E}$ is just $\mathrm{Rep}(\mathrm{GL}(2))$ as before. Let now $\mathcal{E}$ be trivial. Then, as was noted above, we have the isomorphism

$$\pi^{-1}(\mathcal{E}) \simeq (\mathbb{V} \times \mathbb{V}[-1])/\mathrm{GL}(\mathbb{V}),$$

where $\mathbb{V}$ is a two-dimensional vector space. By Koszul duality, the category $\mathrm{IndCoh}((\mathbb{V} \times \mathbb{V}[-1])/\mathrm{GL}(\mathbb{V}))$ is equivalent to the derived category of $\mathrm{GL}(\mathbb{V})$-equivariant dg-modules over $\mathbf{O}_{\mathbb{V}\times\mathbb{V}^*[2]}$. On the other hand, the sought-for fiber $\mathrm{IndCoh}(\mathcal{W}_2^{\mathrm{triv}})_{\mathcal{E}}$ is equivalent to the derived category of $\mathrm{GL}(\mathbb{V})$-equivariant dg-

---

[2] Here $\pi^{-1}(\mathcal{E})$ should be understood in dg-sense.

modules over $\mathbf{O}_{\mathbb{V}\times\mathbb{V}^*[2]}$ which are set-theoretically supported on $\mathcal{Z}_{\mathbb{V}} \subset \mathbb{V} \times \mathbb{V}^*[2]$ consisting of pairs (v, v^*) with $v^*(v) = 0$. We shall denote this category by $\mathrm{IndCoh}_{\mathcal{Z}_{\mathbb{V}}}((\mathbb{V} \times \mathbb{V}[-1])/\mathrm{GL}(\mathbb{V}))$.

Now, any $\mathcal{E}$ as above defines a character D-module $\mathcal{L}$ on $\mathcal{K}^\times$, i.e., a rank 1 local system endowed with an isomorphism $m^*\mathcal{L} \simeq \mathcal{L} \boxtimes \mathcal{L}$ (here $m\colon \mathcal{K}^\times \times \mathcal{K}^\times \to \mathcal{K}^\times$ is the multiplication map) satisfying the standard associativity constraint. Under this correspondence, trivial $\mathcal{E}$ corresponds to trivial $\mathcal{L}$, i.e., $\mathcal{L}$ isomorphic to $\mathbf{O}_{\mathcal{K}^\times}$ (note that $\mathcal{L}$ is trivial if and only if it is trivial when restricted to $\mathcal{O}^\times$). Given any $\mathcal{L}$ as above, and a category $\mathcal{C}$ with a strong action of $\mathcal{K}^\times$, it makes sense to consider the category of $(\mathcal{K}^\times, \mathcal{L})$-equivariant objects in $\mathcal{C}$. When $\mathcal{L}$ is trivial, this is just the category of $\mathcal{K}^\times$-equivariant objects.

Thus, the following result is exactly the "fiberwise version" of Conjecture 1.6:

Theorem 1.8 *Let $\mathcal{K}^\times$ act on $\mathrm{Gr}_{\mathrm{GL}(2)}$ by means of the map η. Then*

(1) *Let $\mathcal{L}$ be a non-trivial character D-module on $\mathcal{K}^\times$. Then the category of $(\mathcal{K}^\times, \mathcal{L})$-equivariant D-modules on* $\mathrm{Gr}_{\mathrm{GL}(2)}$ *is equivalent to* $\mathrm{Rep}(\mathrm{GL}(2))$.
(2) *Let $D^b_{\mathcal{K}^\times}(\mathrm{Gr}_{\mathrm{GL}(2)})$ denote the full subcategory of the derived category of $\mathcal{K}^\times$-equivariant D-modules on $\mathrm{Gr}_{\mathrm{GL}(2)}$ whose restriction to any connected component of $\mathrm{Gr}_{\mathrm{GL}(2)}$ is a bounded complex whose cohomology D-modules have finite-dimensional support and are coherent. Then $D^b_{\mathcal{K}^\times}(\mathrm{Gr}_{\mathrm{GL}(2)})$ is equivalent to* $\mathrm{Coh}((\mathbb{V} \times \mathbb{V}[-1])/\mathrm{GL}(\mathbb{V}))$ *(here $\mathbb{V}$ is again a two-dimensional vector space over $\mathbb{C}$).*
(3) *Let $D_{\mathcal{K}^\times}(\mathrm{Gr}_{\mathrm{GL}(2)})$ denote the derived category of $\mathcal{K}^\times$-equivariant D-modules on $\mathrm{Gr}_{\mathrm{GL}(2)}$. Then an object of $D_{\mathcal{K}^\times}(\mathrm{Gr}_{\mathrm{GL}(2)})$ is compact if and only if*

   (a) It lies in $D^b_{\mathcal{K}^\times}(\mathrm{Gr}_{\mathrm{GL}(2)})$;
   (b) Its image under the equivalence (2) lies in $\mathrm{Coh}_{\mathcal{Z}_{\mathbb{V}}}((\mathbb{V} \times \mathbb{V}[-1])/\mathrm{GL}(\mathbb{V}))$.

   In particular, the equivalence (ii) extends to the equivalence between $D_{\mathcal{K}^\times}(\mathrm{Gr}_{\mathrm{GL}(2)})$ and $\mathrm{IndCoh}_{\mathcal{Z}_{\mathbb{V}}}((\mathbb{V} \times \mathbb{V}[-1])/\mathrm{GL}(\mathbb{V}))$.

The rest of the paper is devoted to the proof of Theorem 1.8.

Remarks The fact that usually not all objects of the bounded equivariant derived category of D-modules (or constructible sheaves) are compact was first observed and studied by V. Drinfeld and D. Gaitsgory, cf. [DG13]. Also the reader should compare the last two assertions of Theorem 1.8 with, respectively, Theorem 12.3.3 and Corollary 12.5.5 of [AG15].

2 Proof of Theorem 1.8(1)

2.1 *Sketch of the Proof*

In what follows, we denote by $\Lambda = \mathbb{Z} \oplus \mathbb{Z}$ the coweight lattice of GL(2) and by

$$\Lambda^+ = \{(a, b) \in \Lambda | \, a \geq b\}$$

the cone of dominant coweights. Fix now a non-trivial character D-module $\mathcal{L}$ on $\mathcal{K}^\times$. We claim that in order to prove Theorem 1.8(1), it is enough to construct an embedding $\iota_{\mathcal{L}}$ from Λ^+ into the set of $\mathcal{K}^\times$-orbits on $\mathrm{Gr}_{\mathrm{GL}(2)}$ such that the following three properties hold:

(i) A $\mathcal{K}^\times$-orbit on $\mathrm{Gr}_{\mathrm{GL}(2)}$ supports a $(\mathcal{K}^\times, \mathcal{L})$-equivariant D-module if and only if it lies in the image of $\iota_{\mathcal{L}}$.
In what follows, for every $\lambda \in \Lambda^+$, let us denote by $\mathcal{F}_!^\lambda$ and $\mathcal{F}_*^\lambda$ the ! and $*$-extensions to all of $\mathrm{Gr}_{\mathrm{GL}(2)}$ of the corresponding irreducible $(\mathcal{K}^\times, \mathcal{L})$-equivariant D-module on the orbit $\iota_{\mathcal{L}}(\lambda)$.
(ii) For any $\lambda \in \Lambda^+$, we have

$$\mathcal{F}_!^0 \star \mathrm{IC}^\lambda \simeq \mathcal{F}_!^\lambda; \quad \mathcal{F}_*^0 \star \mathrm{IC}^\lambda \simeq \mathcal{F}_*^\lambda.$$

(iii) The natural morphism $\mathcal{F}_!^0 \to \mathcal{F}_*^0$ is an isomorphism.

Indeed, (ii) and (iii) together imply that the map $\mathcal{F}_!^\lambda \to \mathcal{F}_*^\lambda$ is an isomorphism for any λ. Hence, the category of $(\mathcal{K}^\times, \mathcal{L})$-equivariant D-modules on $\mathrm{Gr}_{\mathrm{GL}(2)}$ is semi-simple with simple objects $\mathcal{F}^\lambda := \mathcal{F}_!^\lambda \simeq \mathcal{F}_*^\lambda$. Now (ii) implies that the functor $\mathcal{S} \mapsto \mathcal{F}^0 \star \mathcal{S}$ from $D\text{-mod}_{\mathrm{GL}(2,\mathcal{O})}(\mathrm{Gr}_{\mathrm{GL}(2)})$ to the (abelian) category of $(\mathcal{K}^\times, \mathcal{L})$-equivariant D-modules on $\mathrm{Gr}_{\mathrm{GL}(2)}$ is an equivalence which is exactly what we had to prove.

So, it remains to define the map $\iota_{\mathcal{L}}$ and to check the properties (i)–(iii).

2.2 *The Map* $\iota_{\mathcal{L}}$

There exists unique $k > 0$ such that $\mathcal{L}$ is pulled back from $\mathcal{O}^\times/1 + z^k\mathcal{O}$ but not pulled back from $\mathcal{O}^\times/1 + z^{k-1}\mathcal{O}$. The corresponding map $\iota_{\mathcal{L}}$ will only depend on k which will be fixed till the end of this section. To simplify the notation, we shall simply write Y_λ for the $\mathcal{K}^\times$-orbit of $z^{\iota_{\mathcal{L}}(\lambda)}$. Also we set X_λ to be the intersection of Y_λ with $\mathrm{Gr}_{\mathrm{SL}(2)}$.

Let $\lambda = (n_1, n_2)$ with $n_1 \geq n_2$. Then we set Y_λ to be the $\mathcal{K}^\times$-orbit of the (image in $\mathrm{Gr}_{\mathrm{GL}(2)}$ of the) matrix

$$\begin{pmatrix} 1 & z^{-k-n} \\ 0 & 1 \end{pmatrix} \cdot \begin{pmatrix} z^{-n_2} & 0 \\ 0 & z^{n_2} \end{pmatrix}$$

Here $n = n_1 + n_2$.

2.3 *Proof of (i)*

It is enough to deal with $\mathcal{O}^\times$-orbits on $\mathrm{Gr}_{\mathrm{SL}(2)}$ instead of $\mathcal{K}^\times$-orbits on $\mathrm{Gr}_{\mathrm{GL}(2)}$. Such orbits are in one-to-one correspondence with pairs $(m, l) \in \mathbb{Z} \times \mathbb{Z}$ with $l - 2m \leq 0$; the $\mathcal{O}^\times$-orbit corresponding to a given (m, l) is the orbit of the matrix

$$\begin{pmatrix} 1 & z^{l} \\ 0 & 1 \end{pmatrix} \cdot \begin{pmatrix} z^{m} & 0 \\ 0 & z^{-m} \end{pmatrix}$$

The stabilizer of the above point in $\mathcal{O}^\times$ is $1 + z^{2m-l}\mathcal{O}$. Hence, this orbit supports a $(\mathcal{O}^\times, \mathcal{L})$-equivariant D-module if and only if $2m - l \geq k$. This is exactly the condition that there exists a pair $(n_1, n_2) \in \mathbb{Z} \times \mathbb{Z}$ such that $n_1 \geq n_2$ satisfying the equations

$$l = -k - n, \quad m = -n_2.$$

2.4 *Proof of (ii)*

Let us compute the convolution of $\mathcal{F}^0_!$ with IC^λ where $\lambda = (n_1, n_2)$ (the corresponding calculation for $\mathcal{F}^0_*$ is completely analogous). We need to show the following two things:

(1) The $*$-restriction $\mathcal{F}^0_! \star \mathrm{IC}^\lambda$ to X_λ is equal to IC-sheaf of X_λ;
(2) The $*$-restriction $\mathcal{F}^0_! \star \mathrm{IC}^\lambda$ to any $\mathcal{O}^*$-orbit on $\mathrm{Gr}_{\mathrm{SL}(2)}$ different from X_λ is equal to 0.

For this, it is enough to compute the stalk of $\mathcal{F}^0_! \star \mathrm{IC}^\lambda$ at any point of the form

$$g = \begin{pmatrix} 1 & z^{l} \\ 0 & 1 \end{pmatrix} \cdot \begin{pmatrix} z^{m} & 0 \\ 0 & z^{-m} \end{pmatrix}.$$

Let us fix $\lambda = (n_1, n_2)$, m, l, and k, and let

$$Z = \{x \in X_0 |\, x^{-1}g \in \overline{\mathrm{Gr}}^{\lambda}_{\mathrm{GL}(2)}\}.$$

Let i denote the natural map from Z to $X_0 \simeq \mathcal{O}^*/1+z^k\mathcal{O}$. Then the above stalk is equal to $H_c^*(Z, i^*\mathcal{L}[\dim X_0 + \dim \mathrm{Gr}^{\lambda}_{\mathrm{GL}(2)}])$. We can assume that x is of the form

$$x = \begin{pmatrix} z^{-n} & az^{-n-k} \\ 0 & 1 \end{pmatrix}$$

where $a \in \mathcal{O}^{\times}$. Then

$$x^{-1}g = \begin{pmatrix} z^{n+m} & z^{n+l-m} - az^{-k-m} \\ 0 & z^{-m} \end{pmatrix}.$$

This matrix defines a point in $\overline{\mathrm{Gr}}^{\lambda}_{\mathrm{GL}(2)}$ if $n+m, -m \geq n_2$ and $z^{-m}(z^{n+l} - az^{-k}) \in z^{n_2}\mathcal{O}$. Let $a = \sum a_i z_i$. We see that if $-m > n_2$, then changing a_{k-1} does not affect the above conditions; so, "integrating out" a_{k-1} first, we see that $H_c^*(Z, i^*\mathcal{L}[\dim X_0 + \dim \mathrm{Gr}^{\lambda}_{\mathrm{GL}(2)}]) = 0$. Assume now that $-m = n_2$. Then unless $n + l = -k$, the above equations have no solutions; hence, the sought-for stalk is again 0. The case $-m = n_2, n+l = -k$ is precisely the case $g \in X_\lambda$. In this case, we must have $a_0 = 1$ and $a_j = 0$ for $0 < j < k$. So Z consists of just one point and $H_c^*(Z, \mathbb{C}[\dim X_0 + \dim \mathrm{Gr}^{\lambda}_{\mathrm{GL}(2)}] = \mathbb{C}[\dim X_\lambda]$ (since it is easy to see that $\dim X_0 + \dim \mathrm{Gr}^{\lambda}_{\mathrm{GL}(2)} = \dim X_\lambda$).

2.5 *Proof of (iii)*

It follows from the discussion in the beginning of Sect. 2.3 that

(a) If an $\mathcal{O}^{\times}$-orbit X on $\mathrm{Gr}_{\mathrm{SL}(2)}$ carries a non-zero $(\mathcal{O}^{\times}, \mathcal{L})$-equivariant sheaf, then $\dim X \geq k$;
(b) $\dim X_0 = k$.

It follows from (b) that $\overline{X}_0 \backslash X$ is a union of $\mathcal{O}^{\times}$-orbits of dimension $< k$. Thus, (a) implies that the natural morphism $\mathcal{F}^0_! \to \mathcal{F}^0_*$ is an isomorphism.

3 Proof of Theorem 1.8(2)

In this section, we prove the second assertion of Theorem 1.8. It is in fact a mild variation on the proof of the derived geometric Satake equivalence (cf. [BF08]).

3.1 Reduction to SL(2)

We are supposed to study the derived category of $\mathcal{K}^\times$-equivariant D-modules on $\mathrm{Gr}_{\mathrm{GL}(2)}$. We claim that it is the same as the derived category of $\mathcal{O}^\times$-equivariant D-modules on $\mathrm{Gr}_{\mathrm{SL}(2)}$ (here $\mathcal{O}^\times$ is embedded into $\mathrm{SL}(2, \mathcal{K})$ via the identification of the standard Cartan subgroup of SL(2) with $\mathbb{G}_m$). Indeed, we have $\mathcal{K}^\times = \mathcal{O}^\times \times \mathbb{Z}$. The last factor acts simply transitively on the set of connected components of $\mathrm{Gr}_{\mathrm{GL}(2)}$, and the first factor preserves every connected component. Hence, a $\mathcal{K}^\times$-equivariant D-module on $\mathrm{Gr}_{\mathrm{GL}(2)}$ is the same as an $\mathcal{O}^\times$-equivariant D-module on the connected component of 1, which is equal to $\mathrm{Gr}_{\mathrm{SL}(2)}$. The reader must be warned that the action of $\mathcal{O}^\times$ on $\mathrm{Gr}_{\mathrm{SL}(2)}$ coming from our usual $\mathcal{K}^\times$-action on $\mathrm{Gr}_{\mathrm{GL}(2)}$ is not the same as the action coming from the Cartan torus of SL(2), but the latter is obtained from the former by means of the map $x \mapsto x^2$ which doesn't change the equivariant derived category.

For the remainder of this section, we shall write Gr instead of $\mathrm{Gr}_{\mathrm{SL}(2)}$.

3.2 Koszul Duality

We let $D_{\mathcal{O}^\times}(\mathrm{Gr})$ denote the corresponding equivariant derived category; since orbits of $\mathcal{O}^\times$ on Gr are parameterized by discrete set, we can work with constructible sheaves instead of D-modules.

We let $D^b_{\mathcal{O}^\times}(\mathrm{Gr})$ denote the bounded derived category of $\mathcal{O}^\times$-equivariant constructible sheaves on Gr. Recall that we need to show the following:

Theorem 3.3 $D^b_{\mathcal{O}^\times}(\mathrm{Gr}) \simeq \mathrm{Coh}((\mathbb{V} \times \mathbb{V}^*[2])/\mathrm{GL}(\mathbb{V}))$.

3.4 Equivariant Cohomology

Let $\lambda \in \mathbb{Z}_+, \mu \in \mathbb{Z}$. We are going to think about λ as a dominant coweight of PGL(2) and about μ as an arbitrary coweight of PGL(2). Let us also assume that $\lambda - \mu \in 2\mathbb{Z}$. Then we define $\mathcal{F}^{\lambda,\mu}$ to be the IC-sheaf of $z^\mu\overline{\mathrm{Gr}}^\lambda$ (note that because λ and μ have the same parity, it follows that $z^\mu\overline{\mathrm{Gr}}^\lambda \subset \mathrm{Gr}_{\mathrm{SL}(2)} = \mathrm{Gr}$). This is an object of $D^b_{\mathcal{O}^\times}(\mathrm{Gr})$. We would like to describe $H^*_{\mathcal{O}^\times}(\mathrm{Gr}, \mathcal{F}^{\lambda,\mu})$ as a module over $H^*_{\mathcal{O}^\times}(\mathrm{Gr}, \mathbb{C})$.

First, let us describe $H^*_{\mathcal{O}^\times}(\mathrm{Gr}, \mathbb{C})$. Namely, let **Det** denote the standard determinant line bundle on Gr. Then we have

$$H^*_{\mathcal{O}^\times}(\mathrm{Gr}, \mathbb{C}) = \mathbb{C}[\mathbf{a}, \mathbf{c}]$$

where $\mathbf{a}$ is the standard generator of $H^*_{\mathcal{O}^\times}(\mathrm{pt}) = H^*_{\mathbb{C}^\times}(\mathrm{pt})$ and $\mathbf{c} = c_1(\mathbf{Det})$ (equivariant first Chern class).

We can now describe $H^*_{\mathcal{O}^\times}(\mathrm{Gr}, \mathcal{F}^{\lambda,\mu})$.

Proposition 3.5 *Let $V(\lambda)$ denote the irreducible representation of* SL(2) *with highest weight λ (it has dimension $\lambda + 1$). Let $\pi_\lambda \colon \mathfrak{sl}_2 \to \mathrm{End}(V(\lambda))$ denote the corresponding map. Then the $H^*_{\mathcal{O}^\times}(\mathrm{Gr}, \mathbb{C}) = \mathbb{C}[\mathbf{a}, \mathbf{c}]$-module $H^*_{\mathcal{O}^\times}(\mathrm{Gr}, \mathcal{F}^{\lambda,\mu})$ is isomorphic to $\mathbb{C}[\mathbf{a}] \otimes V(\lambda)$ where*

(a) $\mathbf{c}$ *acts by*

$$\pi_\lambda \begin{pmatrix} 0 & 1 \\ \mathbf{a}^2 & 0 \end{pmatrix} + \mu\mathbf{a}. \tag{1}$$

(b) The grading on $\mathbb{C}[\mathbf{a}] \otimes V(\lambda)$ is equal to the tensor product of the standard grading on $\mathbb{C}[\mathbf{a}]$ (recall that $\mathbf{a}$ has degree 2) and the grading on $V(\lambda)$ by eigenvalues of h (here we use the standard basis (e, h, f) of the Lie algebra of SL(2)*). Note that the endomorphism of $\mathbb{C}[\mathbf{a}] \otimes V(\lambda)$ given by the element 1 is homogeneous of degree 2 with respect to this grading.*

Proof This statement is well-known when $\mu = 0$. To prove it for general μ, it is enough to show that $c_1((z^\mu)^*\mathbf{Det}) = \mathbf{c} + \mu\mathbf{a}$. It is enough to check this equality after restricting to every $\mathcal{O}^\times$-fixed point on Gr where it is obvious. □

Let us slightly reformulate this answer. Given λ and μ as above, let $V(\lambda, \mu)$ denote the (unique) irreducible representation of GL(2), such that its restriction to SL(2) is isomorphic to $V(\lambda)$ and its central character is given by μ (note that such a representation exists precisely when $\lambda - \mu \in 2\mathbb{Z}$). In what follows, we shall regard it as a graded vector space, where the grading as before is given by the eigenvalues of $h \in \mathfrak{sl}_2$. Let $\pi_{\lambda,\mu} \colon \mathfrak{gl}_2 \to \mathrm{End}(V(\lambda, \mu))$ denote the corresponding map. Then (1) is equal to

$$\pi_{\lambda,\mu} \begin{pmatrix} \mathbf{a} & 1 \\ \mathbf{a}^2 & \mathbf{a} \end{pmatrix}. \tag{2}$$

Let us make yet another reformulation of the answer. Let

$$S(\mathbf{a}) = \begin{pmatrix} \mathbf{a} & 1 \\ \mathbf{a}^2 & \mathbf{a} \end{pmatrix}, \quad T(\mathbf{a}) = \begin{pmatrix} 0 & 1 \\ 0 & 2\mathbf{a} \end{pmatrix}$$

Then $T(\mathbf{a}) = g(\mathbf{a})^{-1} S(\mathbf{a}) g(\mathbf{a})$ where

$$g(\mathbf{a}) = \begin{pmatrix} 1 & 0 \\ -\mathbf{a} & 1 \end{pmatrix}.$$

Hence, we get the following equivalent version of Proposition 3.5:

Proposition 3.6 *The $H^*_{\mathcal{O}^\times}(\mathrm{Gr}, \mathbb{C}) = \mathbb{C}[\mathbf{a}, \mathbf{c}]$-module $H^*_{\mathcal{O}^\times}(\mathrm{Gr}, \mathcal{F}^{\lambda,\mu})$ is isomorphic to $\mathbb{C}[\mathbf{a}] \otimes V(\lambda, \mu)$ where $\mathbf{c}$ acts by $\pi_{\lambda,\mu}(T(\mathbf{a}))$.*

3.7 The Functor

We can now describe the functor $F\colon D^b_{\mathcal{O}^\times}(\mathrm{Gr}) \to \mathrm{Coh}((\mathbb{V} \times \mathbb{V}^*[2])/\mathrm{GL}(2))$. Namely, it has the property that

$$F(\mathcal{F}^{\lambda,\mu}) = \mathcal{O}_{\mathbb{V}\times\mathbb{V}^*[2]} \otimes V(\lambda, \mu)$$

where the group GL(2) acts on the RHS diagonally. We claim that in order to check existence of F, it is enough to construct isomorphisms

$$\begin{aligned} &\mathrm{Ext}_{D^b_{\mathcal{O}^\times}(\mathrm{Gr})}(\mathcal{F}^{\lambda,\mu}, \mathcal{F}^{\lambda',\mu'}) \simeq \\ &\mathrm{Ext}_{\mathbf{O}_{\mathbb{V}\times\mathbb{V}^*[2]}\rtimes \mathrm{GL}(2)}(\mathbf{O}_{\mathbb{V}\times\mathbb{V}^*[2]} \otimes V(\lambda, \mu), \mathbf{O}_{\mathbb{V}\times\mathbb{V}^*[2]} \otimes V(\lambda', \mu')) \end{aligned} \quad (3)$$

for any (λ, μ) and (λ', μ') as above (these isomorphisms must be compatible with compositions). Indeed, if we have such isomorphisms, then a word-by-word repetition of the arguments of [BF08, Section 6] constructs the functor F (and also proves that it is an equivalence).

3.8 Computing Ext's

The next result allows us to compute Ext's between $\mathcal{O}^\times$-equivariant IC-sheaves on Gr; it is analogous to a theorem of V. Ginzburg from [Gin91], but we do not know how to prove it by any general argument.

Proposition 3.9

$$\mathrm{Ext}_{D^b_{\mathcal{O}^\times}(\mathrm{Gr})}(\mathcal{F}^{\lambda,\mu}, \mathcal{F}^{\lambda',\mu'}) = \mathrm{Hom}_{H^*_{\mathcal{O}^\times}(\mathrm{Gr},\mathbb{C})}(H^*_{\mathcal{O}^\times}(\mathrm{Gr}, \mathcal{F}^{\lambda,\mu}), H^*_{\mathcal{O}^\times}(\mathrm{Gr}, \mathcal{F}^{\lambda',\mu'})). \quad (4)$$

Here we use the following convention: when we write Hom between two graded modules over a graded ring, we consider all *homomorphisms (not just those that preserve the grading).*

Proof Obviously, we have a map from the LHS of (4) to the RHS of (4). First, we claim that this map is injective. For this, it is enough to show the following:

(1) Both sides are free modules over $H^*_{\mathcal{O}^\times}(\mathrm{pt})$;
(2) The map in question becomes an isomorphism after tensoring with the field of fractions of $H^*_{\mathcal{O}^\times}(\mathrm{pt})$.

□

The first assertion is known to follow from the fact that the corresponding non-equivariant Ext's and cohomologies are pure (which follows from the fact that these are Ext's between pure sheaves on a projective variety). The second assertion follows from localization theorem since the set of fixed points of $\mathbb{C}^\times \subset \mathcal{O}^\times$ in the closure of any $\mathcal{O}^\times$-orbit on Gr is finite.

Now let us show that the above map is surjective. It follows from Proposition 3.6 that $H^*_{\mathcal{O}^\times}(\mathrm{Gr}, \mathcal{F}^{\lambda,\mu})$ is a cyclic $H^*_{\mathcal{O}^\times}(\mathrm{Gr}, \mathbb{C}) = \mathbb{C}[\mathbf{a}, \mathbf{c}]$-module generated by one vector $v_{\lambda,\mu}$ of degree $-\lambda$ whose annihilator is generated by the element

$$\prod_{i=0}^{\lambda}(\mathbf{c} - \mathbf{a}(2i + \mu - \lambda)). \tag{5}$$

Let now (λ, μ) and (λ', μ') be as in (4). Let $S(\lambda, \mu)$ be the set $\{\mu - \lambda, \mu - \lambda + 2, \cdots, \lambda\}$ (respectively, let $S(\lambda', \mu') = \{\mu' - \lambda', \mu - \lambda + 2, \cdots, \lambda'\}$). Let k be the cardinality of $S(\lambda, \mu) \cap S(\lambda', \mu')$. Then the RHS of (4) is

(a) equal to 0 if $k = 0$;
(b) generated by one element of degree $2(\lambda' + 1 - k)$ whose annihilator in $\mathbb{C}[\mathbf{a}, \mathbf{c}]$ is generated by $\prod\limits_{i \in S(\lambda,\mu)\cap S(\lambda',\mu')} (\mathbf{c} - \mathbf{a}i)$ for $k > 0$.

We now want to compare this to the LHS of (4). Let $\overline{\mathrm{Gr}}^{\lambda,\mu}$ denote the support of $\mathcal{F}^{\lambda,\mu}$. Since $\mathcal{F}^{\lambda,\mu}$ (resp. $\mathcal{F}^{\lambda',\mu'}$) is the constant sheaf on $\overline{\mathrm{Gr}}^{\lambda,\mu}$ (resp. on $\overline{\mathrm{Gr}}^{\lambda',\mu'}$) shifted by λ (resp. by λ'), it follows that the LHS of (4) is equal to $H^*_{\mathcal{O}^*}(\overline{\mathrm{Gr}}^{\lambda,\mu} \cap \overline{\mathrm{Gr}}^{\lambda',\mu'}, \mathbb{C})[\lambda' - \lambda]$. Thus, Proposition 3.9 follows from the following:

Lemma 3.10

(1) $\overline{\mathrm{Gr}}^{\lambda,\mu} \cap \overline{\mathrm{Gr}}^{\lambda',\mu'} = \emptyset$ *if* $k = 0$.
(2) $\overline{\mathrm{Gr}}^{\lambda,\mu} \cap \overline{\mathrm{Gr}}^{\lambda',\mu'} = \overline{\mathrm{Gr}}^{\lambda'',\mu''}$, *where* λ'', μ'' *are such that* $S(\lambda, \mu) \cap S(\lambda', \mu') = S(\lambda'', \mu'')$ *(for* $k > 0$*).*

Proof The assignment $\mu \mapsto z^\mu$ defines a bijection between $2\mathbb{Z}$ and $\mathrm{Gr}^{\mathbb{C}^\times}$. Any closed $\mathcal{O}^*$-invariant subset of Gr is uniquely determined by its intersection with $\mathrm{Gr}^{\mathbb{C}^\times} = 2\mathbb{Z}$. It is easy to see that $\overline{\mathrm{Gr}}^{\lambda,\mu} \cap \mathrm{Gr}^{\mathbb{C}^\times} = S(\lambda, \mu)$; hence, the lemma follows.

The proposition is proved. □

3.11 The End of the Proof

We need to construct an isomorphism between the RHS of (3) and the RHS of (4). Note that the latter is equal to Hom between two explicit modules over the ring $\mathbb{C}[\mathbf{a},\mathbf{c}]$ over the polynomial ring in two variables of degree 2. We would like to rewrite the former in a similar way. For this, let us do the following.

First, let P denote the stabilizer of the vector $(1, 0)$ in $\mathbb{V}$. Then we claim that

$$\mathrm{Ext}_{\mathbf{O}_{\mathbb{V}\times\mathbb{V}^*[2]}\rtimes\mathrm{GL}(2)}(\mathbf{O}_{\mathbb{V}\times\mathbb{V}^*[2]}\otimes V(\lambda,\mu),\mathbf{O}_{\mathbb{V}\times\mathbb{V}^*[2]}\otimes V(\lambda',\mu')) =$$

$$\mathrm{Hom}_{\mathbf{O}_{\mathbb{V}^*[2]}\rtimes P}(\mathbf{O}_{\mathbb{V}^*[2]}\otimes V(\lambda,\mu),\mathbf{O}_{\mathbb{V}^*[2]}\otimes V(\lambda',\mu')).$$

Indeed, since we are computing Hom's between free modules, we can replace $\mathbb{V}$ by $\mathbb{V}\backslash\{0\}$. Since GL(2) acts transitively on the latter with P being the stabilizer of one element, we obtain the above isomorphism.

Now we would like to describe a functor from the category of P-equivariant coherent sheaves on $\mathbb{V}^*[2]$ to the category of graded modules over $\mathbb{C}[\mathbf{a},\mathbf{c}]$ which is fully faithful on free modules. The category of P-equivariant coherent sheaves on $\mathbb{V}^*[2]$ can be thought of as the category of P-equivariant graded modules over $\mathbb{C}[x, y]$ where x and y both have degree 2. The group P consists of matrices

$$g = \begin{pmatrix} 1 & \alpha \\ 0 & \beta \end{pmatrix}. \tag{6}$$

Such a matrix acts on a vector (x, y) by means of $(g^t)^{-1}$ (here g^t stands for the transposed matrix). Thus, the Lie algebra of P consists of matrices of the form

$$A = \begin{pmatrix} 0 & u \\ 0 & v \end{pmatrix}$$

and $A(x, y) = (0, -ux - vy)$.

Let us take a module M as above, and let us restrict it to the line $y = -1$, i.e., consider the quotient $M/(y+1)M$. This quotient is endowed with a natural action of $\mathbb{C}^\times$ which comes from the $\mathbb{C}^\times$-action on M coming from the grading on M and the action coming from the embedding $\mathbb{C}^\times \hookrightarrow P$ corresponding to matrices as in (6) with $\alpha = 0$. We would like to extend this to a structure of a graded $\mathbb{C}[\mathbf{a},\mathbf{c}]$-module on it.

The action of $\mathbf{a}$ just comes from the action of $x/2$ on M. The action of $\mathbf{c}$ is characterized by the property that its action on the fiber over the point $(x, -1) = (2\mathbf{a}, -1)$ is given by the action of the matrix

$$\begin{pmatrix} 0 & 1 \\ 0 & 2\mathbf{a} \end{pmatrix} \in \mathrm{Lie}(P). \tag{7}$$

This makes sense because this matrix kills the vector $(2\mathbf{a}, -1)$ and hence the corresponding one-parametric subgroup (and hence also its Lie algebra) acts on the fiber of any P-equivariant coherent sheaf over $(2\mathbf{a}, -1)$.

Let us denote the resulting functor from P-equivariant coherent sheaves on $\mathbb{V}^*[2]$ to graded $\mathbb{C}[\mathbf{a}, \mathbf{c}]$-modules by $\widetilde{F}$. It follows from Proposition 3.6 that this functor sends the module $\mathbf{O}_{\mathbb{V}^*[2]} \otimes V(\lambda, \mu)$ to $H^*_{\mathcal{O}^\times}(\mathrm{Gr}, \mathcal{F}^{\lambda,\mu})$. To finish the proof, it remains to show that $\widetilde{F}$ is fully faithful on free modules. This immediately follows from the following two (easy) statements:

(1) $P \cdot \{(x, -1)\} = \mathbb{V}^* \backslash \{0\}$;
(2) The stabilizer of the point $(2\mathbf{a}, -1)$ in P is equal to the one-parametric subgroup generated by the matrix (7).

3.12 Abelian Equivalence

We would like to conclude this section with a variant of Theorem 1.8(2) which in particular will give rise to certain equivalence of abelian categories (this is not strictly speaking needed for the purposes of this paper, but it is important for some future work). Namely, first of all, we claim that the category $\mathrm{Coh}((\mathbb{V} \times \mathbb{V}[-1])/\mathrm{GL}(\mathbb{V}))$ is equivalent to the derived category of $\mathrm{GL}(\mathbb{V})$-equivariant finitely generated modules over the algebra $\Lambda(\mathbb{V}) \otimes \Lambda(\mathbb{V}^*)$. Indeed, $\mathrm{Coh}((\mathbb{V} \times \mathbb{V}[-1])/\mathrm{GL}(\mathbb{V}))$ is the derived category of $\mathrm{GL}(\mathbb{V})$-equivariant dg-modules over $\mathrm{Sym}(\mathbb{V}^*) \otimes \Lambda(\mathbb{V}^*[-1])$ (considered as a dg-algebra with trivial differential).[3]

Let now M be any $\mathrm{GL}(\mathbb{V})$-equivariant dg-module over $\mathrm{Sym}(\mathbb{V}^*) \otimes \Lambda(\mathbb{V}^*[-1])$. Define a new grading of M which is equal to the sum of the old grading and the grading coming from the action of the center of $\mathrm{GL}(\mathbb{V})$. This makes it into a $\mathrm{GL}(\mathbb{V})$-equivariant dg-module over $\mathrm{Sym}(\mathbb{V}^*[1]) \otimes \Lambda(\mathbb{V}^*)$. By applying Koszul duality with respect to the first factor, we can now associate to M a finitely generated $\mathrm{GL}(\mathbb{V})$-equivariant module over $\Lambda(\mathbb{V}) \otimes \Lambda(\mathbb{V}^*)$. It is easy to see that this procedure defines an equivalence between the derived category of $\mathrm{GL}(\mathbb{V})$-equivariant dg-modules over $\mathrm{Sym}(\mathbb{V}^*) \otimes \Lambda(\mathbb{V}^*[-1])$ and the derived category of $\mathrm{GL}(\mathbb{V})$-equivariant modules over $\Lambda(\mathbb{V}) \otimes \Lambda(\mathbb{V}^*)$. The advantage of the latter model is that it comes equipped with an obvious t-structure, whose heart is the abelian category of $\mathrm{GL}(\mathbb{V})$-equivariant modules over the algebra $\Lambda(\mathbb{V}) \otimes \Lambda(\mathbb{V}^*)$.

On the other hand, the category $D^b_{\mathcal{K}^\times}(\mathrm{Gr}_{\mathrm{GL}(2)})$ also has an obvious t-structure whose heart can be identified with the category $\mathrm{Perv}_{\mathcal{K}^\times}(\mathrm{Gr}_{\mathrm{GL}(2)})$ of $\mathcal{K}^\times$-equivariant

[3] Here when we write $\Lambda(W[d])$ (for a vector space W and $d \in \mathbb{Z}$), we just mean the dg-algebra with trivial differential which is equal to the exterior algebra generated by elements of W which have homological degree $-d$, i.e., we are NOT using the "super-notation" here with respect to the homological degree. Same goes for the notation $\mathrm{Sym}(W[d])$.

perverse sheaves on $\mathrm{Gr}_{\mathrm{GL}(2)}$ (the latter category is the same as $\mathrm{Perv}_{\mathcal{O}^\times}(\mathrm{Gr}_{\mathrm{SL}(2)})$ which is just the full subcategory of the category of perverse sheaves (with finite-dimensional support) on $\mathrm{Gr}_{\mathrm{SL}(2)}$ which are constant along $\mathcal{O}^\times$-orbits).

The following statement is an easy corollary of the proof of Theorem 1.8(2); we leave the details to the reader.

Theorem 3.13 *The equivalence between $D^b_{\mathcal{K}^\times}(\mathrm{Gr}_{\mathrm{GL}(2)})$ and the derived category of* $\mathrm{GL}(\mathbb{V})$*-equivariant finitely generated modules over $\Lambda(\mathbb{V}) \otimes \Lambda(\mathbb{V}^*)$ (obtained by combining Theorem 1.8(2) and the equivalence described in the beginning of this subsection) preserves the above t-structures. In particular, the category* $\mathrm{Perv}_{\mathcal{K}^\times}(\mathrm{Gr}_{\mathrm{GL}(2)})$ *is equivalent to the abelian category of* $\mathrm{GL}(\mathbb{V})$*-equivariant finitely generated modules over the algebra $\Lambda(\mathbb{V}) \otimes \Lambda(\mathbb{V}^*)$.*

4 Proof of Theorem 1.8(3)

4.1 Compact Objects in D-mod$_{\mathbf{H}}(X)$

Let X be a scheme of finite type over $\mathbb{C}$. Let also $\mathbf{H}$ be a pro-algebraic group over $\mathbb{C}$ acting on X; we assume that $\mathbf{H}$ has a normal pro-unipotent subgroup with finite-dimensional quotient. As before, we denote by $D\text{-mod}_{\mathbf{H}}(X)$ the derived category of strongly $\mathbf{H}$-equivariant D-modules on X. We also denote by $D^b_{\mathbf{H}}(X)$ its full subcategory consisting of bounded complexes with coherent cohomology. We would like to get a characterization of compact objects in $D\text{-mod}_{\mathbf{H}}(X)$ (under some additional assumptions). This question is studied in detail in [DG13]. The following lemma is an easy consequence of the results of *loc. cit.*:

Lemma 4.2

(1) *Assume that $\mathcal{F} \in D\text{-mod}_{\mathbf{H}}(X)$ is compact. Then $\mathcal{F} \in D^b_{\mathbf{H}}(X)$.*
(2) *Assume that $\mathcal{F} \in D\text{-mod}_{\mathbf{H}}(X)$ is compact. Then its equivariant de Rham cohomology $H^*_{\mathbf{H}}(X, \mathcal{F})$ is finite-dimensional (i.e., it is a bounded complex of vector spaces with finite-dimensional cohomology).*
(3) *Assume that $X =$ pt. Then conditions (1) and (2) above are also sufficient for compactness.*
(4) *Let $\mathbf{H} = \mathbb{C}^\times \times \mathbf{H}^0$ where $\mathbf{H}^0$ is (pro)unipotent. Then $\mathcal{F} \in D^b_{\mathbf{H}}(X)$ is compact if and only if for any embedding $i_x \colon \{x\} \to X$ of $\mathbb{C}^\times$-fixed point x in X, the object $i_x^! \mathcal{F}$ is a compact object of $D\text{-mod}_{\mathbb{C}^\times}(\mathrm{pt})$.*

4.3 The Cohomology Functor

In view of assertion (2) of Lemma 4.2, we would like to describe what happens to the functor of equivariant de Rham cohomology under the equivalence constructed in Sect. 3. Let us denote this equivalence by Φ (this is a functor from $D^b_{\mathcal{O}^\times}(\mathrm{Gr})$ to $\mathrm{Coh}((\mathbb{V}\times\mathbb{V}^*[2])/\mathrm{GL}(\mathbb{V}))$.

Let us consider the closed dg-subscheme $\mathbb{S}$ of $\mathbb{V}\times\mathbb{V}^*[2]$ consisting of pairs (v, v^*) where $v=(1,0)$ and v^* is of the form $(x,-1)$. Then we claim the following:

Lemma 4.4 *We have canonical isomorphism*

$$H^*_{\mathcal{O}^\times}(\mathrm{Gr},\mathcal{F})\simeq \mathcal{F}|_{\mathbb{S}} \tag{8}$$

for any $\mathcal{F}\in D^b_{\mathcal{O}^\times}(\mathrm{Gr})$. Here the grading on the RHS of (8) is defined in the same way as in Sect. 3.11.

The proof follows immediately from the construction of the functor Φ described in Sect. 3.

4.5 Compact Objects in $\mathcal{D}_{\mathcal{O}^\times}(\mathrm{Gr})$

Let us now go back to the proof of Theorem 1.8(3). We want to show that an object $\mathcal{F}$ in $\mathcal{D}_{\mathcal{O}^\times}(\mathrm{Gr})$ is compact if and only if it is a bounded complex of coherent D-modules (which in this case is the same as a bounded complex of constructible sheaves) and $\Phi(\mathcal{F})$ is supported on $\mathcal{Z}_{\mathbb{V}}$. Let us first show the "only if" direction. According to assertion (2) of Lemma 4.2, compactness of $\mathcal{F}$ implies that $H^*_{\mathcal{O}^\times}(\mathrm{Gr},\mathcal{F})$ is finite-dimensional. This condition is equivalent to the condition $\dim \mathrm{supp}(\Phi(\mathcal{F}))\cap\mathbb{S}=0$; here, we regard both $\mathrm{supp}(\Phi(\mathcal{F}))$ and $\mathbb{S}$ as closed subvarieties of $\mathbb{V}\times\mathbb{V}^*$ (i.e., we disregard the cohomological grading on the second factor). However, the fact that $\Phi(\mathcal{F})$ is actually an object of $\mathrm{Coh}((\mathbb{V}\times\mathbb{V}^*[2])/\mathrm{GL}(\mathbb{V}))$ implies that $\mathrm{supp}(\Phi(\mathcal{F}))$ is

(a) $\mathrm{GL}(\mathbb{V})$-invariant.
(b) $\mathbb{C}^\times$-invariant where the $\mathbb{C}^\times$-action on $\mathbb{V}\times\mathbb{V}^*$ comes from dilating the second factor.

It is easy to see that a closed subvariety of $\mathbb{V}\times\mathbb{V}^*$ which satisfies conditions (a) and (b) above has zero-dimensional intersection with $\mathbb{S}$ if and only if it is contained in $\mathcal{Z}_{\mathbb{V}}$, which finishes the proof of the "only if" direction.

4.6 End of the Proof

To prove the "if" direction, we are going to use the fourth assertion of Lemma 4.2 (note that $\mathcal{O}^\times$ is a product of $\mathbb{C}^\times$ and a pro-unipotent group). Let us assume that $\mathrm{supp}(\Phi(\mathcal{F}))\subset\mathcal{Z}_{\mathbb{V}}$. Combining the third and fourth assertions we see that (using the notation of Sect. 3), we just need to check that for any even integer μ, we have

$$\dim \mathrm{Ext}^*(\mathcal{F}^{0,\mu}, \mathcal{F}) < \infty \tag{9}$$

(here we compute Ext in the equivariant derived category). Indeed, the sheaves $\mathcal{F}^{0,\mu}$ are exactly the sky-scraper sheaves at the $\mathbb{C}^\times$-fixed points in Gr.

First of all, we claim that it is enough to assume that $\mu = 0$. Indeed, we have

$$\mathrm{Ext}^*(\mathcal{F}^{0,\mu}, \mathcal{F}) = \mathrm{Ext}^*(\mathcal{F}^{0,0}, (z^{-\mu})^*\mathcal{F})$$

and $\Phi((z^{-\mu})^*\mathcal{F}) = \Phi(\mathcal{F}) \otimes V(0, -\mu)$; hence, if $\Phi(\mathcal{F})$ is supported inside $\mathcal{Z}_{\mathbb{V}}$, then the same is true for $\Phi((z^{-\mu})^*\mathcal{F}))$.

Now, since $\Phi(\mathcal{F}^{0,0}) = \mathbf{O}_{\mathbb{V}\times\mathbb{V}^*[2]}$, it follows that

$$\mathrm{RHom}(\mathcal{F}^{0,0}, \mathcal{F}) = \Phi(\mathcal{F})^{\mathrm{GL}(\mathbb{V})}.$$

To show that the RHS of the above equation has finite-dimensional cohomology (assuming that $\Phi(\mathcal{F})$ is supported inside $\mathcal{Z}_{\mathbb{V}}$), it is enough to show $\mathbf{O}_{\mathcal{Z}_{\mathbb{V}}}^{\mathrm{GL}(\mathbb{V})}$ is finite-dimensional (since $\Phi(\mathcal{F})$ is a finite extension of quotients of $\mathbf{O}_{\mathcal{Z}_{\mathbb{V}}}$). This immediately follows from the fact that $\mathbf{O}_{\mathbb{V}\times\mathbb{V}^*}^{\mathrm{GL}(\mathbb{V})} = \mathbb{C}[v^*(v)]$ which is obvious (here we regard $v^*(v)$ as a function $\mathbb{V} \times \mathbb{V}^* \to \mathbb{C}$).

Acknowledgments This paper resulted from numerous conversations of the first-named author with D. Gaitsgory and S. Raskin which took place during the workshop "Vertex algebras, factorization algebras and applications" at IPMU in July 2018. The authors thank both D. Gaitsgory and S. Raskin for their patient explanations and the organizers of the workshop for hospitality and for providing this opportunity. We would also like to thank Roman Bezrukavnikov for help with some technical details of the proof. M.F. was partially funded within the framework of the HSE University Basic Research Program and the Russian Academic Excellence Project '5-100'.

References

[AG15] D. Arinkin and D. Gaitsgory, *Singular support of coherent sheaves, and the geometric Langlands conjecture*, Selecta Math. (N.S.) **21** (2015), no. 1, 1–199.

[BF08] R. Bezrukavnikov, M. Finkelberg, *Equivariant Satake category and Kostant-Whittaker reduction*, Moscow Math. J. **8** (2008), no. 1, 39–72.

[BF19] A. Braverman and M. Finkelberg, *Coulomb branches of 3-dimensional gauge theories and related structures*, Lecture Notes in Mathematics **2248** (2019), 1–52.

[DG13] V. Drinfeld and D. Gaitsgory, *On some finiteness questions for algebraic stacks*, Geom. Funct. Anal. **23** (2013), no. 1, 149–294.

[Gai17] D. Gaitsgory, *Progrès récents dans la théorie de Langlands géométrique*, Séminaire Bourbaki, Astérisque **390** (2017), exp. no. 1109, 139–168; arXiv:1606.09462.

[Gin91] V. Ginzburg, *Perverse sheaves and $\mathbb{C}^*$-actions*, J. Amer. Math. Soc. **4** no. 3 (1991), 483–490.

[Lau] G. Laumon, *Transformation de Fourier généralisée*, alg-geom/9603004.

The Semi-infinite Intersection Cohomology Sheaf-II: The Ran Space Version

Dennis Gaitsgory

To Sasha Beilinson and Vitya Ginzburg

Contents

D. Gaitsgory (✉)
Harvard University, Department of Mathematics, Cambridge, MA, USA
e-mail: gaitsgde@math.edu

V. Baranovsky et al. (eds.), *Representation Theory and Algebraic Geometry*, Trends in Mathematics, https://doi.org/10.1007/978-3-030-82007-7_6

1 Introduction

1.1 What Are Trying to Do?

1.1.1

This paper is a sequel of [Ga1]. In *loc. cit.*, an attempt was made to construct a certain object, denoted $\mathrm{IC}^{\frac{\infty}{2}}$, in the (derived) category $\mathrm{Shv}(\mathrm{Gr}_G)$ of sheaves on the affine Grassmannian, whose existence had been predicted by G. Lusztig.

Notionally, $\mathrm{IC}^{\frac{\infty}{2}}$ is supposed to be the intersection cohomology complex on the closure $\overline{S}^0$ of the unit $N(\!(t)\!)$-orbit $S^0 \subset \mathrm{Gr}_G$. Its stalks are supposed to be given by periodic Kazhdan–Lusztig polynomials. Ideally, one would want the construction of $\mathrm{IC}^{\frac{\infty}{2}}$ to have the following properties:

- It should be local, i.e., only depend on the formal disc, where we are thinking of Gr_G as $G(\!(t)\!)/G[\![t]\!]$.

- When our formal disc is the formal neighborhood of a point x in a global curve X, then $\mathrm{IC}^{\frac{\infty}{2}}$ should be the pullback along the map $\overline{S}^0 \to \overline{\mathrm{Bun}}_N$ of the intersection cohomology sheaf of $\overline{\mathrm{Bun}}_N$, where the latter is Drinfeld's relative compactification of the stack of G-bundles equipped with a reduction to N (which is an algebraic stack locally of finite type, so $\mathrm{IC}_{\overline{\mathrm{Bun}}_N}$ is well defined).

The construction in [Ga1] indeed produced such an object, but with the following substantial drawback: in *loc. cit.*, $\mathrm{IC}^{\frac{\infty}{2}}$ was given by an *ad hoc* procedure; namely, it was written as a certain explicit direct limit. In particular, $\mathrm{IC}^{\frac{\infty}{2}}$ was *not* the middle extension of the constant[1] sheaf on S^0 with respect to the natural t-structure on $\mathrm{Shv}(\mathrm{Gr}_G)$ (however, $\mathrm{IC}^{\frac{\infty}{2}}$ does belong to the heart of this t-structure).

1.1.2

In this paper, we will construct a variant of $\mathrm{IC}^{\frac{\infty}{2}}$, denoted $\mathrm{IC}^{\frac{\infty}{2}}_{\mathrm{Ran}}$, closely related to $\mathrm{IC}^{\frac{\infty}{2}}$, which is actually given by the procedure of middle extension in a certain t-structure.

Namely, instead of the single copy of the affine Grassmannian Gr_G, we will consider its Ran space version, denoted $\mathrm{Gr}_{G,\mathrm{Ran}}$. We will equip the corresponding category $\mathrm{Shv}(\mathrm{Gr}_{G,\mathrm{Ran}})$ with a t-structure, and we will define $\mathrm{IC}^{\frac{\infty}{2}}_{\mathrm{Ran}}$ as the middle extension of the dualizing sheaf on $S^0_{\mathrm{Ran}} \subset \mathrm{Gr}_{G,\mathrm{Ran}}$.

Remark 1.1.3 Technically, the Ran space is attached to a smooth (but not necessarily complete) curve X, and one may think that this somewhat compromises the locality property of the construction of $\mathrm{IC}^{\frac{\infty}{2}}_{\mathrm{Ran}}$. However, if one day a formalism becomes available for working with the Ran space of a formal disc, the construction of $\mathrm{IC}^{\frac{\infty}{2}}_{\mathrm{Ran}}$ will become purely local.

1.1.4

For a fixed point $x \in X$, we have the embedding

$$\mathrm{Gr}_G \simeq \{x\} \underset{\mathrm{Ran}}{\times} \mathrm{Gr}_{G,\mathrm{Ran}} \hookrightarrow \mathrm{Gr}_{G,\mathrm{Ran}},$$

and we will show that the restriction of $\mathrm{IC}^{\frac{\infty}{2}}_{\mathrm{Ran}}$ along this map recovers $\mathrm{IC}^{\frac{\infty}{2}}$ from [Ga1].

[1] Technically, not constant but rather dualizing.

1.1.5

Our $\mathrm{IC}^{\frac{\infty}{2}}_{\mathrm{Ran}}$ retains the relation to $\mathrm{IC}_{\overline{\mathrm{Bun}}_N}$. Namely, we have a natural map

$$\overline{S}^0_{\mathrm{Ran}} \to \overline{\mathrm{Bun}}_N$$

and we will prove that $\mathrm{IC}^{\frac{\infty}{2}}_{\mathrm{Ran}}$ identifies with the pullback of $\mathrm{IC}_{\overline{\mathrm{Bun}}_N}$ along this map.

In particular, this implies the isomorphism

$$\mathrm{IC}^{\frac{\infty}{2}} \simeq \mathrm{IC}_{\overline{\mathrm{Bun}}_N}|_{\overline{S}^0},$$

which had been established in [Ga1] by a different method.

1.1.6

To summarize, we can say that we still do not know how to intrinsically characterize $\mathrm{IC}^{\frac{\infty}{2}}$ on an individual Gr_G as an intersection cohomology sheaf, but we can do it, once we allow the point of the curve to move along the Ran space.

But *ce n'est pas grave*: as was argued in [Ga1, Sect. 0.4], our $\mathrm{IC}^{\frac{\infty}{2}}_{\mathrm{Ran}}$, equipped with its factorization structure, is perhaps a more fundamental object than the original $\mathrm{IC}^{\frac{\infty}{2}}$.

1.2 What Is Done in This Paper?

The main constructions and results of this paper can be summarized as follows:

1.2.1

We define the *semi-infinite* category on the Ran version of the affine Grassmannian, denoted $\mathrm{SI}_{\mathrm{Ran}}$, and equip it with a t-structure. This is largely parallel to [Ga1].

We define $\mathrm{IC}^{\frac{\infty}{2}}_{\mathrm{Ran}} \in \mathrm{SI}_{\mathrm{Ran}}$ as the middle extension of the dualizing sheaf on the stratum $S^0_{\mathrm{Ran}} \subset \mathrm{Gr}_{G,\mathrm{Ran}}$. (We will also show that the corresponding !- and *-extensions both belong to the heart of the t-structure, see Proposition 3.2.2; this contrasts with the situation for $\mathrm{IC}^{\frac{\infty}{2}}$, see [Ga1, Theorem 1.5.5]).

We describe explicitly the !- and *-restrictions of $\mathrm{IC}^{\frac{\infty}{2}}_{\mathrm{Ran}}$ to the strata $S^\lambda_{\mathrm{Ran}} \subset \overline{S}^0_{\mathrm{Ran}} \subset \mathrm{Gr}_{G,\mathrm{Ran}}$ (here λ is an element of Λ^{neg}, the negative span of positive simple coroots), see Theorem 3.4.5. These descriptions are given in terms of the combinatorics of the Langlands dual Lie algebra: more precisely, in terms of the *factorization algebras* attached to $\mathcal{O}(\check{N})$ and $U(\check{\mathfrak{n}}^-)$.

We give an explicit presentation of $\text{IC}^{\frac{\infty}{2}}_{\text{Ran}}$ as a colimit (parallel to the *definition* of $\text{IC}^{\frac{\infty}{2}}$ in [Ga1]), see Theorem 3.7.2. This implies the identification $\text{IC}^{\frac{\infty}{2}}_{\text{Ran}}|_{\text{Gr}_G} \simeq \text{IC}^{\frac{\infty}{2}}$, where $\text{IC}^{\frac{\infty}{2}} \in \text{Shv}(\text{Gr}_G)$ is the object constructed in [Ga1].

1.2.2

We show that $\text{IC}^{\frac{\infty}{2}}_{\text{Ran}}$ identifies canonically (up to a cohomological shift by $[d]$, $d = \dim(\text{Bun}_N)$) with the pullback of $\text{IC}_{\overline{\text{Bun}}_N}$ along the map

$$\overline{S}^0 \to \overline{\text{Bun}}_N, \tag{1.1}$$

see Theorem 4.3.3.

In fact, we show that the above pullback functor is t-exact (up to the shift by $[d]$), when restricted to the subcategory $\text{SI}^{\leq 0}_{\text{glob}} \subset \text{Shv}(\overline{\text{Bun}}_N)$ that consists of objects equivariant with respect to the action of the adelic N, see Corollary 4.6.7.

The proof of this t-exactness property is based on applying Braden's theorem to $\text{Gr}_{G,\text{Ran}}$ and the *Zastava space*.

We note that, unlike [Ga1], the resulting proof of the isomorphism

$$\text{IC}_{\overline{\text{Bun}}_N}|_{\overline{S}^0_{\text{Ran}}}[d] \simeq \text{IC}^{\frac{\infty}{2}}_{\text{Ran}} \tag{1.2}$$

does not use the computation of $\text{IC}_{\overline{\text{Bun}}_N}$ from [BFGM], but rather reproves it.

As an aside we prove an important geometric fact that the map (1.1) is *universally homologically contractible* (=the pullback functor along any base change of this map is fully faithful), see Theorem 4.4.4.

1.2.3

We show that $\text{IC}^{\frac{\infty}{2}}$ has a *unitality* property: it stays invariant under the operation of "throwing in" more points in Ran without altering the G-bundle.

We use the unitality property of $\text{IC}^{\frac{\infty}{2}}$ to equip it with a *factorization structure*.

1.2.4

We show that $\text{IC}^{\frac{\infty}{2}}_{\text{Ran}}$ has an eigen-property with respect to the action of the Hecke functors for G and T, see Theorem 6.5.7.

In the course of the proof of this theorem, we give yet another characterization of $\text{IC}^{\frac{\infty}{2}}_{\text{Ran}}$ (which works for $\text{IC}^{\frac{\infty}{2}}$ as well):

We show that the δ-function $\delta_{1_{\mathrm{Gr,Ran}}} \in \mathrm{Shv}(\mathrm{Gr}_{G,\mathrm{Ran}})$ on the unit section $\mathrm{Ran} \to \mathrm{Gr}_{G,\mathrm{Ran}}$ possesses a natural *Drinfeld–Plucker* structure with respect to the Hecke actions of G and T (see Sect. 6.4 for what this means), and that $\mathrm{IC}^{\frac{\infty}{2}}_{\mathrm{Ran}}$ can be obtained from $\delta_{1_{\mathrm{Gr,Ran}}}$ by applying the functor from the Drinfeld–Plücker category to the graded Hecke category, left adjoint to the tautological forgetful functor (see Sect. 6.5).

Finally, we establish the compatibility of the isomorphism (1.2) with the Hecke eigen-structures on $\mathrm{IC}^{\frac{\infty}{2}}_{\mathrm{Ran}}$ and $\mathrm{IC}_{\overline{\mathrm{Bun}}_N}$, respectively (see Theorem 7.3.5).

1.3 Organization

1.3.1

In Sect. 2, we recall the definition of the Ran space Ran, the Ran version of the affine Grassmannian $\mathrm{Gr}_{G,\mathrm{Ran}}$, and the stratification of the closure $\overline{S}^0_{\mathrm{Ran}}$ of the adelic N-orbit S^0_{Ran} by locally closed substacks S^λ_{Ran}.

We define the semi-infinite category $\mathrm{SI}_{\mathrm{Ran}}$ and study the standard functors that link it to the corresponding categories on the strata.

1.3.2

In Sect. 3, we define the t-structure on $\mathrm{SI}^{\leq 0}_{\mathrm{Ran}}$ and our main object of study, $\mathrm{IC}^{\frac{\infty}{2}}_{\mathrm{Ran}}$.

We state Theorem 3.4.5 that describes the *- and !- restrictions of $\mathrm{IC}^{\frac{\infty}{2}}_{\mathrm{Ran}}$ to the strata S^λ_{Ran}. The proof of the statement concerning *-restrictions will be given in this same section (it will result from Theorem 3.7.2 mentioned below). The proof of the statement concerning !-restrictions will be given in Sect. 4.

We state and prove Theorem 3.7.2 that gives a presentation of $\mathrm{IC}^{\frac{\infty}{2}}_{\mathrm{Ran}}$ as a colimit.

1.3.3

In Sect. 4, we recall the definition of Drinfeld's relative compactification $\overline{\mathrm{Bun}}_N$.

We define the global semi-infinite category $\mathrm{SI}^{\leq 0}_{\mathrm{glob}} \subset \mathrm{Shv}(\overline{\mathrm{Bun}}_N)$. We prove that the pullback functor along (1.1), viewed as a functor

$$\mathrm{SI}^{\leq 0}_{\mathrm{glob}} \to \mathrm{SI}^{\leq 0}_{\mathrm{Ran}},$$

is t-exact (up to the shift by $[d]$). From here we deduce the identification (1.2), which is Theorem 4.3.3.

We also state Theorem 4.4.4, whose proof is given in Appendix 7.4.4.

1.3.4

In Sect. 5, we introduce the notion of *unital* subcategory inside $\text{Shv}(\text{Gr}_{G,\text{Ran}})$, $\text{Shv}(\overline{S}^0_{\text{Ran}})$, and $\text{SI}^{\leq 0}_{\text{Ran}}$, and we show that $\text{IC}^{\frac{\infty}{2}}$ belongs to $\text{SI}^{\leq 0}_{\text{Ran}}$.

We use this property of $\text{IC}^{\frac{\infty}{2}}$ to equip it with a factorization structure.

1.3.5

In Sect. 6, we establish the Hecke eigen-property of $\text{IC}^{\frac{\infty}{2}}_{\text{Ran}}$. In the process of doing so, we discuss the formalism of *lax central objects* and *Drinfeld–Plücker* structures, and their relation to the Hecke eigen-structures.

In Sect. 7, we prove the compatibility between the eigen-property of $\text{IC}^{\frac{\infty}{2}}_{\text{Ran}}$ and that of $\text{IC}_{\overline{\text{Bun}}_N}$.

1.4 Background, Conventions, and Notation

The notations and conventions in this follow closely those of [Ga1]. Here is a summary:

1.4.1

This paper uses higher category theory. It appears already in the definition of our basic object of study, the *category of sheaves* on the Ran Grassmannian, $\text{Gr}_{G,\text{Ran}}$.

Thus, we will assume that the reader is familiar with the basics of higher categories and higher algebra. The fundamental reference is [Lu1, Lu2], but shorter expositions (or user guides) exist as well, for example, the first chapter of [GR].

1.4.2

Our algebraic geometry happens over an arbitrary algebraically closed ground field k. Our geometric objects are classical (i.e., this paper *does not* need derived algebraic geometry).

We let $\text{Sch}^{\text{aff}}_{\text{ft}}$ denote the category of (classical) affine schemes of finite type over k.

By a prestack (locally of finite type), we mean an arbitrary functor

$$(\text{Sch}^{\text{aff}}_{\text{ft}})^{\text{op}} \to \text{Groupoids} \tag{1.3}$$

(we do not need to consider higher groupoids).

We let $\text{PreSk}_{\text{lft}}$ denote the category of such prestacks. It contains $\text{Sch}^{\text{aff}}_{\text{ft}}$ via the Yoneda embedding. All other types of geometric objects (schemes, algebraic stacks, ind-schemes) are prestacks with some specific *properties* (but *not additional pieces of structure*).

1.4.3

We let G be a connected reductive group over k. We fix a Borel subgroup $B \subset G$ and the opposite Borel $B^- \subset G$. Let $N \subset B$ and $N^- \subset B^-$ denote their respective unipotent radicals.

Set $T = B \cap B^-$; this is a Cartan subgroup of G. We use it to identify the quotients

$$B/N \simeq T \simeq B^-/N^-.$$

We let Λ denote the coweight lattice of G, i.e., the lattice of cocharacters of T. We let $\Lambda^{\text{pos}} \subset \Lambda$ denote the sub-monoid consisting of linear combinations of positive simple roots with non-negative integral coefficients. We let $\Lambda^+ \subset \Lambda$ denote the sub-monoid of *dominant coweights*.

1.4.4

While our geometry happens over a field k, the representation-theoretic categories that we study are *DG categories* over another field, denoted e (assumed algebraically closed and of characteristic 0). For a crash course on DG categories, the reader is referred to [GR, Chapter 1, Sect. 10].

All our DG categories are assumed presentable. When considering functors, we will only consider functors that preserve colimits. We denote the ∞-category of DG categories by DGCat. It carries a symmetric monoidal structure (i.e., one can consider tensor products of DG categories). The unit object is the DG category of complexes of e-vector spaces, denoted Vect.

We will use the notion of t-structure on a DG category. Given a t-structure on $\mathcal{C}$, we will denote by $\mathcal{C}^{\leq 0}$ the corresponding subcategory of cohomologically connective objects, and by $\mathcal{C}^{>0}$ its right orthogonal. We let $\mathcal{C}^{\heartsuit}$ denote the heart $\mathcal{C}^{\leq 0} \cap \mathcal{C}^{\geq 0}$.

1.4.5

The source of DG categories will be a *sheaf theory*, which is a functor

$$\text{Shv} : (\text{Sch}^{\text{aff}}_{\text{ft}})^{\text{op}} \to \text{DGCat}, \quad Y \mapsto \text{Shv}(Y).$$

For a morphism of affine schemes $f : Y_0 \to Y_1$, the corresponding functor

$$\mathrm{Shv}(Y_1) \to \mathrm{Shv}(Y_0)$$

is the !-pullback $f^!$.

We will work with the following particular examples sheaf theories:

(i) We take $\mathsf{e} = \overline{\mathbb{Q}}_\ell$, and we take $\mathrm{Shv}(Y)$ to be the ind-completion of the (small) DG category of constructible $\overline{\mathbb{Q}}_\ell$-sheaves.
(ii) When $k = \mathbb{C}$ and e arbitrary, we take $\mathrm{Shv}(Y)$ to be the ind-completion of the (small) DG category of constructible e-sheaves on $Y(\mathbb{C})$ in the analytic topology.
(iii) If k has characteristic 0, we take $\mathsf{e} = k$ and we take $\mathrm{Shv}(Y)$ to be the DG category of holonomic D-modules on S.
(iv) If k has characteristic 0, we take $\mathsf{e} = k$ and we take $\mathrm{Shv}(Y)$ to be the DG category of D-modules on Y.

We will refer to examples (i), (ii), and (iii) as *a constructible* sheaf theories.

In the constructible case, the functor $f^!$ always has a left adjoint, denoted $f_!$. In example (iv), this is not the case. However, the partially defined left adjoint $f_!$ is defined on holonomic objects. It is defined on the entire category if f is proper.

1.4.6 Sheaves on Prestacks

We apply the procedure of right Kan extension along the embedding

$$(\mathrm{Sch}^{\mathrm{aff}}_{\mathrm{ft}})^{\mathrm{op}} \hookrightarrow (\mathrm{PreStk}_{\mathrm{lft}})^{\mathrm{op}}$$

to the functor Shv and thus obtain a functor (denoted by the same symbol)

$$\mathrm{Shv} : (\mathrm{PreStk}_{\mathrm{lft}})^{\mathrm{op}} \to \mathrm{DGCat}\,.$$

By definition, for $\mathcal{Y} \in \mathrm{PreStk}_{\mathrm{lft}}$, we have

$$\mathrm{Shv}(\mathcal{Y}) = \lim_{S \in \mathrm{Sch}^{\mathrm{aff}}_{\mathrm{ft}},\, y:S \to \mathcal{Y}} \mathrm{Shv}(S), \tag{1.4}$$

where the transition functors in the formation of the limit are the !-pullbacks.[2]

For a map of prestacks $f : \mathcal{Y}_0 \to \mathcal{Y}_1$, we thus have a well-defined pullback functor

[2]Note that even though the index category (i.e., $(\mathrm{Sch}^{\mathrm{aff}}_{\mathrm{ft}})_{/\mathcal{Y}}$) is ordinary, the above limit is formed in the ∞-category DGCat. This is how ∞-categories appear in this paper.

$$f^! : \mathrm{Shv}(\mathcal{Y}_1) \to \mathrm{Shv}(\mathcal{Y}_0).$$

We denote by $\omega_{\mathcal{Y}}$ the dualizing sheaf on $\mathcal{Y}$, i.e., the pullback of

$$\mathsf{e} \in \mathrm{Vect} \simeq \mathrm{Shv}(\mathrm{pt})$$

along the tautological map $\mathcal{Y} \to \mathrm{pt}$.

1.4.7

We let X be a smooth, connected (but not necessarily proper) curve over k. Whenever we need X to be proper, we will explicitly say so.

1.4.8

This paper is closely related to the geometric Langlands theory, and the geometry of the Langlands dual group $\check{G}$ makes it appearance.

By definition, $\check{G}$ is a reductive group over e and geometric objects constructed out of $\check{G}$ give rise to e-linear DG categories by considering quasi-coherent (resp., ind-coherent) sheaves on them.

The most basic example of such a category is

$$\mathrm{QCoh}(\mathrm{pt}/\check{G}) =: \mathrm{Rep}(\check{G}).$$

2 The Ran Version of the Semi-infinite Category

In this section, we extend the definition of the semi-infinite category given in [Ga1] from the affine Grassmannian $\mathrm{Gr}_{G,x}$ corresponding to a fixed point $x \in X$ to the Ran version, denoted $\mathrm{Gr}_{G,\mathrm{Ran}}$.

2.1 The Ran Grassmannian

2.1.1

We recall that the Ran space of X, denoted Ran, is the prestack that assigns to an affine test scheme Y the set of finite non-empty subsets

$$\mathcal{I} \subset \mathrm{Hom}(Y, X).$$

One can explicitly exhibit Ran as a colimit (in PreStk) of schemes:

$$\mathrm{Ran} \simeq \underset{I}{\mathrm{colim}}\, X^I,$$

where the colimit is taken over the category opposite to the category $\mathrm{Fin}^{\mathrm{surj}}$ of finite non-empty sets and surjective maps, where to a map $\phi : I \twoheadrightarrow J$ we assign the corresponding diagonal embedding

$$X^J \overset{\Delta_\phi}{\hookrightarrow} X^I.$$

This description implies, in particular, that if X is proper, then Ran is pseudo-proper as a prestack (see Sect. A.2.4 for what it means).

Another key feature of Ran is that it is homologically contractible (see Sect. A.1.8 for what this means).

2.1.2

We will consider the Ran version of the affine Grassmannian, denoted $\mathrm{Gr}_{G,\mathrm{Ran}}$, defined as follows.

It assigns to an affine test scheme Y, the set of triples $(\mathcal{I}, \mathcal{P}_G, \alpha)$, where $\mathcal{I}$ is a Y-point of Ran, $\mathcal{P}_G$ is a G-bundle on $Y \times X$, and α is a trivialization of $\mathcal{P}_G$ on the open subset of $Y \times X$ equal to the complement of the union $\Gamma_\mathcal{I}$ of the graphs of the maps $Y \to X$ that comprise $\mathcal{I}$.

The projection $\mathrm{Gr}_{G,\mathrm{Ran}} \to \mathrm{Ran}$ is pseudo-proper.

We will also consider the Ran versions of the loop and arc groups (ind)-schemes, denoted

$$\mathfrak{L}^+(G)_{\mathrm{Ran}} \subset \mathfrak{L}(G)_{\mathrm{Ran}}.$$

The Ran Grassmannian $\mathrm{Gr}_{G,\mathrm{Ran}}$ identifies with the étale (equivalently, fppf) sheafification of the prestack quotient $\mathfrak{L}(G)_{\mathrm{Ran}}/\mathfrak{L}^+(G)_{\mathrm{Ran}}$.

2.1.3

For a fixed finite non-empty set I, we denote

$$\mathrm{Gr}_{G,I} := X^I \underset{\mathrm{Ran}}{\times} \mathrm{Gr}_{G,\mathrm{Ran}},\ \mathfrak{L}(G)_I := X^I \underset{\mathrm{Ran}}{\times} \mathfrak{L}(G)_{\mathrm{Ran}},\ \mathfrak{L}^+(G)_I := X^I \underset{\mathrm{Ran}}{\times} \mathfrak{L}^+(G)_{\mathrm{Ran}}.$$

For a map of finite sets $\phi : I \twoheadrightarrow J$, we will denote by Δ_ϕ the corresponding map $\mathrm{Gr}_{G,J} \to \mathrm{Gr}_{G,I}$, so that we have the Cartesian square:

$$\begin{array}{ccc} \mathrm{Gr}_{G,J} & \xrightarrow{\Delta_\phi} & \mathrm{Gr}_{G,I} \\ \downarrow & & \downarrow \\ X^J & \xrightarrow{\Delta_\phi} & X^I. \end{array}$$

2.1.4

We introduce also the following closed (resp., locally closed) subfunctors

$$S^0_{\mathrm{Ran}} \subset \overline{S}^0_{\mathrm{Ran}} \subset \mathrm{Gr}_{G,\mathrm{Ran}}.$$

Namely, for an affine test scheme Y, a Y-point $(\mathcal{I}, \mathcal{P}_G, \alpha)$ belongs to $\overline{S}^0_{\mathrm{Ran}}$ if for every dominant weight $\check{\lambda}$, the composite meromorphic map of vector bundles on $Y \times X$

$$\mathcal{O} \to \mathcal{V}^{\check{\lambda}}_{\mathcal{P}^0_G} \xrightarrow{\alpha} \mathcal{V}^{\check{\lambda}}_{\mathcal{P}_G} \tag{2.1}$$

is regular. In the above formula, the notations are as follows:

- $\mathcal{V}^{\check{\lambda}}$ denotes the Weyl module over G with highest weight $\check{\lambda}$.
- $\mathcal{V}^{\check{\lambda}}_{\mathcal{P}_G}$ (resp., $\mathcal{V}^{\check{\lambda}}_{\mathcal{P}^0_G}$) denotes the vector bundles associated with $\mathcal{V}^{\check{\lambda}}$ and the G-bundle $\mathcal{P}_G$ (resp., the trivial G-bundle $\mathcal{P}^0_G$).
- $\mathcal{O} \to \mathcal{V}^{\check{\lambda}}_{\mathcal{P}^0_G}$ is the map coming from the highest weight vector in $\mathcal{V}^{\check{\lambda}}$.

A point as above belongs to S^0_{Ran} if the above composite map is an injective bundle map (i.e., the cokernel is flat over $Y \times X$).

2.2 *The Semi-infinite Category*

2.2.1

Since $\mathrm{Gr}_{G,\mathrm{Ran}}$ a prestack locally of finite type, we have a well-defined category

$$\mathrm{Shv}(\mathrm{Gr}_{G,\mathrm{Ran}}).$$

We have

$$\mathrm{Shv}(\mathrm{Gr}_{G,\mathrm{Ran}}) := \lim_{I} \mathrm{Shv}(\mathrm{Gr}_{G,I}),$$

where the limit is formed using the !-pullback functors.

2.2.2

Although the group ind-scheme $\mathfrak{L}(N)_{\text{Ran}}$ is *not* locally of finite type, we have a well-defined full subcategory

$$\text{SI}_{\text{Ran}} := \text{Shv}(\text{Gr}_{G,\text{Ran}})^{\mathfrak{L}(N)_{\text{Ran}}} \subset \text{Shv}(\text{Gr}_{G,\text{Ran}}).$$

Namely, for every fixed finite non-empty set I, we consider the full subcategory

$$\text{SI}_I := \text{Shv}(\text{Gr}_{G,I})^{\mathfrak{L}(N)_I} \subset \text{Shv}(\text{Gr}_{G,I}),$$

defined as in [Ga1, Sect. 1.2].

We say that the object of $\text{Shv}(\text{Gr}_{G,\text{Ran}})$ belongs to $\text{Shv}(\text{Gr}_{G,\text{Ran}})^{\mathfrak{L}(N)_{\text{Ran}}}$ if its restriction to any $\text{Gr}_{G,I}$ yields an object of $\text{Shv}(\text{Gr}_{G,I})^{\mathfrak{L}(N)_I}$. By construction, we have an equivalence

$$\text{SI}_{\text{Ran}} := \lim_I \text{SI}_I .$$

2.2.3

Let $\text{SI}^{\leq 0}_{\text{Ran}} \subset \text{SI}_{\text{Ran}}$ be the full subcategory consisting of objects supported on $\overline{S}^0_{\text{Ran}}$, i.e.,

$$\text{SI}^{\leq 0}_{\text{Ran}} = \text{SI}_{\text{Ran}} \cap \text{Shv}(\overline{S}^0_{\text{Ran}}),$$

while

$$\text{Shv}(\overline{S}^0_{\text{Ran}}) \simeq \lim_I \text{Shv}(\overline{S}^0_I).$$

2.3 *Stratification*

In order to study the structure of $\text{SI}^{\leq 0}_{\text{Ran}}$, we will now describe a certain natural stratification of $\overline{S}^0_{\text{Ran}}$, whose open stratum will be S^0_{Ran}.

2.3.1

For $\lambda \in \Lambda^{\text{neg}}$, let X^λ denote the corresponding partially symmetrized power of X. That is, if

$$\lambda = \underset{i}{\Sigma}\,(-n_i)\cdot\alpha_i, \quad n_i \geq 0$$

where α_i are simple positive coroots, then

$$X^\lambda = \underset{i}{\Pi}\, X^{(n_i)}.$$

In other words, Y-points of X^λ are effective Λ^{neg}-valued divisors on X.
For $\lambda = 0$, we by definition have $X^0 = \text{pt}$.

2.3.2

Let

$$(X^\lambda \times \text{Ran})^{\subset} \subset \text{Ran} \times X^\lambda$$

be the ind-closed subfunctor, whose S-points are pairs $(\mathcal{I}, D)$ for which the support of the divisor D is *set-theoretically* supported on the union of the graphs of the maps $S \to X$ that comprise $\mathcal{I}$.

In other words,

$$(X^\lambda \times \text{Ran})^{\subset} = \underset{I}{\text{colim}}\,(X^\lambda \times X^I)^{\subset},$$

where

$$(X^\lambda \times X^I)^{\subset} \subset X^I \times X^\lambda$$

is the formal completion of the corresponding incidence subvariety.

For future use, we note:

Lemma 2.3.3 *The map*

$$\text{pr}^\lambda_{\text{Ran}} : (X^\lambda \times \text{Ran})^{\subset} \to X^\lambda$$

is universally homologically contractible.[3]

The proof in the case when X is proper will be given in Sect. A.2.8. For the proof in the general case, see Remark 5.1.3.

Corollary 2.3.4 *The pullback functor*

$$(\text{pr}^\lambda_{\text{Ran}})^! : \text{Shv}(X^\lambda) \to \text{Shv}((X^\lambda \times \text{Ran})^{\subset})$$

is fully faithful.

[3]The notion of universal local contractibility is recalled in Sect. A.1.8.

2.3.5

We let

$$\overline{S}^{\lambda}_{\text{Ran}} \subset (X^{\lambda} \times \text{Ran})^{\subset} \underset{\text{Ran}}{\times} \text{Gr}_{G,\text{Ran}}$$

be the closed subfunctor defined by the following condition:

An S-point $(\mathcal{I}, D, \mathcal{P}_G, \alpha)$ of the fiber product $(X^{\lambda} \times \text{Ran})^{\subset} \underset{\text{Ran}}{\times} \text{Gr}_{G,\text{Ran}}$ belongs to $\overline{S}^{\lambda}_{\text{Ran}}$ if for every dominant weight $\check{\lambda}$ the map (2.1) extends to a regular map

$$\mathcal{O}(-\check{\lambda}(D)) \to \mathcal{V}^{\check{\lambda}}_{\mathcal{P}_G}. \tag{2.2}$$

We denote by $\bar{\mathbf{i}}^{\lambda}$ the composite map

$$\overline{S}^{\lambda}_{\text{Ran}} \to (X^{\lambda} \times \text{Ran})^{\subset} \underset{\text{Ran}}{\times} \text{Gr}_{G,\text{Ran}} \to \text{Gr}_{G,\text{Ran}}.$$

This map is proper, and its image is contained in $\overline{S}^{0}_{\text{Ran}}$.

Note that for $\lambda = 0$, the map $\bar{\mathbf{i}}^{0}$ is the identity map on $\overline{S}^{0}_{\text{Ran}}$.

Let $\overline{p}^{\lambda}_{\text{Ran}}$ denote the projection

$$\overline{S}^{\lambda}_{\text{Ran}} \to (X^{\lambda} \times \text{Ran})^{\subset}.$$

2.3.6

We define the open subfunctor

$$S^{\lambda}_{\text{Ran}} \subset \overline{S}^{\lambda}_{\text{Ran}}$$

to correspond to those quadruples $(\mathcal{I}, D, \mathcal{P}_G, \alpha)$ for which the map (2.2) is an injective bundle map (i.e., the cokernel is flat over $Y \times X$).

The projection

$$p^{\lambda}_{\text{Ran}} := \overline{p}^{\lambda}_{\text{Ran}}|_{S^{\lambda}_{\text{Ran}}} : S^{\lambda}_{\text{Ran}} \to (X^{\lambda} \times \text{Ran})^{\subset} \tag{2.3}$$

admits a canonically defined section

$$s^{\lambda}_{\text{Ran}} : (X^{\lambda} \times \text{Ran})^{\subset} \to S^{\lambda}_{\text{Ran}}. \tag{2.4}$$

Namely, it sends $(\mathcal{I}, D)$ to the quadruple $(\mathcal{I}, D, \mathcal{P}_G, \alpha)$, where $\mathcal{P}_G$ is the G-bundle induced from the T-bundle $\mathcal{P}_T := \mathcal{P}^0_T(D)$, and α is the trivialization of $\mathcal{P}_G$ induced by the tautological trivialization of $\mathcal{P}_T$ away from the support of D.

2.3.7

We let

$$\mathbf{j}^\lambda : S^\lambda_{\mathrm{Ran}} \hookrightarrow \overline{S}^\lambda_{\mathrm{Ran}}, \quad \mathbf{i}^\lambda = \overline{\mathbf{i}}^\lambda \circ \mathbf{j}^\lambda : S^\lambda_{\mathrm{Ran}} \to \mathrm{Gr}_{G,\mathrm{Ran}}$$

denote the resulting maps.

For a fixed finite non-empty set I, we obtain the corresponding subfunctors

$$\overline{S}^\lambda_I \subset (X^\lambda \times X^I)^{\subset} \underset{X^I}{\times} \mathrm{Gr}_{G,I}$$

and

$$S^\lambda_I \subset \overline{S}^\lambda_I,$$

and maps, denoted by the same symbols $\mathbf{j}^\lambda$, $\overline{\mathbf{i}}^\lambda$, $\mathbf{i}^\lambda$. Let p^λ_I (resp., $\overline{p}^\lambda_I$) denote the resulting map from S^λ_I (resp., $\overline{S}^\lambda_I$) to $(X^\lambda \times X^I)^{\subset}$.

Let s^λ_I denote the resulting section

$$s^\lambda_I : (X^\lambda \times X^I)^{\subset} \to S^\lambda_I.$$

2.3.8

The following results easily from the definitions:

Lemma 2.3.9 *The maps*

$$\mathbf{i}^\lambda : S^\lambda_{\mathrm{Ran}} \to \overline{S}^0_{\mathrm{Ran}} \text{ and } S^\lambda_I \to \overline{S}^0_I$$

are locally closed embeddings. Every field-valued point of $\overline{S}^0_{\mathrm{Ran}}$ *(resp.,* $\overline{S}^0_I$*) belongs to the image of exactly one such map.*

2.3.10

In what follows, we will denote by

$$\mathrm{SI}^{\leq\lambda}_{\mathrm{Ran}} \subset \mathrm{Shv}(\overline{S}^\lambda_{\mathrm{Ran}}) \text{ and } \mathrm{SI}^{=\lambda}_{\mathrm{Ran}} \subset \mathrm{Shv}(S^\lambda_{\mathrm{Ran}}) \tag{2.5}$$

and also

$$\mathrm{SI}^{\leq\lambda}_I \subset \mathrm{Shv}(\overline{S}^\lambda_I) \text{ and } \mathrm{SI}^{=\lambda}_I \subset \mathrm{Shv}(S^\lambda_I),$$

the corresponding full subcategories.

2.4 *The Category on a Single Stratum*

2.4.1

We have the following explicit description of the category on each stratum separately:

Proposition 2.4.2 *Pullback along the map* $p^{\lambda}_{\mathrm{Ran}}$ *of* (2.3) *defines an equivalence*

$$\mathrm{Shv}((X^{\lambda} \times \mathrm{Ran})^{\subset}) \to \mathrm{SI}^{=\lambda}_{\mathrm{Ran}}.$$

The inverse equivalence is given by restriction to the section $s^{\lambda}_{\mathrm{Ran}}$ *of* (2.4), *and similarly for* Ran *replaced by* X^I *for an individual* I.

2.4.3 Proof of Proposition 2.4.2

Follows from the fact that the action of the group ind-scheme

$$(X^{\lambda} \times \mathrm{Ran})^{\subset} \underset{\mathrm{Ran}}{\times} \mathfrak{L}(N)_{\mathrm{Ran}}$$

on $S^{\lambda}_{\mathrm{Ran}}$ is transitive along the fibers of the map (2.3), with the stabilizer of the section $s^{\lambda}_{\mathrm{Ran}}$ being a pro-unipotent group scheme over $(X^{\lambda} \times \mathrm{Ran})^{\subset}$.

□

2.5 *Interaction Between the Strata*

2.5.1

Consider the subcategories (2.5). The maps $\mathbf{j}^{\lambda}$, $\bar{\mathbf{i}}^{\lambda}$ and $\mathbf{i}^{\lambda}$ induce functors

$$(\bar{\mathbf{i}}^{\lambda})_! := (\bar{\mathbf{i}}^{\lambda})_* : \mathrm{SI}^{\leq\lambda}_{\mathrm{Ran}} \to \mathrm{SI}^{\leq 0}_{\mathrm{Ran}},$$

$$(\bar{\mathbf{i}}^{\lambda})^! : \mathrm{SI}^{\leq 0}_{\mathrm{Ran}} \to \mathrm{SI}^{\leq\lambda}_{\mathrm{Ran}};$$

$$(\mathbf{j}^{\lambda})^* := (\mathbf{j}^{\lambda})^! : \mathrm{SI}^{\leq\lambda}_{\mathrm{Ran}} \to \mathrm{SI}^{=\lambda}_{\mathrm{Ran}};$$

$$(\mathbf{j}^{\lambda})_* : \mathrm{SI}^{=\lambda}_{\mathrm{Ran}} \to \mathrm{SI}^{\leq\lambda}_{\mathrm{Ran}};$$

$$(\mathbf{i}^{\lambda})^! \simeq (\mathbf{j}^{\lambda})^! \circ (\bar{\mathbf{i}}^{\lambda})^! : \mathrm{SI}^{\leq 0}_{\mathrm{Ran}} \to \mathrm{SI}^{=\lambda}_{\mathrm{Ran}};$$

$$(\mathbf{i}^{\lambda})_* \simeq (\bar{\mathbf{i}}^{\lambda})_* \circ (\mathbf{j}^{\lambda})_* : \mathrm{SI}^{=\lambda}_{\mathrm{Ran}} \to \mathrm{SI}^{\leq 0}_{\mathrm{Ran}}.$$

The same applies to Ran replaced by X^I for a fixed finite non-empty set I.

2.5.2

In Sect. 2.7, we will prove:

Proposition 2.5.3

(a) *For a fixed finite set I, the left adjoint of*

$$(\mathbf{i}^{\lambda})_* : \mathrm{SI}_I^{=\lambda} \to \mathrm{SI}_I^{\leq 0}$$

is defined as a functor

$$\mathrm{SI}_I^{\leq 0} \to \mathrm{SI}_I^{=\lambda},$$

to be denoted by $(\mathbf{i}^{\lambda})^*$.

(b) *For* $\mathcal{F} \in \mathrm{SI}_I^{\leq 0}$ *and* $\mathcal{F}' \in \mathrm{Shv}(X^I)$, *the map*[4]

$$(\mathbf{i}^{\lambda})^*((\overline{p}_I^0)^!(\mathcal{F}') \overset{!}{\otimes} \mathcal{F}) \to (p_I^{\lambda})^!(\mathcal{F}'|_{(X^{\lambda}\times X^I)^{\subset}} \overset{!}{\otimes} (\mathbf{i}^{\lambda})^*(\mathcal{F}))$$

is an isomorphism.

(c) *For a map of finite sets $\phi : I \twoheadrightarrow J$, the natural transformation*

$$(\mathbf{i}^{\lambda})^* \circ (\Delta_{\phi})^! \to (\Delta_{\phi})^! \circ (\mathbf{i}^{\lambda})^*, \quad \mathrm{SI}_I^{\leq 0} \rightrightarrows \mathrm{SI}_J^{=\lambda}$$

is an isomorphism.

Remark 2.5.4 Let $\mathcal{F} \in \mathrm{SI}_I^{\leq 0}$, be such that the *partially defined* left adjoint $(\mathbf{i}^{\lambda})^*$ of

$$(\mathbf{i}^{\lambda})_* : \mathrm{Shv}(S_I^{\lambda}) \to \mathrm{Shv}(\overline{S}_I^0) \tag{2.6}$$

is defined on $\mathcal{F}$, viewed as an object of $\mathrm{Shv}(\overline{S}_I^0)$.

(Note that the condition of point (a') of Proposition 2.5.3 is always satisfied in the context of constructible sheaves. In the context of D-modules, it is satisfied if, for example, $\mathcal{F}$ is ind-holonomic.)

Then it follows formally that the resulting object of $\mathrm{Shv}(S_I^{\lambda})$ equals

$$(\mathbf{i}^{\lambda})^*(\mathcal{F}) \in \mathrm{SI}_I^{=\lambda} \subset \mathrm{Shv}(S_I^{\lambda}),$$

where $(\mathbf{i}^{\lambda})^*$ is understood in the sense of point (a) of Proposition 2.5.3.

In other words, for such $\mathcal{F}$, the notation $(\mathbf{i}^{\lambda})^*(\mathcal{F})$ is unambiguous.

A similar remark applies to the functor $(\mathbf{i}^{\lambda})_!$ studied in Corollary 2.5.6 below.

[4]In the formula below $-|_{(X^{\lambda}\times X^I)^{\subset}}$ denotes the !-pullback along the projection $(X^{\lambda} \times X^I)^{\subset} \to X^I$.

2.5.3

From Proposition 2.5.3, by a formal Cousin argument, we obtain:

Corollary 2.5.6

(a) *For a fixed finite set I, the left adjoint of*

$$(\mathbf{i}^\lambda)^! : \mathrm{SI}_I^{\leq 0} \to \mathrm{SI}_I^{=\lambda}$$

is defined as a functor

$$\mathrm{SI}_I^{=\lambda} \to \mathrm{SI}_I^{\leq 0},$$

to be denoted $(\mathbf{i}^\lambda)_!$.

(b) *For* $\mathcal{F} \in \mathrm{SI}_I^{=\lambda}$ *and* $\mathcal{F}' \in \mathrm{Shv}(X^I)$*, the map*

$$(\mathbf{i}^\lambda)_!((p_I^\lambda)^!(\mathcal{F}'|_{(X^\lambda \times X^I)^\subset})) \overset{!}{\otimes} \mathcal{F}) \to (\overline{p}_I^0)^!(\mathcal{F}') \overset{!}{\otimes} (\mathbf{i}^\lambda)_!(\mathcal{F})$$

is an isomorphism.

(c) *For a map of finite sets $\phi : I \twoheadrightarrow J$, the natural transformation*

$$(\mathbf{i}^\lambda)_! \circ (\Delta_\phi)^! \to (\Delta_\phi)^! \circ (\mathbf{i}^\lambda)_!, \quad \mathrm{SI}_I^{=\lambda} \rightrightarrows \mathrm{SI}_J^{\leq 0}.$$

2.5.7

Passing to the limit over $I \in \mathrm{Fin}^{\mathrm{surj}}$, we obtain:

Corollary 2.5.8

(a) *The left adjoint of*

$$(\mathbf{i}^\lambda)_* : \mathrm{SI}_{\mathrm{Ran}}^{=\lambda} \to \mathrm{SI}_{\mathrm{Ran}}^{\leq 0}$$

is defined as a functor

$$\mathrm{SI}_{\mathrm{Ran}}^{\leq 0} \to \mathrm{SI}_{\mathrm{Ran}}^{=\lambda},$$

to be denoted by $(\mathbf{i}^\lambda)^*$.

(b) *The left adjoint of*

$$(\mathbf{i}^\lambda)^! : \mathrm{SI}_{\mathrm{Ran}}^{\leq 0} \to \mathrm{SI}_{\mathrm{Ran}}^{=\lambda}$$

is defined as a functor

$$\mathrm{SI}^{=\lambda}_{\mathrm{Ran}} \to \mathrm{SI}^{\leq 0}_{\mathrm{Ran}},$$

to be denoted $(\mathbf{i}^\lambda)_!$.

(c) *For* $\mathcal{F} \in \mathrm{SI}^{\leq 0}_{\mathrm{Ran}}$ *and* $\mathcal{F}' \in \mathrm{Shv}(\mathrm{Ran})$, *the map*

$$(\mathbf{i}^\lambda)^*((\overline{p}^0_{\mathrm{Ran}})^!(\mathcal{F}') \overset{!}{\otimes} \mathcal{F}) \to (p^\lambda_{\mathrm{Ran}})^!(\mathcal{F}'|_{(X^\lambda \times \mathrm{Ran})^{\subset}}) \overset{!}{\otimes} (\mathbf{i}^\lambda)^*(\mathcal{F})$$

is an isomorphism.

(d) *For* $\mathcal{F} \in \mathrm{SI}^{=\lambda}_{\mathrm{Ran}}$ *and* $\mathcal{F}' \in \mathrm{Shv}(\mathrm{Ran})$, *the map*

$$(\mathbf{i}^\lambda)_!((p^\lambda_{\mathrm{Ran}})^!(\mathcal{F}'|_{(X^\lambda \times \mathrm{Ran})^{\subset}}) \overset{!}{\otimes} \mathcal{F}) \to (\overline{p}^0_{\mathrm{Ran}})^!(\mathcal{F}') \overset{!}{\otimes} (\mathbf{i}^\lambda)_!(\mathcal{F})$$

is an isomorphism.

Remark 2.5.9 A slight variation of the proof of Proposition 2.5.3 shows that the assertions of Corollary 2.5.8 remain valid for $\mathbf{i}^\lambda$ replaced by $\bar{\mathbf{i}}^\lambda$. Similarly, the assertion of Corollary 2.5.6 remains valid for $\mathbf{i}^\lambda$ replaced by $\mathbf{j}^\lambda$, and the same is true for their Ran variants.

2.6 An Aside: The ULA Property

Consider the object

$$(\mathbf{j}^0)_!(\omega_{S^0_I}) \in \mathrm{SI}^{\leq 0}_I \subset \mathrm{Shv}(\overline{S}^0_I).$$

Here $(\mathbf{j}^0)_!$ is understood as the (partially defined) left adjoint of

$$(\mathbf{j}^0)^! : \mathrm{Shv}(\overline{S}^0_I) \to (\mathbf{j}^0)_!(\omega_{S^0_I});$$

it is always defined in constructible contexts; in the context of D-modules, it is defined since $\omega_{S^0_I}$ is ind-holonomic.

We will now formulate a certain strong acyclicity property of the above object that it enjoys with respect to the projection

$$\overline{p}^0_I : \overline{S}^0_I \to X^I.$$

2.6.1

Let Y be a scheme, and let $\mathcal{C}$ be a DG category equipped with an action of the $\mathrm{Shv}(Y)$, viewed as a monoidal category with respect to $\overset{!}{\otimes}$.

We shall say that an object $c \in \mathcal{C}$ is ULA with respect to Y if for any compact $\mathcal{F} \in \mathrm{Shv}(Y)^c$, and any $c' \in \mathcal{C}$, the map

$$\mathrm{Hom}_{\mathcal{C}}(\mathcal{F} \overset{!}{\otimes} c, c') \to \mathrm{Hom}(\mathbb{D}(\mathcal{F}) \overset{!}{\otimes} \mathcal{F} \overset{!}{\otimes} c, \mathbb{D}(\mathcal{F}) \overset{!}{\otimes} c') \to \mathrm{Hom}(\mathsf{e}_Y \overset{!}{\otimes} c, \mathbb{D}(\mathcal{F}) \overset{!}{\otimes} c')$$

is an isomorphism.

In the above formula, $\mathbb{D}(-)$ denotes the Verdier duality anti-equivalence of $\mathrm{Shv}(Y)^c$,

$$(\mathrm{Shv}(Y)^c)^{\mathrm{op}} \to \mathrm{Shv}(Y)^c,$$

and e_Y is the "constant sheaf" on Y, i.e., $\mathsf{e}_Y := \mathbb{D}(\omega_Y)$. Note that when Y is smooth of dimension d, we have $\mathsf{e}_Y \simeq \omega_Y[-2d]$.

2.6.2

We regard $\mathrm{Shv}(\overline{S}_I^0)$ as tensored over $\mathrm{Shv}(X^I)$ via

$$\mathcal{F} \in \mathrm{Shv}(X^I),\ \mathcal{F}' \in \mathrm{Shv}(\overline{S}_I^0) \mapsto (\overline{p}_I^0)^!(\mathcal{F}) \overset{!}{\otimes} \mathcal{F}'.$$

We claim:

Proposition 2.6.3 *The object $(\mathbf{j}^0)_!(\omega_{S_I^0}) \in \mathrm{Shv}(\overline{S}_I^0)$ is ULA with respect to X^I.*

Proof For $\mathcal{F} \in \mathrm{Shv}(X^I)$ and $\mathcal{F}' \in \mathrm{Shv}(\overline{S}_I^0)$, we have a commutative square

$$\begin{array}{ccc}
\mathrm{Hom}((p_I^0)^!(\mathcal{F}), (\mathbf{j}^0)^!(\mathcal{F}')) & \longrightarrow & \mathrm{Hom}((p_I^0)^!(\mathsf{e}_{X^I}), (p_I^0)^!(\mathbb{D}(\mathcal{F})) \overset{!}{\otimes} (\mathbf{j}^0)^!(\mathcal{F}')) \\
\sim\downarrow & & \downarrow\sim \\
\mathrm{Hom}((\mathbf{j}^0)_! \circ (p_I^0)^!(\mathcal{F}), \mathcal{F}') & & \mathrm{Hom}((\mathbf{j}^0)_! \circ (p_I^0)^!(\mathsf{e}_{X^I}), (\overline{p}_I^0)^!(\mathbb{D}(\mathcal{F})) \overset{!}{\otimes} \mathcal{F}') \\
\uparrow & & \uparrow \\
\mathrm{Hom}((\overline{p}_I^0)^!(\mathcal{F}) \overset{!}{\otimes} (\mathbf{j}^0)_!(\omega_{S_I^0}), \mathcal{F}') & \longrightarrow & \mathrm{Hom}((\overline{p}_I^0)^!(\mathsf{e}_{X^I}) \overset{!}{\otimes} (\mathbf{j}^0)_!(\omega_{S_I^0}), (\overline{p}_I^0)^!(\mathbb{D}(\mathcal{F})) \overset{!}{\otimes} \mathcal{F}').
\end{array}$$

In this diagram, the lower vertical arrows are isomorphisms by Corollary 2.5.6(b). The top horizontal arrow is an isomorphism because S_I^0 can be exhibited as a union of closed subschemes, each being smooth over X^I. (Indeed, write $\mathfrak{L}(N)_I$ as a union of group subschemes N_I^α pro-smooth over X^I; then S_I^0 is a union of the quotients $N_I^\alpha/\mathfrak{L}^+(N)_I$.)

Hence, the bottom horizontal arrow is an isomorphism, as required.

□

2.7 An Application of Braden's Theorem

In this subsection, we will prove Proposition 2.5.3.

2.7.1

Let

$$S_I^{-,\lambda} \overset{\mathbf{j}^{-,\lambda}}{\hookrightarrow} \overline{S}_I^{-,\lambda} \overset{\mathbf{i}^{-\lambda}}{\to} \mathrm{Gr}_{G,I}$$

be the objects defined in the same way as their counterparts

$$S_I^{\lambda} \overset{\mathbf{j}^{\lambda}}{\hookrightarrow} \overline{S}_I^{\lambda} \overset{\mathbf{i}^{\lambda}}{\to} \mathrm{Gr}_{G,I},$$

but where we replace N by N^-.

Choose a regular dominant coweight $\mathbb{G}_m \to T$. It gives rise to an action of $\mathbb{G}_m$ on $\overline{S}_I^0$ along the fibers of the projection $\overline{p}_I^0$. We have:

Lemma 2.7.2 *The attracting (resp., repelling) locus of the above $\mathbb{G}_m$ action identifies with*

$$\underset{\lambda\in\Lambda^{\mathrm{neg}}}{\sqcup} S_I^{\lambda} \textit{ and } \underset{\lambda\in\Lambda^{\mathrm{neg}}}{\sqcup} S_I^{-,\lambda},$$

respectively. The fixed-point locus identifies with

$$\underset{\lambda\in\Lambda^{\mathrm{neg}}}{\sqcup} s_I^{\lambda}((X^\lambda \times X^I)^{\subset}).$$

2.7.3

Let us now prove point (a) of Proposition 2.5.3.[5]

By Proposition 2.4.2, it suffices to show that the functor

$$\mathrm{Shv}((X^\lambda \times X^I)^{\subset}) \overset{(p_I^\lambda)^!}{\longrightarrow} \mathrm{SI}_I^{=\lambda} \overset{(\mathbf{i}^\lambda)_*}{\longrightarrow} \mathrm{SI}_I^{\leq 0}$$

admits a left adjoint.

For this, it suffices to show that the *partially defined* left adjoint to

[5]We are grateful to Lin Chen for pointing out a mistake in the statement of Proposition 2.5.3 in the previous version of the paper. The corrected argument is due to him.

$$\mathrm{Shv}((X^\lambda \times X^I)^{\subset}) \xrightarrow{(p_I^\lambda)^!} \mathrm{Shv}(S_I^\lambda) \xrightarrow{(\mathbf{i}^\lambda)_*} \mathrm{Shv}(\overline{S}_I^0)$$

is defined on objects that belong to $\mathrm{SI}_I^{\leq 0} \subset \mathrm{Shv}(\overline{S}_I^0)$.

It is easy to see that every object in $\mathrm{Shv}(\overline{S}_I^0)$ is $\mathbb{G}_m$-monodromic. Now, the result follows from Braden's theorem[6] (see [DrGa, Theorem 3.3.4]), combined with Lemma 2.7.2.

2.7.4

Note that Braden's theorem also implies the existence of a canonical isomorphism

$$(s_I^\lambda)^! \circ (\mathbf{i}^\lambda)^* \simeq (p_I^{-,\lambda})_* \circ (\mathbf{i}^{-,\lambda})^!. \tag{2.7}$$

This implies point (b) of Proposition 2.5.3 by base change.

Point (c) is a formal corollary of point (b).

□

Remark 2.7.5 For future use, we note that (2.7) and Proposition 2.5.3(c) imply that an analogous formula holds over the Ran space:

$$(s_{\mathrm{Ran}}^\lambda)^! \circ (\mathbf{i}^\lambda)^*(\mathcal{F}) \simeq (p_{\mathrm{Ran}}^{-,\lambda})_* \circ (\mathbf{i}^{-,\lambda})^!(\mathcal{F}), \quad \mathcal{F} \in \mathrm{SI}_{\mathrm{Ran}}^{\leq 0}. \tag{2.8}$$

3 The t-Structure and the Semi-infinite IC Sheaf

In this section, we define a t-structure on $\mathrm{SI}_{\mathrm{Ran}}^{\leq 0}$ and define the main object of study in this paper— the Ran version of the semi-infinite intersection cohomology sheaf, denoted $\mathrm{IC}_{\mathrm{Ran}}^{\frac{\infty}{2}}$.

We will also give a presentation of $\mathrm{IC}_{\mathrm{Ran}}^{\frac{\infty}{2}}$ as a colimit and describe explicitly its *- and !-restrictions to the strata S_{Ran}^λ.

3.1 The t-Structure on the Semi-infinite Category

3.1.1

We introduce a t-structure on the category $\mathrm{Shv}((X^\lambda \times \mathrm{Ran})^{\subset})$ as follows.

We declare an object $\mathcal{F} \in \mathrm{Shv}((X^\lambda \times \mathrm{Ran})^{\subset})$ to be *connective* if

[6] Braden's theorem extends from schemes to ind-schemes by an easy colimit argument.

$$\mathrm{Hom}(\mathcal{F}, (\mathrm{pr}^{\lambda}_{\mathrm{Ran}})^!(\mathcal{F}')) = 0$$

for all $\mathcal{F}' \in \mathrm{Shv}(X^{\lambda})$ that are *strictly coconnective* (in the perverse t-structure).

Remark 3.1.2 The above t-structure on $\mathrm{Shv}((X^{\lambda} \times \mathrm{Ran})^{\subset})$ is quite pathological in that it is *non-local*, see also Remark 3.1.7.

3.1.3

By construction, the functor

$$(\mathrm{pr}^{\lambda}_{\mathrm{Ran}})^! : \mathrm{Shv}(X^{\lambda}) \to \mathrm{Shv}((X^{\lambda} \times \mathrm{Ran})^{\subset})$$

is left t-exact.

However, we claim:

Proposition 3.1.4 *The functor* $(\mathrm{pr}^{\lambda}_{\mathrm{Ran}})^! : \mathrm{Shv}(X^{\lambda}) \to \mathrm{Shv}((X^{\lambda} \times \mathrm{Ran})^{\subset})$ *is t-exact.*

Proof Follows from Corollary 2.3.4. □

3.1.5

We define a t-structure on $\mathrm{SI}^{=\lambda}_{\mathrm{Ran}}$ as follows. We declare an object $\mathcal{F} \in \mathrm{SI}^{=\lambda}_{\mathrm{Ran}}$ to be connective/coconnective if

$$(s^{\lambda}_{\mathrm{Ran}})^!(\mathcal{F})[\langle\lambda, 2\check{\rho}\rangle] \in \mathrm{Shv}((X^{\lambda} \times \mathrm{Ran})^{\subset})$$

is connective/coconnective.

In other words, this t-structure is transferred from $\mathrm{Shv}((X^{\lambda} \times \mathrm{Ran})^{\subset})$ via the equivalences

$$(s^{\lambda}_{\mathrm{Ran}})^! : \mathrm{SI}^{=\lambda}_{\mathrm{Ran}} \to \mathrm{Shv}((X^{\lambda} \times \mathrm{Ran})^{\subset}) : (p^{\lambda}_{\mathrm{Ran}})^!$$

of Proposition 2.4.2, up to the cohomological shift $[\langle\lambda, 2\check{\rho}\rangle]$.

3.1.6

We define a t-structure on $\mathrm{SI}^{\leq 0}_{\mathrm{Ran}}$ by declaring that an object $\mathcal{F}$ is coconnective if

$$(\mathbf{i}^{\lambda})^!(\mathcal{F}) \in \mathrm{SI}^{=\lambda}_{\mathrm{Ran}}$$

is coconnective in the above t-structure.

Remark 3.1.7 The above t-structure on $\mathrm{SI}^{\leq 0}_{\mathrm{Ran}}$ is a somewhat artificial construct, since the t-structure on the individual strata

$$\mathrm{SI}^{=\lambda}_{\mathrm{Ran}} \simeq \mathrm{Shv}((X^\lambda \times \mathrm{Ran})^{\subset})$$

was transferred from a pathological t-structure on $\mathrm{Shv}(X^\lambda \times \mathrm{Ran})^{\subset}$, see Remark 3.1.2.

This drawback will be cured in Sect. 5.4: we will single out a (naturally defined) full subcategory

$$\mathrm{SI}^{\leq 0}_{\mathrm{Ran,unital}} \subset \mathrm{SI}^{\leq 0}_{\mathrm{Ran}},$$

such that for each stratum S^λ_{Ran}, the functor $(\mathrm{pr}^\lambda_{\mathrm{Ran}} \circ p^\lambda_{\mathrm{Ran}})^!$ defines an *equivalence*

$$\mathrm{Shv}(X^\lambda) \to \mathrm{SI}^{=\lambda}_{\mathrm{Ran,unital}}.$$

3.1.8

By construction, the subcategory of connective objects in $\mathrm{SI}^{\leq 0}_{\mathrm{Ran}}$ is generated under colimits by objects of the form

$$(\mathbf{i}^\lambda)_! \circ (p^\lambda_{\mathrm{Ran}})^!(\mathcal{F})[-\langle \lambda, 2\check{\rho}\rangle], \quad \lambda \in \Lambda^{\mathrm{neg}} \tag{3.1}$$

where $\mathcal{F}$ is a connective object of $\mathrm{Shv}((X^\lambda \times \mathrm{Ran})^{\subset})$.

We claim:

Lemma 3.1.9 *An object $\mathcal{F} \in \mathrm{SI}^{\leq 0}_{\mathrm{Ran}}$ is connective if and only if $(\mathbf{i}^\lambda)^*(\mathcal{F}) \in \mathrm{SI}^{\leq \lambda}_{\mathrm{Ran}}$ is connective for every $\lambda \in \Lambda^{\mathrm{neg}}$.*

Proof It is clear that for objects of the form (3.1), their *-pullback to any S^λ_{Ran} is connective. Hence, the same is true for any connective object of $\mathrm{SI}^{\leq 0}_{\mathrm{Ran}}$.

Vice versa, let $0 \neq \mathcal{F}$ be a strictly coconnective object of $\mathrm{SI}^{\leq 0}_{\mathrm{Ran}}$. We need to show that if all $(\mathbf{i}^\lambda)^*(\mathcal{F})$ are connective, then $\mathcal{F} = 0$. Let λ be the largest element such that $(\mathbf{i}^\lambda)^!(\mathcal{F}) \neq 0$. On the one hand, since $\mathcal{F}$ is strictly coconnective, $(\mathbf{i}^\lambda)^!(\mathcal{F})$ is strictly coconnective. On the other hand, by the maximality of λ, we have

$$(\mathbf{i}^\lambda)^!(\mathcal{F}) \simeq (\mathbf{i}^\lambda)^*(\mathcal{F}),$$

and the assertion follows.

□

3.2 Definition of the Semi-infinite IC Sheaf

When considering the semi-infinite IC sheaf, we will assume that G is semi-simple and simply connected.

3.2.1

By construction, the object $(\mathbf{i}^\lambda)_!(\omega_{S^\lambda_{\mathrm{Ran}}})[-\langle\lambda, 2\check{\rho}\rangle]$ (resp., $(\mathbf{i}^\lambda)_*(\omega_{S^\lambda_{\mathrm{Ran}}})[-\langle\lambda, 2\check{\rho}\rangle]$) of $\mathrm{SI}^{\leq 0}_{\mathrm{Ran}}$ is connective (resp., coconnective).

However, in Sect. 4.6.10, we will prove:

Proposition 3.2.2 *The objects*

$$(\mathbf{i}^\lambda)_!(\omega_{S^\lambda_{\mathrm{Ran}}})[-\langle\lambda, 2\check{\rho}\rangle] \text{ and } (\mathbf{i}^\lambda)_*(\omega_{S^\lambda_{\mathrm{Ran}}})[-\langle\lambda, 2\check{\rho}\rangle]$$

both belong to $(\mathrm{SI}^{\leq 0}_{\mathrm{Ran}})^\heartsuit$.

3.2.3

Consider the canonical map

$$(\mathbf{i}^0)_!(\omega_{S^0_{\mathrm{Ran}}}) \to (\mathbf{i}^0)_*(\omega_{S^0_{\mathrm{Ran}}}).$$

According to Proposition 3.2.2, both sides belong to $(\mathrm{SI}^{\leq 0}_{\mathrm{Ran}})^\heartsuit$. We let

$$\mathrm{IC}^{\frac{\infty}{2}}_{\mathrm{Ran}} \in (\mathrm{SI}^{\leq 0}_{\mathrm{Ran}})^\heartsuit$$

denote the image of this map.

The above object is the main object of study of this paper.

3.2.4

Our goal in this section and the next is to describe $\mathrm{IC}^{\frac{\infty}{2}}_{\mathrm{Ran}}$ as explicitly as possible. Specifically, we will do the following:

- We will describe the !- and *- restrictions of $\mathrm{IC}^{\frac{\infty}{2}}_{\mathrm{Ran}}$ to the strata S^λ_{Ran} (see Theorem 3.4.5).
- We will exhibit the values of $\mathrm{IC}^{\frac{\infty}{2}}_{\mathrm{Ran}}$ in $\mathrm{Shv}(\mathrm{Gr}_{G,I})$ explicitly as colimits (see Theorem 3.7.2).

- We will relate $\mathrm{IC}^{\frac{\infty}{2}}_{\mathrm{Ran}}$ to the intersection cohomology sheaf of Drinfeld's compactification $\overline{\mathrm{Bun}}_N$ (see Theorem 4.3.3).

3.3 Digression: From Commutative Algebras to Factorization Algebras

Let A be a commutative algebra, graded by Λ^{neg} with $A(0) \simeq \mathsf{e}$. To A, we can attach an object

$$\mathrm{Fact}^{\mathrm{alg}}(A)_{X^\lambda} \in \mathrm{Shv}(X^\lambda),$$

characterized by the property that its !-fiber at a divisor

$$\sum_k \lambda_k \cdot x_k \in X^\lambda, \quad 0 \neq \lambda_k \in \Lambda^{\mathrm{neg}}, \quad \sum_k \lambda_k = \lambda, \quad k' \neq k'' \Rightarrow x_{k'} \neq x_{k''}$$

equals $\bigotimes\limits_k A(\lambda_k)$.

In the present subsection, we recall this construction.

3.3.1

Consider the category TwArr_λ whose objects are diagrams

$$\Lambda^{\mathrm{neg}} - 0 \overset{\underline{\lambda}}{\longleftarrow} J \overset{\phi}{\twoheadrightarrow} K, \quad \sum_{j \in J} \underline{\lambda}(j) = \lambda, \tag{3.2}$$

where I and J are finite non-empty sets. A morphism between two such objects is a diagram

$$\begin{array}{ccccc} \Lambda^{\mathrm{neg}} - 0 & \overset{\underline{\lambda}_1}{\longleftarrow} & J_1 & \overset{\phi_1}{\longrightarrow} & K_1 \\ \mathrm{id}\downarrow & & \psi_J \downarrow & & \uparrow \psi_K \\ \Lambda^{\mathrm{neg}} - 0 & \overset{\underline{\lambda}_2}{\longleftarrow} & J_2 & \overset{\phi_2}{\longrightarrow} & K_2, \end{array}$$

where:

- The right square commutes.
- The maps ψ_J and ψ_K are surjective.
- $\underline{\lambda}_2(j_2) = \sum\limits_{j_1 \in \psi_J^{-1}(j_2)} \underline{\lambda}_1(j_1)$.

3.3.2

The algebra A defines a functor

$$\mathrm{TwArr}(A) : \mathrm{TwArr}_{\lambda} \to \mathrm{Shv}(X^{\lambda}),$$

constructed as follows.

For an object (3.2), let $\Delta_{K,\lambda}$ denote the map $X^K \to X^{\lambda}$ that sends a point $\{x_k, k \in K\} \in X^K$ to the divisor

$$\sum_{k\in K} \Big(\sum_{j\in\phi^{-1}(k)} \underline{\lambda}(j)\Big) \cdot x_k \in X^{\lambda}.$$

Then the value of $\mathrm{TwArr}(A)$ on (3.2) is

$$(\Delta_{K,\lambda})_*(\omega_{X^K}) \bigotimes \left(\underset{j\in J}{\otimes} A(\lambda_j)\right),$$

where $\lambda_j = \underline{\lambda}(j)$.

The structure of functor on $\mathrm{TwArr}(A)$ is provided by the commutative algebra structure on A.

3.3.3

We set

$$\mathrm{Fact}^{\mathrm{alg}}(A)_{X^{\lambda}} := \underset{\mathrm{TwArr}_{\lambda}}{\mathrm{colim}}\ \mathrm{TwArr}(A) \in \mathrm{Shv}(X^{\lambda}).$$

3.3.4 An Example

Let V be a Λ^{neg}-graded vector space with $V(0) = 0$. Let us take $A = \mathrm{Sym}(V)$. In this case, $\mathrm{Fact}^{\mathrm{alg}}(A)_{X^{\lambda}}$ can be explicitly described as follows:

It is the direct sum over all ways to write λ as a sum

$$\lambda = \sum_k n_k \cdot \lambda_k, \quad n_k > 0,\ \lambda_k \in \Lambda^{\mathrm{neg}} - 0$$

of the direct images of

$$\bigotimes_k (\omega_X \otimes V(\lambda_k))^{(n_k)}$$

along the maps

$$\prod_k X^{(n_k)} \to X^\lambda,$$

where $(-)^{(n)}$ denotes the n-th symmetric power of a given local system.

3.3.5

Dually, if A is a co-commutative co-algebra, graded by Λ^{neg} with $A(0) \simeq \mathsf{e}$, then to A we attach an object $\mathrm{Fact}^{\mathrm{coalg}}(A)_{X^\lambda} \in \mathrm{Shv}(X^\lambda)$ characterized by the property that its *-fiber at a divisor

$$\sum_k \lambda_k \cdot x_k \in X^\lambda, \quad 0 \neq \lambda_k \in \Lambda^{\mathrm{neg}}, \quad \sum_k \lambda_k = \lambda, \quad k' \neq k'' \Rightarrow x_{k'} \neq x_{k''}$$

equals $\bigotimes_k A(\lambda_k)$.

If all the graded components of A are finite-dimensional, we can view the dual $A^\vee$ of A as a Λ^{neg}-graded algebra, and we have

$$\mathbb{D}(\mathrm{Fact}^{\mathrm{coalg}}(A)_{X^\lambda}) \simeq \mathrm{Fact}^{\mathrm{alg}}(A^\vee)_{X^\lambda}, \tag{3.3}$$

where we remind that $\mathbb{D}$ stands for the Verdier duality functor.

3.4 Restriction of $\mathrm{IC}^{\frac{\infty}{2}}_{\mathrm{Ran}}$ to Strata

3.4.1

We apply the construction of Sect. 3.3 to A being the (classical) algebra $\mathcal{O}(\check{N})$ (resp., co-algebra $U(\check{\mathfrak{n}}^-)$).

Thus, we obtain the objects

$$\mathrm{Fact}^{\mathrm{alg}}(\mathcal{O}(\check{N}))_{X^\lambda} \text{ and } \mathrm{Fact}^{\mathrm{coalg}}(U(\check{\mathfrak{n}}^-))_{X^\lambda}$$

in $\mathrm{Shv}(X^\lambda)$.

Note also that $U(\check{\mathfrak{n}}^-)$ is the graded dual of $\mathcal{O}(\check{N})$, and so the objects $\mathrm{Fact}^{\mathrm{alg}}(\mathcal{O}(\check{N}))_{X^\lambda}$ and $\mathrm{Fact}^{\mathrm{coalg}}(U(\check{\mathfrak{n}}^-))_{X^\lambda}$ are Verdier dual to each other, see (3.3).

3.4.2

From the construction, it follows that for $\lambda \neq 0$,

$$\mathrm{Fact}^{\mathrm{alg}}(\mathcal{O}(\check{N}))_{X^\lambda} \in \mathrm{Shv}(X^\lambda)^{<0},$$

and hence,

$$\mathrm{Fact}^{\mathrm{coalg}}(U(\check{\mathfrak{n}}^-))_{X^\lambda} \in \mathrm{Shv}(X^\lambda)^{>0}.$$

Remark 3.4.3 Note that by the PBW theorem, when viewed as a co-commutative co-algebra, $U(\check{\mathfrak{n}}^-)$ is canonically identified with $\mathrm{Sym}(\check{\mathfrak{n}}^-)$; in this paper, we will not use the algebra structure on $U(\check{\mathfrak{n}}^-)$, which allows to distinguish it from $\mathrm{Sym}(\check{\mathfrak{n}}^-)$.

Dually, when viewed just as a commutative algebra (i.e., ignoring the Hopf algebra structure), $\mathcal{O}(\check{N})$ is canonically identified with $\mathrm{Sym}(\check{\mathfrak{n}}^*)$. So $\mathrm{Fact}^{\mathrm{alg}}(\mathcal{O}(\check{N}))_{X^\lambda}$ falls into the paradigm of Example 3.3.4.

3.4.4

In Sect. 4.9, we will prove:

Theorem 3.4.5 *The objects*

$$(\mathbf{i}^\lambda)^!(\mathrm{IC}^{\frac{\infty}{2}}_{\mathrm{Ran}}) \text{ and } (\mathbf{i}^\lambda)^*(\mathrm{IC}^{\frac{\infty}{2}}_{\mathrm{Ran}})$$

of $\mathrm{Shv}(S^\lambda_{\mathrm{Ran}})$ *identify with the !-pullback along*

$$S^\lambda_{\mathrm{Ran}} \xrightarrow{p^\lambda_{\mathrm{Ran}}} (X^\lambda \times \mathrm{Ran})^{\subset} \xrightarrow{\mathrm{pr}^\lambda_{\mathrm{Ran}}} X^\lambda$$

of $\mathrm{Fact}^{\mathrm{coalg}}(U(\check{\mathfrak{n}}^-))_{X^\lambda}[-\langle\lambda, 2\check{\rho}\rangle]$ *and* $\mathrm{Fact}^{\mathrm{alg}}(\mathcal{O}(\check{N}))_{X^\lambda}[-\langle\lambda, 2\check{\rho}\rangle]$*, respectively.*

3.5 Digression: Categories over the Ran Space

We will now discuss a variant of the construction in Sect. 3.3 that attaches to a symmetric monoidal category $\mathcal{A}$ a category spread over the Ran space, denoted $\mathrm{Fact}^{\mathrm{alg}}(\mathcal{A})_{\mathrm{Ran}}$.

3.5.1

Consider the category TwArr whose objects are

$$I \overset{\phi}{\twoheadrightarrow} J, \tag{3.4}$$

where I and J are finite non-empty sets. A morphism between two such objects is is a commutative diagram

$$\begin{array}{ccc} J_1 & \xrightarrow{\phi_1} & K_1 \\ {\scriptstyle\psi_J}\downarrow & & \uparrow{\scriptstyle\psi_K} \\ J_2 & \xrightarrow{\phi_2} & K_2, \end{array} \tag{3.5}$$

where the maps ψ_J and ψ_K are surjective.

3.5.2

To $\mathcal{A}$, we attach a functor

$$\text{TwArr}(\mathcal{A}) : \text{TwArr} \to \text{DGCat}$$

by sending an object (3.4) to $\text{Shv}(X^K) \otimes \mathcal{A}^{\otimes J}$, and a morphism (3.5) to a functor comprised of

$$\text{Shv}(X^{K_1}) \stackrel{(\Delta_{\psi_J})_*}{\longrightarrow} \text{Shv}(X^{K_2})$$

and the functor

$$\mathcal{A}^{\otimes J_1} \to \mathcal{A}^{\otimes J_2},$$

given by the symmetric monoidal structure on $\mathcal{A}$.

3.5.3

We set

$$\text{Fact}^{\text{alg}}(\mathcal{A})_{\text{Ran}} := \underset{\text{TwArr}}{\text{colim}}\ \text{TwArr}(\mathcal{A}) \in \text{DGCat}\,. \tag{3.6}$$

3.5.4

Let us consider some examples:

(i) Let $\mathcal{A} = \text{Vect}$. Then $\text{Fact}^{\text{alg}}(\mathcal{A})_{\text{Ran}} \simeq \text{Shv}(\text{Ran})$.
(ii) Let $\mathcal{A}$ be the (non-unital) symmetric monoidal category consisting of vector spaces graded by the semi-group $\Lambda^{\text{neg}} - 0$. We have a canonical functor

$$\text{Fact}^{\text{alg}}(\mathcal{A})_{\text{Ran}} \to \text{Shv}(\underset{\lambda \in \Lambda^{\text{neg}}-0}{\sqcup} X^{\lambda}), \tag{3.7}$$

and it follows from [Ga2, Lemma 7.4.11(d)] that this functor is an equivalence.

3.5.5

Similarly, if $\mathcal{A}$ is a symmetric co-monoidal category, we can form the limit of the corresponding functor

$$\text{TwArr}(\mathcal{A}) : \text{TwArr}^{\text{op}} \to \text{DGCat}$$

and obtain a category that we denote by $\text{Fact}^{\text{coalg}}(\mathcal{A})_{\text{Ran}}$.

3.5.6

Recall that whenever we have a diagram of categories

$$t \mapsto \mathcal{C}_t$$

indexed by some category T, then

$$\underset{t \in T}{\text{colim}}\, \mathcal{C}_t$$

is canonically equivalent to

$$\lim_{t \in T^{\text{op}}} \mathcal{C}_t,$$

where the transition functors are given by the right adjoints of those in the original family.

3.5.7

Let $\mathcal{A}$ be again a symmetric monoidal category. Applying the observation of Sect. 3.5.6 to the colimit (3.6), we obtain that $\text{Fact}^{\text{alg}}(\mathcal{A})_{\text{Ran}}$ can also be written as a limit.

Assume now that $\mathcal{A}$ is such that the functor

$$\mathcal{A} \to \mathcal{A} \otimes \mathcal{A},$$

right adjoint to the tensor product operation, is continuous. In this case, the above tensor co-product operation makes $\mathcal{A}$ into a symmetric co-monoidal category, and we can form $\text{Fact}^{\text{coalg}}(\mathcal{A})_{\text{Ran}}$.

We note however that the *limit* presentation of $\text{Fact}^{\text{alg}}(\mathcal{A})_{\text{Ran}}$ tautologically coincides with the limit defining $\text{Fact}^{\text{coalg}}(\mathcal{A})_{\text{Ran}}$. That is, we have a canonical equivalence:

$$\mathrm{Fact}^{\mathrm{alg}}(\mathcal{A})_{\mathrm{Ran}} \simeq \mathrm{Fact}^{\mathrm{coalg}}(\mathcal{A})_{\mathrm{Ran}}.$$

Hence, in what follows, we will sometimes write simply $\mathrm{Fact}(\mathcal{A})_{\mathrm{Ran}}$, having both of the above realizations in mind.

3.5.8

Let I be a fixed finite non-empty set. The above constructions have a variant, where instead of TwArr we use its variant, denoted $\mathrm{TwArr}_{I/}$, whose objects are commutative diagrams

$$I \twoheadrightarrow J \overset{\phi}{\twoheadrightarrow} K,$$

and whose morphisms are commutative diagrams

$$\begin{array}{ccccc} I & \longrightarrow & J_1 & \xrightarrow{\phi_1} & K_1 \\ {\scriptstyle \mathrm{id}}\downarrow & & {\scriptstyle \psi_J}\downarrow & & \uparrow{\scriptstyle \psi_K} \\ I & \longrightarrow & J_2 & \xrightarrow{\phi_2} & K_2. \end{array}$$

We denote the resulting category by $\mathrm{Fact}^{\mathrm{alg}}(\mathcal{A})_I$, i.e.,

$$\mathrm{Fact}^{\mathrm{alg}}(\mathcal{A})_I := \underset{\mathrm{TwArr}_{I/}}{\mathrm{colim}}\ \mathrm{TwArr}(\mathcal{A})|_{\mathrm{TwArr}_{I/}}.$$

3.5.9

For a surjective morphism $\phi : I_1 \twoheadrightarrow I_2$, we have the corresponding functor

$$\mathrm{TwArr}_{I_2/} \to \mathrm{TwArr}_{I_1/},$$

which induces a functor

$$(\Delta_\phi)_* : \mathrm{Fact}^{\mathrm{alg}}(\mathcal{A})_{I_2} \to \mathrm{Fact}^{\mathrm{alg}}(\mathcal{A})_{I_1}. \tag{3.8}$$

An easy cofinality argument shows that the resulting functor

$$\underset{I}{\mathrm{colim}}\ \mathrm{Fact}^{\mathrm{alg}}(\mathcal{A})_I \to \mathrm{Fact}^{\mathrm{alg}}(\mathcal{A})_{\mathrm{Ran}} \tag{3.9}$$

is an equivalence.

3.5.10

Note also that push-out defines a functor

$$\mathrm{TwArr}_{I_1/} \to \mathrm{TwArr}_{I_2/},$$

and we have a natural transformation from the composite

$$\mathrm{TwArr}_{I_1/} \to \mathrm{TwArr} \overset{\mathrm{TwArr}(\mathcal{A})}{\longrightarrow} \mathrm{DGCat}$$

to the composite

$$\mathrm{TwArr}_{I_1/} \to \mathrm{TwArr}_{I_2/} \to \mathrm{TwArr} \overset{\mathrm{TwArr}(\mathcal{A})}{\longrightarrow} \mathrm{DGCat},$$

inducing a functor

$$\mathrm{Fact}^{\mathrm{alg}}(\mathcal{A})_{I_1} := \mathrm{Fact}^{\mathrm{alg}}(\mathcal{A})_{I_1} \to \mathrm{Fact}^{\mathrm{alg}}(\mathcal{A})_{I_2} =: \mathrm{Fact}^{\mathrm{alg}}(\mathcal{A})_{I_2},$$

to be denoted $(\Delta_\phi)^!$.

By unwinding the constructions, it follows that the above functor $(\Delta_\phi)^!$ is the right adjoint of the functor $(\Delta_\phi)_*$ of (3.8).

In particular, by Sect. 3.5.6, we can also write

$$\mathrm{Fact}^{\mathrm{alg}}(\mathcal{A})_{\mathrm{Ran}} \simeq \lim_{I} \mathrm{Fact}^{\mathrm{alg}}(\mathcal{A})_I, \tag{3.10}$$

where the limit is taken with respect to the functors $(\Delta_\phi)^!$.

3.6 *Presentation of* $\mathrm{IC}^{\frac{\infty}{2}}$ *as a Colimit*

Consider the symmetric monoidal category $\mathrm{Rep}(\check{G})$.

3.6.1

For a fixed finite non-empty set I and a map $\underline{\lambda} : I \to \Lambda^+$, we consider the following object of $\mathrm{Fact}(\mathrm{Rep}(\check{G}))_I$, denoted $V^{\underline{\lambda}}$.

Informally, $V^{\underline{\lambda}}$ is designed so its !-fiber at a point

$$I \to X, \quad I = \bigsqcup_k I_k, \ I_k \mapsto x_k, \quad x_{k'} \neq x_{k''}$$

is

$$\bigotimes_k V^{\lambda_k} \in \operatorname{Rep}(\check{G})^{\otimes k}, \quad \lambda_k = \underset{i \in I_k}{\Sigma} \underline{\lambda}(i),$$

where for $\lambda \in \Lambda^+$, we denote by V^λ the corresponding irreducible highest weight representation of $\check{G}$, normalized so that its highest weight line is identified with e.

3.6.2

In terms of the presentation of $\operatorname{Fact}(\operatorname{Rep}(\check{G}))_I$ as a colimit

$$\operatorname{Fact}(\operatorname{Rep}(\check{G}))_I = \underset{\operatorname{TwArr}_{I/}}{\operatorname{colim}} \operatorname{TwArr}(\operatorname{Rep}(\check{G})),$$

the object $V^{\underline{\lambda}}$ corresponds to the colimit over $\operatorname{TwArr}_{I/}$ of the functor

$$\operatorname{TwArr}_{I/} \to \operatorname{Fact}(\operatorname{Rep}(\check{G}))_I$$

that sends

$$(I \twoheadrightarrow J \twoheadrightarrow K) \in \operatorname{TwArr}_{I/} \rightsquigarrow V^{\underline{\lambda}}_{I \twoheadrightarrow J \twoheadrightarrow K} \in \operatorname{Shv}(X^K) \otimes \operatorname{Rep}(\check{G})^{\otimes J} \to \operatorname{Fact}(\operatorname{Rep}(\check{G}))_I, \tag{3.11}$$

where

$$V^{\underline{\lambda}}_{I \twoheadrightarrow J \twoheadrightarrow K} = \omega_{X^K} \bigotimes \left(\underset{j \in J}{\otimes} V^{\lambda_j} \right), \quad \lambda_j = \underset{i \in I, i \mapsto j}{\Sigma} \underline{\lambda}(i).$$

The structure of a functor $\operatorname{TwArr}_{I/} \to \operatorname{Fact}(\operatorname{Rep}(\check{G}))_I$ on (3.11) is given by the Plücker maps

$$\underset{i}{\otimes} V^{\lambda_i} \to V^\lambda, \quad \lambda = \underset{i}{\Sigma} \lambda_i.$$

3.6.3

Denote

$$\operatorname{Sph}_{G,I} := \operatorname{Shv}(\mathfrak{L}^+(G)_I \backslash \operatorname{Gr}_{G,I}) \text{ and } \operatorname{Sph}_{G,\operatorname{Ran}} := \operatorname{Shv}(\mathfrak{L}^+(G)_{\operatorname{Ran}} \backslash \operatorname{Gr}_{G,\operatorname{Ran}}).$$

Consider the symmetric monoidal category $\operatorname{Rep}(\check{G})$. Geometric Satake defines functors

$$\operatorname{Sat}_{G,I} : \operatorname{Fact}(\operatorname{Rep}(\check{G}))_I \to \operatorname{Sph}_{G,I}$$

that glue to a functor

$$\mathrm{Sat}_{G,\mathrm{Ran}} : \mathrm{Fact}(\mathrm{Rep}(\check{G}))_{\mathrm{Ran}} \to \mathrm{Sph}_{G,\mathrm{Ran}}\,.$$

3.6.4

Consider the object

$$\mathrm{Sat}_{G,I}(V^{\underline{\lambda}}) \in \mathrm{Sph}_{G,I}\,.$$

The element $\underline{\lambda}$ gives rise to a section

$$s_I^{-,\underline{\lambda}} : X^I \to \mathrm{Gr}_{T,I} \subset \mathrm{Gr}_{G,I}.$$

Denote

$$\delta_{-\underline{\lambda}} := (s_I^{-,\underline{\lambda}})_!(\omega_{X^I}) \in \mathrm{Shv}(\mathrm{Gr}_{G,I}).$$

Consider the object

$$\delta_{-\underline{\lambda}} \star \mathrm{Sat}_{G,I}(V^{\underline{\lambda}})[\langle \lambda, 2\check{\rho}\rangle] \in \mathrm{Shv}(\mathrm{Gr}_{G,I}).$$

In the above formula, $\lambda = \sum\limits_{i\in I} \underline{\lambda}(i)$, and $\star$ denotes the (right) convolution action of $\mathrm{Sph}_{G,I}$ on $\mathrm{Shv}(\mathrm{Gr}_{G,I})$.

3.6.5

Consider now the set $\mathrm{Maps}(I, \Lambda^+)$ of maps

$$\underline{\lambda} : I \to \Lambda^+.$$

We equip it with a partial order by declaring

$$\underline{\lambda}_1 \leq \underline{\lambda}_2 \Leftrightarrow \underline{\lambda}_2(i) - \underline{\lambda}_1(i) \in \Lambda^+,\ \forall i \in I.$$

The assignment

$$\underline{\lambda} \mapsto \delta_{-\underline{\lambda}} \star \mathrm{Sat}_{G,I}(V^{\underline{\lambda}})[\langle \lambda, 2\check{\rho}\rangle] \in \mathrm{Shv}(\mathrm{Gr}_{G,I})$$

has a structure of a functor

$$\text{Maps}(I, \Lambda^+) \to \text{Shv}(\text{Gr}_{G,I}),$$

see Sects. 6.4.6 and 6.5.2.

Set

$$\text{IC}_I^{\frac{\infty}{2}} := \underset{\underline{\lambda} \in \text{Maps}(I,\Lambda^+)}{\text{colim}} \delta_{-\underline{\lambda}} \star \text{Sat}_{G,I}(V^{\underline{\lambda}})[\langle \lambda, 2\check{\rho} \rangle] \in \text{Shv}(\text{Gr}_{G,I}).$$

As in [Ga1, Proposition 2.3.7(a,b,c)], one shows:

Lemma 3.6.6 *The object* $\text{IC}_I^{\frac{\infty}{2}}$ *has the following properties:*

(a) *It is supported on* $\overline{S}_I^0$.
(b) *It belongs to* $\text{SI}_I^{\leq 0} = \text{Shv}(\overline{S}_I^0)^{\mathfrak{L}(N)_I} \subset \text{Shv}(\overline{S}_I^0)$.
(c) *Its restriction to* S_I^0 *is identified with* $\omega_{S_I^0}$.

3.6.7

For a surjective map

$$\phi : I_2 \twoheadrightarrow I_1$$

and the corresponding map

$$\Delta_\phi : \text{Gr}_{G,I_1} \to \text{Gr}_{G,I_2},$$

we have a canonical identification

$$(\Delta_\phi)^!(\text{IC}_{I_2}^{\frac{\infty}{2}}) \simeq \text{IC}_{I_1}^{\frac{\infty}{2}}.$$

One endows this system of isomorphisms with a homotopy-coherent system of compatibilities, thus making the assignment

$$I \mapsto \text{IC}_I^{\frac{\infty}{2}}$$

into an object of $\text{SI}_{\text{Ran}}^{\leq 0}$, see Sect. 6.4.8.

We denote this object by $'\text{IC}_{\text{Ran}}^{\frac{\infty}{2}}$. By Lemma 3.6.6(c), we have a canonical identification

$$'\text{IC}_{\text{Ran}}^{\frac{\infty}{2}}|_{S_{\text{Ran}}^0} \simeq \omega_{S_{\text{Ran}}^0}. \tag{3.12}$$

3.6.8

Fix a point $x \in X$ and consider the restriction of $'\mathrm{IC}^{\frac{\infty}{2}}_{\mathrm{Ran}}$ along the map

$$\mathrm{Gr}_{G,x} \simeq \{x\} \underset{\mathrm{Ran}}{\times} \mathrm{Gr}_{G,\mathrm{Ran}} \to \mathrm{Gr}_{G,\mathrm{Ran}}.$$

It follows from the construction that this restriction identifies canonically with the object

$$\mathrm{IC}^{\frac{\infty}{2}}_{x} \in \mathrm{Shv}(\mathrm{Gr}_{G,x}),$$

constructed in [Ga1, Sect. 2.3].

3.7 *Presentation of* $\mathbf{IC}^{\frac{\infty}{2}}_{\mathbf{Ran}}$ *as a Colimit*

3.7.1

The rest of this section will be devoted to the proof of the following result:

Theorem 3.7.2 *There exists a unique isomorphism* $'\mathrm{IC}^{\frac{\infty}{2}}_{\mathrm{Ran}} \simeq \mathrm{IC}^{\frac{\infty}{2}}_{\mathrm{Ran}}$, *extending the identification*

$$'\mathrm{IC}^{\frac{\infty}{2}}_{\mathrm{Ran}}|_{S^0_{\mathrm{Ran}}} \overset{(3.12)}{\simeq} \omega_{S^0_{\mathrm{Ran}}} \simeq \mathrm{IC}^{\frac{\infty}{2}}_{\mathrm{Ran}}|_{S^0_{\mathrm{Ran}}}.$$

The proof of Theorem 3.7.2 will amount to the combination of the following two assertions:

Proposition 3.7.3 *For* $\mu \in \Lambda^{\mathrm{neg}}$, *the object*

$$(\mathbf{i}^{\mu})^{*}('\mathrm{IC}^{\frac{\infty}{2}}_{\mathrm{Ran}}) \in \mathrm{SI}^{=\mu}_{\mathrm{Ran}}$$

identifies canonically with the !-pullback along

$$S^{\mu}_{\mathrm{Ran}} \overset{p^{\mu}_{\mathrm{Ran}}}{\longrightarrow} (X^{\mu} \times \mathrm{Ran})^{\subset} \overset{\mathrm{pr}^{\mu}_{\mathrm{Ran}}}{\longrightarrow} X^{\mu}$$

of $\mathrm{Fact}^{\mathrm{alg}}(\mathcal{O}(\check{N}))_{X^{\mu}}[-\langle \mu, 2\check{\rho}\rangle]$.

Proposition 3.7.4 *For* $0 \neq \mu \in \Lambda^{\mathrm{neg}}$, *the object*

$$(\mathbf{i}^{\mu})^{!}('\mathrm{IC}^{\frac{\infty}{2}}_{\mathrm{Ran}})[\langle \mu, 2\check{\rho}\rangle] \in \mathrm{SI}^{=\mu}_{\mathrm{Ran}}$$

is a pullback along $\mathrm{pr}^{\mu}_{\mathrm{Ran}} \circ p^{\mu}_{\mathrm{Ran}}$ *of an object of* $\mathrm{Shv}(X^{\mu})$ *that is strictly coconnective.*

Proof of Theorem 3.7.2 Modulo the Propositions By the definition of the t-structure on $\mathrm{SI}^{\leq 0}_{\mathrm{Ran}}$ and Lemma 3.1.9, it suffices to show that for $\mu \in \Lambda^{\mathrm{neg}} - 0$, the !-restriction (resp., *-restriction) of $'\mathrm{IC}^{\frac{\infty}{2}}_{\mathrm{Ran}}$ to S^{μ}_{Ran} is cohomologically > 0 (resp., < 0).

Now, Proposition 3.7.4 (resp., Proposition 3.7.3) implies the required cohomological estimate by Proposition 3.1.4. □

Remark 3.7.5 Note that Theorem 3.7.2 and Proposition 3.7.3 imply the assertion of Theorem 3.4.5 about the *-fibers.

We will use this observation in the sequel, for the proof of the assertion of Theorem 3.4.5 about the !-fibers.

3.7.6

Let us assume Theorem 3.7.2 for a moment. As a corollary, and taking into account Sect. 3.6.8, we obtain:

Corollary 3.7.7 *The restriction of* $\mathrm{IC}^{\frac{\infty}{2}}_{\mathrm{Ran}}$ *along the map*

$$\mathrm{Gr}_{G,x} \simeq \{x\} \underset{\mathrm{Ran}}{\times} \mathrm{Gr}_{G,\mathrm{Ran}} \to \mathrm{Gr}_{G,\mathrm{Ran}}$$

identifies canonically with the object $\mathrm{IC}^{\frac{\infty}{2}}_{x} \in \mathrm{Shv}(\mathrm{Gr}_{G,x})$ *of* [Ga1, Sect. 2.3].

3.7.8

Before we proceed with the proof of Propositions 3.7.3 and 3.7.4, let us make the following observation concerning the object $'\mathrm{IC}^{\frac{\infty}{2}}_{\mathrm{Ran}}$ (it will be used in the proof of 3.7.4):

By construction,

$$'\mathrm{IC}^{\frac{\infty}{2}}_{\mathrm{Ran}} \in \mathrm{Shv}(\overline{S}^{0}_{\mathrm{Ran}})$$

has the following *factorization property* with respect to Ran:

Let $(\mathrm{Ran} \times \mathrm{Ran})_{\mathrm{disj}}$ denote the *disjoint locus*. That is, for an affine test scheme Y,

$$\mathrm{Hom}(Y, (\mathrm{Ran} \times \mathrm{Ran})_{\mathrm{disj}}) \subset \mathrm{Hom}(Y, \mathrm{Ran}) \times \mathrm{Hom}(Y, \mathrm{Ran})$$

consists of those pairs $\mathcal{I}_1, \mathcal{I}_2 \in \mathrm{Hom}(Y, X)$, for which for every $i_1 \in I_1$ and $i_2 \in I_2$, the corresponding two maps $Y \rightrightarrows X$ have non-intersecting images.

It is well known that we have a canonical isomorphism

$$(\mathrm{Gr}_{G,\mathrm{Ran}}\times\mathrm{Gr}_{G,\mathrm{Ran}}) \underset{\mathrm{Ran}\times\mathrm{Ran}}{\times} (\mathrm{Ran}\times\mathrm{Ran})_{\mathrm{disj}} \simeq \mathrm{Gr}_{G,\mathrm{Ran}} \underset{\mathrm{Ran}}{\times} (\mathrm{Ran}\times\mathrm{Ran})_{\mathrm{disj}}, \tag{3.13}$$

where

$$(\mathrm{Ran}\times\mathrm{Ran})_{\mathrm{disj}} \to \mathrm{Ran}\times\mathrm{Ran} \to \mathrm{Ran}$$

is the map

$$\mathcal{I}_1, \mathcal{I}_2 \mapsto \mathcal{I}_1 \cup \mathcal{I}_2.$$

Then, in terms of the identification (3.13), we have a canonical isomorphism

$$({}'\mathrm{IC}^{\frac{\infty}{2}}_{\mathrm{Ran}} \boxtimes {}'\mathrm{IC}^{\frac{\infty}{2}}_{\mathrm{Ran}})|_{(\mathrm{Gr}_{G,\mathrm{Ran}}\times\mathrm{Gr}_{G,\mathrm{Ran}}) \underset{\mathrm{Ran}\times\mathrm{Ran}}{\times} (\mathrm{Ran}\times\mathrm{Ran})_{\mathrm{disj}}} \simeq$$

$$\simeq {}'\mathrm{IC}^{\frac{\infty}{2}}_{\mathrm{Ran}}|\mathrm{Gr}_{G,\mathrm{Ran}} \underset{\mathrm{Ran}}{\times} (\mathrm{Ran}\times\mathrm{Ran})_{\mathrm{disj}}. \tag{3.14}$$

3.8 *Description of the *-Restriction to Strata*

The goal of this subsection is to prove Proposition 3.7.3.

3.8.1

We will compute

$$(\mathbf{i}^\mu)^*({}'\mathrm{IC}^{\frac{\infty}{2}}_I) \in \mathrm{SI}^{\leq 0}_I$$

for each individual finite non-empty set I and obtain the !-pullback of

$$(\mathrm{pr}^\mu_{\mathrm{Ran}} \circ p^\mu_{\mathrm{Ran}})^!(\mathrm{Fact}^{\mathrm{alg}}(\mathcal{O}(\check{N}))_{X^\mu})[-\langle\mu, 2\check{\rho}\rangle]$$

along $S^\mu_I \to S^\mu_{\mathrm{Ran}}$.

Thus, by Proposition 2.4.2, we need to construct an identification

$$(p^\mu_I)_! \circ (\mathbf{i}^\mu)^*({}'\mathrm{IC}^{\frac{\infty}{2}}_I)[\langle\mu, 2\check{\rho}\rangle] \simeq (\mathrm{pr}^\mu_I)^!(\mathrm{Fact}^{\mathrm{alg}}(\mathcal{O}(\check{N}))_{X^\mu}), \tag{3.15}$$

where pr^μ_I denotes the map

$$(X^\mu \times X^I)^{\subset} \to X^\mu.$$

3.8.2

We will compute

$$(p_I^\mu)_! \circ (\mathbf{i}^\mu)^*(\delta_{-\underline{\lambda}} \star \mathrm{Sat}_{G,I}(V^{\underline{\lambda}}))[\langle \lambda + \mu, 2\check{\rho} \rangle] \in \mathrm{Shv}((X^\mu \times X^I)^\subset) \tag{3.16}$$

for each individual $\underline{\lambda} : I \to \Lambda^+$ with $\lambda = \sum_{i \in I} \underline{\lambda}(i)$.

Namely, we will show that (3.16) identifies with the following object, denoted

$$V^{\underline{\lambda}}(\underline{\lambda} + \mu) \in \mathrm{Shv}((X^\mu \times X^I)^\subset),$$

described below.

3.8.3

Before we give the definition of $V^{\underline{\lambda}}(\underline{\lambda} + \mu)$, let us describe what its !-fibers are. Fix a point of $(X^\mu \times X^I)^\subset$. By definition, a datum of such a point consists of:

- A partition $I = \sqcup_k I_k$
- A collection of distinct points x_k of x
- An assignment $x_k \mapsto \mu_k \in \Lambda^{\mathrm{neg}} - 0$, so that $\sum_k \mu_k = \mu$

Then the !-fiber of $V^{\underline{\lambda}}(\underline{\lambda} + \mu)$ at a such a point is

$$\bigotimes_k V^{\lambda_k}(\lambda_k + \mu_k),$$

where $\lambda_k = \sum_{i \in I_k} \underline{\lambda}(i)$, and where $V(\nu)$ denotes the ν-weight space in a $\check{G}$-representation V.

3.8.4

Consider the category, denoted $\mathrm{TwArr}_{\mu, I/}$, whose objects are commutative diagrams

$$\begin{array}{ccccc}
I & \xrightarrow{\upsilon} & J & \xrightarrow{\psi} & K \\
 & & {\scriptstyle \phi_J}\downarrow & & \downarrow{\scriptstyle \phi_K} \\
 & & \widetilde{J} & \xrightarrow{\widetilde{\psi}} & \widetilde{K} \\
 & & {\scriptstyle \phi'_J}\uparrow & & \uparrow{\scriptstyle \phi'_K} \\
\Lambda^{\mathrm{neg}} & \xleftarrow{\underline{\mu}} & \widetilde{J}' & \xrightarrow{\widetilde{\psi}'} & \widetilde{K}',
\end{array}$$

where the maps $\upsilon, \psi, \widetilde{\psi}, \widetilde{\psi}', \phi_J, \phi_K$ are surjective (but ϕ'_J and ϕ'_K are not necessarily so), and

$$\sum_{\tilde{j}' \in \widetilde{J}'} \underline{\mu}(\tilde{j}') = \mu.$$

Morphisms in this category are defined by the same principle as in TwArr_μ and $\mathrm{TwArr}_{I/}$ introduced earlier, i.e., the sets J, $\widetilde{J}$, $\widetilde{J}'$ map forward and the sets K, $\widetilde{K}$, $\widetilde{K}'$ map backward.

Let $\Delta_{\widetilde{K},I,\lambda}$ denote the map

$$X^{\widetilde{K}} \to X^\mu \times X^I,$$

comprised of

$$\Delta_{\phi_K \circ \psi \circ \upsilon} : X^{\widetilde{K}} \to X^I$$

and

$$X^{\widetilde{K}} \overset{\Delta_{\phi'_K}}{\longrightarrow} X^{\widetilde{K}'} \overset{\Delta_{\widetilde{K}',\mu}}{\longrightarrow} X^\mu.$$

We let $V^{\underline{\lambda}}(\underline{\lambda} + \mu)$ be the colimit over $\mathrm{TwArr}_{\mu,I/}$ of the objects

$$(\Delta_{\widetilde{K},I,\lambda})_*(\omega_{X^{\widetilde{K}}}) \bigotimes \left(\underset{\tilde{j} \in \widetilde{J}}{\otimes} V^{\lambda_{\tilde{j}}}(\lambda_{\tilde{j}} + \mu_{\tilde{j}}) \right),$$

where

$$\lambda_{\tilde{j}} = \sum_{i \in I, i \mapsto \tilde{j}} \underline{\lambda}(i) \text{ and } \mu_{\tilde{j}} = \sum_{\tilde{j}' \in \widetilde{J}', \tilde{j}' \mapsto \tilde{j}} \underline{\mu}(\tilde{j}').$$

3.8.5

Applying Braden's theorem (see Sect. 2.7.1), we obtain a canonical isomorphism

$$(p_I^\mu)_! \circ (\mathbf{i}^\mu)^*(\delta_{-\underline{\lambda}} \star \mathrm{Sat}_{G,I}(V^{\underline{\lambda}})) \simeq (p_I^{-,\mu})_* \circ (\mathbf{i}^{-,\mu})^!(\delta_{-\underline{\lambda}} \star \mathrm{Sat}_{G,I}(V^{\underline{\lambda}})).$$

The key property of the geometric Satake functor $\mathrm{Sat}_{G,I}$ for $I = \{1\}$ is that for $V' \in \mathrm{Rep}(\check{G})$ and $\mu' \in \Lambda$

$$(p_{\{1\}}^{-,\mu'})_* \circ (\mathbf{i}^{-,\mu'})^!(\mathrm{Sat}_{G,\{1\}}(V'))[\langle \mu, 2\check{\rho} \rangle] \simeq \omega_X \otimes V'(\mu').$$

Unwinding the construction of its multi-point version $\mathrm{Sat}_{G,I}$, we obtain a canonical isomorphism

$$(p_I^{-,\mu})_* \circ (\mathbf{i}^{-,\mu})^!(\delta_{-\underline{\lambda}} \star \mathrm{Sat}_{G,I}(V^{\underline{\lambda}}))[\langle \lambda + \mu, 2\check{\rho} \rangle] \simeq V^{\underline{\lambda}}(\underline{\lambda} + \mu),$$

giving rise to the desired expression for (3.16).

3.8.6

To finish the proof of Proposition 3.7.3, we have to show that

$$\underset{\underline{\lambda} \in \mathrm{Maps}(I, \Lambda^+)}{\mathrm{colim}} V^{\underline{\lambda}}(\underline{\lambda} + \mu)$$

identifies canonically with $(\mathrm{pr}_I^{\mu})^!(\mathrm{Fact}^{\mathrm{alg}}(\mathcal{O}(\check{N}))_{X^\lambda})$.

Indeed, this follows from the fact that we have a canonical identification

$$\underset{\lambda \in \Lambda^+}{\mathrm{colim}}\, V^{\lambda}(\lambda + \mu) \simeq \mathcal{O}(\check{N})(\mu),$$

where Λ^+ is endowed with the order relation

$$\lambda_1 \leq \lambda_2 \Leftrightarrow \lambda_2 - \lambda_1 \in \Lambda^+.$$

3.9 *Proof of Coconnectivity*

In this subsection, we will prove Proposition 3.7.4, thereby completing the proof of Theorem 3.7.2.

3.9.1

Consider the diagonal stratification of X^μ. For each parameter β of the stratification, let X^μ_β denote the corresponding stratum, and denote by

$$(X^\mu_\beta \times \mathrm{Ran})^{\subset} := X^\mu_\beta \underset{X^\mu}{\times} (X^\mu \times \mathrm{Ran})^{\subset}) \overset{\iota_\beta}{\hookrightarrow} (X^\mu \times \mathrm{Ran})^{\subset}$$

and

$$(X^\mu_\beta \times \mathrm{Ran})^{\subset} \overset{\mathrm{pr}^\mu_{\mathrm{Ran},\beta}}{\longrightarrow} X^\mu_\beta$$

the resulting maps.

Let $\mathcal{F}^\mu \in \mathrm{Shv}((X^\mu \times \mathrm{Ran})^\subset)$ be such that

$$(\mathbf{i}^\mu)^!({}'\mathrm{IC}^{\frac{\infty}{2}}_{\mathrm{Ran}}) \simeq (p^\mu_{\mathrm{Ran}})^!(\mathcal{F}^\mu).$$

Since the functor $(\mathrm{pr}^\mu_{\mathrm{Ran}})^!$ is left t-exact and fully faithful (the latter by Corollary 2.3.4), in order to prove Proposition 3.7.4, it suffices to show that each

$$(\iota_\beta)^! \circ \mathcal{F}^\mu \in \mathrm{Shv}((X^\mu_\beta \times \mathrm{Ran})^\subset)$$

is of the form

$$(\mathrm{pr}^\mu_{\mathrm{Ran},\beta})^!(\mathcal{F}^\mu_\beta),$$

where $\mathcal{F}^\mu_\beta \in \mathrm{Shv}(X^\mu_\beta)$ is such that $\mathcal{F}^\mu_\beta[\langle \mu, 2\check{\rho}\rangle]$ is strictly coconnective.

3.9.2

By the factorization property of ${}'\mathrm{IC}^{\frac{\infty}{2}}_{\mathrm{Ran}}$ (see (3.14)), it suffices to prove the above assertion for β corresponding to the main diagonal $X \to X^\mu$. Denote the corresponding stratum in $(X^\mu \times \mathrm{Ran})^\subset$ by

$$(X \times \mathrm{Ran})^\subset.$$

Denote the corresponding map $\mathrm{pr}^\mu_{\mathrm{Ran},\beta}$ by

$$\mathrm{pr}^\mu_{(X\times\mathrm{Ran})^\subset} : (X \times \mathrm{Ran})^\subset \to X.$$

Denote the restriction of the section

$$s^\mu_{\mathrm{Ran}} : (X^\mu \times \mathrm{Ran})^\subset \to S^\mu_{\mathrm{Ran}}$$

to this stratum by $s^\mu_{(X\times\mathrm{Ran})^\subset}$.

We claim that

$$(s^\mu_{(X\times\mathrm{Ran})^\subset})^!({}'\mathrm{IC}^{\frac{\infty}{2}}_{\mathrm{Ran}}) \simeq (\mathrm{pr}^\mu_{(X\times\mathrm{Ran})^\subset})^!(\omega_X) \otimes \mathsf{W}_\mu[-\langle \mu, 2\check{\rho}\rangle],$$

where $\mathsf{W}_\mu \in \mathrm{Vect}$ lives in cohomological degrees ≥ 2.

Remark 3.9.3 One can show that there is a canonical identification

$$\mathsf{W}_\mu \simeq \mathrm{Sym}(\check{\mathfrak{n}}^-[-2])(\mu),$$

where $\check{\mathfrak{n}}$ is the unipotent radical of the Langlands dual Lie algebra. In fact, such an identification would follow once we prove Theorem 3.4.5 for the !-restrictions.

3.9.4

In fact, we claim that for every I, we have

$$(s^{\mu}_{(X\times X^I)^{\subset}})^!('\mathrm{IC}^{\frac{\infty}{2}}_I) \simeq (\mathrm{pr}^{\mu}_{(X\times X^I)^{\subset}})^!(\omega_X)\otimes \mathsf{W}_\mu[-\langle \mu, 2\check{\rho}\rangle],$$

where

$$\mathrm{pr}^{\mu}_{(X\times X^I)^{\subset}} := \mathrm{pr}^{\mu}_{(X\times \mathrm{Ran})^{\subset}}|_{(X\times X^I)^{\subset}}.$$

3.9.5

Indeed, it follows from the definitions that for any $\underline{\lambda}: I \to \Lambda^+$,

$$(s^{\mu}_{(X\times X^I)^{\subset}})^!(\delta_{-\underline{\lambda}} \star \mathrm{Sat}_{G,I}(V^{\underline{\lambda}}))[\langle \lambda, 2\check{\rho}\rangle] \simeq (\mathrm{pr}^{\mu}_{(X\times X^I)^{\subset}})^!(\omega_X)\otimes W_{\lambda,\mu}[-\langle \mu, 2\check{\rho}\rangle],$$

where $W_{\lambda,\mu}$ is the cohomogically graded vector space such that

$$\mathrm{Sat}(V^\lambda)|_{\mathrm{Gr}^{\lambda+\mu}_G} \simeq \mathrm{IC}_{\mathrm{Gr}^{\lambda+\mu}_G} \otimes W_{\lambda,\mu}, \quad \mathsf{W}_{\lambda,\mu} \in \mathrm{Vect},$$

where $-|_-$ means !-restriction. By parity vanishing, $\mathsf{W}_{\lambda,\mu}$ lives in cohomological degrees ≥ 2.

Finally,

$$\mathsf{W}_\mu = \underset{\lambda\in\Lambda^+}{\mathrm{colim}}\, W_{\lambda,\mu},$$

and the cohomological estimate holds for W_μ because the poset Λ^+ is filtered (here we use the assumption that G is semi-simple and simply connected).

4 The Semi-infinite IC Sheaf and Drinfeld's Compactification

In this section, we will express $\mathrm{IC}^{\frac{\infty}{2}}_{\mathrm{Ran}}$ in terms of an *actual intersection cohomology sheaf*, i.e., one arising in finite-dimensional algebraic geometry (technically, on an algebraic stack locally of finite type).

Throughout this section, the curve X is assumed to be proper.

4.1 Drinfeld's Compactification

4.1.1

Let $\overline{\mathrm{Bun}}_B$ Drinfeld's relative compactification of the stack Bun_B along the fibers of the map $\mathrm{Bun}_B \to \mathrm{Bun}_G$.

That is, $\overline{\mathrm{Bun}}_B$ is the algebraic stack that classifies triples $(\mathcal{P}_G, \mathcal{P}_T, \kappa)$, where:

(i) $\mathcal{P}_G$ is a G-bundle on X.
(ii) $\mathcal{P}_T$ is a T-bundle on X.
(iii) κ is a *Plücker* data, i.e., a system of non-zero maps

$$\kappa^{\check{\lambda}} : \check{\lambda}(\mathcal{P}_T) \to \mathcal{V}^{\check{\lambda}}_{\mathcal{P}_G},$$

(here $\mathcal{V}^{\check{\lambda}}$ denotes the Weyl module with highest weight $\check{\lambda} \in \check{\Lambda}^+$) that satisfy Plücker relations, i.e., for $\check{\lambda}_1$ and $\check{\lambda}_2$ the diagram

$$\begin{array}{ccc} \check{\lambda}_1(\mathcal{P}_T) \otimes \check{\lambda}_2(\mathcal{P}_T) & \xrightarrow{\kappa^{\check{\lambda}_1} \otimes \kappa^{\check{\lambda}_2}} & \mathcal{V}^{\check{\lambda}_1}_{\mathcal{P}_G} \otimes \mathcal{V}^{\check{\lambda}_2}_{\mathcal{P}_G} \\ \sim\uparrow & & \uparrow \\ (\check{\lambda}_1 + \check{\lambda}_2)(\mathcal{P}_T) & \xrightarrow{\kappa^{\check{\lambda}_1+\check{\lambda}_2}} & \mathcal{V}^{\check{\lambda}_1+\check{\lambda}_2}_{\mathcal{P}_G} \end{array}$$

must commute.

The open substack

$$\mathrm{Bun}_B \overset{\mathbf{j}_{\mathrm{glob}}}{\hookrightarrow} \overline{\mathrm{Bun}}_B \tag{4.1}$$

corresponds to the condition that the maps $\kappa^{\check{\lambda}}$ be injective bundle maps.

We denote by $\overline{\mathsf{p}}$ (resp., $\overline{\mathsf{q}}$) resulting map from $\overline{\mathrm{Bun}}_B$ to Bun_G (resp., Bun_T), which sends $(\mathcal{P}_G, \mathcal{P}_T, \kappa)$ to $\mathcal{P}_G$ (resp., $\mathcal{P}_T$).

Its restriction to $\mathrm{Bun}_B \subset \overline{\mathrm{Bun}}_B$ is the usual map $\mathsf{q} : \mathrm{Bun}_B \to \mathrm{Bun}_G$ (resp., $\mathsf{q} : \mathrm{Bun}_B \to \mathrm{Bun}_T$) induced by the map of groups $B \to G$ (resp., $B \to T$).

4.1.2

For $\lambda \in \Lambda^{\mathrm{neg}}$, we let $\mathbf{i}^{\lambda}_{\mathrm{glob}}$ denote the map

$$\overline{\mathrm{Bun}}_B^{\leq\lambda} := \overline{\mathrm{Bun}}_B \times X^{\lambda} \to \overline{\mathrm{Bun}}_B,$$

given by

$$(\mathcal{P}_G, \mathcal{P}_T, \kappa, D) \mapsto (\mathcal{P}'_G, \mathcal{P}'_T, \kappa')$$

with $\mathcal{P}'_G = \mathcal{P}_G$, $\mathcal{P}'_T = \mathcal{P}_T(D)$, and κ' given by precomposing κ with the natural maps

$$\check{\lambda}(\mathcal{P}'_T) = \check{\lambda}(\mathcal{P}_T)(\check{\lambda}(D)) \hookrightarrow \check{\lambda}(\mathcal{P}_T).$$

It is known that $\bar{\mathbf{i}}^{\lambda}_{\text{glob}}$ is a finite morphism.

4.1.3

Let $\mathbf{j}^{\lambda}_{\text{glob}}$ denote the open embedding

$$\overline{\text{Bun}}_B^{=\lambda} := \text{Bun}_B \times X^{\lambda} \hookrightarrow \overline{\text{Bun}}_B \times X^{\lambda} =: \overline{\text{Bun}}_B^{\leq\lambda}.$$

Denote

$$\mathbf{i}^{\lambda}_{\text{glob}} = \bar{\mathbf{i}}^{\lambda}_{\text{glob}} \circ \mathbf{j}^{\lambda}_{\text{glob}}.$$

Note that by definition $\mathbf{i}^{0}_{\text{glob}} = \mathbf{j}^{0}_{\text{glob}} = \mathbf{j}_{\text{glob}}$ is the embedding (4.1).
The following is known:

Lemma 4.1.4 *The maps $\mathbf{i}^{\lambda}_{\text{glob}}$ are locally closed embeddings. Every field-valued point of $\overline{\text{Bun}}_B$ belongs to the image of exactly one such map.*

4.2 The Global Semi-infinite Category

4.2.1

Denote

$$\overline{\text{Bun}}_N := \overline{\text{Bun}}_B \underset{\text{Bun}_T}{\times} \text{pt}, \ \overline{\text{Bun}}_N^{\leq\lambda} := \overline{\text{Bun}}_B^{\leq\lambda} \underset{\text{Bun}_T}{\times} \text{pt}, \ \overline{\text{Bun}}_N^{=\lambda} := \overline{\text{Bun}}_B^{=\lambda} \underset{\text{Bun}_T}{\times} \text{pt},$$

where $\text{pt} \to \text{Bun}_T$ corresponds to the trivial bundle and the map $\overline{\text{Bun}}_B^{\leq\lambda} \to \text{Bun}_T$ is

$$\overline{\text{Bun}}_B^{\leq\lambda} = \overline{\text{Bun}}_B \times X^{\lambda} \xrightarrow{\bar{\mathfrak{q}}\times\text{id}} \text{Bun}_T \times X^{\lambda} \xrightarrow{\text{id}\times\text{AJ}} \text{Bun}_T \times \text{Bun}_T \xrightarrow{\text{mult}} \text{Bun}_T,$$

where AJ is the Abel–Jacobi map,

$$D \mapsto \mathcal{O}(D).$$

In particular,

$$\overline{\mathrm{Bun}}_N^{=\lambda} \simeq \mathrm{Bun}_B \underset{\mathrm{Bun}_T}{\times} X^\lambda,$$

where $X^\lambda \to \mathrm{Bun}_T$ is the composition of the map AJ and inversion on Bun_T.

We will denote by the same symbols the corresponding maps

$$\bar{\mathbf{i}}^\lambda_{\mathrm{glob}} : \overline{\mathrm{Bun}}_N^{\leq\lambda} \to \overline{\mathrm{Bun}}_N,\ \mathbf{j}^\lambda_{\mathrm{glob}} : \overline{\mathrm{Bun}}_N^{=\lambda} \to \overline{\mathrm{Bun}}_N^{\leq\lambda},\ \mathbf{i}^\lambda_{\mathrm{glob}} : \overline{\mathrm{Bun}}_N^{=\lambda} \to \overline{\mathrm{Bun}}_N.$$

Denote by $p^\lambda_{\mathrm{glob}}$ (resp., $\overline{p}^\lambda_{\mathrm{glob}}$) the projection from $\overline{\mathrm{Bun}}_N^{=\lambda}$ (resp., $\overline{\mathrm{Bun}}_N^{\leq\lambda}$) to X^λ.

4.2.2

We define

$$\mathrm{SI}^{\leq 0}_{\mathrm{glob}} \subset \mathrm{Shv}(\overline{\mathrm{Bun}}_N) \tag{4.2}$$

to be the full subcategory defined by the following condition:

An object $\mathcal{F} \in \mathrm{Shv}(\overline{\mathrm{Bun}}_N)$ belongs to $\mathrm{SI}^{\leq 0}_{\mathrm{glob}}$ if and only if for every $\lambda \in \Lambda^{\mathrm{neg}}$, the object

$$(\mathbf{i}^\lambda_{\mathrm{glob}})^!(\mathcal{F}) \in \mathrm{Shv}(\overline{\mathrm{Bun}}_N^{=\lambda})$$

belongs to the full subcategory

$$\mathrm{SI}^{=\lambda}_{\mathrm{glob}} \subset \mathrm{Shv}(\overline{\mathrm{Bun}}_N^{=\lambda}), \tag{4.3}$$

equal by definition to the essential image of the pullback functor

$$(p^\lambda_{\mathrm{glob}})^! : \mathrm{Shv}(X^\lambda) \to \mathrm{Shv}(\overline{\mathrm{Bun}}_N^{=\lambda}).$$

We note that the above pullback functor is fully faithful, since the map $p^\lambda_{\mathrm{glob}}$, being a base change of $\mathrm{Bun}_B \to \mathrm{Bun}_T$, is smooth with homologically contractible fibers.

4.2.3

Proceeding as in [Ga4, Sects. 4.5–4.7], one shows that the full subcategory (4.2) (resp., (4.3)) can also be defined by an equivariance condition with respect to a certain pro-unipotent groupoid.

In particular, the embedding (4.2) (resp., (4.3)) admits a *right* adjoint,[7] to be denoted $\mathrm{Av}_*^{\mathrm{SI}}$.

4.2.4

The functors

$$(\mathbf{i}^{\lambda}_{\mathrm{glob}})^! : \mathrm{Shv}(\overline{\mathrm{Bun}}_N) \to \mathrm{Shv}(\overline{\mathrm{Bun}}_N^{=\lambda}) \tag{4.4}$$

and

$$(\mathbf{i}^{\lambda}_{\mathrm{glob}})_* : \mathrm{Shv}(\overline{\mathrm{Bun}}_N^{=\lambda}) \to \mathrm{Shv}(\overline{\mathrm{Bun}}_N) \tag{4.5}$$

induce (same-named) functors

$$(\mathbf{i}^{\lambda}_{\mathrm{glob}})^! : \mathrm{SI}^{\leq 0}_{\mathrm{glob}} \to \mathrm{SI}^{=\lambda}_{\mathrm{glob}} \tag{4.6}$$

and

$$(\mathbf{i}^{\lambda}_{\mathrm{glob}})_* : \mathrm{SI}^{=\lambda}_{\mathrm{glob}} \to \mathrm{SI}^{\leq 0}_{\mathrm{glob}}\,. \tag{4.7}$$

Moreover, the diagram

$$\begin{array}{ccc} \mathrm{SI}^{\leq 0}_{\mathrm{glob}} & \xleftarrow{\mathrm{Av}_*^{\mathrm{SI}}} & \mathrm{Shv}(\overline{\mathrm{Bun}}_N) \\ {\scriptstyle (\mathbf{i}^{\lambda}_{\mathrm{glob}})^!}\downarrow & & \downarrow{\scriptstyle (\mathbf{i}^{\lambda}_{\mathrm{glob}})^!} \\ \mathrm{SI}^{=\lambda}_{\mathrm{glob}} & \xleftarrow{\mathrm{Av}_*^{\mathrm{SI}}} & \mathrm{Shv}(\overline{\mathrm{Bun}}_N^{=\lambda}), \end{array} \tag{4.8}$$

and similarly for $(\mathbf{i}^{\lambda}_{\mathrm{glob}})_*$.

Proposition 4.2.5

(a) *The partially defined left adjoint* $(\mathbf{i}^{\lambda}_{\mathrm{glob}})_!$ *of* (4.4)

$$(\mathbf{i}^{\lambda}_{\mathrm{glob}})^! : \mathrm{Shv}(\overline{\mathrm{Bun}}_N) \to \mathrm{Shv}(\overline{\mathrm{Bun}}_N^{=\lambda})$$

is defined on

[7] The corresponding assertion would be *false* for the corresponding embedding $\mathrm{SI}^{\leq 0}_{\mathrm{Ran}} \subset \mathrm{Shv}(\overline{S}^0_{\mathrm{Ran}})$; this is a geometric counterpart of the fact that the local field is not compact, while the quotient of adeles by principal adeles is compact.

$$\mathrm{SI}^{=\lambda}_{\mathrm{glob}} \subset \mathrm{Shv}(\overline{\mathrm{Bun}}_N^{=\lambda}).$$

(b) *The resulting functor*

$$\mathrm{SI}^{=\lambda}_{\mathrm{glob}} \to \mathrm{Shv}(\overline{\mathrm{Bun}}_N)$$

takes values in

$$\mathrm{SI}^{\leq 0}_{\mathrm{glob}} \subset \mathrm{Shv}(\overline{\mathrm{Bun}}_N)$$

and thus provides a left adjoint to (4.6).

Proof To prove point (a), it suffices to do so for the map $\mathbf{j}^{\lambda}_{\mathrm{glob}}$. We claim that the objects

$$(\mathbf{j}^{\lambda}_{\mathrm{glob}})_!(\omega_{\overline{\mathrm{Bun}}_N^{=\lambda}}) \in \mathrm{Shv}(\overline{\mathrm{Bun}}_N^{\leq\lambda}))$$

and

$$\omega_{\overline{\mathrm{Bun}}_N^{=\lambda}} \in \mathrm{Shv}(\overline{\mathrm{Bun}}_N^{=\lambda}))$$

are ULA with respect to the maps $\overline{p}^{\lambda}_{\mathrm{glob}}$ and $p^{\lambda}_{\mathrm{glob}}$, respectively. This would imply that for $\mathcal{F} \in \mathrm{Shv}(X^{\lambda})$, we have

$$(\mathbf{j}^{\lambda}_{\mathrm{glob}})_! \circ (p^{\lambda}_{\mathrm{glob}})^!(\mathcal{F}) \simeq (\overline{p}^{\lambda}_{\mathrm{glob}})^!(\mathcal{F}),$$

in particular, giving a formula for the left-hand side as an object of $\mathrm{Shv}(\overline{\mathrm{Bun}}_N^{\leq\lambda}))$.

To prove the required ULA property, it suffices to do so for the embedding

$$\mathbf{j}_{\mathrm{glob}} : \mathrm{Bun}_B \hookrightarrow \overline{\mathrm{Bun}}_B,$$

in which case this is the assertion of [BG1, Theorem 5.1.5].

Point (b) follows from the commutativity of the diagram (4.8) by passing to left adjoints.

□

By a Cousin argument, it follows formally from Proposition 4.2.5 that the partially defined functor $(\mathbf{i}^{\lambda}_{\mathrm{glob}})^*$, left adjoint to (4.5), is defined on $\mathrm{SI}^{\leq 0}_{\mathrm{glob}} \subset \mathrm{Shv}(\overline{\mathrm{Bun}}_N)$ and takes values in $\mathrm{SI}^{=\lambda}_{\mathrm{glob}} \subset \mathrm{Shv}(\overline{\mathrm{Bun}}_N^{=\lambda})$, thus providing a left adjoint to (4.7).

4.2.6

The embeddings

$$\mathrm{SI}_{\mathrm{glob}}^{=\lambda} \hookrightarrow \mathrm{Shv}(\overline{\mathrm{Bun}}_N^{=\lambda}) \text{ and } \mathrm{SI}_{\mathrm{glob}}^{\leq 0} \hookrightarrow \mathrm{Shv}(\overline{\mathrm{Bun}}_N)$$

are compatible with the t-structure on the target categories. This follows from the fact that the right adjoints $\mathrm{Av}_*^{\mathrm{SI}}$ (see Sect. 4.2.3) are right t-exact.

Hence, the categories $\mathrm{SI}_{\mathrm{glob}}^{=\lambda}$ and $\mathrm{SI}_{\mathrm{glob}}^{\leq 0}$ acquire t-structures. By construction, an object $\mathcal{F} \in \mathrm{SI}_{\mathrm{glob}}^{\leq 0}$ is connective (resp., coconnective) if and only if $(\mathbf{i}_{\mathrm{glob}}^{\lambda})^*(\mathcal{F})$ (resp., $(\mathbf{i}_{\mathrm{glob}}^{\lambda})^!(\mathcal{F})$ is connective (resp., coconnective) for every $\lambda \in \Lambda^{\mathrm{neg}}$.

4.2.7

We will denote by

$$\mathrm{IC}_{\mathrm{glob}}^{\frac{\infty}{2}} \in (\mathrm{SI}_{\mathrm{glob}}^{\leq 0})^{\heartsuit}$$

the minimal extension of $\mathrm{IC}_{\mathrm{Bun}_N} \in (\mathrm{SI}_{\mathrm{glob}}^{=0})^{\heartsuit}$ along $\mathbf{j}_{\mathrm{glob}}^{0}$.

4.3 Local vs. Global Compatibility for the Semi-infinite IC Sheaf

4.3.1

For every finite set I we have a canonically defined map

$$\pi_I : \overline{S}_I^0 \to \overline{\mathrm{Bun}}_N.$$

Together these maps combine to a map

$$\pi_{\mathrm{Ran}} : \overline{S}_{\mathrm{Ran}}^0 \to \overline{\mathrm{Bun}}_N.$$

4.3.2

Let $d = \dim(\mathrm{Bun}_N) = (g-1) \cdot \dim(N)$. The main result of this section is:

Theorem 4.3.3 *There exists an (unique) isomorphism*

$$(\pi_{\mathrm{Ran}})^!(\mathrm{IC}_{\mathrm{glob}}^{\frac{\infty}{2}})[d] = \mathrm{IC}_{\mathrm{Ran}}^{\frac{\infty}{2}},$$

extending the tautological identification over Bun_N.

4.3.4

The next few subsections are devoted to the proof of this theorem. Modulo auxiliary assertions, the proof will be given in Sect. 4.6.8.

4.4 The Local vs. Global Compatibility for the Semi-infinite Category

This subsection contains some preparatory material for the proof of Theorem 4.3.3.

4.4.1

First, we observe:

Lemma 4.4.2 *For every λ, we have a commutative diagram*

$$\begin{array}{ccc} \overline{S}^{\leq\lambda}_{\mathrm{Ran}} & \xrightarrow{\bar{\mathbf{i}}^{\lambda}} & \overline{S}^{0}_{\mathrm{Ran}} \\ \downarrow & & \downarrow \pi_{\mathrm{Ran}} \\ \overline{\mathrm{Bun}}_N^{\leq\lambda} & \xrightarrow{\bar{\mathbf{i}}^{\lambda}_{\mathrm{glob}}} & \overline{\mathrm{Bun}}_N. \end{array}$$

The corresponding diagram

$$\begin{array}{ccc} S^{=\lambda}_{\mathrm{Ran}} & \xrightarrow{\mathbf{i}^{\lambda}} & \overline{S}^{0}_{\mathrm{Ran}} \\ \pi^{\lambda}_{\mathrm{Ran}} \downarrow & & \downarrow \pi_{\mathrm{Ran}} \\ \overline{\mathrm{Bun}}_N^{=\lambda} & \xrightarrow{\mathbf{i}^{\lambda}_{\mathrm{glob}}} & \overline{\mathrm{Bun}}_N \end{array} \tag{4.9}$$

is Cartesian, and we have a commutative diagram

$$\begin{array}{ccc} S^{=\lambda}_{\mathrm{Ran}} & \xrightarrow{p^{\lambda}_{\mathrm{Ran}}} & (X^{\lambda} \times \mathrm{Ran})^{\subset} \\ \pi^{\lambda}_{\mathrm{Ran}} \downarrow & & \downarrow \mathrm{pr}^{\lambda}_{\mathrm{Ran}} \\ \overline{\mathrm{Bun}}_N^{=\lambda} & \xrightarrow{p^{\lambda}_{\mathrm{glob}}} & X^{\lambda}. \end{array}$$

The assertions parallel to those in the above lemma hold for Ran replaced by X^I for an individual finite set I.

4.4.3

The following assertion is not necessary for the needs of this paper, but we will prove it for the sake of completeness (see Sect. A.1.11):

Theorem 4.4.4 *The functor*

$$(\pi_{\text{Ran}})^! : \text{Shv}(\overline{\text{Bun}}_N) \to \text{Shv}(\overline{S}^0_{\text{Ran}})$$

is fully faithful.

When working with an individual stratum, a stronger assertion is true (to be proved in Sect. 4.5): Consider the map

$$(p^\lambda_{\text{Ran}} \times \pi^\lambda_{\text{Ran}}) : S^\lambda_{\text{Ran}} \to (X^\lambda \times \text{Ran})^{\subset} \underset{X^\lambda}{\times} \overline{\text{Bun}}_N^{=\lambda}.$$

Proposition 4.4.5 *The functor*

$$(p^\lambda_{\text{Ran}} \times \pi^\lambda_{\text{Ran}})^! : \text{Shv}((X^\lambda \times \text{Ran})^{\subset} \underset{X^\lambda}{\times} \overline{\text{Bun}}_N^{=\lambda}) \to \text{Shv}(S^\lambda_{\text{Ran}})$$

is fully faithful.

Combining with Lemma 2.3.3, we obtain:

Corollary 4.4.6 *The functor*

$$(\pi^\lambda_{\text{Ran}})^! : \text{Shv}(\overline{\text{Bun}}_N^{=\lambda}) \to \text{Shv}(S^\lambda_{\text{Ran}})$$

is fully faithful.

4.4.7

Next we claim:

Proposition 4.4.8 *For every finite set I, the functor*

$$(\pi_I)^! : \text{Shv}(\overline{\text{Bun}}_N) \to \text{Shv}(\overline{S}^0_I)$$

sends $\text{SI}^{\leq 0}_{\text{glob}}$ *to* $\text{SI}^{\leq 0}_I$.

Proof Note that an object $\mathcal{F} \in \text{Shv}(\overline{S}^0_I)$ belongs to $\text{SI}^{\leq 0}_I$ if and only if $(\mathbf{i}^\lambda)^!(\mathcal{F})$ belongs to $\text{SI}^{=\lambda}_I$ for every λ. Now the result follows from the identification

$$\text{pr}^\lambda_I \circ p^\lambda_I = p^\lambda_{\text{glob}} \circ \pi^\lambda_I.$$

□

We will now deduce:

Corollary 4.4.9 *An object of* $\mathrm{Shv}(\overline{\mathrm{Bun}}_N)$ *belongs to* $\mathrm{SI}^{\leq 0}_{\mathrm{glob}}$ if and only if *its pullback under* $(\pi_{\mathrm{Ran}})^!$ *belongs to* $\mathrm{SI}^{\leq 0}_{\mathrm{Ran}} \subset \mathrm{Shv}(\overline{S}^0_{\mathrm{Ran}})$.

Proof The "only if" direction is the content of Proposition 4.4.8.

For the "if" direction, we need to show that if an object $\mathcal{F} \in \mathrm{Shv}(\overline{\mathrm{Bun}}_N^{=\lambda})$ is such that

$$(\pi^\lambda_{\mathrm{Ran}})^!(\mathcal{F}) \simeq (p^\lambda_{\mathrm{Ran}})^!(\mathcal{F}')$$

for some $\mathcal{F}' \in \mathrm{Shv}((X^\lambda \times \mathrm{Ran})^\subset)$, then $\mathcal{F}$ is the pullback of an object in $\mathrm{Shv}(X^\lambda)$ along $p^\lambda_{\mathrm{glob}}$.

By Proposition 4.4.5, in the diagram

$$\begin{array}{ccc}
S^\lambda_{\mathrm{Ran}} & & \\
\Big\downarrow{\scriptstyle p^\lambda_{\mathrm{Ran}} \times \pi^\lambda_{\mathrm{Ran}}} & & \\
(X^\lambda \times \mathrm{Ran})^\subset \underset{X^\lambda}{\times} \overline{\mathrm{Bun}}_N^{=\lambda} & \xrightarrow{\mathrm{id}_{(X^\lambda \times \mathrm{Ran})^\subset} \times p^\lambda_{\mathrm{glob}}} & (X^\lambda \times \mathrm{Ran})^\subset \\
\Big\downarrow{\scriptstyle \mathrm{pr}^\lambda_{\mathrm{Ran}} \times \mathrm{id}_{\overline{\mathrm{Bun}}_N^{=\lambda}}} & & \Big\downarrow{\scriptstyle \mathrm{pr}^\lambda_{\mathrm{Ran}}} \\
\overline{\mathrm{Bun}}_N^{=\lambda} & \xrightarrow{p^\lambda_{\mathrm{glob}}} & X^\lambda
\end{array}$$

we have

$$\mathcal{F} \overset{\text{Lemma 2.3.3}}{\simeq} (\mathrm{pr}^\lambda_{\mathrm{Ran}} \times \mathrm{id}_{\overline{\mathrm{Bun}}_N^{=\lambda}})_! \circ (\mathrm{pr}^\lambda_{\mathrm{Ran}} \times \mathrm{id}_{\overline{\mathrm{Bun}}_N^{=\lambda}})^!(\mathcal{F}) \overset{\text{Proposition 4.4.5}}{\simeq}$$

$$\simeq (\mathrm{pr}^\lambda_{\mathrm{Ran}} \times \mathrm{id}_{\overline{\mathrm{Bun}}_N^{=\lambda}})_! \circ (p^\lambda_{\mathrm{Ran}} \times \pi^\lambda_{\mathrm{Ran}})_! \circ (p^\lambda_{\mathrm{Ran}} \times \pi^\lambda_{\mathrm{Ran}})^! \circ (\mathrm{pr}^\lambda_{\mathrm{Ran}} \times \mathrm{id}_{\overline{\mathrm{Bun}}_N^{=\lambda}})^!(\mathcal{F}) \simeq$$

$$\simeq (\mathrm{pr}^\lambda_{\mathrm{Ran}} \times \mathrm{id}_{\overline{\mathrm{Bun}}_N^{=\lambda}})_! \circ (p^\lambda_{\mathrm{Ran}} \times \pi^\lambda_{\mathrm{Ran}})_! \circ (\pi^\lambda_{\mathrm{Ran}})^!(\mathcal{F}) \overset{\text{assumption}}{\simeq}$$

$$\simeq (\mathrm{pr}^\lambda_{\mathrm{Ran}} \times \mathrm{id}_{\overline{\mathrm{Bun}}_N^{=\lambda}})_! \circ (p^\lambda_{\mathrm{Ran}} \times \pi^\lambda_{\mathrm{Ran}})_! \circ (p^\lambda_{\mathrm{Ran}})^!(\mathcal{F}') \simeq$$

$$\simeq (\mathrm{pr}^\lambda_{\mathrm{Ran}} \times \mathrm{id}_{\overline{\mathrm{Bun}}_N^{=\lambda}})_! \circ (p^\lambda_{\mathrm{Ran}} \times \pi^\lambda_{\mathrm{Ran}})_! \circ (p^\lambda_{\mathrm{Ran}} \times \pi^\lambda_{\mathrm{Ran}})^! \circ (\mathrm{id}_{(X^\lambda \times \mathrm{Ran})^\subset} \times p^\lambda_{\mathrm{glob}})^!(\mathcal{F}') \simeq$$

$$\overset{\text{Proposition 4.4.5}}{\simeq} (\mathrm{pr}^\lambda_{\mathrm{Ran}} \times \mathrm{id}_{\overline{\mathrm{Bun}}_N^{=\lambda}})_! \circ (\mathrm{id}_{(X^\lambda \times \mathrm{Ran})^\subset} \times p^\lambda_{\mathrm{glob}})^!(\mathcal{F}') \simeq (p^\lambda_{\mathrm{glob}})^! \circ (\mathrm{pr}^\lambda_{\mathrm{Ran}})_!(\mathcal{F}'),$$

as required (the last isomorphism is base change, which holds due to the fact that the map $\mathrm{pr}^{\lambda}_{\mathrm{Ran}}$ is pseudo-proper[8]). □

4.5 Proof of Proposition 4.4.5

4.5.1

Consider the morphism

$$(p^{\lambda}_{\mathrm{Ran}} \times \pi^{\lambda}_{\mathrm{Ran}}) : S^{\lambda}_{\mathrm{Ran}} \to (X^{\lambda} \times \mathrm{Ran})^{\subset} \underset{X^{\lambda}}{\times} \overline{\mathrm{Bun}}_N^{=\lambda}.$$

A point of $S^{\lambda}_{\mathrm{Ran}}$ is the following data:

(i) A B-bundle $\mathcal{P}_B$ on X (denote by $\mathcal{P}_T$ the induced T-bundle)
(ii) A Λ^{neg}-valued divisor D on X (we denote by $\mathcal{O}(D)$ the corresponding T-bundle)
(iii) An identification $\mathcal{P}_T \simeq \mathcal{O}(D)$
(iv) A finite non-empty set $\mathcal{J}$ of points of X that contains the support of D
(v) A trivialization α of $\mathcal{P}_B$ away from $\mathcal{J}$, such that the induced trivialization of $\mathcal{P}_T|_{X-\mathcal{J}}$ agrees with the tautological trivialization of $\mathcal{O}(D)|_{X-\mathcal{J}}$

4.5.2

The map $(p^{\lambda}_{\mathrm{Ran}} \times \pi^{\lambda}_{\mathrm{Ran}})$ amounts to forgetting the data of (v) above. It is clear that for an affine test scheme Y and a Y-point of

$$(X^{\lambda} \times \mathrm{Ran})^{\subset} \underset{X^{\lambda}}{\times} \overline{\mathrm{Bun}}_N^{=\lambda},$$

the set of its lifts to a Y-point of $S^{\lambda}_{\mathrm{Ran}}$ is non-empty and is a torsor for the group

$$\mathrm{Maps}(Y \times X - \Gamma_{\mathcal{J}}, N).$$

For a given Y and $\mathcal{J} \subset \mathrm{Maps}(Y, X)$, let $\mathbf{Maps}_Y(X - \mathcal{J}, N)$ be the prestack over Y that assigns to $Y' \to Y$ the set of maps

$$\mathrm{Maps}(Y' \times X - (Y' \underset{Y}{\times} \Gamma_{\mathcal{J}}), N).$$

[8] See Sect. A.2.4, where this notion is recalled.

Thus, it suffices to show that the projection $\mathbf{Maps}_Y(X-\mathfrak{I}, N) \to Y$ is *universally homologically contractible*, see Sect. A.1.8 for what this means.

4.5.3

Since N is unipotent, it is isomorphic to $\mathbb{A}^m$, where $m = \dim(N)$. Hence, it suffices to show that the map

$$\mathbf{Maps}_Y(X - \mathfrak{I}, \mathbb{A}^1) \to Y$$

is universally homologically contractible.

However, the latter is clear: the prestack $\mathbf{Maps}_Y(X - \mathfrak{I}, \mathbb{A}^1)$ is isomorphic to the ind-scheme $\mathbb{A}^\infty \times Y$, where

$$\mathbb{A}^\infty \simeq \underset{n}{\operatorname{colim}}\, \mathbb{A}^n.$$

□

4.6 The Key Isomorphism

4.6.1

The base-change isomorphism

$$(\pi_I)^! \circ (\mathbf{i}^\lambda_{\text{glob}})_* \simeq (\mathbf{i}^\lambda)_* \circ (\pi_I)^!$$

in the diagram (4.9) gives rise to a natural transformation

$$(\mathbf{i}^\lambda)^* \circ (\pi_I)^! \to (\pi_I^\lambda)^! \circ (\mathbf{i}^\lambda_{\text{glob}})^* \tag{4.10}$$

as functors

$$\mathrm{SI}^{\leq 0}_{\text{glob}} \rightrightarrows \mathrm{SI}^{=\lambda}_I,$$

see Proposition 2.5.3(a) for the notation $(\mathbf{i}^\lambda)^*$.

4.6.2

In Sect. 4.7, we will prove:

Proposition 4.6.3 *The natural transformation* (4.10) *is an isomorphism.*

We will now deduce some corollaries of Proposition 4.6.3; these will easily imply Theorem 4.3.3, see Sect. 4.6.8.

First, combining Proposition 4.6.3 with Proposition 2.5.3(c), we obtain:

Corollary 4.6.4 *The natural transformation*

$$(\mathbf{i}^\lambda)^* \circ (\pi_{\mathrm{Ran}})^! \to (\pi^\lambda_{\mathrm{Ran}})^! \circ (\mathbf{i}^\lambda_{\mathrm{glob}})^*$$

as functors

$$\mathrm{SI}^{\leq 0}_{\mathrm{glob}} \rightrightarrows \mathrm{SI}^{=\lambda}_{\mathrm{Ran}}$$

is an isomorphism.

Next, by a Cousin argument, from Proposition 4.6.3, we formally obtain:

Corollary 4.6.5 *The natural transformation*

$$(\mathbf{i}^\lambda)_! \circ (\pi^\lambda_I)^! \to (\pi_I)^! \circ (\mathbf{i}^\lambda_{\mathrm{glob}})_!,$$

arising by adjunction from

$$(\pi^\lambda_I)^! \circ (\mathbf{i}^\lambda_{\mathrm{glob}})^! \simeq (\mathbf{i}^\lambda)^! \circ (\pi_I)^!,$$

is an isomorphism of functors

$$\mathrm{SI}^{=\lambda}_{\mathrm{glob}} \rightrightarrows \mathrm{SI}^{\leq 0}_I .$$

Combining Corollary 4.6.5 with Corollary 2.5.6(c), we obtain:

Corollary 4.6.6 *The natural transformation*

$$(\mathbf{i}^\lambda)_! \circ (\pi^\lambda_{\mathrm{Ran}})^! \to (\pi_{\mathrm{Ran}})^! \circ (\mathbf{i}^\lambda_{\mathrm{glob}})_!$$

is an isomorphism of functors

$$\mathrm{SI}^{=\lambda}_{\mathrm{glob}} \rightrightarrows \mathrm{SI}^{\leq 0}_{\mathrm{Ran}} .$$

Finally, we claim:

Corollary 4.6.7 *The functor*

$$\pi^![d] : \mathrm{SI}^{\leq 0}_{\mathrm{glob}} \to \mathrm{SI}^{\leq 0}_{\mathrm{Ran}}$$

is t-exact.

Proof This follows from Corollary 4.6.4, combined with the (tautological) isomorphism

$$(\mathbf{i}^\lambda)^! \circ (\pi_{\mathrm{Ran}})^! \simeq (\pi^\lambda_{\mathrm{Ran}})^! \circ (\mathbf{i}^\lambda_{\mathrm{glob}})^!.$$

□

4.6.8

Note that Corollary 4.6.7 immediately implies Theorem 4.3.3.

Remark 4.6.9 In Sect. 7.4, we will present another construction of the map in one direction

$$\mathrm{IC}^{\frac{\infty}{2}}_{\mathrm{Ran}} \to \pi^!(\mathrm{IC}^{\frac{\infty}{2}}_{\mathrm{glob}})[d],$$

where we will realize $\mathrm{IC}^{\frac{\infty}{2}}_{\mathrm{Ran}}$ as $'\mathrm{IC}^{\frac{\infty}{2}}_{\mathrm{Ran}}$.

4.6.10

Let us now prove Proposition 3.2.2.

Proof By Corollary 4.6.7, it suffices to show that the objects

$$(\mathbf{i}^{\lambda}_{\mathrm{glob}})_!(\mathrm{IC}_{\overline{\mathrm{Bun}}_N^{=\lambda}}) \text{ and } (\mathbf{i}^{\lambda}_{\mathrm{glob}})_*(\mathrm{IC}_{\overline{\mathrm{Bun}}_N^{=\lambda}})$$

belong to the heart of the t-structure (i.e., are perverse sheaves on $\overline{\mathrm{Bun}}_N$).

We claim that the morphism $\mathbf{i}^{\lambda}_{\mathrm{glob}}$ is affine, which would imply that the functors $(\mathbf{i}^{\lambda}_{\mathrm{glob}})_!$ and $(\mathbf{i}^{\lambda}_{\mathrm{glob}})_*$ are t-exact.

Indeed, $\mathbf{i}^{\lambda}_{\mathrm{glob}}$ is the base change of the morphism

$$\mathbf{i}^{\lambda}_{\mathrm{glob}} : \mathrm{Bun}_B \times X^{\lambda} \to \overline{\mathrm{Bun}}_B,$$

which we claim to be affine.

Indeed, $\mathbf{i}^{\lambda}_{\mathrm{glob}} = \bar{\mathbf{i}}^{\lambda}_{\mathrm{glob}} \circ \mathbf{j}^{\lambda}_{\mathrm{glob}}$, where $\bar{\mathbf{i}}^{\lambda}_{\mathrm{glob}}$ is a finite morphism, and $\mathbf{j}^{\lambda}_{\mathrm{glob}}$ is known to be an affine open embedding (see [FGV, Proposition 3.3.1]). □

4.7 Proof of Proposition 4.6.3

4.7.1

Let $\mathcal{F}$ be an object of $\mathrm{SI}^{\leq 0}_{\mathrm{glob}}$. We need to show that the map

$$(s^{\lambda}_{I})^! \circ (\mathbf{i}^{\lambda})^* \circ (\pi_I)^!(\mathcal{F}) \simeq (s^{\lambda}_{I})^! \circ (\pi^{\lambda}_{I})^! \circ (\mathbf{i}^{\lambda}_{\mathrm{glob}})^*(\mathcal{F}) \quad (4.11)$$

is an isomorphism.

4.7.2

We first rewrite the left-hand side in (4.11).

As a first step, we note that by (2.7), we have

$$(s_I^{\lambda})^! \circ (\mathbf{i}^{\lambda})^* \circ (\pi_I)^!(\mathcal{F}) \simeq (p_I^{-,\lambda})_* \circ (\mathbf{i}^{-,\lambda})^! \circ (\pi_I)^!(\mathcal{F}). \tag{4.12}$$

4.7.3

For $\lambda \in \Lambda^{\mathrm{neg}}$, let $\mathcal{Z}^{\lambda}$ be the Zastava space, i.e., this is the open substack of

$$\overline{\mathrm{Bun}}_N \underset{\mathrm{Bun}_G}{\times} \mathrm{Bun}_{B^-}^{-\lambda},$$

corresponding to the condition that the B^--reduction and the generalized N-reduction of a given G-bundle are generically transversal.

Let $\mathfrak{q}$ denote the forgetful map $\mathcal{Z}^{\lambda} \to \overline{\mathrm{Bun}}_N$. Let $\mathfrak{p}$ denote the projection

$$\mathcal{Z}^{\lambda} \to X^{\lambda},$$

and let $\mathfrak{s}$ denote its section

$$X^{\lambda} \to \mathcal{Z}^{\lambda}.$$

4.7.4

Note that we have a canonical identification

$$(X^{\lambda} \times X^I)^{\subset} \underset{X^{\lambda}}{\times} \mathcal{Z}^{\lambda} \simeq \overline{S}_I^0 \cap S_I^{-,\lambda}, \tag{4.13}$$

so that the projection

$$(\mathrm{id}_{(X^{\lambda} \times X^I)^{\subset}} \times \mathfrak{p}) : (X^{\lambda} \times X^I)^{\subset} \underset{X^{\lambda}}{\times} \mathcal{Z}^{\lambda} \to (X^{\lambda} \times X^I)^{\subset}$$

identifies with

$$\overline{S}_I^0 \cap S_I^{-,\lambda} \to S_I^{-,\lambda} \xrightarrow{p_I^{-,\lambda}} (X^{\lambda} \times X^I)^{\subset}.$$

4.7.5

Hence, the right-hand side in (4.12) can be rewritten as

$$(\mathrm{id}_{(X^{\lambda} \times X^I)^{\subset}} \times \mathfrak{p})_* \circ (\mathrm{pr}_I^{\lambda} \times \mathrm{id}_{\mathcal{Z}^{\lambda}})^! \circ \mathfrak{q}^!(\mathcal{F}), \tag{4.14}$$

where the maps are as shown in the diagram

$$\begin{array}{ccccc}
(X^\lambda \times X^I)^{\subset} \underset{X^\lambda}{\times} \mathcal{Z}^\lambda & \xrightarrow{\mathrm{pr}^\lambda_I \times \mathrm{id}_{\mathcal{Z}^\lambda}} & \mathcal{Z}^\lambda & \xrightarrow{\mathfrak{q}} & \overline{\mathrm{Bun}}_N \\
{\scriptstyle \mathrm{id}_{(X^\lambda \times X^I)^{\subset}} \times \mathfrak{p}}\downarrow & & \downarrow{\scriptstyle \mathfrak{p}} & & \\
(X^\lambda \times X^I)^{\subset} & \xrightarrow{\mathrm{pr}^\lambda_I} & X^\lambda. & &
\end{array}$$

By base change, we rewrite (4.14) as

$$(\mathrm{pr}^\lambda_I)^! \circ \mathfrak{p}_* \circ \mathfrak{q}^!(\mathcal{F}). \tag{4.15}$$

4.7.6

The adjoint action of T on N defines an action of T on $\overline{\mathrm{Bun}}_N$. It is easy to see that every object of $\mathrm{SI}^{\leq 0}$ is monodromic for this action. Hence, the same is true for its pullback to $\mathcal{Z}^\lambda$.

Choose a dominant coweight as in Sect. 2.7.1. Applying the contraction principle for the action of $\mathbb{G}_m$ along the fibers of $\mathfrak{p}$ (see [DrGa, Proposition 3.2.2]), we rewrite (4.15) as

$$(\mathrm{pr}^\lambda_I)^! \circ \mathfrak{s}^* \circ \mathfrak{q}^!(\mathcal{F}). \tag{4.16}$$

To summarize, we have rewritten the left-hand side in (4.11) as (4.16).

4.7.7

We now rewrite the right-hand side in (4.11).

Note that we have a Cartesian diagram

$$\begin{array}{ccc}
X^\lambda & \xrightarrow{\mathfrak{s}} & \mathcal{Z}^\lambda \\
{\scriptstyle \mathfrak{q}^\lambda}\downarrow & & \downarrow{\scriptstyle \mathfrak{q}} \\
\overline{\mathrm{Bun}}_N^{=\lambda} & \xrightarrow{\mathbf{i}^\lambda_{\mathrm{glob}}} & \overline{\mathrm{Bun}}_N,
\end{array} \tag{4.17}$$

where the map $\mathfrak{q}^\lambda$ is given by

$$X^\lambda \simeq X^\lambda \underset{\mathrm{Bun}_T}{\times} \mathrm{Bun}_T \to X^\lambda \underset{\mathrm{Bun}_T}{\times} \mathrm{Bun}_B \simeq \overline{\mathrm{Bun}}_N^{=\lambda}.$$

Note also that the map

$$(X^\lambda \times X^I)^\subset \xrightarrow{s_I^\lambda} S_I^\lambda \xrightarrow{\pi_I^\lambda} \overline{\mathrm{Bun}}_N^{=\lambda}$$

identifies with

$$(X^\lambda \times X^I)^\subset \xrightarrow{\mathrm{pr}_I^\lambda} X^\lambda \xrightarrow{\mathfrak{q}^\lambda} \overline{\mathrm{Bun}}_N^{=\lambda}.$$

Hence, the right-hand side in (4.11) identifies with

$$(\mathrm{pr}_I^\lambda)^! \circ (\mathfrak{q}^\lambda)^! \circ (\mathbf{i}_{\mathrm{glob}}^\lambda)^*(\mathcal{F}). \tag{4.18}$$

4.7.8

Unwinding the identifications, we obtain that the map in (4.11) is induced by the natural transformation

$$\mathfrak{s}^* \circ \mathfrak{q}^! \to (\mathfrak{q}^\lambda)^! \circ (\mathbf{i}_{\mathrm{glob}}^\lambda)^*, \tag{4.19}$$

coming from the Cartesian square (4.17).

Thus, it suffices to show that the natural transformation (4.19) is an isomorphism, when evaluated on objects from $\mathrm{SI}_{\mathrm{glob}}^{\leq 0}$.

However, the latter is done by repeating the argument of [Ga1, Sect. 3.9]:

We first consider the case when $-\lambda$ is *sufficiently dominant*, in which case the morphism $\mathfrak{q}$ is smooth, being the base change of $\mathrm{Bun}_{B^-}^{-\lambda} \to \mathrm{Bun}_G$. In this case, the fact that (4.19) is an isomorphism follows by smoothness.

Then we reduce the case of a general λ to one above using the factorization property of $\mathcal{Z}^\lambda$.

□

4.7.9

Thus, we have completed the proof of Proposition 4.6.3 and hence also of Theorem 4.3.3.

4.8 Relation to the IC Sheaf on Zastava Spaces

4.8.1

Recall the Zastava spaces

$$\mathcal{Z}^\lambda \subset \overline{\mathrm{Bun}}_N \underset{\mathrm{Bun}_G}{\times} \mathrm{Bun}_{B^-}^{-\lambda},$$

introduced in Sect. 4.7.3.

Let $\overset{\circ}{\mathcal{Z}}^\lambda \subset \mathcal{Z}^\lambda$ denote the open subscheme equal to

$$\mathrm{Bun}_N \underset{\overline{\mathrm{Bun}}_N}{\times} \mathcal{Z}^\lambda.$$

4.8.2

Note now that the identification (4.13) gives rise to a map

$$\mathrm{q}' : (X^\lambda \times \mathrm{Ran})^{\subset} \underset{X^\lambda}{\times} \mathcal{Z}^\lambda \to \overline{S}^0_{\mathrm{Ran}}. \tag{4.20}$$

Let $\mathrm{pr}^\lambda_{\mathrm{Ran}} \times \mathrm{id}_{\mathcal{Z}^\lambda}$ denote the projection

$$(X^\lambda \times \mathrm{Ran})^{\subset} \underset{X^\lambda}{\times} \mathcal{Z}^\lambda \to \mathcal{Z}^\lambda.$$

We claim:

Proposition 4.8.3 *There exists a canonical isomorphism*

$$(\mathrm{pr}^\lambda_{\mathrm{Ran}} \times \mathrm{id}_{\mathcal{Z}^\lambda})^!(\mathrm{IC}_{\mathcal{Z}^\lambda}) \simeq (\mathrm{q}')^!(\mathrm{IC}^{\frac{\infty}{2}})[\langle \lambda, 2\check{\rho} \rangle],$$

extending the tautological identification of the restriction of either side to

$$(X^\lambda \times \mathrm{Ran})^{\subset} \underset{X^\lambda}{\times} \overset{\circ}{\mathcal{Z}}^\lambda$$

with $\omega_{(X^\lambda \times \mathrm{Ran})^{\subset} \underset{X^\lambda}{\times} \overset{\circ}{\mathcal{Z}}^\lambda}[\langle \lambda, 2\check{\rho} \rangle]$.

4.8.4 Proof of Proposition 4.8.3

We have a commutative diagram

$$\begin{array}{ccc} (X^\lambda \times \mathrm{Ran})^{\subset} \underset{X^\lambda}{\times} \mathcal{Z}^\lambda & \xrightarrow{\mathrm{q}'} & \overline{S}^0_{\mathrm{Ran}} \\ {\scriptstyle \mathrm{pr}^\lambda_{\mathrm{Ran}} \times \mathrm{id}_{\mathcal{Z}^\lambda}} \downarrow & & \downarrow {\scriptstyle \pi_{\mathrm{Ran}}} \\ \mathcal{Z}^\lambda & \xrightarrow{\mathrm{q}} & \overline{\mathrm{Bun}}_N. \end{array}$$

According to [Ga1, Prop. 3.6.5(a)], we have a canonical isomorphism

$$\mathfrak{q}^!(\mathrm{IC}^{\frac{\infty}{2}}_{\mathrm{glob}})[(g-1)\cdot \dim(N) + \langle \lambda, 2\check{\rho}\rangle] \simeq \mathrm{IC}_{\mathcal{Z}^\lambda}.$$

Now the assertion follows from Theorem 4.3.3.

□

4.9 Computation of Fibers

In this subsection, we will prove Theorem 3.4.5. One possible proof follows from the description of the objects

$$(\mathbf{i}^\lambda_{\mathrm{glob}})^!(\mathrm{IC}^{\frac{\infty}{2}}_{\mathrm{glob}}) \text{ and } (\mathbf{i}^\lambda_{\mathrm{glob}})^*(\mathrm{IC}^{\frac{\infty}{2}}_{\mathrm{glob}})$$

in [BG2, Proposition 4.4], combined with Theorem 4.3.3.

Instead, we will actually reprove [BG2, Proposition 4.4], see Theorem 4.9.3 below, using our Theorem 4.3.3.

Remark 4.9.1 Let us add a clarification on the order of the argument proving Theorems 3.4.5 and 4.9.3:

(1) In Sect. 3.6, we defined the object $'\mathrm{IC}^{\frac{\infty}{2}}_{\mathrm{Ran}}$.
(2) In Proposition 3.7.4, we showed that the !-restrictions of $'\mathrm{IC}^{\frac{\infty}{2}}_{\mathrm{Ran}}$ to the strata S^λ_{Ran} are strictly coconnective.
(3) In Proposition 3.7.3, we calculated the *-restrictions of $'\mathrm{IC}^{\frac{\infty}{2}}_{\mathrm{Ran}}$ to the @strata S^λ_{Ran} and showed that they are isomorphic to (the pullbacks of) $\mathrm{Fact}^{\mathrm{alg}}(\mathcal{O}(\check{N}))_{X^\lambda}[-\langle \lambda, 2\check{\rho}\rangle]$; in particular, they are strictly connective.
(4) Points (2) and (3) imply that $'\mathrm{IC}^{\frac{\infty}{2}}_{\mathrm{Ran}}$ is isomorphic to $\mathrm{IC}^{\frac{\infty}{2}}_{\mathrm{Ran}}$.
(5) Points (3) and (4) imply that the *-restrictions of $\mathrm{IC}^{\frac{\infty}{2}}_{\mathrm{Ran}}$ to the strata S^λ_{Ran} are isomorphic to (the pullbacks of) $\mathrm{Fact}^{\mathrm{alg}}(\mathcal{O}(\check{N}))_{X^\lambda}[-\langle \lambda, 2\check{\rho}\rangle]$, thus proving the part of Theorem 3.4.5 about *-restrictions.
(6) Point (5) above, combined with Theorem 4.3.3 and Corollary 4.6.4, will imply Theorem 4.9.3(a) (see below).
(7) Point (a) of Theorem 4.9.3 will imply point (b) by a duality argument (see below).
(8) Point (b) of Theorem 4.9.3 will imply the assertion of Theorem 3.4.5 about !-restrictions (see below).

4.9.2

We first prove:

Theorem 4.9.3

(a) $(\mathbf{i}^{\lambda}_{\text{glob}})^{*}(\text{IC}^{\frac{\infty}{2}}_{\text{glob}}) \simeq (p^{\lambda}_{\text{glob}})^{!}(\text{Fact}^{\text{alg}}(\mathcal{O}(\check{N}))_{X^{\lambda}})[-d - \langle \lambda, 2\check{\rho} \rangle]$.

(b) $(\mathbf{i}^{\lambda}_{\text{glob}})^{!}(\text{IC}^{\frac{\infty}{2}}_{\text{glob}}) \simeq (p^{\lambda}_{\text{glob}})^{!}(\text{Fact}^{\text{coalg}}(U(\check{\mathfrak{n}}^{-}))_{X^{\lambda}})[-d - \langle \lambda, 2\check{\rho} \rangle]$.

Proof We first prove point (a). Let $\mathcal{F}^{\lambda} \in \text{Shv}(X^{\lambda})$ be such that

$$(\mathbf{i}^{\lambda}_{\text{glob}})^{*}(\text{IC}^{\frac{\infty}{2}}_{\text{glob}}) \simeq (p^{\lambda}_{\text{glob}})^{!}(\mathcal{F}^{\lambda})[-d - \langle \lambda, 2\check{\rho} \rangle]. \tag{4.21}$$

We will show that

$$\mathcal{F}^{\lambda} \simeq \text{Fact}^{\text{alg}}(\mathcal{O}(\check{N}))_{X^{\lambda}}.$$

Applying $(\pi^{\lambda}_{\text{Ran}})^{!}$ to both sides in (4.21), we obtain

$$\begin{aligned}(\pi^{\lambda}_{\text{Ran}})^{!} \circ (\mathbf{i}^{\lambda}_{\text{glob}})^{*}(\text{IC}^{\frac{\infty}{2}}_{\text{glob}}) &\simeq \\ \simeq (\pi^{\lambda}_{\text{Ran}})^{!} \circ (p^{\lambda}_{\text{glob}})^{!}(\mathcal{F}^{\lambda})[-d - \langle \lambda, 2\check{\rho} \rangle] &\simeq (p^{\lambda}_{\text{Ran}})^{!} \circ (\text{pr}^{\lambda}_{\text{Ran}})^{!}(\mathcal{F}^{\lambda})[-d - \langle \lambda, 2\check{\rho} \rangle]\end{aligned} \tag{4.22}$$

By Corollary 4.6.4 and Theorem 4.3.3, we have

$$(\pi^{\lambda}_{\text{Ran}})^{!} \circ (\mathbf{i}^{\lambda}_{\text{glob}})^{*}(\text{IC}^{\frac{\infty}{2}}_{\text{glob}}) \simeq (\mathbf{i}^{\lambda})^{*}(\text{IC}^{\frac{\infty}{2}}_{\text{Ran}})[-d].$$

Further, by Remark 3.7.5, we have

$$(\mathbf{i}^{\lambda})^{*}(\text{IC}^{\frac{\infty}{2}}_{\text{Ran}}) \simeq (p^{\lambda}_{\text{Ran}})^{!} \circ (\text{pr}^{\lambda}_{\text{Ran}})^{!}(\text{Fact}^{\text{alg}}(\mathcal{O}(\check{N}))_{X^{\lambda}})[-\langle \lambda, 2\check{\rho} \rangle].$$

Combining with (4.22), we obtain

$$(p^{\lambda}_{\text{Ran}})^{!} \circ (\text{pr}^{\lambda}_{\text{Ran}})^{!}(\mathcal{F}^{\lambda}) \simeq (p^{\lambda}_{\text{Ran}})^{!} \circ (\text{pr}^{\lambda}_{\text{Ran}})^{!}(\text{Fact}^{\text{alg}}(\mathcal{O}(\check{N}))_{X^{\lambda}}).$$

Since the functor $(p^{\lambda}_{\text{Ran}})^{!} \circ (\text{pr}^{\lambda}_{\text{Ran}})^{!}$ is fully faithful, we obtain the desired

$$\mathcal{F}^{\lambda} \simeq \text{Fact}^{\text{alg}}(\mathcal{O}(\check{N}))_{X^{\lambda}},$$

proving point (a).

Since $\text{IC}^{\frac{\infty}{2}}_{\text{glob}}$ is Verdier self-dual, and using the fact that

$$\mathbb{D}(\text{Fact}^{\text{coalg}}(U(\check{\mathfrak{n}}^{-}))_{X^{\lambda}}) \simeq \text{Fact}^{\text{alg}}(\mathcal{O}(\check{N}))_{X^{\lambda}},$$

from the isomorphism of point (a), we obtain

$$(\mathbf{i}^\lambda_{\mathrm{glob}})^!(\mathrm{IC}^{\frac{\infty}{2}}_{\mathrm{glob}}) \simeq (p^\lambda_{\mathrm{glob}})^*(\mathrm{Fact}^{\mathrm{coalg}}(U(\check{\mathfrak{n}}^-))_{X^\lambda})[d + \langle \lambda, 2\check{\rho}\rangle] \simeq$$
$$\simeq (p^\lambda_{\mathrm{glob}})^!(\mathrm{Fact}^{\mathrm{coalg}}(U(\check{\mathfrak{n}}^-))_{X^\lambda})[-d - \langle \lambda, 2\check{\rho}\rangle],$$

the latter isomorphism because $p^\lambda_{\mathrm{glob}}$ is smooth of relative dimension $d + \langle \lambda, 2\check{\rho}\rangle$. This proves point (b).

□

4.9.4

Let us now prove Theorem 3.4.5.

Proof By Remark 3.7.5, it remains to prove the assertion regarding $(\mathbf{i}^\lambda)^!(\mathrm{IC}^{\frac{\infty}{2}}_{\mathrm{Ran}})$.

Let $\mathcal{G}^\lambda \in \mathrm{Shv}((X^\lambda \times \mathrm{Ran})^{\subset})$ be such that

$$(\mathbf{i}^\lambda)^!(\mathrm{IC}^{\frac{\infty}{2}}_{\mathrm{Ran}}) \simeq (p^\lambda_{\mathrm{Ran}})^!(\mathcal{G}^\lambda)[-\langle \lambda, 2\check{\rho}\rangle].$$

Let us show that

$$\mathcal{G}^\lambda \simeq (\mathrm{pr}^\lambda_{\mathrm{Ran}})^!(\mathrm{Fact}^{\mathrm{coalg}}(U(\check{\mathfrak{n}}^-))_{X^\lambda}).$$

Indeed, by Theorem 4.3.3 and Theorem 4.9.3(b), we have

$$(p^\lambda_{\mathrm{Ran}})^!(\mathcal{G}^\lambda)[-\langle \lambda, 2\check{\rho}\rangle] = (\mathbf{i}^\lambda)^!(\mathrm{IC}^{\frac{\infty}{2}}_{\mathrm{Ran}}) \simeq (\mathbf{i}^\lambda)^! \circ (\pi_{\mathrm{Ran}})^!(\mathrm{IC}^{\frac{\infty}{2}}_{\mathrm{glob}})[d] \simeq$$
$$\simeq (\pi^\lambda_{\mathrm{Ran}})^! \circ (\mathbf{i}^\lambda_{\mathrm{glob}})^!(\mathrm{IC}^{\frac{\infty}{2}}_{\mathrm{glob}})[d] \simeq (\pi^\lambda_{\mathrm{Ran}})^! \circ (p^\lambda_{\mathrm{glob}})^!(\mathrm{Fact}^{\mathrm{coalg}}(U(\check{\mathfrak{n}}^-))_{X^\lambda})[-\langle \lambda, 2\check{\rho}\rangle] \simeq$$
$$\simeq (p^\lambda_{\mathrm{Ran}})^! \circ (\mathrm{pr}^\lambda_{\mathrm{Ran}})^!(\mathrm{Fact}^{\mathrm{coalg}}(U(\check{\mathfrak{n}}^-))_{X^\lambda})[-\langle \lambda, 2\check{\rho}\rangle].$$

Since $(p^\lambda_{\mathrm{Ran}})^!$ is fully faithful, this gives the desired isomorphism.

□

5 Unital Structure and Factorization

The goal of this section is to explore an additional property of $\mathrm{IC}^{\frac{\infty}{2}}$, which we will refer to as *unitality*. It has to do with the following additional structure on $\mathrm{Gr}_{G,\mathrm{Ran}}$: one can "throw in" more points in Ran without altering the G-bundle.

The unital property of $\mathrm{IC}^{\frac{\infty}{2}}$ will allow us to construct on it a *factorization structure*.

5.1 Unital Structure on the Affine Grassmannian

In this subsection, we introduce the geometric structure on $\mathrm{Gr}_{G,\mathrm{Ran}}$ that would allow us to talk about *unitality*.

5.1.1

Let $(\mathrm{Ran} \times \mathrm{Ran})^{\subset}$ be the following subfunctor of $\mathrm{Ran} \times \mathrm{Ran}$: for an affine test scheme Y, the set $\mathrm{Hom}(Y, (\mathrm{Ran} \times \mathrm{Ran})^{\subset})$ consists of those

$$\mathfrak{I}, \mathfrak{I}' \subset \mathrm{Hom}(Y, X)$$

for which

$$\mathfrak{I} \subseteq \mathfrak{I}' \subset \mathrm{Hom}(Y, X). \tag{5.1}$$

The diagonal map

$$\Delta_{\mathrm{Ran}} : \mathrm{Ran} \to \mathrm{Ran} \times \mathrm{Ran}$$

factors through a map $\mathrm{Ran} \to (\mathrm{Ran} \times \mathrm{Ran})^{\subset}$, which, by a slight abuse of notation, we denote by the same symbol Δ_{Ran}.

There are two obvious projections

$$\mathsf{ob}_{\mathrm{small}}, \mathsf{ob}_{\mathrm{big}} : (\mathrm{Ran} \times \mathrm{Ran})^{\subset} \to \mathrm{Ran}$$

that send a point (5.1) to

$$\mathfrak{I} \subset \mathrm{Hom}(Y, X) \text{ and } \mathfrak{I}' \subset \mathrm{Hom}(Y, X),$$

respectively.

We have

$$\mathsf{ob}_{\mathrm{small}} \circ \Delta_{\mathrm{Ran}} = \mathrm{id} \text{ and } \mathsf{ob}_{\mathrm{big}} \circ \Delta_{\mathrm{Ran}} = \mathrm{id}\,.$$

For future use, we note:

Lemma 5.1.2 *The map* $\mathsf{ob}_{\mathrm{small}}$ *is universally homologically contractible.*

Remark 5.1.3 One proof of Lemma 5.1.2 can be obtained by mimicking the argument in Sect. A.2.8. We will now give a different argument, which does *not* use the properness of X (we note that the argument below can also be used to give an alternative proof of Lemma 2.3.3, see Proposition 5.2.7 below).

Proof Let Y be an affine scheme and let us be given a Y-point $\mathcal{J} \subset \mathrm{Hom}(Y, X)$ of Ran. We need to show that the pullback functor

$$\mathrm{Shv}(Y) \to \mathrm{Shv}(Y \underset{\mathrm{Ran}}{\times} (\mathrm{Ran} \times \mathrm{Ran})^{\subset})$$

is fully faithful, where the map $(\mathrm{Ran} \times \mathrm{Ran})^{\subset} \to \mathrm{Ran}$ is $\mathsf{ob}_{\mathrm{small}}$.

To show this, it suffices to show that the map $\mathsf{ob}_{\mathrm{small}}$ can be obtained as a *retract* of a map that is universally homologically contractible. We let this other map be the projection

$$\mathrm{Ran} \times \mathrm{Ran} \to \mathrm{Ran}, \quad (\mathcal{I}_1, \mathcal{I}_2) \mapsto \mathcal{I}_1.$$

It is universally homologically contractible because the Ran space is homologically contractible (i.e., universally homologically contractible over pt).

We realize $(\mathrm{Ran} \times \mathrm{Ran})^{\subset} \to \mathrm{Ran}$ as a retract of $\mathrm{Ran} \times \mathrm{Ran} \to \mathrm{Ran}$ as follows. The map

$$(\mathrm{Ran} \times \mathrm{Ran})^{\subset} \to \mathrm{Ran} \times \mathrm{Ran}$$

sends

$$(\mathcal{I} \subset \mathcal{I}') \mapsto (\mathcal{I}, \mathcal{I}').$$

The retraction $\mathrm{Ran} \times \mathrm{Ran} \to (\mathrm{Ran} \times \mathrm{Ran})^{\subset}$ sends

$$(\mathcal{I}_1, \mathcal{I}_2) \mapsto (\mathcal{I}_1 \subseteq \mathcal{I}_1 \cup \mathcal{I}_2).$$

□

5.1.4

Consider the fiber product

$$\mathrm{Gr}_{G,(\mathrm{Ran}\times\mathrm{Ran})^{\subset}} := \mathrm{Gr}_{G,\mathrm{Ran}} \underset{\mathrm{Ran}}{\times} (\mathrm{Ran} \times \mathrm{Ran})^{\subset},$$

where the map $(\mathrm{Ran} \times \mathrm{Ran})^{\subset} \to \mathrm{Ran}$ is $\mathsf{ob}_{\mathrm{small}}$. By a slight abuse of notation, we will denote by the same symbol $\mathsf{ob}_{\mathrm{small}}$ the projection

$$\mathrm{Gr}_{G,(\mathrm{Ran}\times\mathrm{Ran})^{\subset}} \to \mathrm{Gr}_{G,\mathrm{Ran}}.$$

Note, however, that we have another map, denoted

$$\mathsf{ob}_{\mathrm{big}} : \mathrm{Gr}_{G,(\mathrm{Ran}\times\mathrm{Ran})^{\subset}} \to \mathrm{Gr}_{G,\mathrm{Ran}}$$

that makes the following diagram commute:

$$\begin{array}{ccc} \mathrm{Gr}_{G,(\mathrm{Ran}\times\mathrm{Ran})^{\subset}} & \xrightarrow{\mathsf{ob}_{\mathrm{big}}} & \mathrm{Gr}_{G,\mathrm{Ran}} \\ \downarrow & & \downarrow \\ (\mathrm{Ran}\times\mathrm{Ran})^{\subset} & \xrightarrow{\mathsf{ob}_{\mathrm{big}}} & \mathrm{Ran}. \end{array} \tag{5.2}$$

Namely, it sends a quadruple $(\mathcal{I} \subseteq \mathcal{I}', \mathcal{P}_G, \alpha)$ to $(\mathcal{I}', \mathcal{P}_G, \alpha')$, where α is a trivialization of $\mathcal{P}_G$ on the complement of $\Gamma_{\mathcal{I}}$ and α' is the restriction of α to the complement of $\Gamma_{\mathcal{I}'}$.

Warning Note, however, that the diagram (5.2) is *not* Cartesian.

Denote by Δ_{Ran} the natural map

$$\mathrm{Gr}_{G,\mathrm{Ran}} \to \mathrm{Gr}_{G,(\mathrm{Ran}\times\mathrm{Ran})^{\subset}}.$$

We have

$$\mathsf{ob}_{\mathrm{small}} \circ \Delta_{\mathrm{Ran}} \simeq \mathrm{id} \text{ and } \mathsf{ob}_{\mathrm{big}} \circ \Delta_{\mathrm{Ran}} \simeq \mathrm{id}\,.$$

5.1.5

We shall say that an object

$$\mathcal{F} \in \mathrm{Shv}(\mathrm{Gr}_{G,\mathrm{Ran}})$$

is *unital* if *there exists* an isomorphism

$$\mathsf{ob}^!_{\mathrm{small}}(\mathcal{F}) \simeq \mathsf{ob}^!_{\mathrm{big}}(\mathcal{F})$$

for which the composition

$$\mathcal{F} \simeq \Delta^!_{\mathrm{Ran}} \circ \mathsf{ob}^!_{\mathrm{small}}(\mathcal{F}) \simeq \Delta^!_{\mathrm{Ran}} \circ \mathsf{ob}^!_{\mathrm{big}}(\mathcal{F}) \simeq \mathcal{F}$$

is the identity map.

Note that it follows from Lemma 5.1.2 that if such an isomorphism exists, then it is unique.

5.1.6

Let

$$\mathrm{Shv}(\mathrm{Gr}_{G,\mathrm{Ran}})_{\mathrm{unital}} \subset \mathrm{Shv}(\mathrm{Gr}_{G,\mathrm{Ran}})$$

be the full subcategory formed by unital objects.

From Lemma 5.1.2, we obtain:

Corollary 5.1.7 *The subcategory* $\mathrm{Shv}(\mathrm{Gr}_{G,\mathrm{Ran}})_{\mathrm{unital}} \subset \mathrm{Shv}(\mathrm{Gr}_{G,\mathrm{Ran}})$ *is closed under colimits.*

In particular, we obtain that $\mathrm{Shv}(\mathrm{Gr}_{G,\mathrm{Ran}})_{\mathrm{unital}}$ is a (cocomplete) DG subcategory of $\mathrm{Shv}(\mathrm{Gr}_{G,\mathrm{Ran}})$.

5.1.8

Set:

$$\mathrm{SI}_{\mathrm{Ran},\mathrm{unital}} := \mathrm{SI}_{\mathrm{Ran}} \cap \mathrm{Shv}(\mathrm{Gr}_{G,\mathrm{Ran}})_{\mathrm{unital}} \subset \mathrm{Shv}(\mathrm{Gr}_{G,\mathrm{Ran}}).$$

Our next goal is to characterize $\mathrm{SI}_{\mathrm{Ran},\mathrm{unital}}$ more explicitly as a full subcategory of $\mathrm{SI}_{\mathrm{Ran}}$.

5.2 *Unital Structure on the Strata*

In this subsection, we will extend the discussion of Sect. 5.1 from $\mathrm{Gr}_{\mathrm{Ran}}$ to the prestacks S^λ_{Ran} and $(X^\lambda \times \mathrm{Ran})^{\subset}$.

We will see that the unital subcategory of $\mathrm{Shv}((X^\lambda \times \mathrm{Ran})^{\subset})$ is actually equivalent to $\mathrm{Shv}(X^\lambda)$.

5.2.1

For a fixed λ, consider the functors

$$S^\lambda_{\mathrm{Ran}} \hookrightarrow \overline{S}^\lambda_{\mathrm{Ran}} \to \mathrm{Gr}_{G,\mathrm{Ran}},$$

and consider the corresponding diagram of prestacks

$$\begin{array}{ccccc} S^\lambda_{(\mathrm{Ran}\times\mathrm{Ran})^{\subset}} & \xrightarrow{\mathbf{j}^\lambda} & \overline{S}^\lambda_{(\mathrm{Ran}\times\mathrm{Ran})^{\subset}} & \xrightarrow{\overline{\mathbf{i}}^\lambda} & \mathrm{Gr}_{G,(\mathrm{Ran}\times\mathrm{Ran})^{\subset}} \\ {\scriptstyle \mathrm{ob}_{\mathrm{big}}}\downarrow & & {\scriptstyle \mathrm{ob}_{\mathrm{big}}}\downarrow & & \downarrow{\scriptstyle \mathrm{ob}_{\mathrm{big}}} \\ S^\lambda_{\mathrm{Ran}} & \xrightarrow{\mathbf{j}^\lambda} & \overline{S}^\lambda_{\mathrm{Ran}} & \xrightarrow{\overline{\mathbf{i}}^\lambda} & \mathrm{Gr}_{G,\mathrm{Ran}}. \end{array}$$

The discussion in Sect. 5.1 applies to the present situation as well. In particular, we obtain the *full* subcategories

$$\mathrm{Shv}(\overline{S}^{\lambda}_{\mathrm{Ran}})_{\mathrm{unital}} \subset \mathrm{Shv}(\overline{S}^{\lambda}_{\mathrm{Ran}}) \text{ and } \mathrm{Shv}(S^{\lambda}_{\mathrm{Ran}})_{\mathrm{unital}} \subset \mathrm{Shv}(S^{\lambda}_{\mathrm{Ran}})$$

as well as

$$\mathrm{SI}^{\leq\lambda}_{\mathrm{Ran,unital}} \subset \mathrm{SI}^{\leq\lambda} \text{ and } \mathrm{SI}^{=\lambda}_{\mathrm{Ran,unital}} \subset \mathrm{SI}^{=\lambda}.$$

It is clear that the functors $(\overline{\mathbf{i}}^{\lambda})^!$, $(\mathbf{j}^{\lambda})^!$ and $(\overline{\mathbf{i}}^{\lambda})_*$, $(\mathbf{j}^{\lambda})_*$ send the corresponding unital subcategories to one another. In particular, from Lemma 5.1.2, we obtain:

Corollary 5.2.2 *An object $\mathcal{F} \in \mathrm{SI}^{\leq 0}_{\mathrm{Ran}}$ belongs to $\mathrm{SI}^{\leq 0}_{\mathrm{Ran,unital}}$ if and only if $(\mathbf{i}^{\lambda})^!(\mathcal{F})$ belongs to $\mathrm{SI}^{=\lambda}_{\mathrm{Ran,unital}}$ for all λ.*

Finally, from (2.8), one obtains:

Corollary 5.2.3

(a) *The functor $(\mathbf{i}^{\lambda})^* : \mathrm{SI}^{\leq 0}_{\mathrm{Ran}} \to \mathrm{SI}^{=\lambda}_{\mathrm{Ran}}$ sends $\mathrm{SI}^{\leq 0}_{\mathrm{Ran,unital}}$ to $\mathrm{SI}^{=\lambda}_{\mathrm{Ran,unital}}$.*
(b) *The functor $(\mathbf{i}^{\lambda})_! : \mathrm{SI}^{=\lambda}_{\mathrm{Ran}} \to \mathrm{SI}^{\leq 0}_{\mathrm{Ran}}$ sends $\mathrm{SI}^{=\lambda}_{\mathrm{Ran,unital}}$ to $\mathrm{SI}^{\leq 0}_{\mathrm{Ran,unital}}$.*

5.2.4

For a fixed λ, consider the prestack

$$(X^{\lambda} \times \mathrm{Ran} \times \mathrm{Ran})^{\subset} := (X^{\lambda} \times \mathrm{Ran})^{\subset} \underset{\mathrm{Ran}}{\times} (\mathrm{Ran} \times \mathrm{Ran})^{\subset},$$

where the map $(\mathrm{Ran} \times \mathrm{Ran})^{\subset} \to \mathrm{Ran}$ is $\mathsf{ob}_{\mathrm{small}}$. By a slight abuse of notation, we will denote by the same symbol $\mathsf{ob}_{\mathrm{small}}$ the projection

$$(X^{\lambda} \times \mathrm{Ran} \times \mathrm{Ran})^{\subset} \to (X^{\lambda} \times \mathrm{Ran})^{\subset}, \quad (D, \mathcal{I}, \mathcal{I}') \mapsto (D, \mathcal{I}).$$

Let us denote by $\mathsf{ob}_{\mathrm{big}}$ the map

$$(X^{\lambda} \times \mathrm{Ran} \times \mathrm{Ran})^{\subset} \to (X^{\lambda} \times \mathrm{Ran})^{\subset}, \quad (D, \mathcal{I}, \mathcal{I}') \mapsto (D, \mathcal{I}').$$

Using this map, we define a full subcategory

$$\mathrm{Shv}((X^{\lambda} \times \mathrm{Ran})^{\subset})_{\mathrm{unital}} \subset \mathrm{Shv}((X^{\lambda} \times \mathrm{Ran})^{\subset}).$$

From Proposition 2.4.2, we obtain:

Corollary 5.2.5 *The equivalence*

$$(p^{\lambda}_{\mathrm{Ran}})^! : \mathrm{Shv}((X^{\lambda} \times \mathrm{Ran})^{\subset}) \to \mathrm{SI}^{=\lambda}_{\mathrm{Ran}}$$

restricts to an equivalence

$$\mathrm{Shv}((X^\lambda \times \mathrm{Ran})^{\subset})_{\mathrm{unital}} \to \mathrm{SI}^{=\lambda}_{\mathrm{Ran},\mathrm{unital}}.$$

5.2.6

We now claim:

Proposition 5.2.7 *The pullback functor*

$$(\mathrm{pr}^\lambda_{\mathrm{Ran}})^! : \mathrm{Shv}(X^\lambda) \to \mathrm{Shv}((X^\lambda \times \mathrm{Ran})^{\subset})$$

defines an equivalence

$$\mathrm{Shv}(X^\lambda) \xrightarrow{\sim} \mathrm{Shv}((X^\lambda \times \mathrm{Ran})^{\subset})_{\mathrm{unital}}.$$

Proof The fact that the functor $(\mathrm{pr}^\lambda_{\mathrm{Ran}})^!$ sends $\mathrm{Shv}(X^\lambda)$ to $\mathrm{Shv}((X^\lambda \times \mathrm{Ran})^{\subset})_{\mathrm{unital}}$ is immediate from the definition.[9]

Choose a finite set I so that we have a surjective symmetrization map $\mathrm{sym}^{I\to\lambda} : X^I \to X^\lambda$. Since the map $\mathrm{sym}^{I\to\lambda}$ is finite and surjective, it satisfies descent for $\mathrm{Shv}(-)$. So it is sufficient to prove the assertion of the proposition when the original map

$$\mathrm{pr}^\lambda_{\mathrm{Ran}} : (X^\lambda \times \mathrm{Ran})^{\subset} \to X^\lambda$$

is base-changed by the Čech nerve of the map $X^I \to X^\lambda$.

We will prove that the pullback functor

$$(\mathrm{pr}^I_{\mathrm{Ran}})^! : \mathrm{Shv}(X^I) \to \mathrm{Shv}((X^I \times \mathrm{Ran})^{\subset})$$

defines an equivalence onto $\mathrm{Shv}((X^I \times \mathrm{Ran})^{\subset})_{\mathrm{unital}}$. That is, we will prove the assertion for the 0-simplices of the Čech nerve; for higher simplices, the proof is the same.

Note that the map

$$\mathrm{pr}^I_{\mathrm{Ran}} : (X^I \times \mathrm{Ran})^{\subset} \to X^I$$

admits a section, denoted r^I. Namely, for an affine test scheme Y and a Y-point of X^I, which is a map $I \to \mathrm{Hom}(Y, X)$, we assign its image, denoted $\mathcal{J}$ in $\mathrm{Hom}(Y, X)$.

[9] Note also that the fully faithfulness of $(\mathrm{pr}^\lambda_{\mathrm{Ran}})^!$ has been already stated in Lemma 2.3.3; however, the argument given below will give an alternative proof of this fact.

Pullback with respect to r^I defines a functor $\mathrm{Shv}((X^I \times \mathrm{Ran})^{\subset}) \to \mathrm{Shv}(X^I)$. We claim that the restriction of $(r^I)^!$ to $\mathrm{Shv}((X^I \times \mathrm{Ran})^{\subset})_{\mathrm{unital}}$ is a functor inverse to $(\mathrm{pr}^I_{\mathrm{Ran}})^!$.

Indeed, the fact that $(r^I)^! \circ (\mathrm{pr}^I_{\mathrm{Ran}})^! \simeq \mathrm{Id}$ is obvious. To construct an isomorphism

$$(\mathrm{pr}^I_{\mathrm{Ran}})^! \circ (r^I)^!|_{\mathrm{Shv}((X^I \times \mathrm{Ran})^{\subset})_{\mathrm{unital}}} \simeq \mathrm{Id},$$

we note that there exist canonically defined maps

$$''r^I ; 'r^I : (X^I \times \mathrm{Ran})^{\subset} \to (X^I \times \mathrm{Ran} \times \mathrm{Ran})^{\subset}$$

such that

$$\mathsf{ob}_{\mathrm{big}} \circ 'r^I = \mathsf{ob}_{\mathrm{big}} \circ ''r^I,$$

while

$$\mathsf{ob}_{\mathrm{small}} \circ 'r^I = \mathrm{id} \text{ and } \mathsf{ob}_{\mathrm{small}} \circ ''r^I = r^I \circ \mathrm{pr}^I_{\mathrm{Ran}}.$$

The maps $'r^I ; ''r^I$ are given by sending a pair $(x, \mathcal{J}')$ to

$$(x, \mathcal{J}', \mathcal{J} \cup \mathcal{J}') \text{ and } (x, \mathcal{J}, \mathcal{J} \cup \mathcal{J}'),$$

respectively, where $x \in \mathrm{Hom}(Y, X^I)$, and $\mathcal{J}$ denotes the image of the resulting map $I \to \mathrm{Hom}(Y, X)$.

□

5.3 *Local-to-Global Comparison, Revisited*

Once we have defined the category $\mathrm{SI}^{\leq 0}_{\mathrm{Ran,unital}}$, we can sharpen the assertion of Theorem 4.4.4, by directly comparing the global semi-infinite category and the unital Ran version of the local one.

5.3.1

We claim:

Theorem 5.3.2 *The pullback functor*

$$(\pi_{\mathrm{Ran}})^! : \mathrm{SI}^{\leq 0}_{\mathrm{glob}} \to \mathrm{SI}^{\leq 0}_{\mathrm{Ran}}$$

defines an equivalence onto $\mathrm{SI}^{\leq 0}_{\mathrm{Ran,unital}}$.

The rest of this subsection is devoted to the proof of this theorem.

5.3.3

First off, it is clear that the essential image of the functor

$$(\pi_{\mathrm{Ran}})^! : \mathrm{Shv}(\overline{\mathrm{Bun}}_N) \to \mathrm{Shv}(\overline{S}^{\leq 0}_{\mathrm{Ran}})$$

belongs to the full subcategory

$$\mathrm{Shv}(\overline{S}^{\leq 0}_{\mathrm{Ran}})_{\mathrm{unital}} \subset \mathrm{Shv}(\overline{S}^{\leq 0}_{\mathrm{Ran}}).$$

Indeed, this follows from the fact that the following diagram commutes:

$$\begin{array}{ccc} \mathrm{Gr}_{G,(\mathrm{Ran}\times\mathrm{Ran})^{\subset}} & \xrightarrow{\mathrm{ob}_{\mathrm{small}}} & \mathrm{Gr}_{G,\mathrm{Ran}} \\ {\scriptstyle \mathrm{ob}_{\mathrm{big}}}\downarrow & & \downarrow{\scriptstyle \pi_{\mathrm{Ran}}} \\ \mathrm{Gr}_{G,\mathrm{Ran}} & \xrightarrow{\pi_{\mathrm{Ran}}} & \overline{\mathrm{Bun}}_N . \end{array}$$

5.3.4

Second, the fact that the functor in question is fully faithful follows from Theorem 4.4.4.

Thus, it remains to show that the functor

$$(\pi_{\mathrm{Ran}})^! : \mathrm{SI}^{\leq 0}_{\mathrm{glob}} \to \mathrm{SI}^{\leq 0}_{\mathrm{Ran,unital}}$$

is essentially surjective.

Taking into account Corollary 4.6.6, it suffices to show that the functor

$$(\pi^{\lambda}_{\mathrm{Ran}})^! : \mathrm{SI}^{=\lambda}_{\mathrm{glob}} \to \mathrm{SI}^{=\lambda}_{\mathrm{Ran}}$$

defines an equivalence onto $\mathrm{SI}^{=\lambda}_{\mathrm{Ran,unital}} \subset \mathrm{SI}^{=\lambda}_{\mathrm{Ran}}$.

However, this follows from Corollary 5.2.5 and Proposition 5.2.7 using the commutative diagram

$$\begin{array}{ccc} S^\lambda_{\mathrm{Ran}} & \xrightarrow{p^\lambda_{\mathrm{Ran}}} & (X^\lambda \times \mathrm{Ran})^\subset \\ {\scriptstyle \pi^\lambda_{\mathrm{Ran}}}\downarrow & & \downarrow{\scriptstyle \mathrm{pr}^\lambda_{\mathrm{Ran}}} \\ \overline{\mathrm{Bun}}_N^{=\lambda} & \longrightarrow & X^\lambda. \end{array}$$

5.4 The t-Structure on the Unital Category

In this subsection, we will show that the t-structure on $\mathrm{SI}^{\leq 0}_{\mathrm{Ran}}$ restricts to a t-structure on $\mathrm{SI}^{\leq 0}_{\mathrm{Ran,unital}}$.

5.4.1

We define a t-structure on $\mathrm{SI}^{=\lambda}_{\mathrm{Ran,unital}}$ by transferring the (perverse) t-structure on $\mathrm{Shv}(X^\lambda)$ via the equivalences

$$\mathrm{Shv}(X^\lambda) \overset{(\mathrm{pr}^\lambda_{\mathrm{Ran}})^!}{\longrightarrow} \mathrm{Shv}((X^\lambda \times \mathrm{Ran})^\subset)_{\mathrm{unital}} \overset{(p^\lambda_{\mathrm{Ran}})^!}{\longrightarrow} \mathrm{SI}^{=\lambda}_{\mathrm{Ran,unital}}$$

and applying the shift $[\langle \lambda, 2\check{\rho}\rangle]$.

5.4.2

We define a t-structure on $\mathrm{SI}^{\leq 0}_{\mathrm{Ran,unital}}$ by declaring that an object $\mathcal{F}$ is *coconnective* if

$$(\mathbf{i}^\lambda)^!(\mathcal{F}) \in \mathrm{SI}^{=\lambda}_{\mathrm{Ran,unital}}$$

is *coconnective* for each λ.

As in Lemma 3.1.9, one shows that an object $\mathcal{F} \in \mathrm{SI}^{\leq 0}_{\mathrm{Ran,unital}}$ is *connective* if and only if

$$(\mathbf{i}^\lambda)^*(\mathcal{F}) \in \mathrm{SI}^{=\lambda}_{\mathrm{Ran,unital}}$$

is *connective* for each λ.

From here, we obtain:

Corollary 5.4.3 *The inclusion* $\mathrm{SI}^{\leq 0}_{\mathrm{Ran,unital}} \hookrightarrow \mathrm{SI}^{\leq 0}_{\mathrm{Ran}}$ *is compatible with t-structures (i.e., is t-exact).*

5.4.4

We now claim:

Proposition 5.4.5 *The object* $\mathrm{IC}^{\frac{\infty}{2}} \in (\mathrm{SI}^{\leq 0}_{\mathrm{Ran}})^{\heartsuit}$ *belongs to* $(\mathrm{SI}^{\leq 0}_{\mathrm{Ran,unital}})^{\heartsuit}$.

Proof The assertion follows from Corollary 5.4.3 and the fact that both

$$(\mathbf{i}^0)_!(\omega_{S^0_{\mathrm{Ran}}}) \text{ and } (\mathbf{i}^0)_*(\omega_{S^0_{\mathrm{Ran}}})$$

belong to $\mathrm{SI}^{\leq 0}_{\mathrm{Ran,unital}}$. □

5.5 Comparison with IC on Zastava Spaces, Continued

Recall the isomorphism

$$(\mathrm{pr}^{\lambda}_{\mathrm{Ran}} \times \mathrm{id}_{\mathcal{Z}^{\lambda}})^!(\mathrm{IC}_{\mathcal{Z}^{\lambda}}) \simeq (\mathfrak{q}')^!(\mathrm{IC}^{\frac{\infty}{2}})[\langle\lambda, 2\check{\rho}\rangle]$$

established in Proposition 4.8.3.

In this subsection, we will sharpen this assertion by showing that it is *uniquely* characterized by the property that its restriction to the open substack

$$(X^{\lambda} \times \mathrm{Ran})^{\subset} \underset{X^{\lambda}}{\times} \overset{\circ}{\mathcal{Z}}{}^{\lambda} \subset (X^{\lambda} \times \mathrm{Ran})^{\subset} \underset{X^{\lambda}}{\times} \mathcal{Z}^{\lambda}$$

is the tautological identification of both sides with the dualizing sheaf.

5.5.1

First off, we note that the recipe in Sect. 5.2 allows to introduce a full subcategory

$$\mathrm{Shv}((X^{\lambda} \times \mathrm{Ran})^{\subset} \underset{X^{\lambda}}{\times} \mathcal{Z}^{\lambda})_{\mathrm{unital}} \subset \mathrm{Shv}((X^{\lambda} \times \mathrm{Ran})^{\subset} \underset{X^{\lambda}}{\times} \mathcal{Z}^{\lambda}),$$

and the functor $(\mathfrak{q}')^!$ (see (4.20)) sends

$$\mathrm{SI}^{\leq 0}_{\mathrm{unital}} \to \mathrm{Shv}((X^{\lambda} \times \mathrm{Ran})^{\subset} \underset{X^{\lambda}}{\times} \mathcal{Z}^{\lambda})_{\mathrm{unital}}.$$

Moreover, an analog of Proposition 5.2.7 applies, and the functor $(\mathrm{pr}^{\lambda}_{\mathrm{Ran}} \times \mathrm{id}_{\mathcal{Z}^{\lambda}})^!$ defines an equivalence

$$\mathrm{Shv}(\mathcal{Z}^{\lambda}) \xrightarrow{\sim} \mathrm{Shv}((X^{\lambda} \times \mathrm{Ran})^{\subset} \underset{X^{\lambda}}{\times} \mathcal{Z}^{\lambda})_{\mathrm{unital}}. \tag{5.3}$$

5.5.2

We define a t-structure on $\mathrm{Shv}((X^\lambda \times \mathrm{Ran})^{\subset} \underset{X^\lambda}{\times} \mathcal{Z}^\lambda)_{\mathrm{unital}}$ by transferring the t-structure on $\mathrm{Shv}(\mathcal{Z}^\lambda)$ via the equivalence of (5.3).

In particular, we obtain that the object

$$(\mathrm{pr}^\lambda_{\mathrm{Ran}} \times \mathrm{id}_{\mathcal{Z}^\lambda})^!(\mathrm{IC}_{\mathcal{Z}^\lambda}) \in \mathrm{Shv}((X^\lambda \times \mathrm{Ran})^{\subset} \underset{X^\lambda}{\times} \mathcal{Z}^\lambda)_{\mathrm{unital}}$$

lies in the heart of the t-structure and is the minimal extension of

$$(\mathrm{pr}^\lambda_{\mathrm{Ran}} \times \mathrm{id}_{\mathcal{Z}^\lambda})^!(\mathrm{IC}_{\mathcal{Z}^\lambda})|_{(X^\lambda \times \mathrm{Ran})^{\subset} \underset{X^\lambda}{\times} \overset{\circ}{\mathcal{Z}}{}^\lambda} \in \mathrm{Shv}((X^\lambda \times \mathrm{Ran})^{\subset} \underset{X^\lambda}{\times} \overset{\circ}{\mathcal{Z}}{}^\lambda)_{\mathrm{unital}}.$$

5.5.3

Hence, we obtain:

Corollary 5.5.4 *The isomorphism*

$$(\mathrm{pr}^\lambda_{\mathrm{Ran}} \times \mathrm{id}_{\mathcal{Z}^\lambda})^!(\mathrm{IC}_{\mathcal{Z}^\lambda}) \simeq (\mathfrak{q}')^!(\mathrm{IC}^{\frac{\infty}{2}})[\langle \lambda, 2\check{\rho}\rangle]$$

of Proposition 4.8.3 is uniquely characterized by the property that it extends the tautological isomorphism over $(X^\lambda \times \mathrm{Ran})^{\subset} \underset{X^\lambda}{\times} \overset{\circ}{\mathcal{Z}}{}^\lambda$.

5.6 *Factorization Structure on* $\mathrm{IC}^{\frac{\infty}{2}}$

We now arrive to the key point of this section. We will show that *unitality* allows one to construct the factorization structure on the semi-infinite cohomology sheaf $\mathrm{IC}^{\frac{\infty}{2}}$.

5.6.1

Recall that identification

$$(\mathrm{Gr}_{G,\mathrm{Ran}} \times \mathrm{Gr}_{G,\mathrm{Ran}}) \underset{\mathrm{Ran}\times\mathrm{Ran}}{\times} (\mathrm{Ran}\times\mathrm{Ran})_{\mathrm{disj}} \simeq \mathrm{Gr}_{G,\mathrm{Ran}} \underset{\mathrm{Ran}}{\times} (\mathrm{Ran}\times\mathrm{Ran})_{\mathrm{disj}} \quad (5.4)$$

of (3.13).

Our current goal is to show that, with respect to this identification, we have a canonical isomorphism

$$(\mathrm{IC}^{\frac{\infty}{2}} \boxtimes \mathrm{IC}^{\frac{\infty}{2}})_{(\mathrm{Gr}_{G,\mathrm{Ran}} \times \mathrm{Gr}_{G,\mathrm{Ran}}) \underset{\mathrm{Ran}\times\mathrm{Ran}}{\times} (\mathrm{Ran}\times\mathrm{Ran})_{\mathrm{disj}}} \simeq \mathrm{IC}^{\frac{\infty}{2}}_{\mathrm{Gr}_{G,\mathrm{Ran}} \underset{\mathrm{Ran}}{\times} (\mathrm{Ran}\times\mathrm{Ran})_{\mathrm{disj}}}. \tag{5.5}$$

Note that we already know that such an isomorphism takes place, due to the identification

$$('\mathrm{IC}^{\frac{\infty}{2}} \boxtimes {'\mathrm{IC}}^{\frac{\infty}{2}})_{(\mathrm{Gr}_{G,\mathrm{Ran}} \times \mathrm{Gr}_{G,\mathrm{Ran}}) \underset{\mathrm{Ran}\times\mathrm{Ran}}{\times} (\mathrm{Ran}\times\mathrm{Ran})_{\mathrm{disj}}} \simeq {'\mathrm{IC}}^{\frac{\infty}{2}}_{\mathrm{Gr}_{G,\mathrm{Ran}} \underset{\mathrm{Ran}}{\times} (\mathrm{Ran}\times\mathrm{Ran})_{\mathrm{disj}}}$$

of (3.14) and the isomorphism

$$\mathrm{IC}^{\frac{\infty}{2}} \simeq {'\mathrm{IC}}^{\frac{\infty}{2}}. \tag{5.6}$$

However, we would like to present a different construction of the isomorphism (5.5). It will be based on "abstract" t-structure considerations rather the identification of $\mathrm{IC}^{\frac{\infty}{2}}$ with the (explicitly constructed) object $'\mathrm{IC}^{\frac{\infty}{2}}$.

5.6.2

Let $\mathrm{Ran}^{\subset,\bullet}$ be the simplicial prestack whose prestack of n-simplices $\mathrm{Ran}^{\subset,n}$ attaches to an affine test scheme Y the set of

$$\mathcal{I}_0 \subseteq \ldots \subseteq \mathcal{I}_n \subset \mathrm{Hom}(Y, X).$$

Let

$$(\mathrm{Ran}^{\subset,\bullet} \times \mathrm{Ran}^{\subset,\bullet})_{\mathrm{disj}} \subset \mathrm{Ran}^{\subset,\bullet} \times \mathrm{Ran}^{\subset,\bullet}$$

be an open simplicial sub-prestack equal to

$$(\mathrm{Ran}^{\subset,\bullet} \times \mathrm{Ran}^{\subset,\bullet}) \underset{\mathrm{Ran}\times\mathrm{Ran}}{\times} (\mathrm{Ran} \times \mathrm{Ran})_{\mathrm{disj}},$$

where the map

$$\mathrm{Ran}^{\subset,\bullet} \times \mathrm{Ran}^{\subset,\bullet} \to \mathrm{Ran} \times \mathrm{Ran}$$

sends

$$(\mathcal{I}'_0 \subseteq \ldots \subseteq \mathcal{I}'_n, \mathcal{I}''_0 \subseteq \ldots \subseteq \mathcal{I}''_n) \mapsto (\mathcal{I}'_n, \mathcal{I}''_n).$$

Consider the simplicial prestack

$$(\mathrm{Gr}_{G,\mathrm{Ran}} \times \mathrm{Gr}_{G,\mathrm{Ran}}) \underset{\mathrm{Ran}\times\mathrm{Ran}}{\times} (\mathrm{Ran}^{\subseteq,\bullet} \times \mathrm{Ran}^{\subseteq,\bullet})_{\mathrm{disj}},$$

where the map

$$(\mathrm{Ran}^{\subseteq,\bullet} \times \mathrm{Ran}^{\subseteq,\bullet})_{\mathrm{disj}} \to (\mathrm{Ran} \times \mathrm{Ran})$$

sends

$$(\mathcal{I}_0' \subseteq \ldots \subseteq \mathcal{I}_n', \mathcal{I}_0'' \subseteq \ldots \subseteq \mathcal{I}_n'') \mapsto (\mathcal{I}_0', \mathcal{I}_0''). \tag{5.7}$$

Note also that the identification (5.4) extends to an identification of simplicial prestacks

$$(\mathrm{Gr}_{G,\mathrm{Ran}}\times\mathrm{Gr}_{G,\mathrm{Ran}}) \underset{\mathrm{Ran}\times\mathrm{Ran}}{\times} (\mathrm{Ran}^{\subseteq,\bullet}\times\mathrm{Ran}^{\subseteq,\bullet})_{\mathrm{disj}} \simeq \mathrm{Gr}_{G,\mathrm{Ran}} \underset{\mathrm{Ran}}{\times} (\mathrm{Ran}^{\subseteq,\bullet}\times\mathrm{Ran}^{\subseteq,\bullet})_{\mathrm{disj}}, \tag{5.8}$$

where the map

$$(\mathrm{Ran}^{\subseteq,\bullet} \times \mathrm{Ran}^{\subseteq,\bullet})_{\mathrm{disj}} \to (\mathrm{Ran} \times \mathrm{Ran})$$

is again (5.7), and the map

$$(\mathrm{Ran}^{\subseteq,\bullet} \times \mathrm{Ran}^{\subseteq,\bullet})_{\mathrm{disj}} \to \mathrm{Ran}$$

is

$$(\mathcal{I}_0' \subseteq \ldots \subseteq \mathcal{I}_n', \mathcal{I}_0'' \subseteq \ldots \subseteq \mathcal{I}_n'') \mapsto (\mathcal{I}_0' \cup \mathcal{I}_0'').$$

We define

$$\begin{aligned}&\mathrm{Shv}((\mathrm{Gr}_{G,\mathrm{Ran}} \times \mathrm{Gr}_{G,\mathrm{Ran}}) \underset{\mathrm{Ran}\times\mathrm{Ran}}{\times} (\mathrm{Ran} \times \mathrm{Ran})_{\mathrm{disj}})_{\mathrm{unital}} := \\ &\quad = \mathrm{Tot}\left(\mathrm{Shv}((\mathrm{Gr}_{G,\mathrm{Ran}} \times \mathrm{Gr}_{G,\mathrm{Ran}}) \underset{\mathrm{Ran}\times\mathrm{Ran}}{\times} (\mathrm{Ran}^{\subseteq,\bullet} \times \mathrm{Ran}^{\subseteq,\bullet})_{\mathrm{disj}})\right).\end{aligned}$$

Warning Unlike the case of the functor $\mathrm{Shv}(\mathrm{Gr}_{G,\mathrm{Ran}})_{\mathrm{unital}} \to \mathrm{Shv}(\mathrm{Gr}_{G,\mathrm{Ran}})$, it is *no longer* true that the functor of restriction to 0-simplices

$$\begin{aligned}&\mathrm{Shv}((\mathrm{Gr}_{G,\mathrm{Ran}} \times \mathrm{Gr}_{G,\mathrm{Ran}}) \underset{\mathrm{Ran}\times\mathrm{Ran}}{\times} (\mathrm{Ran} \times \mathrm{Ran})_{\mathrm{disj}})_{\mathrm{unital}} \to \\ &\qquad \to \mathrm{Shv}((\mathrm{Gr}_{G,\mathrm{Ran}} \times \mathrm{Gr}_{G,\mathrm{Ran}}) \underset{\mathrm{Ran}\times\mathrm{Ran}}{\times} (\mathrm{Ran} \times \mathrm{Ran})_{\mathrm{disj}})\end{aligned}$$

is fully faithful.

5.6.3

Proceeding as in Sect. 2.2, we define a full subcategory

$$\mathrm{SI}_{(\mathrm{Ran}\times\mathrm{Ran})_{\mathrm{disj}}} \subset \mathrm{Shv}((\mathrm{Gr}_{G,\mathrm{Ran}} \times \mathrm{Gr}_{G,\mathrm{Ran}}) \underset{\mathrm{Ran}\times\mathrm{Ran}}{\times} (\mathrm{Ran}\times\mathrm{Ran})_{\mathrm{disj}})$$

and a full subcategory

$$\mathrm{SI}_{(\mathrm{Ran}\times\mathrm{Ran})_{\mathrm{disj}},\mathrm{unital}} \subset \mathrm{Shv}((\mathrm{Gr}_{G,\mathrm{Ran}} \times \mathrm{Gr}_{G,\mathrm{Ran}}) \underset{\mathrm{Ran}\times\mathrm{Ran}}{\times} (\mathrm{Ran}\times\mathrm{Ran})_{\mathrm{disj}})_{\mathrm{unital}}.$$

5.6.4

It is clear that if $\mathcal{F}_1, \mathcal{F}_2$ are objects in $\mathrm{SI}_{\mathrm{Ran},\mathrm{unital}}$, then

$$(\mathcal{F}_1 \boxtimes \mathcal{F}_2)|_{(\mathrm{Gr}_{G,\mathrm{Ran}}\times\mathrm{Gr}_{G,\mathrm{Ran}}) \underset{\mathrm{Ran}\times\mathrm{Ran}}{\times} (\mathrm{Ran}\times\mathrm{Ran})_{\mathrm{disj}}} \in \mathrm{SI}_{(\mathrm{Ran}\times\mathrm{Ran})_{\mathrm{disj}}}$$

naturally upgrades to an object of $\mathrm{SI}_{(\mathrm{Ran}\times\mathrm{Ran})_{\mathrm{disj}},\mathrm{unital}}$.

Similarly, it is clear that if $\mathcal{F}$ is an object of $\mathrm{SI}_{\mathrm{Ran},\mathrm{unital}}$, then

$$\mathcal{F}|_{\mathrm{Gr}_{G,\mathrm{Ran}} \underset{\mathrm{Ran}}{\times} (\mathrm{Ran}^{\subset}\times\mathrm{Ran}^{\subset})_{\mathrm{disj}}} \in \mathrm{SI}_{(\mathrm{Ran}\times\mathrm{Ran})_{\mathrm{disj}}}$$

naturally upgrades to an object of $\mathrm{SI}_{(\mathrm{Ran}\times\mathrm{Ran})_{\mathrm{disj}},\mathrm{unital}}$.

In particular, we obtain that both sides in (5.5) are naturally objects of $\mathrm{SI}_{(\mathrm{Ran}\times\mathrm{Ran})_{\mathrm{disj}},\mathrm{unital}}$.

5.6.5

Similar definitions apply to $\mathrm{Gr}_{G,\mathrm{Ran}} \times \mathrm{Gr}_{G,\mathrm{Ran}}$ replaced by $\overline{S}^0_{\mathrm{Ran}} \times \overline{S}^0_{\mathrm{Ran}}$ and also by

$$S^{\lambda}_{\mathrm{Ran}} \times S^{\mu}_{\mathrm{Ran}}$$

for a pair of elements $\lambda, \mu \in \Lambda$. Denote the resulting categories by

$$\mathrm{SI}^{\leq 0}_{(\mathrm{Ran}\times\mathrm{Ran})_{\mathrm{disj}},\mathrm{unital}} \text{ and } \mathrm{SI}^{=\lambda,\mu}_{(\mathrm{Ran}\times\mathrm{Ran})_{\mathrm{disj}},\mathrm{unital}},$$

respectively.

As in Corollary 5.2.5, we have:

Corollary 5.6.6 *The pullback functor along the map* $p^{\lambda,\mu}_{(\mathrm{Ran}\times\mathrm{Ran})_{\mathrm{disj}}}$

$$(S^{\lambda}_{\mathrm{Ran}} \times S^{\mu}_{\mathrm{Ran}}) \underset{\mathrm{Ran}\times\mathrm{Ran}}{\times} (\mathrm{Ran} \times \mathrm{Ran})_{\mathrm{disj}} \to ((X^{\lambda} \times \mathrm{Ran})^{\subset} \times (X^{\mu} \times \mathrm{Ran})^{\subset}) \underset{\mathrm{Ran}\times\mathrm{Ran}}{\times} (\mathrm{Ran} \times \mathrm{Ran})_{\mathrm{disj}}$$

defines equivalences

$$\mathrm{Shv}(((X^{\lambda} \times \mathrm{Ran})^{\subset} \times (X^{\mu} \times \mathrm{Ran})^{\subset}) \underset{\mathrm{Ran}\times\mathrm{Ran}}{\times} (\mathrm{Ran} \times \mathrm{Ran})_{\mathrm{disj}}) \to \mathrm{SI}^{=\lambda,\mu}_{(\mathrm{Ran}\times\mathrm{Ran})_{\mathrm{disj}}}$$

and

$$\mathrm{Shv}(((X^{\lambda} \times \mathrm{Ran})^{\subset} \times (X^{\mu} \times \mathrm{Ran})^{\subset}) \underset{\mathrm{Ran}\times\mathrm{Ran}}{\times} (\mathrm{Ran} \times \mathrm{Ran})_{\mathrm{disj}})_{\mathrm{unital}} \to \mathrm{SI}^{=\lambda,\mu}_{(\mathrm{Ran}\times\mathrm{Ran})_{\mathrm{disj}},\mathrm{unital}}.$$

In addition, by repeating the argument of Proposition 5.2.7, one shows:

Proposition 5.6.7 *The pullback functor along the map* $\mathrm{pr}^{\lambda,\mu}_{(\mathrm{Ran}\times\mathrm{Ran})_{\mathrm{disj}}}$

$$((X^{\lambda} \times \mathrm{Ran})^{\subset} \times (X^{\mu} \times \mathrm{Ran})^{\subset}) \underset{\mathrm{Ran}\times\mathrm{Ran}}{\times} (\mathrm{Ran} \times \mathrm{Ran})_{\mathrm{disj}} \to (X^{\lambda} \times X^{\mu})_{\mathrm{disj}}$$

defines an equivalence

$$\mathrm{Shv}((X^{\lambda} \times X^{\mu})_{\mathrm{disj}}) \to \mathrm{Shv}(((X^{\lambda} \times \mathrm{Ran})^{\subset} \times (X^{\mu} \times \mathrm{Ran})^{\subset}) \underset{\mathrm{Ran}\times\mathrm{Ran}}{\times} (\mathrm{Ran} \times \mathrm{Ran})_{\mathrm{disj}})_{\mathrm{unital}}.$$

5.6.8

Using Corollary 5.6.6 and Proposition 5.6.7, proceeding as in Sect. 5.4, we define a t-structure on the categories $\mathrm{SI}^{=\lambda,\mu}_{(\mathrm{Ran}\times\mathrm{Ran})_{\mathrm{disj}},\mathrm{unital}}$ and $\mathrm{SI}^{\leq 0}_{(\mathrm{Ran}\times\mathrm{Ran})_{\mathrm{disj}},\mathrm{unital}}$.

It is clear that in the situation of Sect. 5.6.4, if $\mathcal{F}_1, \mathcal{F}_2 \in \mathrm{SI}^{\leq 0}_{\mathrm{Ran},\mathrm{unital}}$ (resp., $\mathcal{F} \in \mathrm{SI}^{\leq 0}_{\mathrm{Ran},\mathrm{unital}}$) are connective/coconnective, then so are the corresponding objects

$$(\mathcal{F}_1 \boxtimes \mathcal{F}_2)|_{(\mathrm{Gr}_{G,\mathrm{Ran}}\times\mathrm{Gr}_{G,\mathrm{Ran}}) \underset{\mathrm{Ran}\times\mathrm{Ran}}{\times} (\mathrm{Ran}\times\mathrm{Ran})_{\mathrm{disj}}} \in \mathrm{SI}^{\leq 0}_{(\mathrm{Ran}\times\mathrm{Ran})_{\mathrm{disj}},\mathrm{unital}}$$

and

$$\mathcal{F}|_{\mathrm{Gr}_{G,\mathrm{Ran}} \underset{\mathrm{Ran}}{\times} (\mathrm{Ran}^{\subset}\times\mathrm{Ran}^{\subset})_{\mathrm{disj}}} \in \mathrm{SI}^{\leq 0}_{(\mathrm{Ran}\times\mathrm{Ran})_{\mathrm{disj}},\mathrm{unital}}.$$

connective/coconnective.

This implies that both sides in (5.5) are minimal extensions of the object

$$\omega_{(S^0_{\mathrm{Ran}}\times S^0_{\mathrm{Ran}}) \underset{\mathrm{Ran}\times\mathrm{Ran}}{\times} (\mathrm{Ran}\times\mathrm{Ran})_{\mathrm{disj}}} \in \mathrm{SI}^{=0,0}_{(\mathrm{Ran}\times\mathrm{Ran})_{\mathrm{disj}},\mathrm{unital}}.$$

This implies the sought-for canonical isomorphism (5.5).

5.7 *Factorization and Zastava Spaces*

In this subsection, we will establish the compatibility of the factorization structure on $\mathrm{IC}^{\frac{\infty}{2}}$ given by (5.5) and the factorization property of the IC sheaf on Zastava spaces.

5.7.1

Recall again the Zastava spaces $\mathcal{Z}^\lambda$.

According to [BFGM, Prop. 2.4], we have canonical isomorphisms

$$(\mathcal{Z}^\lambda \times \mathcal{Z}^\mu) \underset{X^\lambda \times X^\mu}{\times} (X^\lambda \times X^\mu)_{\mathrm{disj}} \simeq \mathcal{Z}^{\lambda+\mu} \underset{X^{\lambda+\mu}}{\times} (X^\lambda \times X^\mu)_{\mathrm{disj}}. \tag{5.9}$$

Since the composite map

$$(X^\lambda \times X^\mu)_{\mathrm{disj}} \to X^\lambda \times X^\mu \to X^{\lambda+\mu}$$

is étale, we have a canonical isomorphism

$$(\mathrm{IC}_{\mathcal{Z}^\lambda} \boxtimes \mathrm{IC}_{\mathcal{Z}^\mu})|_{(\mathcal{Z}^\lambda \times \mathcal{Z}^\mu) \underset{X^\lambda \times X^\mu}{\times} (X^\lambda \times X^\mu)_{\mathrm{disj}}} \simeq \mathrm{IC}_{\mathcal{Z}^{\lambda+\mu}}|_{\mathcal{Z}^{\lambda+\mu} \underset{X^{\lambda+\mu}}{\times} (X^\lambda \times X^\mu)_{\mathrm{disj}}}. \tag{5.10}$$

5.7.2

Note that we have an identification

$$\begin{aligned}\left(((X^\lambda \times \mathrm{Ran})^\subset \underset{X^\lambda}{\times} \mathcal{Z}^\lambda) \times ((X^\mu \times \mathrm{Ran})^\subset \underset{X^\mu}{\times} \mathcal{Z}^\mu)\right) \underset{\mathrm{Ran}\times\mathrm{Ran}}{\times} (\mathrm{Ran}\times\mathrm{Ran})_{\mathrm{disj}} \simeq \\ \simeq \left((X^{\lambda+\mu} \times \mathrm{Ran})^\subset \underset{X^{\lambda+\mu}}{\times} \mathcal{Z}^{\lambda+\mu}\right) \underset{(X^{\lambda+\mu}\times\mathrm{Ran})^\subset}{\times} \left((X^\lambda \times \mathrm{Ran})^\subset \times (X^\mu \times \mathrm{Ran})^\subset\right)_{\mathrm{disj}},\end{aligned} \tag{5.11}$$

where

$$\begin{aligned}\left((X^\lambda \times \mathrm{Ran})^\subset \times (X^\mu \times \mathrm{Ran})^\subset\right)_{\mathrm{disj}} := \\ = \left(((X^\lambda \times \mathrm{Ran})^\subset \times (X^\mu \times \mathrm{Ran})^\subset) \underset{\mathrm{Ran}\times\mathrm{Ran}}{\times} (\mathrm{Ran}\times\mathrm{Ran})_{\mathrm{disj}}\right).\end{aligned}$$

Consider the maps

$$\left(((X^\lambda \times \text{Ran})^{\subset} \underset{X^\lambda}{\times} \mathcal{Z}^\lambda) \times ((X^\mu \times \text{Ran})^{\subset} \underset{X^\mu}{\times} \mathcal{Z}^\mu)\right) \underset{\text{Ran}\times\text{Ran}}{\times} (\text{Ran} \times \text{Ran})_{\text{disj}} \to$$
$$\to (\overline{S}^0_{\text{Ran}} \times \overline{S}^0_{\text{Ran}}) \underset{\text{Ran}\times\text{Ran}}{\times} (\text{Ran} \times \text{Ran})_{\text{disj}}$$

and

$$\left((X^{\lambda+\mu} \times \text{Ran})^{\subset} \underset{X^{\lambda+\mu}}{\times} \mathcal{Z}^{\lambda+\mu}\right) \underset{(X^{\lambda+\mu}\times\text{Ran})^{\subset}}{\times} \left((X^\lambda \times \text{Ran})^{\subset} \times (X^\mu \times \text{Ran})^{\subset}\right)_{\text{disj}} \to$$
$$\to \overline{S}^0_{\text{Ran}} \underset{\text{Ran}}{\times} (\text{Ran} \times \text{Ran})_{\text{disj}}.$$

They are compatible with respect to the identifications (5.11) and

$$(\overline{S}^0_{\text{Ran}} \times \overline{S}^0_{\text{Ran}}) \underset{\text{Ran}\times\text{Ran}}{\times} (\text{Ran} \times \text{Ran})_{\text{disj}} \simeq \overline{S}^0_{\text{Ran}} \underset{\text{Ran}}{\times} (\text{Ran} \times \text{Ran})_{\text{disj}}. \tag{5.12}$$

Hence, from (5.9), we obtain an isomorphism:

$$\text{IC}^{\frac{\infty}{2}} \boxtimes \text{IC}^{\frac{\infty}{2}} |_{\left(((X^\lambda\times\text{Ran})^{\subset} \underset{X^\lambda}{\times} \mathcal{Z}^\lambda)\times((X^\mu\times\text{Ran})^{\subset} \underset{X^\mu}{\times} \mathcal{Z}^\mu)\right) \underset{\text{Ran}\times\text{Ran}}{\times} (\text{Ran}\times\text{Ran})_{\text{disj}}} \simeq$$
$$\simeq$$
$$\simeq \text{IC}^{\frac{\infty}{2}} |_{\left((X^{\lambda+\mu}\times\text{Ran})^{\subset} \underset{X^{\lambda+\mu}}{\times} \mathcal{Z}^{\lambda+\mu}\right) \underset{(X^{\lambda+\mu}\times\text{Ran})^{\subset}}{\times} ((X^\lambda\times\text{Ran})^{\subset}\times(X^\mu\times\text{Ran})^{\subset})_{\text{disj}}} \tag{5.13}$$

5.7.3

Consider now the maps

$$\left(((X^\lambda \times \text{Ran})^{\subset} \underset{X^\lambda}{\times} \mathcal{Z}^\lambda) \times ((X^\mu \times \text{Ran})^{\subset} \underset{X^\mu}{\times} \mathcal{Z}^\mu)\right) \underset{\text{Ran}\times\text{Ran}}{\times} (\text{Ran} \times \text{Ran})_{\text{disj}} \to$$
$$\to (\mathcal{Z}^\lambda \times \mathcal{Z}^\mu) \underset{X^\lambda\times X^\mu}{\times} (X^\lambda \times X^\mu)_{\text{disj}}$$

and

$$\left((X^{\lambda+\mu} \times \text{Ran})^{\subset} \underset{X^{\lambda+\mu}}{\times} \mathcal{Z}^{\lambda+\mu}\right) \underset{(X^{\lambda+\mu}\times\text{Ran})^{\subset}}{\times} \left((X^\lambda \times \text{Ran})^{\subset} \times (X^\mu \times \text{Ran})^{\subset}\right)_{\text{disj}} \to$$
$$\to \mathcal{Z}^{\lambda+\mu} \underset{X^{\lambda+\mu}}{\times} (X^\lambda \times X^\mu)_{\text{disj}}.$$

They are compatible with respect to the identifications (5.11) and (5.9). Hence, from (5.10), we obtain the isomorphism

$$\mathrm{IC}_{\mathcal{Z}^\lambda} \boxtimes \mathrm{IC}_{\mathcal{Z}^\mu} |_{\left(((X^\lambda \times \mathrm{Ran})^{\subset} \underset{X^\lambda}{\times} \mathcal{Z}^\lambda) \times ((X^\mu \times \mathrm{Ran})^{\subset} \underset{X^\mu}{\times} \mathcal{Z}^\mu)\right) \underset{\mathrm{Ran}\times\mathrm{Ran}}{\times} (\mathrm{Ran}\times\mathrm{Ran})_{\mathrm{disj}}} \simeq$$
$$\simeq \mathrm{IC}_{\mathcal{Z}^{\lambda+\mu}} |_{\left((X^{\lambda+\mu} \times \mathrm{Ran})^{\subset} \underset{X^{\lambda+\mu}}{\times} \mathcal{Z}^{\lambda+\mu}\right) \underset{(X^{\lambda+\mu}\times \mathrm{Ran})^{\subset}}{\times} ((X^\lambda \times \mathrm{Ran})^{\subset} \times (X^\mu \times \mathrm{Ran})^{\subset})_{\mathrm{disj}}}. \tag{5.14}$$

5.7.4

We claim:

Proposition 5.7.5 *The isomorphisms* (5.13) *and* (5.14) *are compatible with respect to the isomorphisms*

$$(\mathrm{pr}^\lambda_{\mathrm{Ran}} \times \mathrm{id}_{\mathcal{Z}^\lambda})^!(\mathrm{IC}_{Z^\lambda}) \simeq (\mathrm{q}')^!(\mathrm{IC}^{\frac{\infty}{2}})[\langle\lambda, 2\check{\rho}\rangle], \ (\mathrm{pr}^\mu_{\mathrm{Ran}} \times \mathrm{id}_{\mathcal{Z}^\mu})^!(\mathrm{IC}_{Z^\mu}) \simeq (\mathrm{q}')^!(\mathrm{IC}^{\frac{\infty}{2}})[\langle\mu, 2\check{\rho}\rangle]$$

and

$$(\mathrm{pr}^{\lambda+\mu}_{\mathrm{Ran}} \times \mathrm{id}_{\mathcal{Z}^{\lambda+\mu}})^!(\mathrm{IC}_{Z^{\lambda+\mu}}) \simeq (\mathrm{q}')^!(\mathrm{IC}^{\frac{\infty}{2}})[\langle\lambda+\mu, 2\check{\rho}\rangle]$$

of Proposition 4.8.3.

Proof By mimicking the procedure in Sect. 5.6.4, we introduce the category

$$\mathrm{Shv}\left(\left(((X^\lambda \times \mathrm{Ran})^{\subset} \underset{X^\lambda}{\times} \mathcal{Z}^\lambda) \times ((X^\mu \times \mathrm{Ran})^{\subset} \underset{X^\mu}{\times} \mathcal{Z}^\mu)\right) \underset{\mathrm{Ran}\times\mathrm{Ran}}{\times} (\mathrm{Ran}\times\mathrm{Ran})_{\mathrm{disj}}\right)_{\mathrm{unital}}, \tag{5.15}$$

and we show that the object

$$\mathrm{IC}_{\mathcal{Z}^\lambda} \boxtimes \mathrm{IC}_{\mathcal{Z}^\mu} |_{\left(((X^\lambda \times \mathrm{Ran})^{\subset} \underset{X^\lambda}{\times} \mathcal{Z}^\lambda) \times ((X^\mu \times \mathrm{Ran})^{\subset} \underset{X^\mu}{\times} \mathcal{Z}^\mu)\right) \underset{\mathrm{Ran}\times\mathrm{Ran}}{\times} (\mathrm{Ran}\times\mathrm{Ran})_{\mathrm{disj}}} \in$$
$$\mathrm{Shv}\left(\left(((X^\lambda \times \mathrm{Ran})^{\subset} \underset{X^\lambda}{\times} \mathcal{Z}^\lambda) \times ((X^\mu \times \mathrm{Ran})^{\subset} \underset{X^\mu}{\times} \mathcal{Z}^\mu)\right) \underset{\mathrm{Ran}\times\mathrm{Ran}}{\times} (\mathrm{Ran}\times\mathrm{Ran})_{\mathrm{disj}}\right)$$

naturally upgrades to an object of (5.15).

Furthermore, by mimicking the procedure in Sect. 5.6.8, we introduce a t-structure on (5.15) and we show that the above object

$$\mathrm{IC}_{\mathcal{Z}^\lambda}\boxtimes\mathrm{IC}_{\mathcal{Z}^\mu}|_{\left(((X^\lambda\times\mathrm{Ran})^{\subset}\underset{X^\lambda}{\times}\mathcal{Z}^\lambda)\times((X^\mu\times\mathrm{Ran})^{\subset}\underset{X^\mu}{\times}\mathcal{Z}^\mu)\right)\underset{\mathrm{Ran}\times\mathrm{Ran}}{\times}(\mathrm{Ran}\times\mathrm{Ran})_{\mathrm{disj}}}\in$$

$$\mathrm{Shv}\left(\left(((X^\lambda\times\mathrm{Ran})^{\subset}\underset{X^\lambda}{\times}\mathcal{Z}^\lambda)\times((X^\mu\times\mathrm{Ran})^{\subset}\underset{X^\mu}{\times}\mathcal{Z}^\mu)\right)\underset{\mathrm{Ran}\times\mathrm{Ran}}{\times}(\mathrm{Ran}\times\mathrm{Ran})_{\mathrm{disj}}\right)_{\mathrm{unital}}$$

is the minimal extension of its restriction to

$$\left(((X^\lambda\times\mathrm{Ran})^{\subset}\underset{X^\lambda}{\times}\overset{\circ}{\mathcal{Z}}{}^\lambda)\times((X^\mu\times\mathrm{Ran})^{\subset}\underset{X^\mu}{\times}\overset{\circ}{\mathcal{Z}}{}^\mu)\right)\underset{\mathrm{Ran}\times\mathrm{Ran}}{\times}(\mathrm{Ran}\times\mathrm{Ran})_{\mathrm{disj}}. \tag{5.16}$$

Now the compatibility stated in Sect. 5.7.5 follows from the fact that it does so after restriction to (5.16).

□

6 The Hecke Property of the Semi-infinite IC Sheaf

The goal of this section is to show that the object $\mathrm{IC}^{\frac{\infty}{2}}_{\mathrm{Ran}}$ that we have constructed satisfies the (appropriately formulated) Hecke eigen-property.

6.1 Pointwise Hecke Property

6.1.1

Consider the category $\mathrm{Shv}(\mathfrak{L}^+(T)_{\mathrm{Ran}}\backslash\mathrm{Gr}_{G,\mathrm{Ran}})$, i.e., we impose the structure of equivariance with respect to group scheme of arcs into T over the base prestack Ran.

The action of $\mathfrak{L}(T)_{\mathrm{Ran}}$ on $\mathrm{Gr}_{G,\mathrm{Ran}}$ by left multiplication defines an action of $\mathrm{Sph}_{T,\mathrm{Ran}}$ on $\mathrm{Shv}(\mathfrak{L}^+(T)_{\mathrm{Ran}}\backslash\mathrm{Gr}_{G,\mathrm{Ran}})$.

We consider $\mathrm{Shv}(\mathfrak{L}^+(T)_{\mathrm{Ran}}\backslash\mathrm{Gr}_{G,\mathrm{Ran}})$ as acted on by the monoidal category $\mathrm{Sph}_{G,\mathrm{Ran}}$ on the right by convolutions.

This action commutes with the left action of $\mathrm{Sph}_{T,\mathrm{Ran}}$.

6.1.2

Since $\mathfrak{L}(T)_{\mathrm{Ran}}$ normalizes $\mathfrak{L}(N)_{\mathrm{Ran}}$, the category

$$(\mathrm{SI}_{\mathrm{Ran}})^{\mathfrak{L}^+(T)_{\mathrm{Ran}}}:=\mathrm{Shv}(\mathrm{Gr}_{G,\mathrm{Ran}})^{\mathfrak{L}^+(T)_{\mathrm{Ran}}\cdot\mathfrak{L}(N)_{\mathrm{Ran}}}$$

inherits an action of $\mathrm{Sph}_{T,\mathrm{Ran}}$ and a commuting $\mathrm{Sph}_{G,\mathrm{Ran}}$-action.

Working with this version of the semi-infinite category, we can define a t-structure on it in the same way as for

$$\mathrm{SI}_{\mathrm{Ran}} := \mathrm{Shv}(\mathrm{Gr}_{G,\mathrm{Ran}})^{\mathfrak{L}(N)_{\mathrm{Ran}}},$$

so that the forgetful functor

$$(\mathrm{SI}_{\mathrm{Ran}})^{\mathfrak{L}^+(T)_{\mathrm{Ran}}} \to \mathrm{SI}_{\mathrm{Ran}}$$

is t-exact.

Thus, we obtain that the object $\mathrm{IC}^{\infty}_{\mathrm{Ran}} \in \mathrm{SI}_{\mathrm{Ran}} \subset \mathrm{Shv}(\mathrm{Gr}_{G,\mathrm{Ran}})$ naturally lifts to an object of

$$(\mathrm{SI}_{\mathrm{Ran}})^{\mathfrak{L}^+(T)_{\mathrm{Ran}}} := \mathrm{Shv}(\mathrm{Gr}_{G,\mathrm{Ran}})^{\mathfrak{L}(N)_{\mathrm{Ran}}\cdot\mathfrak{L}^+(T)_{\mathrm{Ran}}} \subset \mathrm{Shv}(\mathfrak{L}^+(T)_{\mathrm{Ran}}\backslash\mathrm{Gr}_{G,\mathrm{Ran}});$$

by a slight abuse of notation we denote it by the symbol $\mathrm{IC}^{\infty}_{\mathrm{Ran}}$.

6.1.3

Fix a point x. Let Ran_x be the version of the Ran space with x as a marked point. By definition, for an affine test scheme Y, the set $\mathrm{Hom}(Y, \mathrm{Ran}_x)$ consists of finite subsets

$$\mathfrak{I} \subset \mathrm{Hom}(Y, X)$$

equipped with distinguished element $* \in \mathfrak{I}$ such that the corresponding map $Y \to X$ is

$$X \to \mathrm{pt} \xrightarrow{x} X.$$

6.1.4

We have the natural forgetful map $\mathrm{Ran}_x \to \mathrm{Ran}$, and we can use it to base change all the objects considered above.

In particular, we consider the prestack

$$\mathrm{Gr}_{G,\mathrm{Ran}_x} := \mathrm{Gr}_{G,\mathrm{Ran}} \underset{\mathrm{Ran}}{\times} \mathrm{Ran}_x,$$

the category

$$\mathrm{Shv}(\mathfrak{L}^+(T)_{\mathrm{Ran}_x}\backslash\mathrm{Gr}_{G,\mathrm{Ran}_x}),$$

acted on by

$$\mathrm{Sph}_{G,\mathrm{Ran}_x} := \mathrm{Shv}(\mathfrak{L}^+(G)_{\mathrm{Ran}_x} \backslash \mathrm{Gr}_{G,\mathrm{Ran}_x}) \text{ and } \mathrm{Sph}_{T,\mathrm{Ran}_x} := \mathrm{Shv}(\mathfrak{L}^+(T)_{\mathrm{Ran}_x} \backslash \mathrm{Gr}_{T,\mathrm{Ran}_x}),$$

etc.

We can consider the corresponding object

$$\mathrm{IC}^{\infty}_{\mathrm{Ran}_x} \in \mathrm{Shv}(\mathrm{Gr}_{G,\mathrm{Ran}_x})^{\mathfrak{L}(N)_{\mathrm{Ran}_x} \cdot \mathfrak{L}^+(T)_{\mathrm{Ran}_x}} \subset \mathrm{Shv}(\mathfrak{L}^+(T)_{\mathrm{Ran}_x} \backslash \mathrm{Gr}_{G,\mathrm{Ran}_x}),$$

equal to the !-pullback of $\mathrm{IC}^{\infty}_{\mathrm{Ran}}$ along the projection $\mathrm{Gr}_{G,\mathrm{Ran}_x} \to \mathrm{Gr}_{G,\mathrm{Ran}}$.

6.1.5

Note that we have a tautologically defined map

$$\mathrm{Ran}_x \times \mathrm{Gr}_{G,x} \to \mathrm{Gr}_{G,\mathrm{Ran}_x}. \tag{6.1}$$

From (6.1), we obtain a canonically defined monoidal functor

$$\mathrm{Sph}_{G,x} \to \mathrm{Sph}_{G,\mathrm{Ran}_x}.$$

Composing with the geometric Satake functor

$$\mathrm{Sat}_{G,x} : \mathrm{Rep}(\check{G}) \to \mathrm{Sph}_{G,x},$$

we obtain a monoidal functor

$$\mathrm{Sph}_{G,x} \to \mathrm{Sph}_{G,\mathrm{Ran}_x}.$$

We modify the geometric Satake functor for T by applying the cohomological shift by $[-\langle\lambda, 2\check{\rho}\rangle]$ on $\mathsf{e}^{\lambda} \in \mathrm{Rep}(\check{T})$. Denote the resulting functor by

$$\mathrm{Sat}'_{T,x} : \mathrm{Rep}(\check{T}) \to \mathrm{Sph}_{T,x}.$$

Precomposing with

$$\mathrm{Sph}_{T,x} \to \mathrm{Sph}_{T,\mathrm{Ran}_x},$$

we obtain a monoidal functor

$$\mathrm{Rep}(\check{T}) \to \mathrm{Sph}_{T,\mathrm{Ran}_x}.$$

6.1.6

Thus, we obtain that $\text{Shv}(\mathfrak{L}^+(T)_{\text{Ran}_x}\backslash \text{Gr}_{G,\text{Ran}_x})$ is a bimodule category for $(\text{Rep}(\check{T}), \text{Rep}(\check{G}))$. In this case, we can talk about the category of *graded Hecke objects* in $\text{Shv}(\mathfrak{L}^+(T)_{\text{Ran}_x}\backslash \text{Gr}_{G,\text{Ran}_x})$, denoted

$$\text{Hecke}_{\check{G},\check{T}}(\text{Shv}(\mathfrak{L}^+(T)_{\text{Ran}_x}\backslash \text{Gr}_{G,\text{Ran}_x})),$$

see [Ga1, Sect. 4.3.5], and also Sect. 6.4.1 below.

These are objects $\mathcal{F} \in \text{Shv}(\mathfrak{L}^+(T)_{\text{Ran}_x}\backslash \text{Gr}_{G,\text{Ran}_x})$, equipped with a system of isomorphisms

$$\mathcal{F} \star \text{Sat}_{G,x}(V) \overset{\phi(V,\mathcal{F})}{\longrightarrow} \text{Sat}'_{T,x}(\text{Res}^{\check{G}}_{\check{T}}(V)) \star \mathcal{F}, \quad V \in \text{Rep}(\check{G})$$

that are compatible with the monoidal structure on $\text{Rep}(\check{G})$ in the sense that the diagrams

$$\begin{array}{ccc}
\mathcal{F} \star \text{Sat}_{G,x}(V_1) \star \text{Sat}_{G,x}(V_2) & \xrightarrow{\phi(V_1,\mathcal{F})} & \text{Sat}'_{T,x}(\text{Res}^{\check{G}}_{\check{T}}(V_1)) \star \mathcal{F} \star \text{Sat}_{G,x}(V_2) \\
\sim\downarrow & & \downarrow\phi(V_2,\mathcal{F}) \\
\mathcal{F} \star \text{Sat}_{G,x}(V_1 \otimes V_2) & \longrightarrow & \text{Sat}'_{T,x}(\text{Res}^{\check{G}}_{\check{T}}(V_1)) \star \text{Sat}'_{T,x}(\text{Res}^{\check{G}}_{\check{T}}(V_2)) \star \mathcal{F} \\
\phi(V_1 \otimes V_2,\mathcal{F})\downarrow & & \downarrow\sim \\
\text{Sat}'_{T,x}(\text{Res}^{\check{G}}_{\check{T}}(V_1 \otimes V_2)) \star \mathcal{F} & \overset{\sim}{\longrightarrow} & \text{Sat}'_{T,x}(\text{Res}^{\check{G}}_{\check{T}}(V_1) \otimes \text{Res}^{\check{G}}_{\check{T}}(V_2)) \star \mathcal{F},
\end{array}$$

along with a coherent system of higher compatibilities.

6.1.7

We will prove:

Theorem-Construction 6.1.8 *The object* $\text{IC}^{\infty}_{\text{Ran}_x} \in \text{Shv}(\mathfrak{L}^+(T)_{\text{Ran}_x}\backslash \text{Gr}_{G,\text{Ran}_x})$ *naturally lifts to an object of* $\text{Hecke}_{\check{G},\check{T}}(\text{Shv}(\mathfrak{L}^+(T)_{\text{Ran}_x}\backslash \text{Gr}_{G,\text{Ran}_x}))$.

Several remarks are in order.

Remark 6.1.9 In the proof of Theorem 6.1.8, the object $\text{IC}^{\infty}_{\text{Ran}}$ will come in its incarnation as $'\text{IC}^{\infty}_{\text{Ran}}$, constructed in Sect. 3.6.

Remark 6.1.10 Consider the restriction

$$\text{IC}^{\infty}_x \simeq \text{IC}^{\infty}_{\text{Ran}_x}|_{\text{Gr}_{G,x}}.$$

The Hecke structure on $\mathrm{IC}^{\infty}_{\mathrm{Ran}_x}$ induces one on IC^{∞}_x. It will follow from the construction and [Ga1, Sect. 6.2.5] that the resulting Hecke structure on IC^{∞}_x coincides with one constructed in [Ga1, Sect. 5.1].

Remark 6.1.11 In order to prove Theorem 6.1.8, we will need to consider the Hecke action of $\mathrm{Rep}(\check{G})$ on $\mathrm{Shv}(\mathfrak{L}^+(T)_{\mathrm{Ran}}\backslash \mathrm{Gr}_{G,\mathrm{Ran}})$ over the entire Ran space. The next few subsections are devoted to setting up the corresponding formalism.

6.2 *Categories over the Ran Space, Continued*

6.2.1

Recall the construction

$$\mathcal{A} \rightsquigarrow \mathrm{Fact}^{\mathrm{alg}}(\mathcal{A})_I \tag{6.2}$$

of Sect. 3.5, viewed as a functor $\mathrm{DGCat}^{\mathrm{SymMon}} \to \mathrm{Shv}(X^I)$-**mod**.

Note that the functor (6.2) has a natural right-lax symmetric monoidal structure, i.e., we have the natural transformation

$$\mathrm{Fact}^{\mathrm{alg}}(\mathcal{A}')_I \underset{\mathrm{Shv}(X^I)}{\otimes} \mathrm{Fact}^{\mathrm{alg}}(\mathcal{A}'')_I \to \mathrm{Fact}^{\mathrm{alg}}(\mathcal{A}' \otimes \mathcal{A}'')_I.$$

In particular, since any $\mathcal{A} \in \mathrm{DGCat}^{\mathrm{SymMon}}$ can be viewed as an object in $\mathrm{ComAlg}(\mathrm{DGCat}^{\mathrm{SymMon}})$, we obtain that $\mathrm{Fact}^{\mathrm{alg}}(\mathcal{A})_I$ itself acquires a structure of symmetric monoidal category.

6.2.2

For a surjection of finite sets $\phi : I_1 \twoheadrightarrow I_2$, the corresponding functor

$$(\Delta_\phi)^! : \mathrm{Fact}^{\mathrm{alg}}(\mathcal{A})_{I_1} \to \mathrm{Fact}^{\mathrm{alg}}(\mathcal{A})_{I_2} \tag{6.3}$$

(see Sect. 3.5.10) is naturally symmetric monoidal. In particular, we obtain that

$$\mathrm{Fact}^{\mathrm{alg}}(\mathcal{A})_{\mathrm{Ran}} \simeq \lim_{I} \mathrm{Fact}(\mathcal{A})_I$$

(see (3.10)) acquires a natural symmetric monoidal structure.

6.2.3

Let $\mathcal{A}' \to \mathcal{A}''$ be a right-lax symmetric monoidal functor. The functor (6.2) gives rise to a right-lax symmetric monoidal functor

$$\mathrm{Fact}^{\mathrm{alg}}(\mathcal{A}')_I \to \mathrm{Fact}^{\mathrm{alg}}(\mathcal{A}'')_I,$$

compatible with the restriction functors (6.3). Varying I, we obtain a right-lax symmetric monoidal functor

$$\mathrm{Fact}^{\mathrm{alg}}(\mathcal{A}')_{\mathrm{Ran}} \to \mathrm{Fact}^{\mathrm{alg}}(\mathcal{A}'')_{\mathrm{Ran}}.$$

In particular, a commutative algebra object A in $\mathcal{A}$, viewed as a right-lax symmetric monoidal functor Vect $\to \mathcal{A}$, gives rise to a commutative algebra

$$\mathrm{Fact}^{\mathrm{alg}}(A)_I \in \mathrm{Fact}^{\mathrm{alg}}(\mathcal{A})_I.$$

These algebra a objects are compatible under the restriction functors (6.3). Varying I, we obtain a commutative algebra object

$$\mathrm{Fact}^{\mathrm{alg}}(A)_{\mathrm{Ran}} \in \mathrm{Fact}(\mathcal{A})_{\mathrm{Ran}}.$$

6.2.4 Examples

Let us consider the two examples of $\mathcal{A}$ from Sect. 3.5.4:

(i) Let $\mathcal{A} = \mathrm{Vect}$. We obtain that to $A \in \mathrm{ComAlg(Vect)}$ we can canonically assign an object $\mathrm{Fact}^{\mathrm{alg}}(A)_{\mathrm{Ran}} \in \mathrm{Shv(Ran)}$.
(ii) Let $\mathcal{A}$ be the category of $\Lambda^{\mathrm{neg}}-0$-graded vector spaces. Note that a commutative algebra A in $\mathcal{A}$ is the same as a commutative Λ^{neg}-algebra with $A(0) = k$. On the one hand, the construction of Sect. 3.3 assigns to such an A a collection of objects

$$\mathrm{Fact}^{\mathrm{alg}}(A)_{X^\lambda} \in \mathrm{Shv}(X^\lambda), \quad \lambda \in \Lambda^{\mathrm{neg}} - 0.$$

On the other hand, we have the above object

$$\mathrm{Fact}^{\mathrm{alg}}(A)_{\mathrm{Ran}} \in \mathrm{Fact}^{\mathrm{alg}}(\mathcal{A})_{\mathrm{Ran}}.$$

By unwinding the constructions, we obtain that these two objects match up under the equivalence (3.7).

6.3 Digression: Right-Lax Central Structures

6.3.1

Let $\mathcal{A}$ and $\mathcal{A}'$ be symmetric monoidal categories, and let $\mathcal{C}$ be a $(\mathcal{A}', \mathcal{A})$-bimodule category. Let $F : \mathcal{A} \to \mathcal{A}'$ be a right-lax symmetric monoidal functor.

A right-lax central structure on an object $c \in \mathcal{C}$ with respect to F is a system of maps

$$F(a) \otimes c \overset{\phi(a,c)}{\longrightarrow} c \otimes a, \quad a \in \mathcal{A}$$

that make the diagrams

$$\begin{array}{ccc} F(a_1) \otimes (F(a_2) \otimes c) & \xrightarrow{\phi(a_2,c)} & F(a_1) \otimes (c \otimes a_2) \\ \sim\big\downarrow & & \big\downarrow\sim \\ (F(a_1) \otimes F(a_2)) \otimes c & & (F(a_1) \otimes c) \otimes a_2 \\ \big\downarrow & & \big\downarrow\phi(a_1,c) \\ F(a_1 \otimes a_2) \otimes c & & (c \otimes a_1) \otimes a_2 \\ \phi(a_1 \otimes a_2, c)\big\downarrow & & \big\downarrow\sim \\ c \otimes (a_1 \otimes a_2) & \xrightarrow{\text{id}} & c \otimes (a_1 \otimes a_2) \end{array}$$

commute, along with a coherent system of higher compatibilities.

Denote the category of objects of $\mathcal{C}$ equipped with a right-lax central structure on an object with respect to F by $Z_F(\mathcal{C})$.

6.3.2

From now on, we will assume that $\mathcal{A}$ is rigid (see [GR, Chapter 1, Sect. 9.1] for what this means).

If $\mathcal{A}$ is compactly generated, this condition is equivalent to requiring that the class of compact objects in $\mathcal{A}$ coincides with the class of objects that are dualizable with respect to the symmetric monoidal structure on $\mathcal{A}$.

6.3.3

Assume for a moment that F is strict (i.e., is a genuine symmetric monoidal functor). We have:

Lemma 6.3.4 *If $c \in Z_F(\mathcal{C})$, then the morphisms $\phi(a, c)$ are isomorphisms.*

In other words, this lemma says that if F is genuine, then any right-lax central structure is a genuine central structure (under the assumption that $\mathcal{A}$ is rigid).

6.3.5

Let $R_{\mathcal{A}} \in \mathcal{A} \otimes \mathcal{A}$ be the (commutative) algebra object, obtained by applying the right adjoint

$$\mathcal{A} \to \mathcal{A} \otimes \mathcal{A}$$

of the monoidal operation $\mathcal{A} \otimes \mathcal{A} \to \mathcal{A}$, to the unit object $\mathbf{1}_{\mathcal{A}} \in \mathcal{A}$.

Consider the (commutative) algebra object

$$R^F_{\mathcal{A}} := (F \otimes \mathrm{id})(R_{\mathcal{A}}) \in \mathcal{A}' \otimes \mathcal{A}.$$

We have:

Lemma 6.3.6 *A datum of right-lax central structure on an object $c \in \mathcal{C}$ is equivalent to upgrading c to an object of $R^F_{\mathcal{A}}$-*mod$(\mathcal{C})$.

6.3.7

Let F' be another right-lax symmetric monoidal functor, and let $F \to F'$ be a right-lax symmetric monoidal natural transformation. Restriction defines a functor

$$Z_{F'}(\mathcal{C}) \to Z_F(\mathcal{C}). \tag{6.4}$$

In addition, we have a homomorphism of commutative algebra objects in $\mathcal{A}' \otimes \mathcal{A}$

$$R^F_{\mathcal{A}} \to R^{F'}_{\mathcal{A}}.$$

It is easy to see that with respect to the equivalence of Lemma 6.3.6, the diagram

$$\begin{array}{ccc} Z_{F'}(\mathcal{C}) & \longrightarrow & Z_F(\mathcal{C}) \\ \sim\downarrow & & \downarrow\sim \\ R^{F'}_{\mathcal{A}}\text{-mod}(\mathcal{C}) & \longrightarrow & R^F_{\mathcal{A}}\text{-mod}(\mathcal{C}) \end{array}$$

commutes, where the bottom arrow is given by restriction.

In particular, we obtain that the functor (6.4) admits a left adjoint, given by

$$R^{F'}_{\mathcal{A}} \underset{R^F_{\mathcal{A}}}{\otimes} -.$$

6.3.8

We now modify our context, and we let $\mathcal{C}$ be a module category for

$$\text{Fact}^{\text{alg}}(\mathcal{A}' \otimes \mathcal{A})_I.$$

We have the corresponding category of right-lax central objects, denoted by the same symbol $Z_F(\mathcal{C})$, which can be identified with

$$\text{Fact}^{\text{alg}}(R^F_{\mathcal{A}})_I\text{-mod}(\mathcal{C}).$$

For a right-lax symmetric monoidal natural transformation $F \to F'$, the left adjoint to the restriction functor $Z_{F'}(\mathcal{C}) \to Z_F(\mathcal{C})$ is given by

$$\text{Fact}^{\text{alg}}(R^{F'}_{\mathcal{A}})_I \underset{\text{Fact}^{\text{alg}}(R^F_{\mathcal{A}})_I}{\otimes} -. \tag{6.5}$$

6.3.9

Let

$$I \rightsquigarrow \mathcal{C}_I, \quad I \in \text{Fin}^{\text{surj}}$$

be a compatible family of module categories over $\text{Fact}(\mathcal{A}' \otimes \mathcal{A})_I$.

Set

$$\mathcal{C}_{\text{Ran}} := \lim_{I \in \text{Fin}^{\text{surj}}} \mathcal{C}_I.$$

We can thus talk about an object $c \in \mathcal{C}_{\text{Ran}}$ being equipped with a right-lax central structure with respect to F. Denote the corresponding category of right-lax central objects by $Z_F(\mathcal{C}_{\text{Ran}})$.

The functors (6.5) provide a left adjoint to the forgetful functor

$$Z_{F'}(\mathcal{C}_{\text{Ran}}) \to Z_F(\mathcal{C}_{\text{Ran}}).$$

This follows from the fact that for a surjective map of finite sets $\phi : I_1 \twoheadrightarrow I_2$, the natural transformation in the diagram

$$\begin{array}{ccc}
Z_F(\mathcal{C}_{I_1}) & \xrightarrow{\Delta^!_\phi} & Z_F(\mathcal{C}_{I_2}) \\
{\scriptstyle \mathrm{Fact}^{\mathrm{alg}}(R^{F'}_{\mathcal{A}})_{I_1} \underset{\mathrm{Fact}^{\mathrm{alg}}(R^{F}_{\mathcal{A}})_{I_1}}{\otimes} -}\downarrow & & \downarrow{\scriptstyle \mathrm{Fact}^{\mathrm{alg}}(R^{F'}_{\mathcal{A}})_{I_2} \underset{\mathrm{Fact}^{\mathrm{alg}}(R^{F}_{\mathcal{A}})_{I_2}}{\otimes} -} \\
Z_{F'}(\mathcal{C}_{I_1}) & \xrightarrow{\Delta^!_\phi} & Z_{F'}(\mathcal{C}_{I_2})
\end{array}$$

is an isomorphism.

6.4 Hecke and Drinfeld–Plücker Structures

We will be interested in the following particular cases of the above situation.[10]

6.4.1

Take $\mathcal{A} = \mathrm{Rep}(\check{G})$ and $\mathcal{A}' = \mathrm{Rep}(\check{T})$ with F' being given by restriction along $\check{T} \to \check{G}$. We denote the corresponding category $Z_{F'}(\mathcal{C})$ by

$$\mathrm{Hecke}_{\check{G},\check{T}}(\mathcal{C}).$$

By Lemma 6.3.4, its objects are $c \in \mathcal{C}$, equipped with a system of *iso*morphisms

$$\mathrm{Res}^{\check{G}}_{\check{T}}(V) \otimes c \simeq c \otimes V, \quad V \in \mathrm{Rep}(\check{G}),$$

compatible with tensor products of the V's.

For this reason, we call a (right-lax) central structure on an object of $\mathcal{C}$ in this case a *graded Hecke structure.*

Equivalently, these are objects of $\mathcal{C}$ equipped with an action of the algebra

$$R^{F'}_{\mathcal{A}} := (\mathrm{Res}(\check{G}_{\check{T}}) \otimes \mathrm{id})(R_{\check{G}}),$$

where $R_{\check{G}} \in \mathrm{Rep}(\check{G}) \otimes \mathrm{Rep}(\check{G})$ is the regular representation.

6.4.2

Let us now take $\mathcal{A} = \mathrm{Rep}(\check{G})$ and $\mathcal{A}' = \mathrm{Rep}(\check{T})$, but the functor F is given by the *non-derived* functor of $\check{N}$-invariants

[10]The formalism described in this subsection (as well as the term) was suggested by S. Raskin.

$$V^\lambda \mapsto V^\lambda(\lambda) = \mathsf{e}^\lambda.$$

The corresponding algebra object

$$R^F_{\mathcal{A}} \in \mathrm{Rep}(\check{T}) \otimes \mathrm{Rep}(\check{G})$$

is $\mathcal{O}(\overline{\check{N}\backslash\check{G}})$, where $\overline{\check{N}\backslash\check{G}}$ is the base affine space of $\check{G}$, viewed as acted on the left by $\check{T}$ and on the right by $\check{G}$.

We denote the corresponding category $Z_F(\mathcal{C})$ by

$$\mathrm{DrPl}(\mathcal{C}).$$

By definition, its objects are $c \in \mathcal{C}$, equipped with a collection of maps

$$\mathsf{e}^\lambda \otimes c \overset{\phi(\lambda,c)}{\longrightarrow} c \otimes V^\lambda$$

that make the diagrams

$$\begin{array}{ccc}
\mathsf{e}^\lambda \otimes (\mathsf{e}^\mu \otimes c) & \xrightarrow{\phi(\mu,c)} & \mathsf{e}^\lambda \otimes (c \otimes V^\mu) \\
\sim\downarrow & & \downarrow\sim \\
(\mathsf{e}^\lambda \otimes \mathsf{e}^\mu) \otimes c & & (\mathsf{e}^\lambda \otimes c) \otimes V^\mu \\
\sim\downarrow & & \downarrow\phi(\lambda,c) \\
\mathsf{e}^{\lambda+\mu} \otimes c & & (c \otimes V^\lambda) \otimes V^\mu \\
\phi(\lambda+\mu,c)\downarrow & & \downarrow\sim \\
c \otimes V^{\lambda+\mu} & \longrightarrow & c \otimes (V^\lambda \otimes V^\mu)
\end{array}$$

commute, along with a coherent system of higher compatibilities.

We will call a right-lax central structure on an object of $\mathcal{C}$ in this case a *Drinfeld–Plücker* structure.

6.4.3

We have a right-lax symmetric monoidal natural transformation $F \to F'$,

$$\mathsf{e}^\lambda \to \mathrm{Res}^{\check{G}}_{\check{T}}(V^\lambda).$$

The corresponding morphism of commutative algebra objects in $\mathrm{Rep}(\check{T}) \otimes \mathrm{Rep}(\check{G})$ is given by pullback along the projection map

$$\check{G} \to \overline{\check{N}\backslash\check{G}}.$$

Consider the forgetful functor

$$\mathrm{Res}_{\mathrm{DrPl}}^{\mathrm{Hecke}_{\check{G},\check{T}}} : \mathrm{Hecke}_{\check{G},\check{T}}(\mathcal{C}) \to \mathrm{DrPl}(\mathcal{C}),$$

and its left adjoint

$$\mathrm{Ind}_{\mathrm{DrPl}}^{\mathrm{Hecke}_{\check{G},\check{T}}} : \mathrm{DrPl}(\mathcal{C}) \to \mathrm{Hecke}_{\check{G},\check{T}}(\mathcal{C}).$$

6.4.4

Let us now recall the statement of [Ga1, Proposition 6.2.4] that describes the composition

$$\mathrm{DrPl}(\mathcal{C}) \overset{\mathrm{Ind}_{\mathrm{DrPl}}^{\mathrm{Hecke}_{\check{G},\check{T}}}}{\longrightarrow} \mathrm{Hecke}_{\check{G},\check{T}}(\mathcal{C}) \to \mathcal{C}, \tag{6.6}$$

where the second arrow is the forgetful functor.

Given an object $c \in \mathrm{DrPl}(\mathcal{C})$, the construction of [Ga1, Sect. 2.7] defines a functor $\Lambda^+ \to \mathcal{C}$, which at the level of objects sends $\lambda \in \Lambda^+$ to

$$\mathsf{e}^{-\lambda} \otimes c \otimes V^\lambda.$$

The assertion [Ga1, Proposition 6.2.4] says that the value of (6.6) on the above c is canonically identified with

$$\underset{\lambda\in\Lambda^+}{\mathrm{colim}}\, \mathsf{e}^{-\lambda} \otimes c \otimes V^\lambda.$$

6.4.5

We now place ourselves in the context of Sect. 6.3.8. Let $\mathcal{C}$ be a module category for

$$\mathrm{Fact}^{\mathrm{alg}}(\mathrm{Rep}(\check{T}) \otimes \mathrm{Rep}(\check{G}))_I.$$

We denote the corresponding categories $Z_{F'}(\mathcal{C})$ and $Z_F(\mathcal{C})$ by $\mathrm{Hecke}_{\check{G},\check{T}}(\mathcal{C})$ and $\mathrm{DrPl}(\mathcal{C})$, respectively.

Let $c \in \mathcal{C}$ be an object of $Z_F(\mathcal{C})$. We wish to describe the value on c of the composite functor

$$\mathrm{DrPl}(\mathcal{C}) \overset{\mathrm{Ind}_{\mathrm{DrPl}}^{\mathrm{Hecke}_{\check{G},\check{T}}}}{\longrightarrow} \mathrm{Hecke}_{\check{G},\check{T}}(\mathcal{C}) \to \mathcal{C} \tag{6.7}$$

6.4.6

For $\underline{\lambda} \in \mathrm{Maps}(I, \Lambda^+)$, recall the object $V^{\underline{\lambda}} \in \mathrm{Fact}(\mathrm{Rep}(\check{G}))_I$, see Sect. 3.6.1. Similarly, we have the object

$$\mathsf{e}^{\underline{\lambda}} \in \mathrm{Fact}(\mathrm{Rep}(\check{T}))_I.$$

The construction of [Ga1, Sect.2.7] defines on the assignment

$$\underline{\lambda} \mapsto \mathsf{e}^{-\underline{\lambda}} \otimes c \otimes V^{\underline{\lambda}}$$

a structure of a functor

$$\mathrm{Maps}(I, \Lambda^+) \to \mathcal{C}.$$

Generalizing [Ga1, Proposition 6.2.4], one shows:

Proposition 6.4.7 *The value of the composite functor* (6.7) *on* $c \in \mathrm{DrPl}(\mathcal{C})$ *identifies canonically with*

$$\underset{\underline{\lambda} \in \mathrm{Maps}(I,\Lambda^+)}{\mathrm{colim}} \; \underline{\lambda} \mapsto \mathsf{e}^{-\underline{\lambda}} \otimes c \otimes V^{\underline{\lambda}}.$$

6.4.8

Let $I \rightsquigarrow \mathcal{C}_I$ be as in Sect. 6.3.9. Consider the corresponding categories $\mathrm{DrPl}(\mathcal{C}_{\mathrm{Ran}})$ and $\mathrm{Hecke}_{\check{G},\check{T}}(\mathcal{C}_{\mathrm{Ran}})$.

The compatibility of the functors $\mathrm{Ind}_{\mathrm{DrPl}}^{\mathrm{Hecke}_{\check{G},\check{T}}}$ for surjections of finite sets gives rise to a well-defined functor

$$\mathrm{Ind}_{\mathrm{DrPl}}^{\mathrm{Hecke}_{\check{G},\check{T}}} : \mathrm{DrPl}(\mathcal{C}_{\mathrm{Ran}}) \to \mathrm{Hecke}_{\check{G},\check{T}}(\mathcal{C}_{\mathrm{Ran}}),$$

left adjoint to the restriction functor.

For $c \in \mathrm{DrPl}(\mathcal{C}_{\mathrm{Ran}})$, the value of the composite functor

$$\mathrm{DrPl}(\mathcal{C}) \overset{\mathrm{Ind}_{\mathrm{DrPl}}^{\mathrm{Hecke}_{\check{G},\check{T}}}}{\longrightarrow} \mathrm{Hecke}_{\check{G},\check{T}}(\mathcal{C}) \to \mathcal{C} \to \mathcal{C}_I$$

is given by

$$\underset{\underline{\lambda}\in \mathrm{Maps}(I,\Lambda^+)}{\mathrm{colim}} \mathsf{e}^{-\underline{\lambda}} \otimes c_I \otimes V^{\underline{\lambda}},$$

where c_I is the value of c in $\mathcal{C}_I$.

6.5 The Hecke Property-Enhanced Statement

6.5.1

The key property of the geometric Satake functor

$$\mathrm{Sat}_{G,I} : \mathrm{Fact}^{\mathrm{alg}}(\mathrm{Rep}(\check{G}))_I \to \mathrm{Sph}_{G,I}$$

is that it is has a natural monoidal structure.

The same applies to the modified geometric Satake functor $\mathrm{Sat}'_{T,I}$ for T.

Thus, we obtain that the category $\mathrm{Shv}(\mathfrak{L}^+(T)_I\backslash \mathrm{Gr}_{G,I})$ is as acted on by the monoidal category $\mathrm{Fact}^{\mathrm{alg}}(\mathrm{Rep}(\check{T}) \otimes \mathrm{Rep}(\check{G}))_I$.

These actions are compatible under surjective maps of finite sets $I_1 \twoheadrightarrow I_2$.

6.5.2

Consider the object

$$\delta_{1_{\mathrm{Gr}},I} := (s_I)_!(\omega_{X^I}) \in \mathrm{Shv}(\mathfrak{L}^+(T)_I\backslash \mathrm{Gr}_{G,I}),$$

where $s_I : X^I \to \mathrm{Gr}_{G,I}$ is the unit section.

It follows from the construction of the functor $\mathrm{Sat}_{G,I}$ that $\delta_{\underline{0},I}$ lifts canonically to an object of

$$\mathrm{DrPl}(\mathrm{Shv}(\mathfrak{L}^+(T)_I\backslash \mathrm{Gr}_{G,I})).$$

6.5.3

Consider the corresponding object

$$\mathrm{Ind}_{\mathrm{DrPl}}^{\mathrm{Hecke}_{\check{G},\check{T}}}(\delta_{1_{\mathrm{Gr}},I}) \in \mathrm{Hecke}_{\check{G},\check{T}}(\mathrm{Shv}(\mathfrak{L}^+(T)_I\backslash \mathrm{Gr}_{G,I})).$$

It follows from Proposition 6.4.7 that its image under the forgetful functor

$$\mathrm{Hecke}_{\check{G},\check{T}}(\mathrm{Shv}(\mathfrak{L}^+(T)_I\backslash \mathrm{Gr}_{G,I})) \to \mathrm{Shv}(\mathfrak{L}^+(T)_I\backslash \mathrm{Gr}_{G,I}) \to \mathrm{Shv}(\mathrm{Gr}_{G,I})$$

identifies canonically with the object $\mathrm{IC}_I^{\frac{\infty}{2}}$, constructed in Sect. 3.6.5.

6.5.4

Consider now the object

$$\delta_{1_{\mathrm{Gr}},\mathrm{Ran}} := (s_{\mathrm{Ran}})_!(\omega_{\mathrm{Ran}}) \in \mathrm{Shv}(\mathfrak{L}^+(T)_{\mathrm{Ran}}\backslash \mathrm{Gr}_{G,\mathrm{Ran}}),$$

where $s_{\mathrm{Ran}} : \mathrm{Ran} \to \mathrm{Gr}_{G,\mathrm{Ran}}$ is the unit section.

It naturally lifts to an object of

$$\mathrm{DrPl}(\mathrm{Shv}(\mathfrak{L}^+(T)_{\mathrm{Ran}}\backslash \mathrm{Gr}_{G,\mathrm{Ran}})).$$

Consider the corresponding object

$$\mathrm{Ind}_{\mathrm{DrPl}}^{\mathrm{Hecke}_{\check{G},\check{T}}}(\delta_{1_{\mathrm{Gr}},\mathrm{Ran}}) \in \mathrm{Hecke}_{\check{G},\check{T}}(\mathrm{Shv}(\mathfrak{L}^+(T)_{\mathrm{Ran}}\backslash \mathrm{Gr}_{G,\mathrm{Ran}})).$$

By Sect. 6.4.8, the image of $\mathrm{Ind}_{\mathrm{DrPl}}^{\mathrm{Hecke}_{\check{G},\check{T}}}(\delta_{1_{\mathrm{Gr}},\mathrm{Ran}})$ under the forgetful functor

$$\mathrm{Hecke}_{\check{G},\check{T}}(\mathrm{Shv}(\mathfrak{L}^+(T)_{\mathrm{Ran}}\backslash \mathrm{Gr}_{G,\mathrm{Ran}})) \to \mathrm{Shv}(\mathfrak{L}^+(T)_{\mathrm{Ran}}\backslash \mathrm{Gr}_{G,\mathrm{Ran}}) \to \mathrm{Shv}(\mathrm{Gr}_{G,\mathrm{Ran}})$$

identifies canonically with the object $'\mathrm{IC}_{\mathrm{Ran}}^{\frac{\infty}{2}}$, constructed in Sect. 3.6.7.

Remark 6.5.5 The latter could be used to define on the assignment

$$I \rightsquigarrow \mathrm{IC}_I^{\frac{\infty}{2}}$$

a homotopy-coherent system of compatibilities as I varies over $\mathrm{Fin}^{\mathrm{surj}}$.

6.5.6

Using the isomorphism

$$'\mathrm{IC}_{\mathrm{Ran}}^{\frac{\infty}{2}} \simeq \mathrm{IC}_{\mathrm{Ran}}^{\frac{\infty}{2}}$$

of Theorem 3.7.2, we thus obtain a lift of $\mathrm{IC}_{\mathrm{Ran}}^{\frac{\infty}{2}}$ to an object of $\mathrm{Hecke}_{\check{G},\check{T}}$ $(\mathrm{Shv}(\mathfrak{L}^+(T)_{\mathrm{Ran}}\backslash \mathrm{Gr}_{G,\mathrm{Ran}}))$.

Summarizing, we obtain:

Theorem 6.5.7 *The object* $\mathrm{IC}^{\frac{\infty}{2}}_{\mathrm{Ran}} \in \mathrm{Shv}(\mathfrak{L}^+(T)_{\mathrm{Ran}}\backslash \mathrm{Gr}_{G,\mathrm{Ran}}))$ *naturally lifts to an object of* $\mathrm{Hecke}_{\check{G},\check{T}}(\mathrm{Shv}(\mathfrak{L}^+(T)_{\mathrm{Ran}}\backslash \mathrm{Gr}_{G,\mathrm{Ran}}))$.

6.6 *Recovering the Pointwise Hecke Structure*

In this subsection, we will finally complete the proof of Theorem 6.1.8.

6.6.1

The constructions in Sects. 6.2–6.4 carry over to the situation when Ran is replaced by Ran_x. From Theorem 6.5.7, we obtain that the object

$$\mathrm{IC}^{\frac{\infty}{2}}_{\mathrm{Ran}_x} \in \mathrm{Shv}(\mathfrak{L}^+(T)_{\mathrm{Ran}_x}\backslash \mathrm{Gr}_{G,\mathrm{Ran}_x}))$$

naturally lifts to an object of $\mathrm{Hecke}_{\check{G},\check{T}}(\mathrm{Shv}(\mathfrak{L}^+(T)_{\mathrm{Ran}_x}\backslash \mathrm{Gr}_{G,\mathrm{Ran}_x}))$.

6.6.2

Now, we have a symmetric monoidal functor

$$\mathrm{Rep}(\check{T}) \otimes \mathrm{Rep}(\check{G}) \to \mathrm{Fact}(\mathrm{Rep}(\check{T}) \otimes \mathrm{Rep}(\check{G}))_{\mathrm{Ran}_x}.$$

Restricting, we obtain that $\mathrm{IC}^{\frac{\infty}{2}}_{\mathrm{Ran}_x}$ lifts to an object of $\mathrm{Hecke}_{\check{G},\check{T}}(\mathrm{Shv}(\mathfrak{L}^+(T)_{\mathrm{Ran}_x} \backslash \mathrm{Gr}_{G,\mathrm{Ran}_x}))$, as stated in Theorem 6.1.8.

7 Local vs. Global Compatibility of the Hecke Structure

In this section, we will establish a compatibility between the Hecke structure on $\mathrm{IC}^{\frac{\infty}{2}}_{\mathrm{Ran}}$ constructed in the previous section and the corresponding structure on $\mathrm{IC}^{\frac{\infty}{2}}_{\mathrm{glob}}$ established in [BG1].

7.1 The Relative Version of the Ran Grassmannian

7.1.1

We introduce a relative version of the prestack $\mathrm{Gr}_{G,\mathrm{Ran}}$ over Bun_T, denoted $\mathrm{Gr}_{G,\mathrm{Ran}} \widetilde{\times} \mathrm{Bun}_T$, as follows.

Let $(\mathrm{Ran} \times \mathrm{Bun}_T)^{\mathrm{level}}$ be the prestack that classifies the data of $(\mathcal{P}_T, \mathcal{I}, \beta)$, where:

(i) $\mathcal{I}$ is a finite non-empty collection of points on X.
(ii) $\mathcal{P}_T$ is a T-bundle on X.
(iii) β is a trivialization of $\mathcal{P}_T$ on the formal neighborhood of $\Gamma_{\mathcal{I}}$.

The prestack $(\mathrm{Ran} \times \mathrm{Bun}_T)^{\mathrm{level}}$ is acted on by $\mathfrak{L}(T)_{\mathrm{Ran}}$, and the map

$$(\mathrm{Ran} \times \mathrm{Bun}_T)^{\mathrm{level}} \to \mathrm{Bun}_T \times \mathrm{Ran}$$

is a $\mathfrak{L}^+(T)_{\mathrm{Ran}}$-torsor, locally trivial in the étale (in fact, even Zariski, since T is a torus) topology.

We set

$$\mathrm{Gr}_{G,\mathrm{Ran}} \widetilde{\times} \mathrm{Bun}_T := \mathfrak{L}^+(T)_{\mathrm{Ran}} \backslash \left(\mathrm{Gr}_{G,\mathrm{Ran}} \underset{\mathrm{Ran}}{\times} (\mathrm{Ran} \times \mathrm{Bun}_T)^{\mathrm{level}} \right).$$

We have a tautological projection

$$r : \mathrm{Gr}_{G,\mathrm{Ran}} \widetilde{\times} \mathrm{Bun}_T \to \mathfrak{L}^+(T)_{\mathrm{Ran}} \backslash \mathrm{Gr}_{G,\mathrm{Ran}}.$$

7.1.2

The right action of the groupoid

$$\mathfrak{L}^+(G)_{\mathrm{Ran}} \backslash \mathfrak{L}(G)_{\mathrm{Ran}} / \mathfrak{L}^+(G)_{\mathrm{Ran}} \tag{7.1}$$

on $\mathrm{Gr}_{G,\mathrm{Ran}}$ naturally lifts to an action on $\mathrm{Gr}_{G,\mathrm{Ran}} \widetilde{\times} \mathrm{Bun}_T$, in a way compatible with the projection r.

In addition, by construction, we have an action of the groupoid

$$\mathfrak{L}^+(T)_{\mathrm{Ran}} \backslash \mathfrak{L}(T)_{\mathrm{Ran}} / \mathfrak{L}^+(T)_{\mathrm{Ran}} \tag{7.2}$$

on $\mathrm{Gr}_{G,\mathrm{Ran}} \widetilde{\times} \mathrm{Bun}_T$, also compatible with the projection r.

In particular, we obtain that $\mathrm{Shv}(\mathrm{Gr}_{G,\mathrm{Ran}} \widetilde{\times} \mathrm{Bun}_T)$ is a bimodule category for $(\mathrm{Sph}_{T,\mathrm{Ran}}, \mathrm{Sph}_{G,\mathrm{Ran}})$, and hence for $(\mathrm{Fact}(\mathrm{Rep}(\check{T})_{\mathrm{Ran}}, \mathrm{Fact}(\mathrm{Rep}(\check{G}))_{\mathrm{Ran}})$, via the Geometric Satake functor, where we use the functor $\mathrm{Sat}'_{T,\mathrm{Ran}}$ to map

$$\text{Fact}^{\text{alg}}(\text{Rep}(\check{T}))_{\text{Ran}} \to \text{Sph}_{T,\text{Ran}}\,.$$

Base-changing along $X^I \to \text{Ran}$, we obtain a compatible family of module categories for $(\text{Fact}^{\text{alg}}(\text{Rep}(\check{T})_I, \text{Fact}^{\text{alg}}(\text{Rep}(\check{G}))_I)$, for $I \in \text{Fin}^{\text{surj}}$.

7.1.3

Denote:

$$\text{IC}^{\frac{\infty}{2}}_{\text{Ran},\text{Bun}_T} := r^!(\text{IC}^{\frac{\infty}{2}}_{\text{Ran}}).$$

From Theorem 6.5.7, we obtain that $\text{IC}^{\frac{\infty}{2}}_{\text{Ran},\text{Bun}_T}$ naturally lifts to an object of

$$\text{Hecke}_{\check{G},\check{T}}(\text{Shv}(\text{Gr}_{G,\text{Ran}} \tilde{\times} \text{Bun}_T));$$

moreover, we have

$$\text{IC}^{\frac{\infty}{2}}_{\text{Ran},\text{Bun}_T} \simeq \text{Ind}^{\text{Hecke}_{\check{G},\check{T}}}_{\text{DrPl}}(\delta_{1_{\text{Gr}},\text{Ran},\text{Bun}_T}), \tag{7.3}$$

where

$$\delta_{1_{\text{Gr}},\text{Ran},\text{Bun}_T} = (s_{\text{Ran},\text{Bun}_T})_!(\omega_{\text{Ran}\times\text{Bun}_T}),$$

and where $s_{\text{Ran},\text{Bun}_T}$ is the unit section

$$\text{Ran} \times \text{Bun}_T \to \text{Gr}_{G,\text{Ran}} \tilde{\times} \text{Bun}_T\,.$$

7.2 *Hecke Property in the Global Setting*

7.2.1

Consider the stack $\overline{\text{Bun}}_B$, and consider its version

$$(\overline{\text{Bun}}_B \times \text{Ran})_{\text{poles}}$$

defined as follows:

A point of $(\overline{\text{Bun}}_B \times \text{Ran})_{\text{poles}}$ is a quadruple $(\mathcal{P}_G, \mathcal{P}_T, \kappa, \mathcal{I})$, where:

(i) $\mathcal{P}_G$ is a G-bundle on X.
(ii) $\mathcal{P}_T$ is a T-bundle on X.
(iii) $\mathcal{I}$ is a finite non-empty collection of points on X.

(iv) κ is a datum of maps

$$\kappa^{\check{\lambda}} : \check{\lambda}(\mathcal{P}_T) \to \mathcal{V}^{\check{\lambda}}_{\mathcal{P}_G}$$

that are allowed to have poles on $\Gamma_{\mathcal{J}}$ and that satisfy the Plücker relations.

Note that we have a closed embedding

$$\overline{\mathrm{Bun}}_B \times \mathrm{Ran} \hookrightarrow (\overline{\mathrm{Bun}}_B \times \mathrm{Ran})_{\mathrm{poles}},$$

corresponding to the condition that the maps $\kappa^{\check{\lambda}}$ have no poles.

7.2.2

Hecke modifications of the G-bundle (resp., T-bundle) define a right (resp., left) action of the groupoid (7.1) (resp., (7.2)) on $(\overline{\mathrm{Bun}}_B \times \mathrm{Ran})_{\mathrm{poles}}$.

In particular, the category $\mathrm{Shv}((\overline{\mathrm{Bun}}_B \times \mathrm{Ran})_{\mathrm{poles}})$ acquires a natural structure of bimodule category for $(\mathrm{Sph}_{T,\mathrm{Ran}}, \mathrm{Sph}_{G,\mathrm{Ran}})$, and hence for $(\mathrm{Fact}^{\mathrm{alg}}(\mathrm{Rep}(\check{T}))_{\mathrm{Ran}}, \mathrm{Fact}^{\mathrm{alg}}(\mathrm{Rep}(\check{G}))_{\mathrm{Ran}})$.

Base-changing along $X^I \to \mathrm{Ran}$, we obtain a compatible family of module categories for $(\mathrm{Fact}^{\mathrm{alg}}(\mathrm{Rep}(\check{T}))_I, \mathrm{Fact}^{\mathrm{alg}}(\mathrm{Rep}(\check{G}))_I)$, for $I \in \mathrm{Fin}^{\mathrm{surj}}$.

7.2.3

Denote

$$\mathrm{IC}^{\frac{\infty}{2}}_{\mathrm{glob},\mathrm{Bun}_T} := \mathrm{IC}_{\overline{\mathrm{Bun}}_B} \boxtimes \omega_{\mathrm{Ran}} \subset \mathrm{Shv}((\overline{\mathrm{Bun}}_B \times \mathrm{Ran})_{\mathrm{poles}}).$$

The following assertion is (essentially) established in [BG1, Theorem 3.1.4]:

Theorem 7.2.4 *The object* $\mathrm{IC}^{\frac{\infty}{2}}_{\mathrm{glob},\mathrm{Bun}_T}$ *naturally lifts to an object of the category*

$$\mathrm{Hecke}_{\check{G},\check{T}}(\mathrm{Shv}((\overline{\mathrm{Bun}}_B \times \mathrm{Ran})_{\mathrm{poles}})).$$

7.3 Local vs. Global Compatibility

7.3.1

Note now that the map

$$\pi_{\text{Ran}} : \overline{S}^0_{\text{Ran}} \to \overline{\text{Bun}}_N$$

naturally extends to a map

$$\pi_{\text{Ran},\text{Bun}_T} : \text{Gr}_{G,\text{Ran}} \widetilde{\times} \text{Bun}_T \to (\overline{\text{Bun}}_B \times \text{Ran})_{\text{poles}}.$$

We consider the functor

$$(\pi_{\text{Ran},\text{Bun}_T})'^! : \text{Shv}((\overline{\text{Bun}}_B \times \text{Ran})_{\text{poles}}) \to \text{Shv}(\text{Gr}_{G,\text{Ran}} \widetilde{\times} \text{Bun}_T)$$

obtained from $(\pi_{\text{Ran},\text{Bun}_T})^!$ by applying the shift by $[d - \langle\lambda, 2\check{\rho}\rangle]$ over the connected component Bun_T^λ of Bun_T.

A relative version of the calculation performed in the proof of Theorem 4.3.3 shows:

Theorem 7.3.2 *There exists a canonical isomorphism in* $\text{Shv}(\text{Gr}_{G,\text{Ran}} \widetilde{\times} \text{Bun}_T)$

$$(\pi_{\text{Ran},\text{Bun}_T})'^!(\text{IC}^{\frac{\infty}{2}}_{\text{glob},\text{Bun}_T}) \simeq \text{IC}^{\frac{\infty}{2}}_{\text{Ran},\text{Bun}_T}.$$

7.3.3

The map r is compatible with the actions of the groupoids (7.1) and (7.2). In particular, the pullback functor

$$(\pi_{\text{Ran},\text{Bun}_T})^! : \text{Shv}((\overline{\text{Bun}}_B \times \text{Ran})_{\text{poles}}) \to \text{Shv}(\text{Gr}_{G,\text{Ran}} \widetilde{\times} \text{Bun}_T)$$

is a map of bimodule categories for $(\text{Sph}_{T,\text{Ran}}, \text{Sph}_{G,\text{Ran}})$.

Hence, we obtain that the functor $(\pi_{\text{Ran},\text{Bun}_T})'^!$ can be thought of as a map of bimodule categories for $(\text{Fact}^{\text{alg}}(\text{Rep}(\check{T}))_{\text{Ran}}, \text{Fact}^{\text{alg}}((\text{Rep}(\check{G}))_{\text{Ran}})$.

7.3.4

We are now ready to state the main result of this section:

Theorem 7.3.5 *The isomorphism* $(\pi_{\text{Ran},\text{Bun}_T})'^!(\text{IC}^{\frac{\infty}{2}}_{\text{glob},\text{Bun}_T}) \simeq \text{IC}^{\frac{\infty}{2}}_{\text{Ran},\text{Bun}_T}$ *of Theorem 7.3.2 canonically lifts to an isomorphism of objects of* $\text{Hecke}_{\check{G},\check{T}}(\text{Shv}(\text{Gr}_{G,\text{Ran}} \widetilde{\times} \text{Bun}_T))$.

7.4 *Proof of Theorem 7.3.5*

7.4.1

Consider the tautological map

$$\delta_{1_{\mathrm{Gr}},\mathrm{Ran},\mathrm{Bun}_T} \to \mathrm{Ind}_{\mathrm{DrPl}}^{\mathrm{Hecke}_{\check{G},\check{T}}}(\delta_{1_{\mathrm{Gr}},\mathrm{Ran},\mathrm{Bun}_T}). \tag{7.4}$$

Under the isomorphism

$$\mathrm{Ind}_{\mathrm{DrPl}}^{\mathrm{Hecke}_{\check{G},\check{T}}}(\delta_{1_{\mathrm{Gr}},\mathrm{Ran},\mathrm{Bun}_T}) \simeq \mathrm{IC}^{\frac{\infty}{2}}_{\mathrm{Ran},\mathrm{Bun}_T}$$

of (7.3), this map corresponds to the map

$$\delta_{1_{\mathrm{Gr}},\mathrm{Ran},\mathrm{Bun}_T} \to \mathrm{IC}^{\frac{\infty}{2}}_{\mathrm{Ran},\mathrm{Bun}_T}, \tag{7.5}$$

arising, by the $((s_{\mathrm{Ran},\mathrm{Bun}_T})_!, (s_{\mathrm{Ran},\mathrm{Bun}_T})^!)$ adjunction, from the isomorphism

$$\omega_{\mathrm{Ran}\times\mathrm{Bun}_T} \to (s_{\mathrm{Ran},\mathrm{Bun}_T})^!(\mathrm{IC}^{\frac{\infty}{2}}_{\mathrm{Ran},\mathrm{Bun}_T}).$$

7.4.2

Consider the composite

$$\delta_{1_{\mathrm{Gr}},\mathrm{Ran},\mathrm{Bun}_T} \to \mathrm{Ind}_{\mathrm{DrPl}}^{\mathrm{Hecke}_{\check{G},\check{T}}}(\delta_{1_{\mathrm{Gr}},\mathrm{Ran},\mathrm{Bun}_T}) \simeq \mathrm{IC}^{\frac{\infty}{2}}_{\mathrm{Ran},\mathrm{Bun}_T} \to (\pi_{\mathrm{Ran},\mathrm{Bun}_T})'^!(\mathrm{IC}^{\frac{\infty}{2}}_{\mathrm{glob},\mathrm{Bun}_T}). \tag{7.6}$$

We obtain that the data on the morphism

$$\mathrm{IC}^{\frac{\infty}{2}}_{\mathrm{Ran},\mathrm{Bun}_T} \to (\pi_{\mathrm{Ran},\mathrm{Bun}_T})'^!(\mathrm{IC}^{\frac{\infty}{2}}_{\mathrm{glob},\mathrm{Bun}_T})$$

of a map of objects of $\mathrm{Hecke}_{\check{G},\check{T}}(\mathrm{Shv}(\mathrm{Gr}_{G,\mathrm{Ran}} \tilde{\times} \mathrm{Bun}_T))$ is equivalent to the data on (7.6) of a map of objects of $\mathrm{DrPl}(\mathrm{Shv}(\mathrm{Gr}_{G,\mathrm{Ran}} \tilde{\times} \mathrm{Bun}_T))$.

7.4.3

The map (7.6) can be explicitly described as follows. By the $((s_{\mathrm{Ran},\mathrm{Bun}_T})_!, (s_{\mathrm{Ran},\mathrm{Bun}_T})^!)$ adjunction, it corresponds to the (iso)morphism

$$\omega_{\mathrm{Ran}\times\mathrm{Bun}_T} \to (s_{\mathrm{Ran},\mathrm{Bun}_T})^! \circ (\pi_{\mathrm{Ran},\mathrm{Bun}_T})'^!(\mathrm{IC}^{\frac{\infty}{2}}_{\mathrm{glob},\mathrm{Bun}_T}) \tag{7.7}$$

constructed as follows:

We note that the map

$$\pi_{\mathrm{Ran},\mathrm{Bun}_T} \circ s_{\mathrm{Ran},\mathrm{Bun}_T} : \mathrm{Ran} \times \mathrm{Bun}_T \to (\overline{\mathrm{Bun}}_B \times \mathrm{Ran})_{\mathrm{poles}}$$

factors as

$$\mathrm{Ran} \times \mathrm{Bun}_T \to \mathrm{Ran} \times \mathrm{Bun}_B \to \mathrm{Ran} \times \overline{\mathrm{Bun}}_B \to (\overline{\mathrm{Bun}}_B \times \mathrm{Ran})_{\mathrm{poles}}.$$

Now, the map (7.7) is the natural isomorphism coming from the identification

$$\mathrm{IC}^{\frac{\infty}{2}}_{\mathrm{glob},\mathrm{Bun}_T}|_{\mathrm{Ran}\times\mathrm{Bun}_B^{\lambda}}[d - \langle\lambda, 2\check{\rho}\rangle] \simeq \omega_{\mathrm{Ran}\times\mathrm{Bun}_B^{\lambda}}.$$

7.4.4

Now, by unwinding the construction of the Hecke structure on $\mathrm{IC}^{\frac{\infty}{2}}_{\mathrm{glob},\mathrm{Bun}_T}$ in [BG1, Theorem 3.1.4], one shows that the map (7.6) indeed canonically lifts to a map in $\mathrm{DrPl}(\mathrm{Shv}(\mathrm{Gr}_{G,\mathrm{Ran}} \tilde{\times} \mathrm{Bun}_T))$.

□

Appendix A: Proof of Theorem 4.4.4

With future applications in mind, we will prove a generalization of Theorem 4.4.4. The proof is a paraphrase of the theory developed in [Bar].

Throughout this appendix, the curve X will be assumed proper.

A.1 The Space of G-Bundles with a Generic Reduction

A.1.1

Let Y be a test affine scheme. We shall say that an open subset of $Y \times X$ is a *domain* if it is dense in every fiber of the projection $Y \times X \to X$. Note that the intersection of two domains is again a domain.

Observe that for $\mathcal{I} \in \mathrm{Maps}(Y, \mathrm{Ran})$, the subscheme $Y \times X - \Gamma_{\mathcal{I}}$ is a domain.

A.1.2

Let $\mathrm{Bun}_{G\text{-gen}}$ be the prestack that assigns to an affine test scheme Y the groupoid, whose objects are pairs:

(i) A domain $U \subset Y \times X$
(ii) A G-bundle $\mathcal{P}_G$ defined on U.

An (iso)morphism between two such points is by definition an isomorphism of G-bundles defined over a *sub-domain* of the intersection of their respective domains of definition.

Remark A.1.3 In particular, given $(\mathcal{P}_G, U)$, if $U' \subset U$ is a sub-domain, then the points $(\mathcal{P}_G, U)$ and $(\mathcal{P}_G|_{U'}, U')$ are canonically isomorphic. Hence, in the definition of $\mathrm{Bun}_{G\text{-gen}}$, we can combine points (i) and (ii) into:

(i') A G-bundle $\mathcal{P}_G$ defined over *some* domain in $Y \times X$.

A.1.4

Let $H \to G$ be a homomorphism of algebraic groups. Consider the prestack

$$\mathrm{Bun}_{H\text{-gen}} \underset{\mathrm{Bun}_{G\text{-gen}}}{\times} \mathrm{Bun}_G .$$

By definition, for a test affine scheme Y, its groupoid of Y-points has as objects triples:

(i) A G-bundle $\mathcal{P}_G$ on $Y \times X$
(ii) A domain $U \subset Y \times X$
(iii) A reduction β of $\mathcal{P}_G$ to H defined over $U \subset Y \times X$

An (iso)morphism between two such points is by definition an isomorphism of G-bundles, compatible with the reductions *over the intersection of the corresponding domains*.

Remark A.1.5 As in Remark A.1.3 above, we can combine (ii) and (iii) into:

(ii') A reduction β of $\mathcal{P}_G$ to H defined over *some* domain in $Y \times X$.

A.1.6

For $H = \{1\}$, we will use the notation

$$\mathrm{Gr}_{G,\mathrm{gen}} := \mathrm{pt} \underset{\mathrm{Bun}_{G\text{-gen}}}{\times} \mathrm{Bun}_G .$$

By definition, for an affine test scheme Y, the set $\operatorname{Maps}(Y, \operatorname{Gr}_{G,\text{gen}})$ consists of pairs $(\mathcal{P}_G, \alpha)$, where $\mathcal{P}_G$ is a G-bundle on $Y \times X$, and α is a trivialization of $\mathcal{P}_G$ defined on *some* domain in $Y \times X$.

A.1.7

We have a canonically defined map

$$\operatorname{Gr}_{G,\text{gen}} \to \operatorname{Bun}_{H\text{-gen}} \underset{\operatorname{Bun}_{G\text{-gen}}}{\times} \operatorname{Bun}_G,$$

obtained by base change along $\operatorname{Bun}_G \to \operatorname{Bun}_{G\text{-gen}}$ from the map

$$\text{pt} \to \operatorname{Bun}_{H\text{-gen}}.$$

In addition, we have a canonical map

$$\operatorname{Gr}_{G,\text{Ran}} \to \operatorname{Gr}_{G,\text{gen}}.$$

Composing, we obtain a map

$$\operatorname{Gr}_{G,\text{Ran}} \to \operatorname{Bun}_{H\text{-gen}} \underset{\operatorname{Bun}_{G\text{-gen}}}{\times} \operatorname{Bun}_G. \tag{A.8}$$

A.1.8

We recall the following definition from [Ga2, Sect. 2.5.1]:

A map between prestacks $\mathcal{X}_1 \to \mathcal{X}_2$ is said to be *universally homologically contractible* if for any affine test scheme Y and a map $Y \to \mathcal{X}_2$, the !-pullback functor

$$\operatorname{Shv}(Y) \to \operatorname{Shv}(Y \underset{\mathcal{X}_2}{\times} \mathcal{X}_1)$$

is fully faithful.

If this happens, a formal argument shows that for any prestack $\mathcal{Y}$ and a map $\mathcal{Y} \to \mathcal{X}_2$, the !-pullback functor

$$\operatorname{Shv}(\mathcal{Y}) \to \operatorname{Shv}(\mathcal{Y} \underset{\mathcal{X}_2}{\times} \mathcal{X}_1)$$

is also fully faithful. In particular, the pullback functor

$$f^! : \operatorname{Shv}(\mathcal{X}_2) \to \operatorname{Shv}(\mathcal{X}_1)$$

is fully faithful.

We shall call a prestack $\mathcal{X}$ *homologically contractible* if the map $\mathcal{X} \to \mathrm{pt}$ induces a fully faithful embedding

$$\mathrm{Vect} \to \mathrm{Shv}(\mathcal{Y});$$

this is equivalent to the trace map

$$\mathrm{C}_\bullet(\mathcal{Y}) := \mathrm{C}^\bullet_c(\mathcal{Y}, \omega_\mathcal{Y}) \to \mathsf{e}$$

being an isomorphism. It is not difficult to see that this condition implies a stronger one, namely, that $\mathcal{X} \to \mathrm{pt}$ is universally homologically contractible.

A.1.9

The goal of this section is to prove:

Theorem A.1.10 *Assume that H is connected. Then the map* (A.8) *is universally homologically contractible.*

A.1.11

Let us show how Theorem A.1.10 implies Theorem 4.4.4. We take $H = N$. Note that there is a canonically defined map (in fact, a closed embedding)

$$\overline{\mathrm{Bun}}_N \to \mathrm{Bun}_{N\text{-gen}} \underset{\mathrm{Bun}_{G\text{-gen}}}{\times} \mathrm{Bun}_G .$$

Indeed, a Y-point of $\mathrm{Bun}_{N\text{-gen}} \underset{\mathrm{Bun}_{G\text{-gen}}}{\times} \mathrm{Bun}_G$ can be thought of as a data of $(\mathcal{P}_G, \kappa)$, where $\mathcal{P}_G$ is a G-bundle on $Y \times X$, and κ is a system of bundle maps

$$\kappa^{\check{\lambda}} : \mathcal{O}_X \to \mathcal{V}^{\check{\lambda}}_{\mathcal{P}_G}, \quad \check{\lambda} \in \check{\Lambda}^+$$

defined over some domain $U \subset T \times X$, and satisfying the Plücker relations.

Such a point belongs to $\overline{\mathrm{Bun}}_N$ if and only if the maps $\kappa^{\check{\lambda}}$ extend to regular maps on all of $Y \times X$.

Finally, we note that we have a Cartesian square:

$$\begin{array}{ccc}
\overline{S}^0_{\mathrm{Ran}} & \longrightarrow & \mathrm{Gr}_{G,\mathrm{Ran}} \\
\downarrow & & \downarrow \\
\overline{\mathrm{Bun}_N} & \longrightarrow & \mathrm{Bun}_{N\text{-gen}} \underset{\mathrm{Bun}_{G\text{-gen}}}{\times} \mathrm{Bun}_G .
\end{array}$$

A.2 Toward the Proof of Theorem A.1.10

A.2.1

The assertion of Theorem A.1.10 is obtained as a combination of the following two statements:

Proposition A.2.2 *The map* $\mathrm{Gr}_{G,\mathrm{Ran}} \to \mathrm{Gr}_{G,\mathrm{gen}}$ *is universally homologically contractible.*

Theorem A.2.3 *Let H be connected. Then the map* $\mathrm{pt} \to \mathrm{Bun}_{H\text{-gen}}$ *is universally homologically contractible.*

A.2.4

Let us recall the notion of what it means for a map of prestacks $\mathcal{X}_1 \to \mathcal{X}_2$ to be *pseudo-proper* (cf. [Ga2, Sect. 1.5]):

We shall say that a prestack $\mathcal{X}$ over an affine scheme Y is *pseudo-proper* if it can be represented as a colimit of schemes proper over Y.

We shall say that a map of prestacks $f : \mathcal{Y}_1 \to \mathcal{Y}_2$ is *pseudo-proper* if for any affine test scheme Y and a map $Y \to \mathcal{X}_2$, the map

$$Y \underset{\mathcal{X}_2}{\times} \mathcal{X}_1 \to Y$$

is pseudo-proper.

In *loc.cit.*, it is shown that if f is pseudo-proper, the functor $f_!$, left adjoint to $f^!$, is defined and satisfies base change against !-pullbacks and the projection formula with the $\overset{!}{\otimes}$ tensor product.

From here, we obtain:

Lemma A.2.5 *Let $\mathcal{X}_1 \to \mathcal{X}_2$ be pseudo-proper. Then it is universally homologically contractible if and only if its fibers over field-valued points (potentially, after extending the ground field) are homologically contractible.*

A.2.6 Interlude: The Relative Ran Space

Let $\mathcal{I}_0$ be a finite subset of k-points of X. We define the relative Ran space $\mathrm{Ran}^{\supset \mathcal{I}_0}$ as follows:

For an affine test scheme Y, the set of Y-points of $\mathrm{Ran}^{\supset \mathcal{I}_0}$ consists of finite non-empty subsets

$$\mathcal{I} \subset \mathrm{Hom}(Y, X),$$

such that $Y \times \mathcal{I}_0$ is set-theoretically contained in $\Gamma_{\mathcal{I}}$.

We claim:

Proposition A.2.7 *The prestack* $\mathrm{Ran}^{\supset \mathcal{I}_0}$ *is homologically contractible.*

The proof repeats the proof of the homological contractibility of Ran, see [Ga3, Appendix].

A.2.8 Proof of Lemma 2.3.3 for X Proper

If X is proper, Ran is pseudo-proper. Hence, in this case, the map p^λ_{Ran} is pseudo-proper. Therefore, by Lemma A.2.5, it suffices to show that the fibers of p^λ_{Ran} (over field-valued points) are homologically contractible.

For a given field-valued point $D \in X^\lambda$, let $\mathcal{I}_0 \subset X$ be its support. The fiber of p^λ_{Ran} identifies with $\mathrm{Ran}^{\supset \mathcal{I}_0}$.

Now the assertion follows from Proposition A.2.7.

□

A.2.9 Proof of Proposition A.2.2

It is easy to see that the map $\mathrm{Gr}_{G,\mathrm{Ran}} \to \mathrm{Gr}_{G,\mathrm{gen}}$ is pseudo-proper. Hence, by Lemma A.2.5, it suffices to see that its fibers over field-valued points are homologically contractible.

For a given (field-valued) point of $\mathrm{Gr}_{G,\mathrm{gen}}$, let $U \subset X$ be the maximal open subset over which α is defined. Let $\mathcal{I}_0$ be its set-theoretic complement. Then

$$\mathrm{pt} \underset{\mathrm{Gr}_{G,\mathrm{gen}}}{\times} \mathrm{Gr}_{G,\mathrm{Ran}}$$

identifies with $\mathrm{Ran}^{\supset \mathcal{I}_0}$.

Now the required assertion follows from Proposition A.2.7.

□

A.3 Proof of Theorem A.2.3

A.3.1

Let $\mathrm{Bun}_{H\text{-gen,triv}}$ be the prestack, whose value on an affine test scheme Y is the full subgroupoid of $\mathrm{Maps}(Y, \mathrm{Bun}_{H\text{-gen}})$ consisting of objects isomorphic to the trivial one. In other words, this is the essential image of the functor

$$* = \mathrm{Maps}(Y, \mathrm{pt}) \to \mathrm{Maps}(Y, \mathrm{Bun}_{H\text{-gen}}).$$

The assertion of Theorem A.2.3 is obtained as a combination of the following two statements:

Theorem A.3.2 *For H connected, the map* $\mathrm{pt} \to \mathrm{Bun}_{H\text{-gen,triv}}$ *is universally homologically contractible.*

Theorem A.3.3 *The map* $\mathrm{Bun}_{H\text{-gen,triv}} \to \mathrm{Bun}_{H\text{-gen}}$ *is universally homologically contractible.*

A.3.4 Proof of Theorem A.3.2

Let $\underline{\mathrm{Maps}}(X, H)_{\mathrm{gen}}$ be the group prestack that attaches to an affine test scheme Y the group of maps from a domain in $Y \times X$ to H. By definition

$$\mathrm{Bun}_{H\text{-gen,triv}} \simeq B(\underline{\mathrm{Maps}}(X, H)_{\mathrm{gen}}).$$

Hence, in order to prove Theorem A.3.2, it suffices to show that the prestack $\underline{\mathrm{Maps}}(X, H)_{\mathrm{gen}}$ is homologically contractible. However, this is essentially what is proved in [Ga3, Theorem 1.8.2]:

In order to formally deduce the homological contractibility of $\underline{\mathrm{Maps}}(X, H)_{\mathrm{gen}}$ from *loc. cit.*, we argue as follows:

Let $\underline{\mathrm{Maps}}(X, H)_{\mathrm{Ran}}$ be the prestack that assigns to an affine test scheme Y the set of pairs $(\mathfrak{I}, h)$, where $\mathfrak{I}$ is a finite non-empty subset in $\mathrm{Hom}(Y, X)$ and h is a map

$$(Y \times X - \Gamma_{\mathfrak{I}}) \to H.$$

We have a tautologically defined map

$$\underline{\mathrm{Maps}}(X, H)_{\mathrm{Ran}} \to \underline{\mathrm{Maps}}(X, H)_{\mathrm{gen}},$$

and as in Proposition A.2.2, we show that this map is universally homologically contractible.

Now, the assertion of [Ga2, Theorem 1.8.2] is precisely that for H connected, the prestack $\underline{\mathrm{Maps}}(X, H)_{\mathrm{Ran}}$ is homologically contractible.

□

A.3.5

The remainder of this section is devoted to the proof of Theorem A.3.3. Write

$$1 \to H_u \to H \to H_r \to 1,$$

where H_u is the unipotent radical of H and H_r is the reductive quotient.

We factor the map $\mathrm{Bun}_{H\text{-gen,triv}} \to \mathrm{Bun}_{H\text{-gen}}$ as

$$\mathrm{Bun}_{H\text{-gen,triv}} \to \mathrm{Bun}_{H_r\text{-gen,triv}} \underset{\mathrm{Bun}_{H_r\text{-gen}}}{\times} \mathrm{Bun}_{H\text{-gen}} \to \mathrm{Bun}_{H\text{-gen}}.$$

We will prove that the maps

$$\mathrm{Bun}_{H\text{-gen,triv}} \to \mathrm{Bun}_{H_r\text{-gen,triv}} \underset{\mathrm{Bun}_{H_r\text{-gen}}}{\times} \mathrm{Bun}_{H\text{-gen}} \qquad \text{(A.9)}$$

and

$$\mathrm{Bun}_{H_r\text{-gen,triv}} \to \mathrm{Bun}_{H_r\text{-gen}} \qquad \text{(A.10)}$$

are universally homologically contractible, which would imply the assertion of Theorem A.3.3.

Remark A.3.6 Note that in the applications for the present paper, we have $H = N$, so we do not actually need to consider (A.10).

A.3.7

In order to prove the universal homological contractibility property of (A.9), we can base change with respect to the (value-wise surjective) map $\mathrm{pt} \to \mathrm{Bun}_{H_r\text{-gen,triv}}$. We obtain a map

$$\mathrm{Bun}_{H_u\text{-gen,triv}} \to \mathrm{Bun}_{H_u\text{-gen}},$$

and the statement that (A.9) is universally homologically contractible amounts to the statement of Theorem A.3.3 for H unipotent.

However, we claim that for H unipotent, the map $\mathrm{Bun}_{H\text{-gen,triv}} \to \mathrm{Bun}_{H\text{-gen}}$ is actually an isomorphism. Indeed, every H-bundle is (non-canonically) trivial over a domain that is affine.

A.3.8

Let us observe that the statement that (A.10) is universally homologically contractible is equivalent to the statement of Theorem A.3.3 for H reductive. Hence, for the rest of the argument, H will be assumed reductive.

A.4 Proof of Theorem A.3.3 for H Reductive

A.4.1

In order to prove that

$$\mathrm{Bun}_{H\text{-gen,triv}} \to \mathrm{Bun}_{H\text{-gen}}$$

is universally homologically contractible, it suffices to show that it becomes an isomorphism after localization in the h-topology. (We recall that h-covers include fppf covers as well as maps that are proper and surjective at the level of k-points.)

Since (A.10) is a value-wise monomorphism, it suffices to show that it is a surjection in the h-topology.

A.4.2

Consider the Cartesian square

$$\begin{array}{ccc} \mathrm{Bun}_{H\text{-gen,triv}} \underset{\mathrm{Bun}_{H\text{-gen}}}{\times} \mathrm{Bun}_H & \longrightarrow & \mathrm{Bun}_H \\ \downarrow & & \downarrow \\ \mathrm{Bun}_{H\text{-gen,triv}} & \longrightarrow & \mathrm{Bun}_{H\text{-gen}}. \end{array}$$

It suffices to show that both maps

$$\mathrm{Bun}_{H\text{-gen,triv}} \underset{\mathrm{Bun}_{H\text{-gen}}}{\times} \mathrm{Bun}_H \to \mathrm{Bun}_H \tag{A.11}$$

and

$$\mathrm{Bun}_H \to \mathrm{Bun}_{H\text{-gen}} \tag{A.12}$$

are h-surjections.

A.4.3

The fact that map (A.11) is an h-surjection follows from [DS]; in fact the main theorem of *loc.cit.* asserts that this map is an fppf surjection.

A.4.4

Let us show that (A.12) is an h-surjection.

Fix a Y-point $(\mathcal{P}_G, U)$ of $\mathrm{Bun}_{H\text{-gen}}$ for an affine test scheme Y. The fiber product

$$Y \underset{\mathrm{Bun}_{H\text{-gen}}}{\times} \mathrm{Bun}_H$$

is a prestack that assigns to $Y' \to Y$ the set of extensions of the G-bundle $\mathcal{P}_G|_{Y' \underset{Y}{\times} U}$ to all of $Y' \times X$.

It is easy to see that this prestack is (ind)representable by an ind-scheme, ind-proper over Y. Hence, it is enough to show that the map

$$Y \underset{\mathrm{Bun}_{H\text{-gen}}}{\times} \mathrm{Bun}_H \to Y$$

is surjective at the level of k-points.

However, the latter means that any H-bundle on open subset of X can be extended to all of X, which is well known.

Acknowledgments The author would like to thank S. Raskin for his suggestion to consider the formalism of Drinfeld–Plücker structures, as well as numerous stimulating discussions.

The author is grateful to M. Finkelberg for igniting his interest in the semi-infinite IC sheaf, which has been on author's mind for some 20 years now.

The author is grateful to J. Campbell and L. Chen for pointing out important mistakes in an earlier version of the paper.

The author is supported by NSF grant DMS-1063470. He has also received support from ERC grant 669655.

References

[Bar] J. Barlev, *Moduli spaces of generic data*, arXiv:1204.3469.

[BFGM] A. Braverman, M. Finkelberg, D. Gaitsgory and I. Mirkovic, *Intersection cohomology of Drinfeld compactifications*, Selecta Math. (N.S.) **8** (2002), 381–418.

[BG1] A. Braverman and D. Gaitsgory, *Geometric Eisenstein series*, Invent. Math. **150** (2002), 287–84.

[BG2] A. Braverman and D. Gaitsgory, *Deformations of local systems and Eisenstein series*, GAFA **17** (2008), 1788–1850.

[DrGa] V. Drinfeld and D. Gaitsgory, *On a theorem of Braden*, Transformation groups **19** no. 2 (2014), 313–358.

[DS] V. Drinfeld and C. Simpson, *B-Structures on G-bundles and Local Triviality*, Mathematical Research Letters **2** (1995), 823–829.

[FGV] E. Frenkel, D. Gaitsgory and K. Vilonen, *Whittaker patterns in the geometry of moduli space of bundles on curves*, Annals of Math. **153** (2001), no. 3, 699–748.

[Ga1] D. Gaitsgory, *The semi-infinite intersection cohomology sheaf*, arXiv:1703.04199, to appear in AIM.

[Ga2] D. Gaitsgory, *The Atiyah-Bott formula for the cohomology of the moduli space of bundles on a curve*, arXiv:1505.02331.

[Ga3] D. Gaitsgory, *Contractibility of the space of rational maps*, Invent. Math. **191** (2013), 91–196.

[Ga4] D. Gaitsgory, *The local and global versions of the Whittaker category*, arXiv: 1811.02468.

[GR] D. Gaitsgory and N. Rozenblyum, *A study in derived algebraic geometry, volume I*, AMS (2017).

[Lu1] J. Lurie *Higher Topos Theory*, Princeton University Press (2009).

[Lu2] J. Lurie *Higher Algebra*, available at: http://math.harvard.edu/~lurie

A Topological Approach to Soergel Theory

Roman Bezrukavnikov and Simon Riche

To Sasha Beĭlinson and Vitya Ginzburg with gratitude and admiration.

Contents

R.B. was partially supported by the NSF grant DMS-1601953. This project has received funding from the European Research Council (ERC) under the European Union's Horizon 2020 research and innovation program (grant agreement No 677147).

R. Bezrukavnikov (✉)
Department of Mathematics, Massachusetts Institute of Technology, Cambridge, MA, USA
e-mail: bezrukav@math.mit.edu

S. Riche
CNRS, LMBP, Université Clermont Auvergne, Clermont-Ferrand, France
e-mail: simon.riche@uca.fr

V. Baranovsky et al. (eds.), *Representation Theory and Algebraic Geometry*, Trends in Mathematics, https://doi.org/10.1007/978-3-030-82007-7_7

1 Introduction

1.1 Soergel Theory

In [So2], Soergel developed a new approach to the study of the principal block $\mathscr{O}_0$ of the Bernstein–Gelfand–Gelfand category $\mathscr{O}$ of a complex semisimple Lie algebra $\mathfrak{g}$ (with a fixed Borel subalgebra $\mathfrak{b}$ and Cartan subalgebra $\mathfrak{h} \subset \mathfrak{b}$). Namely, let P be the projective cover of the unique simple object in $\mathcal{O}_0$ with antidominant highest weight (in other words, of the unique simple Verma module). Then Soergel establishes the following results:

(1) (Endomorphismensatz) There exists a canonical algebra isomorphism

$$\mathrm{S}(\mathfrak{h})/\langle \mathrm{S}(\mathfrak{h})^W_+ \rangle \xrightarrow{\sim} \mathrm{End}(P).$$

where W is the Weyl group of $(\mathfrak{g}, \mathfrak{h})$, $\mathrm{S}(\mathfrak{h})$ is the symmetric algebra of $\mathfrak{h}$, and $\langle \mathrm{S}(\mathfrak{h})^W_+ \rangle$ is the ideal generated by homogeneous W-invariant elements of positive degree.

(2) (Struktursatz) The functor $\mathbb{V} := \mathrm{Hom}_{\mathcal{O}_0}(P, -)$ is fully faithful on projective objects; in other words, for any projective objects Q, Q' this functor induces an isomorphism

$$\mathrm{Hom}_{\mathcal{O}_0}(Q, Q') \xrightarrow{\sim} \mathrm{Hom}_{\mathrm{End}(P)}(\mathbb{V}(Q), \mathbb{V}(Q')).$$

(3) The essential image of the restriction of $\mathbb{V}$ to projective objects in $\mathcal{O}_0$ is the subcategory generated by the trivial module $\mathbb{C}$ under the operations of (repeatedly) applying the functors $\mathrm{S}(\mathfrak{h}) \otimes_{\mathrm{S}(\mathfrak{h})^s} -$ with s is a simple reflection and taking direct sums and direct summands.

Taken together, these results allow him to describe the category of projective objects in $\mathcal{O}_0$, and hence the category $\mathcal{O}_0$ itself, in terms of commutative algebra ("Soergel modules"). On the other hand, Soergel relates these modules to cohomology of Bruhat-constructible simple perverse sheaves on the Langlands dual flag variety opening the way to the ideas of *Koszul duality* further developed in his celebrated work with Beĭlinson and Ginzburg [BGS]. Another celebrated application of these ideas is Soergel's new proof of the Kazhdan–Lusztig conjecture [KL] proved earlier by Beĭlinson–Bernstein and Brylinsky–Kashiwara.

1.2 Geometric Version

If G is the semisimple complex algebraic group of adjoint type whose Lie algebra is $\mathfrak{g}$, and if $B \subset G$ is the Borel subgroup whose Lie algebra is $\mathfrak{b}$, then combining the Beĭlinson–Bernstein localization theory [BB] and an equivalence due to Soergel [So1] one obtains that the category $\mathcal{O}_0$ is equivalent to the category $\mathsf{Perv}_U(G/B, \mathbb{C})$ of U-equivariant (equivalently, B-constructible) $\mathbb{C}$-perverse sheaves on the flag variety G/B, where U is the unipotent radical of B (see, e.g., [BGS, Proposition 3.5.2]). Under this equivalence, the simple Verma module corresponds to the skyscraper sheaf at the base point B/B. The main goal of the present paper is to develop a geometric approach to the results in Sect. 1.1, purely in the framework of perverse sheaves, and moreover valid in the setting where the coefficients can be in an arbitrary field $\Bbbk$ (of possibly positive characteristic) instead of $\mathbb{C}$.

In fact, a geometric proof of the Struktursatz (stated for coefficients of characteristic 0, but in fact valid in the general case) was already found by Beĭlinson, the first author and Mirković in [BBM]. One of the main themes of the latter paper, which is fundamental in our approach too, is an idea introduced by Beĭlinson–Ginzburg in [BG], namely, that it is easier (but equivalent) to work with *tilting* objects in $\mathscr{O}_0$ (or its geometric counterparts) rather than with projective objects. Our main contribution is generalization of the Endomorphismensatz to the present setting; then the description of the essential image of the functor $\mathbb{V}$ follows by rather standard methods.

1.3 Monodromy

So we fix a field $\Bbbk$, and consider the category $\mathsf{Perv}_U(G/B, \Bbbk)$ of U-equivariant $\Bbbk$-perverse sheaves on the complex variety G/B. This category has a natural highest weight structure, with weight poset the Weyl group W, and as in the characteristic-0 setting, the projective cover of the skyscraper sheaf at B/B is also the tilting object associated with the longest element w_0 in W; we will therefore denote it $\mathscr{T}_{w_0}$. Our first task is then to describe the $\Bbbk$-algebra $\mathrm{End}_{\mathsf{Perv}_U(G/B,\Bbbk)}(\mathscr{T}_{w_0})$.

In the representation-theoretic context studied by Soergel (see Sect. 1.1), the morphism $\mathrm{S}(\mathfrak{h})/\langle \mathrm{S}(\mathfrak{h})^W_+ \rangle \xrightarrow{\sim} \mathrm{End}(P)$ is obtained from the action of the center of the enveloping algebra $\mathcal{U}\mathfrak{g}$ on P. It has been known for a long time (see, e.g., [BGS, §4.6] or [BBM, Footnote 8 on p. 556]) that from the geometric point of view, this morphism can be obtained via the logarithm of monodromy for the action of T on G. But of course, the logarithm will not make sense over an arbitrary field $\Bbbk$; therefore, what we consider here is the monodromy itself, which defines an algebra morphism

$$\varphi_{\mathscr{T}_{w_0}} : \Bbbk[X_*(T)] \to \mathrm{End}(\mathscr{T}_{w_0}).$$

We then need to show that:

(1) The morphism $\varphi_{\mathscr{T}_{w_0}}$ factors through the quotient $\Bbbk[X_*(T)]/\langle \Bbbk[X_*(T)]^W_+ \rangle$, where $\langle \Bbbk[X_*(T)]^W_+ \rangle$, is the ideal generated by W-invariant elements in the kernel of the natural augmentation morphism $\Bbbk[X_*(T)] \to \Bbbk$.
(2) The resulting morphism $\Bbbk[X_*(T)]/\langle \Bbbk[X_*(T)]^W_+ \rangle \to \mathrm{End}(\mathscr{T}_{w_0})$ is an isomorphism.

1.4 Free-Monodromic Deformation

To prove these claims, we need the second main ingredient of our approach, namely, the "completed category" defined by Yun in [BY, Appendix A]. This category (which is constructed using certain pro-objects in the derived category of sheaves on G/U) is a triangulated category endowed with a t-structure, which we will denote $\widehat{D}_U((G/U)/\!\!/T, \Bbbk)$, and which contains certain objects whose monodromy is "free unipotent." Killing this monodromy provides a functor to the U-equivariant derived category $D^{\mathrm{b}}_U(G/B, \Bbbk)$. The tilting objects in $\mathsf{Perv}_U(G/B, \Bbbk)$ admit "lifts" (or "deformations") to this category, and we can in particular consider the lift $\widehat{\mathscr{T}}_{w_0}$ of $\mathscr{T}_{w_0}$. Now, the algebra $\mathrm{End}_{\widehat{D}_U((G/U)/\!\!/T,\Bbbk)}(\widehat{\mathscr{T}}_{w_0})$ admits *two* morphisms from (the completion $\Bbbk[X_*(T)]^\wedge$ with respect to the augmentation ideal of) $\Bbbk[X_*(T)]$ coming from the monodromy for the left and the right actions of T on G/U, and moreover, we have a canonical isomorphism

$$\mathrm{End}(\mathscr{T}_{w_0}) \cong \mathrm{End}(\widehat{\mathscr{T}}_{w_0}) \otimes_{\Bbbk[X_*(T)]^\wedge} \Bbbk.$$

Hence, what we have to prove transforms into the following claims:

(1) The monodromy morphism $\Bbbk[X_*(T)]^\wedge \otimes_\Bbbk \Bbbk[X_*(T)]^\wedge \to \mathrm{End}(\widehat{\mathscr{T}}_{w_0})$ factors through $\Bbbk[X_*(T)]^\wedge \otimes_{(\Bbbk[X_*(T)]^\wedge)^W} \Bbbk[X_*(T)]^\wedge$.
(2) The resulting morphism $\Bbbk[X_*(T)]^\wedge \otimes_{(\Bbbk[X_*(T)]^\wedge)^W} \Bbbk[X_*(T)]^\wedge \to \mathrm{End}(\widehat{\mathscr{T}}_{w_0})$ is an isomorphism.

1.5 Identification of $\mathbf{End}(\widehat{\mathscr{T}}_{w_0})$

One of the main advantages of working with the category $\widehat{D}_U((G/U)/\!\!/T, \Bbbk)$ rather than with $D^{\mathrm{b}}_U(G/B, \Bbbk)$ is that the natural lifts $(\widehat{\Delta}_w : w \in W)$ of the standard perverse sheaves satisfy $\mathrm{Hom}(\widehat{\Delta}_x, \widehat{\Delta}_y) = 0$ if $x \neq y$. This implies that the functor of "taking the associated graded for the standard filtration" is faithful, and we obtain an injective algebra morphism

$$\mathrm{End}(\widehat{\mathscr{T}}_{w_0}) \to \mathrm{End}(\mathrm{gr}(\widehat{\mathscr{T}}_{w_0})). \tag{1.1}$$

Now, we have $\mathrm{gr}(\widehat{\mathscr{T}}_{w_0}) \cong \bigoplus_{w \in W} \widehat{\Delta}_w$ so that the right-hand side identifies with $\bigoplus_{w \in W} \Bbbk[X_*(T)]^\wedge$. To conclude, it remains to identify the image of (1.1); for this, we use some algebraic results due to Kostant–Kumar [KK] (in their study of the K-theory of flag varieties) and Andersen–Jantzen–Soergel [AJS].

1.6 The Functor $\mathbb{V}$

Once we have identified $\mathrm{End}(\mathscr{T}_{w_0})$ and $\mathrm{End}(\widehat{\mathscr{T}}_{w_0})$, we can consider the functor

$$\mathbb{V} := \mathrm{Hom}(\mathscr{T}_{w_0}, -) : \mathsf{Perv}_U(G/B, \Bbbk) \to \mathsf{Mod}(\Bbbk[X_*(T)]/\langle \Bbbk[X_*(T)]^W_+ \rangle)$$

and its version $\widehat{\mathbb{V}}$ for free-monodromic perverse sheaves. As explained in Sect. 1.2, a short argument from [BBM] shows that these functors are fully faithful on tilting objects. To conclude our study, we need to identify their essential image. The main step for this is to show that $\widehat{\mathbb{V}}$ is monoidal. (Here, the monoidal structure on tilting objects is given by a "convolution" construction, and the monoidal structure on modules over $\Bbbk[X_*(T)]^\wedge \otimes_{(\Bbbk[X_*(T)]^\wedge)^W} \Bbbk[X_*(T)]^\wedge$ is given by tensor product over $\Bbbk[X_*(T)]^\wedge$.) Adapting an argument of [BY], we show that there exists an isomorphism of bifunctors

$$\widehat{\mathbb{V}}(- \mathbin{\widehat{\star}} -) \cong \widehat{\mathbb{V}}(-) \otimes_{\Bbbk[X_*(T)]^\wedge} \widehat{\mathbb{V}}(-). \tag{1.2}$$

However, constructing a *monoidal* structure (i.e., an isomorphism compatible with the relevant structures) is a bit harder. In fact, we construct such a structure in the similar context of étale sheaves on the analogue of G/B (or G/U) over an algebraically closed field of positive characteristic, using a "Whittaker-type" construction. We then deduce the similar claim in the classical topology over $\mathbb{C}$ using the general formalism explained in [BBD, §6.1].

With this at hand, we obtain a description of the monoidal triangulated category $(\widehat{D}_U((G/U)/\!\!/T, \Bbbk), \widehat{\star})$ and its module category $D^{\mathrm{b}}_U(G/B, \Bbbk)$ in terms of coherent sheaves on the formal neighborhood of the point $(1, 1)$ in $T^\vee_\Bbbk \times_{(T^\vee_\Bbbk)/W} T^\vee_\Bbbk$ and on the fiber of the quotient morphism $T^\vee_\Bbbk \to (T^\vee_\Bbbk)/W$ over the image of 1, respectively (where $T^\vee_\Bbbk$ is the split $\Bbbk$-torus which is Langlands dual to T); see Theorem 11.9.

Remark 1.1 Identification of the essential image of $\mathbb{V}$ and $\widehat{\mathbb{V}}$ does not require monoidal structure on $\widehat{\mathbb{V}}$: An isomorphism as in (1.2) would be sufficient. However, description of the monoidal structure provides a stronger statement.

1.7 Some Remarks

We conclude this introduction with a few remarks.

As explained in Sect. 1.3, in the present paper, we work with the group algebra $\Bbbk[X_*(T)]$ and not with the symmetric algebra $\mathrm{S}(\Bbbk \otimes_{\mathbb{Z}} X_*(T))$ as one might have expected from the known characteristic-0 setting. However, one can check (see, e.g., [AR2, Proposition 5.5]) that if $\mathrm{char}(\Bbbk)$ is very good for G, then there exists a W-equivariant algebra isomorphism between the completions of $\Bbbk[X_*(T)]$ and $\mathrm{S}(\Bbbk \otimes_{\mathbb{Z}} X_*(T))$ with respect to their natural augmentation ideals. (In the characteristic-0 setting, there exists a canonical choice of identification, given by the logarithm; in positive characteristic, there exists no "preferred" isomorphism.) Therefore, fixing such an isomorphism, under this assumption, our results can also be stated in terms of $\mathrm{S}(\Bbbk \otimes_{\mathbb{Z}} X_*(T))$. An important observation in [So2, BGS] is that the identification between $\mathrm{End}(P)$ and the coinvariant algebra allows one to define a grading on $\mathrm{End}(P)$ and then to define a "graded version" of $\mathscr{O}_0$. This graded version can be realized geometrically via mixed perverse sheaves (either in the sense of Deligne, see [BGS], or in a more elementary sense constructed using semisimple complexes, see [AR1, AR3]). When $\mathrm{char}(\Bbbk)$ is not very good, the algebra $\Bbbk[X_*(T)]/\langle \Bbbk[X_*(T)]^W_+ \rangle$ does not admit an obvious grading; we do not know how to interpret this, and the relation with the corresponding category of "mixed perverse sheaves" constructed in [AR3]. (In very good characteristic, this category indeed provides a "graded version" of $\mathsf{Perv}_U(G/B, \Bbbk)$, as proved in [AR2, AR3].)

As explained already, in the case of characteristic-0 coefficients, our results are equivalent to those of Soergel in [So2]. They are also proved by geometric means in this case in [BY]. In the case of very good characteristic, these methods were extended in [AR2] (except for the consideration of the free-monodromic objects). The method we follow here is completely general (in particular, new in bad characteristic), more direct (since it does not involve Koszul duality), and more canonical (since it does not rely on any choice of identification relating $\mathrm{S}(\Bbbk \otimes_{\mathbb{Z}} X_*(T))$ and $\Bbbk[X_*(T)]$).

In the complex coefficients setting, the category $\mathsf{Perv}_U(G/B, \mathbb{C})$ has a representation-theoretic interpretation, in terms of the category $\mathscr{O}_0$. It also admits a representation-theoretic description in the case when $\mathrm{char}(\Bbbk)$ is bigger than the Coxeter number of G, in terms of Soergel's modular category $\mathscr{O}$ [So3]. This fact was first proved in [AR2, Theorem 2.4]; it can also be deduced more directly by comparing the results of [So3] and those of the present paper.

1.8 Contents

The paper starts with a detailed review of the construction of Yun's "completed category" (see [BY, Appendix A]) in Sects. 2–5. More precisely, we adapt his constructions (performed initially for étale $\mathbb{Q}_\ell$-complexes) to the setting of sheaves on complex algebraic varieties, with coefficients in an arbitrary field. This adaptation does not require new ideas, but since the wording in [BY] is quite dense, we reproduce most proofs and propose alternative arguments in a few cases.

Starting from Sect. 6, we concentrate on the case of the flag variety. We start by constructing the "associated graded" functor. Then in Sect. 7, we review the construction of the convolution product on $\widehat{D}_U((G/U)/\!\!/T, \Bbbk)$ (again, mainly following Yun). In Sect. 8, we recall some algebraic results of Kostant–Kumar, and we apply all of this to prove our "Endomorphismensatz" in Sect. 9. In Sect. 10, we explain how to adapt our constructions in the setting of étale sheaves, and in Sect. 11, we study the functors $\mathbb{V}$ and $\widehat{\mathbb{V}}$. Finally, in Sect. 12, we take the opportunity to correct the proof of a technical lemma in [AB].

Part I: Reminder on Completed Categories
We fix a field $\Bbbk$.

2 Monodromy

2.1 Construction

We consider a complex algebraic torus A and an A-torsor[1] $\pi : X \to Y$. We then denote by $D^{\mathrm{b}}_c(X /\!\!/ A, \Bbbk)$ the full triangulated subcategory of $D^{\mathrm{b}}_c(X, \Bbbk)$ generated by the essential image of the functor $\pi^* : D^{\mathrm{b}}_c(Y, \Bbbk) \to D^{\mathrm{b}}_c(X, \Bbbk)$.

Fix some $\lambda \in X_*(A)$. We then set

$$\theta_\lambda : \begin{cases} \mathbb{C} \times X \to & X \\ (z, x) \mapsto \lambda(\exp(z)) \cdot x \end{cases}.$$

We will also denote by $\mathrm{pr} : \mathbb{C} \times X \to X$ the projection.

The following claims follow from the considerations in [Ve, §9].

Lemma 2.1

(1) *For any $\mathscr{F}$ in $D^{\mathrm{b}}_c(X /\!\!/ A, \Bbbk)$, there exists a unique morphism*

$$\iota^\lambda_{\mathscr{F}} : \theta^*_\lambda(\mathscr{F}) \to \mathrm{pr}^*(\mathscr{F})$$

whose restriction to $\{0\} \times X$ is $\mathrm{id}_{\mathscr{F}}$. Moreover, $\iota^\lambda_{\mathscr{F}}$ is an isomorphism.

(2) *If $\mathscr{F}, \mathscr{G}$ are in $D^{\mathrm{b}}_c(X /\!\!/ A, \Bbbk)$ and $f : \mathscr{F} \to \mathscr{G}$ is a morphism, then the following diagram commutes:*

$$\begin{array}{ccc} \theta^*_\lambda(\mathscr{F}) & \xrightarrow{\iota^\lambda_{\mathscr{F}}} & \mathrm{pr}^*(\mathscr{F}) \\ {\scriptstyle \theta^*_\lambda(f)} \downarrow & & \downarrow {\scriptstyle \mathrm{pr}^*(f)} \\ \theta^*_\lambda(\mathscr{G}) & \xrightarrow{\iota^\lambda_{\mathscr{G}}} & \mathrm{pr}^*(\mathscr{G}). \end{array}$$

Sketch of Proof The essential ingredient of the proof is the (obvious) fact that the functor pr^* is fully faithful so that its essential image is a triangulated subcategory of $D^{\mathrm{b}}_c(\mathbb{C} \times X, \Bbbk)$. We see that for any $\mathscr{G}$ in $D^{\mathrm{b}}_c(Y, \Bbbk)$, the object $\theta^*_\lambda \pi^*(\mathscr{G})$ belongs to this essential image; hence, for any $\mathscr{F}$ in $D^{\mathrm{b}}_c(X /\!\!/ A, \Bbbk)$ the object $\theta^*_\lambda(\mathscr{F})$ is isomorphic to $\mathrm{pr}^*(\mathscr{F}')$ for some $\mathscr{F}'$ in $D^{\mathrm{b}}_c(X, \Bbbk)$. Restricting to $\{0\} \times X$, we

[1]All the torsors we will encounter in the present paper will be locally trivial for the Zariski topology.

obtain an isomorphism $f : \mathscr{F} \xrightarrow{\sim} \mathscr{F}'$, and we can define $\iota^{\lambda}_{\mathscr{F}}$ as the composition $\theta^*_\lambda(\mathscr{F}) \xrightarrow{\sim} \mathrm{pr}^*(\mathscr{F}') \xrightarrow{\mathrm{pr}^*(f^{-1})} \mathrm{pr}^*(\mathscr{F})$. □

Using this lemma and restricting $\iota^{\lambda}_{\mathscr{F}}$ to $\{2\mathbf{i}\pi\} \times X$, we obtain an automorphism $\varphi^{\lambda}_{\mathscr{F}}$ of $\mathscr{F}$. This automorphism satisfies the property that if $\mathscr{F}, \mathscr{G}$ are in $D^{\mathrm{b}}_c(X /\!\!/ A, \Bbbk)$ and $f : \mathscr{F} \to \mathscr{G}$ is a morphism, then $\varphi^{\lambda}_{\mathscr{G}} \circ f = f \circ \varphi^{\lambda}_{\mathscr{F}}$.

For any $\mathscr{F}$ in $D^{\mathrm{b}}_c(X /\!\!/ A, \Bbbk)$, the automorphism $\varphi^{\lambda}_{\mathscr{F}}$ is unipotent. (In fact, this automorphism is the identity if $\mathscr{F}$ belongs to the essential image of π^*, and the category $D^{\mathrm{b}}_c(X /\!\!/ A, \Bbbk)$ is generated by such objects.) Moreover, if $\lambda, \mu \in X_*(A)$, we have

$$\varphi^{\lambda \cdot \mu}_{\mathscr{F}} = \varphi^{\lambda}_{\mathscr{F}} \circ \varphi^{\mu}_{\mathscr{F}}.$$

In other words, the assignment $\lambda \mapsto \varphi^{\lambda}_{\mathscr{F}}$ defines a group morphism

$$X_*(A) \to \mathrm{Aut}(\mathscr{F}). \tag{2.1}$$

We now set

$$R_A := \Bbbk[X_*(A)].$$

The group morphism (2.1) induces a $\Bbbk$-algebra morphism

$$\varphi_{\mathscr{F}} : R_A \to \mathrm{End}(\mathscr{F}).$$

Since each $\varphi_{\mathscr{F}}(\lambda)$ is unipotent, this morphism factors through an algebra morphism

$$\varphi^{\wedge}_{\mathscr{F}} : R^{\wedge}_A \to \mathrm{End}(\mathscr{F}),$$

where $R^{\wedge}_A$ is the completion of R_A with respect to the maximal ideal $\mathfrak{m}_A$ given by the kernel of the algebra map $\varepsilon_A : R_A \to \Bbbk$ sending each $\lambda \in X_*(A)$ to 1. This construction is functorial, in the sense that it makes $D^{\mathrm{b}}_c(X /\!\!/ A, \Bbbk)$ an $R^{\wedge}_A$-linear category. (Here, the $R^{\wedge}_A$-action on $\mathrm{Hom}_{D^{\mathrm{b}}_c(X /\!\!/ A, \Bbbk)}(\mathscr{F}, \mathscr{G})$ is given by $r \cdot f = f \circ \varphi^{\wedge}_{\mathscr{F}}(r) = \varphi^{\wedge}_{\mathscr{G}}(r) \circ f$.)

Remark 2.2 Geometrically, we have $R_A = \mathcal{O}(A^{\vee}_{\Bbbk})$, where $A^{\vee}_{\Bbbk}$ is the $\Bbbk$-torus such that $X^*(A^{\vee}_{\Bbbk}) = X_*(A)$, and $R^{\wedge}_A$ identifies with the algebra of functions on the formal neighborhood of 1 in $A^{\vee}_{\Bbbk}$. Note that any choice of trivialization $A \xrightarrow{\sim} (\mathbb{C}^{\times})^r$ provides isomorphisms

$$R_A \cong \Bbbk[y_1^{\pm 1}, \cdots, y_r^{\pm 1}] \quad \text{and} \quad R^{\wedge}_A \cong \Bbbk[[x_1, \cdots, x_r]]$$

(where $x_i = y_i - 1$).

2.2 Basic Properties

We denote by $\varepsilon_A^\wedge : R_A^\wedge \to \Bbbk$ the continuous morphism which extends ε_A.

Lemma 2.3 *For any $\mathscr{F}$ in $D_c^{\mathrm{b}}(X /\!\!/ A, \Bbbk)$ and $x \in R_A^\wedge$, we have*

$$\pi_!(\varphi_{\mathscr{F}}^\wedge(x)) = \varepsilon_A^\wedge(x) \cdot \mathrm{id}_{\pi_! \mathscr{F}}.$$

Proof Let $\lambda \in X_*(A)$, and let $p : \mathbb{C} \times Y \to Y$ be the projection. Then both of the following squares are Cartesian:

$$\begin{array}{ccc} \mathbb{C} \times X & \xrightarrow{\theta_\lambda} & X \\ {\scriptstyle \mathrm{id}_{\mathbb{C}} \times \pi} \downarrow & & \downarrow {\scriptstyle \pi} \\ \mathbb{C} \times Y & \xrightarrow{p} & Y, \end{array} \qquad \begin{array}{ccc} \mathbb{C} \times X & \xrightarrow{\mathrm{pr}} & X \\ {\scriptstyle \mathrm{id}_{\mathbb{C}} \times \pi} \downarrow & & \downarrow {\scriptstyle \pi} \\ \mathbb{C} \times Y & \xrightarrow{p} & Y. \end{array}$$

By the base change theorem, we deduce canonical isomorphisms

$$(\mathrm{id}_{\mathbb{C}} \times \pi)_! \theta_\lambda^*(\mathscr{F}) \cong p^* \pi_!(\mathscr{F}), \qquad (\mathrm{id}_{\mathbb{C}} \times \pi)_! \mathrm{pr}^*(\mathscr{F}) \cong p^* \pi_!(\mathscr{F}).$$

Under these isomorphisms, the map $(\mathrm{id}_{\mathbb{C}} \times \pi)_! \iota_{\mathscr{F}}^\lambda$ identifies with an endomorphism of $p^* \pi_!(\mathscr{F})$. Now, the functor p^* is fully faithful; hence, this morphism must be of the form $p^*(f)$ for f an endomorphism of $\pi_! \mathscr{F}$. Restricting to $\{0\} \times Y$, we see that $f = \mathrm{id}_{\pi_! \mathscr{F}}$. Hence, the restriction of $(\mathrm{id}_{\mathbb{C}} \times \pi)_! \iota_{\mathscr{F}}^\lambda$ to $\{2\mathbf{i}\pi\} \times Y$ is also the identity. But this morphism identifies with $\pi_!(\varphi_{\mathscr{F}}^\lambda)$, which completes the proof. □

We now consider a second A-torsor $\pi' : X' \to Y'$, and an A-equivariant morphism $f : X \to X'$. The following claims follow easily from the definitions.

Lemma 2.4

(1) *The functors $f^!$ and f^* induce functors*

$$f^!, f^* : D_c^{\mathrm{b}}(X' /\!\!/ A, \Bbbk) \to D_c^{\mathrm{b}}(X /\!\!/ A, \Bbbk).$$

Moreover, for any $\mathscr{F}$ in $D_c^{\mathrm{b}}(X' /\!\!/ A, \Bbbk)$ and $r \in R_A^\wedge$, we have

$$\varphi_{f^! \mathscr{F}}^\wedge(r) = f^!(\varphi_{\mathscr{F}}^\wedge(r)), \quad \varphi_{f^* \mathscr{F}}^\wedge(r) = f^*(\varphi_{\mathscr{F}}^\wedge(r)).$$

(2) *The functors $f_!$ and f_* induce functors*

$$f_!, f_* : D_c^{\mathrm{b}}(X /\!\!/ A, \Bbbk) \to D_c^{\mathrm{b}}(X' /\!\!/ A, \Bbbk).$$

Moreover, for any $\mathscr{F}$ in $D_c^{\mathrm{b}}(X /\!\!/ A, \Bbbk)$ and $r \in R_A^\wedge$, we have

$$\varphi^{\wedge}_{f_!\mathscr{F}}(r) = f_!(\varphi^{\wedge}_{\mathscr{F}}(r)), \quad \varphi^{\wedge}_{f_*\mathscr{F}}(r) = f_*(\varphi^{\wedge}_{\mathscr{F}}(r)).$$

Finally, we consider a second torus A', and an injective morphism $\phi : A' \to A$. Of course, in this setting, we can consider X either as an A-torsor or as an A'-torsor, and $D^{\mathrm{b}}_c(X /\!\!/ A, \Bbbk)$ is a full subcategory in $D^{\mathrm{b}}_c(X /\!\!/ A', \Bbbk)$. In particular, for $\mathscr{F}$ in $D^{\mathrm{b}}_c(X /\!\!/ A, \Bbbk)$, we can consider the morphism $\varphi^{\wedge}_{\mathscr{F}}$ both for the action of A (in which case we will denote it $\varphi^{\wedge}_{\mathscr{F},A}$) and for the action of A' (in which case we will denote it $\varphi^{\wedge}_{\mathscr{F},A'}$). Once again, the following lemma immediately follows from the definitions.

Lemma 2.5 *For $\mathscr{F}$ in $D^{\mathrm{b}}_c(X /\!\!/ A, \Bbbk)$, the morphism $\varphi^{\wedge}_{\mathscr{F},A'}$ is the composition of $\varphi^{\wedge}_{\mathscr{F},A}$ with the morphism $R^{\wedge}_{A'} \to R^{\wedge}_{A}$ induced by ϕ.*

2.3 Monodromy and Equivariance

For simplicity, in this subsection, we assume that $A = \mathbb{C}^{\times}$. We denote by $a, p : A \times X \to X$ the action and projection maps, respectively. Recall that a perverse sheaf $\mathscr{F}$ in $D^{\mathrm{b}}_c(X, \Bbbk)$ is said to be A-equivariant if $a^*(\mathscr{F}) \cong p^*(\mathscr{F})$. (See [BR, Appendix A] for the equivalence with other "classical" definitions.)

Lemma 2.6 *Let $\mathscr{F}$ be a perverse sheaf in $D^{\mathrm{b}}_c(X /\!\!/ A, \Bbbk)$. Then $\mathscr{F}$ is A-equivariant iff the morphism $\varphi^{\wedge}_{\mathscr{F}}$ factors through $\varepsilon^{\wedge}_{A}$.*

Proof If $\mathscr{F}$ is equivariant, then there exists an isomorphism $a^*(\mathscr{F}) \xrightarrow{\sim} p^*(\mathscr{F})$ whose restriction to $\{1\} \times X$ is the identity. For $\lambda \in X_*(A)$, pulling back under the morphism $\mathbb{C} \times X \to A \times X$ given by $(z, x) \mapsto (\lambda(\exp(z)), x)$, we obtain the morphism $\iota^{\lambda}_{\mathscr{F}}$ of Lemma 2.1, whose restriction to $\{2\mathbf{i}\pi\} \times X$ is therefore the identity.

Conversely, assume that $\varphi^{\wedge}_{\mathscr{F}}$ factors through ε_A. Let $\lambda : \mathbb{C}^{\times} \to A$ be the tautological cocharacter, and let $f : \mathbb{C} \times X \to \mathbb{C} \times X$ be the map defined by $f(z, x) = (z + 2\mathbf{i}\pi, x)$. Then $f^*(\iota^{\lambda}_{\mathscr{F}})$ is a morphism $\theta^*_{\lambda}(\mathscr{F}) \to \mathrm{pr}^*(\mathscr{F})$ whose restriction to $\{0\} \times X$ is, by assumption, the identity of $\mathscr{F}$. Therefore, by the unicity claim in Lemma 2.1, we have $f^*(\iota^{\lambda}_{\mathscr{F}}) = \iota^{\lambda}_{\mathscr{F}}$.

Now, we explain how to construct an isomorphism $\eta : a^*(\mathscr{F}) \xrightarrow{\sim} p^*(\mathscr{F})$. Recall (see [BBD, Corollaire 2.1.22]) that since we consider (shifts of) perverse sheaves, such an isomorphism can be constructed locally; more concretely, if we set $U_1 = \mathbb{C} \smallsetminus \mathbb{R}_{\geq 0}$ and $U_2 = \mathbb{C} \smallsetminus \mathbb{R}_{\leq 0}$, then to construct η, it suffices to construct isomorphisms on $U_1 \times X$ and $U_2 \times X$, which coincide on $(U_1 \cap U_2) \times X$. The map $\mathbb{C} \times X \to A \times X$ given by $(z, x) \mapsto (\lambda(\exp(z)), x)$ restricts to homeomorphisms between $\{z \in \mathbb{C} \mid \Im(z) \in (0, 2\pi)\} \times X$ and $U_1 \times X$ and between $\{z \in \mathbb{C} \mid \Im(z) \in (-\pi, \pi)\} \times X$ and $U_2 \times X$. Therefore, we can obtain the isomorphisms on $U_1 \times X$ and $U_2 \times X$ by simply restricting $\iota^{\lambda}_{\mathscr{F}}$ to these open subsets. The intersection $U_1 \cap U_2$ has two connected components: $U_+ = \{z \in \mathbb{C} \mid \Im(z) > 0\}$

and $U_- = \{z \in \mathbb{C} \mid \Im(z) < 0\}$. Our two isomorphisms coincide on $U_+ \times X$ by definition, and they coincide on $U_- \times X$ because of the equality $f^*(\iota^\lambda_{\mathscr{F}}) = \iota^\lambda_{\mathscr{F}}$ justified above. Hence, they indeed glue to an isomorphism $\eta : a^*(\mathscr{F}) \xrightarrow{\sim} p^*(\mathscr{F})$, which finishes the proof. □

Remark 2.7

(1) Our proof of Lemma 2.6 can easily be adapted to the case of a general torus; we leave the details to interested readers.
(2) In [Ve], Verdier defines (by the exact same procedure) monodromy for a more general class of objects in $D^{\mathrm{b}}_c(X, \Bbbk)$, called the *monodromic complexes*, namely, those complexes $\mathscr{F}$ such that the restriction of $\mathscr{H}^i(\mathscr{F})$ to each A-orbit is locally constant for any $i \in \mathbb{Z}$. As was suggested to one of us by J. Bernstein, one can give an alternative definition of the category $D^{\mathrm{b}}_c(X /\!\!/ A, \Bbbk)$ as the category of monodromic complexes $\mathscr{F}$ (in this sense) such that the monodromy morphism $\varphi_{\mathscr{F}} : R_A \to \mathrm{End}(\mathscr{F})$ is unipotent, that is, factors through $R_A/\mathfrak{m}_A^n$ for some n. Indeed, it is clear that our category $D^{\mathrm{b}}_c(X /\!\!/ A, \Bbbk)$ is included in the latter category. Now, if $\mathscr{F}$ is monodromic with unipotent monodromy, then $\mathscr{F}$ is an extension of its perverse cohomology objects, which have the same property; hence, we can assume that $\mathscr{F}$ is perverse. Then one can consider the (finite) filtration

$$\mathscr{F} \supset \sum_{x \in \mathfrak{m}_A} \mathrm{Im}(x) \supset \sum_{x \in \mathfrak{m}_A^2} \mathrm{Im}(x) \supset \cdots .$$

Each subquotient in this filtration is a perverse sheaf with trivial monodromy and hence belongs to the essential image of π^* by (the general version of) Lemma 2.6.

3 Completed Category

3.1 Definition

As in Sect. 2, we consider a complex torus A of rank r and an A-torsor $\pi : X \to Y$. We also assume we are given a finite algebraic stratification

$$Y = \bigsqcup_{s \in \mathcal{S}} Y_s$$

where each Y_s is isomorphic to an affine space, and such that for any $s \in \mathcal{S}$, the restriction $\pi_s : \pi^{-1}(Y_s) \to Y_s$ is a trivial A-torsor. We set

$$\pi_\dagger := \pi_![r], \qquad \pi^\dagger := \pi^![-r] \cong \pi^*[r].$$

Then $(\pi_\dagger, \pi^\dagger)$ is an adjoint pair, and $\pi^\dagger$ is t-exact with respect to the perverse t-structures.

We denote by $D^{\mathrm{b}}_{\mathcal{S}}(Y, \Bbbk)$ the $\mathcal{S}$-constructible derived category of $\Bbbk$-sheaves on Y, and by $D^{\mathrm{b}}_{\mathcal{S}}(X /\!\!/ A, \Bbbk)$ the full triangulated subcategory of $D^{\mathrm{b}}_{c}(X, \Bbbk)$ generated by the essential image of the restriction of $\pi^\dagger$ to $D^{\mathrm{b}}_{\mathcal{S}}(Y, \Bbbk)$.

Definition 3.1 The category $\widehat{D}_{\mathcal{S}}(X /\!\!/ A, \Bbbk)$ is defined as the full subcategory of the category of pro-objects[2] in $D^{\mathrm{b}}_{\mathcal{S}}(X /\!\!/ A, \Bbbk)$ consisting of the objects "$\varprojlim$"$\mathscr{F}_n$ which are:

- π-constant, that is, such that the pro-object "$\varprojlim$"$\pi_\dagger(\mathscr{F}_n)$ in $D^{\mathrm{b}}_{\mathcal{S}}(Y, \Bbbk)$ is isomorphic to an object of $D^{\mathrm{b}}_{\mathcal{S}}(Y, \Bbbk)$
- Uniformly bounded in degrees, that is, isomorphic to a pro-object "$\varprojlim$"$\mathscr{F}'_n$ such that each $\mathscr{F}'_n$ belongs to $D^{[a,b]}_{\mathcal{S}}(X /\!\!/ A, \Bbbk)$ for some $a, b \in \mathbb{Z}$ (independent of n).

The morphisms in this category can be described as

$$\mathrm{Hom}_{\widehat{D}_{\mathcal{S}}(X /\!\!/ A, \Bbbk)}(\text{"}\varprojlim\text{"}\mathscr{F}_n, \text{"}\varprojlim\text{"}\mathscr{G}_n) = \varprojlim_n \varinjlim_m \mathrm{Hom}_{D^{\mathrm{b}}_{\mathcal{S}}(X /\!\!/ A, \Bbbk)}(\mathscr{F}_m, \mathscr{G}_n). \tag{3.1}$$

According to [BY, Theorem A.3.2], the category $\widehat{D}_{\mathcal{S}}(X /\!\!/ A, \Bbbk)$ has a natural triangulated structure, for which the distinguished triangles are the triangles isomorphic to those of the form

$$\text{"}\varprojlim\text{"}\mathscr{F}_n \xrightarrow{\text{"}\varprojlim\text{"}f_n} \text{"}\varprojlim\text{"}\mathscr{G}_n \xrightarrow{\text{"}\varprojlim\text{"}g_n} \text{"}\varprojlim\text{"}\mathscr{H}_n \xrightarrow{\text{"}\varprojlim\text{"}h_n} \text{"}\varprojlim\text{"}\mathscr{F}_n[1]$$

obtained from projective systems of distinguished triangles

$$\mathscr{F}_n \xrightarrow{f_n} \mathscr{G}_n \xrightarrow{g_n} \mathscr{H}_n \xrightarrow{h_n} \mathscr{F}_n[1]$$

in $D^{\mathrm{b}}_{\mathcal{S}}(X /\!\!/ A, \Bbbk)$. By definition, the functor $\pi_\dagger$ induces a functor

$$\widehat{D}_{\mathcal{S}}(X /\!\!/ A, \Bbbk) \to D^{\mathrm{b}}_{\mathcal{S}}(Y, \Bbbk),$$

which will also be denoted $\pi_\dagger$. From the proof of [BY, Theorem A.3.2], we see that this functor is triangulated.

The monodromy construction from Sect. 2 makes the category $\widehat{D}_{\mathcal{S}}(X /\!\!/ A, \Bbbk)$ an $R^\wedge_A$-linear category. More precisely, for any object $\mathscr{F} = \text{"}\varprojlim\text{"}\mathscr{F}_n$ in $\widehat{D}_{\mathcal{S}}(X /\!\!/ A, \Bbbk)$, we have

$$\mathrm{End}(\mathscr{F}) = \varprojlim_n \varinjlim_m \mathrm{Hom}_{D^{\mathrm{b}}_{\mathcal{S}}(X /\!\!/ A, \Bbbk)}(\mathscr{F}_m, \mathscr{F}_n).$$

[2] All our pro-objects are tacitly parametrized by $\mathbb{Z}_{\geq 0}$ (with its standard order).

See (3.1). We have a natural algebra morphism $R_A \to \mathrm{End}(\mathscr{F})$, sending $r \in R_A$ to $(\varphi_{\mathscr{F}_n}(r))_n$. Since each $\varphi_{\mathscr{F}_n}$ factors through a quotient $R_A/\mathfrak{m}_A^N$ for some N (depending on n), this morphism extends to a morphism $\varphi^{\wedge}_{\mathscr{F}} : R_A^{\wedge} \to \mathrm{End}(\mathscr{F})$. As in Sect. 2.1, this construction provides an $R_A^{\wedge}$-linear structure on $\widehat{D}_{\mathcal{S}}(X /\!\!/ A, \Bbbk)$.

All the familiar functors (in particular, the pushforward and pullback functors associated with morphisms of A-torsors) induce functors between the appropriate completed categories, which will be denoted similarly; for details, the reader might consult [BY, Proposition A.3.3 and Corollary A.3.4].

Remark 3.2 As explained in [BY, Remark A.2.3], there exists a filtered triangulated category $\widehat{D}^F_{\mathcal{S}}(X /\!\!/ A, \Bbbk)$ over $\widehat{D}_{\mathcal{S}}(X /\!\!/ A, \Bbbk)$ in the sense of [Be, Definition A.1(c)]. Namely, consider a filtered triangulated category $D^F_{\mathcal{S}}(X /\!\!/ A, \Bbbk)$ over $D^{\mathrm{b}}_{\mathcal{S}}(X /\!\!/ A, \Bbbk)$ (constructed, e.g., following [Be, Example A.2]). Then one can take as $\widehat{D}^F_{\mathcal{S}}(X /\!\!/ A, \Bbbk)$ the category of pro-objects "$\varprojlim$"$\mathscr{F}_n$ in $D^F_{\mathcal{S}}(X /\!\!/ A, \Bbbk)$ such that the filtrations on the objects $\mathscr{F}_n$ are uniformly bounded and such that "$\varprojlim$"$\mathrm{gr}^F_i(\mathscr{F}_n)$ belongs to $\widehat{D}_{\mathcal{S}}(X /\!\!/ A, \Bbbk)$ for any $i \in \mathbb{Z}$.

3.2 The Free-Monodromic Local System

Let us consider the special case $X = A$ (with its natural action) and $Y = \mathrm{pt}$. Let us choose as a generator of the fundamental group $\pi_1(\mathbb{C}^\times)$ the anticlockwise loop $\gamma : t \in [0, 1] \mapsto \exp(2\mathbf{i}\pi t)$. Then we obtain a group isomorphism

$$X_*(A) \xrightarrow{\sim} \pi_1(A) \tag{3.2}$$

by sending $\lambda \in X_*(A)$ to the class of the loop $t \mapsto \lambda(\gamma(t))$. (Here, our fundamental groups are taken with the neutral element as base point.) Of course, the category of $\Bbbk$-local systems on A is equivalent to the category of finite-dimension $\Bbbk$-representations of $\pi_1(A)$. Via the isomorphism (3.2), we thus obtain an equivalence between the category of $\Bbbk$-local systems on A and that of finite-dimensional R_A-modules. The Serre subcategory consisting of local systems which are extensions of copies of the constant local system $\underline{\Bbbk}_A$ then identifies with the category of finite-dimensional R_A-modules annihilated by a power of $\mathfrak{m}_A$ or equivalently with the category of finite-dimensional $R_A^{\wedge}$-modules annihilated by a power of $\mathfrak{m}_A^{\wedge} := \mathfrak{m}_A R_A^{\wedge}$. The latter category will be denoted $\mathsf{Mod}^{\mathrm{nil}}(R_A^{\wedge})$.

For any $n \in \mathbb{Z}_{\geq 0}$, we denote by $\mathscr{L}_{A,n}$ the local system on A corresponding to the R_A-module $R_A/\mathfrak{m}_A^{n+1}$. Then we have natural surjections $\mathscr{L}_{A,n+1} \to \mathscr{L}_{A,n}$; hence, we can define $\widehat{\mathscr{L}}_A$ as the pro-object "$\varprojlim$"$\mathscr{L}_{A,n}$. It is clear that this pro-object is uniformly bounded. It is easily seen that it is also π-constant; in fact, the surjections $\mathscr{L}_{A,n} \twoheadrightarrow \mathscr{L}_{A,0} = \underline{\Bbbk}_A$ induce an isomorphism

$$\text{"}\varprojlim\text{"}\pi_!(\mathscr{L}_{A,n}) \xrightarrow{\sim} \mathscr{H}^{2r}(\pi_!\mathscr{L}_{A,0})[-2r] = \Bbbk[-2r].$$

In particular, this shows that $\widehat{\mathscr{L}}_A$ defines an object of $\widehat{D}(A /\!\!/ A, \Bbbk)$, which satisfies

$$\pi_{\dagger}(\widehat{\mathscr{L}}_A) \cong \Bbbk[-r]. \tag{3.3}$$

(The stratification of $Y = \mathrm{pt}$ we consider here is the obvious one.)

Remark 3.3 Choose a trivialization $A \xrightarrow{\sim} (\mathbb{C}^{\times})^r$. Then we obtain an isomorphism $R_A \cong (R_{\mathbb{C}^\times})^{\otimes r}$; see Remark 2.2. For any $n \geq 0$, we have

$$\mathfrak{m}_A^{n \cdot r} \subset \mathfrak{m}_{\mathbb{C}^\times}^n \otimes (R_{\mathbb{C}^\times})^{\otimes(r-1)} + R_{\mathbb{C}^\times} \otimes \mathfrak{m}_{\mathbb{C}^\times}^n \otimes (R_{\mathbb{C}^\times})^{\otimes(r-2)} + \cdots + (R_{\mathbb{C}^\times})^{\otimes(r-1)} \otimes \mathfrak{m}_{\mathbb{C}^\times}^n \subset \mathfrak{m}_A^n,$$

hence an isomorphism

$$\widehat{\mathscr{L}}_A \xrightarrow{\sim} \text{“}\varprojlim\text{”}\big(\mathscr{L}_{\mathbb{C}^\times, n}\big)^{\boxtimes r}. \tag{3.4}$$

The definition of $\widehat{\mathscr{L}}_A$ given above is much more canonical, but the description as the right-hand side in (3.4) is sometimes useful to reduce the proofs to the case $r = 1$.

3.3 “Averaging” with the Free-Monodromic Local System

In this subsection, for simplicity, we assume that $\ell := \mathrm{char}(\Bbbk)$ is positive. We will prove a technical lemma that will allow us later to prove that in the flag variety, setting the convolution product admits a unit (see Lemma 7.6). A reader ready to accept (or ignore) this question might skip this subsection.

We denote by $a : A \times X \to X$ the action morphism.

Lemma 3.4 *For any $\mathscr{F}$ in $D^{\mathrm{b}}_{\mathscr{S}}(X /\!\!/ A, \Bbbk)$, there exists a canonical isomorphism*

$$a_!\big(\widehat{\mathscr{L}}_A \boxtimes \mathscr{F}\big) \cong \mathscr{F}[-2r].$$

Proof We first want to construct a morphism of functors $a_!\big(\widehat{\mathscr{L}}_A \boxtimes -\big) \to \mathrm{id}[-2r]$. For this, by adjunction, it suffices to construct a morphism of functors

$$\big(\widehat{\mathscr{L}}_A \boxtimes -\big) \to a^![-2r]. \tag{3.5}$$

For any $s \geq 0$, we denote by $[s] : A \to A$ the morphism sending z to z^{ℓ^s}, and set $a_s := a \circ ([s] \times \mathrm{id}_X)$. Since any unipotent matrix M with coefficients in $\Bbbk$ satisfies $M^{\ell^s} = 1$ for $s \gg 0$, we see that for $\mathscr{F}$ in $D^{\mathrm{b}}_{\mathscr{S}}(X /\!\!/ A, \Bbbk)$, for $s \gg 0$, all the cohomology objects of $(a_s)^* \mathscr{F}$ are constant on the fibers of the projection to X. In fact, the techniques of [Ve, §5] show that for any such $\mathscr{F}$ and for $s \gg 0$, there exists an isomorphism $f_s^{\mathscr{F}} : (a_s)^*\mathscr{F} \xrightarrow{\sim} p^*(\mathscr{F})$ whose restriction to $\{1\} \times X$ is the identity. Moreover, these morphisms are essentially unique in the sense that given s, s' such that $f_s^{\mathscr{F}}$ and $f_{s'}^{\mathscr{F}}$ are defined, for $t \gg s, s'$, we have

$$([t-s] \times \mathrm{id}_X)^* f_s^{\mathscr{F}} = ([t-s'] \times \mathrm{id}_X)^* f_{s'}^{\mathscr{F}},$$

and functorial in the sense that if $u : \mathscr{F} \to \mathscr{G}$ is a morphism, then for $s \gg 0$, the diagram

$$\begin{array}{ccc} (a_s)^*\mathscr{F} & \xrightarrow[\sim]{f_s^{\mathscr{F}}} & p^*\mathscr{F} \\ {\scriptstyle (a_s)^*u}\downarrow & & \downarrow{\scriptstyle p^*u} \\ (a_s)^*\mathscr{G} & \xrightarrow[\sim]{f_s^{\mathscr{G}}} & p^*\mathscr{G} \end{array}$$

commutes.

Now, fix $\mathscr{F}$ in $D^{\mathrm{b}}_{\mathcal{S}}(X /\!\!/ A, \Bbbk)$. For $s \gg 0$, we have the morphism

$$\begin{aligned}(f_s^{\mathscr{F}})^{-1} \in \mathrm{Hom}\big(p^*(\mathscr{F}), (a_s)^*\mathscr{F}\big) &= \mathrm{Hom}\big(p^*(\mathscr{F}), (a_s)^!\mathscr{F}[-2r]\big) \\ &\cong \mathrm{Hom}\big(([s] \times \mathrm{id}_X)_! p^*(\mathscr{F}), a^!(\mathscr{F})[-2r]\big).\end{aligned}$$

The "essential unicity" claimed above implies that these morphisms define a canonical element in

$$\varinjlim_s \mathrm{Hom}\big(([s]\times \mathrm{id}_X)_! p^*(\mathscr{F}), a^!(\mathscr{F})[-2r]\big) = \mathrm{Hom}\left(\big(\text{“}\varprojlim_s\text{”}[s]_!\underline{\Bbbk}\big) \boxtimes \mathscr{F}, a^!\mathscr{F}[-2r]\right).$$

Now, we observe that $[s]_!\underline{\Bbbk} = \mathscr{L}_{A,\ell^s}$ so that "$\varprojlim_s$"$[s]_!\underline{\Bbbk} \cong \widehat{\mathscr{L}}_A$, and we deduce the wished-for morphism (3.5). (The functoriality of our morphism follows from the "functoriality" of the morphisms $f_s^{\mathscr{F}}$ claimed above.)

To conclude the proof, it remains to show that the morphism $a_!\big(\widehat{\mathscr{L}}_A \boxtimes \mathscr{F}\big) \to \mathscr{F}$ is an isomorphism for any $\mathscr{F}$ in $D^{\mathrm{b}}_{\mathcal{S}}(X /\!\!/ A, \Bbbk)$. By the 5-lemma and the definition of this category, it suffices to do so in case $\mathscr{F} = \pi^\dagger\mathscr{G}$ for some $\mathscr{G}$ in $D^{\mathrm{b}}_{\mathcal{S}}(Y, \Bbbk)$. In this case, the morphism $f_t^{\mathscr{F}}$ is defined for any $t \geq 0$ and can be chosen as the obvious isomorphism

$$(a_t)^*\mathscr{F} = (a_t)^*\pi^*\mathscr{G}[-r] = (\pi \circ a_t)^*\mathscr{G}[-r] = (\pi \circ p)^*\mathscr{G}[-r] = p^*\mathscr{F}.$$

Then under the identification

$$a_!\big(\widehat{\mathscr{L}}_A \boxtimes \mathscr{F}\big) = \pi^\dagger(p_Y)_!\big(\widehat{\mathscr{L}}_A \boxtimes \mathscr{G}\big) = \pi^\dagger\big((\pi')_!(\widehat{\mathscr{L}}_A) \boxtimes \mathscr{G}\big),$$

where $p_Y : A \times Y \to Y$ and $\pi' : A \to \mathrm{pt}$ are the projections, our morphism is induced by the isomorphism $(\pi')_!(\widehat{\mathscr{L}}_A) \cong \Bbbk[-2r]$ from Sect. 3.2. This concludes the proof. □

4 The Case of the Trivial Torsor

In this section, we study the category $\widehat{D}_{\mathcal{S}}(X /\!\!/ A, \Bbbk)$ in the special case $X = A$.

4.1 Description of $\widehat{D}(A /\!\!/ A, \Bbbk)$ in Terms of Pro-complexes of $R_A^\wedge$-Modules

As explained in Sect. 3.2, every object of $\mathsf{Mod}^{\mathrm{nil}}(R_A^\wedge)$ defines a sheaf on A; this assignment therefore defines a functor $D^{\mathrm{b}}\mathsf{Mod}^{\mathrm{nil}}(R_A^\wedge) \to D_c^{\mathrm{b}}(A, \Bbbk)$, which clearly takes values in $D^{\mathrm{b}}(A /\!\!/ A, \Bbbk)$. We will denote by

$$\Phi_A : D^{\mathrm{b}}\mathsf{Mod}^{\mathrm{nil}}(R_A^\wedge) \to D^{\mathrm{b}}(A /\!\!/ A, \Bbbk)$$

the composition of this functor with the shift of complexes by r to the left (where r is the rank of A). In this way, Φ_A is t-exact if $D^{\mathrm{b}}(A /\!\!/ A, \Bbbk)$ is equipped with the perverse t-structure.

Lemma 4.1 *The functor Φ_A is an equivalence of triangulated categories.*

Proof If we denote by $\Bbbk$ the $R_A^\wedge$-module $R_A^\wedge/\mathfrak{m}_A^\wedge$, then it is clear that $\Phi_A(\Bbbk) = \underline{\Bbbk}_A[r]$. We claim that Φ_A induces an isomorphism

$$\bigoplus_{n \in \mathbb{Z}} \mathrm{Hom}_{D^{\mathrm{b}}\mathsf{Mod}^{\mathrm{nil}}(R_A^\wedge)}(\Bbbk, \Bbbk[n]) \xrightarrow{\sim} \bigoplus_{n \in \mathbb{Z}} \mathrm{Hom}_{D^{\mathrm{b}}(A /\!\!/ A, \Bbbk)}(\underline{\Bbbk}_A[r], \underline{\Bbbk}_A[r+n]).$$

Here, the right-hand side identifies with $\mathsf{H}^\bullet(A; \Bbbk)$.

Choosing a trivialization of A, we reduce the claim to the case $r = 1$, that is, $A = \mathbb{G}_{\mathrm{m}}$ (see Remark 2.2). In this case, the left-hand side has dimension 2, with a basis consisting of $\mathrm{id} : \Bbbk \to \Bbbk$ and the natural extension

$$\Bbbk = \mathfrak{m}_{\mathbb{C}^\times}^\wedge/(\mathfrak{m}_{\mathbb{C}^\times}^\wedge)^2 \hookrightarrow R_{\mathbb{C}^\times}^\wedge/(\mathfrak{m}_{\mathbb{C}^\times}^\wedge)^2 \twoheadrightarrow R_{\mathbb{C}^\times}^\wedge/\mathfrak{m}_{\mathbb{C}^\times}^\wedge = \Bbbk.$$

It is clear that $\Phi_{\mathbb{C}^\times}$ identifies this space with $\mathsf{H}^\bullet(\mathbb{C}^\times; \Bbbk)$, and the claim is proved.

Since the object $\Bbbk$, resp. the object $\underline{\Bbbk}_A[r]$ generates the triangulated category $D^{\mathrm{b}}\mathsf{Mod}^{\mathrm{nil}}(R_A^\wedge)$, resp. $D^{\mathrm{b}}(A /\!\!/ A, \Bbbk)$, this claim and Beĭlinson's lemma imply that Φ_A indeed is an equivalence of categories. □

The category $D^{\mathrm{b}}_c(\mathrm{pt}, \Bbbk)$ identifies with $D^{\mathrm{b}}\mathsf{Vect}^{\mathrm{fd}}_{\Bbbk}$, where $\mathsf{Vect}^{\mathrm{fd}}_{\Bbbk}$ is the category of finite-dimensional $\Bbbk$-vector spaces. Under this identification, the functor $\pi^{\dagger}$ corresponds to the composition of Φ_A with the restriction-of-scalars functor associated with the natural surjection $R^{\wedge}_A \twoheadrightarrow \Bbbk$. By adjunction, we deduce an isomorphism

$$\pi_{\dagger} \circ \Phi_A \cong \Bbbk \overset{L}{\otimes}_{R^{\wedge}_A} (-).$$

In view of these identifications, the category $\widehat{D}(A /\!\!/ A, \Bbbk)$ is therefore equivalent to the category $\widehat{D}(R^{\wedge}_A)$ of pro-objects "$\varprojlim$"M_n in $D^{\mathrm{b}}\mathsf{Mod}^{\mathrm{nil}}(R^{\wedge}_A)$ which are uniformly bounded and such that the object

$$\text{“}\varprojlim\text{”}\Bbbk \overset{L}{\otimes}_{R^{\wedge}_A} M_n$$

is isomorphic to an object of $D^{\mathrm{b}}\mathsf{Vect}^{\mathrm{fd}}_{\Bbbk}$. We use this equivalence to transport the triangulated structure on $\widehat{D}(A /\!\!/ A, \Bbbk)$ to $\widehat{D}(R^{\wedge}_A)$.

4.2 Some Results on Pro-complexes of $R^{\wedge}_A$-Modules

We now consider

$$\widehat{\mathcal{L}}_A := \text{“}\varprojlim\text{”} R^{\wedge}_A/(\mathfrak{m}^{\wedge}_A)^{n+1},$$

a pro-object in the category $D^{\mathrm{b}}\mathsf{Mod}^{\mathrm{nil}}(R^{\wedge}_A)$.

Lemma 4.2 *For any M in $D^{\mathrm{b}}\mathsf{Mod}^{\mathrm{nil}}(R^{\wedge}_A)$, there exists a canonical isomorphism*

$$\mathrm{Hom}(\widehat{\mathcal{L}}_A, M) \cong \mathsf{H}^0(M)$$

(where morphisms are taken in the category of pro-objects in $D^{\mathrm{b}}\mathsf{Mod}^{\mathrm{nil}}(R^{\wedge}_A)$).

Proof By dévissage, it is sufficient to prove this claim when M is concentrated in a certain degree k, that is, $M = N[-k]$ for some N in $\mathsf{Mod}^{\mathrm{nil}}(R^{\wedge}_A)$. By definition, we have

$$\mathrm{Hom}(\widehat{\mathcal{L}}_A, N[-k]) = \varinjlim_n \mathrm{Ext}^{-k}_{R^{\wedge}_A}(R^{\wedge}_A/(\mathfrak{m}^{\wedge}_A)^{n+1}, N).$$

If $k = 0$, it is easily seen that the right-hand side identifies with N. Now, if $k \neq 0$, we use the fact that the natural functor from $D^{\mathrm{b}}\mathsf{Mod}^{\mathrm{nil}}(R^{\wedge}_A)$ to the bounded derived category of $R^{\wedge}_A$-modules is fully faithful (see, e.g., [Or, Lemma 2.1]), which implies that any morphism $f : R^{\wedge}_A/(\mathfrak{m}^{\wedge}_A)^{n+1} \to N[-k]$ is the image of a morphism in the category $D^{\mathrm{b}}\mathsf{Mod}(R^{\wedge}_A/(\mathfrak{m}^{\wedge}_A)^{m+1})$ for $m \gg 0$. Then the image of f in $\mathrm{Ext}^{-k}_{R^\wedge_A}(R^\wedge_A/(\mathfrak{m}^\wedge_A)^{m+1}, N)$ vanishes, since it is the image of a morphism in $\mathrm{Hom}_{D^b\mathsf{Mod}(R^\wedge_A/(\mathfrak{m}^\wedge_A)^{m+1})}(R^\wedge_A/(\mathfrak{m}^\wedge_A)^{m+1}, N[-k]) = 0$. □

As a consequence of this lemma, one obtains in particular an isomorphism

$$\text{“}\varprojlim\text{”}\Bbbk \overset{L}{\otimes}_{R^\wedge_A} R^\wedge_A/(\mathfrak{m}^\wedge_A)^{n+1} \cong \Bbbk \tag{4.1}$$

in the category of pro-objects in $D^b\mathsf{Vect}^{\mathrm{fd}}_{\Bbbk}$. This shows that $\widehat{L}_A$ belongs to $\widehat{D}(R^\wedge_A)$. (Of course, this property also follows from the fact that this object is the image of $\widehat{\mathscr{L}}_A[r]$ under the equivalence considered in Sect. 4.1.)

Lemma 4.3 *Let* “$\varprojlim$”M_n *be an object of* $\widehat{D}(R^\wedge_A)$, *and assume that the object* “$\varprojlim$”$\Bbbk \otimes^L_{R^\wedge_A} M_n$ *belongs to* $D^{\leq 0}\mathsf{Vect}^{\mathrm{fd}}_{\Bbbk}$. *Then the obvious morphism*

$$\text{“}\varprojlim\text{”}\tau_{\leq 0}M_n \to \text{“}\varprojlim\text{”}M_n$$

is an isomorphism in the category of pro-objects in $D^b\mathsf{Mod}^{\mathrm{nil}}(R^\wedge_A)$, *where* $\tau_{\leq 0}$ *is the usual truncation functor for complexes of* $R^\wedge_A$*-modules.*

Proof By uniform boundedness, we can assume that each complex M_n belongs to $D^{\leq d}\mathsf{Mod}^{\mathrm{nil}}(R^\wedge_A)$ for some $d \in \mathbb{Z}$. If $d \leq 0$, then there is nothing to prove. Hence, we assume that $d > 0$. We will prove that in this case, the pro-object “$\varprojlim$”$H^d(M_n)$ is isomorphic to 0. Since filtrant direct limits are exact, this will show that for any X in $D^b\mathsf{Mod}^{\mathrm{nil}}(R^\wedge_A)$, the morphism

$$\varinjlim \mathrm{Hom}(M_n, X) \to \varinjlim \mathrm{Hom}(\tau_{<d}M_n, X)$$

is an isomorphism and hence that the morphism of pro-objects

$$\text{“}\varprojlim\text{”}\tau_{<d}M_n \to \text{“}\varprojlim\text{”}M_n$$

is an isomorphism. Of course, this property is sufficient to conclude.

We observe that the pro-object

$$\text{“}\varprojlim\text{”}\Bbbk \otimes_{R^\wedge_A} H^d(M_n) = \text{“}\varprojlim\text{”}H^d\left(\Bbbk \overset{L}{\otimes}_{R^\wedge_A} M_n\right) = H^d\left(\text{“}\varprojlim\text{”}\Bbbk \overset{L}{\otimes}_{R^\wedge_A} M_n\right)$$

in the category $\mathsf{Vect}^{\mathrm{fd}}_{\Bbbk}$ vanishes. Hence for any fixed n, for $m \gg n$, the map $\Bbbk \otimes_{R^\wedge_A} H^d(M_m) \to \Bbbk \otimes_{R^\wedge_A} H^d(M_n)$ vanishes, or in other words, the map $H^d(M_m) \to H^d(M_n)$ takes values in $\mathfrak{m}^\wedge_A \cdot H^d(M_n)$. Since $H^d(M_n)$ is annihilated by $(\mathfrak{m}^\wedge_A)^q$ for some q, this implies that the map $H^d(M_m) \to H^d(M_n)$ vanishes for $m \gg 0$. Clearly, this implies that “$\varprojlim$”$H^d(M_n) \cong 0$ and concludes the proof. □

Lemma 4.4 *The object* $\widehat{L}_A$ *generates* $\widehat{D}(R^\wedge_A)$ *as a triangulated category.*

Proof We will prove, by induction on the length of the shortest interval $I \subset \mathbb{Z}$, such that "$\varprojlim$"$\Bbbk \otimes^L_{R^\wedge_A} M_n$ belongs to $D^I\mathsf{Vect}^{\mathrm{fd}}_\Bbbk$ and that any object "$\varprojlim$"M_n of $\widehat{D}(R^\wedge_A)$ belongs to the triangulated subcategory generated by $\widehat{L}_A$.

First, assume that $I = \varnothing$. Then for any X in $D^{\mathrm{b}}\mathsf{Vect}^{\mathrm{fd}}_\Bbbk$, we have

$$0 = \varinjlim_n \mathrm{Hom}_{D^{\mathrm{b}}\mathsf{Vect}^{\mathrm{fd}}_\Bbbk}(\Bbbk \overset{L}{\otimes}_{R^\wedge_A} M_n, X) \cong \varinjlim_n \mathrm{Hom}_{D^{\mathrm{b}}\mathsf{Mod}^{\mathrm{nil}}(R^\wedge_A)}(M_n, X).$$

Since the essential image of $D^{\mathrm{b}}\mathsf{Vect}^{\mathrm{fd}}_\Bbbk$ generates $D^{\mathrm{b}}\mathsf{Mod}^{\mathrm{nil}}(R^\wedge_A)$ as a triangulated category, and since filtrant direct limits are exact, it follows that

$$\varinjlim_n \mathrm{Hom}_{D^{\mathrm{b}}\mathsf{Mod}^{\mathrm{nil}}(R^\wedge_A)}(M_n, X) = 0$$

for any X in $D^{\mathrm{b}}\mathsf{Mod}^{\mathrm{nil}}(R^\wedge_A)$. By definition, this implies that "$\varprojlim$"$M_n = 0$, proving the claim in this case.

Now, we assume that $I \neq \varnothing$. Shifting complexes if necessary, we can assume that $I = [-d, 0]$ for some $d \in \mathbb{Z}_{\geq 0}$. Using Lemma 4.3, we can then assume that each M_n belongs to $D^{\leq 0}\mathsf{Mod}^{\mathrm{nil}}(R^\wedge_A)$. Set

$$V := H^0\left(\text{“}\varprojlim\text{”}\Bbbk \overset{L}{\otimes}_{R^\wedge_A} M_n\right) = \text{“}\varprojlim\text{”}H^0(\Bbbk \overset{L}{\otimes}_{R^\wedge_A} M_n) = \text{“}\varprojlim\text{”}\Bbbk \otimes_{R^\wedge_A} H^0(M_n).$$

Then V is a finite-dimensional $\Bbbk$-vector space, and id_V defines an element in

$$\mathrm{Hom}_\Bbbk\left(V, \text{“}\varprojlim\text{”}\Bbbk \otimes_{R^\wedge_A} H^0(M_n)\right) = \varprojlim \mathrm{Hom}_\Bbbk\left(V, \Bbbk \otimes_{R^\wedge_A} H^0(M_n)\right).$$

Consider the object

$$\mathcal{V} := \text{“}\varprojlim\text{”}\left(R^\wedge_A/(\mathfrak{m}^\wedge_A)^{n+1} \otimes_\Bbbk V\right)$$

in $\widehat{D}(R^\wedge_A)$. (Of course, $\mathcal{V}$ is isomorphic to a direct sum of copies of $\widehat{L}_A$.) Then by Lemma 4.2, we have

$$\mathrm{Hom}_{\widehat{D}(R^\wedge_A)}(\mathcal{V}, \text{“}\varprojlim\text{”}M_m) = \varprojlim_m \mathrm{Hom}_{\widehat{D}(R^\wedge_A)}(\mathcal{V}, M_m) \cong \varprojlim_m \mathrm{Hom}_\Bbbk(V, H^0(M_m)).$$

Now, for any m, we have a surjection

$$H^0(M_m) \twoheadrightarrow \Bbbk \otimes_{R^\wedge_A} H^0(M_m),$$

which induces a surjection

$$\mathrm{Hom}_\Bbbk(V, H^0(M_m)) \twoheadrightarrow \mathrm{Hom}_\Bbbk\left(V, \Bbbk \otimes_{R^\wedge_A} H^0(M_m)\right).$$

Each vector space $\ker\big(\mathrm{Hom}_{\Bbbk}(V, H^0(M_m)) \to \mathrm{Hom}_{\Bbbk}(V, \Bbbk \otimes_{R_A^\wedge} H^0(M_m))\big)$ is finite-dimensional; therefore, the projective system formed by these spaces satisfies the Mittag–Leffler condition. This implies that the map

$$\varprojlim_m \mathrm{Hom}_{\Bbbk}(V, H^0(M_m)) \to \varprojlim_m \mathrm{Hom}_{\Bbbk}\left(V, \Bbbk \otimes_{R_A^\wedge} H^0(M_n)\right) \tag{4.2}$$

is surjective (see, e.g., [KS1, Proposition 1.12.3]).

Let now $f : \mathcal{V} \to$ "$\varprojlim$"M_n be a morphism whose image in the right-hand side of (4.2) is id_V. By definition (and in view of (4.1)), the morphism

$$\Bbbk \overset{L}{\otimes}_{R_A^\wedge} f : \Bbbk \overset{L}{\otimes}_{R_A^\wedge} \mathcal{V} \to \Bbbk \overset{L}{\otimes}_{R_A^\wedge} \text{“}\varprojlim\text{”} M_n$$

induces an isomorphism in degree-0 cohomology. Hence, the cone C of f (in the triangulated category $\widehat{D}(R_A^\wedge)$) is such that $\Bbbk \otimes^L_{R_A^\wedge} C$ belongs to $D^{[-d,-1]}\mathsf{Vect}^{\mathrm{fd}}_{\Bbbk}$. By induction, this objects belongs to the triangulated subcategory of $\widehat{D}(R_A^\wedge)$ generated by $\widehat{L}_A$. Then the distinguished triangle

$$\mathcal{V} \to \text{“}\varprojlim\text{”} M_n \to C \xrightarrow{[1]}$$

shows that "$\varprojlim$"M_n also belongs to this subcategory, which finishes the proof. □

4.3 Description of $\widehat{D}(A /\!\!/ A, \Bbbk)$ in Terms of Complexes of $R_A^\wedge$-Modules

Recall that the algebra $R_A^\wedge$ is isomorphic to an algebra of formal power series in r indeterminates; see Remark 2.2. In particular, this shows that this algebra is local, Noetherian, and of finite global dimension. We will denote by $\mathsf{Mod}^{\mathrm{fg}}(R_A^\wedge)$ the category of finitely generated $R_A^\wedge$-modules.

Proposition 4.5 *There exists a natural equivalence of triangulated categories*

$$D^{\mathrm{b}}\mathsf{Mod}^{\mathrm{fg}}(R_A^\wedge) \xrightarrow{\sim} \widehat{D}(R_A^\wedge).$$

Proof We consider the functor φ from $D^{\mathrm{b}}\mathsf{Mod}^{\mathrm{fg}}(R_A^\wedge)$ to the category of pro-objects in $D^{\mathrm{b}}\mathsf{Mod}^{\mathrm{nil}}(R_A^\wedge)$ sending a complex M to

$$\varphi(M) := \text{“}\varprojlim\text{”}\left(R_A^\wedge/(\mathfrak{m}_A^\wedge)^{n+1} \overset{L}{\otimes}_{R_A^\wedge} M\right).$$

Since $R_A^\wedge$ is local, Noetherian, and of finite global dimension, any object in the category $D^{\mathrm{b}}\mathsf{Mod}^{\mathrm{fg}}(R_A^\wedge)$ is isomorphic to a bounded complex of free $R_A^\wedge$-modules. It is clear that the image of such a complex belongs to $\widehat{D}(R_A^\wedge)$; hence, φ takes values in $\widehat{D}(R_A^\wedge)$. Once this is established, it is clear that this functor is triangulated.

By Lemma 4.2, for $k \in \mathbb{Z}$, we have

$$\operatorname{Hom}_{\widehat{D}(R_A^\wedge)}(\widehat{L}_A, \widehat{L}_A[k]) = \varprojlim_n \operatorname{Hom}_{\widehat{D}(R_A^\wedge)}(\widehat{L}_A, R_A^\wedge/(\mathfrak{m}_A^\wedge)^{n+1}[k])$$
$$\cong \begin{cases} R_A^\wedge & \text{if } k = 0; \\ 0 & \text{otherwise.} \end{cases} \tag{4.3}$$

Hence, φ induces an isomorphism

$$\operatorname{Hom}_{D^{\mathrm{b}}\mathsf{Mod}^{\mathrm{fg}}(R_A^\wedge)}(R_A^\wedge, R_A^\wedge[k]) \xrightarrow{\sim} \operatorname{Hom}_{\widehat{D}(R_A^\wedge)}(\widehat{L}_A, \widehat{L}_A[k]).$$

Since the object $R_A^\wedge$, resp. $\widehat{L}_A$, generates $D^{\mathrm{b}}\mathsf{Mod}^{\mathrm{fg}}(R_A^\wedge)$, resp. $\widehat{D}(R_A^\wedge)$, as a triangulated category (see Lemma 4.4), this observation and Beĭlinson's lemma imply that φ is an equivalence of categories. □

Combining Proposition 4.5 and the considerations of Sect. 4.1, we finally obtain the following result.

Corollary 4.6 *There exists a canonical equivalence of triangulated categories*

$$D^{\mathrm{b}}\mathsf{Mod}^{\mathrm{fg}}(R_A^\wedge) \xrightarrow{\sim} \widehat{D}(A /\!\!/ A, \Bbbk)$$

sending the free module $R_A^\wedge$ to $\widehat{\mathscr{L}}_A[r]$.

From (4.1), we see that the equivalence of Proposition 4.5 intertwines the functors $\Bbbk \otimes^L_{R_A^\wedge} (-)$ on both sides. Therefore, under the equivalence of Corollary 4.6, the functor $\Bbbk \otimes^L_{R_A^\wedge} (-)$ on the left-hand side corresponds to the functor $\pi_\dagger$ on the right-hand side. (Here, $\pi : A \to \mathrm{pt}$ is the unique map, and we identify the categories $D^{\mathrm{b}}_c(\mathrm{pt}, \Bbbk)$ and $D^{\mathrm{b}}\mathsf{Vect}^{\mathrm{fd}}_{\Bbbk}$ as in Sect. 4.1.)

5 The Perverse t-Structure

5.1 Recollement

We now come back to the setting of Sect. 3.1. If $Z \subset Y$ is a locally closed union of strata, and if we denote by $h : \pi^{-1}(Z) \to X$ the embedding, in view of the results recalled in Sect. 3.1, the functors $h_!$, h_*, $h^!$, and h^* induce triangulated functors

$$h_!, h_* : \widehat{D}_{\mathcal{S}}(\pi^{-1}(Z) /\!\!/ A, \Bbbk) \to \widehat{D}_{\mathcal{S}}(X /\!\!/ A, \Bbbk),$$
$$h^*, h^! : \widehat{D}_{\mathcal{S}}(X /\!\!/ A, \Bbbk) \to \widehat{D}_{\mathcal{S}}(\pi^{-1}(Z) /\!\!/ A, \Bbbk)$$

which satisfy the usual adjunction and fully faithfulness properties. (Here, following standard conventions, we write $\widehat{D}_{\mathcal{S}}(\pi^{-1}(Z) /\!\!/ A, \Bbbk)$ for $\widehat{D}_{\mathcal{T}}(\pi^{-1}(Z) /\!\!/ A, \Bbbk)$ where $\mathcal{T} = \{s \in \mathcal{S} \mid Y_s \subset Z\}$.) If $\pi_Z : \pi^{-1}(Z) \to Z$ is the restriction of π, and if $\overline{h} : Z \to Y$ is the embedding, then the arguments of the proof of [BY, Corollary A.3.4] show that we have canonical isomorphisms

$$(\pi_Z)_\dagger \circ h_? \cong \overline{h}_? \circ \pi_\dagger, \quad (\pi_Z)_\dagger \circ h^? \cong \overline{h}^? \circ \pi_\dagger \tag{5.1}$$

for $? \in \{!, *\}$.

In particular, if Z is closed with $U := Y \smallsetminus Z$, it is open complement, and if we denote the corresponding embeddings by $i : \pi^{-1}(Z) \to X$ and $j : \pi^{-1}(U) \to X$, then we obtain a recollement diagram

$$\widehat{D}_{\mathcal{S}}(\pi^{-1}(Z) /\!\!/ A, \Bbbk) \underset{i^!}{\overset{i^*}{\leftleftarrows}} \xrightarrow{i_*} \widehat{D}_{\mathcal{S}}(X /\!\!/ A, \Bbbk) \underset{j_*}{\overset{j_!}{\leftleftarrows}} \xrightarrow{j^*} \widehat{D}_{\mathcal{S}}(\pi^{-1}(U) /\!\!/ A, \Bbbk)$$

in the sense of [BBD].

5.2 *Definition of the Perverse t-Structure*

Let us choose, for any $s \in \mathcal{S}$, an A-equivariant map $p_s : X_s \to A$, where $X_s := \pi^{-1}(Y_s)$. (Such a map exists by assumption.) Then the functor $(p_s)^*[\dim(Y_s)] \cong (p_s)^![-\dim(Y_s)]$ induces an equivalence of triangulated categories

$$\widehat{D}(A /\!\!/ A, \Bbbk) \xrightarrow{\sim} \widehat{D}_{\mathcal{S}}(X_s /\!\!/ A, \Bbbk).$$

Composing with the equivalence of Corollary 4.6, we deduce an equivalence of categories

$$D^{\mathrm{b}}\mathsf{Mod}^{\mathrm{fg}}(R_A^\wedge) \xrightarrow{\sim} \widehat{D}_{\mathcal{S}}(X_s /\!\!/ A, \Bbbk). \tag{5.2}$$

The transport, via this equivalence, of the tautological t-structure on $D^{\mathrm{b}}\mathsf{Mod}^{\mathrm{fg}}(R_A^\wedge)$, will be called the *perverse* t-structure and will be denoted

$$\left({}^p\widehat{D}_{\mathcal{S}}(X_s /\!\!/ A, \Bbbk)^{\leq 0}, {}^p\widehat{D}_{\mathcal{S}}(X_s /\!\!/ A, \Bbbk)^{\geq 0}\right).$$

Using the recollement formalism from Sect. 5.1, by gluing these t-structures, we obtain a t-structure on $\widehat{D}_{\mathcal{S}}(X /\!\!/ A, \Bbbk)$, which we also call the perverse t-structure. More precisely, for any $s \in \mathcal{S}$, we denote by $j_s : X_s \to X$ the embedding. Then the full subcategory ${}^p\widehat{D}_{\mathcal{S}}(X /\!\!/ A, \Bbbk)^{\leq 0}$ consists of the objects $\mathscr{F}$ such that $j_s^* \mathscr{F}$ belongs to ${}^p\widehat{D}_{\mathcal{S}}(X_s /\!\!/ A, \Bbbk)^{\leq 0}$ for any s, and the full subcategory ${}^p\widehat{D}_{\mathcal{S}}(X /\!\!/ A, \Bbbk)^{\geq 0}$ consists of the objects $\mathscr{F}$ such that $j_s^! \mathscr{F}$ belongs to ${}^p\widehat{D}_{\mathcal{S}}(X_s /\!\!/ A, \Bbbk)^{\geq 0}$ for any s.

The heart of the perverse t-structure will be denoted $\widehat{P}_{\mathcal{S}}(X /\!\!/ A, \Bbbk)$, and an object of $\widehat{D}_{\mathcal{S}}(X /\!\!/ A, \Bbbk)$ will be called perverse if it belongs to this heart.

Remark 5.1 By construction, there exists an obvious fully faithful triangulated functor $D^{\mathrm{b}}_{\mathcal{S}}(X /\!\!/ A, \Bbbk) \to \widehat{D}_{\mathcal{S}}(X /\!\!/ A, \Bbbk)$. The essential image of this functor consists of the objects $\widehat{\mathscr{F}}$ such that the monodromy morphism $R_A^\wedge \to \mathrm{End}(\widehat{\mathscr{F}})$ factors through some quotient $R_A^\wedge/(\mathfrak{m}_A^\wedge)^n = R_A/\mathfrak{m}_A^n$. In fact, it is clear that the objects in the essential image of our functor satisfy this property. For the converse statement, using the fact that this essential image is a triangulated subcategory and the usual recollement triangles, we reduce the proof to the case X has only one stratum. Then the equivalence (5.2) allows to translate the question in terms of complexes of $R_A^\wedge$-modules. Using once again the triangulated structure (and the result quoted in the proof of Lemma 4.2), one can then assume that the complex is concentrated in one degree; in this case, the claim is obvious.

The following (well-known) claim will be needed for certain proofs below.

Lemma 5.2 *Let M be in $D^{\mathrm{b}}\mathsf{Mod}^{\mathrm{fg}}(R_A^\wedge)$, and assume that $\Bbbk \otimes^L_{R_A^\wedge} M$ is concentrated in nonnegative degrees. Then M is isomorphic to a complex of free $R_A^\wedge$-modules with nonzero terms in nonnegative degrees only.*

Proof Since $R_A^\wedge$ is local and of finite global dimensional, M is isomorphic to a bounded complex $N^\bullet$ of free R_A-modules. Let n be the smallest integer with $N^n \neq 0$. If $n < 0$, then our assumption implies that the morphism $\Bbbk \otimes_{R_A^\wedge} N^n \to \Bbbk \otimes_{R_A^\wedge} N^{n+1}$ is injective. Then by the Nakayama lemma, the map $N^n \to N^{n+1}$ is a split embedding, and choosing a (free) complement to its image in N^{n+1}, we see that M isomorphic to a complex of free $R_A^\wedge$-modules concentrated in degrees $\geq n+1$. Repeating this procedure if necessary, we obtain the desired claim. □

Lemma 5.3 *Let $\mathscr{F}$ in $\widehat{D}_{\mathcal{S}}(X /\!\!/ A, \Bbbk)$.*

(1) *If $\pi_\dagger \mathscr{F}$ is perverse, then $\mathscr{F}$ is perverse.*
(2) *If $\pi_\dagger \mathscr{F} = 0$, then $\mathscr{F} = 0$.*
(3) *If $\mathscr{F}$ is perverse and ${}^p\mathscr{H}^0(\pi_\dagger \mathscr{F}) = 0$, then $\mathscr{F} = 0$.*

Proof (1) The shifted pullback functor associated with the projection $Y_s \to \mathrm{pt}$ induces a (perverse) t-exact equivalence between $D^{\mathrm{b}}\mathsf{Vect}^{\mathrm{fd}}_{\Bbbk}$ and $D^{\mathrm{b}}_{\mathcal{S}}(Y_s, \Bbbk)$. Under this equivalence and (5.2), the functor $(\pi_s)_\dagger$ corresponds to the functor $\Bbbk \otimes^L_{R_A^\wedge} (-)$ (see the comments after Corollary 4.6). In view of the isomorphisms (5.1), this reduces the lemma to the claim that if an object M of $D^{\mathrm{b}}\mathsf{Mod}^{\mathrm{fg}}(R_A^\wedge)$ satisfies $\mathsf{H}^k(\Bbbk \otimes^L_{R_A^\wedge} M) = 0$ for all $k > 0$, resp. for all $k < 0$, then we have $\mathsf{H}^k(M) = 0$ for all $k > 0$, resp. for all $k < 0$. This claim is a standard consequence of the Nakayama lemma, resp. follows from Lemma 5.2.

The proof of parts (2), (3) are similar to that of (1); details are left to the reader. □

5.3 Standard and Costandard Perverse Sheaves

For any $s \in \mathscr{S}$, we denote by $i_s : Y_s \to Y$ the embedding, and consider the objects

$$\Delta_s := (i_s)_! \underline{\Bbbk}_{Y_s}[\dim Y_s], \quad \nabla_s := (i_s)_* \underline{\Bbbk}_{Y_s}[\dim Y_s].$$

We will also set

$$\widehat{\mathscr{L}}_{A,s} := (p_s)^* \widehat{\mathscr{L}}_A,$$

and consider the objects

$$\widehat{\Delta}_s := (j_s)_! \widehat{\mathscr{L}}_{A,s}[\dim X_s], \quad \widehat{\nabla}_s := (j_s)_* \widehat{\mathscr{L}}_{A,s}[\dim X_s]$$

in $\widehat{D}_{\mathcal{S}}(X /\!\!/ A, \Bbbk)$. In view of (5.1) and (3.3), we have canonical isomorphisms

$$\pi_{\dagger} \widehat{\Delta}_s \cong \Delta_s, \quad \pi_{\dagger} \widehat{\nabla}_s \cong \nabla_s. \tag{5.3}$$

We also have isomorphisms of $R_A^\wedge$-modules

$$\mathrm{Hom}_{\widehat{D}_{\mathcal{S}}(X /\!\!/ A, \Bbbk)} \left(\widehat{\Delta}_s, \widehat{\nabla}_t[k] \right) \cong \begin{cases} R_A^\wedge & \text{if } s = t \text{ and } k = 0; \\ 0 & \text{otherwise.} \end{cases} \tag{5.4}$$

Our map i_s is an affine morphism so that the objects Δ_s and ∇_s are perverse sheaves on Y. By Lemma 5.3, (1) and (5.3), this implies that the objects $\widehat{\Delta}_s$ and $\widehat{\nabla}_s$ are perverse too.

Lemma 5.4

(1) *The triangulated category $\widehat{D}_{\mathcal{S}}(X /\!\!/ A, \Bbbk)$ is generated by the objects $\widehat{\Delta}_s$ for $s \in \mathcal{S}$, as well as by the objects $\widehat{\nabla}_s$ for $s \in \mathcal{S}$.*

(2) *For any $s \in \mathcal{S}$, the monodromy morphism $\varphi_{\widehat{\Delta}_s}^\wedge$ induces an isomorphism*

$$R_A^\wedge \xrightarrow{\sim} \mathrm{Hom}_{\widehat{D}_{\mathcal{S}}(X /\!\!/ A, \Bbbk)}(\widehat{\Delta}_s, \widehat{\Delta}_s).$$

Moreover, any nonzero endomorphism of $\widehat{\Delta}_s$ is injective.

Proof Property (1) follows from the equivalences (5.2) and the gluing formalism. And in (2), the isomorphism follows from the equivalence (5.2) and the fact that $(j_s)_!$ is fully faithful.

Now, let $x \in R_A^\wedge \smallsetminus \{0\}$, and consider the induced endomorphism $\varphi_{\widehat{\Delta}_s}^\wedge(x)$. Let $\mathscr{C}$ be the cone of this morphism; then we need to show that $\mathscr{C}$ is concentrated in nonnegative perverse degrees, or in other words that for any $t \in \mathcal{S}$, the complex $j_t^! \mathscr{C}$ belongs to ${}^p\widehat{D}_{\mathcal{S}}(X_t /\!\!/ A, \Bbbk)^{\geq 0}$. Fix $t \in \mathcal{S}$, and denote by M the inverse image of the complex $j_t^! \widehat{\Delta}_s$ under the equivalence (5.2) (for the stratum labelled by t); then the inverse image of $j_t^! \mathscr{C}$ is the cone of the endomorphism of M induced by the action of x.

Using (5.3), we see that $(\pi_t)_\dagger(j_t^! \widehat{\Delta}_s) \cong i_t^! \Delta_s$. Since Δ_s is perverse, this complex is concentrated in nonnegative perverse degrees, which implies that the complex of vector spaces $\Bbbk \otimes^L_{R_A^\wedge} M$ is concentrated in nonnegative degrees. Hence, by Lemma 5.2, M is isomorphic to a complex N of free $R_A^\wedge$-modules with $N^i = 0$ for all $i < 0$. It is clear that the cone of the endomorphism of N induced by the action of x has cohomology only in nonnegative degrees; therefore, the same is true for M, and finally, $j_t^! \mathscr{C}$ indeed belongs to ${}^p\widehat{D}_{\mathcal{S}}(X_t /\!\!/ A, \Bbbk)^{\geq 0}$. □

Corollary 5.5

(1) *For any* $\mathscr{F}, \mathscr{G}$ *in* $\widehat{D}_{\mathcal{S}}(X /\!\!/ A, \Bbbk)$*, the* $R_A^\wedge$*-module*

$$\mathrm{Hom}_{\widehat{D}_{\mathcal{S}}(X /\!\!/ A, \Bbbk)}(\mathscr{F}, \mathscr{G})$$

is finitely generated.

(2) *The category* $\widehat{D}_{\mathcal{S}}(X /\!\!/ A, \Bbbk)$ *is Krull–Schmidt.*

Proof (1) Lemma 5.4 (1) reduces the claim to the special case $\mathscr{F} = \widehat{\Delta}_s, \mathscr{G} = \widehat{\nabla}_t$ for some $s, t \in \mathcal{S}$, which is clear from (5.4).

(2) Since the triangulated category $\widehat{D}_{\mathcal{S}}(X /\!\!/ A, \Bbbk)$ admits a bounded t-structure, it is Karoubian by [LC]. By (1) and [La, Example 23.3], the endomorphism ring of any of its objects is semi-local. By [CYZ, Theorem A.1], this implies that $\widehat{D}_{\mathcal{S}}(X /\!\!/ A, \Bbbk)$ is Krull–Schmidt. □

The standard objects also allow one to describe the perverse t-structure on the category $\widehat{D}_{\mathcal{S}}(X /\!\!/ A, \Bbbk)$, as follows.

Lemma 5.6 *The subcategory* ${}^p\widehat{D}_{\mathcal{S}}(X /\!\!/ A, \Bbbk)^{\leq 0}$ *is generated under extensions by the objects of the form* $\widehat{\Delta}_s[n]$ *for* $s \in \mathcal{S}$ *and* $n \geq 0$.

Proof This claim follows from the yoga of recollement, starting from the observation that the subcategory $D^{\mathrm{b}}\mathsf{Mod}^{\mathrm{fg}}(R_A^\wedge)^{\leq 0}$ is generated under extensions by the objects of the form $R_A^\wedge[n]$ with $n \geq 0$. (Here, we use the fact that $R_A^\wedge$ is local so that any finitely generated projective module is free.) □

Remark 5.7 It is *not* true that the subcategory ${}^p\widehat{D}_{\mathcal{S}}(X /\!\!/ A, \Bbbk)^{\geq 0}$ is generated under extensions by the objects of the form $\widehat{\nabla}_s[n]$ for $s \in \mathcal{S}$ and $n \leq 0$. (This is already false if $Y = \mathrm{pt}$ and $r > 0$.)

Corollary 5.8 *The functor $\pi_\dagger$ is right t-exact with respect to the perverse t-structures.*

Proof This follows from Lemma 5.6 and (5.3). □

5.4 Tilting Perverse Sheaves

It is a standard fact (see, e.g., [BGS]) that under our assumptions, the category $\mathsf{Perv}_{\mathcal{S}}(Y, \Bbbk)$ of $\mathcal{S}$-constructible perverse sheaves on Y is a highest weight category, with weight poset $\mathcal{S}$ (for the order induced by inclusions of closures of strata), standard objects $(\Delta_s : s \in \mathcal{S})$, and costandard objects $(\nabla_s : s \in \mathcal{S})$. Hence, we can consider the tilting objects in this category, that is, those which admit both a filtration with subquotients of the form Δ_s $(s \in \mathcal{S})$ and a filtration with subquotients of the form ∇_s $(s \in \mathcal{S})$. If $\mathscr{F}$ is a tilting object, the number of occurrences of Δ_s, resp. ∇_s, in a filtration of the first kind, resp. second kind, does not depend on the choice of filtration and equals the dimension of $\mathrm{Hom}(\mathscr{F}, \nabla_s)$, resp. $\mathrm{Hom}(\Delta_s, \mathscr{F})$. This number will be denoted $(\mathscr{F} : \Delta_s)$, resp. $(\mathscr{F} : \nabla_s)$. The indecomposable tilting objects are parametrized (up to isomorphism) by $\mathcal{S}$; the object corresponding to s will be denoted $\mathscr{T}_s$.

Similarly, an object $\mathscr{F}$ of $\widehat{P}_{\mathcal{S}}(X /\!\!/ A, \Bbbk)$ will be called tilting if it admits both a filtration with subquotients of the form $\widehat{\Delta}_s$ $(s \in \mathcal{S})$ and a filtration with subquotients of the form $\widehat{\nabla}_s$ $(s \in \mathcal{S})$. From (5.4), we see that the number of occurrences of $\widehat{\Delta}_s$, resp. $\widehat{\nabla}_s$, in a filtration of the first kind, resp. second kind, does not depend on the choice of filtration and equals the rank of $\mathrm{Hom}(\mathscr{F}, \widehat{\nabla}_s)$, resp. $\mathrm{Hom}(\widehat{\Delta}_s, \mathscr{F})$, as an $R_A^\wedge$-module. (These modules are automatically free of finite rank.) This number will be denoted $(\mathscr{F} : \widehat{\Delta}_s)$, resp. $(\mathscr{F} : \widehat{\nabla}_s)$.

It is clear from definitions and (5.3) that if $\mathscr{F}$ is a tilting object in $\widehat{P}_{\mathcal{S}}(X /\!\!/ A, \Bbbk)$, then $\pi_\dagger(\mathscr{F})$ is a tilting perverse sheaf and that moreover,

$$(\pi_\dagger(\mathscr{F}) : \Delta_s) = (\mathscr{F} : \widehat{\Delta}_s), \quad (\pi_\dagger(\mathscr{F}) : \nabla_s) = (\mathscr{F} : \widehat{\nabla}_s). \tag{5.5}$$

Lemma 5.9

(1) *If $\mathscr{F}$ belongs to $\widehat{D}_{\mathcal{S}}(X /\!\!/ A, \Bbbk)$, then $\mathscr{F}$ is a tilting perverse sheaf iff $\pi_\dagger(\mathscr{F})$ is a tilting perverse sheaf.*

(2) *If $\mathscr{F}, \mathscr{G}$ are tilting perverse sheaves in $\widehat{D}_{\mathcal{S}}(X /\!\!/ A, \Bbbk)$, then we have*

$$\mathrm{Hom}_{\widehat{D}_{\mathcal{S}}(X /\!\!/ A, \Bbbk)}(\mathscr{F}, \mathscr{G}[k]) = 0 \quad \text{if } k \neq 0,$$

the $R_A^\wedge$-module $\mathrm{Hom}_{\widehat{D}_{\mathcal{S}}(X /\!\!/ A, \Bbbk)}(\mathscr{F}, \mathscr{G})$ is free of finite rank, and the functor $\pi_\dagger$ induces an isomorphism

$$\Bbbk \otimes_{R_A^\wedge} \mathrm{Hom}_{\widehat{D}_{\mathcal{S}}(X /\!\!/ A, \Bbbk)}(\mathscr{F}, \mathscr{G}) \xrightarrow{\sim} \mathrm{Hom}_{D^{\mathrm{b}}_{\mathcal{S}}(Y, \Bbbk)}(\pi_\dagger \mathscr{F}, \pi_\dagger \mathscr{G}).$$

Proof (1) Using recollement triangles, it is easy to show that $\mathscr{F}$ is a tilting perverse sheaf iff for any $s \in \mathcal{S}$, the objects $j_s^* \mathscr{F}$ and $j_s^! \mathscr{F}$ are direct sums of copies of $\widehat{\mathscr{L}}_{A,s}[\dim X_s]$ (see [BBM] for this point of view in the case of usual tilting perverse sheaves). In turn, this condition is equivalent to the requirement that the inverse images of $j_s^* \mathscr{F}$ and $j_s^! \mathscr{F}$ under the equivalence (5.2) are isomorphic to a free $R_A^\wedge$-module. It is well known that the latter condition is equivalent to the property that the image under $\Bbbk \otimes^L_{R_A^\wedge} (-)$ of these objects is concentrated in degree 0. We deduce that $\mathscr{F}$ is a tilting perverse sheaf iff for any $s \in \mathcal{S}$, the complexes $(\pi_s)_\dagger j_s^* \mathscr{F}$ and $(\pi_s)_\dagger j_s^! \mathscr{F}$ are concentrated in perverse degree 0 (see the proof of Lemma 5.3). Since

$$(\pi_s)_\dagger j_s^* \mathscr{F} \cong i_s^* \pi_\dagger \mathscr{F} \quad \text{and} \quad (\pi_s)_\dagger j_s^! \mathscr{F} \cong i_s^! \pi_\dagger \mathscr{F}$$

by (5.1), we finally obtain that $\mathscr{F}$ is a tilting perverse sheaf iff the object $\mathscr{G} := \pi_\dagger \mathscr{F}$ is such that for any $s \in \mathcal{S}$, the complexes $i_s^* \mathscr{G}$ and $i_s^! \mathscr{G}$ are concentrated in perverse degree 0. This condition is equivalent to the fact that $\mathscr{G}$ is a tilting perverse sheaf; see [BBM], which concludes the proof.

(2) By Lemma 2.3, the morphism

$$\mathrm{Hom}_{\widehat{D}_{\mathcal{S}}(X/\!\!/A,\Bbbk)}(\mathscr{F}, \mathscr{G}) \to \mathrm{Hom}_{D^{\mathrm{b}}_{\mathcal{S}}(Y,\Bbbk)}(\pi_\dagger \mathscr{F}, \pi_\dagger \mathscr{G})$$

induced by $\pi_\dagger$ factors through the quotient $\Bbbk \otimes_{R_A^\wedge} \mathrm{Hom}_{\widehat{D}_{\mathcal{S}}(X/\!\!/A,\Bbbk)}(\mathscr{F}, \mathscr{G})$. Then the desired properties follow from (5.4) and the 5-lemma. □

Remark 5.10 The arguments in the proof of Lemma 5.9(1) show more generally that if $\mathscr{F}$ belongs to $\widehat{D}_{\mathcal{S}}(X/\!\!/A,\Bbbk)$ and if $\pi_\dagger(\mathscr{F})$ is a perverse sheaf admitting a standard filtration, then $\mathscr{F}$ is perverse and admits a filtration with subquotients of the form $\widehat{\Delta}_s$ for $s \in \mathcal{S}$, with $\widehat{\Delta}_s$ occurring as many times as Δ_s occurs in $\pi_\dagger(\mathscr{F})$. Of course, a similar claim holds for costandard filtrations.

We will denote by $\widehat{T}_{\mathcal{S}}(X/\!\!/A,\Bbbk)$ the full subcategory of $\widehat{D}_{\mathcal{S}}(X/\!\!/A,\Bbbk)$ whose objects are the tilting perverse sheaves. Lemma 5.9(2) has the following consequence.

Proposition 5.11 *There exists an equivalence of triangulated categories*

$$K^{\mathrm{b}}\widehat{T}_{\mathcal{S}}(X/\!\!/A,\Bbbk) \xrightarrow{\sim} \widehat{D}_{\mathcal{S}}(X/\!\!/A,\Bbbk).$$

Proof As explained in Remark 3.2, the category $\widehat{D}_{\mathcal{S}}(X/\!\!/A,\Bbbk)$ admits a filtered version. Hence, by [AMRW, Proposition 2.2] (see also [Be, §A.6]), there exists a triangulated functor $K^{\mathrm{b}}\widehat{T}_{\mathcal{S}}(X/\!\!/A,\Bbbk) \to \widehat{D}_{\mathcal{S}}(X/\!\!/A,\Bbbk)$ whose restriction to $\widehat{T}_{\mathcal{S}}(X/\!\!/A,\Bbbk)$ is the natural embedding. The fact that this functor is an equivalence follows from Beĭlinson's lemma. □

5.5 Classification of Tilting Perverse Sheaves

It follows from Corollary 5.5(2) that the category $\widehat{T}_{\mathcal{S}}(X /\!\!/ A, \Bbbk)$ is Krull–Schmidt. To proceed further, we need to classify its indecomposable objects.

The following classification result is proved in [BY, Lemma A.7.3]. Here, we provide a different proof, based on some ideas developed in [RSW] and [AR2, Appendix B]. (These ideas are themselves closely inspired by the methods of [BGS].)

Proposition 5.12 *For any $s \in \mathcal{S}$, there exists a unique (up to isomorphism) object $\widehat{\mathscr{T}}_s$ in $\widehat{D}_{\mathcal{S}}(X /\!\!/ A, \Bbbk)$ such that $\pi_{\dagger}(\widehat{\mathscr{T}}_s) \cong \mathscr{T}_s$. Moreover, $\widehat{\mathscr{T}}_s$ is an indecomposable tilting perverse sheaf, and the assignment $s \mapsto \widehat{\mathscr{T}}_s$ induces a bijection between $\mathcal{S}$ and the set of isomorphism classes of indecomposable tilting objects in $\widehat{D}_{\mathcal{S}}(X /\!\!/ A, \Bbbk)$.*

We begin with the following lemma, where we fix $s \in \mathcal{S}$.

Lemma 5.13 *For any open subset $U \subset \overline{Y_s}$ which is a union of strata, there exists a tilting perverse sheaf in $\widehat{D}_{\mathcal{S}}(\pi^{-1}(U) /\!\!/ A, \Bbbk)$ whose restriction to X_s is $\mathscr{L}_{A,s}[\dim X_s]$.*

Proof We proceed by induction on the number of strata in U, the initial case being when $U = Y_s$ (which is of course obvious).

Consider now a general U as in the statement, and $t \in \mathcal{S}$ such that $Y_t \subset U$ and Y_t is closed in U. Then we set $V := U \smallsetminus Y_t$, and assume (by induction) that we have a suitable object $\widehat{\mathscr{T}}_V$ in $\widehat{D}_{\mathcal{S}}(\pi^{-1}(V) /\!\!/ A, \Bbbk)$. We then denote by $j : V \to U$ the embedding, and consider the object $j_! \widehat{\mathscr{T}}_V$. This object admits a filtration (in the sense of triangulated categories) whose subquotients are standard objects in $\widehat{D}_{\mathcal{S}}(\pi^{-1}(U) /\!\!/ A, \Bbbk)$. In particular, it is perverse. We now consider the $R_A^{\wedge}$-module

$$E := \mathrm{Ext}^1_{\widehat{P}_{\mathcal{S}}(\pi^{-1}(U) /\!\!/ A, \Bbbk)}(\widehat{\Delta}^U_t, j_! \widehat{\mathscr{T}}_V) = \mathrm{Hom}_{\widehat{D}_{\mathcal{S}}(\pi^{-1}(U) /\!\!/ A, \Bbbk)}(\widehat{\Delta}^U_t, j_! \widehat{\mathscr{T}}_V[1])$$

(where $\widehat{\Delta}^U_t$ is the standard object in $\widehat{D}_{\mathcal{S}}(\pi^{-1}(U) /\!\!/ A, \Bbbk)$ associated with t). By Corollary 5.5(1), E is finitely generated as an $R_A^{\wedge}$-module; therefore, we can choose a nonnegative integer n and a surjection $(R_A^{\wedge})^{\oplus n} \twoheadrightarrow E$. This morphism defines an element in

$$\mathrm{Hom}_{R_A^{\wedge}}\left((R_A^{\wedge})^{\oplus n}, E\right) \cong E^{\oplus n} \cong \mathrm{Ext}^1_{\widehat{P}_{\mathcal{S}}(\pi^{-1}(U) /\!\!/ A, \Bbbk)}\left((\widehat{\Delta}^U_t)^{\oplus n}, j_! \widehat{\mathscr{T}}_V\right),$$

and therefore, an extension

$$j_! \widehat{\mathscr{T}}_V \hookrightarrow \widehat{\mathscr{T}}_U \twoheadrightarrow (\widehat{\Delta}^U_t)^{\oplus n} \tag{5.6}$$

in $\widehat{P}_{\mathcal{S}}(\pi^{-1}(U) /\!\!/ A, \Bbbk)$, for some object $\widehat{\mathscr{T}}_U$. It is clear that this object admits a filtration with subquotients of the form $\widehat{\Delta}_u$ ($u \in \mathcal{S}$) and has the appropriate restriction to X_s. Hence, to conclude the proof of the claim, in view of Remark 5.10, it suffices to prove that if $\mathscr{T}_U := (\pi_U)_\dagger \widehat{\mathscr{T}}_U$ (where π_U is the restriction of π to $\pi^{-1}(U)$), then $\mathscr{T}_U$ admits a costandard filtration in the highest weight category $\mathsf{Perv}_{\mathcal{S}}(U, \Bbbk)$ or in other words that

$$\mathrm{Ext}^1_{\mathsf{Perv}_{\mathcal{S}}(U,\Bbbk)}(\Delta_u^U, \mathscr{T}_U) = 0$$

for any $u \in \mathcal{S}$ such that $Y_u \subset U$. (Here, Δ_u^U is the standard perverse sheaf in $\mathsf{Perv}_{\mathcal{S}}(U, \Bbbk)$ associated with u.)

The case $u \neq t$ is easy and left to the reader. We then remark that applying the functor $(\pi_U)_\dagger$ to (5.6), we obtain an exact sequence

$$\overline{\jmath}_! \mathscr{T}_V \hookrightarrow \mathscr{T}_U \twoheadrightarrow (\Delta_t^U)^{\oplus n} \tag{5.7}$$

in $\mathsf{Perv}_{\mathcal{S}}(U, \Bbbk)$, where $\overline{\jmath} : V \to U$ is the embedding and $\mathscr{T}_V := (\pi_V)_\dagger \widehat{\mathscr{T}}_V$ for $\pi_V : \pi^{-1}(V) \to V$ the restriction of π.

We now claim that there exists a canonical isomorphism

$$\Bbbk \otimes_{R_A^\wedge} E \cong \mathrm{Ext}^1_{\mathsf{Perv}_{\mathcal{S}}(U,\Bbbk)}(\Delta_u^U, \overline{\jmath}_! \mathscr{T}_V). \tag{5.8}$$

In fact, using the natural exact sequences

$$\ker \hookrightarrow \overline{\jmath}_! \mathscr{T}_V \twoheadrightarrow \overline{\jmath}_{!*} \mathscr{T}_V, \quad \overline{\jmath}_{!*} \mathscr{T}_V \hookrightarrow \overline{\jmath}_* \mathscr{T}_V \twoheadrightarrow \mathrm{coker}$$

and the fact that $\mathscr{T}_V$ admits a standard filtration, it is easily checked that

$$\mathrm{Ext}^i_{\mathsf{Perv}_{\mathcal{S}}(U,\Bbbk)}(\Delta_u^U, \overline{\jmath}_! \mathscr{T}_V) = 0 \quad \text{for } i \geq 2$$

(see [AR2, Proof of Proposition B.2] for details). If M is the inverse image of $j_t^! j_! \widehat{\mathscr{T}}_V$ under the equivalence (5.2), this means that the complex $\Bbbk \otimes^L_{R_A^\wedge} M$ is concentrated in degrees ≤ 1. This implies that M itself is concentrated in degrees ≤ 1 and that we have a canonical isomorphism $\Bbbk \otimes_{R_A^\wedge} \mathsf{H}^1(M) \cong \mathsf{H}^1(\Bbbk \otimes^L_{R_A^\wedge} M)$. This isomorphism is precisely (5.8).

Once (5.8) is established, we see that our surjection $(R_A^\wedge)^{\oplus n} \twoheadrightarrow E$ induces a surjection $\Bbbk^{\oplus n} \twoheadrightarrow \mathrm{Ext}^1_{\mathsf{Perv}_{\mathcal{S}}(U,\Bbbk)}(\Delta_u^U, \overline{\jmath}_! \mathscr{T}_V)$. Using this fact and considering the long exact sequence obtained by applying the functor $\mathrm{Hom}(\Delta_t^U, -)$ to (5.7), we conclude that $\mathrm{Ext}^1_{\mathsf{Perv}_{\mathcal{S}}(U,\Bbbk)}(\Delta_t^U, \mathscr{T}_U) = 0$, which finishes the proof. □

Proof of Proposition 5.12 By Lemma 5.12, there exists a tilting object $\widehat{\mathscr{T}}_s$ in the category $\widehat{D}_{\mathcal{S}}(X /\!\!/ A, \Bbbk)$ which is supported on $\overline{X}_s$ and whose restriction to X_s is $\widehat{\mathscr{L}}_{A,s}$. Of course, we can (and will) further require that this object is indecomposable. By Lemma 5.9, the object $\pi_\dagger \widehat{\mathscr{T}}_s$ is then a tilting perverse sheaf, and it's endomorphism ring is a quotient of $\mathrm{End}(\widehat{\mathscr{T}}_s)$ and hence is local; in other words, $\pi_\dagger \widehat{\mathscr{T}}_s$ is indecomposable. Since it is supported on $\overline{Y_s}$, and since its restriction to Y_s is $\underline{\Bbbk}_{Y_s}[\dim(Y_s)]$, it follows that $\pi_\dagger(\widehat{\mathscr{T}}_s) \cong \mathscr{T}_s$.

These arguments show more generally that if $\widehat{\mathscr{T}}$ is any indecomposable tilting object in $\widehat{D}_{\mathcal{S}}(X /\!\!/ A, \Bbbk)$, the object $\pi_\dagger(\widehat{\mathscr{T}})$ is isomorphic to $\mathscr{T}_t$ for some $t \in \mathcal{S}$. To conclude the proof, it remains to prove that in this case, we must have $\widehat{\mathscr{T}} \cong \widehat{\mathscr{T}}_t$. By Lemma 5.9(2), the functor $\pi_\dagger$ induces an isomorphism

$$\Bbbk \otimes_{R_A^\wedge} \mathrm{Hom}_{\widehat{D}_{\mathcal{S}}(X /\!\!/ A, \Bbbk)}(\widehat{\mathscr{T}}, \widehat{\mathscr{T}}_t) \xrightarrow{\sim} \mathrm{Hom}_{D^{\mathrm{b}}_{\mathcal{S}}(Y, \Bbbk)}(\pi_\dagger \widehat{\mathscr{T}}, \pi_\dagger \widehat{\mathscr{T}}_t).$$

Hence, there exists a morphism $f : \widehat{\mathscr{T}} \to \widehat{\mathscr{T}}_t$ such that $\pi_\dagger(f)$ is an isomorphism. Then the cone $\mathscr{C}$ of f satisfies $\pi_\dagger(\mathscr{C}) = 0$. By Lemma 5.3(2), this implies that $\mathscr{C} = 0$; hence, that f is an isomorphism. □

Part 2: The Case of Flag Varieties

6 Study of Tilting Perverse Objects

6.1 Notation

From now on, we fix a complex connected reductive algebraic group G and choose a maximal torus and a Borel subgroup $T \subset B \subset G$. We will denote by U the unipotent radical of B and, by W, the Weyl group of (G, T). The choice of B determines a subset $S \subset W$ of simple reflections and a choice of positive roots (such that B is the *negative* Borel subgroup).

We will study further the previous constructions in the special case

$$X = G/U, \quad Y = G/B$$

(with the action of $A = T$ given by $t \cdot gU = gtU$), $\pi : X \to Y$ is the natural projection, and the stratification is

$$Y = \bigsqcup_{w \in W} Y_w \quad \text{with} \quad Y_w := BwB/B.$$

The corresponding categories in this case will be denoted

$$D^{\mathrm{b}}_U(Y, \Bbbk), \quad \widehat{D}_U(X /\!\!/ T, \Bbbk).$$

(Here, $D^{\mathrm{b}}_U(Y, \Bbbk)$ is indeed equivalent to the U-equivariant constructible derived category in the sense of Bernstein–Lunts, which explains the notation.)

Recall that to define the objects $\widehat{\Delta}_w$ and $\widehat{\nabla}_w$, we need to choose a T-equivariant morphism $X_w \to T$, where $X_w = \pi^{-1}(Y_w)$. For this, we choose a lift $\dot{w}$ of w in $N_G(T)$ and consider the subgroup $U_{w^{-1}} \subset U$ defined as in [Sp, Lemma 8.3.5]. Then the map $u \mapsto u\dot{w}B$ induces an isomorphism $U_w \xrightarrow{\sim} Y_w$, and the map $(u, t) \mapsto$

$u\dot{w}tU$ induces an isomorphism $U_w \times T \xrightarrow{\sim} X_w$; see [Sp, Lemma 8.3.6]. We will choose p_w as the composition of the inverse isomorphism with the projection to the T factor.

The category $D^{\mathrm{b}}_U(Y, \Bbbk)$ admits a natural perverse t-structure; its heart will be denoted

$$\mathscr{O} := \mathsf{Perv}_U(Y, \Bbbk).$$

Similarly, the constructions of Sect. 5.2 provide a perverse t-structure on the category $\widehat{D}_U(X /\!\!/ T, \Bbbk)$, whose heart will be denoted

$$\widehat{\mathscr{O}} := \mathsf{Perv}_U(X /\!\!/ T, \Bbbk).$$

6.2 Right and Left Monodromy

By the general formalism of the completed monodromic category (see Sect. 3.1), for any $\mathscr{F}$ in $\widehat{D}_U(X /\!\!/ T, \Bbbk)$, we have an algebra morphism

$$\varphi^{\wedge}_{\mathscr{F}} : R^{\wedge}_T \to \mathrm{End}(\mathscr{F}).$$

Since this monodromy comes from the action of T by right multiplication, we will denote it in this case by $\varphi^{\wedge}_{\mathrm{r},\mathscr{F}}$.

Now, let $a : G \to G/U$ be the projection (a locally trivial fibration, with fibers isomorphic to affine spaces). Then the functor $a^* : D^{\mathrm{b}}_c(Y, \Bbbk) \to D^{\mathrm{b}}_c(G, \Bbbk)$ is fully faithful since $a_* \circ a^* \cong \mathrm{id}$. The triangulated category $D^{\mathrm{b}}_U(X, \Bbbk)$ is generated by the image of the forgetful functor $D^{\mathrm{b}}_B(X, \Bbbk) \to D^{\mathrm{b}}_c(X, \Bbbk)$; therefore, if $\mathscr{G}$ belongs to $D^{\mathrm{b}}_U(X /\!\!/ T, \Bbbk)$, then $a^*(\mathscr{G})$ belongs to the monodromic category $D^{\mathrm{b}}_c(G /\!\!/ T, \Bbbk)$ where T acts on G via $t \cdot g = tg$. Hence, we can consider the morphism $\varphi^{\wedge}_{a^*(\mathscr{G})}$. Since a^* is fully faithful, this morphism can be interpreted as a morphism

$$\varphi^{\wedge}_{\mathrm{l},\mathscr{G}} : R^{\wedge}_T \to \mathrm{End}(\mathscr{G})$$

(where "l" stands for left). Passing to projective limits, we deduce, for any $\mathscr{F}$ in $\widehat{D}_U(X /\!\!/ T, \Bbbk)$, an algebra morphism

$$\varphi^{\wedge}_{\mathrm{l},\mathscr{F}} : R^{\wedge}_T \to \mathrm{End}(\mathscr{F}).$$

Combining these two constructions, we obtain an algebra morphism

$$\varphi^{\wedge}_{\mathrm{lr},\mathscr{F}} : R^{\wedge}_T \otimes_{\Bbbk} R^{\wedge}_T \to \mathrm{End}(\mathscr{F})$$

sending $r \otimes r'$ to $\varphi^{\wedge}_{1,\mathscr{F}}(r) \circ \varphi^{\wedge}_{\mathrm{r},\mathscr{F}}(r') = \varphi^{\wedge}_{\mathrm{r},\mathscr{F}}(r') \circ \varphi^{\wedge}_{1,\mathscr{F}}(r)$.

Lemma 6.1 *For any $w \in W$, the morphism $\varphi^{\wedge}_{\mathrm{r},\widehat{\Delta}_w}$, resp. $\varphi^{\wedge}_{\mathrm{r},\widehat{\nabla}_w}$, is the composition of $\varphi^{\wedge}_{1,\widehat{\Delta}_w}$, resp. $\varphi^{\wedge}_{1,\widehat{\nabla}_w}$, with the automorphism of $R^{\wedge}_T$ induced by w.*

Proof We treat the case of $\widehat{\Delta}_w$; the case of $\widehat{\nabla}_w$ is similar. More precisely, we will prove a similar claim for the monodromy endomorphisms of each object $\Delta^n_w := (j_w)_! p^*_w(\mathscr{L}_{T,n})[\dim X_w]$.

By the base change theorem, we have

$$a^*(\Delta^n_w) \cong (\widetilde{j}_w)_!(p_w \circ a_w)^* \mathscr{L}_{T,n}[\dim X_w],$$

where $\widetilde{j}_w : a^{-1}(X_w) \hookrightarrow G$ is the embedding and $a_w : a^{-1}(X_w) \to X_w$ is the restriction of a. By Lemma 2.4, we deduce that for any $r \in R^{\wedge}_T$, we have

$$a^*\big(\varphi^{\wedge}_{\mathrm{r},\Delta^n_w}(r)\big) = \varphi^{\wedge}_{\mathrm{r},a^*(\Delta^n_w)}(r) = (\widetilde{j}_w)_!(p_w \circ a_w)^* \varphi^{\wedge}_{\mathscr{L}_{T,n}}(r)[\dim X_w], \tag{6.1}$$

where in the first two terms, we consider the monodromy operation with respect to the action of T on G/U and G by multiplication on the right.

Now, we consider the actions induced by multiplication on the left. It is not difficult to check that

$$(p_w \circ a_w)(t \cdot x) = w^{-1}(t)(p_w \circ a_w)(x)$$

for any $t \in T$ and $x \in a^{-1}(X_w)$. In other words, $p_w \circ a_w$ is T-equivariant when T acts on $a^{-1}(X_w)$ by multiplication on the left, and on T via the natural action twisted by w^{-1}. From this, using the same arguments as above and Lemma 2.5, we deduce that

$$a^*\big(\varphi^{\wedge}_{1,\Delta^n_w}(r)\big) = (\widetilde{j}_w)_!(p_w \circ a_w)^* \varphi^{\wedge}_{\mathscr{L}_{T,n}}(w^{-1}(r))[\dim X_w]. \tag{6.2}$$

Comparing (6.1) and (6.2), and using the fact that a^* is fully faithful, we deduce the desired claim. □

Similar considerations hold for objects in $D^{\mathrm{b}}_U(Y, \Bbbk)$. Below, we will only consider the case of perverse sheaves, so we restrict to this setting. Let $b = \pi \circ a : G \to Y$ be the natural projection, and let $\mathscr{F}$ in $\mathscr{O}$. Then the object $b^*(\mathscr{F})$ belongs to $D^{\mathrm{b}}_c(G /\!\!/ T, \Bbbk)$, where the T-action on G is induced by multiplication on the *left*. Hence, the monodromy construction from Sect. 2 provides a morphism $R^{\wedge}_T \to \mathrm{End}(b^*(\mathscr{F}))$. Now, the functor b^* is fully faithful on perverse sheaves since b is smooth with connected fibers (see [BBD, Proposition 4.2.5]); hence, this morphism can be considered as an algebra morphism

$$\varphi^{\wedge}_{1,\mathscr{F}} : R^{\wedge}_T \to \mathrm{End}(\mathscr{F}).$$

It is clear that if $\mathscr{F}$ belongs to $\widehat{D}_U(X /\!\!/ T, \Bbbk)$ and $\pi_\dagger(\mathscr{F})$ is perverse, the composition

$$R_T^\wedge \xrightarrow{\varphi^\wedge_{1,\mathscr{F}}} \mathrm{End}(\mathscr{F}) \xrightarrow{\pi_\dagger} \mathrm{End}(\pi_\dagger(\mathscr{F}))$$

coincides with $\varphi^\wedge_{1,\pi_\dagger(\mathscr{F})}$.

6.3 The Associated Graded Functor

Let us now fix a total order $\preceq$ on W that refines the Bruhat order. We then denote by $j_{\prec w}$ the embedding of the closed subvariety $\bigsqcup_{y \prec w} X_y$ in X. For any $\widehat{\mathscr{T}}$ in $\widehat{T}_U(X /\!\!/ T, \Bbbk)$, the adjunction morphism

$$\widehat{\mathscr{T}} \to (j_{\prec w})_*(j_{\prec w})^* \widehat{\mathscr{T}}$$

is surjective. If we denote its kernel by $\widehat{\mathscr{T}}_{\succeq w}$, then the family of subobjects of $\widehat{\mathscr{T}}$ given by $(\widehat{\mathscr{T}}_{\succeq w})_{w \in W}$ is an exhaustive filtration on $\widehat{\mathscr{T}}$ indexed by W, endowed with the order opposite to $\preceq$ (meaning that $\widehat{\mathscr{T}}_{\succeq w} \subset \widehat{\mathscr{T}}_{\succeq y}$ if $y \preceq w$). Moreover, if we set

$$\mathrm{gr}_w(\widehat{\mathscr{T}}) := \widehat{\mathscr{T}}_{\succeq w} / \widehat{\mathscr{T}}_{\succeq w'},$$

where w' is the successor of w for $\preceq$, then $\mathrm{gr}_w(\widehat{\mathscr{T}})$ is a direct sum of copies of $\widehat{\Delta}_w$. (Here by convention $\widehat{\mathscr{T}}_{\succeq w'} = 0$ if w has no successor, i.e., if w is the longest element in W.) Since by adjunction we have $\mathrm{Hom}_{\widehat{D}_U(X /\!\!/ T, \Bbbk)}(\widehat{\Delta}_y, \widehat{\Delta}_w) = 0$ if $y \succ w$, we see that if $f : \widehat{\mathscr{T}} \to \widehat{\mathscr{T}}'$ is a morphism in $\widehat{T}_U(X /\!\!/ T, \Bbbk)$, then $f(\widehat{\mathscr{T}}_{\succeq w}) \subset \widehat{\mathscr{T}}'_{\succeq w}$ for any $w \in W$. In other words, the assignment $\widehat{\mathscr{T}} \mapsto \widehat{\mathscr{T}}_{\succeq w}$ is functorial. This allows us to define the functor

$$\mathrm{gr} : \begin{cases} \widehat{T}_U(X /\!\!/ T, \Bbbk) \to & \widehat{P}_U(X /\!\!/ T, \Bbbk) \\ \quad \widehat{\mathscr{T}} & \mapsto \bigoplus_{w \in W} \mathrm{gr}_w(\widehat{\mathscr{T}}) \end{cases}.$$

This functor is clearly additive.

Lemma 6.2 *For any $y, w \in W$ with $y \neq w$, we have*

$$\mathrm{Hom}_{\widehat{D}_U(X /\!\!/ T, \Bbbk)}(\widehat{\Delta}_y, \widehat{\Delta}_w) = 0.$$

Proof Let $f : \widehat{\Delta}_y \to \widehat{\Delta}_w$ be a nonzero morphism. We denote by $\mathscr{F}$ the image of f, and write $f = f_1 \circ f_2$ with $f_2 : \widehat{\Delta}_y \to \mathscr{F}$ the natural surjection and $f_1 : \mathscr{F} \to \widehat{\Delta}_w$ the natural embedding. Then for any $r \in R_T^\wedge$, we have a commutative diagram

$$
\begin{array}{ccccc}
\widehat{\Delta}_y & \xrightarrow{\quad f_2 \quad} & \mathscr{F} & \xrightarrow{\quad f_1 \quad} & \widehat{\Delta}_w \\
{\scriptstyle \varphi^{\wedge}_{\mathrm{l},\widehat{\Delta}_y}(r)} \downarrow \; \downarrow {\scriptstyle \varphi^{\wedge}_{\mathrm{r},\widehat{\Delta}_y}(r)} & & {\scriptstyle \varphi^{\wedge}_{\mathrm{l},\mathscr{F}}(r)} \downarrow \; \downarrow {\scriptstyle \varphi^{\wedge}_{\mathrm{r},\mathscr{F}}(r)} & & {\scriptstyle \varphi^{\wedge}_{\mathrm{l},\widehat{\Delta}_w}(r)} \downarrow \; \downarrow {\scriptstyle \varphi^{\wedge}_{\mathrm{r},\widehat{\Delta}_w}(r)} \\
\widehat{\Delta}_w & \xrightarrow{\quad f_2 \quad} & \mathscr{F} & \xrightarrow{\quad f_1 \quad} & \widehat{\Delta}_w
\end{array}
$$

By Lemma 5.4, if $r \neq 0$, then $\varphi^{\wedge}_{\mathrm{r},\widehat{\Delta}_w}(r)$ is injective. Hence,

$$\varphi^{\wedge}_{\mathrm{r},\mathscr{F}}(r) \text{ is injective (in particular, nonzero) if } r \neq 0. \tag{6.3}$$

On the other hand, using Lemma 6.1, we see that

$$f_1 \circ \varphi^{\wedge}_{\mathrm{r},\mathscr{F}}(r) = \varphi^{\wedge}_{\mathrm{r},\widehat{\Delta}_w}(r) \circ f_1 = \varphi^{\wedge}_{\mathrm{l},\widehat{\Delta}_w}(w(r)) \circ f_1 = f_1 \circ \varphi^{\wedge}_{\mathrm{l},\mathscr{F}}(w(r)),$$

which implies that $\varphi^{\wedge}_{\mathrm{r},\mathscr{F}}(r) = \varphi^{\wedge}_{\mathrm{l},\mathscr{F}}(w(r))$ since f_1 is injective and that

$$\varphi^{\wedge}_{\mathrm{r},\mathscr{F}}(r) \circ f_2 = f_2 \circ \varphi^{\wedge}_{\mathrm{r},\widehat{\Delta}_y}(r) = f_2 \circ \varphi^{\wedge}_{\mathrm{l},\widehat{\Delta}_y}(y(r)) = \varphi^{\wedge}_{\mathrm{l},\mathscr{F}}(y(r)) \circ f_2,$$

which implies that $\varphi^{\wedge}_{\mathrm{r},\mathscr{F}}(r) = \varphi^{\wedge}_{\mathrm{l},\mathscr{F}}(y(r))$ since f_2 is surjective. Comparing these two equations, we deduce that $\varphi^{\wedge}_{\mathrm{r},\mathscr{F}}(r) = \varphi^{\wedge}_{\mathrm{r},\mathscr{F}}(y^{-1}w(r))$, or in other words that

$$\varphi^{\wedge}_{\mathrm{r},\mathscr{F}}(r - y^{-1}w(r)) = 0,$$

for any $r \in R^{\wedge}_T$. In view of (6.3), this implies that $r = y^{-1}w(r)$ for any $r \in R^{\wedge}_T$ and hence that $y = w$. □

As a consequence, we obtain the following claim.

Corollary 6.3 *The functor* gr *is faithful.*

Proof Let $\widehat{\mathscr{T}}, \widehat{\mathscr{T}'}$ be in $\widehat{T}_U(X /\!\!/ T, \Bbbk)$, and let $f : \widehat{\mathscr{T}} \to \widehat{\mathscr{T}'}$ be a nonzero morphism. Let $w \in W$ be an element which is maximal with respect to the property that $f(\widehat{\mathscr{T}}_{\geq w}) \neq 0$. Then f induces a nonzero morphism $\tilde{f}_w : \mathrm{gr}_w(\widehat{\mathscr{T}}) \to \widehat{\mathscr{T}'}$. We have $f(\widehat{\mathscr{T}}_{\geq w}) \subset \widehat{\mathscr{T}'}_{\geq w}$; hence, $\tilde{f}_w$ factors through a nonzero morphism $\mathrm{gr}_w(\widehat{\mathscr{T}}) \to \widehat{\mathscr{T}'}_{\geq w}$. Lemma 6.2 implies that the natural morphism

$$\mathrm{Hom}(\mathrm{gr}_w(\widehat{\mathscr{T}}), \widehat{\mathscr{T}'}_{\geq w}) \to \mathrm{Hom}(\mathrm{gr}_w(\widehat{\mathscr{T}}), \mathrm{gr}_w(\widehat{\mathscr{T}'}))$$

is injective; hence, $\mathrm{gr}_w(f) \neq 0$ so that a fortiori $\mathrm{gr}(f) \neq 0$. □

Note that by functoriality of monodromy, for any $r \in R^{\wedge}_T$, we have

$$\varphi^{\wedge}_{\mathrm{l},\mathrm{gr}(\widehat{\mathscr{T}})}(r) = \mathrm{gr}\big(\varphi^{\wedge}_{\mathrm{l},\widehat{\mathscr{T}}}(r)\big), \qquad \varphi^{\wedge}_{\mathrm{r},\mathrm{gr}(\widehat{\mathscr{T}})}(r) = \mathrm{gr}\big(\varphi^{\wedge}_{\mathrm{r},\widehat{\mathscr{T}}}(r)\big). \tag{6.4}$$

6.4 Monodromy and Coinvariants

Proposition 6.4 *For any $\widehat{\mathcal{T}}$ in $\widehat{T}_U(X /\!\!/ T, \Bbbk)$, the morphism $\varphi^{\wedge}_{\mathrm{lr},\widehat{\mathcal{T}}}$ factors through an algebra morphism*

$$R_T^{\wedge} \otimes_{(R_T^{\wedge})^W} R_T^{\wedge} \to \mathrm{End}(\widehat{\mathcal{T}}).$$

Proof We have to prove that $\varphi^{\wedge}_{\mathrm{l},\widehat{\mathcal{T}}}(r) = \varphi^{\wedge}_{\mathrm{r},\widehat{\mathcal{T}}}(r)$ for any $r \in (R_T^{\wedge})^W$. Since the functor gr is faithful (see Corollary 6.3), for this, it suffices to prove that $\mathrm{gr}(\varphi^{\wedge}_{\mathrm{l},\widehat{\mathcal{T}}}(r)) = \mathrm{gr}(\varphi^{\wedge}_{\mathrm{r},\widehat{\mathcal{T}}}(r))$. This equality follows from (6.4) and Lemma 6.1 since $\mathrm{gr}(\widehat{\mathcal{T}})$ is a direct sum of copies of objects $\widehat{\Delta}_w$. □

6.5 The Case of $\widehat{\mathcal{T}}_s$

In this subsection, we fix a simple reflection s and denote by α the associated simple root.

We consider the closure $\overline{Y_s} = Y_s \sqcup Y_e$. This subvariety of Y is isomorphic to $\mathbb{P}^1$, in such a way that Y_e identifies with $\{0\}$. The structure of the category $\mathsf{Perv}_U(\overline{Y_s}, \Bbbk)$ of $\Bbbk$-perverse sheaves on $\overline{Y_s}$ constructible with respect to the stratification $\overline{Y_s} = Y_s \sqcup Y_e$ is well known: This category admits five indecomposable objects (up to isomorphism):

- Two simple objects $\mathcal{IC}_e$ and $\mathcal{IC}_s$
- Two indecomposable objects of length 2, namely, Δ_s and ∇_s, which fit into nonsplit exact sequences

$$\mathcal{IC}_e \hookrightarrow \Delta_s \twoheadrightarrow \mathcal{IC}_s, \qquad \mathcal{IC}_s \hookrightarrow \nabla_s \twoheadrightarrow \mathcal{IC}_e;$$

- One indecomposable object of length 3, namely, the tilting object $\mathcal{T}_s$, which fits into nonsplit exact sequences

$$\Delta_s \hookrightarrow \mathcal{T}_s \twoheadrightarrow \mathcal{IC}_e, \qquad \mathcal{IC}_e \hookrightarrow \mathcal{T}_s \twoheadrightarrow \nabla_s.$$

We now fix a cocharacter $\lambda : \mathbb{C}^\times \to T$ and consider the full subcategory $\mathsf{Perv}_{\mathbb{C}^\times,U}(\overline{Y_s}, \Bbbk) \subset \mathsf{Perv}_U(\overline{Y_s}, \Bbbk)$ consisting of perverse sheaves which are $\mathbb{C}^\times$-equivariant for the action determined by $z \cdot xB = \lambda(z)xB$.

Lemma 6.5 *If the image of $\langle \lambda, \alpha \rangle$ in $\Bbbk$ is nonzero, then $\mathcal{T}_s$ does not belong to $\mathsf{Perv}_{\mathbb{C}^\times,U}(\overline{Y_s}, \Bbbk)$.*

Proof Let $B^+ \subset G$ be the Borel subgroup opposite to B with respect to T, let U^+ be its unipotent radical, and let $U_s^+ \subset U^+$ be the root subgroup associated with

s. If we set $Y_s^\circ := \overline{Y_s} \smallsetminus \{sB\}$, then the map $u \mapsto u \cdot B$ induces an isomorphism $U_s^+ \xrightarrow{\sim} Y_s^\circ$. In particular, this open subset is $\mathbb{C}^\times$-stable, with an action of $\mathbb{C}^\times$ via the character $\langle \lambda, \alpha \rangle$.

The object $\mathscr{T}_s$ is the unique nonsplit extension of $\mathcal{IC}_e$ by Δ_s in $\mathsf{Perv}_U(\overline{Y_s}, \Bbbk)$; hence, to conclude, it suffices to show that $\mathrm{Ext}^1_{\mathsf{Perv}_{\mathbb{C}^\times,U}(\overline{Y_s},\Bbbk)}(\mathcal{IC}_e, \Delta_s) = 0$ if the image of $\langle \lambda, \alpha \rangle$ in $\Bbbk$ is nonzero. Note that we have

$$\mathrm{Ext}^1_{\mathsf{Perv}_{\mathbb{C}^\times,U}(\overline{Y_s},\Bbbk)}(\mathcal{IC}_e, \Delta_s) = \mathrm{Hom}_{D^{\mathrm{b}}_{\mathbb{C}^\times,U}(\overline{Y_s},\Bbbk)}(\mathcal{IC}_e, \Delta_s[1])$$

where $D^{\mathrm{b}}_{\mathbb{C}^\times,U}(\overline{Y_s}, \Bbbk)$ is the $\mathbb{C}^\times$-equivariant constructible derived category in the sense of Bernstein–Lunts. Let us consider the long exact sequence

$$\mathrm{Hom}_{D^{\mathrm{b}}_{\mathbb{C}^\times,U}(\overline{Y_s},\Bbbk)}(\mathcal{IC}_e, \mathcal{IC}_e[1]) \to \mathrm{Hom}_{D^{\mathrm{b}}_{\mathbb{C}^\times,U}(\overline{Y_s},\Bbbk)}(\mathcal{IC}_e, \Delta_s[1])$$
$$\to \mathrm{Hom}_{D^{\mathrm{b}}_{\mathbb{C}^\times,U}(\overline{Y_s},\Bbbk)}(\mathcal{IC}_e, \mathcal{IC}_s[1]) \to \mathrm{Hom}_{D^{\mathrm{b}}_{\mathbb{C}^\times,U}(\overline{Y_s},\Bbbk)}(\mathcal{IC}_e, \mathcal{IC}_e[2])$$

obtained from the short exact sequence $\mathcal{IC}_e \hookrightarrow \Delta_s \twoheadrightarrow \mathcal{IC}_s$. Here, the first, resp. fourth, term identifies with the degree-1, resp. degree-2, $\mathbb{C}^\times$-equivariant cohomology of the point. In particular, this term vanishes, resp. is canonically isomorphic to $\Bbbk$. Now, we observe that restriction induces an isomorphism

$$\mathrm{Hom}_{D^{\mathrm{b}}_{\mathbb{C}^\times,U}(\overline{Y_s},\Bbbk)}(\mathcal{IC}_e, \mathcal{IC}_s[1]) \xrightarrow{\sim} \mathrm{Hom}_{D^{\mathrm{b}}_{\mathbb{C}^\times,U}(Y_s^\circ,\Bbbk)}(\underline{\Bbbk}_{Y_e}, \underline{\Bbbk}_{Y_s^\circ}[2]).$$

The right-hand side is 1-dimensional, with a basis consisting of the adjunction morphism associated with the embedding $Y_e \hookrightarrow Y_s^\circ$. Moreover, in view of the classical description of the $\mathbb{C}^\times$-equivariant cohomology of the point recalled, for example, in [Lu, §1.10], the map

$$\mathrm{Hom}_{D^{\mathrm{b}}_{\mathbb{C}^\times,U}(\overline{Y_s},\Bbbk)}(\mathcal{IC}_e, \mathcal{IC}_s[1]) \to \mathrm{Hom}_{D^{\mathrm{b}}_{\mathbb{C}^\times,U}(\overline{Y_s},\Bbbk)}(\mathcal{IC}_e, \mathcal{IC}_e[2])$$

considered above identifies with the map $\Bbbk \to \Bbbk$ given by multiplication by $\langle \lambda, \alpha \rangle$. Our assumption is precisely that this map is injective; we deduce that the vector space $\mathrm{Ext}^1_{\mathsf{Perv}_{\mathbb{C}^\times,U}(\overline{Y_s},\Bbbk)}(\mathcal{IC}_e, \Delta_s)$ vanishes, as claimed. □

Corollary 6.6 *Assume that there exists $\lambda \in X_*(T)$ such that the image of $\langle \lambda, \alpha \rangle$ in $\Bbbk$ is nonzero. Then in the special case $\widehat{\mathscr{T}} = \widehat{\mathscr{T}_s}$, the morphism*

$$R_T^\wedge \otimes_{(R_T^\wedge)^W} R_T^\wedge \to \mathrm{End}(\widehat{\mathscr{T}_s})$$

of Proposition 6.4 is surjective.

Proof By Nakayama's lemma and Lemma 5.9(2), it suffices to prove that the morphism

$$\varphi^{\wedge}_{1,\mathcal{T}_s} : R^{\wedge}_{T} \to \mathrm{End}(\mathcal{T}_s)$$

of Sect. 6.2 is surjective. Now we have $\dim(\mathrm{End}(\mathcal{T}_s)) = 2$; hence, for this, it suffices to prove that the image of $\varphi^{\wedge}_{1,\mathcal{T}_s}$ is not reduced to $\Bbbk \cdot \mathrm{id}_{\mathcal{T}_s}$. However, if $\lambda \in X_*(T)$ is such that $\langle \lambda, \alpha \rangle \neq 0$ in $\Bbbk$, then by Lemmas 2.5, 2.6, and 6.5, the automorphism $\varphi^{\wedge}_{1,\mathcal{T}_s}(\lambda)$ is unipotent but not equal to $\mathrm{id}_{\mathcal{T}_s}$; therefore, it does not belong to $\Bbbk \cdot \mathrm{id}_{\mathcal{T}_s}$, and the claim is proved. □

6.6 Properties of $\mathcal{T}_{w_0}$

We finish this section with a reminder of some properties of the category $\mathcal{O}$ which are well known (at least in the case $\mathrm{char}(\Bbbk) = 0$).

The following claim is fundamental. It is proved in [BBM, Lemma in §2.1] under the assumption that $\mathrm{char}(\Bbbk) = 0$, but the arguments apply in full generality.

Lemma 6.7 *For any $w \in W$, the socle of the object Δ_w is $\mathcal{IC}_e$, and all the composition factors of $\Delta_w/\mathrm{soc}(\Delta_w)$ are of the form $\mathcal{IC}_v$ with $v \neq e$. Dually, the top of the object ∇_w is $\mathcal{IC}_e$, and all the composition factors of the kernel of the surjection $\nabla_w \to \mathrm{top}(\nabla_w)$ are of the form $\mathcal{IC}_v$ with $v \neq e$.*

This lemma has the following important consequence.

Corollary 6.8 *If $\mathcal{F}$ is an object of $\mathcal{O}$ which admits a standard filtration, then its socle is a direct sum of copies of $\mathcal{IC}_e$. In other words, any nonzero subobject of $\mathcal{F}$ admits $\mathcal{IC}_e$ as a composition factor. Dually, if $\mathcal{F}$ is an object of $\mathcal{O}$ which admits a costandard filtration, then its top is a direct sum of copies of $\mathcal{IC}_e$. In other words, any nonzero quotient of $\mathcal{F}$ admits $\mathcal{IC}_e$ as a composition factor.*

To finish this section, we recall the main properties of the object $\mathcal{T}_{w_0}$ that we will need in Sect. 9.

Lemma 6.9

(1) *For any $w \in W$, we have $(\mathcal{T}_{w_0} : \Delta_w) = 1$.*
(2) *The object $\mathcal{T}_{w_0}$ is both the projective cover and the injective hull of $\mathcal{IC}_e$ in $\mathcal{O}$.*

Proof Both of these claims are consequences of Lemma 6.7. For details, see [AR2, Lemma 5.25] for (1), and [AR2, Proposition 5.26] for (2). □

Lemma 6.10 *Let s be a simple reflection, and let $\bar{\imath}_s : \overline{Y_s} \to Y$ be the embedding. Then we have $\bar{\imath}_s^*(\mathcal{T}_{w_0}) \cong \mathcal{T}_s$.*

Proof Since $\mathcal{T}_{w_0}$ is tilting (in particular, admits a standard filtration), the object $\bar{\imath}_s^*(\mathcal{T}_{w_0})$ is perverse and admits a standard filtration. More precisely, in view of

Lemma 6.9(1), we have

$$(\overline{\iota}_s^*(\mathscr{T}_{w_0}) : \Delta_e) = (\overline{\iota}_s^*(\mathscr{T}_{w_0}) : \Delta_s) = 1.$$

By the description of the indecomposable objects of $\mathsf{Perv}_U(\overline{Y_s}, \Bbbk)$ recalled in Sect. 6.5, we deduce that $\overline{\iota}_s^*(\mathscr{T}_{w_0})$ is isomorphic to either $\mathscr{T}_s$ or $\Delta_e \oplus \Delta_s$. However, we have

$$\mathrm{Hom}(\overline{\iota}_s^*(\mathscr{T}_{w_0}), \mathcal{IC}_s) = \mathrm{Hom}(\mathscr{T}_{w_0}, \mathcal{IC}_s) = 0$$

by adjunction and Lemma 6.9(2) respectively; hence, this object cannot admit Δ_s as a direct summand. □

7 Convolution

7.1 *Definition*

Let us denote by

$$m : G \times^U X \to X$$

the map defined by $m([g : hU]) = ghU$. If $\mathscr{F}, \mathscr{G}$ belong to $D^{\mathrm{b}}_U(X, \Bbbk)$, there exists a unique object $\mathscr{F} \widetilde{\boxtimes} \mathscr{G}$ in $D^{\mathrm{b}}_U(G \times^U X, \Bbbk)$ whose pullback under the quotient map $G \times X \to G \times^U X$ is $a^*(\mathscr{F}) \boxtimes \mathscr{G}$ (where a is as in Sect. 6.2). We then set

$$\mathscr{F} \star^U \mathscr{G} := m_!(\mathscr{F} \widetilde{\boxtimes} \mathscr{G})[\dim T].$$

This construction defines a functor $D^{\mathrm{b}}_U(X, \Bbbk) \times D^{\mathrm{b}}_U(X, \Bbbk) \to D^{\mathrm{b}}_U(X, \Bbbk)$, which is associative up to (canonical) isomorphism.

Similarly, we denote by

$$m' : G \times^U Y \to Y$$

the map defined by $m([g : hB]) = ghB$. If $\mathscr{F}$ belongs to $D^{\mathrm{b}}_U(X, \Bbbk)$ and $\mathscr{G}$ belongs to $D^{\mathrm{b}}_U(Y, \Bbbk)$, there exists a unique object $\mathscr{F} \widetilde{\boxtimes} \mathscr{G}$ in $D^{\mathrm{b}}_U(G \times^U Y, \Bbbk)$ whose pullback under the quotient map $G \times Y \to G \times^U Y$ is $a^*(\mathscr{F}) \boxtimes \mathscr{G}$. We then set

$$\mathscr{F} \star^U \mathscr{G} := m'_!(\mathscr{F} \widetilde{\boxtimes} \mathscr{G})[\dim T].$$

This construction defines a functor $D^{\mathrm{b}}_U(X, \Bbbk) \times D^{\mathrm{b}}_U(Y, \Bbbk) \to D^{\mathrm{b}}_U(Y, \Bbbk)$, which is compatible with the product $\star^U$ on $D^{\mathrm{b}}_U(X, \Bbbk)$ in the obvious sense.

Remark 7.1 Since the quotient G/U is not proper, there exist two possible conventions to define the convolution product on $D^{\mathrm{b}}_U(X,\Bbbk)$: one involving the functor $m_!$ and one involving the functor m_*. We insist that here, we consider the version with !-pushforward.

It is straightforward (using the base change theorem) to check that for $\mathscr{F},\mathscr{G}$ in $D^{\mathrm{b}}_U(X,\Bbbk)$ and $\mathscr{G}$ in $D^{\mathrm{b}}_U(Y,\Bbbk)$, there exist canonical isomorphisms

$$\pi_!(\mathscr{F}\star^U \mathscr{G}) \cong \mathscr{F}\star^U \pi_!(\mathscr{G}), \tag{7.1}$$

$$\pi^!(\mathscr{F}\star^U \mathscr{G}) \cong \mathscr{F}\star^U \pi^!(\mathscr{G}), \tag{7.2}$$

$$\pi^*(\mathscr{F}\star^U \mathscr{G}) \cong \mathscr{F}\star^U \pi^*(\mathscr{G}). \tag{7.3}$$

Instead of the U-equivariant categories, one can also consider the B-equivariant categories. In particular, very similar considerations lead to the definition of a functor

$$(-)\star^B(-) : D^{\mathrm{b}}_U(Y,\Bbbk)\times D^{\mathrm{b}}_B(Y,\Bbbk)\to D^{\mathrm{b}}_U(Y,\Bbbk).$$

(Here, we do not insert any cohomological shift in the definition. Note also that since G/B is proper, there is no difference between the $*$- and !-versions of convolution.)

We will denote by $\mathrm{For}^B_U : D^{\mathrm{b}}_B(Y,\Bbbk)\to D^{\mathrm{b}}_U(Y,\Bbbk)$ the natural forgetful functor. The following fact is standard.

Lemma 7.2 *For any $\mathscr{F}$ in $D^{\mathrm{b}}_U(X,\Bbbk)$ and $\mathscr{G}$ in $D^{\mathrm{b}}_B(Y,\Bbbk)$, there exists a canonical isomorphism*

$$\mathscr{F}\star^U \mathrm{For}^B_U(\mathscr{G}) \cong \pi_\dagger(\mathscr{F})\star^B \mathscr{G}.$$

7.2 Convolution and Monodromy

Lemma 7.3 *For any $\mathscr{F}$, $\mathscr{G}$ in $D^{\mathrm{b}}_U(X/\!\!/T,\Bbbk)$, the object $\mathscr{F}\star^U\mathscr{G}$ belongs to the subcategory $D^{\mathrm{b}}_U(X/\!\!/T,\Bbbk)$. Moreover, for any $x\in R^{\wedge}_T$, we have*

$$\varphi^{\wedge}_{\mathrm{l},\mathscr{F}\star^U\mathscr{G}}(x) = \varphi^{\wedge}_{\mathrm{l},\mathscr{F}}(x)\star^U \mathrm{id}_{\mathscr{G}},$$

$$\varphi^{\wedge}_{\mathrm{r},\mathscr{F}\star^U\mathscr{G}}(x) = \mathrm{id}_{\mathscr{F}}\star^U \varphi^{\wedge}_{\mathrm{r},\mathscr{G}}(x),$$

$$\varphi^{\wedge}_{\mathrm{r},\mathscr{F}}(x)\star^U \mathrm{id}_{\mathscr{G}} = \mathrm{id}_{\mathscr{F}}\star^U \varphi^{\wedge}_{\mathrm{l},\mathscr{G}}(x).$$

Proof The first claim is clear from (7.3). The proof of the first two isomorphisms is easy and left to the reader. To prove the third one, we write the map m as a composition $m = m_1\circ m_2$ where $m_1 : G\times^B X\to X$ and $m_2 : G\times^U X\to G\times^B X$

are the obvious morphisms. Then we have

$$\mathscr{F} \star^U \mathscr{G} = m_!(\mathscr{F} \widetilde{\boxtimes} \mathscr{G})[\dim T] = (m_1)_!(m_2)_!(\mathscr{F} \widetilde{\boxtimes} \mathscr{G})[\dim T].$$

We consider the action of T on $G \times^U X$ defined by $t \cdot [g : hU] = [gt^{-1} : thU]$. Then $\mathscr{F} \widetilde{\boxtimes} \mathscr{G}$ belongs to $D^{\mathrm{b}}_c\big((G \times^U X) /\!\!/ T, \Bbbk\big)$ for this action, and the corresponding monodromy morphism satisfies

$$\varphi^{\wedge}_{\mathscr{F} \widetilde{\boxtimes} \mathscr{G}}(\lambda) = \varphi^{\wedge}_{\mathscr{F}}(\lambda^{-1}) \widetilde{\boxtimes} \varphi^{\wedge}_{\mathscr{G}}(\lambda)$$

for any $\lambda \in X_*(T)$. (In fact, this equality can be checked after pullback to $G \times G/U$, where it follows from Lemma 2.5.) Now, m_2 is the quotient map for this T-action; hence, Lemma 2.3 implies that

$$(m_2)_! \varphi^{\wedge}_{\mathscr{F} \widetilde{\boxtimes} \mathscr{G}}(\lambda) = \mathrm{id},$$

or in other words that

$$(m_2)_!\big(\varphi^{\wedge}_{\mathscr{F}}(\lambda) \widetilde{\boxtimes} \mathrm{id}_{\mathscr{G}}\big) = (m_2)_!\big(\mathrm{id}_{\mathscr{F}} \widetilde{\boxtimes} \varphi^{\wedge}_{\mathscr{G}}(\lambda)\big).$$

Applying $(m_1)_!$, we deduce the desired equality. □

7.3 Extension to the Completed Category

We now explain how to extend the construction of the convolution product to the framework of the completed category $\widehat{D}_U(X /\!\!/ T, \Bbbk)$.

Lemma 7.4 *Let* "$\varprojlim_n$"$\mathscr{F}_n$ *be an object of* $\widehat{D}_U(X /\!\!/ T, \Bbbk)$. *If* $\mathscr{G}$ *is in* $D^{\mathrm{b}}_U(X /\!\!/ T, \Bbbk)$, *resp. if* $\mathscr{G}$ *is in* $D^{\mathrm{b}}_U(Y, \Bbbk)$, *then the pro-object*

$$\text{“}\varprojlim_n\text{”} \mathscr{F}_n \star^U \mathscr{G}, \quad \textit{resp.} \quad \text{“}\varprojlim_n\text{”} \mathscr{F}_n \star^U \mathscr{G},$$

is representable by an object of $D^{\mathrm{b}}_U(X /\!\!/ T, \Bbbk)$, *resp. of* $D^{\mathrm{b}}_U(Y, \Bbbk)$.

Sketch of Proof This property is proved along the lines of [BY, §4.3]; we sketch the proof in the second case and leave the details and the first case to the reader. If $\mathscr{G}$ is of the form $\mathrm{For}^B_U(\mathscr{G}')$ for some $\mathscr{G}'$ in $D^{\mathrm{b}}_B(Y, \Bbbk)$, then by Lemma 7.2, we have $\mathscr{F}_n \star^U \mathscr{G} \cong \pi_\dagger(\mathscr{F}_n) \star^B \mathscr{G}'$. Hence, the claim follows from the assumption that the pro-object "$\varprojlim_n$"$\pi_\dagger(\mathscr{F}_n)$ is representable. The general case follows since the objects of this form generate $D^{\mathrm{b}}_U(Y, \Bbbk)$ as a triangulated category, using the following observation (which can be checked using the methods of [BY, Appendix A]): Given a projective system of distinguished triangles $\mathscr{A}_n \to \mathscr{B}_n \to \mathscr{C}_n \xrightarrow{[1]}$

in $D^{\mathrm{b}}_U(Y,\Bbbk)$, if the pro-objects "$\varprojlim_n$"$\mathscr{A}_n$ and "$\varprojlim_n$"$\mathscr{B}_n$ are representable, then "$\varprojlim_n$"$\mathscr{C}_n$ is representable too (and this object is a cone of the induced morphism "$\varprojlim_n$"$\mathscr{A}_n \to$ "$\varprojlim_n$"$\mathscr{B}_n$). □

Using Lemma 7.4, we already see that the functor $\star^U : D^{\mathrm{b}}_U(X,\Bbbk) \times D^{\mathrm{b}}_U(Y,\Bbbk) \to D^{\mathrm{b}}_U(Y,\Bbbk)$ induces a functor

$$\widehat{\star} : \widehat{D}_U(X/\!\!/T,\Bbbk) \times D^{\mathrm{b}}_U(Y,\Bbbk) \to D^{\mathrm{b}}_U(Y,\Bbbk). \tag{7.4}$$

Now, let $\mathscr{F} =$ "$\varprojlim_n$"$\mathscr{F}_n$ and $\mathscr{G} =$ "$\varprojlim_m$"$\mathscr{G}_m$ be two objects of $\widehat{D}_U(X/\!\!/T,\Bbbk)$. For any fixed m, by Lemma 7.4, the pro-object "$\varprojlim_n$"$\mathscr{F}_n \star^U \mathscr{G}_m$ is representable by an object of $D^{\mathrm{b}}_U(X/\!\!/T,\Bbbk)$. Therefore, we can consider the pro-object

$$\mathscr{F}\widehat{\star}\mathscr{G} := \text{“}\varprojlim_m\text{”“}\varprojlim_n\text{”}\mathscr{F}_n \star^U \mathscr{G}_m.$$

We claim that this pro-object belongs to $\widehat{D}_U(X/\!\!/T,\Bbbk)$. Indeed, it is clearly uniformly bounded. And using (7.1), we see that

$$\begin{aligned}\text{“}\varprojlim_m\text{”}\pi_\dagger\left(\text{“}\varprojlim_n\text{”}\mathscr{F}_n \star^U \mathscr{G}_m\right) &\cong \text{“}\varprojlim_m\text{”}\left(\text{“}\varprojlim_n\text{”}\mathscr{F}_n \star^U \pi_\dagger(\mathscr{G}_m)\right)\\ &\cong \text{“}\varprojlim_m\text{”}\left(\mathscr{F}\widehat{\star}\pi_\dagger(\mathscr{G}_m)\right) \cong \mathscr{F}\widehat{\star}\left(\text{“}\varprojlim_m\text{”}\pi_\dagger(\mathscr{G}_m)\right).\end{aligned}$$

Since by assumption the pro-object "$\varprojlim_m$"$\pi_\dagger(\mathscr{G}_m)$ is representable, this shows that $\mathscr{F}\widehat{\star}\mathscr{G}$ is π-constant, which finishes the proof of our claim.

Remark 7.5 Let $\mathscr{F}$ and $\mathscr{G}$ be as above. Using similar arguments, one can check that for any fixed $n \geq 0$, the pro-object "$\varprojlim_m$"$\mathscr{F}_n \star^U \mathscr{G}_m$ is representable so that it makes sense to consider the pro-object

$$\text{“}\varprojlim_n\text{”“}\varprojlim_m\text{”}\mathscr{F}_n \star^U \mathscr{G}_m.$$

Using standard results on inverse limits (see, e.g., [KS2, Proposition 2.1.7]), one can show that this pro-object is canonically isomorphic to $\mathscr{F}\widehat{\star}\mathscr{G}$.

This construction provides us with a functor

$$\widehat{\star} : \widehat{D}_U(X/\!\!/T,\Bbbk) \times \widehat{D}_U(X/\!\!/T,\Bbbk) \to \widehat{D}_U(X/\!\!/T,\Bbbk).$$

This functor is associative in the obvious sense and compatible with (7.4) in the sense that for $\mathscr{F},\mathscr{G}$ in $\widehat{D}_U(X/\!\!/T,\Bbbk)$ and $\mathscr{H}$ in $D^{\mathrm{b}}_U(Y,\Bbbk)$, we have canonical isomorphisms

$$(\mathscr{F} \mathbin{\widehat{\star}} \mathscr{G}) \mathbin{\widehat{\star}} \mathscr{H} \cong \mathscr{F} \mathbin{\widehat{\star}} (\mathscr{G} \mathbin{\widehat{\star}} \mathscr{H}), \tag{7.5}$$

$$\pi_{\dagger}(\mathscr{F} \mathbin{\widehat{\star}} \mathscr{G}) \cong \mathscr{F} \mathbin{\widehat{\star}} \pi_{\dagger}(\mathscr{G}). \tag{7.6}$$

The object $\widehat{\Delta}_e = \widehat{\nabla}_e$ is a unit for this product (at least in the case when $\mathrm{char}(\Bbbk) > 0$),[3] as proved in the following lemma.

Lemma 7.6 *Assume that* $\mathrm{char}(\Bbbk) > 0$. *Then for any* $\mathscr{F}$ *in* $\widehat{D}_U(X /\!\!/ T, \Bbbk)$, *there exist canonical isomorphisms*

$$\widehat{\Delta}_e \mathbin{\widehat{\star}} \mathscr{F} \cong \mathscr{F} \cong \mathscr{F} \mathbin{\widehat{\star}} \widehat{\Delta}_e.$$

Proof For any $\mathscr{G}$ in $D^{\mathrm{b}}_U(X /\!\!/ T, \Bbbk)$, we have

$$\widehat{\Delta}_e \mathbin{\widehat{\star}} \mathscr{G} \cong \text{“}\varprojlim_n\text{”} a_!(\mathscr{L}_{T,n} \boxtimes \mathscr{G})[2r],$$

where $a : T \times X \to X$ is the action morphism defined by $a(t, gU) = tgU$. Now, we have canonical identifications

$$D^{\mathrm{b}}_U(G/U, \Bbbk) \cong D^{\mathrm{b}}_{U\times U}(G, \Bbbk) \cong D^{\mathrm{b}}_U(U\backslash G, \Bbbk).$$

Under these identifications, the full subcategory $D^{\mathrm{b}}_U(X /\!\!/ T, \Bbbk) \subset D^{\mathrm{b}}_U(G/U, \Bbbk)$ coincides with the category $D^{\mathrm{b}}_U(U\backslash G /\!\!/ T, \Bbbk)$ defined relative to the T-action on $U\backslash G$ defined by $t \cdot Ug = Utg$ and the stratification of $B\backslash G$ by B-orbits. Hence, Lemma 3.4 provides a canonical isomorphism $\widehat{\Delta}_e \mathbin{\widehat{\star}} \mathscr{G} \cong \mathscr{G}$. Passing to (formal) projective limits, we deduce a similar isomorphism for any $\mathscr{G}$ in $\widehat{D}_U(X /\!\!/ T, \Bbbk)$.

The proof of the isomorphism $\mathscr{F} \cong \mathscr{F} \mathbin{\widehat{\star}} \widehat{\Delta}_e$ follows from similar considerations together with Remark 7.5. □

One can easily check that these constructions provide $D^{\mathrm{b}}_U(X /\!\!/ T, \Bbbk)$ with the structure of a monoidal category (in the case when $\mathrm{char}(\Bbbk) > 0$).

7.4 Convolution of Standard, Costandard, and Tilting Objects

Lemma 7.7

(1) *For any* $w \in W$, *we have* $\widehat{\nabla}_{w^{-1}} \mathbin{\widehat{\star}} \widehat{\Delta}_w \cong \widehat{\Delta}_e$.
(2) *If* $v, w \in W$ *and if* $\ell(vw) = \ell(v) + \ell(w)$, *then we have*

$$\widehat{\Delta}_v \mathbin{\widehat{\star}} \widehat{\Delta}_w \cong \widehat{\Delta}_{vw}, \qquad \widehat{\nabla}_v \mathbin{\widehat{\star}} \widehat{\nabla}_w \cong \widehat{\nabla}_{vw}.$$

[3] This assumption is probably unnecessary. But since this is the setting we are mostly interested in, we will not consider the possible extension of this claim to the characteristic-0 setting.

Proof We prove the first isomorphism in (2); the other claims can be obtained similarly. By (7.6) and (5.3), we have

$$\pi_{\dagger}(\widehat{\Delta}_v \mathbin{\widehat{\star}} \widehat{\Delta}_w) \cong \widehat{\Delta}_v \mathbin{\widehat{\star}} \pi_{\dagger}(\widehat{\Delta}_w) \cong \widehat{\Delta}_v \mathbin{\widehat{\star}} \Delta_w.$$

Since Δ_w is a B-equivariant perverse sheaf, using Lemma 7.2, we deduce that

$$\pi_{\dagger}(\widehat{\Delta}_v \mathbin{\widehat{\star}} \widehat{\Delta}_w) \cong \Delta_v \star^B \Delta_w.$$

Now, it is well known that the right-hand side is isomorphic to Δ_{vw}; see, for example, [BBM, §2.2] or [AR3, Proposition 4.4]. Then the claim follows from Remark 5.10. □

Lemma 7.8 *Let $s \in S$. For any tilting perverse sheaf $\widehat{\mathscr{T}}$ in $\widehat{D}_U(X /\!\!/ T, \Bbbk)$, the object $\widehat{\mathscr{T}}_s \mathbin{\widehat{\star}} \widehat{\mathscr{T}}$ is a tilting perverse sheaf, and for any $w \in W$, we have*

$$(\widehat{\mathscr{T}}_s \mathbin{\widehat{\star}} \widehat{\mathscr{T}} : \widehat{\Delta}_w) = (\widehat{\mathscr{T}} : \widehat{\Delta}_w) + (\widehat{\mathscr{T}} : \widehat{\Delta}_{sw}).$$

Proof We will prove that for any $w \in W$, the object $\widehat{\mathscr{T}}_s \mathbin{\widehat{\star}} \widehat{\Delta}_w$ admits a standard filtration, the multiplicity of $\widehat{\Delta}_v$ being 1 if $v \in \{w, sw\}$, and 0 otherwise. Similar arguments show that $\widehat{\mathscr{T}}_s \mathbin{\widehat{\star}} \widehat{\nabla}_w$ admits a costandard filtration, and the desired claim will follow. First, assume that $sw > w$. Then using the exact sequence $\widehat{\Delta}_s \hookrightarrow \widehat{\mathscr{T}}_s \twoheadrightarrow \widehat{\Delta}_e$ (see Sect. 6.5) and applying $(-) \mathbin{\widehat{\star}} \widehat{\Delta}_w$, we obtain a distinguished triangle

$$\widehat{\Delta}_s \mathbin{\widehat{\star}} \widehat{\Delta}_w \to \widehat{\mathscr{T}}_s \mathbin{\widehat{\star}} \widehat{\Delta}_w \to \widehat{\Delta}_e \mathbin{\widehat{\star}} \widehat{\Delta}_w \xrightarrow{[1]} .$$

Here, Lemma 7.7(2) implies that the first term is isomorphic to $\widehat{\Delta}_{sw}$ and that the third term is isomorphic to $\widehat{\Delta}_w$, which shows the desired property. If now $sw < w$, we use the exact sequence $\widehat{\Delta}_e \hookrightarrow \widehat{\mathscr{T}}_s \twoheadrightarrow \widehat{\nabla}_s$ to obtain a distinguished triangle

$$\widehat{\Delta}_e \mathbin{\widehat{\star}} \widehat{\Delta}_w \to \widehat{\mathscr{T}}_s \mathbin{\widehat{\star}} \widehat{\Delta}_w \to \widehat{\nabla}_s \mathbin{\widehat{\star}} \widehat{\Delta}_w \xrightarrow{[1]} .$$

We conclude as above, using also Lemma 7.7(1) to see that the third term is isomorphic to $\widehat{\Delta}_{sw}$. □

Remark 7.9 One can easily deduce from Lemma 7.8 that the tilting objects in $\widehat{\mathscr{O}}$ are the direct sums of direct summands of objects of the form $\widehat{\mathscr{T}}_{s_1} \mathbin{\widehat{\star}} \cdots \mathbin{\widehat{\star}} \widehat{\mathscr{T}}_{s_r}$ with $s_1, \cdots, s_r \in S$ and moreover that the convolution product of two tilting objects is again a tilting object. Similarly, the tilting objects in $\mathscr{O}$ are the direct sums of direct summands of objects of the form $\widehat{\mathscr{T}}_{s_1} \mathbin{\widehat{\star}} \cdots \mathbin{\widehat{\star}} \widehat{\mathscr{T}}_{s_r} \mathbin{\widehat{\star}} \Delta_e$ with $s_1, \cdots, s_r \in S$, and $\widehat{\mathscr{T}} \mathbin{\widehat{\star}} \mathscr{T}$ is tilting in $\mathscr{O}$ if $\widehat{\mathscr{T}}$ is tilting in $\widehat{\mathscr{O}}$ and $\mathscr{T}$ is tilting in $\mathscr{O}$. In particular, this provides a "Bott–Samelson type" construction of these tilting objects.

Proposition 7.10 *For any $v, w \in W$, we have*

$$\widehat{\Delta}_v \mathbin{\widehat{\star}} \widehat{\mathscr{T}}_{w_0} \mathbin{\widehat{\star}} \widehat{\Delta}_w \cong \widehat{\mathscr{T}}_{w_0}.$$

Proof Of course, it is enough to prove that for $v, w \in W$, we have

$$\widehat{\Delta}_v \,\widehat{\star}\, \widehat{\mathcal{T}}_{w_0} \cong \widehat{\mathcal{T}}_{w_0} \quad \text{and} \quad \widehat{\mathcal{T}}_{w_0} \,\widehat{\star}\, \widehat{\Delta}_w \cong \widehat{\mathcal{T}}_{w_0}.$$

And for this, in view of Proposition 5.12, it suffices to prove that

$$\pi_{\dagger}\big(\widehat{\Delta}_v \,\widehat{\star}\, \widehat{\mathcal{T}}_{w_0}\big) \cong \mathcal{T}_{w_0} \quad \text{and} \quad \pi_{\dagger}\big(\widehat{\mathcal{T}}_{w_0} \,\widehat{\star}\, \widehat{\Delta}_w\big) \cong \mathcal{T}_{w_0}. \tag{7.7}$$

We first prove the second isomorphism in (7.7). By (7.6) and (5.3), we have

$$\pi_{\dagger}\big(\widehat{\mathcal{T}}_{w_0} \,\widehat{\star}\, \widehat{\Delta}_w\big) \cong \widehat{\mathcal{T}}_{w_0} \,\widehat{\star}\, \pi_{\dagger}(\widehat{\Delta}_w) \cong \widehat{\mathcal{T}}_{w_0} \,\widehat{\star}\, \Delta_w.$$

Since Δ_w is a B-equivariant perverse sheaf, using Lemma 7.2, we deduce that

$$\widehat{\mathcal{T}}_{w_0} \,\widehat{\star}\, \Delta_w \cong \mathcal{T}_{w_0} \star^B \Delta_w.$$

Hence, to prove the second isomorphism in (7.7), we only have to prove that

$$\mathcal{T}_{w_0} \star^B \Delta_w \cong \mathcal{T}_{w_0}. \tag{7.8}$$

It is known that any object of the form $\nabla_u \star^B \Delta_v$ is perverse (see, e.g., [AR3, Proposition 4.6] or [ABG, Proposition 8.2.4] for similar claims). In particular, it follows that $\mathcal{T}_{w_0} \star^B \Delta_w$ is perverse. And since $\Delta_w \star^B \nabla_{w^{-1}} \cong \Delta_e$, for any $x \in W$ and $n \in \mathbb{Z}$, we have

$$\mathrm{Hom}_{D^{\mathrm{b}}_U(Y,\Bbbk)}(\mathcal{T}_{w_0} \star^B \Delta_w, \mathcal{IC}_x[n]) \cong \mathrm{Hom}_{D^{\mathrm{b}}_U(Y,\Bbbk)}(\mathcal{T}_{w_0}, \mathcal{IC}_x \star^B \nabla_{w^{-1}}[n]).$$

Now since the realization functor $D^{\mathrm{b}}\mathcal{O} \to D^{\mathrm{b}}_U(Y, \Bbbk)$ is an equivalence of categories (see, e.g., [BGS, Corollary 3.3.2], whose proof applies to any field of coefficients), and in view of Lemma 6.9(2), for $y \in W$ and $m \in \mathbb{Z}$, we have

$$\mathrm{Hom}_{D^{\mathrm{b}}_U(Y,\Bbbk)}(\mathcal{T}_{w_0}, \mathcal{IC}_y[m]) \cong \begin{cases} \Bbbk & \text{if } y = e \text{ and } m = 0; \\ 0 & \text{otherwise.} \end{cases}$$

It is not difficult to see that if $x \neq e$ and if $\mathcal{G}$ belongs to $D^{\mathrm{b}}_B(Y, \Bbbk)$, then all the composition factors of the perverse cohomology objects of $\mathcal{IC}_x \star^B \mathcal{G}$ are of the form $\mathcal{IC}_y$ with $y \neq e$; using also Lemma 6.7, we deduce that

$$\mathrm{Hom}_{D^{\mathrm{b}}_U(Y,\Bbbk)}(\mathcal{T}_{w_0}, \mathcal{IC}_x \star^B \nabla_{w^{-1}}[n]) \cong \begin{cases} \Bbbk & \text{if } x = e \text{ and } n = 0; \\ 0 & \text{otherwise.} \end{cases}$$

It follows that the perverse sheaf $\mathscr{T}_{w_0} \star^B \Delta_w$ is the projective cover of $\mathcal{IC}_e$ and hence that it is isomorphic to $\mathscr{T}_{w_0}$ by Lemma 6.9(2). This finally proves (7.8) and hence also the second isomorphism in (7.7).

We now consider the first isomorphism in (7.7). If $v = e$, then it follows from Lemma 7.7 that $\widehat{\Delta}_e \widehat{\star} \widehat{\mathscr{T}}_{w_0}$ is a tilting perverse sheaf and has the same standard multiplicities as $\widehat{\mathscr{T}}_{w_0}$; therefore, it is isomorphic to $\widehat{\mathscr{T}}_{w_0}$. Now, assume the claim is known for $v \neq w_0$, and choose $s \in S$ such that $vs > v$. By the same arguments as in the proof of Lemma 7.8, we have an exact sequence of perverse sheaves

$$\widehat{\Delta}_{vs} \hookrightarrow \widehat{\Delta}_v \widehat{\star} \widehat{\mathscr{T}}_s \twoheadrightarrow \widehat{\Delta}_v. \tag{7.9}$$

From Lemma 7.8, we deduce that $\widehat{\mathscr{T}}_s \widehat{\star} \mathscr{T}_{w_0} \cong (\mathscr{T}_{w_0})^{\oplus 2}$. Therefore, convolving (7.9) with $\mathscr{T}_{w_0}$ on the right and using induction, we obtain a distinguished triangle

$$\widehat{\Delta}_{vs} \widehat{\star} \mathscr{T}_{w_0} \to (\mathscr{T}_{w_0})^{\oplus 2} \to \mathscr{T}_{w_0} \xrightarrow{[1]}$$

in $D^{\mathrm{b}}_U(Y, \Bbbk)$. As above, the object $\widehat{\Delta}_{vs} \widehat{\star} \mathscr{T}_{w_0}$ is perverse; hence, this triangle is a short exact sequence in $\mathscr{O}$. Since $\mathscr{T}_{w_0}$ is projective (see Lemma 6.9(2)), the surjection $(\mathscr{T}_{w_0})^{\oplus 2} \twoheadrightarrow \mathscr{T}_{w_0}$ must be split, and we finally obtain that $\widehat{\Delta}_{vs} \widehat{\star} \mathscr{T}_{w_0} \cong \mathscr{T}_{w_0}$, as desired. □

8 Variations on Some Results of Kostant–Kumar

From now on, we assume that G is semisimple, of adjoint type. (Of course, this assumption is harmless if one is mainly interested in the category $\mathscr{O}$.) We will denote by $\Phi^\vee$ the coroot system of (G, T), and by $\Phi^\vee_+ \subset \Phi^\vee$ the positive coroots.

8.1 The Pittie–Steinberg Theorem

We set

$$\mathsf{d} = \prod_{\alpha^\vee \in \Phi^\vee_+} (1 - e^{\alpha^\vee}) \quad \in R_T$$

and denote by $\rho^\vee \in X_*(T)$ the half sum of the positive coroots.

The following result is an easy application of the Pittie–Steinberg theorem.

Theorem 8.1 *The $(R^\wedge_T)^W$-module $R^\wedge_T$ is free of rank $\#W$. More precisely, this module admits a basis $(\mathsf{e}_w)_{w \in W}$ such that*

$$\det\bigl((w(\mathsf{e}_v))_{v,w \in W}\bigr) = \bigl((-1)^{|\Phi^\vee_+|} e^{-\rho^\vee} \mathsf{d}\bigr)^{|W|/2}. \tag{8.1}$$

Proof By the Pittie–Steinberg theorem (see [St]), we know that under our assumptions, $\mathbb{Z}[X_*(T)]$ is free over $\mathbb{Z}[X_*(T)]^W$, of rank $\#W$. Moreover, from the proof in [St], one sees that this module admits a basis such that (8.1) holds (see, e.g., [KK, Proof of Theorem 4.4]). Now, there are canonical isomorphisms

$$\Bbbk \otimes_{\mathbb{Z}} \mathbb{Z}[X_*(T)] \xrightarrow{\sim} \Bbbk[X_*(T)], \qquad \Bbbk \otimes_{\mathbb{Z}} \mathbb{Z}[X_*(T)]^W \xrightarrow{\sim} \Bbbk[X_*(T)]^W.$$

(For the second one, we remark that $\mathbb{Z}[X_*(T)]^W$ is a free $\mathbb{Z}$-module, with a basis consisting of the elements $\sum_{\lambda \in \mathbb{O}} e^\lambda$ where $\mathbb{O}$ runs over W-orbits in $X_*(T)$. Since a similar fact holds for $\Bbbk[X_*(T)]^W$, we deduce that the natural morphism $\Bbbk \otimes_{\mathbb{Z}} \mathbb{Z}[X_*(T)]^W \to \Bbbk[X_*(T)]^W$ is indeed an isomorphism.) Hence, R_T is free over $(R_T)^W$, of rank $\#W$, and admits a basis $(\mathsf{e}_w)_{w \in W}$ such that (8.1) holds.

Now, we consider completions. Let $a \in R_T^\wedge$, and write a as the limit of a sequence $(a_n)_{n \geq 0}$ of elements of R_T. For any $n \geq 0$, there exist (unique) elements $(p_w^n)_{w \in W}$ in $(R_T)^W$ such that

$$a_n = \sum_{w \in W} p_w^n \cdot \mathsf{e}_w. \tag{8.2}$$

We claim that each sequence $(p_w^n)_{n \geq 0}$ converges to a certain $p_w \in R_T^\wedge$; then p_w will belong to $(R_T^\wedge)^W$, and we will have $a = \sum_{w \in W} p_w \cdot \mathsf{e}_w$, which will prove that the elements $(\mathsf{e}_w)_{w \in W}$ generate $R_T^\wedge$ over $(R_T^\wedge)^W$.

Consider the matrix $M := (v(\mathsf{e}_w))_{v,w \in W}$, with rows and columns parametrized by W, and coefficients in R_T. Then the equalities (8.2) imply that for any $n \geq 0$, we have

$$(v(a_n))_{v \in W} = M \cdot (p_w^n)_{w \in W}$$

in the space of vectors parametrized by W, and with values in the ring R_T. Now (8.1) shows that M is invertible in the space of matrices with coefficients in the fraction field of R_T and that $\mathsf{d}^{|W|/2} \cdot M^{-1}$ in fact has coefficients in R_T. Moreover, we have

$$(\mathsf{d}^{|W|/2} \cdot p_w^n)_{w \in W} = (\mathsf{d}^{|W|/2} \cdot M^{-1}) \cdot (v(a_n))_{v \in W}. \tag{8.3}$$

From this, we will deduce that each sequence $(p_w^n)_{n \geq 0}$ is Cauchy, which will prove our claim. In fact, by the Artin–Rees lemma (applied to the R_T-modules $\mathsf{d}^{|W|/2} \cdot R_T \subset R_T$ and the ideal $\mathfrak{m}_T$), there exists an integer c such that

$$\mathfrak{m}_T^n \cap \mathsf{d}^{|W|/2} \cdot R_T \subset \mathsf{d}^{|W|/2} \cdot \mathfrak{m}_T^{n-c}$$

for any $n \geq c$. Now, if $k \geq 0$ is fixed, for $n, m \gg 0$, we have $a_n - a_m \in \mathfrak{m}_T^{c+k}$. From (8.3), we deduce that $\mathsf{d}^{|W|/2} \cdot (p_w^n - p_w^m)$ belongs to $\mathfrak{m}_T^{c+k}$ also, hence to $\mathsf{d}^{|W|/2} \cdot \mathfrak{m}_T^k$. Hence, $p_w^n - p_w^m$ belongs to $\mathfrak{m}_T^k$, which finishes the proof of the claim.

To conclude the proof, it remains to check that the elements $(\mathsf{e}_w)_{w\in W}$ are linearly independent over $(R_T^{\wedge})^W$. However, if

$$\sum_{w\in W} p_w \cdot \mathsf{e}_w = 0$$

for some elements p_w in $(R_T^{\wedge})^W$, then as above, we have $M \cdot (p_w)_{w\in W} = 0$. Since M is invertible (as a matrix with coefficients in the fraction field of $R_T^{\wedge}$), it follows that $p_w = 0$ for any $w \in W$. □

Let us note the following consequences of this theorem:

- The $R_T^{\wedge}$-module $R_T^{\wedge} \otimes_{(R_T^{\wedge})^W} R_T^{\wedge}$ is free of rank $\#W$.
- The $\Bbbk$-vector space $R_T^{\wedge}/(R_T^{\wedge})^W_+$ has dimension $\#W$, where $(R_T^{\wedge})^W_+$ is the kernel of the map $(R_T^{\wedge})^W \hookrightarrow R_T^{\wedge} \xrightarrow{\varepsilon_T^{\wedge}} \Bbbk$.

8.2 Some $R_T^{\wedge}$-Modules

In this subsection, we recall some constructions due to Kostant–Kumar [KK] (replacing everywhere the $T^\vee$-equivariant K-theory of the point—where $T^\vee$ is the torus dual to T—by $R_T^{\wedge}$).

We will denote by $Q_T^{\wedge}$ the fraction field of $R_T^{\wedge}$. We then denote by Q_W the smash product of $Q_T^{\wedge}$ and W; in other words Q_W is a $Q_T^{\wedge}$-vector space with a basis $(\delta_w)_{w\in W}$, with the multiplication determined by

$$(a\delta_w) \cdot (b\delta_v) = aw(b)\delta_{wv}.$$

Of course, $(\delta_w)_{w\in W}$ is also a basis for the action of $Q_T^{\wedge}$ given by right multiplication in Q_W. We will denote by ι the anti-involution of Q_W determined by

$$\iota(a) = a, \qquad \iota(\delta_w) = \delta_{w^{-1}}$$

for $a \in Q_T^{\wedge}$ and $w \in W$.

Following [KK], for $s \in S$, we set

$$y_s := (\delta_e + \delta_s)\frac{1}{1-e^{-\alpha_s^\vee}} = \frac{1}{1-e^{-\alpha_s^\vee}}(\delta_e - e^{-\alpha_s^\vee}\delta_s),$$

where $\alpha_s^\vee$ is the simple coroot associated with s. The same computation as for [KK, Proposition 2.4] shows that these elements satisfy the braid relations of W; therefore, by Matsumoto's lemma, for $w \in W$, we can set

$$y_w := y_{s_1} \cdot y_{s_2} \cdot (\cdots) \cdot y_{s_r},$$

where $w = s_1 \cdots s_r$ is any reduced expression. It is clear from definitions that the matrix expressing these elements in the basis $(\delta_w)_{w \in W}$ is upper triangular with respect to the Bruhat order; in particular, $(y_w)_{w \in W}$ is also a $Q_T^\wedge$-basis of Q_W. We set

$$Y_W := \bigoplus_{w \in W} R_T^\wedge \cdot y_w,$$

a free $R_T^\wedge$-module of rank $\#W$. As in [KK, Corollary 2.5], one sees that Y_W is a subring in Q_W and that $(y_w)_{w \in W}$ is also a basis of Y_W as an $R_T^\wedge$-module for the action induced by right multiplication.

We now consider

$$\Omega_W := \mathrm{Hom}_{Q_T^\wedge}(Q_W, Q_T^\wedge),$$

where Q_W is regarded as a $Q_T^\wedge$-vector space for the action by right multiplication. We will regard Ω_W as a $Q_T^\wedge$-vector space via $(a \cdot \psi)(b) = a\psi(b) = \psi(ba)$ for $a \in Q_T^\wedge$ and $b \in Q_W$. We will sometimes identify this vector space with the vector space $\mathsf{Fun}(W, Q_T^\wedge)$ of functions from W to $Q_T^\wedge$, by sending the map ψ to the function $w \mapsto \psi(\delta_w)$.

The space Ω_W admits an action of Q_W (by $Q_T^\wedge$-vector space automorphisms) defined by

$$(y \cdot \psi)(z) = \psi(\iota(y) \cdot z)$$

for $y, z \in Q_W$ and $\psi \in \Omega_W$. (Note that the action of $Q_T^\wedge \cdot \delta_e \subset Q_W$ does *not* coincide with the action of $Q_T^\wedge$ considered above.) Explicitly, we have

$$(y_s \cdot \psi)(\delta_w) = \frac{\psi(\delta_w) - e^{-w^{-1}\alpha_s^\vee}\psi(\delta_{sw})}{1 - e^{-w^{-1}\alpha_s^\vee}}. \tag{8.4}$$

We will be interested in the subspace

$$\Psi_W := \{\psi \in \Omega_W \mid \forall y \in Y_W,\ \psi(\iota(y)) \in R_T^\wedge\}.$$

Of course, this subspace is stable under the action of $R_T^\wedge \subset Q_T^\wedge$. Since Y_W is a subalgebra in Q_W, Ψ_W is also stable under the action of $Y_W \subset Q_W$. Since $(\iota(y_w))_{w \in W}$ is a basis of $\iota(Y_W)$ as a right $R_T^\wedge$-module, Ψ_W is free as an $R_T^\wedge$-module, with a basis $(\psi_w)_{w \in W}$ determined by

$$\psi_w(\iota(y_v)) = \begin{cases} 1 & \text{if } v = w; \\ 0 & \text{otherwise.} \end{cases}$$

The following properties can be checked as in [KK, Proposition 2.22].

Lemma 8.2

(1) *For any* $v, w \in W$, *the element* $\psi_v(\delta_w)$ *belongs to* $R_T^\wedge$ *and vanishes unless* $v \leq w$.

(2) *For any* $w \in W$, *we have*

$$\psi_w(\delta_w) = \prod_{\substack{\alpha^\vee \in \Phi_+^\vee \\ w(\alpha^\vee) \in -\Phi_+^\vee}} (1 - e^{\alpha^\vee}).$$

(3) *For* $w \in W$ *and* $s \in S$, *we have*

$$y_s \cdot \psi_w = \begin{cases} \psi_w + \psi_{sw} & \text{if } sw < w; \\ 0 & \text{otherwise.} \end{cases}$$

In particular, Point (1) in this lemma shows that under the identification of Ω_W with $\mathsf{Fun}(W, Q_T^\wedge)$ considered above, Ψ_W is contained in the subset $\mathsf{Fun}(W, R_T^\wedge)$ of functions taking values in $R_T^\wedge$.

8.3 An Isomorphism of $R_T^\wedge$-Modules

Our goal in this subsection is to relate the algebra $R_T^\wedge \otimes_{(R_T^\wedge)^W} R_T^\wedge$ with the objects introduced in Sect. 8.2. Our proofs are based on "K-theoretic analogues" of some arguments from [AJS, Appendix D].

Below, we will need the following lemma.

Lemma 8.3 *Let* $f \in R_T^\wedge$, *and let* $\alpha^\vee, \beta^\vee$ *be distinct positive coroots. If* $(1 - e^{\alpha^\vee}) \cdot f$ *is divisible (in* $R_T^\wedge$*) by* $1 - e^{\beta^\vee}$, *then* f *is divisible by* $1 - e^{\beta^\vee}$.

Proof Let us first prove the similar claim where $R_T^\wedge$ is replaced by R_T everywhere. For this, we denote by $T_\Bbbk^\vee$ the torus dual to T and consider $\alpha^\vee$ and $\beta^\vee$ as characters of $T_\Bbbk^\vee$. Since $\alpha^\vee$ and $\beta^\vee$ are linearly independent in the $\mathbb{Z}$-module $X_*(T)$, the group morphism

$$(\alpha^\vee, \beta^\vee) : T_\Bbbk^\vee \to (\Bbbk^\times)^2$$

is dominant, hence surjective. It follows that $\dim(\ker(\alpha^\vee) \cap \ker(\beta^\vee)) = \dim(T^\vee) - 2$. Now, R_T is a UFD, and the dimension condition means that $1 - e^{\alpha^\vee}$ and $1 - e^{\beta^\vee}$ have no common prime factor. If $f \in R_T$ and $(1 - e^{\alpha^\vee}) \cdot f$ is divisible by $1 - e^{\beta^\vee}$, each prime factor in the decomposition of $1 - e^{\beta^\vee}$ must appear in f, with at least the same multiplicity. It follows that $1 - e^{\beta^\vee}$ divides f, as desired.

The claim we have just proved can be translated into the fact that the "Koszul complex"

$$0 \to R_T \xrightarrow{f \mapsto ((1-e^{\beta^\vee})f, (1-e^{\alpha^\vee})f)} R_T \oplus R_T \xrightarrow{(g,h) \mapsto (1-e^{\alpha^\vee})g - (1-e^{\beta^\vee})h} R_T \to 0$$

(with nonzero terms in degrees -2, -1, and 0) has no cohomology in degree -1. Since $R_T^\wedge$ is flat over R_T, applying the functor $R_T^\wedge$, we deduce that the complex

$$0 \to R_T^\wedge \xrightarrow{f \mapsto ((1-e^{\beta^\vee})f, (1-e^{\alpha^\vee})f)} R_T^\wedge \oplus R_T^\wedge \xrightarrow{(g,h) \mapsto (1-e^{\alpha^\vee})g - (1-e^{\beta^\vee})h} R_T^\wedge \to 0$$

has no cohomology in degree -1 either, which implies our lemma. □

Theorem 8.4 *The morphism*

$$\tau : R_T^\wedge \otimes_{(R_T^\wedge)^W} R_T^\wedge \to \mathsf{Fun}(W, R_T^\wedge)$$

sending $a \otimes b$ to the function $w \mapsto a \cdot w^{-1}(b)$ is injective. Its image consists of the functions f such that

$$f(w) \equiv f(ws_{\alpha^\vee}) \mod (1 - e^{\alpha^\vee})$$

for any $w \in W$ and any coroot $\alpha^\vee$.

Proof Consider the basis $(\mathsf{e}_w)_{w \in W}$ of $R_T^\wedge$ as an $(R_T^\wedge)^W$-module considered in Theorem 8.1. Then $(1 \otimes \mathsf{e}_w)_{w \in W}$ is a basis of $R_T^\wedge \otimes_{(R_T^\wedge)^W} R_T^\wedge$ as an $R_T^\wedge$-module. Moreover, $\tau(1 \otimes \mathsf{e}_w)$ is the function $v \mapsto v^{-1}(\mathsf{e}_w)$. In view of (8.1), these functions are linearly independent in $\mathsf{Fun}(W, Q_T^\wedge)$. Hence indeed, our map is injective, and its image is (freely) spanned by these functions as an $R_T^\wedge$-module.

Now, let us identify Ψ_W with a subset of $\mathsf{Fun}(W, R_T^\wedge)$ (see Lemma 8.2). We claim that ψ_{w_0} belongs to the image of τ. In fact, this is equivalent to the existence of elements $(p_w)_{w \in W}$ in $R_T^\wedge$ such that

$$\tau\left(\sum_{w \in W} p_w \otimes \mathsf{e}_w\right) = \psi_{w_0},$$

or in other words (using Lemma 8.2(1)–(2)) such that

$$\sum_{w \in W} p_w v(\mathsf{e}_w) = \begin{cases} \mathsf{d} & \text{if } v = w_0; \\ 0 & \text{otherwise.} \end{cases}$$

The arguments above show that there exist unique elements $(p_w)_{w \in W}$ in $Q_T^\wedge$ which satisfy these equalities. As explained in [KK, Proof of Theorem 4.4], these elements in fact belong to R_T, hence in particular to $R_T^\wedge$.

Recall the action of Q_W on Ω_W considered in Sect. 8.2. Using the formula (8.4), one sees that for any $a, b \in R_T^\wedge$ and $s \in S$, we have

$$y_s \cdot \tau(a \otimes b) = \tau \left(a \otimes \frac{b - e^{-\alpha_s^\vee} s(b)}{1 - e^{-\alpha_s^\vee}} \right).$$

In particular, this shows that the image of τ is stable under the operators y_s ($s \in S$). Since (as we have seen above) this image contains ψ_{w_0}, by Lemma 8.2(3), it contains all the elements ψ_w ($w \in W$), hence Ψ_W.

It is clear that any function f in the image of τ satisfies

$$f(w) \equiv f(ws_{\alpha^\vee}) \mod (1 - e^{\alpha^\vee})$$

for any $w \in W$ and any coroot $\alpha^\vee$. To conclude the proof, it only remains to prove that any function which satisfies these conditions is a linear combination of the elements $(\psi_w)_{w \in W}$. For this, we choose a total order on W which extends the Bruhat order and argue by descending induction on the smallest element $w \in W$ such that $f(w) \neq 0$. Fix f, and let w be this smallest element. Then for any positive coroot $\alpha^\vee$ such that $w(\alpha^\vee) \in -\Phi_+^\vee$, we have $ws_{\alpha^\vee} < w$ in the Bruhat order. Hence, $f(ws_{\alpha^\vee}) = 0$, which implies that $f(w)$ is divible by $1 - e^{\alpha^\vee}$. By Lemmas 8.2(2) and 8.3, we deduce that there exists $a \in R_T^\wedge$ such that

$$f(w) = a\psi_w(\delta_w).$$

Then $f - a\psi_w$ vanishes on w and all the elements smaller than w (by definition of w and Lemma 8.2(1)). By induction, we deduce that $f - a\psi_w$ is a linear combination of elements $(\psi_v)_{v \in W}$, which concludes the proof. □

8.4 A Different Description of the Algebra $R_T^\wedge \otimes_{(R_T^\wedge)^W} R_T^\wedge$

The results in this subsection do not play any significant role below; we state them only for completeness.

As in Remark 2.2, the algebra $R_T^\wedge$ identifies with the algebra of functions on the formal neighborhood $\mathsf{FN}_{T_\Bbbk^\vee}(\{1\})$ of the identity in the $\Bbbk$-torus $T_\Bbbk^\vee$ which is Langlands dual to T (considered as a scheme). Hence, $R_T^\wedge \otimes_{(R_T^\wedge)^W} R_T^\wedge$ identifies with the algebra of functions on the fiber product

$$\mathsf{FN}_{T_\Bbbk^\vee}(\{1\}) \times_{(\mathsf{FN}_{T_\Bbbk^\vee}(\{1\}))/W} \mathsf{FN}_{T_\Bbbk^\vee}(\{1\}).$$

On the other hand, consider the formal neighborhood $\mathsf{FN}_{T_{\Bbbk}^{\vee}\times_{(T_{\Bbbk}^{\vee})/W}T_{\Bbbk}^{\vee}}(\{(1,1)\})$ of the base point in $T_{\Bbbk}^{\vee} \times_{(T_{\Bbbk}^{\vee})/W} T_{\Bbbk}^{\vee}$ (again, considered as a scheme). By the universal property of the fiber product, there exists a natural morphism of schemes

$$\mathsf{FN}_{T_{\Bbbk}^{\vee}\times_{(T_{\Bbbk}^{\vee})/W}T_{\Bbbk}^{\vee}}(\{(1,1)\}) \to \mathsf{FN}_{T_{\Bbbk}^{\vee}}(\{1\}) \times_{(\mathsf{FN}_{T_{\Bbbk}^{\vee}}(\{1\}))/W} \mathsf{FN}_{T_{\Bbbk}^{\vee}}(\{1\}). \tag{8.5}$$

Lemma 8.5 *The morphism* (8.5) *is an isomorphism.*

Proof We have to prove that the natural algebra morphism

$$R_T^{\wedge} \otimes_{(R_T^{\wedge})^W} R_T^{\wedge} \to \mathcal{O}(T_{\Bbbk}^{\vee} \times_{(T_{\Bbbk}^{\vee})/W} T_{\Bbbk}^{\vee})^{\wedge} \tag{8.6}$$

is an isomorphism, where the right-hand side is the completion of $\mathcal{O}(T_{\Bbbk}^{\vee} \times_{(T_{\Bbbk}^{\vee})/W} T_{\Bbbk}^{\vee})$ at its natural augmentation ideal J.

Let $I := \ker(\varepsilon_T) \subset R_T$; then we have $J = I \otimes_{(R_T)^W} R_T + R_T \otimes_{(R_T)^W} I$. For any $n \in \mathbb{Z}_{\geq 1}$, we have $J^{2n} \subset I^n \otimes_{(R_T)^W} R_T + R_T \otimes_{(R_T)^W} I^n$. Hence, for any $w \in W$, the morphism $R_T \otimes_{(R_T)^W} R_T \to R_T/I^n$ sending $a \otimes b$ to the class of $a \cdot w^{-1}(b)$ factors through a morphism $(R_T \otimes_{(R_T)^W} R_T)/J^{2n} \to R_T/I^n$. From this observation, it follows that the morphism τ of Theorem 8.4 factors through (8.6), proving that the latter morphism is injective.

On the other hand, from Theorem 8.1, we see that the natural morphism

$$R_T^{\wedge} \otimes_{(R_T)^W} R_T \to R_T^{\wedge} \otimes_{(R_T^{\wedge})^W} R_T^{\wedge}$$

is an isomorphism; hence, $R_T^{\wedge} \otimes_{(R_T^{\wedge})^W} R_T^{\wedge}$ is the completion of $R_T \otimes_{(R_T)^W} R_T$ with respect to the ideal $I \otimes_{(R_T)^W} R_T$. Since $I \otimes_{(R_T)^W} R_T \subset J$, we have for any n a surjection

$$(R_T \otimes_{(R_T)^W} R_T)/(I \otimes_{(R_T)^W} R_T)^n \twoheadrightarrow (R_T \otimes_{(R_T)^W} R_T)/J^n.$$

Since these algebras are finite-dimensional, passing to inverse limits, we deduce that (8.6) is surjective (by the Mittag–Leffler condition), which finishes the proof. □

9 Endomorphismensatz

9.1 Statement and Strategy of Proof

Our goal in this section is to prove the following theorem, which constitutes the main result of this article.

Theorem 9.1 *In the case $\widehat{\mathscr{T}} = \widehat{\mathscr{T}}_{w_0}$, the monodromy morphism of Proposition 6.4 is an algebra isomorphism*

$$R_T^\wedge \otimes_{(R_T^\wedge)^W} R_T^\wedge \xrightarrow{\sim} \mathrm{End}(\widehat{\mathscr{T}}_{w_0}).$$

Let us note the following consequence, which does not involve the completed category.

Corollary 9.2 *The morphism $\varphi^\wedge_{1,\mathscr{T}_{w_0}}$ of Sect. 6.2 induces an algebra isomorphism*

$$R_T/(R_T)^W_+ \xrightarrow{\sim} \mathrm{End}(\mathscr{T}_{w_0}),$$

where $(R_T)^W_+$ is the kernel of the composition $(R_T)^W \hookrightarrow R_T \xrightarrow{\varepsilon_T} \Bbbk$.

Proof Theorem 9.1 and Lemma 5.9(2) imply that monodromy induces an algebra isomorphism

$$R_T^\wedge/(R_T^\wedge)^W_+ \xrightarrow{\sim} \mathrm{End}(\mathscr{T}_{w_0}),$$

where $(R_T^\wedge)^W_+$ is the kernel of the composition $(R_T^\wedge)^W \hookrightarrow R_T^\wedge \xrightarrow{\varepsilon_T^\wedge} \Bbbk$. Hence, to conclude, it suffices to prove that the morphism

$$R_T/(R_T)^W_+ \to R_T^\wedge/(R_T^\wedge)^W_+$$

induced by the inclusion $R_T \hookrightarrow R_T^\wedge$ is an isomorphism. However, this morphism is easily seen to be injective. Since (by Theorem 8.1 and its proof) both sides have dimension $\#W$, the desired claim follows. □

In order to prove Theorem 9.1, we first remark that by Lemma 6.9(1) and (5.5), we have

$$\mathrm{gr}_w(\widehat{\mathscr{T}}_{w_0}) \cong \widehat{\Delta}_w$$

for any $w \in W$. We fix such isomorphisms, which provides an isomorphism

$$\mathrm{gr}(\widehat{\mathscr{T}}_{w_0}) \cong \bigoplus_{w \in W} \widehat{\Delta}_w.$$

By Lemma 5.4(2), the right monodromy morphism induces an isomorphism

$$R_T^\wedge \xrightarrow{\sim} \mathrm{End}(\widehat{\Delta}_w)$$

for any $w \in W$. Taking also Lemma 6.2 into account, we deduce an algebra isomorphism

$$\mathrm{End}\big(\mathrm{gr}(\widehat{\mathscr{T}}_{w_0})\big) \cong \bigoplus_{w \in W} R_T^\wedge. \tag{9.1}$$

We now consider the morphisms

$$R_T^\wedge \otimes_{(R_T^\wedge)^W} R_T^\wedge \xrightarrow{\sim} R_T^\wedge \otimes_{(R_T^\wedge)^W} R_T^\wedge \to \mathrm{End}(\widehat{\mathscr{T}}_{w_0}) \to \bigoplus_{w \in W} R_T^\wedge, \tag{9.2}$$

where:

- The first arrow is given by $a \otimes b \mapsto b \otimes a$.
- The second arrow is the morphism from Proposition 6.4.
- The third arrow is induced by the functor gr, taking into account the isomorphism (9.1).

By (6.4) and Lemma 6.1, the composition of the morphisms in (9.2) is the morphism τ of Theorem 8.4, if we identify $\bigoplus_{w \in W} R_T^\wedge$ with $\mathsf{Fun}(W, R_T^\wedge)$ in the obvious way. In particular, this composition is injective, which proves that the morphism

$$R_T^\wedge \otimes_{(R_T^\wedge)^W} R_T^\wedge \to \mathrm{End}(\widehat{\mathscr{T}}_{w_0})$$

from Theorem 9.1 is injective. We also deduce (using Theorem 8.4) that the image of the third morphism in (9.2) contains the subset of vectors $(a_w)_{w \in W}$ such that

$$a_{ws_{\alpha^\vee}} \equiv a_w \mod (1 - e^{\alpha^\vee}) \tag{9.3}$$

for any coroot $\alpha^\vee$. Below, we will prove the following claim.

Proposition 9.3 *If* $(a_y)_{y \in W}$ *belongs to the image of the third morphism in* (9.2), *then we have* (9.3) *for any* $w \in W$ *and any coroot* $\alpha^\vee$.

This proposition will complete the proof of Theorem 9.1. Indeed, from Corollary 6.3, we know that the third arrow in (9.2) is injective. The discussion above shows that its image coincides with the image of its composition with the second arrow in (9.2). Hence, this second arrow (i.e., the morphism from Theorem 9.1) is surjective.

9.2 A Special Case

In this subsection, we will prove that if $(a_y)_{y \in W}$ belongs to the image of third morphism in (9.2), then (9.3) holds when $w = e$ and $\alpha^\vee$ is a simple coroot. We will denote by α the (simple) root associated with $\alpha^\vee$. To simplify notation, we set $s := s_{\alpha^\vee}$.

We will denote by $\overline{\jmath}_s$ the (closed) embedding of $\pi^{-1}(\overline{Y_s}) = X_s \sqcup X_e$ in X.

Lemma 9.4 *We have $\overline{J}_s^*(\widehat{\mathcal{T}}_{w_0}) \cong \widehat{\mathcal{T}}_s$. Moreover, the morphism*

$$\mathrm{gr}_w(\widehat{\mathcal{T}}_{w_0}) \to \mathrm{gr}_w(\widehat{\mathcal{T}}_s)$$

induced by the adjunction morphism $\widehat{\mathcal{T}}_{w_0} \to (\overline{J}_s)_* \overline{J}_s^* \widehat{\mathcal{T}}_{w_0} = \widehat{\mathcal{T}}_s$ is an isomorphism if $w \in \{e, s\}$, and 0 otherwise.

Proof Since $\widehat{\mathcal{T}}_{w_0}$ is tilting (in particular, admits a standard filtration), it is clear that the adjunction morphism $\widehat{\mathcal{T}}_{w_0} \to (\overline{J}_s)_*(\overline{J}_s)^* \widehat{\mathcal{T}}_{w_0}$ is surjective and that the induced morphism $\mathrm{gr}_w(\widehat{\mathcal{T}}_{w_0}) \to \mathrm{gr}_w\big((\overline{J}_s)_* \overline{J}_s^* \widehat{\mathcal{T}}_{w_0}\big)$ is an isomorphism if $w \in \{e, s\}$, and 0 otherwise. Hence, it suffices to prove the isomorphism $\overline{J}_s^*(\widehat{\mathcal{T}}_{w_0}) \cong \widehat{\mathcal{T}}_s$. However, if we still denote by π the morphism $\pi^{-1}(\overline{Y_s}) \to \overline{Y_s}$ induced by π, we have

$$\pi_\dagger(\overline{J}_s^* \widehat{\mathcal{T}}_{w_0}) \cong \overline{\imath}_s^* \pi_\dagger(\widehat{\mathcal{T}}_{w_0}) = \overline{\imath}_s^* \mathcal{T}_{w_0}$$

where $\overline{\imath}_s : \overline{Y_s} \to Y$ is the embedding (see (5.1)). By Lemma 6.10, it follows that

$$\pi_\dagger(\overline{J}_s^* \widehat{\mathcal{T}}_{w_0}) \cong \mathcal{T}_s.$$

We deduce the desired isomorphism, in view of Proposition 5.12. □

Remark 9.5 The objects $\widehat{\mathcal{T}}_w$ are not canonical; they can be chosen only up to isomorphism. (This does not affect Theorem 9.1, since monodromy commutes with any morphism and hence is invariant under conjugation in the obvious sense.) However, the proof of Lemma 9.4 shows that once $\widehat{\mathcal{T}}_{w_0}$ is chosen, the object $\widehat{\mathcal{T}}_s$ (for any $s \in S$) can be defined canonically as $(\overline{J}_s)_* \overline{J}_s^* \widehat{\mathcal{T}}_{w_0}$.

From Lemma 9.4, we deduce that the composition

$$\mathrm{End}(\widehat{\mathcal{T}}_{w_0}) \to \bigoplus_{w \in W} R_T^\wedge \xrightarrow{(a_w)_{w \in W} \mapsto (a_e, a_s)} R_T^\wedge \oplus R_T^\wedge$$

(where the first arrow is the third morphism in (9.2)) factors as the composition

$$\mathrm{End}(\widehat{\mathcal{T}}_{w_0}) \xrightarrow{\overline{J}_s^*} \mathrm{End}(\widehat{\mathcal{T}}_s) \xrightarrow{\mathrm{gr}} R_T^\wedge \oplus R_T^\wedge. \tag{9.4}$$

Now, by Corollary 6.6, the morphism

$$R_T^\wedge \otimes_{(R_T^\wedge)^W} R_T^\wedge \to \mathrm{End}(\widehat{\mathcal{T}}_s)$$

of Proposition 6.4 is surjective, and its composition with the second arrow in (9.4) identifies with the morphism

$$a \otimes b \mapsto (a \cdot b, s(a) \cdot b)$$

(see Lemma 6.1). Since $ab \equiv s(a)b \mod (1 - e^{\alpha^\vee})$, this proves that if $(a_y)_{y\in W}$ belongs to the image of the third morphism in (9.2), then indeed we have $a_e \equiv a_s \mod (1 - e^{\alpha^\vee})$.

9.3 The General Case

In this subsection, we deduce Proposition 9.3 from the special case considered in Sect. 9.2. The main idea will be the following: recall diagram (9.2). We have natural actions of $W \times W$ on the first, second, and fourth terms in this diagram, respectively, defined by

$$\vartheta^{(1)}_{(w,v)}(a \otimes b) = w(a) \otimes v(b), \quad \vartheta^{(2)}_{(w,v)}(a \otimes b) = v(a) \otimes w(b),$$
$$\big(\vartheta^{(4)}_{(w,v)}(f)\big)(x) = w(f(v^{-1}xw))$$

for $w, v \in W$, $a, b \in R_T^\wedge$, $f \in \mathsf{Fun}(W, R_T^\wedge)$, $x \in W$. It is easily seen that the first arrow and the composition of the second and third arrows are equivariant with respect to these actions. We will now define an action of $W \times W$ on $\mathrm{End}(\widehat{\mathscr{T}}_{w_0})$ that makes the whole diagram (9.2) equivariant. This will imply that the image of the third morphism in this diagram is stable under this $(W \times W)$-action.

For $w, v \in W$, we denote by $\vartheta^{(3)}_{(w,v)}$ the automorphism of $\mathrm{End}(\widehat{\mathscr{T}}_{w_0})$ defined as the composition

$$\mathrm{End}(\widehat{\mathscr{T}}_{w_0}) \xrightarrow{\sim} \mathrm{End}(\widehat{\Delta}_v \widehat{\star} \widehat{\mathscr{T}}_{w_0} \widehat{\star} \widehat{\Delta}_{w^{-1}}) \xrightarrow{\sim} \mathrm{End}(\widehat{\mathscr{T}}_{w_0})$$

where the first arrow is induced by the functor $\widehat{\Delta}_v \widehat{\star} (-) \widehat{\star} \widehat{\Delta}_{w^{-1}}$ and the second arrow is induced by any choice of isomorphism as in Proposition 7.10.

Lemma 9.6 *For any $w, v \in W$, the automorphism $\vartheta^{(3)}_{(w,v)}$ does not depend on the choice of isomorphism as in Proposition 7.10. Moreover, these isomorphisms define an action of $W \times W$ on* $\mathrm{End}(\widehat{\mathscr{T}}_{w_0})$*, and the second and third arrows in* (9.2) *are equivariant with respect to this action and the ones defined above.*

Proof First, we claim that the second morphism in (9.2) intertwines the automorphisms $\vartheta^{(2)}_{(w,v)}$ and $\vartheta^{(3)}_{(w,v)}$. For this, we remark that the image under this morphism of $a \otimes b$ is $\varphi^\wedge_{\mathrm{l},\widehat{\mathscr{T}}_{w_0}}(a) \circ \varphi^\wedge_{\mathrm{r},\widehat{\mathscr{T}}_{w_0}}(b)$. Now, we have

$$\begin{aligned}
\mathrm{id}_{\widehat{\Delta}_v} \widehat{\star} \left(\varphi^\wedge_{\mathrm{l},\widehat{\mathscr{T}}_{w_0}}(a) \circ \varphi^\wedge_{\mathrm{r},\widehat{\mathscr{T}}_{w_0}}(b)\right) \widehat{\star} \mathrm{id}_{\widehat{\Delta}_{w^{-1}}} &= \varphi^\wedge_{\mathrm{r},\widehat{\Delta}_v}(a) \widehat{\star} \mathrm{id}_{\widehat{\mathscr{T}}_{w_0}} \widehat{\star} \varphi^\wedge_{\mathrm{l},\widehat{\Delta}_{w^{-1}}}(b) \\
&= \varphi^\wedge_{\mathrm{l},\widehat{\Delta}_v}(v(a)) \widehat{\star} \mathrm{id}_{\widehat{\mathscr{T}}_{w_0}} \widehat{\star} \varphi^\wedge_{\mathrm{r},\widehat{\Delta}_{w^{-1}}}(w(b)) \\
&= \varphi^\wedge_{\mathrm{l},\widehat{\Delta}_v \widehat{\star} \widehat{\mathscr{T}}_{w_0} \widehat{\star} \widehat{\Delta}_{w^{-1}}}(v(a)) \circ \varphi^\wedge_{\mathrm{r},\widehat{\Delta}_v \widehat{\star} \widehat{\mathscr{T}}_{w_0} \widehat{\star} \widehat{\Delta}_{w^{-1}}}(w(b)),
\end{aligned}$$

where the first and third equalities follow from Lemma 7.3, and the second one from Lemma 6.1. Now, by functoriality of monodromy, the conjugate of this automorphism with any choice of isomorphism $\widehat{\Delta}_v \,\widehat{\star}\, \widehat{\mathcal{T}}_{w_0} \,\widehat{\star}\, \widehat{\Delta}_{w^{-1}} \xrightarrow{\sim} \widehat{\mathcal{T}}_{w_0}$ is $\varphi^{\wedge}_{1,\widehat{\mathcal{T}}_{w_0}}(v(a)) \circ \varphi^{\wedge}_{\mathrm{r},\widehat{\mathcal{T}}_{w_0}}(w(b))$, which concludes the proof of our claim.

We have already remarked that all the $R_T^{\wedge}$-modules appearing in (9.2) are free of rank $\#W$ (see in particular Lemma 5.9(2) and Theorem 8.1). Moreover, from the proof of Theorem 8.4, we see that the image under the functor $Q_T^{\wedge} \otimes_{R_T^{\wedge}} -$ (where, as in Sect. 8.2, $Q_T^{\wedge}$ is the fraction field of $R_T^{\wedge}$) of the composition of the three arrows in this diagram is an isomorphism. Hence, the same property holds for any of the maps in this diagram. Since the composition of the second and third maps intertwines $\vartheta^{(2)}_{(w,v)}$ and $\vartheta^{(4)}_{(w,v)}$, and since the second map intertwines $\vartheta^{(2)}_{(w,v)}$ and $\vartheta^{(3)}_{(w,v)}$, we deduce that the third map intertwines $\vartheta^{(3)}_{(w,v)}$ and $\vartheta^{(4)}_{(w,v)}$. Since this map is injective, from this property, we see that $\vartheta^{(3)}_{(w,v)}$ does not depend on the choice of isomorphism as in Proposition 7.10 and that these isomorphisms define an action of $W \times W$ on $\mathrm{End}(\widehat{\mathcal{T}}_{w_0})$. □

Proof of Proposition 9.3 First, we assume that $\alpha^\vee$ is a simple coroot. Then since $\vartheta^{(4)}_{(e,w^{-1})}((a_y)_{y\in W})$ also belongs to the image of the third map in (9.2), by the special case considered in Sect. 9.2, we must have

$$a_w \equiv a_{ws_{\alpha^\vee}} \mod (1 - e^{\alpha^\vee}),$$

as desired.

Now, we consider the general case. We choose $v \in W$ such that $\beta^\vee := v(\alpha^\vee)$ is a simple coroot. To prove that $a_w \equiv a_{ws_{\alpha^\vee}} \mod (1 - e^{\alpha^\vee})$, we only have to prove that

$$v(a_w) \equiv v(a_{ws_{\alpha^\vee}}) \mod (1 - e^{\beta^\vee}).$$

However, since $ws_{\alpha^\vee} = wv^{-1}s_{\beta^\vee}v$, this fact follows from the observation that $\vartheta^{(4)}_{(v,e)}((a_y)_{y\in W})$ also belongs to the image of the third map in (9.2), and the case of simple coroots treated above (applied with "w" replaced by wv^{-1}). □

10 Variant: The étale Setting

All the constructions we have considered so far have counterparts in the world of étale sheaves, which we briefly review in this section. Here, we need to assume that $\Bbbk$ is a finite field and will denote its characteristic by ℓ.

10.1 Completed Derived Categories

We choose an algebraically closed field $\mathbb{F}$ of characteristic $p \neq \ell$. Instead of considering a *complex* connected reductive group, one can consider a connected reductive group $\mathbf{G}$ over $\mathbb{F}$, a Borel subgroup $\mathbf{B} \subset \mathbf{G}$, and a maximal torus $\mathbf{T} \subset \mathbf{B}$. Then we denote by $\mathbf{U}$ the unipotent radical of $\mathbf{B}$, and we set $\mathbf{X} := \mathbf{G}/\mathbf{U}$, $\mathbf{Y} := \mathbf{G}/\mathbf{B}$. We will denote by $D_c^{\mathrm{b,et}}(\mathbf{X}, \Bbbk)$ and $D_c^{\mathrm{b,et}}(\mathbf{Y}, \Bbbk)$ the bounded constructible derived categories of étale $\Bbbk$-sheaves on $\mathbf{X}$ and $\mathbf{Y}$, respectively. Then one can define the subcategory $D_{\mathbf{U}}^{\mathrm{b,et}}(\mathbf{Y}, \Bbbk) \subset D_c^{\mathrm{b,et}}(\mathbf{Y}, \Bbbk)$ as the $\mathbf{U}$-equivariant[4] derived category of $\mathbf{Y}$ and out of that define the associated categories $D_{\mathbf{U}}^{\mathrm{b,et}}(\mathbf{X} /\!\!/ \mathbf{T}, \Bbbk)$ and $\widehat{D}_{\mathbf{U}}^{\mathrm{et}}(\mathbf{X} /\!\!/ \mathbf{T}, \Bbbk)$ exactly as above.

In this setting, the monodromy construction (see Sect. 2) is a bit more subtle, but the required work has been done by Verdier [Ve]. Namely, we start by choosing once and for all a topological generator $(x_n)_{n \geq 0}$ of the pro-finite group

$$\varprojlim_n \{x \in \mathbb{F} \mid x^{\ell^n} = 1\}$$

(where the transition maps are given by $x \mapsto x^{\ell}$). As in the proof of Lemma 3.4, we denote, for $n \geq 0$, by $[n] : \mathbf{T} \to \mathbf{T}$ the morphism $z \mapsto z^{\ell^n}$, and set $a_n := a \circ ([n] \times \mathrm{id}_{\mathbf{X}})$, where $a : \mathbf{T} \times \mathbf{X} \to \mathbf{X}$ is the action morphism. Then given $\mathscr{F}$ in $D_{\mathbf{U}}^{\mathrm{b,et}}(\mathbf{X} /\!\!/ \mathbf{T}, \Bbbk)$, for $n \gg 0$, there exists an isomorphism

$$f_n^{\mathscr{F}} : (a_n)^* \mathscr{F} \xrightarrow{\sim} p^* \mathscr{F}$$

whose restriction to $\{1\} \times \mathbf{X}$ is the identity. Moreover, these isomorphisms are essentially unique and functorial in the same sense as in the proof of Lemma 3.4; see [Ve, Proposition 5.1]. Given $\lambda \in X_*(\mathbf{T})$, restricting the isomorphism $f_n^{\mathscr{F}}$ to $\{\lambda(x_n)\} \times \mathbf{X}$ (for $n \gg 0$) provides a canonical automorphism of $\mathscr{F}$, which by definition is $\varphi_{\mathscr{F}}^{\lambda}$. Starting with these automorphisms, one obtains the morphism $\varphi_{\mathscr{F}}^{\wedge}$, which still satisfies the properties of Sect. 2.2.

Lemma 2.6 continues to hold in this setting, but its proof has to be adapted to the new definition of monodromy. Note that when $\mathscr{F}$ is a perverse sheaf, the morphisms $f_n^{\mathscr{F}}$ are unique when they exist; in other words, they are determined by the condition that their restriction to $\{1\} \times \mathbf{X}$ is the identity. So if $\mathscr{F}$ is as in Lemma 2.6, there exists n and an isomorphism $f_n^{\mathscr{F}} : (a_n)^* \mathscr{F} \xrightarrow{\sim} p^* \mathscr{F}$ whose restriction to $\{1\} \times \mathbf{X}$ is the identity. The fact that the monodromy is trivial means that the restriction of $f_n^{\mathscr{F}}$ to

[4]Recall that in the étale setting, the $\mathbf{U}$-equivariant and $\mathbf{B}$-constructible derived categories are different if $p > 0$, due to the existence of nonconstant local systems on affine spaces. Here $D_{\mathbf{U}}^{\mathrm{b,et}}(\mathbf{Y}, \Bbbk)$ is the full triangulated subcategory of $D_c^{\mathrm{b,et}}(\mathbf{Y}, \Bbbk)$ generated by pushforwards of *constant* local systems on strata.

$\{x_n\} \times \mathbf{X}$ is the identity also. Hence, the pullback of $f_n^{\mathscr{F}}$ under the automorphism of $\mathbb{G}_m \times \mathbf{X}$ sending (z, x) to (zx_n, x) is also an isomorphism $(a_n)^* \mathscr{F} \xrightarrow{\sim} p^* \mathscr{F}$ whose restriction to $\{1\} \times \mathbf{X}$ is the identity; therefore, this isomorphism coincides with $f_n^{\mathscr{F}}$. Now, the morphism $[n] \times \mathrm{id}_{\mathbf{X}}$ is étale since $p \neq \ell$, and our observation amounts to saying that the morphism $f_n^{\mathscr{F}}$ satisfies the property that its pullbacks under both projections $(\mathbb{G}_m \times \mathbf{X}) \times_{(\mathbb{G}_m \times \mathbf{X})} (\mathbb{G}_m \times \mathbf{X}) \to \mathbb{G}_m \times \mathbf{X}$ (where the fiber product is taken with respect to the morphism $[n] \times \mathbf{X}$ on both sides) coincide. Since perverse sheaves form a stack for the étale topology (see [BBD, §2.2.19]), it follows that this morphism descends to an isomorphism $a^* \mathscr{F} \xrightarrow{\sim} p^* \mathscr{F}$; in other words, $\mathscr{F}$ is a $\mathbb{G}_m$-equivariant perverse sheaf.

Next, the étale fundamental group $\pi_1^{\mathrm{et}}(\mathbb{G}_m)$ of $\mathbb{G}_m$ is more complicated than $\pi_1(\mathbb{C}^\times)$. However, the étale covers $[n] : \mathbb{G}_m \to \mathbb{G}_m$ define a surjective morphism

$$\pi_1^{\mathrm{et}}(\mathbb{G}_m) \twoheadrightarrow \varprojlim_n \{x \in \mathbb{F} \mid x^{\ell^n} = 1\}.$$

Recall that we have fixed a topological generator of the right-hand side; this allows us to identify this group with $\varprojlim_n \mathbb{Z}/\ell^n\mathbb{Z}$. We have a natural isomorphism

$$X_*(\mathbf{T}) \otimes_{\mathbb{Z}} \pi_1^{\mathrm{et}}(\mathbb{G}_m) \xrightarrow{\sim} \pi_1^{\mathrm{et}}(\mathbf{T}),$$

hence a natural surjection

$$\pi_1^{\mathrm{et}}(\mathbf{T}) \to X_*(\mathbf{T}) \otimes_{\mathbb{Z}} \left(\varprojlim_n \mathbb{Z}/\ell^n\mathbb{Z} \right).$$

For $n \geq 0$, one can then consider the quotient $R_{\mathbf{T}}/\mathfrak{m}_{\mathbf{T}}^{\ell^n}$, with its natural action of $X_*(\mathbf{T})$. This action factors through an action of $X_*(\mathbf{T}) \otimes_{\mathbb{Z}} \mathbb{Z}/\ell^n\mathbb{Z}$; hence, it defines an action of $X_*(\mathbf{T}) \otimes_{\mathbb{Z}} \left(\varprojlim_n \mathbb{Z}/\ell^n\mathbb{Z} \right)$. By pullback, we deduce a finite-dimensional continuous $\pi_1^{\mathrm{et}}(\mathbf{T})$-module; the corresponding $\Bbbk$-local system on $\mathbf{T}$ will be denoted $\mathscr{L}_{\mathbf{T},n}^{\mathrm{et}}$. Then we can define the pro-unipotent local system as

$$\widehat{\mathscr{L}}_{\mathbf{T}}^{\mathrm{et}} = \text{“}\varprojlim_n\text{”} \mathscr{L}_{\mathbf{T},n}^{\mathrm{et}}.$$

Using this object as a replacement for $\widehat{\mathscr{L}}_{\mathbf{T}}$, all the constructions of Sects. 4–5 carry over to the present context, with identical proofs.

10.2 Soergel's Endomorphismensatz

Once the formalism of completed categories is in place, all the considerations of Sects. 6–7 carry over also. This allows one to extend the results of Sect. 9, in particular Theorem 9.1 and Corollary 9.2, to the étale setting (assuming that $\mathbf{G}$ is semisimple, of adjoint type).

10.3 Whittaker Derived Category

The main point of introducing the étale variant is that one can combine our considerations with the following "Whittaker-type" construction. Here, we have to assume that there exists a primitive p-th root of unity in $\mathbb{F}$; we will fix once and for all a choice of such a root.

Let $\mathbf{U}^+$ be the unipotent radical of the Borel subgroup of $\mathbf{G}$ opposite to $\mathbf{B}$ with respect to $\mathbf{T}$, and choose for any s an isomorphism between the root subgroup of $\mathbf{G}$ associated with the simple root corresponding to s and the additive group $\mathbb{G}_{\mathrm{a}}$. (Here, we assume that the roots of $\mathbf{B}$ are the *negative* roots.) We deduce an isomorphism $\mathbf{U}^+/[\mathbf{U}^+, \mathbf{U}^+] \cong (\mathbb{G}_{\mathrm{a}})^S$. Composing with the addition map $(\mathbb{G}_{\mathrm{a}})^S \to \mathbb{G}_{\mathrm{a}}$, we deduce a "nondegenerate" morphism $\chi : \mathbf{U}^+ \to \mathbb{G}_{\mathrm{a}}$. Our choice of primitive p-th root of unity determines an Artin–Schreier local system on $\mathbb{G}_{\mathrm{m}}$, whose pullback to $\mathbf{U}^+$ will be denoted $\mathscr{L}_\chi$. Then we can define the "Whittaker" derived category $D^{\mathrm{b,et}}_{\mathrm{Wh}}(\mathbf{Y}, \Bbbk)$ as the full subcategory of $D^{\mathrm{b,et}}_c(\mathbf{Y}, \Bbbk)$ consisting of $(\mathbf{U}^+, \mathscr{L}_\chi)$-equivariant objects. (See, e.g., [AR2, Appendix A] for a reminder on the construction of this category.) If $j : \mathbf{U}^+\mathbf{B}/\mathbf{B} \hookrightarrow \mathbf{Y}$ is the (open) embedding then, for any $\mathscr{F}$ in $D^{\mathrm{b,et}}_{\mathrm{Wh}}(\mathbf{Y}, \Bbbk)$, adjunction provides isomorphisms

$$j_! j^* \mathscr{F} \xrightarrow{\sim} \mathscr{F} \xrightarrow{\sim} j_* j^* \mathscr{F}.$$

Next, we can define the corresponding category $D^{\mathrm{b,et}}_{\mathrm{Wh}}(\mathbf{X} /\!\!/ \mathbf{T}, \Bbbk)$ as the triangulated subcategory generated by the objects of the form $\pi^\dagger \mathscr{F}$ with $\mathscr{F}$ in $D^{\mathrm{b,et}}_{\mathrm{Wh}}(\mathbf{Y}, \Bbbk)$ and deduce a completed category $\widehat{D}^{\mathrm{et}}_{\mathrm{Wh}}(\mathbf{X} /\!\!/ \mathbf{T}, \Bbbk)$. If $\widehat{\jmath} : \pi^{-1}(\mathbf{U}^+\mathbf{B}/\mathbf{B}) \hookrightarrow \mathbf{X}$ is the embedding, then for any object $\mathscr{F}$ in $\widehat{D}^{\mathrm{et}}_{\mathrm{Wh}}(\mathbf{X} /\!\!/ \mathbf{T}, \Bbbk)$, adjunction provides isomorphisms

$$\widehat{\jmath}_! \widehat{\jmath}^* \mathscr{F} \xrightarrow{\sim} \mathscr{F} \xrightarrow{\sim} \widehat{\jmath}_* \widehat{\jmath}^* \mathscr{F}.$$

In particular, using the obvious projection $\pi^{-1}(\mathbf{U}^+\mathbf{B}/\mathbf{B}) = \mathbf{U}^+\mathbf{B}/\mathbf{U} \cong \mathbf{U}^+ \times \mathbf{T} \to \mathbf{T}$, we obtain a canonical equivalence of triangulated categories

$$D^{\mathrm{b}}\mathsf{Mod}^{\mathrm{fg}}(R^\wedge_{\mathbf{T}}) \xrightarrow{\sim} \widehat{D}^{\mathrm{et}}_{\mathrm{Wh}}(\mathbf{X} /\!\!/ \mathbf{T}, \Bbbk). \tag{10.1}$$

The image of the free rank-1 $R^{\wedge}_{\mathbf{T}}$-module is the standard object $\widehat{\Delta}_\chi$ constructed as in Sect. 5.3 (with respect to the orbit $\mathbf{U}^+\mathbf{B}/\mathbf{B} \subset \mathbf{X}$). This object is canonically isomorphic to the corresponding costandard object $\widehat{\nabla}_\chi$. Transporting the tautological t-structure along the equivalence (10.1), we obtain a t-structure on $\widehat{D}^{\mathrm{et}}_{\mathrm{Wh}}(\mathbf{X}/\!\!/\mathbf{T}, \Bbbk)$ which we will call the perverse t-structure and whose heart will be denoted $\widehat{P}^{\mathrm{et}}_{\mathrm{Wh}}(\mathbf{X}/\!\!/\mathbf{T}, \Bbbk)$.

The categories $D^{\mathrm{b,et}}_{\mathrm{Wh}}(\mathbf{Y}, \Bbbk)$, $D^{\mathrm{b,et}}_{\mathrm{Wh}}(\mathbf{X}/\!\!/\mathbf{T}, \Bbbk)$, and $\widehat{D}^{\mathrm{et}}_{\mathrm{Wh}}(\mathbf{X}/\!\!/\mathbf{T}, \Bbbk)$ are related to the categories $D^{\mathrm{b,et}}_{\mathbf{U}}(\mathbf{Y}, \Bbbk)$, $D^{\mathrm{b,et}}_{\mathbf{U}}(\mathbf{X}/\!\!/\mathbf{T}, \Bbbk)$, and $\widehat{D}^{\mathrm{et}}_{\mathbf{U}}(\mathbf{X}/\!\!/\mathbf{T}, \Bbbk)$ in several ways. First, the convolution construction of Sect. 7 defines a right action of the monoidal category $\big(\widehat{D}^{\mathrm{et}}_{\mathbf{U}}(\mathbf{X}/\!\!/\mathbf{T}, \Bbbk), \widehat{\star}\big)$ on $\widehat{D}^{\mathrm{et}}_{\mathrm{Wh}}(\mathbf{X}/\!\!/\mathbf{T}, \Bbbk)$; the corresponding bifunctor will again be denoted $\widehat{\star}$. Next, we have "averaging" functors $D^{\mathrm{b,et}}_{\mathbf{U}}(\mathbf{Y}, \Bbbk) \to D^{\mathrm{b,et}}_{\mathrm{Wh}}(\mathbf{Y}, \Bbbk)$ and $D^{\mathrm{b,et}}_{\mathbf{U}}(\mathbf{X}/\!\!/\mathbf{T}, \Bbbk) \to D^{\mathrm{b,et}}_{\mathrm{Wh}}(\mathbf{X}/\!\!/\mathbf{T}, \Bbbk)$, sending a complex $\mathscr{F}$ to $(a_{\mathbf{U}^+})_!(\mathscr{L}_\chi \boxtimes \mathscr{F})[\dim \mathbf{U}^+]$, where $a_{\mathbf{U}^+} : \mathbf{U}^+ \times \mathbf{Y} \to \mathbf{Y}$ and $a_{\mathbf{U}^+} : \mathbf{U}^+ \times \mathbf{X} \to \mathbf{X}$ are the natural morphisms. Standard arguments (see [BBM, BY]) show that $(a_{\mathbf{U}^+})_!$ can be replaced by $(a_{\mathbf{U}^+})_*$ in this formula without changing the functor up to isomorphism. These functors will be denoted Av_χ; then we have canonical isomorphisms

$$\mathsf{Av}_\chi \circ \pi_\dagger \cong \pi_\dagger \circ \mathsf{Av}_\chi, \qquad \mathsf{Av}_\chi \circ \pi^\dagger \cong \pi^\dagger \circ \mathsf{Av}_\chi.$$

In particular, we obtain an induced functor

$$\mathsf{Av}_\chi : \widehat{D}^{\mathrm{et}}_{\mathbf{U}}(\mathbf{X}/\!\!/\mathbf{T}, \Bbbk) \to \widehat{D}^{\mathrm{et}}_{\mathrm{Wh}}(\mathbf{X}/\!\!/\mathbf{T}, \Bbbk).$$

By construction, this functor satisfies

$$\mathsf{Av}_\chi(\widehat{\Delta}_e) = \widehat{\Delta}_\chi.$$

We also have averaging functors in the other direction, defined in terms of the action morphisms $a_{\mathbf{U}} : \mathbf{U} \times \mathbf{Y} \to \mathbf{Y}$ and $a_{\mathbf{U}} : \mathbf{U} \times \mathbf{X} \to \mathbf{X}$ and the constant local system on $\mathbf{U}$. This time, the versions with $*$- and !-pushforwards are different and will be denoted $\mathsf{Av}^{\mathbf{U}}_*$ and $\mathsf{Av}^{\mathbf{U}}_!$. Here also, we have isomorphisms

$$\mathsf{Av}^{\mathbf{U}}_? \circ \pi_\dagger \cong \pi_\dagger \circ \mathsf{Av}^{\mathbf{U}}_?, \qquad \mathsf{Av}^{\mathbf{U}}_? \circ \pi^\dagger \cong \pi^\dagger \circ \mathsf{Av}^{\mathbf{U}}_?$$

for $? \in \{*, !\}$ (see the arguments in [BY, Proof of Corollary A.3.4] for the first isomorphism in the case $? = *$). Hence, we deduce induced functors

$$\mathsf{Av}^{\mathbf{U}}_! : \widehat{D}^{\mathrm{et}}_{\mathrm{Wh}}(\mathbf{X}/\!\!/\mathbf{T}, \Bbbk) \to \widehat{D}^{\mathrm{et}}_{\mathbf{U}}(\mathbf{X}/\!\!/\mathbf{T}, \Bbbk), \quad \mathsf{Av}^{\mathbf{U}}_* : \widehat{D}^{\mathrm{et}}_{\mathrm{Wh}}(\mathbf{X}/\!\!/\mathbf{T}, \Bbbk) \to \widehat{D}^{\mathrm{et}}_{\mathbf{U}}(\mathbf{X}/\!\!/\mathbf{T}, \Bbbk).$$

Standard arguments (see, e.g., [BY, Lemma 4.4.5] or [AR2, Lemma 5.15]) show that the pairs $(\mathsf{Av}^{\mathbf{U}}_!, \mathsf{Av}_\chi)$ and $(\mathsf{Av}_\chi, \mathsf{Av}^{\mathbf{U}}_*)$ form adjoint pairs of functors.

10.4 Geometric Construction of $\widehat{\mathscr{T}}_{w_0}$

The Whittaker constructions of Sect. 10.3 allow us in particular to give a concrete and explicit description of the objects $\widehat{\mathscr{T}}_{w_0}$ and $\mathscr{T}_{w_0}$, as follows.

Lemma 10.1 *There exist isomorphisms*

$$\mathscr{T}_{w_0} \cong \mathsf{Av}_!^{\mathbf{U}} \circ \mathsf{Av}_\chi(\Delta_e) \cong \mathsf{Av}_*^{\mathbf{U}} \circ \mathsf{Av}_\chi(\Delta_e), \quad \widehat{\mathscr{T}}_{w_0} \cong \mathsf{Av}_!^{\mathbf{U}} \circ \mathsf{Av}_\chi(\widehat{\Delta}_e) \cong \mathsf{Av}_*^{\mathbf{U}} \circ \mathsf{Av}_\chi(\widehat{\Delta}_e).$$

Proof Since the averaging functors commute with $\pi_\dagger$, in view of the characterization of $\widehat{\mathscr{T}}_{w_0}$ in Proposition 5.12, it is sufficient to prove the isomorphisms $\mathscr{T}_{w_0} \cong \mathsf{Av}_!^{\mathbf{U}} \circ \mathsf{Av}_\chi(\Delta_e) \cong \mathsf{Av}_*^{\mathbf{U}} \circ \mathsf{Av}_\chi(\Delta_e)$. This follows from standard arguments, showing that $\mathsf{Av}_!^{\mathbf{U}} \circ \mathsf{Av}_\chi(\Delta_e)$ is the projective cover of $\mathcal{IC}_e$ and that $\mathsf{Av}_*^{\mathbf{U}} \circ \mathsf{Av}_\chi(\Delta_e)$ is the injective hull of $\mathcal{IC}_e$ and then using Lemma 6.9(2); see [BY, Lemma 4.4.11] or [AR2, Lemma 5.18] for details. □

Remark 10.2 As explained above, Lemma 10.1 provides a canonical representative for the object $\widehat{\mathscr{T}}_{w_0}$ (in the present étale setting). In view of Remark 9.5, the objects $\widehat{\mathscr{T}}_s$ with $s \in S$ are then also canonically defined.

11 Soergel Theory

In this section, we use Theorem 9.1 and Corollary 9.2 to obtain a description of tilting objects in $\mathscr{O}$ and $\widehat{\mathscr{O}}$ in terms of some kinds of Soergel bimodules. For simplicity, we assume that $\Bbbk$ is a finite field. (This assumption does not play any role in Sects. 11.1 and 11.2.)

In Sects. 11.1 and 11.2, we work either in the "classical" setting of Sects. 6–9 or in the étale setting of Sect. 10. (For simplicity, we do not distinguish the two cases and use the notation of Sects. 6–9.) Then in Sect. 11.3, we consider a construction that is available only in the étale setting, and in Sect. 11.4, we explain how to extend the results obtained using this construction to the classical setting. Finally, in Sect. 11.5, we derive an explicit description of the categories of tilting objects in $\mathscr{O}$ and $\widehat{\mathscr{O}}$.

11.1 The Functor $\mathbb{V}$

We fix a representative $\widehat{\mathscr{T}}_{w_0}$ and set $\mathscr{T}_{w_0} := \pi_\dagger(\widehat{\mathscr{T}}_{w_0})$ (so that $\mathscr{T}_{w_0}$ is as above the indecomposable tilting object in $\mathscr{O}$ associated with w_0, but now chosen in a slightly more specific way).

Thanks to Theorem 9.1 and Corollary 9.2, respectively, we have isomorphisms

$$R_T^{\wedge} \otimes_{(R_T^{\wedge})^W} R_T^{\wedge} \xrightarrow{\sim} \mathrm{End}(\widehat{\mathcal{T}}_{w_0}), \quad R_T/(R_T)_+^W \xrightarrow{\sim} \mathrm{End}(\mathcal{T}_{w_0})$$

so that we can consider the functors

$$\widehat{\mathbb{V}} : \widehat{\mathcal{O}} \to \mathsf{Mof}(R_T^{\wedge} \otimes_{(R_T^{\wedge})^W} R_T^{\wedge})$$
$$\mathbb{V} : \mathcal{O} \to \mathsf{Mof}(R_T/(R_T)_+^W)$$

(where for A, a Noetherian ring, we denote by $\mathsf{Mof}(A)$ the abelian category of finitely generated left A-modules) defined by

$$\widehat{\mathbb{V}}(\widehat{\mathcal{F}}) = \mathrm{Hom}(\widehat{\mathcal{T}}_{w_0}, \widehat{\mathcal{F}}), \quad \mathbb{V}(\mathcal{F}) = \mathrm{Hom}(\mathcal{T}_{w_0}, \mathcal{F}).$$

Here, the fact that $\mathbb{V}$ takes values in $\mathsf{Mof}(R_T/(R_T)_+^W)$ is obvious, while for $\widehat{\mathbb{V}}$, the corresponding property follows from Corollary 5.5(1). If $\widehat{\mathcal{T}}$ is a tilting object in $\widehat{\mathcal{O}}$, then by Lemma 5.9(2), we have a canonical isomorphism

$$\Bbbk \otimes_{R_T^{\wedge}} \widehat{\mathbb{V}}(\widehat{\mathcal{T}}) \cong \mathbb{V}(\pi_{\dagger}\widehat{\mathcal{T}}), \tag{11.1}$$

where the tensor product is taken with respect to the action of the right copy of $R_T^{\wedge}$.

Remark 11.1 Lemma 8.5 shows that the category $\mathsf{Mof}(R_T^{\wedge} \otimes_{(R_T^{\wedge})^W} R_T^{\wedge})$ can be described more geometrically as the category of coherent sheaves on the formal neighborhood of the point $(1, 1)$ in $T_{\Bbbk}^{\vee} \times_{(T_{\Bbbk}^{\vee})^W} T_{\Bbbk}^{\vee}$ (considered as a scheme). Similarly, the category $\mathsf{Mof}(R_T/(R_T)_+^W)$ is the category of coherent sheaves on the fiber of the quotient morphism $T_{\Bbbk}^{\vee} \to (T_{\Bbbk}^{\vee})/W$ over the image of 1. In these terms, the monoidal structure on $\mathsf{Mof}(R_T^{\wedge} \otimes_{(R_T^{\wedge})^W} R_T^{\wedge})$ considered in Sect. 11.3 below can be described as a convolution product.

These functors are "fully faithful on tilting objects" in the following sense.

Proposition 11.2 *For any tilting perverse sheaves $\widehat{\mathcal{T}}, \widehat{\mathcal{T}}'$ in $\widehat{\mathcal{O}}$, the functor $\widehat{\mathbb{V}}$ induces an isomorphism*

$$\mathrm{Hom}_{\widehat{\mathcal{O}}}(\widehat{\mathcal{T}}, \widehat{\mathcal{T}}') \xrightarrow{\sim} \mathrm{Hom}_{R_T^{\wedge} \otimes_{(R_T^{\wedge})^W} R_T^{\wedge}}\big(\widehat{\mathbb{V}}(\widehat{\mathcal{T}}), \widehat{\mathbb{V}}(\widehat{\mathcal{T}}')\big).$$

Similarly, for any tilting perverse sheaves $\mathcal{T}$, $\mathcal{T}'$ in $\mathcal{O}$, the functor $\mathbb{V}$ induces an isomorphism

$$\mathrm{Hom}_{\mathcal{O}}(\mathcal{T}, \mathcal{T}') \xrightarrow{\sim} \mathrm{Hom}_{R_T/(R_T)_+^W}\big(\mathbb{V}(\mathcal{T}), \mathbb{V}(\mathcal{T}')\big).$$

Proof The second case is treated in [BBM, §2.1]. Here, we prove both cases using a closely related argument explained in [BY, §4.7].

We start with the case of the functor $\mathbb{V}$. We remark that this functor admits a left adjoint $\gamma : \mathsf{Mof}(R_T/(R_T)^W_+) \to \mathscr{O}$ defined by $\gamma(M) = \mathscr{T}_{w_0} \otimes_{R_T/(R_T)^W_+} M$. More concretely, if M is written as the cokernel of a map $f : (R_T/(R_T)^W_+)^{\oplus n} \to (R_T/(R_T)^W_+)^{\oplus m}$, then in view of the isomorphism $R_T/(R_T)^W_+ \xrightarrow{\sim} \mathrm{End}(\mathscr{T}_{w_0})$, the map f defines a morphism $(\mathscr{T}_{w_0})^{\oplus n} \to (\mathscr{T}_{w_0})^{\oplus m}$, whose cokernel is $\gamma(M)$. From this description and using the exactness of $\mathbb{V}$ (see Lemma 6.9(2)), we see that the adjunction morphism $\mathrm{id} \to \mathbb{V} \circ \gamma$ is an isomorphism.

We now assume that $\mathscr{T}$ is a tilting perverse sheaf and consider the adjunction morphism

$$\gamma(\mathbb{V}(\mathscr{T})) \to \mathscr{T}. \tag{11.2}$$

The image of this morphism under $\mathbb{V}$ is an isomorphism since its composition with the (invertible) adjunction morphism $\mathrm{id} \to \mathbb{V} \circ \gamma$ applied to $\mathbb{V}(\mathscr{T})$ is $\mathrm{id}_{\mathbb{V}(\mathscr{T})}$. Hence, its kernel and cokernel are killed by $\mathbb{V}$; in other words, they do not admit $\mathcal{IC}_e$ as a composition factor. In view of Corollary 6.8, this shows that the cokernel of this morphism vanishes, that is, that (11.2) is surjective. Moreover, if $\mathscr{T}'$ is another tilting object in $\mathscr{O}$, then the kernel of this morphism does not admit any nonzero morphism to $\mathscr{T}'$, again by Corollary 6.8. Hence, the induced morphism

$$\mathrm{Hom}(\mathscr{T}, \mathscr{T}') \to \mathrm{Hom}(\gamma(\mathbb{V}(\mathscr{T})), \mathscr{T}')$$

is an isomorphism, which finishes the proof in this case.

Now, we consider the case of $\widehat{\mathbb{V}}$. As for $\mathbb{V}$, this functor admits a left adjoint

$$\widehat{\gamma} : \mathsf{Mof}(R_T^\wedge \otimes_{(R_T^\wedge)^W} R_T^\wedge) \to \widehat{\mathscr{O}}$$

defined by $\widehat{\gamma}(M) = \widehat{\mathscr{T}}_{w_0} \otimes_{R_T^\wedge \otimes_{(R_T^\wedge)^W} R_T^\wedge} M$; in more concrete terms, if M is the cokernel of a map $(R_T^\wedge \otimes_{(R_T^\wedge)^W} R_T^\wedge)^{\oplus n} \to (R_T^\wedge \otimes_{(R_T^\wedge)^W} R_T^\wedge)^{\oplus m}$, then $\widehat{\gamma}(M)$ is the cokernel of the corresponding map $(\widehat{\mathscr{T}}_{w_0})^{\oplus n} \to (\widehat{\mathscr{T}}_{w_0})^{\oplus m}$. From this description and the fact that the functor ${}^p\mathscr{H}^0 \circ \pi_\dagger$ is right exact (see Corollary 5.8), we see that for any M in $\mathsf{Mof}(R_T^\wedge \otimes_{(R_T^\wedge)^W} R_T^\wedge)$, we have

$${}^p\mathscr{H}^0(\pi_\dagger(\widehat{\gamma}(M))) \cong \gamma\big(\Bbbk \otimes_{R_T^\wedge} M\big).$$

Moreover, if $\widehat{\mathscr{T}}$ is a tilting object in $\widehat{\mathscr{O}}$, under this identification and that in (11.1), applying ${}^p\mathscr{H}^0 \circ \pi_\dagger$ to the adjunction morphism

$$\widehat{\gamma}(\widehat{\mathbb{V}}(\widehat{\mathscr{T}})) \to \widehat{\mathscr{T}} \tag{11.3}$$

we recover the adjunction morphism (11.2) for $\mathscr{T} = \pi_\dagger(\widehat{\mathscr{T}})$. Since the latter map is known to be surjective, this shows that the cokernel of (11.3) is killed by ${}^p\mathscr{H}^0 \circ \pi_\dagger$ and hence, in view of Lemma 5.3(3), that this morphism is surjective.

Let now $\widehat{\mathscr{K}}$ be the kernel of (11.3). To conclude the proof, it now suffices to prove that $\mathrm{Hom}_{\widehat{\mathscr{O}}}(\widehat{\mathscr{K}}, \widehat{\mathscr{T}}) = 0$ for any tilting object $\widehat{\mathscr{T}}$ in $\widehat{\mathscr{O}}$. For this, it suffices to prove that $\mathrm{Hom}_{\widehat{\mathscr{O}}}(\widehat{\mathscr{K}}, \widehat{\Delta}_w) = 0$ for any $w \in W$. And finally, by the description of morphisms as in (3.1) and since each local system $\mathscr{L}_{A,n}$ is an extension of copies of the trivial local system, for this, it suffices to prove that

$$\mathrm{Hom}_{\widehat{\mathscr{O}}}(\widehat{\mathscr{K}}, \pi^{\dagger}\Delta_w) = 0$$

for any $w \in W$.

By adjunction and right-exactness of $\pi_{\dagger}$ (see Corollary 5.8), we have

$$\mathrm{Hom}_{\widehat{\mathscr{O}}}(\widehat{\mathscr{K}}, \pi^{\dagger}\Delta_w) \cong \mathrm{Hom}_{D^{\mathrm{b}}_U(Y,\Bbbk)}(\pi_{\dagger}\widehat{\mathscr{K}}, \Delta_w) \cong \mathrm{Hom}_{\mathscr{O}}({}^p\mathscr{H}^0(\pi_{\dagger}\widehat{\mathscr{K}}), \Delta_w).$$

Now, the remarks above (and the observation that ${}^p\mathscr{H}^{-1}(\pi_{\dagger}\widehat{\mathscr{T}}) = 0$) show that ${}^p\mathscr{H}^0(\pi_{\dagger}\widehat{\mathscr{K}})$ is the kernel of the morphism (11.2) for $\mathscr{T} = \pi_{\dagger}(\widehat{\mathscr{T}})$. In particular, this object does not admit $\mathcal{IC}_e$ as a composition factor; by Lemma 6.7, this implies that $\mathrm{Hom}_{\mathscr{O}}({}^p\mathscr{H}^0(\pi_{\dagger}\widehat{\mathscr{K}}), \Delta_w) = 0$ and finishes the proof. □

We also observe the following consequence of Proposition 11.2, following [BBM].

Corollary 11.3 *For any projective perverse sheaves $\mathscr{P}$, $\mathscr{P}'$ in $\mathscr{O}$, the functor $\mathbb{V}$ induces an isomorphism*

$$\mathrm{Hom}_{\mathscr{O}}(\mathscr{P}, \mathscr{P}') \xrightarrow{\sim} \mathrm{Hom}_{R_T/(R_T)^W_+}\big(\mathbb{V}(\mathscr{P}), \mathbb{V}(\mathscr{P}')\big).$$

Proof It is well known that the functor

$$(-) \star^B \Delta_{w_0} : D^{\mathrm{b}}_U(Y,\Bbbk) \to D^{\mathrm{b}}_U(Y,\Bbbk)$$

is an equivalence of triangulated categories which restricts to an equivalence between tilting and projective objects in $\mathscr{O}$; see [BBM] or [AR2]. The inverse equivalence is the functor

$$(-) \star^B \nabla_{w_0} : D^{\mathrm{b}}_U(Y,\Bbbk) \to D^{\mathrm{b}}_U(Y,\Bbbk).$$

Therefore, we have

$$\mathbb{V}(\mathscr{P}) = \mathrm{Hom}(\mathscr{T}_{w_0}, \mathscr{P}) \cong \mathrm{Hom}(\mathscr{T}_{w_0} \star^B \nabla_{w_0}, \mathscr{P} \star^B \nabla_{w_0}) \cong \mathbb{V}(\mathscr{P} \star^B \nabla_{w_0})$$

since $\mathscr{T}_{w_0} \star^B \nabla_{w_0} \cong \mathscr{T}_{w_0}$; see (7.8). In other words, we have constructed an isomorphism between the restriction of $\mathbb{V}$ to the subcategory $\mathsf{Proj}(\mathscr{O})$ of projective objects in $\mathscr{O}$ and the composition

$$\mathsf{Proj}(\mathscr{O}) \xrightarrow[\sim]{(-)\star^B\nabla_{w_0}} \mathsf{Tilt}(\mathscr{O}) \xrightarrow{\mathbb{V}} \mathsf{Mof}(R_T/(R_T)_+^W),$$

where $\mathsf{Tilt}(\mathscr{O})$ is the category of tilting objects in $\mathscr{O}$. Hence, the desired claim follows from Proposition 11.2. □

11.2 Image of $\widehat{\mathscr{T}}_s$

Let us fix $s \in S$. Recall (see Remark 9.5) that since we have chosen a representative for $\widehat{\mathscr{T}}_{w_0}$, we have a canonical representative for $\widehat{\mathscr{T}}_s$. In the following lemma, we denote by $(R_T^\wedge)^s$ the s-invariants in $R_T^\wedge$.

Lemma 11.4 *There exists a canonical isomorphism*

$$R_T^\wedge \otimes_{(R_T^\wedge)^s} R_T^\wedge \xrightarrow{\sim} \widehat{\mathbb{V}}(\widehat{\mathscr{T}}_s).$$

Proof Recall that $\widehat{\mathscr{T}}_s = (\overline{J}_s)_* \overline{J}_s^* \widehat{\mathscr{T}}_{w_0}$; hence, by adjunction, we have

$$\widehat{\mathbb{V}}(\widehat{\mathscr{T}}_s) = \mathrm{Hom}(\widehat{\mathscr{T}}_{w_0}, \widehat{\mathscr{T}}_s) \cong \mathrm{End}(\widehat{\mathscr{T}}_s).$$

By Proposition 6.4 (applied to the Levi subgroup of G containing T associated with s), the morphism

$$R_T^\wedge \otimes_{\Bbbk} R_T^\wedge \to \mathrm{End}(\widehat{\mathscr{T}}_s)$$

induced by monodromy factors through a morphism $R_T^\wedge \otimes_{(R_T^\wedge)^s} R_T^\wedge \to \mathrm{End}(\widehat{\mathscr{T}}_s)$, and by Corollary 6.6, this morphism is surjective. Now, under our assumptions, $R_T^\wedge$ is free of rank 2 over $(R_T^\wedge)^s$. (In fact, if $\delta^\vee \in X_*(T)$ is a cocharacter such that $\langle \delta^\vee, \alpha_s \rangle = 1$, then $\{1, \delta^\vee\}$ is a basis of this module.) Hence, $R_T^\wedge \otimes_{(R_T^\wedge)^s} R_T^\wedge$ is free of rank 2 as an $R_T^\wedge$-module. Since $\mathrm{End}(\widehat{\mathscr{T}}_s)$ also has this property (see Lemma 5.9(2)), this morphism must be an isomorphism. □

11.3 Monoidal Structure: étale Setting

In this subsection, we consider the setting of Sect. 10. In this case, in view of Lemma 10.1, we have a canonical choice for the object $\widehat{\mathscr{T}}_{w_0}$; this is the choice we consider.

We will denote by

$$\widehat{T}_{\mathbf{U}}^{\mathrm{et}}(\mathbf{X} /\!\!/ \mathbf{T}, \Bbbk)$$

the category of tilting perverse sheaves in $\widehat{D}^{\mathrm{et}}_{\mathbf{U}}(\mathbf{X}/\!\!/\mathbf{T}, \Bbbk)$. By Remark 7.9, this subcategory is stable under the convolution product $\widehat{\star}$; moreover, it contains the unit object $\widehat{\Delta}_e$ (see Lemma 7.6); hence, it has a natural structure of monoidal category.

In the following proposition, we consider the monoidal structure on the category $\mathsf{Mof}(R^{\wedge}_{\mathbf{T}} \otimes_{(R^{\wedge}_{\mathbf{T}})^W} R^{\wedge}_{\mathbf{T}})$ given by $(M, N) \mapsto M \otimes_{R^{\wedge}_{\mathbf{T}}} N$, where the tensor product is defined with respect to the action of the second copy of $R^{\wedge}_{\mathbf{T}}$ on M and the first copy on N, and the action of $R^{\wedge}_{\mathbf{T}} \otimes_{(R^{\wedge}_{\mathbf{T}})^W} R^{\wedge}_{\mathbf{T}}$ on $M \otimes_{R^{\wedge}_{\mathbf{T}}} N$ is induced by the action of the first copy of $R^{\wedge}_{\mathbf{T}}$ on M and the second copy of $R^{\wedge}_{\mathbf{T}}$ on N.

Proposition 11.5 *The functor* $\widehat{\mathbb{V}} : \widehat{T}^{\mathrm{et}}_{\mathbf{U}}(\mathbf{X}/\!\!/\mathbf{T}, \Bbbk) \to \mathsf{Mof}(R^{\wedge}_{\mathbf{T}} \otimes_{(R^{\wedge}_{\mathbf{T}})^W} R^{\wedge}_{\mathbf{T}})$ *has a canonical monoidal structure.*

Proof Recall from Sect. 10.3 the category $\widehat{D}^{\mathrm{et}}_{\mathrm{Wh}}(\mathbf{X}/\!\!/\mathbf{T}, \Bbbk)$, the object $\widehat{\Delta}_\chi$, the equivalence

$$\Upsilon : D^{\mathrm{b}}\mathsf{Mod}^{\mathrm{fg}}(R^{\wedge}_{\mathbf{T}}) \xrightarrow{\sim} \widehat{D}^{\mathrm{et}}_{\mathrm{Wh}}(\mathbf{X}/\!\!/\mathbf{T}, \Bbbk)$$

from (10.1), and the functor $\mathsf{Av}_\chi : \widehat{D}^{\mathrm{et}}_{\mathbf{U}}(\mathbf{X}/\!\!/\mathbf{T}, \Bbbk) \to \widehat{D}^{\mathrm{et}}_{\mathrm{Wh}}(\mathbf{X}/\!\!/\mathbf{T}, \Bbbk)$. We also have a right action of the monoidal category $\widehat{D}^{\mathrm{et}}_{\mathbf{U}}(\mathbf{X}/\!\!/\mathbf{T}, \Bbbk)$ on $\widehat{D}^{\mathrm{et}}_{\mathrm{Wh}}(\mathbf{X}/\!\!/\mathbf{T}, \Bbbk)$, denoted again $\widehat{\star}$.

Let us denote by $\widehat{T}^{\mathrm{et}}_{\mathrm{Wh}}(\mathbf{X}/\!\!/\mathbf{T}, \Bbbk)$ the full subcategory of $\widehat{D}^{\mathrm{et}}_{\mathrm{Wh}}(\mathbf{X}/\!\!/\mathbf{T}, \Bbbk)$ whose objects are the direct sums of copies of $\widehat{\Delta}_\chi$ or equivalently the image under Υ of the category of free $R^{\wedge}_{\mathbf{T}}$-modules. We claim that for $\widehat{\mathscr{T}}$ in $\widehat{T}^{\mathrm{et}}_{\mathbf{U}}(\mathbf{X}/\!\!/\mathbf{T}, \Bbbk)$, the functor

$$(-)\,\widehat{\star}\,\widehat{\mathscr{T}} : \widehat{D}^{\mathrm{et}}_{\mathrm{Wh}}(\mathbf{X}/\!\!/\mathbf{T}, \Bbbk) \to \widehat{D}^{\mathrm{et}}_{\mathrm{Wh}}(\mathbf{X}/\!\!/\mathbf{T}, \Bbbk) \tag{11.4}$$

stabilizes the subcategory $\widehat{T}^{\mathrm{et}}_{\mathrm{Wh}}(\mathbf{X}/\!\!/\mathbf{T}, \Bbbk)$. In fact, to prove this, it suffices to prove that $\widehat{\Delta}_\chi \,\widehat{\star}\, \widehat{\mathscr{T}}$ belongs to $\widehat{T}^{\mathrm{et}}_{\mathrm{Wh}}(\mathbf{X}/\!\!/\mathbf{T}, \Bbbk)$. But we have $\widehat{\Delta}_\chi \,\widehat{\star}\, \widehat{\mathscr{T}} \cong \mathsf{Av}_\chi(\widehat{\mathscr{T}})$, and

$$\begin{aligned} \mathsf{H}^\bullet(\Upsilon^{-1}(\mathsf{Av}_\chi(\widehat{\mathscr{T}}))) &\cong \mathrm{Hom}^\bullet_{\widehat{D}^{\mathrm{et}}_{\mathrm{Wh}}(\mathbf{X}/\!\!/\mathbf{T}, \Bbbk)}(\widehat{\Delta}_\chi, \mathsf{Av}_\chi(\widehat{\mathscr{T}})) \\ &\cong \mathrm{Hom}^\bullet_{\widehat{D}^{\mathrm{et}}_{\mathbf{U}}(\mathbf{X}/\!\!/\mathbf{T}, \Bbbk)}(\mathsf{Av}^{\mathbf{U}}_!(\widehat{\Delta}_\chi), \widehat{\mathscr{T}}) \cong \mathrm{Hom}^\bullet_{\widehat{D}^{\mathrm{et}}_{\mathbf{U}}(\mathbf{X}/\!\!/\mathbf{T}, \Bbbk)}(\widehat{\mathscr{T}}_{w_0}, \widehat{\mathscr{T}}) \end{aligned}$$

where the second isomorphism uses adjunction, and the third one uses Lemma 10.1. Now, the right-hand side is concentrated in degree 0 and free over $R^{\wedge}_{\mathbf{T}}$ by Lemma 5.9(2). Hence, $\mathsf{Av}_\chi(\widehat{\mathscr{T}})$ is indeed a direct sum of copies of $\widehat{\Delta}_\chi$.

The claim we have just proved shows in particular that the functor (11.4) is right exact for the perverse t-structure. Hence, the functor

$$\mathsf{Mof}(R^{\wedge}_{\mathbf{T}}) \underset{\sim}{\xrightarrow{\Upsilon}} \widehat{P}^{\mathrm{et}}_{\mathrm{Wh}}(\mathbf{X}/\!\!/\mathbf{T}, \Bbbk) \xrightarrow{{}^p\!\mathscr{H}^0(-\,\widehat{\star}\,\widehat{\mathscr{T}})} \widehat{P}^{\mathrm{et}}_{\mathrm{Wh}}(\mathbf{X}/\!\!/\mathbf{T}, \Bbbk) \underset{\sim}{\xrightarrow{\Upsilon^{-1}}} \mathsf{Mof}(R^{\wedge}_{\mathbf{T}})$$

is right exact and therefore representable by the $R_{\mathbf{T}}^{\wedge}$-bimodule

$$\Upsilon^{-1}\left({}^{p}\mathscr{H}^{0}\left(\Upsilon(R_{\mathbf{T}}^{\wedge})\widehat{\star}\widehat{\mathscr{T}}\right)\right)=\widehat{\mathbb{V}}(\widehat{\mathscr{T}}).$$

In the case $\widehat{\mathscr{T}}=\widehat{\Delta}_e$, since the functor $(-)\widehat{\star}\widehat{\Delta}_e$ is canonically isomorphic to the identity functor, we must have a canonical isomorphism $\widehat{\mathbb{V}}(\widehat{\Delta}_e)\cong R_{\mathbf{T}}^{\wedge}$ (which can of course also been seen directly). And if $\widehat{\mathscr{T}},\widehat{\mathscr{T}}'$ belong to $\widehat{T}_{\mathbf{U}}^{\mathrm{et}}(\mathbf{X}/\!\!/\mathbf{T},\Bbbk)$, since the functor constructed as above from $\widehat{\mathscr{T}}\widehat{\star}\widehat{\mathscr{T}}'$ is canonically isomorphic to the composition of the functors associated with $\widehat{\mathscr{T}}$and with $\widehat{\mathscr{T}}'$, respectively, we obtain a canonical isomorphism

$$\widehat{\mathbb{V}}(\widehat{\mathscr{T}}\widehat{\star}\widehat{\mathscr{T}}')\cong\widehat{\mathbb{V}}(\widehat{\mathscr{T}})\otimes_{R_{\mathbf{T}}^{\wedge}}\widehat{\mathbb{V}}(\widehat{\mathscr{T}}').$$

It is easy to check that these isomorphisms are compatible with the associativity and unit constraints and hence define a monoidal structure on $\widehat{\mathbb{V}}$. □

11.4 Monoidal Structure: Classical Setting

In this subsection, we consider the "classical" setting of Sects. 6–9. Here, we do not have (at present) a counterpart of the Whittaker category, but an analogue of Proposition 11.5 can be obtained from general principles. For this, we have to assume that $\Bbbk$ contains a primitive p-th root of unity for some prime number $p\neq\ell$; we fix a choice of p and of a primitive root.

Proposition 11.6 *There exists a choice of object $\widehat{\mathscr{T}}_{w_0}$ such that the functor $\widehat{\mathbb{V}}$: $\widehat{T}_U(X/\!\!/T,\Bbbk)\to\mathsf{Mof}(R_T^{\wedge}\otimes_{(R_T^{\wedge})^W}R_T^{\wedge})$ admits a monoidal structure.*

Proof We follow the procedure of [BBD, §6.1] to deduce the result in the classical topology (over $\mathbb{C}$) from its étale counterpart (over an algebraically closed field of characteristic p).

Let $\mathbf{G}_{\mathbb{Z}}$ be split connected reductive group over $\mathbb{Z}$ such that $\mathrm{Spec}(\mathbb{C})\times_{\mathrm{Spec}(\mathbb{Z})}\mathbf{G}_{\mathbb{Z}}$ is isomorphic to G, and let $\mathbf{B}_{\mathbb{Z}}$ be a Borel subgroup of $\mathbf{G}_{\mathbb{Z}}$ and $\mathbf{T}_{\mathbb{Z}}\subset\mathbf{B}_{\mathbb{Z}}$ be a (split) maximal torus; then we can assume that $B=\mathrm{Spec}(\mathbb{C})\times_{\mathrm{Spec}(\mathbb{Z})}\mathbf{B}_{\mathbb{Z}}$ and $T=\mathrm{Spec}(\mathbb{C})\times_{\mathrm{Spec}(\mathbb{Z})}\mathbf{T}_{\mathbb{Z}}$. Let also $\mathbf{U}_{\mathbb{Z}}$ be the unipotent radical of $\mathbf{B}_{\mathbb{Z}}$ so that $U=\mathrm{Spec}(\mathbb{C})\times_{\mathrm{Spec}(\mathbb{Z})}\mathbf{U}_{\mathbb{Z}}$; then we can set $\mathbf{X}_{\mathbb{Z}}:=\mathbf{G}_{\mathbb{Z}}/\mathbf{U}_{\mathbb{Z}}$, $\mathbf{Y}_{\mathbb{Z}}:=\mathbf{G}_{\mathbb{Z}}/\mathbf{B}_{\mathbb{Z}}$, which provides versions of X and Y over $\mathbb{Z}$. We set $\mathbf{X}_{\mathbb{C}}:=\mathrm{Spec}(\mathbb{C})\times_{\mathrm{Spec}(\mathbb{Z})}\mathbf{X}_{\mathbb{Z}}$, $\mathbf{Y}_{\mathbb{C}}:=\mathrm{Spec}(\mathbb{C})\times_{\mathrm{Spec}(\mathbb{Z})}\mathbf{Y}_{\mathbb{Z}}$; of course, these varieties coincide with X and Y, but we change notation to emphasize the fact that we now consider them as schemes (with the Zariski topology) rather than topological spaces (with the classical topology). If $\mathbf{U}_{\mathbb{C}}=\mathrm{Spec}(\mathbb{C})\times_{\mathrm{Spec}(\mathbb{Z})}\mathbf{U}_{\mathbb{Z}}$ and $\mathbf{T}_{\mathbb{C}}=\mathrm{Spec}(\mathbb{C})\times_{\mathrm{Spec}(\mathbb{Z})}\mathbf{T}_{\mathbb{Z}}$, we can consider the categories $D^{\mathrm{b,et}}_{\mathbf{U}_{\mathbb{C}}}(\mathbf{Y}_{\mathbb{C}},\Bbbk)$ and $D^{\mathrm{b,et}}_{\mathbf{U}_{\mathbb{C}}}(\mathbf{X}_{\mathbb{C}}/\!\!/\mathbf{T}_{\mathbb{C}},\Bbbk)$ defined using étale sheaves (but now over a complex scheme) as in Sect. 10. The general results recalled in [BBD, §6.1.2] provide canonical equivalences of categories

$$D^{\mathrm{b,et}}_{\mathbf{U}_{\mathbb{C}}}(\mathbf{Y}_{\mathbb{C}}, \Bbbk) \cong D^{\mathrm{b}}_{U}(Y, \Bbbk), \qquad D^{\mathrm{b,et}}_{\mathbf{U}_{\mathbb{C}}}(\mathbf{X}_{\mathbb{C}} /\!\!/ \mathbf{T}_{\mathbb{C}}, \Bbbk) \cong D^{\mathrm{b}}_{U}(X /\!\!/ T, \Bbbk)$$

which commute (in the obvious sense) with pullback and pushforward functors.

Now, choose an algebraically closed field $\mathbb{F}$ whose characteristic is p, and a strictly Henselian discrete valuation ring $\mathfrak{R} \subset \mathbb{C}$ whose residue field is $\mathbb{F}$. Then we can consider the base changes of $\mathbf{G}_{\mathbb{Z}}$, $\mathbf{B}_{\mathbb{Z}}$, etc. to $\mathfrak{R}$ or $\mathbb{F}$, which we will denote by the same letter with a subscript $\mathfrak{R}$ or $\mathbb{F}$. We can consider the versions of the categories considered above for $\mathbf{X}_{\mathfrak{R}}$ and $\mathbf{Y}_{\mathfrak{R}}$ instead of $\mathbf{X}_{\mathbb{C}}$ and $\mathbf{Y}_{\mathbb{C}}$; the results explained in [BBD, §§6.1.8–6.1.9] (see also [Mi, Corollary VI.4.20 and Remark VI.4.21]) guarantee that pullback along the natural morphisms

$$\mathbf{Y}_{\mathbb{C}} \longrightarrow \mathbf{Y}_{\mathfrak{R}} \longleftarrow \mathbf{Y}_{\mathbb{F}} \quad \text{and} \quad \mathbf{X}_{\mathbb{C}} \longrightarrow \mathbf{X}_{\mathfrak{R}} \longleftarrow \mathbf{X}_{\mathbb{F}}$$

induces equivalences of triangulated categories

$$D^{\mathrm{b,et}}_{\mathbf{U}_{\mathbb{C}}}(\mathbf{Y}_{\mathbb{C}}, \Bbbk) \xleftarrow{\sim} D^{\mathrm{b,et}}_{\mathbf{U}_{\mathfrak{R}}}(\mathbf{Y}_{\mathfrak{R}}, \Bbbk) \xrightarrow{\sim} D^{\mathrm{b,et}}_{\mathbf{U}_{\mathbb{F}}}(\mathbf{Y}_{\mathbb{F}}, \Bbbk)$$

and

$$D^{\mathrm{b,et}}_{\mathbf{U}_{\mathbb{C}}}(\mathbf{X}_{\mathbb{C}} /\!\!/ \mathbf{T}_{\mathbb{C}}, \Bbbk) \xleftarrow{\sim} D^{\mathrm{b,et}}_{\mathbf{U}_{\mathfrak{R}}}(\mathbf{X}_{\mathfrak{R}} /\!\!/ \mathbf{T}_{\mathfrak{R}}, \Bbbk) \xrightarrow{\sim} D^{\mathrm{b,et}}_{\mathbf{U}_{\mathbb{F}}}(\mathbf{X}_{\mathbb{F}} /\!\!/ \mathbf{T}_{\mathbb{F}}, \Bbbk).$$

Combining these two constructions, we obtain an equivalence of categories

$$\widehat{T}_U(X /\!\!/ T, \Bbbk) \xrightarrow{\sim} \widehat{T}^{\mathrm{et}}_{\mathbf{U}_{\mathbb{F}}}(\mathbf{X}_{\mathbb{F}} /\!\!/ \mathbf{T}_{\mathbb{F}}, \Bbbk) \tag{11.5}$$

which is easily seen to be monoidal. Let us denote by $\widehat{\mathscr{T}}^{\mathrm{et}}_{w_0}$ the object of the category $\widehat{T}^{\mathrm{et}}_{\mathbf{U}_{\mathbb{F}}}(\mathbf{X}_{\mathbb{F}} /\!\!/ \mathbf{T}_{\mathbb{F}}, \Bbbk)$ considered in Sect. 11.3; then Proposition 11.5 provides us with a coalgebra structure on $\widehat{\mathscr{T}}^{\mathrm{et}}_{w_0}$ (in the monoidal category $(\widehat{T}^{\mathrm{et}}_{\mathbf{U}_{\mathbb{F}}}(\mathbf{X}_{\mathbb{F}} /\!\!/ \mathbf{T}_{\mathbb{F}}, \Bbbk), \widehat{\star})$). If we choose the object $\widehat{\mathscr{T}}_{w_0}$ as the inverse image of $\widehat{\mathscr{T}}^{\mathrm{et}}_{w_0}$ under (11.5), then the coalgebra structure on $\widehat{\mathscr{T}}^{\mathrm{et}}_{w_0}$ induces a coalgebra structure on $\widehat{\mathscr{T}}_{w_0}$. Given such a structure, it is not difficult (see, e.g., [BY, Proposition 4.6.4 and its proof]) to construct a monoidal structure on the associated functor $\widehat{\mathbb{V}}$. □

Remark 11.7 One can obtain a result weaker than Proposition 11.6 without using the comparison with étale sheaves. Namely, choose an identification $(i_e)_* i_e^* \widehat{\mathscr{T}}_{w_0} \cong \widehat{\Delta}_e$. Then by adjunction, we deduce a morphism $\xi : \widehat{\mathscr{T}}_{w_0} \to \widehat{\Delta}_e$, which itself induces a morphism

$$\xi \,\widehat{\star}\, \xi : \widehat{\mathscr{T}}_{w_0} \,\widehat{\star}\, \widehat{\mathscr{T}}_{w_0} \to \widehat{\Delta}_e \,\widehat{\star}\, \widehat{\Delta}_e = \widehat{\Delta}_e.$$

One can show (following, e.g., the ideas in [BY, Proof of Proposition 4.6.4]) that there exists a morphism $\eta : \widehat{\mathscr{T}}_{w_0} \to \widehat{\mathscr{T}}_{w_0} \mathbin{\widehat{\star}} \widehat{\mathscr{T}}_{w_0}$ which makes the diagram

$$\begin{array}{ccc} \widehat{\mathscr{T}}_{w_0} & \xrightarrow{\eta} & \widehat{\mathscr{T}}_{w_0} \mathbin{\widehat{\star}} \widehat{\mathscr{T}}_{w_0} \\ & {\scriptstyle\xi}\searrow \quad \swarrow{\scriptstyle \xi \widehat{\star} \xi} & \\ & \widehat{\Delta}_e & \end{array}$$

commutative and that moreover for any such η, the morphism of bifunctors

$$\widehat{\mathbb{V}}(-) \otimes_{R_T^\wedge} \widehat{\mathbb{V}}(-) \to \widehat{\mathbb{V}}(- \mathbin{\widehat{\star}} -)$$

sending $f \otimes g$ to $(f \mathbin{\widehat{\star}} g) \circ \eta$ is an isomorphism of functors. However, to make sure that this isomorphism induces a monoidal structure, we would have to choose η such that $(\eta \mathbin{\widehat{\star}} \mathrm{id}) \circ \eta = (\mathrm{id} \mathbin{\widehat{\star}} \eta) \circ \eta$. We do not know how to make such a choice.

11.5 Soergel Theory

In this subsection, we work either in the classical or in the étale setting (but use the notation from Sects. 6–9).

With Proposition 11.2, Lemma 11.4, and Proposition 11.5 (or Proposition 11.6) at hand, one can obtain a very explicit description of the categories $\widehat{T}_U(X /\!\!/ T, \Bbbk)$ and $\mathsf{Tilt}(\mathscr{O})$, as follows.

Theorem 11.8

(1) *The functor $\widehat{\mathbb{V}}$ induces an equivalence of monoidal categories between $\widehat{T}_U(X /\!\!/ T, \Bbbk)$ and the full subcategory $\mathsf{SMof}(R_T^\wedge \otimes_{(R_T^\wedge)^W} R_T^\wedge)$ of $\mathsf{Mof}(R_T^\wedge \otimes_{(R_T^\wedge)^W} R_T^\wedge)$ generated under direct sums, direct summands, and tensor products, by the objects $R_T^\wedge$ and $R_T^\wedge \otimes_{(R_T^\wedge)^s} R_T^\wedge$ with $s \in S$.*

(2) *The functor $\mathbb{V}$ induces an equivalence of categories between $\mathsf{Tilt}(\mathscr{O})$ and the full subcategory $\mathsf{SMof}(R_T^\wedge)$ of $\mathsf{Mof}(R_T^\wedge)$ generated under direct sums, direct summands, and application of functors $R_T^\wedge \otimes_{(R_T^\wedge)^s} -$ (with $s \in S$) by the trivial module $\Bbbk$.*

(3) *These equivalences are compatible in the sense that the diagram*

$$\begin{array}{ccc} \widehat{T}_U(X /\!\!/ T, \Bbbk) & \xrightarrow[\sim]{\widehat{\mathbb{V}}} & \mathsf{SMof}(R_T^\wedge \otimes_{(R_T^\wedge)^W} R_T^\wedge) \\ {\scriptstyle \pi_\dagger}\downarrow & & \downarrow{\scriptstyle -\otimes_{R_T^\wedge} \Bbbk} \\ \mathsf{Tilt}(\mathscr{O}) & \xrightarrow[\sim]{\mathbb{V}} & \mathsf{SMof}(R_T^\wedge) \end{array}$$

commutes (up to canonical isomorphism) and that the convolution action of $\widehat{T}_U(X /\!\!/ T, \Bbbk)$ *on* $\mathsf{Tilt}(\mathscr{O})$ *identifies with the action induced by the action of* $\mathsf{Mof}(R_T^\wedge \otimes_{(R_T^\wedge)^W} R_T^\wedge)$ *on* $\mathsf{Mof}(R_T^\wedge)$ *by tensor product over* $R_T^\wedge$.

Proof The theorem follows from the results quoted above and Remark 7.9. □

One can also state similar results for triangulated categories.

Theorem 11.9 *There exist canonical equivalences of monoidal triangulated categories*

$$K^{\mathrm{b}}\mathsf{SMof}(R_T^\wedge \otimes_{(R_T^\wedge)^W} R_T^\wedge) \xrightarrow{\sim} \widehat{D}_U(X /\!\!/ T, \Bbbk),$$

$$K^{\mathrm{b}}\mathsf{SMof}(R_T^\wedge) \xrightarrow{\sim} D_U^{\mathrm{b}}(Y, \Bbbk).$$

These equivalences are compatible in a sense similar to that in Theorem 11.8.

Proof The first equivalence follows from Proposition 5.11 and Theorem 11.8(1). (The fact that the equivalence of Proposition 5.11 is monoidal in our setting follows from standard arguments; see [Be, Lemma A.7.1] or [AMRW, Proposition 2.3].) The second equivalence and their compatibilities follow from similar arguments. □

Remark 11.10

(1) Using Remark 5.1, from the description of the category $\widehat{D}_U(X /\!\!/ T, \Bbbk)$ given in Theorem 11.9, one can deduce a description of the category $D_U^{\mathrm{b}}(X /\!\!/ T, \Bbbk)$ in algebraic terms.
(2) Using Theorem 11.8 and the known structure of the additive categories $\widehat{T}_U(X /\!\!/ T, \Bbbk)$ and $\mathsf{Tilt}(\mathscr{O})$, one obtains some sort of "multiplicative variant" of the theory of Soergel modules and bimodules (see [So4]) in our present setting. It might be interesting to understand if such a theory can be developed algebraically and in bigger generality.

Finally, following [BBM], from our results, we deduce the following description of the category $\mathscr{O}$. Here, for $\underline{w} = (s_1, \cdots, s_r)$, a sequence of elements of S, we set

$$\mathsf{B}(\underline{w}) = R_T^\wedge \otimes_{(R_T^\wedge)^{s_1}} \cdots \otimes_{(R_T^\wedge)^{s_{r-1}}} R_T^\wedge \otimes_{(R_T^\wedge)^{s_r}} \Bbbk.$$

Theorem 11.11 *Choose, for any* $w \in W$, *a reduced expression* $\underline{w}$ *for* w. *Then there exists an equivalence of categories between* $\mathscr{O}$ *and the category* $\mathsf{Mof}(A)$, *where*

$$A = \left(\mathrm{End}_{R_T^\wedge}\Bigl(\bigoplus_{w \in W} \mathsf{B}(\underline{w}) \Bigr) \right)^{\mathrm{op}}.$$

Proof For $\underline{v} = (s_1, \cdots, s_r)$, a sequence of elements of S, we set

$$\mathscr{T}(\underline{v}) = \widehat{\mathscr{T}}_{s_1} \mathbin{\widehat{\star}} \cdots \mathbin{\widehat{\star}} \widehat{\mathscr{T}}_{s_r} \mathbin{\widehat{\star}} \Delta_e.$$

Then by Corollary 11.3 and its proof, the object

$$\mathscr{P} := \bigoplus_{\underline{w} \in W} \mathscr{T}(\underline{w}) \star^B \Delta_{w_0}$$

is a projective generator of $\mathscr{O}$, and we have $\mathrm{End}(\mathscr{P}) \cong A^{\mathrm{op}}$. Then the claim follows from general and well-known result; see, for example, [Ba, Exercise on p. 55]. □

12 Erratum to [AB]

In this section, we use the above results to correct an error found in the proof of [AB, Lemma 5][5] and generalize that statement to arbitrary coefficients. The new proof below follows the strategy suggested in [AB, Remark 3]. The statement of [AB, Lemma 5] involves an *affine* flag variety, but it readily reduces to Lemma 12.1 below restricted to the special case of characteristic zero coefficients.

As in Sect. 10, we consider a connected reductive algebraic group $\mathbf{G}$ over an algebraically closed field $\mathbb{F}$ of characteristic $p \neq \ell$ and choose a Borel subgroup $\mathbf{B} \subset \mathbf{G}$ and a maximal torus $\mathbf{T} \subset \mathbf{B}$. Fixing the same data as in Sect. 10.3, we can consider the standard perverse sheaf $\Delta_\chi := \mathsf{Av}_\chi(\Delta_e)$. (Note that the natural morphism $\Delta_\chi \to \nabla_\chi := \mathsf{Av}_\chi(\nabla_e)$ is an isomorphism.) In Sect. 10.3, we have considered the averaging functors $\mathsf{Av}_!^{\mathbf{U}}$ and $\mathsf{Av}_*^{\mathbf{U}}$. We can similarly define the functors

$$\mathsf{Av}_!^{\mathbf{B}} := (a_{\mathbf{B}})_!\big(\underline{\Bbbk}_{\mathbf{B}} \boxtimes -\big)[\dim \mathbf{B}], \quad \mathsf{Av}_*^{\mathbf{B}} := (a_{\mathbf{B}})_*\big(\underline{\Bbbk}_{\mathbf{B}} \boxtimes -\big)[\dim \mathbf{B}],$$

from $D^{\mathrm{b,et}}_{\mathrm{Wh}}(\mathbf{Y}, \Bbbk)$ to the $\mathbf{B}$-equivariant derived category $D^{\mathrm{b,et}}_{\mathbf{B}}(\mathbf{Y}, \Bbbk)$, where $a_{\mathbf{B}} : \mathbf{B} \times \mathbf{Y} \to \mathbf{Y}$ is the action morphism.

In the following lemma, we denote by $\Phi \subset X^*(\mathbf{T})$ the root system of $(\mathbf{G}, \mathbf{T})$, and by $\mathbb{Z}\Phi$ the lattice generated by Φ.

Lemma 12.1 *The* $\mathbf{B}$*-equivariant complex* $\mathsf{Av}_*^{\mathbf{B}}(\Delta_\chi)$ *is concentrated in perverse degrees* $\geq -\dim(\mathbf{T})$. *Moreover, if* $X^*(\mathbf{T})/\mathbb{Z}\Phi$ *has no torsion, then we have*

$${}^p\mathscr{H}^{-\dim \mathbf{T}}\big(\mathsf{Av}_*^{\mathbf{B}}(\Delta_\chi)\big) \cong \Delta_{w_0}.$$

Proof Using Verdier duality, this statement is equivalent to the fact that $\mathsf{Av}_!^{\mathbf{B}}(\Delta_\chi)$ is concentrated in perverse degrees $\leq \dim(\mathbf{T})$ and that if $X^*(\mathbf{T})/\mathbb{Z}\Phi$ has no torsion,

[5] Namely, it is claimed in this proof that the complex denoted "C" is concentrated in positive perverse degrees. But the arguments given there only imply that its *negative* perverse cohomology objects vanish.

then we have ${}^p\mathscr{H}^{\dim \mathbf{T}}\big(\mathsf{Av}_!^{\mathbf{B}}(\Delta_\chi)\big) \cong \nabla_{w_0}$. This is the statement we will actually prove.

We have

$$\mathsf{Av}_!^{\mathbf{B}} \cong {}^!\mathrm{Ind}_{\mathbf{U}}^{\mathbf{B}} \circ \mathsf{Av}_!^{\mathbf{U}},$$

where ${}^!\mathrm{Ind}_{\mathbf{U}}^{\mathbf{B}} : D^{\mathrm{b,et}}_{\mathbf{U}}(\mathbf{Y}, \Bbbk) \to D^{\mathrm{b,et}}_{\mathbf{B}}(\mathbf{Y}, \Bbbk)$ is the functor sending $\mathscr{F}$ to

$$(a'_{\mathbf{B}})_!(\underline{\Bbbk}_{\mathbf{B}/\mathbf{U}} \mathbin{\widetilde{\boxtimes}} \mathscr{F})[\dim(\mathbf{B}/\mathbf{U})].$$

(Here, $a'_{\mathbf{B}} : \mathbf{B} \times^{\mathbf{U}} \mathbf{Y} \to \mathbf{Y}$ is the natural map, and $\widetilde{\boxtimes}$ is the twisted external product.) Using Lemma 10.1, we deduce that

$$\mathsf{Av}_!^{\mathbf{B}}(\Delta_\chi) \cong {}^!\mathrm{Ind}_{\mathbf{U}}^{\mathbf{B}}(\mathscr{T}_{w_0}).$$

It is clear that for any $\mathbf{B}$-equivariant perverse sheaf $\mathscr{F}$ on $\mathbf{Y}$, the complex ${}^!\mathrm{Ind}_{\mathbf{U}}^{\mathbf{B}}(\mathscr{F})$ is concentrated in perverse degrees between 0 and $\dim(\mathbf{T})$. Hence, the same claim holds for any extension of such perverse sheaves, that is, for any $\mathbf{U}$-equivariant perverse sheaf; thus, the first claim of the lemma is proved. Now, the functor ${}^!\mathrm{Ind}_{\mathbf{U}}^{\mathbf{B}}$ is left adjoint to $\mathrm{For}_{\mathbf{U}}^{\mathbf{B}}[\dim(\mathbf{B}/\mathbf{U})]$, where $\mathrm{For}_{\mathbf{U}}^{\mathbf{B}} : D^{\mathrm{b,et}}_{\mathbf{B}}(\mathbf{Y}, \Bbbk) \to D^{\mathrm{b,et}}_{\mathbf{U}}(\mathbf{Y}, \Bbbk)$ is the forgetful functor. Using this fact, it is not difficult to check that for any $\mathbf{U}$-equivariant perverse sheaf $\mathscr{F}$ on $\mathbf{Y}$, the perverse sheaf

$${}^p\mathscr{H}^{\dim \mathbf{T}}\big({}^!\mathrm{Ind}_{\mathbf{U}}^{\mathbf{B}}(\mathscr{F})\big)$$

is characterized as the largest $\mathbf{B}$-equivariant quotient of $\mathscr{F}$.

To conclude the proof, it remains to prove that if $X^*(\mathbf{T})/\mathbb{Z}\Phi$ has no torsion, then ∇_{w_0} is the largest $\mathbf{B}$-equivariant quotient of $\mathscr{T}_{w_0}$. Now, $\mathscr{T}_{w_0}$ has a costandard filtration, whose last term is ∇_{w_0}; therefore, there exists a surjection $\mathscr{T}_{w_0} \twoheadrightarrow \nabla_{w_0}$ (which is unique up to scalar). Since ∇_{w_0} is $\mathbf{B}$-equivariant, we deduce that this map factors as a composition

$$\mathscr{T}_{w_0} \twoheadrightarrow {}^p\mathscr{H}^{\dim \mathbf{T}}\big({}^!\mathrm{Ind}_{\mathbf{U}}^{\mathbf{B}}(\mathscr{T}_{w_0})\big) \twoheadrightarrow \nabla_{w_0}.$$

The kernel of the second map here is the image of the kernel of our surjection $\mathscr{T}_{w_0} \twoheadrightarrow \nabla_{w_0}$. Since the latter admits a costandard filtration, in view of Lemma 6.8, if the former is nonzero, then it admits $\mathcal{IC}_e$ as a composition factor; in other words, the vector space

$$\mathrm{Hom}\big(\mathscr{T}_{w_0}, {}^p\mathscr{H}^{\dim \mathbf{T}}({}^!\mathrm{Ind}_{\mathbf{U}}^{\mathbf{B}}(\mathscr{T}_{w_0}))\big)$$

has dimension at least 2.

On the other hand, we have a surjection

$$\mathrm{Hom}(\mathscr{T}_{w_0}, \mathscr{T}_{w_0}) \twoheadrightarrow \mathrm{Hom}\big(\mathscr{T}_{w_0}, {}^p\mathscr{H}^{\dim \mathbf{T}}({}^!\mathrm{Ind}_{\mathbf{U}}^{\mathbf{B}}(\mathscr{T}_{w_0}))\big).$$

Our assumption on $\mathbf{G}$ means that the quotient morphism $\mathbf{G} \twoheadrightarrow \mathbf{G}/Z(\mathbf{G})$ (where $Z(\mathbf{G})$ is the center of $\mathbf{G}$) induces a surjection $X_*(\mathbf{T}) \twoheadrightarrow X_*(\mathbf{T}/Z(\mathbf{G}))$. Applying Corollary 9.2 to $\mathbf{G}/Z(\mathbf{G})$, we obtain that monodromy induces a surjection

$$R_{\mathbf{T}} \twoheadrightarrow \mathrm{Hom}(\mathscr{T}_{w_0}, \mathscr{T}_{w_0}).$$

Since ${}^p\mathscr{H}^{\dim \mathbf{T}}({}^!\mathrm{Ind}_{\mathbf{U}}^{\mathbf{B}}(\mathscr{T}_{w_0}))$ is $\mathbf{B}$-equivariant, the composition

$$R_{\mathbf{T}} \twoheadrightarrow \mathrm{Hom}(\mathscr{T}_{w_0}, \mathscr{T}_{w_0}) \twoheadrightarrow \mathrm{Hom}\big(\mathscr{T}_{w_0}, {}^p\mathscr{H}^{\dim \mathbf{T}}({}^!\mathrm{Ind}_{\mathbf{U}}^{\mathbf{B}}(\mathscr{T}_{w_0}))\big)$$

factors through $\varepsilon_{\mathbf{T}}$, proving that the rightmost term has dimension at most 1. This condition prevents the kernel of the surjection ${}^p\mathscr{H}^{\dim \mathbf{T}}\big({}^!\mathrm{Ind}_{\mathbf{U}}^{\mathbf{B}}(\mathscr{T}_{w_0})\big) \twoheadrightarrow \nabla_{w_0}$ to be nonzero, which concludes the proof. □

Remark 12.2

(1) Using the remarks in Sect. 1.7, one can show that another setting in which the second claim in Lemma 12.1 holds is when ℓ is very good for $\mathbf{G}$ hence, in particular, when $\ell = 0$. (Note that under this assumption, $X^*(\mathbf{T})/\mathbb{Z}\Phi$ has no ℓ-torsion; see [He, §2.10].)
(2) Replacing the proof of [AB, Lemma 5] by the proof given above, one can check that all the results of [AB, §2] (hence, in particular, [AB, Proposition 2]) extend in a straightforward way to positive-characteristic coefficients.

Acknowledgments Part of the work on this paper was done while the second author was a member of the Freiburg Institute for Advanced Studies, as part of the Research Focus "Cohomology in Algebraic Geometry and Representation Theory" led by A. Huber–Klawitter, S. Kebekus and W. Soergel.

We thank Geordie Williamson for useful discussions on the subject of this paper (in particular for the suggestion to compare with the K-theory of the flag variety), Pramod Achar for useful comments, and Valentin Gouttard for spotting several typos and minor errors.

References

[AMRW] P. Achar, S. Makisumi, S. Riche, and G. Williamson, *Koszul duality for Kac-Moody groups and characters of tilting modules*, J. Amer. Math. Soc. **32** (2019), 261–310.

[AR1] P. Achar and S. Riche, *Koszul duality and semisimplicity of Frobenius*, Ann. Inst. Fourier **63** (2013), 1511–1612.

[AR2] P. Achar and S. Riche, *Modular perverse sheaves on flag varieties I: tilting and parity sheaves*, with a joint appendix with G. Williamson, Ann. Sci. Éc. Norm. Supér. **49** (2016), 325–370.

[AR3] P. Achar and S. Riche, *Modular perverse sheaves on flag varieties II: Koszul duality and formality*, Duke Math. J. **165** (2016), 161–215.

[AJS] H. H. Andersen, J. C. Jantzen, and W. Soergel, *Representations of quantum groups at a p-th root of unity and of semisimple groups in characteristic p: independence of p*, Astérisque **220** (1994), 1–321.

[AB] S. Arkhipov and R. Bezrukavnikov, *Perverse sheaves on affine flags and Langlands dual group*, with an appendix by R. Bezrukavnikov and I Mirković, Israel J. Math **170** (2009), 135–183.

[ABG] S. Arkhipov, R. Bezrukavnikov, and V. Ginzburg, *Quantum groups, the loop Grassmannian, and the Springer resolution*, J. Amer. Math. Soc. **17** (2004), 595–678.

[Ba] H. Bass, *Algebraic K-theory*, W. A. Benjamin, 1968.

[BR] P. Baumann and S. Riche, *Notes on the geometric Satake equivalence*, in *Relative Aspects in Representation Theory, Langlands Functoriality and Automorphic Forms, CIRM Jean-Morlet Chair, Spring 2016* (V. Heiermann, D. Prasad, Eds.), 1–134, Lecture Notes in Math. 2221, Springer, 2018.

[Be] A. Beĭlinson, *On the derived category of perverse sheaves*, in *K-theory, arithmetic and geometry (Moscow, 1984–1986)*, 27–41, Lecture Notes in Math. 1289, Springer-Verlag, 1987.

[BB] A. Beĭlinson and J. Bernstein, *Localisation de $\mathfrak{g}$-modules*, C. R. Acad. Sci. Paris Sér. I Math. **292** (1981), 15–18.

[BBD] A. Beĭlinson, J. Bernstein, and P. Deligne, *Faisceaux pervers*, in *Analyse et topologie sur les espaces singuliers, I (Luminy, 1981)*, Astérisque **100** (1982), 5–171.

[BBM] A. Beĭlinson, R. Bezrukavnikov, and I. Mirković, *Tilting exercises*, Mosc. Math. J. **4** (2004), 547–557, 782.

[BG] A. Beĭlinson and V. Ginzburg, *Wall-crossing functors and $\mathcal{D}$-modules*, Represent. Theory **3** (1999), 1–31.

[BGS] A. Beĭlinson, V. Ginzburg, and W. Soergel, *Koszul duality patterns in representation theory*, J. Amer. Math. Soc. **9** (1996), 473–527.

[BY] R. Bezrukavnikov and Z. Yun, *On Koszul duality for Kac–Moody groups*, Represent. Theory **17** (2013), 1–98.

[CYZ] X. W. Chen, Y. Ye, and P. Zhang, *Algebras of derived dimension zero*, Comm. Algebra **36** (2008), 1–10.

[He] S. Herpel, *On the smoothness of centralizers in reductive groups*, Trans. Amer. Math. Soc. **365** (2013), 3753–3774.

[KS1] M. Kashiwara and P. Schapira, *Sheaves on manifolds*, Grundlehren der Mathematischen Wissenschaften 292, Springer, 1990.

[KS2] M. Kashiwara and P. Schapira, *Categories and sheaves*, Grundlehren der Mathematischen Wissenschaften 332, Springer, 2006.

[KL] D. Kazhdan and G. Lusztig, *Representations of Coxeter groups and Hecke algebras*, Invent. Math. **53** (1979), 165–184.

[KK] B. Kostant and S. Kumar, *T-equivariant K-theory of generalized flag varieties*, J. Differential Geom. **32** (1990), 549–603.

[La] T. Y. Lam, *A First course in noncommutative rings*, second edition, Graduate Texts in Mathematics 131, Springer, 2001.

[LC] J. Le and X.-W. Chen, *Karoubianness of a triangulated category*, J. Algebra **310** (2007), 452–457.

[Lu] G. Lusztig, *Cuspidal local systems and graded Hecke algebras. I.* Inst. Hautes Études Sci. Publ. Math. **67** (1988), 145–202.

[Mi] J. S. Milne, *Étale cohomology*, Princeton Mathematical Series 33, Princeton University Press, 1980.

[Or] D. Orlov, *Formal completions and idempotent completions of triangulated categories of singularities*, Adv. in Math. **226** (2011), 206–217.

[RSW] S. Riche, W. Soergel, and G. Williamson, *Modular Koszul duality*, Compos. Math. **150** (2014), 273–332.

[So1] W. Soergel, *Équivalences de certaines catégories de* $\mathfrak{g}$*-modules*, C. R. Acad. Sci. Paris Sér. I Math. **303** (1986), 725–728.

[So2] W. Soergel, *Kategorie* $\mathcal{O}$*, perverse Garben und Moduln über den Koinvarianten zur Weylgruppe*, J. Amer. Math. Soc. **3** (1990), 421–445.

[So3] W. Soergel, *On the relation between intersection cohomology and representation theory in positive characteristic*, in *Commutative algebra, homological algebra and representation theory (Catania/Genoa/Rome, 1998)*, J. Pure Appl. Algebra **152** (2000), 311–335.

[So4] W. Soergel, *Kazhdan–Lusztig-Polynome und unzerlegbare Bimoduln über Polynomringen*, J. Inst. Math. Jussieu **6** (2007), 501–525.

[Sp] T. A. Springer, *Linear algebraic groups, Second edition*, Progress in Mathematics 9, Birkhäuser Boston, 1998.

[St] R. Steinberg, *On a theorem of Pittie*, Topology **14** (1975), 173–177.

[Ve] J.-L. Verdier, *Spécialisation de faisceaux et monodromie modérée*, in *Analysis and topology on singular spaces, II, III (Luminy, 1981)*, 332–364, Astérisque **101**, 1983.

Part III
Varieties Associated to Quivers and Relations to Representation Theory and Symplectic Geometry

Loop Grassmannians of Quivers and Affine Quantum Groups

Ivan Mirković, Yaping Yang, and Gufang Zhao

Contents

I. Mirković (✉)
University of Massachusetts, Department of Mathematics and Statistics, Amherst, MA, USA
e-mail: mirkovic@math.umass.edu

Y. Yang
School of Mathematics and Statistics, The University of Melbourne, Parkville, VIC, Australia
e-mail: yaping.yang1@unimelb.edu.au

G. Zhao
Institute of Science and Technology Austria, Klosterneuburg, Austria

School of Mathematics and Statistics, The University of Melbourne, Parkville, VIC, Australia
e-mail: gufangz@unimelb.edu.au

V. Baranovsky et al. (eds.), *Representation Theory and Algebraic Geometry*, Trends in Mathematics, https://doi.org/10.1007/978-3-030-82007-7_8

I.M. is very happy for opportunity to mention just a few transformative effects of personalities of Sasha Beilinson and Vitya Ginzburg. I.M.'s understanding of possibilities of being a mathematician have been upturned through Bernstein's talk at Park City and Beilinson's talks in Boston. A part of the magic was that mathematics was alive, high on ideas, and low on ownership, and each talk would open for thinking some topic in mathematics, almost regardless of one's preparation. Before meeting Ginzburg, I.M. has come to view him as a smarter twin brother in mathematical tastes. Of biggest influence on I.M. was Ginzburg's paper on loop Grassmannians that offered a new kind of mathematics, orchestrated by an explosion of geometric ideas.

1 Introduction

For a semisimple algebraic group G of ADE type, the corresponding quiver Q is used to study representations of G, its loop group $G((t))$, and their quantum versions. Here, we reconstruct from Q the loop Grassmannian $\mathcal{G}(G)$ of G. This construction produces a candidate for the loop Grassmannian Kac-Moody extensions $\mathfrak{g}^{\mathrm{aff}}$ of loop Lie algebras $\mathfrak{g}((t))$ and should provide a tool to study its representations.

1.1 Loop Grassmannians Associated to Quivers

An advantage of the quiver approach is that it works in large generality. It provides a "loop Grassmannian" $\mathcal{G}^P_{\mathcal{D}}(Q, A)$ associated to the data of an arbitrary quiver Q, a cohomology theory A, a poset P, and a torus $\mathcal{D}$ of dilations. Intuitively, a quiver Q should provide a "grouplike" object $G(Q, A)$ though at the moment, we only see objects that should correspond to (quantization of) its affinization.

A cohomology theory A gives a "cohomological schematization" functor $\mathfrak{A}(X) \overset{\text{def}}{=} \mathrm{Spec}[A(X)]$ which assigns to a space X the affine scheme $\mathfrak{A}(X)$ over the ring of constants of theory A.[1] For us, this simplifies stacks (spaces with much symmetry) to classical geometry. It takes the moduli of lines, i.e., the classifying space $\mathbb{B}\mathbb{G}_m$, to a curve $\mathbb{G} = \mathfrak{A}(\mathbb{B}\mathbb{G}_m)$. Next, it turns the moduli $\mathcal{V}$ of finite-dimensional vector spaces into the space of configurations on the curve $\mathbb{G}$, i.e., the Hilbert scheme of points $\mathcal{H}_{\mathbb{G}} = \sqcup_n \mathbb{G}^{(n)}$ of $\mathbb{G}$.

[1] When A is de Rham cohomology, then $\mathfrak{A}_X$ can be viewed as an affinization of the de Rham space X_{dR} of X.

This configuration space is then used as the setting for the Beilinson-Drinfeld version of the "loop Grassmannian" $\mathcal{G}(Q, A)$ that intuitively corresponds to a (yet undefined) group $G(Q, A)$.

Finally, we construct the space $\mathcal{G}_{\mathcal{D}}(Q, A)$ which should be the quantum loop Grassmannian of the group $G(Q, A)$. Here, one adds quantization by letting a torus $\mathcal{D}$ act on the extension correspondence for representations of quivers (and its cotangent stack). At this level, there is a well-defined related "group theoretic" object, the "affine quantum group," that was constructed in [YZ16] and is denoted here by $U_{\mathcal{D}}(Q, A)$.

1.2 Construction

Since we avoid the group $G(Q, A)$ and its affinization, the construction is less standard. We will argue that it is of "homological nature."

The geometric ingredient is the technique of *local projective spaces* from [M17]. This refers to the notion of *I-colored local vector bundles* over a curve C, i.e., vector bundles over the I-colored configuration space $\mathcal{H}_{C\times I}$ (the moduli of finite flat subschemes of $C \times I$), that are in a certain sense "local with respect to C."

We actually start with a local line bundle L over $\mathcal{H}_{C\times I}$, and we induce it using a finite poset P to a local sheaf $Ind^P(\mathsf{L})$ over $\mathcal{H}_{C\times I}$. The corresponding projective scheme $\mathbf{P}(Ind^P\mathsf{L})$ is the projective spectrum of it symmetric algebra. The ("projective") *zastava space* $Z^P(\mathsf{L})$ is its "local part" $\mathbf{P}^{loc}[Ind^P\mathsf{L}]$. Its fibers are obtained as collisions of products of fibers at colored points $ai \in \mathcal{H}_{C\times I}$ (for a point $a \in C$ and a color $i \in I$). The collision happens inside the projective scheme $\mathbf{P}[Ind^P(\mathsf{L})]$, and the "rules of collision" are specified by the locality structure on the line bundle L.

The local line bundles L on $\mathcal{H}_{C\times I}$ are classified by symmetric bilinear forms $\mathcal{Q}$ on $\mathbb{Z}[I]$, and then the zastava space is denoted $Z^P(I, \mathcal{Q})$. However, we will concentrate on the case when $\mathcal{Q}$ is associated to a quiver Q (with vertices I). In this case, the topological ingredient of our construction is an explicit reconstruction of L from the quiver Q as the Thom line bundle of the moduli of representations of a quiver Q. This topological construction allows adding to the data a choice of a cohomology theory A, to get the zastava space $Z^P(Q, A)$. Moreover, we can upgrade to the *quantum version* $Z^P_{\mathcal{D}}(Q, A)$ by replacing the moduli of representations with the cotangent stack of the moduli of extensions of such representations and by switching on an action of a certain small torus $\mathcal{D}$ from [YZ16] on this cotangent stack.

Finally, we get the loop Grassmannian $\mathcal{G}^P_{\mathcal{D}}(Q, A)$ as a certain union of fibers of these zastava spaces.

Remark The classical loop Grassmannians of reductive groups are recovered when the poset P is a point (then P is omitted from notation). In that case, the fiber of $Z^P(\mathsf{L})$ at any colored point is $\mathbb{P}^1$.[2]

1.3 "Quantum Nature" of $\mathcal{G}^P_{\mathcal{D}}(Q, A)$

The loop Grassmannian $\mathcal{G}(G)$ of a semisimple group G is a partial flag variety of G^{aff} so it has a known quantum version which is a noncommutative geometric object. For the $\mathcal{G}_{\mathcal{D}}(Q, A)$ construction, this corresponds to the case when A is the K-theory. However, our incarnation $\mathcal{G}_{\mathcal{D}}(Q, A)$ is an object in standard geometry, and the hidden noncommutativity manifests in its Beilinson-Drinfeld form, i.e., when $\mathcal{G}_{\mathcal{D}}(Q, A)$ is extended to lie over a configuration space. The configuration space is necessarily ordered ("noncommutative"), i.e., $\mathcal{H}^n_{C\times I} = (C \times I)^{(n)}$ is replaced by $(C \times I)^n$. This has more connected components, but this increase is ameliorated by a nonstandard feature, a meromorphic braiding relating different connected components of the configuration moduli.

We expect to have more explicit descriptions of $\mathcal{G}^P_{\mathcal{D}}(Q, A)$ in terms of the graded algebra of sections of line bundles $\mathcal{O}(m)$ or in terms of the equation for the embedding into the projective space corresponding to sections of $\mathcal{O}(1)$.[3]In this paper, we only do some preparatory steps toward identifying the cases of $\mathcal{G}^P_{\mathcal{D}}(Q, A)$ with the classical loop Grassmannian of reductive groups.

This paper is related to the work of Z. Dong [D18] that studies the relation between the Mirković-Vilonen cycles in loop Grassmannians and the quiver Grassmannian of representations of the preprojective algebra (see Sect. 3.2.4).

1.4 Contents

In Sect. 2, we recall the method of generalized cohomology theories. Section 3 covers relevant aspects of classical loop Grassmannians and how to rebuild and generalize these in a "homological" way, i.e., by turning the notion of locality into a construction. In Sect. 4, we find a realization of these ideas in the setting of quivers by constructing local line bundles on configuration spaces from representations of quivers. Finally, in Sect. 5, we get quantum generalization of the notion of local

[2]Here, we do not pay attention to a choice of P, but when $P = (1 < \cdots < m)$, the fibers at colored points are $\mathbb{P}^m$ and $\mathcal{G}^{1<\cdots<m}_{\mathcal{D}}(Q, A)$ should "correspond to level m" in the sense that the sections of the standard line bundle $\mathcal{O}(1)$ on this object should be the same as the sections of $\mathcal{O}(m)$ in the case when $P = \text{pt}$.

[3]These embedding equations should be integrable hierarchies of differential equations indexed by Q, P and A since this is true in the classical case of $\mathcal{G}(G)$.

line bundles and of the corresponding loop Grassmannians using dilations on the cotangent bundle of moduli of extensions of representations of a quiver.

Appendix "Loop Grassmannians with a Condition" completes the description of Cartan fixed points in intersections of closures of semi-infinite orbits in loop Grassmannians (Proposition 3.2.3). This observation was the starting point for our generalization $\mathcal{G}(Q, A)$ of loop Grassmannians $\mathcal{G}(G)$. Appendix "Appendix 2: Calculation of Thom Line Bundles from [YZ17]" compares computations of Thom line bundles of convolution diagrams in Sect. 4.4 and in [YZ17].

2 Recollections on Cohomology Theories

2.1 Equivariant-Oriented Cohomology Theories

An *oriented cohomology theory* is a contravariant functor A that takes spaces X to graded commutative rings $A(X)$ and has certain properties such as the proper direct image.[4]For us, an oriented cohomological theory A can be either a topological cohomology theory or an algebraic cohomology theory. In the first case, the "spaces" are topological spaces, and we will use the ones that are given by complex algebraic varieties. In the second case, the "spaces" mean schemes over a given base ring $\Bbbk$.

Here, we list some of the common properties of such theories A that we will use. First, A extends canonically to pairs of spaces $A(X, Y)$ for $Y \subseteq X$. In particular, we get cohomology $A_Y(X) \stackrel{\text{def}}{=} A(X, X - Y)$ of X with supports in Y. Such theory A is functorial under flat pullbacks and proper push forwards with usual properties (homotopy invariance, projection formula, base change, and the projective bundle formula [LM07, Lev15]).

Also, such A has an equivariant version $A_{\mathsf{G}}(X)$ defined as $\lim\limits_{\leftarrow} A(\mathcal{X}_i)$ for ind-systems of approximations $\mathcal{X}_i$ of the stack $\mathsf{G}\backslash X$, For this reason, it is consistent to denote $A_{\mathsf{G}}(X)$ symbolically as $A(\mathsf{G}\backslash X)$ even if we do not really extend A to category of stacks.

The basic invariants of A are the commutative ring of constants $R = A(\text{pt})$ and the one-dimensional formal group $\mathbb{G}$ over R with a choice of a coordinate $\mathfrak{l}$ on $\mathbb{G}$ (called *orientation* of theory A).

The geometric form of the theory A is the functor $\mathfrak{A}$ from spaces to affine R-schemes defined by $\mathfrak{A}(X) = \text{Spec}(A(X))$. The G-equivariant version is again denoted by the index G, and it yields indschemes $\mathfrak{A}_{\mathsf{G}}(X) = \text{Spec}(A_{\mathsf{G}}(X))$, also denoted $\mathfrak{A}(\mathsf{G}\backslash X)$, that lie above $\mathfrak{A}_G \stackrel{\text{def}}{=} \mathfrak{A}_{\mathsf{G}}(\text{pt})$. For instance, the formal group $\mathbb{G}$ associated to A is $\mathfrak{A}_{\mathbb{G}_m}$ (approximations of $\mathbb{B}\mathbb{G}_m$ are given by $\mathbb{P}^\infty$, the ind-system of finite projective spaces).

[4]While the grading of a cohomology theory is fundamental, we will disregard it in this paper.

We denote $A_G = A_G(\mathrm{pt})$ and $\mathfrak{A}_G = \mathfrak{A}_G(\mathrm{pt})$. For a torus T, let $X^*(\mathsf{T})$, $X_*(\mathsf{T})$ be the dual lattices of characters and cocharacters of T and then $\mathfrak{A}_\mathsf{T} = X_*(\mathsf{T})\otimes_\mathbb{Z}\mathbb{G}$. For a reductive group G with a Cartan T and Weyl group W, $\mathfrak{A}_G$ is the categorical quotient $\mathfrak{A}_\mathsf{T}//W$. For instance, for the Cartan $\mathsf{T} = (\mathbb{G}_m)^n$ in GL_n, the Weyl group is the symmetric group $\mathfrak{S}_n$, and one has $\mathfrak{A}_\mathsf{T} = \mathbb{G}^n$ while $\mathfrak{A}_{GL_n} = \mathbb{G}^{(n)}$ is the symmetric power $\mathbb{G}^n//\mathfrak{S}_n$ of $\mathbb{G}$.

Remarks

(0) In the case when $\mathbb{G}$ is the germ of an algebraic group $\mathbb{G}_{alg}$, the equivariant A-cohomology has a refinement which gives indschemes over $\mathbb{G}_{alg}$. All of our results extend to this setting, and we will abuse the notation by allowing $\mathbb{G}$ to stand either for the formal group or for this algebraic group. For simplicity, our formulations will assume that $\mathbb{G}_{alg}$ is affine—the adjustment for the non-affine case is clear from the paper [YZ17] on elliptic curves (then $\mathbb{G}$ is an elliptic curve and $\mathfrak{A}(X)$ is affine over $\mathbb{G}$ rather than affine). Either version satisfies equivariant localization.

(1) For *algebraic* oriented cohomology theories, the basic reference is [LM07, Chapter 2] (one can also use [CZZ14, § 2] and [ZZ14, § 5.1]).[5] Here, cohomology theory is defined on smooth schemes over a given base ring $\mathbb{k}$. However, such cohomology theory A then extends (with a shift in degrees) under the formalism of *oriented Borel-Moore homology* to schemes over $\mathbb{k}$ that are of finite type and separable.[6]

2.2 *Thom Line Bundles*

When V is a G-equivariant vector bundle over X, the equivariant cohomology of V supported in the zero section $\Theta_\mathsf{G}(V) \overset{\text{def}}{=} A_\mathsf{G}(V, V-X)$ is known to be a line bundle over $\mathfrak{A}_\mathsf{G}(X)$, i.e., a rank one locally free module over $A_G(X)$, called the *Thom line bundle* of V. Moreover, this is an ideal sheaf of an effective divisor in $\mathfrak{A}_\mathsf{G}(X)$ called the *Thom divisor of* V (see section 2.1 in [GKV95]). As usual, one can think of this as the Thom line bundle $\Theta(\mathsf{G}\backslash V)$ over $\mathfrak{A}(\mathsf{G}\backslash X)$ for the vector bundle $\mathsf{G}\backslash V$ over $\mathsf{G}\backslash X$.

[5] The terminology of "algebraic cohomology" is also used by Panin-Smirnov for a refinement of the formalism in which the theory is bigraded (to adequately encode the example of motivic cohomology). We will not be concerned with this version.

[6] What is called Borel-Moore homology here is not quite what this means in classical topology; however, this is just a choice of terminology since the A-setting does contain the precise analogue of Borel-Moore homology. For instance, for smooth X, the more appropriate version would be $BM_A(X) = \Theta_A(TX)^{-1}$ in terms of the Thom bundle which is defined next.

Lemma

(a) *Let* $\mathsf{V} \to \mathsf{X}$ *be a vector bundle equivariant for a reductive group* G *with a Cartan* T*. Then* $\Theta_\mathsf{T}(\mathsf{V})$ *is the pullback of* $\Theta_\mathsf{G}(\mathsf{V})$ *by a flat map* $\mathfrak{A}_\mathsf{T} \to \mathfrak{A}_\mathsf{G}$*, and* $\Theta_\mathsf{T}(\mathsf{V})$ *determines* $\Theta_\mathsf{G}(\mathsf{V})$*.*

(b) *For a cohomology theory A and a character* η *of a torus* T*, if* η *is trivial, then* $\Theta_\mathsf{T}(\eta) = \mathcal{O}_{\mathfrak{A}_\mathsf{T}}$*, and otherwise,* $\Theta_\mathsf{T}(\eta)$ *is the ideal sheaf of the (Thom) divisor* $\mathrm{Ker}(\eta) \subseteq \mathsf{T}$*.*

(c) *For an extension of vector bundles* $0 \to V' \to V \to V'' \to 0$*, one has* $\Theta_\mathsf{G}(V) \cong \Theta_\mathsf{G}(V') \otimes \Theta_\mathsf{G}(V'')$*. So* Θ_G *is defined on the K-group of* G*-equivariant vector bundles over X.*

Proof These are proved in [GKV95, § 2.1]. See also [ZZ15, Proposition 3.13].

2.2.1 $\Theta_\mathsf{G}(\mathsf{V})$ for a Representation V of G

This is the case when X is a point. We can write $\Theta_\mathsf{G}(\mathsf{V})$ in terms of the character $ch(V)$. First for a torus T and $\mathfrak{A}_\mathsf{T} \stackrel{\text{def}}{=} \mathfrak{A}_\mathsf{T}(\mathrm{pt})$, there is a unique homomorphism $\mathfrak{l} : (\mathbb{N}[X^*(\mathsf{T})], +) \to (A_\mathsf{T}, \cdot)$ such that for any character χ of T, the function $\mathfrak{l}_\chi$ is the composition $\mathfrak{A}_\mathsf{T} \xrightarrow{\mathfrak{A}_\chi} \mathfrak{A}_{\mathbb{G}_m} \xrightarrow{\mathfrak{l}} \mathbb{A}^1$. Now, for a reductive group G with a Cartan T, this restricts to a homomorphism $\mathfrak{l}$ from $(\mathbb{N}[X^*(\mathsf{T})]^W, +)$ to $(A_\mathsf{G}, \cdot)$. Then the ideal $\Theta_\mathsf{G}(\mathsf{V})$ in functions on $\mathfrak{A}_\mathsf{G} = [X_*(\mathsf{T}) \otimes \mathbb{G}]//W$ is generated by the function $\mathfrak{l}_{ch(\mathsf{V})}$ on $\mathfrak{A}_\mathsf{G}$. (By the preceding lemmas, it suffices to check this when G is a torus, which in turn can be reduced to the case when V is one dimensional and $\mathsf{T} = \mathbb{G}_m$.)

2.2.2 Thom Line Bundles $\Theta(f)$ of Maps f

For a map of smooth spaces $f : \mathsf{X} \to \mathsf{Y}$, we have the tangent complex $T(f) = [T\mathsf{X} \to f^*T\mathsf{Y}]_{-1,0}$ on X and in degrees $-1, 0$. The line bundle $\Theta(f) = \Theta(T(f))$ on $\mathfrak{A}(\mathsf{X})$ is defined as the value of Θ on the corresponding virtual vector bundle $f^*T\mathsf{Y} - T\mathsf{X}$.

3 Loop Grassmannians and Local Spaces

In Sect. 3.1, we recall loop Grassmannians and in Sect. 3.2, we check the description of T-fixed points in intersections $(\overline{S_0} \cap \overline{S^-_{-\alpha}})^T$ of closures of semi-infinite orbits in a loop Grassmannian. This is used in the "homological" reconstruction and generalization of loop Grassmannians in Sect. 3.4, which is itself based on the formalism of *local spaces* from Sect. 3.3.

3.1 Loop Grassmannians

We start with the standard loop Grassmannians $\underline{\mathcal{G}}(G)$. Let $\Bbbk$ be a commutative ring, and let $\mathcal{O} = \Bbbk[[z]] \subseteq \mathcal{K} = \Bbbk((z))$ be the Taylor and Laurent series over $\Bbbk$. These are functions on the indscheme d (the formal disc) and its punctured version $d^* = d - 0$. For an algebraic group scheme G, we denote by $G_\mathcal{O} \subseteq G_\mathcal{K}$ its *disc* group scheme and *loop* group indscheme over $\Bbbk$, the points over a $\Bbbk$-algebra $\Bbbk'$ are $G_\mathcal{O}(\Bbbk') = G(\Bbbk'[[z]])$ and $G_\mathcal{K}(\Bbbk') = G\big(\Bbbk'((z))\big)$. The standard loop Grassmannian is the indscheme given by the quotient $\underline{\mathcal{G}}(G) = G_\mathcal{K}/G_\mathcal{O}$.

We also notice that $\mathcal{O}_- = \Bbbk[z^{-1}]$ defines group indscheme $G_{\mathcal{O}_-} \subseteq G_\mathcal{K}$. The congruence subgroups $K_\pm(G)$ are the kernels of evaluations $G_\mathcal{O} \to G$ and $G_{\mathcal{O}_-} \to G$ at $z = 0$ and $z = \infty$.

3.1.1 Global (Beilinson-Drinfeld) Loop Grassmannians $\mathcal{G}_{\mathcal{H}_C}(G)$

Let C be a smooth curve. For a finite subscheme $D \subseteq C$, the first G-cohomology $H^1_D(C, G)$ of C with the support at D is the moduli indscheme of pairs $(\mathcal{T}, \tau)$ of a G-torsor $\mathcal{T}$ over C and its section τ over $C - D$. As D varies in the Hilbert scheme of points $\mathcal{H}_C$, one assembles these into an indscheme $\mathcal{G}(G) = \mathcal{G}_{\mathcal{H}_C}(G)$ over $\mathcal{H}_C$, with fibers $\mathcal{G}_{\mathcal{H}_C}(G)_D = H^1_D(C, G)$.

A point of a smooth curve $c \in C$ defines a "smooth formal curve" $\widehat{c}$, and we equally get $\mathcal{G}_{\mathcal{H}_{\widehat{c}}}(G) \to \mathcal{H}_{\widehat{c}}$, which is a restriction of $\mathcal{G}_{\mathcal{H}_C}(G) \to \mathcal{H}_C$. Moreover, a choice of a formal coordinate z at c identifies the fiber at c with the standard loop Grassmannian

$$\mathcal{G}_{\mathcal{H}_{\widehat{c}}}(G)_c \xrightarrow[\cong]{} \underline{\mathcal{G}}(G).$$

One can also think of it the compactly supported cohomology $H^1_c(\widehat{c}, G)$.[7]

3.1.2 The Abel-Jacobi Map

Recall that on a smooth curve C (hence also for $C = d$), $\mathcal{H}_C$ is a commutative monoid for addition of effective divisors. Moreover, the *Abel-Jacobi map* is a map of monoids:

$$\mathrm{AJ}_C : \mathcal{H}_C \to \underline{\mathcal{G}}(\mathbb{G}_m), \quad \mathrm{AJ}_C(D) \overset{\text{def}}{=} \mathcal{O}_C(-D)).$$

[7]By *compactly supported cohomology* of X, we mean the cohomology of a compactification $\overline{X}$ trivialized on the formal neighborhood of the boundary of X in $\overline{X}$.

Precisely, $AJ_C(D)$ consists of $D \in \mathcal{H}_C$, the $\mathbb{G}_m$-torsor corresponding to the line bundle $\mathcal{O}_C(-D)$, and the canonical trivialization 1 of $\mathcal{O}_C(-D)$ off D (which we often omit).

Remark

(0) There are two simple ways to realize all of $\underline{\mathcal{G}}(\mathbb{G}_m)$. One fixes the trivialization fixed at 1 as above, and the other uses the trivial torsor $G \times d$. The transition involves a minus since for any $f \in \mathcal{K}^*$, one has f^{-1} : $(f\mathcal{O}_C, 1) \xrightarrow[\cong]{} (\mathcal{O}_C, f^{-1})$.

Lemma ([CC81] (see also [M17])) *The inclusion $d \subseteq \mathcal{H}_d$ makes $\mathcal{H}_d$ into the commutative monoid affine indscheme freely generated by the formal disc d. The Abel-Jacobi map $d \to \underline{\mathcal{G}}(\mathbb{G}_m)$ makes $\underline{\mathcal{G}}(\mathbb{G}_m)$ into the commutative group indscheme freely generated by d.* □

Remarks

(1) In [M17], this is viewed as interpretation of $\underline{\mathcal{G}}(\mathbb{G}_m)$ as homology $\mathbb{H}_*(d)$ of d for a certain conjectural cohomology theory $\mathbb{H}$.[8] The above interpretations of $\underline{\mathcal{G}}(\mathbb{G}_m)$ as both homology and the compactly supported cohomology (see Sect. 3.1) can then be viewed as local Poincaré duality in algebraic geometry.
(2) A formal coordinate z on d gives a correspondence of subschemes $D \in \mathcal{H}_d$ and monic polynomials χ_D in $\mathbb{k}[z]$ with nilpotent coefficients, such that χ_D is an equation of D. This gives a lift of the Abel-Jacobi map that embeds $\mathcal{H}_d$ into $\mathbb{G}_{m,\mathcal{K}}$ by sending $D \in \mathcal{H}_d^n$ to $\chi_D{}^{-1}$. For instance, for $n \in \mathbb{N}$, the divisor $n[0] \overset{\text{def}}{=} \{z^n = 0\}$ goes to $z^{-n} \in \mathbb{G}_{m,\mathcal{K}}$.
(3) The group indscheme $\mathbb{G}_{m,\mathcal{K}}$ has a factorization $\mathbb{G}_m \times z^{\mathbb{Z}} \times K_+(\mathbb{G}_m) \times K_-(\mathbb{G}_m)$ where the points of congruence subgroups are $K_+(\mathbb{G}_m)(\mathbb{k}') = 1 + z\mathbb{k}'[[z]]$ and, $K_-(\mathbb{G}_m)(\mathbb{k}')$ is the invertible part of $1 + z^{-1}\mathbb{k}'[z^{-1}]$, i.e., the part where the coefficients are nilpotent [CC81].

3.2 The T-Fixed Points in Semi-infinite Varieties $\overline{S_\alpha}$

3.2.1 Tori

Let us restate the remarks in Sect. 3.1.2 in the generality of split tori $T \cong X_*(T) \otimes_{\mathbb{Z}} \mathbb{G}_m$. First, $\lambda \in X_*(T)$ gives $\mathcal{K}^* \to T_{\mathcal{K}}$, and the image of a coordinate z on the disc is denoted z^λ. This gives isomorphisms $X_*(T) \xrightarrow[\cong]{} \pi_0(T_{\mathcal{K}})$ and $X_*(T) \xrightarrow[\cong]{} \underline{\mathcal{G}}(T)_{reduced} \xrightarrow[\cong]{} \pi_0(\underline{\mathcal{G}}(T))$. For $\lambda \in X_*(T)$, we denote

[8] In general, the derived version of homology $\mathbb{H}_*(X)$ should be the free abelian commutative group object in derived stacks freely generated by X.

$L_\lambda \overset{\text{def}}{=} z^{-\lambda} T_{\mathcal{O}} \in \underline{\mathcal{G}}(T)$ (independently of z), meaning a trivial T-torsor on d with a section $z^{-\lambda}$ on d^*. The corresponding connected component $\underline{\mathcal{G}}(T)_\lambda$ of $\underline{\mathcal{G}}(T)$ is a $K_-(T)$-torsor.

3.2.2 The "Semi-Infinite" Orbits $S_\lambda^\pm$

Now let G be reductive with a Cartan T. Then the T-fixed point subscheme $\underline{\mathcal{G}}(G)^T$ is $\underline{\mathcal{G}}(T)$. A choice of opposite Borel subgroups $B^\pm = TN^\pm$ yields orbits $S_\lambda^\pm \overset{\text{def}}{=} N_{\mathcal{K}}^\pm L_\lambda$ indexed by $\lambda \in X_*(T)$ (we often omit the super index $+$). If G is semisimple, then $\underline{\mathcal{G}}(G)$ is reduced, and these orbits provide two stratifications of $\underline{\mathcal{G}}(G)$. The following is well known:

Lemma ([MV07]) *For $\lambda, \mu \in X_*(T)$, the following are equivalent: (0) $\overline{S_\lambda} \ni \mu$, (i) $\overline{S_\lambda} \supseteq S_\mu$, (ii) S_λ meets S_μ^-, and (iii) $\lambda \geq \mu$ (in the sense that $\lambda - \mu$ lies in the cone $\check{Q}^+$ generated by the coroots $\check{\alpha}$ dual to roots α in N).* □

Example The loop Grassmannian of $G = SL_2$ is identified with the space $\mathcal{L}$ of *lattices* in $\mathcal{K}^2 = \mathcal{K}e_1 \oplus \mathcal{K}e_2$ (the $\mathcal{O}$-submodules that lie between two submodules of form $z^n\mathcal{O}^2$) and have volume zero. Here., $vol(L) \overset{\text{def}}{=} \dim(L/z^n\mathcal{O}^2) - \dim(\mathcal{O}^2/z^n\mathcal{O}^2)$ for $n >> 0$. For the standard Borel subgroup $B = TN$, we have $N = \left(\begin{smallmatrix} 1 & * \\ 0 & 1 \end{smallmatrix}\right)$, and the coroot $\check{\alpha}$ in N is $\check{\alpha}(a) = \begin{pmatrix} a & 0 \\ 0 & a^{-1} \end{pmatrix} \in T$. Then $X_*(T) = \mathbb{Z}\check{\alpha}$ and $L_{n\check{\alpha}} = z^{-n\check{\alpha}} L_0$ is the lattice generated by two vectors $\langle z^{-n}e_1, z^n e_2 \rangle$. For a lattice $L \in \mathcal{L}$, one has $L \in S_{n\check{\alpha}}$ if $L \cap \mathcal{K}e_1 = z^{-n}\mathcal{O}e_1$ and $L \in \overline{S_{n\check{\alpha}}}$ if $L \ni z^{-n}e_1$.

3.2.3 The T-Fixed Points

Now, let $G \supseteq B = TN$ be semisimple with a simply connected cover $G_{sc} \supseteq T_{sc}$. If α_i, $i \in I$, are simple roots in N, then $\prod_{i \in I} \check{\alpha}_i : \mathbb{G}_m^I \xrightarrow[\cong]{} T_{sc}$ defines the *Abel-Jacobi embedding* AJ_d^G as the composition $\mathcal{H}_{d \times I} \hookrightarrow \underline{\mathcal{G}}(\mathbb{G}_m{}^I) \xrightarrow[\cong]{} \underline{\mathcal{G}}(T_{sc}) \subseteq \underline{\mathcal{G}}(G_{sc}) \subseteq \underline{\mathcal{G}}(G)$, where $D = (D_i)_I \mapsto (\mathcal{O}_C(-D_i))_I \mapsto (\check{\alpha}_i[\mathcal{O}_C(-D_i)])_I$. In particular, for $\alpha = \sum_I a_i i \in \mathbb{N}[I]$,

$$\mathrm{AJ}^G(\alpha[0]) = (\check{\alpha}_i[\mathcal{O}_C(-a_i[0]])_I = (\check{\alpha}_i[z^{a_i}\mathcal{O}_C)])_I \overset{\text{Remark 3.1.2.0}}{=} z^{-\alpha} L_0 = L_\alpha.$$

The following has been announced in [M17].

Proposition

(a) *The image of the Abel-Jacobi embedding $AJ_d^G : \mathcal{H}_{d \times I} \hookrightarrow \underline{\mathcal{G}}(G)$ is the fixed point sub-indscheme $(\overline{S_0})^T$.*

(b) *For* $\alpha \in \mathbb{N}[I]$, *this identifies the connected component* $\mathcal{H}^{\alpha}_{d\times I} = \prod_{i\in I} \mathcal{H}^{\alpha_i}_d$ *of* $\mathcal{H}_{d\times I}$ *with the intersection of* $\overline{S_0}$ *with the connected component* $\underline{\mathcal{G}}(T)_\alpha$ *of* $\underline{\mathcal{G}}(T)$.
(c) *Also, the moduli* $\mathcal{H}_{\alpha[0]}\subseteq\mathcal{H}_{d\times I}$ *of all subschemes of the finite flat colored scheme* $\alpha[0]$ *is identified with* $(\overline{S_0} \cap \overline{S^-_\alpha})^T$.[9]

Proof We start with the proof for $G = SL_2$. Parts (a-b) claim that $\overline{S_0}$ meets the connected component $\underline{\mathcal{G}}(T)_{p\check{\alpha}}$ of $\underline{\mathcal{G}}(T)$ if and only if $p \geq 0$, and then the intersection is $\mathrm{AJ}^G_d(\mathcal{H}^p_d)$.

The points of the negative congruence subgroup $K_-(\mathbb{G}_m)\subseteq\mathbb{G}_{m,\mathcal{K}}$ are the comonic polynomials $Q = 1+q_1z^{-1}+\cdots+q_sz^{-s}$ in z^{-1} with nilpotent coefficients. Now, the isomorphism $K_-(\mathbb{G}_m) \xrightarrow{\cong} \underline{\mathcal{G}}(T)_{p\check{\alpha}}$ is given by $Q\mapsto \check{\alpha}(Q)L_p$ which is the lattice $\langle Q^{-1}z^{-p}e_1, Qz^pe_2\rangle$. According to the example in Sect. 3.2.2, this is in $\overline{S_0}$ if $(Qz^p)^{-1}\mathcal{O} \ni z^0$. This means that $z^pQ \in \mathcal{O}$, i.e., that the powers of z^{-1} in Q are $\leq p$. Such z^pQ form all monic polynomials in z of degree p with nilpotent coefficients. So all such $\check{\alpha}(Qz^p)L_0$ form exactly $\mathrm{AJ}^G_d(\mathcal{H}^p_d)$.

For part (c), the example in Sect. 3.2.2 says that $\overline{S^-_{m\check{\alpha}}}$ consists of lattices L that contain z^me_2. Now, for $D \in \mathcal{H}^p_d$ with a monic equation $P \in \Bbbk[z]$, $\mathrm{AJ}^G_d(D) = \check{\alpha}(P)L_0 = \langle P^{-1}e_1, Pe_2\rangle$ lies in $\overline{S^-_m}$ if $P\mathcal{O} \ni z^m$, i.e., polynomial P divides z^m. This is equivalent to $D\subseteq m[0]$.

The proof in the general case is postponed to the Appendix "Proof of the Proposition 3.2.3".

Remark The connected component of $\underline{\mathcal{G}}(G)$ is $\underline{\mathcal{G}}(G_{sc})$, so the spaces $\overline{S_0}\supseteq \overline{S_0} \cap \overline{S^-_{-\alpha}}$ and their T-fixed points do not depend on the center of G.

3.2.4 The Kamnitzer-Knutson Program of Reconstructing MV Cycles

Here, we restate the proposition and recall one of the origins of this paper. Consider a simply laced semisimple Lie algebra $\mathfrak{g}$ and its adjoint group G. In [BK10], the irreducible components C of the variety Λ of representations of the *preprojective algebra* Π of a Dynkin quiver Q of G are put into a canonical bijection with certain irreducible subschemes X_{C} of the corresponding loop Grassmannian $\underline{\mathcal{G}}(G)$, called *MV cycles* [MV07].

For any representation $\dot{V}$ of the preprojective algebra Π, the moduli $\mathrm{Gr}_\Pi(\dot{V})$ of all Π-submodules of $\dot{V}$ is called the *quiver Grassmannian* of $\dot{V}$.

Conjecture ([M17]) For any irreducible component C of Λ, and a sufficiently generic representation $\dot{V}$ in C, the cohomology of its quiver Grassmannian $\mathrm{Gr}_\Pi(\dot{V})$

[9] So the connected components of $\overline{S_0}^T$ are $(\overline{S_0} \cap \overline{S^-_{-\alpha}}) \cap \underline{\mathcal{G}}(T)_{-\beta}$, for $0 \leq \beta \leq \alpha$, identified with the moduli $\mathcal{H}^\beta_{\alpha[0]}$ of length β subschemes of $\alpha[0]$.

is the ring of functions on the subscheme X_{C}^T of points in the corresponding MV cycle X_{C} in $\underline{\mathcal{G}}(G)$ that are fixed by a Cartan subgroup T of G.

The grading on cohomology corresponds to the action of loop rotations on X_{C}^T.

Remarks

(0) This is a version of a conjecture of Kamnitzer and Knutson on equality of dimensions of cohomology $H^*[Gr_\Pi(\dot{V})]$ and of sections of the line bundle $\mathcal{O}(1)$ over the MV cycle X_{C}.
(1) Zhijie Dong has constructed a map in one direction in this conjecture [D18].
(2) The form of this conjecture is alike the Hikita conjecture in the symplectic duality framework.
(3) The MV cycles are defined as irreducible components of intersections in $\underline{\mathcal{G}}(G)$ of closures of semi-infinite orbits $\overline{S_0} \cap \overline{S_{-\alpha}^-}$ for $\alpha \in \mathbb{N}[I]$. Proposition 3.2.3.c will imply the following simplified version of the conjecture that replaces on the loop Grassmannian side the individual MV cycles with the intersections $\overline{S_0} \cap \overline{S_{-\alpha}^-}$, and on the quiver side, it degenerates the operators in the representation of Π to zero:

Corollary *Let $\alpha \in \mathbb{N}[I]$. Then $(\overline{S_0} \cap \overline{S_{-\alpha}^-})^T$ is the spectrum of cohomology of the quiver Grassmannian $\mathrm{Gr}_\Pi(\dot{V})$, where $\dot{V}$ is the zero representation of Π of dimension α. Also, the grading on cohomology corresponds to the action of loop rotations on the fixed point subscheme of the loop Grassmannian.*

Proof For $\alpha = \sum \alpha_i i$, we have $\dot{V} = \oplus_{i\in I} V_i$ with $\dim(V_i) = \alpha_i$. The quiver Grassmannian $Gr_\Pi(\dot{V})$ is then the product $\prod_{i\in I} Gr(V_i)$ of total Grassmannians of components V_i.

Since $(\overline{S_0} \cap \overline{S_{-\alpha}^-})^T = \mathcal{H}_{\alpha[0]} = \prod_{i\in I} \mathcal{H}_{\alpha_i[0]}$ by Proposition 3.2.3.c, it remains to notice that $H^*(Gr_p(n))$ can be calculated by Carell's theorem as functions on the fixed point subscheme $Gr_p(n)^e$ of a regular nilpotent e on $\Bbbk^n$. If we realize $\Bbbk^n$ and e as $\mathcal{O}(n[0])$ and the operator of multiplication by z, we see that $Gr_p(n)^e$ is $\mathcal{H}_{n[0]}^p$ (a subspace of $\mathcal{O}(n[0])$ is z-invariant if it is the ideal of a subscheme).

Finally, the degree $2p$ cohomology corresponds to the p-power of z which is the grading by rotations of the disc d.

3.3 Local Spaces Over a Curve

The notion of local spaces has appeared in [M14] as a common framework for the factorization spaces of Beilinson-Drinfeld and the factorizable sheaves of Finkelberg-Schechtman.

3.3.1 Local Spaces

For a set I and a smooth curve C, we consider the Hilbert scheme $\mathcal{H}_{C\times I} \cong (\mathcal{H}_C)^I$ of I-*colored* points of C.[10] Its connected components $\mathcal{H}^{\alpha}_{C\times I} \cong \prod_{i\in I} \mathcal{H}^{\alpha_i}_C$ are given by subschemes of length $\alpha \in \mathbb{N}[I]$. For a space Z over $\mathcal{H}_{C\times I}$, we denote the fiber at $D \in \mathcal{H}_{C\times I}$ by Z_D.

An I-*colored local space* Z over C is a space Z over $\mathcal{H}_{C\times I}$, together with a commutative and associative system of isomorphisms for disjoint $D', D'' \in \mathcal{H}_{C\times I}$

$$\iota_{D',D''} : Z_{D'} \times Z_{D''} \xrightarrow[\cong]{} Z_{D'\sqcup D''}.$$

We have $Z_\emptyset = \text{pt}$. When $\alpha = i \in I$, the connected component $\mathcal{H}^i_{C\times I}$ is $C \times i$. We call the fiber Z_{ai} at $a \in C$ the "i-particle at a," and we think of Z as a *fusion diagram* for these particles.

Example A *factorization space* in the sense of Beilinson and Drinfeld is a local space $Z \to \mathcal{H}_{C\times I}$ such that the fibers Z_D only depend on the formal neighborhood $\hat{D}$ of D in C. These can be viewed as spaces over the *Ran space* $\mathcal{R}_C$, the moduli of finite subsets of C.

Remarks

(1) A *weakly local* structure is the case when the structure maps ι are only embeddings. Any weakly local space Z has its *local part* $Z^{loc} \subseteq Z$ which we define as the least closed local subspace of Z that contains all particles. Explicitly, one first constructs $Z^{loc,reg}$ over $\mathcal{H}^{reg}_{C\times I}$ so that at a discrete $D \in \mathcal{H}_{C\times I}$ the fiber is $\prod_{ai\in D} Z_{ai}$, then Z^{loc} is the closure in Z of $Z^{loc,reg}$.
(2) A *local structure* on a (super) vector bundle V over a local space Z is an associative and commutative system of isomorphisms $V|_{Z_{D'}} \boxtimes V|_{Z_{D'}} \cong V|_{Z_{D'\sqcup D''}}$. By the Segre embedding, its projective bundle $\mathbb{P}(V)$ is a weakly local space. Its local part $\mathbb{P}(V)^{loc}$ is called the *local projective space* $\mathbb{P}^{loc}(V)$ of a local vector bundle V.
(3) The notion of "locality structure" is a version of the Beilinson-Drinfeld "factorization structure" where emphasis is changed slightly to get a tool for producing spaces such as $\mathbb{P}^{loc}(V)$. However, the use of closure makes this construction existential rather than explicit.
(4) One would like to extend this locality mechanism from a smooth curve to a formal disc d, but this requires a supply of disjoint finite flat subschemes of d. This is expected (or known) to be doable in terms of rigid geometry.

[10] One could try replacing a curve by a more general scheme and $\mathcal{H}$ by other notions of powers of a scheme like the Cartesian powers $C^n_C = C^n$.

3.3.2 Classification of Local Line Bundles on the Colored Hilbert Scheme

The following is a simplified version of the classification of factorizable line bundles on the space of colored divisor in Proposition 3.10.7 of [BD04].

Lemma

(a) For a line bundle $\mathsf{M} = (\mathsf{M}_i)_{i\in I}$ *over* $C \times I$*, there is a unique local line bundle* $\widetilde{\mathsf{M}}$ *over* $\mathcal{H}_{C\times I}$ *which agrees with* M *on* $C \times I$ *and whose locality structure maps extend to isomorphisms across the diagonals.*

(b) Isomorphism classes of local line bundles on $\mathcal{H}_{C\times I}$ *are classified by pairs* $(\mathsf{M}, \mathcal{Q})$ *of a line bundle* $\mathsf{M} = (\mathsf{M}_i)_{i\in I}$ *over* $C \times I$ *and a symmetric bilinear form ("sb-form")* $\mathcal{Q}$ *on* $\mathbb{Z}[I]$ *by* $(\mathsf{M}, \mathcal{Q}) \mapsto \widetilde{\mathsf{M}}(\mathcal{Q}\Delta^{\mathcal{H}})$ *where* $\mathcal{Q}\Delta^{\mathcal{H}}$ *is the divisor in* $\mathcal{H}_{C\times I}$ *given by* $\sum_{i\leq j} \mathcal{Q}(i, j)\Delta_{ij}$ *for the discriminant divisors* $\Delta^{\mathcal{H}}_{ij} \subseteq \mathcal{H}_{C\times I}$*.*

Remarks

(0) We also check the same classification for local super line bundles on the "ordered configuration space" $\mathcal{C}_{C\times I} \stackrel{\text{def}}{=} \sqcup_n (C \times I)^n$. A line bundle M on $C \times I$ and an sb-form κ on $\mathbb{Z}[I]$ give the corresponding local super line bundle by the "same" formula $\widetilde{\mathsf{M}}(\kappa\Delta^{\mathcal{C}})$ where the diagonals $\Delta^{\mathcal{C}}_{ij}$ are now in $\mathcal{C}_{C\times I}$ and the parity of M^i is that of $\kappa(i, i)$. Local line bundles L of the form $\widetilde{\mathsf{M}}(-\kappa\Delta^{\mathcal{C}}))$ are characterized by requiring that the locality maps $\iota^{ij} : \mathsf{L}^i \boxtimes \mathsf{L}^j \to \mathsf{L}^{i,j}$ defined on $\mathcal{C}^{i,j}_C = C^2$ and off Δ_C, have vanishing of order $\kappa(i, j)$ along Δ_C.

(1) A local super line bundle $\widetilde{\mathsf{M}}(\kappa\Delta^{\mathcal{C}})$ on $\mathcal{C}_{C\times I}$ descends to $\mathcal{H}_{C\times I}$ if the quadratic form κ is even. (Since the pullback of the diagonal in $\mathcal{H}^{2i}_C = C^{(2)}$ to C^2 is the double of the diagonal divisor, the pullback of $\mathcal{O}_{\mathcal{H}_{C\times I}}(\mathcal{Q}\Delta^{\mathcal{H}})$ from $\mathcal{H}_{C\times I}$ to $\mathcal{C}_{C\times I}$ is $\mathcal{O}_{\mathcal{C}}(\kappa\Delta^{\mathcal{C}})$ where κ is obtained from $\mathcal{Q}$ by doubling the numbers on the diagonal.)

Proof

(a) In the setting of Cartesian powers, the restriction $\widetilde{\mathsf{M}}^{i_1,\dots,i_n}$ of $\widetilde{\mathsf{M}}$ to the connected component $\mathcal{C}^{i_1,\dots,i_n}_{C\times I} = \prod_{k=1}^n (C \times i_k)$ is simply $\boxtimes_{k=1}^n M_{i_k}$.

In the setting of Hilbert powers, consider the tautological bundle $\mathcal{T} \xrightarrow{q} \mathcal{H}_{C\times I}$, the fiber at $D \in \mathcal{H}_{C\times I}$ is the subscheme D of $C \times I$. From $\mathcal{T} \subseteq \mathcal{H}_{C\times I} \times (C \times I) \xrightarrow{pr_2} C \times I$, we have a line bundle $i^* pr_2^* \mathsf{M}$ on $\mathcal{T}$, and we define the line bundle $\widetilde{\mathsf{M}}$ on $\mathcal{H}_{C\times I}$ as its Deligne direct image in line bundles, along the map q. (For an abelian group A, one has the direct image of A-torsors along a finite flat map.)

(b) Let us write the proof in the more general case of Cartesian powers. For any $\mathsf{M}, \mathcal{Q}$, the line bundle $\widetilde{M}(\mathcal{Q}\Delta^{\mathcal{C}})$ on $\mathcal{C}_{C\times I}$ is clearly local. Conversely, let L be any local line bundle and denote $\mathsf{M} = L|_{C\times I}$.

For any $i, j \in I$, the locality isomorphism $L^i \boxtimes L^j \cong L^{i,j}$ is defined on $\mathcal{C}^{ij}_{C\times I} = (C \times i) \times (C \times j)$ minus the diagonal. It extends to an identification over all of $\mathcal{C}^{ij}_{C\times I}$: $L^i \boxtimes L^j \cong L^{i,j}(-\kappa(i, j))$ for a unique integer $\kappa(i, j)$.

Now, local line bundles L and $\mathsf{M}(\kappa\Delta^{\mathcal{C}})$ on $\mathcal{C}_{C\times I}$ have been identified over $\mathcal{C}^{\leq 2}_{C\times I}$. However, an isomorphism over $\mathcal{C}^{\leq 2}_{C\times I}$ extends uniquely to $\mathcal{C}_{C\times I}$ since the the remaining higher incidences have codimension ≥ 2.

3.4 A Generalization $\mathcal{G}^P(I, \mathcal{Q})$ of Loop Grassmannians of Reductive Groups

This is a case of the local projective space construction (Remark 3.3.1(2)), in the setting of a local line bundle L on the configuration space $\mathcal{H}_{C\times I}$ of colored effective divisors on a curve C. In Sect. 3.4.1, we notice that for a semisimple group, G leads to such local line bundle L as a restriction of the line bundle $\mathcal{O}(1)$ on the loop Grassmannian $\underline{\mathcal{G}}(G)$. In Sects. 3.4.2–3.4.4, we will associate to a based symmetric bilinear form $(I, \mathcal{Q})$ and a poset P the corresponding zastava space $Z^P(I, \mathcal{Q})$ and the loop Grassmannian $\mathcal{G}^P(I, \mathcal{Q})$. Finally, in Sect. 3.4.6, we check that this is indeed a generalization of the corresponding spaces for simply connected semisimple groups.

This reconstruction roughly says that the loop Grassmannian $\mathcal{G}(G)$ can be effectively reconstructed from $\mathcal{G}(T)$. Previous results in this direction include [Zhu07], [FK], and [Se]. The key observation is that the equations of the semi-infinite variety $\overline{S_0}$ in the projective space $\mathbb{P}(\Gamma[\overline{S_0}, \mathcal{O}_{\underline{\mathcal{G}}(G)}(1)]^*) = \mathbb{P}(\Gamma[\overline{S_0}^T, \mathsf{L}^*])$ are given by the locality structure on $\mathcal{O}_{\underline{\mathcal{G}}(G)}(1)$. So the locality structure allows us to reconstruct $\overline{S_0}$ and then also $\underline{\mathcal{G}}(G)$ as a certain limit of copies of $\overline{S_0}$.

3.4.1 Local Line Bundles from Loop Grassmannians

If G is simple and simply connected, then $Pic[\underline{\mathcal{G}}(G)] \cong \mathbb{Z}$ with the canonical generator $\mathcal{O}(1)$ (given, for instance, by the divisor which is the complement of the open Bruhat cell $\underline{\mathcal{G}}^0$ in $\underline{\mathcal{G}}(G)$). To study $\mathcal{O}(1)$, we will use *factorizable line bundles* on various versions of loop Grassmannians, and these are defined and compared in [TZ19].

We now choose two relevant versions of the Abel-Jacobi map. For a smooth curve C, define the maps $\mathcal{C}_{C\times I} \xrightarrow{x\mapsto\overline{x}} \mathcal{H}_{C\times I} \xrightarrow{N} \mathcal{H}_C$. For $x = (c_1, i_1, \ldots, c_n, i_n) \in \mathcal{C}^n_{C\times I} = (C \times I)^n$ (so, $c_p \in C$, $i_p \in I$), let $\overline{x} = \sum_{i\in I} ix_i \in \mathcal{H}_{C\times I}$ with $x_i = \sum_{i_p=i} c_p \in \mathcal{H}_C$. Also, for $D_i \in \mathcal{H}_C$, let $N(\sum_{i\in I} iD_i) = \sum_I D_i$. For a semisimple group G, consider the global Abel-Jacobi map

$$\mathsf{AJ}_C^G \overset{\text{def}}{=} \left(\mathcal{C}_{C\times I} \xrightarrow{x\mapsto\overline{x}} \mathcal{H}_{C\times I} \xrightarrow{\mathrm{AJ}_C^G} \mathcal{G}_{\mathcal{H}_C}(G)\right), \quad \mathsf{AJ}_C^G(x) \overset{\text{def}}{=} \mathrm{AJ}_C^G(\overline{x}) = \left(N\overline{x}, \mathcal{O}_C(-\overline{x})\right),$$

with $N\overline{x} \in \mathcal{H}_C$ and $\mathcal{O}_C(-\overline{x}) \in \mathcal{G}_{\mathcal{H}_C}(T)_{N\overline{x}}$. We also consider a version at one point $0 \in C$:

$$\mathsf{AJ}_d^G \overset{\text{def}}{=} \left(\mathcal{C}_{d\times I} \xrightarrow{x\mapsto\overline{x}} \mathcal{H}_{d\times I} \xrightarrow{\mathrm{AJ}_d^G} \underline{\mathcal{G}}(G)\right), \quad \mathsf{AJ}_d^G(x) \overset{\text{def}}{=} \mathrm{AJ}_d^G(\overline{x}) \overset{\text{def}}{=} \mathcal{O}_C(-\overline{x}).$$

Proposition *Let G be simple and simply connected. Then the pullback $(AJ_d^G)^*\mathcal{O}(1)$ of $\mathcal{O}(1)$ via the above Abel-Jacobi map is the line bundle $\mathcal{O}_{\mathcal{C}_{d\times I}}(-\kappa\Delta^{\mathcal{C}})$ on $\mathcal{C}_{d\times I}$ corresponding to the negative of the Cartan form κ on $\mathbb{Z}[I]$. It has a canonical extension to a local line bundle $\mathcal{O}_{\mathcal{C}_{C\times I}}(-\kappa\Delta^{\mathcal{C}})$ on $\mathcal{C}_{C\times I}$.*

Proof We will use the curve $C = \mathbb{A}^1$ which contains the disc $d = \widehat{0}$. Then $\underline{\mathcal{G}}(G)$ is the fiber $\mathcal{G}_{\mathcal{C}_C}(G)_0$ of the Beilinson-Drinfeld Grassmannian $\mathcal{G}_{\mathcal{C}_C}(G)$ at the divisor $0 \in C = \mathcal{C}_C^1$. We use the extension of $\mathcal{O}(1)$ to a factorizable line bundle $\mathcal{O}^{\mathcal{C}}(1)$ on the loop Grassmannian $\mathcal{G}_{\mathcal{C}_C}(G)$. By Proposition 2.5 in [TZ19], this extension is unique once we trivialize $\mathcal{O}(1)$ at the origin of $\underline{\mathcal{G}}(G)$ (this is essentially Section 3.4 in [Zhu16]).

The restriction $\mathcal{L}$ of $\mathcal{O}^{\mathcal{C}}(1)$ to $\mathcal{G}_{\mathcal{C}_C}(T)\subseteq\mathcal{G}_{\mathcal{C}_C}(G)$ inherits the structure of a factorizable line bundle. This will imply that $(\mathsf{AJ}_d^G)^*\mathcal{O}(1)$ is canonically identified with the restriction of $(\mathsf{AJ}_C^G)^*\mathcal{O}^{\mathcal{C}}(1)$ to $\mathcal{H}_{d\times I}\subseteq\mathcal{H}_{C\times I}$. The point is that at $x \in \mathcal{C}_{d\times I}$, the fiber of $(\mathsf{AJ}_d^G)^*\mathcal{O}(1)$ is $\mathcal{L}_{0,\mathcal{O}_C(-\overline{x})}$, and the fiber of $(\mathsf{AJ}_C^G)^*\mathcal{O}^{\mathcal{C}}(1)$ is $\mathcal{L}_{N\overline{x},\mathcal{O}_C(-\overline{x})}$. The canonical identification of these fibers of $\mathcal{L}$ is a special case of the unique descent of any factorizable line bundle on $\mathcal{G}_{\mathcal{C}_{C\times I}}(T)$ to the so-called *rational Grassmannian* $\mathcal{G}_{rat}(T)$, and this was proved in Proposition 1.4 of [TZ19].

We are now interested in the line bundle $(\mathsf{AJ}_C^G)^*\mathcal{O}^{\mathcal{C}}(1)$ on $\mathcal{C}_{C\times I}$ with its structure of a local line bundle that it inherits—by definitions—from the factorization line bundle structure on $\mathcal{O}^{\mathcal{C}}(1)$. To any factorization line bundle L on $\mathcal{G}_{\mathcal{C}_C}(G)$, one associates a quadratic form $q_L \in QF(X_*(T))$ whose symmetric bilinear form $\kappa_L(\lambda,\mu) = q_L(\lambda+\mu) - q_L(\lambda) - q_L(\mu)$ is given by the order of vanishing of the locality structure along the diagonals $\Delta_C\subseteq C^2$ corresponding to the pair λ,μ.

The proof of Proposition 2.5 in [TZ19] checks that if G is simple and $\pi_1(G) = 0$ the quadratic form corresponding to $\mathcal{O}^{\mathcal{C}}(1)$ is the "minimal" integral invariant quadratic form q_m, characterized by $q_m(\check{\alpha}) = 1$ for short coroots $\check{\alpha}$.

Finally, in the simply connected case, this is the Cartan matrix κ in the sense that $\kappa_{ij} \overset{\text{def}}{=} \langle\alpha_i, \check{\alpha}_j\rangle$ equals $(\check{\alpha}_i, \check{\alpha}_j)$ for the sb-form $(-,-)$ given by the quadratic form q_m.

3.4.2 Zastava Spaces of Local Line Bundles

We can induce any local line bundle L on $\mathcal{H}_{C\times I}$ along a poset P. Consider the correspondence

$$(\mathcal{H}_{C\times I})^P \overset{\pi}{\longleftarrow} {}^P\mathcal{H} \overset{\sigma}{\longrightarrow} \mathcal{H}_{C\times I},$$

where the fiber $\sigma^{-1}\{D\}$ of ${}^P\mathcal{H}$ at $D \in \mathcal{H}_{C\times I}$ is the set $\mathrm{Hom}(P, \mathcal{H}_D)$ of maps E of posets, i.e., for $a \le b$ in P, one has $E^a \subseteq E^b \subseteq D$. This gives a sheaf $Ind^P(\mathsf{L}) \overset{\text{def}}{=} \sigma_*\pi^*(\mathsf{L}^{\boxtimes P})$ on $\mathcal{H}_{C\times I}$.

Lemma *$Ind^P(\mathsf{L})$ is a local sheaf on $\mathcal{H}_{C\times I}$. If P is an interval, $[m] = (1 < \cdots < m)$ then $Ind^P(\mathsf{L})$ is a local vector bundle.*[11]

Proof If $D = D^1 \sqcup D^2$ in $\mathcal{H}_{C\times I}$, then a representation of P in D is a pair of representations in D^i's. Sheaf $Ind^P(\mathsf{L})$ is a vector bundle if ${}^P\mathcal{H}$ is flat over $\mathcal{H}_{C\times I}$. If D is a length n subscheme of a smooth curve, then the length of the Hilbert scheme $\mathcal{H}_D$ is 2^n. This gives the case $m = 1$. The general case follows by induction in posets $i < \cdots < m$ as i goes from m to 1.

Define the mth *zastava space* of L as the local projective space

$$Z^m(\mathsf{L}) \overset{\text{def}}{=} \mathbb{P}^{loc}([Ind^{[m]}\mathsf{L}]^*).$$

Remarks

(0) One can extend the construction to zastava spaces $Z^P(\mathsf{L})$ for finite posets P, by replacing the projective space of $Ind^P\mathsf{L}^*$ with the projective spectrum of the symmetric algebra of $Ind^P\mathsf{L}$.
(1) Any symmetric bilinear form $\mathcal{Q}$ on $\mathbb{Z}[I]$ defines a local line bundle $\mathcal{O}(\mathcal{Q}) = \mathcal{O}_{\mathcal{H}_{C\times I}}(\mathcal{Q}\Delta^{\mathcal{H}})$ on $\mathcal{H}_{C\times I}$ (3.3.2), hence also zastava spaces $Z^P(I, \mathcal{Q}) \overset{\text{def}}{=} Z^P(\mathcal{O}(\mathcal{Q}))$.

Example When P is a point, we omit P from notation. Then ${}^P\mathcal{H} \overset{\sigma}{\longrightarrow} \mathcal{H}_{C\times I}$ is the relative Hilbert scheme $\mathcal{H}_{\mathcal{T}/\mathcal{H}_{C\times I}}$ for the tautological bundle $\mathcal{T}/\mathcal{H}_{C\times I}$, i.e., the fiber $\mathcal{T}_D$ at $D \in \mathcal{H}_{C\times I}$ is the Hilbert scheme $\mathcal{H}_D$ of all subschemes of the finite scheme $D \subseteq C \times I$.

If D is a point ai with $a \in C$ and $i \in I$, then $\mathcal{H}_{ai} = \{\emptyset, ai\}$, hence $Ind(\mathsf{L}) = \mathsf{L}_\emptyset \oplus \mathsf{L}_{ai} = \mathbb{k} \oplus \mathsf{L}_{ai}$. So the particle at ai is $\mathbb{P}^1$ with two chosen points. Then one is constructing $Z(\mathsf{L})$ by colliding $\mathbb{P}^1$'s according to the prescription given by the locality structure on the line bundle L.

[11] Flatness fails for the poset $P = (0 < a, b)$.

Similarly, when $P = [m] = \{1 < \cdots < m\}$, then all particles of $Z^m(\mathsf{L}) = Z^{[m]}(\mathsf{L})$ are $\mathbb{P}^m$. Actually, this $\mathbb{P}^m$ is naturally the mth symmetric power of the particle $\mathbb{P}^1$, so $\mathbb{P}^1$ embeds into $\mathbb{P}^m$ by Veronese embedding.

3.4.3 Weak Flatness of Zastavas

We start with a general statement about Kodaira embeddings.

Lemma *Let $\mathcal{L}$ be a relatively very ample line bundle on a projective scheme Y over a base variety S. Let X be a projective subvariety over an open dense $U \subseteq S$. Suppose that there exists a finite flat S-subscheme $F \subseteq Y$ such that over U it lies inside X and the restriction map $\mathcal{L}_{X/U} \to (\mathcal{L}_{F/S})|_U$ is an isomorphism.*[12] *Then the closure $\overline{X}$ of X in Y is a closed subscheme of $\mathbb{P}(\mathcal{L}_F{}^*) \cap Y$ and also $\mathcal{L}_{\overline{X}/S} = \mathcal{L}_{F/S}$ is a vector bundle.*

Proof For a sheaf $\mathcal{E}$, denote by $\mathbf{P}(\mathcal{E})$ the projective spectrum of the symmetric algebra of $\mathcal{E}$. The restriction map $\mathcal{L}_{Y/S} \to \mathcal{L}_{F/S}$ is surjective since $\mathcal{L}$ is very ample for Y/S and F/S is finite flat. Therefore, inside $\mathbf{P}(\mathcal{L}_{Y/F})$, we have both Y and $\mathbf{P}(\mathcal{L}_{F/S})$. Over U, both contain X since $\mathbf{P}(\mathcal{L}_{F/S})|_U = \mathbf{P}\mathcal{L}_{X/U}$, and this contains X since $\mathcal{L}$ is very ample. So $Y \cap \mathbf{P}(\mathcal{L}_{F/S})$ also contains $\overline{X}$.

The restriction $\rho : \mathcal{L}_{\overline{X}/S} \to \mathcal{L}_{F/S}$ is surjective since we have $F \subseteq \overline{X} \subseteq Y$, and $\mathcal{L}_{Y/S} \to \mathcal{L}_{F/S}$ is surjective since $\mathcal{L}$ is very ample for Y/S and F/S is finite flat. Moreover, if S is a variety, then $\mathrm{Ker}(\rho)$ is supported over the boundary $S - U$ of U. If X is a variety, then so is $\overline{X}$. Moreover, $\mathrm{Ker}(\rho) = 0$ since the line bundle $\mathcal{L}$ has no sections on $\overline{X}$ that vanish on X. So $\mathcal{L}_{\overline{X}/S} = \mathcal{L}_{F/S}$, and this is a vector bundle since F/S is finite flat.

Corollary *Recall that the zastava spaces $Z^m(\mathsf{L})$ lie inside the bundle of projective spaces $\mathbb{P}[Ind^{[m]}(\mathsf{L})^*]$ which carries the line bundle $\mathcal{O}(1)$. Then the sheaf $(Z(\mathsf{L})/\mathcal{H}_{C\times I})_*\mathcal{O}(1)$ is the vector bundle $Ind^{[m]}(\mathsf{L})$.*

Proof In the lemma, we choose $S \supseteq U$ as $\mathcal{H}_{C\times I} \supseteq \mathcal{H}^{reg}_{C\times I}$. Let F/S be the map σ in the correspondence $\mathcal{H}_{C\times I} \xleftarrow{p} {}^{[m]}\mathcal{H} \xrightarrow{q} \mathcal{H}_{C\times I}$ from Sect. 3.4.2.

A local line bundle L on S gives a line bundle $\pi^*\mathsf{L}^{\boxtimes m}$ on F, and $(\pi^*\mathsf{L}^{\boxtimes m})_{F/S}$ is just $Ind^{[m]}(\mathsf{L})$. Now, the projective bundle $Y/S \overset{\text{def}}{=} \mathbb{P}(\pi^*\mathsf{L}_{F/S})/S$ carries a very ample line bundle $\mathcal{L} = \mathcal{O}(1)$ and contains $\overline{X} = \mathbb{P}^{loc}([\pi^*\mathsf{L}]_{F/S}) = Z^m(\mathsf{L})$ and $X = \mathbb{P}^{loc,reg}([\pi^*\mathsf{L}]_{F/S}) = Z^{m,loc}(\mathsf{L})$. So the claim $\mathcal{L}_{\overline{X}/S} = \mathcal{L}_{F/S}$ from the lemma means that $(Z^m(L)/\mathcal{H}_{C\times I})_*\mathcal{O}(1)$ equals $({}^{[m]}\mathcal{H}/\mathcal{H}_{C\times I})_*\mathcal{O}(1)$. However, $\mathcal{O}(1)|_{{}^{[m]}\mathcal{H}}$ is just $\pi^*\mathsf{L}^{\boxtimes m}$ (a property of Kodaira embeddings), and $({}^{[m]}\mathcal{H}/\mathcal{H}_{C\times I})_*\pi^*\mathsf{L}^{\boxtimes m}$ is $Ind^{[m]}(\mathsf{L})$.

---

[12] We denote by $\mathcal{L}_{F/S}$ the direct image of $\mathcal{L}|_F$ to S, etc.

3.4.4 Grassmannians from Based Symmetric Bilinear Forms

Let $\mathcal{Q}$ be a quadratic form on $\mathbb{Z}[I]$. Its zastava space $Z^P(I,\mathcal{Q}) \overset{\text{def}}{=} Z^P(\mathcal{O}(\mathcal{Q}))$ defines the *semi-infinite* space $S^P(I,\mathcal{Q})$ over $\mathcal{H}_{C\times I}$, and the fiber at $D \in \mathcal{H}_{C\times I}$ is the colimit (union) of zastava fibers $Z^P(I,\mathcal{Q})_D$ at multiples of D

$$S^P(I,\mathcal{Q})_D = \lim_{\to n} Z^P(I,\mathcal{Q})_{nD}.$$

Finally, $\mathbb{N}[I]$ acts on $S^P(I,\mathcal{Q})$, and the corresponding *loop Grassmannian* is defined as

$$\mathcal{G}^P(I,\mathcal{Q}) \overset{\text{def}}{=} \mathbb{Z}[I] \times_{\mathbb{N}[I]} S^P(I,\mathcal{Q}).$$

3.4.5 Zastava Spaces $Z^\alpha(G)$ for Groups

We will recall these spaces from [FM99] which compares several definitions. The description in terms of the Beilinson-Drinfeld Grassmannian $\mathcal{G}^{BD}(G)$ is from Section 6 of [FM99].

We will be interested in the connected component $\mathcal{G}^{BD}(G_{sc})$ of $\mathcal{G}^{BD}(G)$. For a Borel B, the simple coroots identify $H = B/N$ with ${\mathbb{G}_m}^I$. Therefore, the singularities of a rational section τ of any H-torsor over C define an I-colored divisor D. We say that the *singularity of* τ is the degree $\deg(D) \in \mathbb{Z}[I]$.

For $\alpha \in \check{Q}^+$, we define $S^{BD}(\alpha)\subseteq\mathcal{G}^{BD}(G_{sc})$ so that the fiber $S_D(\alpha)$ at $D \in \mathcal{H}_C$ consists of all pairs $(\mathcal{T},\tau)$ of a G-torsor $\mathcal{T}$ over C and its section τ over $C-D$, such that the B-subtorsor $B\tau\subseteq\mathcal{T}|_{C-D}$ extends to C (necessarily as the closure $\overline{B\tau}$ in $\mathcal{T}$), and that the section τ of the H-torsor $N\backslash\overline{B\tau}$ has singularity α. If $\alpha \in \mathbb{N}[I]$, this means that the divisor D is in $\mathcal{H}^\alpha_{C\times I}$. At a single point $a \in C$, the fiber $S_a(\alpha)$ is the orbit $N_\mathcal{K}L_\alpha$ from Sect. 3.2.3.

Now, a pair of opposite Borels $B^\pm = TN^\pm$ (with $B^+ = B$) gives two semi-infinite stratifications $S^{BD,\pm}(\alpha)$, $\alpha \in \check{Q}^+$, of $\mathcal{G}^{BD}(G_{sc})$. For $\alpha \in \mathbb{N}[I]$, consider a version $\mathbf{Z}^\alpha(G)$ of "zastava" for G which is obtained by pulling back the intersection of closures $\overline{S_0^{BD,+}} \cap \overline{S_\alpha^{BD,-}}$ to the connected component $\mathcal{H}^\alpha_{C\times I}$ of the colored configuration space, via the sum map $\mathcal{H}_{C\times I} \to \mathcal{H}_C$, $(D_i)_{i\in I}\mapsto \sum D_i$. Our "zastava" space $\mathbf{Z} \to \mathcal{H}_{C\times I}$ is local since the singularity of a section τ regular off $D'\sqcup D''$ is the sum of contributions at D' and D''.

When ai is a point in $C \times I = \mathcal{H}^i_{C\times I}$, then the fiber $(\mathbf{Z}^i)_{ai}$ is $\mathbb{P}^1$ with Cartan fixed points $L_0, L_{\check{\alpha}_i}$ (this reduces to $G = SL_2$, and then it is as easily seen from the example in Sect. 3.2.2). Then by locality, the fiber at a regular divisor $D \in \mathcal{H}^{\alpha,reg}_{C\times I}$ is isomorphic to $(\mathbb{P}^1)^D$, and the T-fixed points in $\mathbf{Z}^\alpha_D$ are all L_β with $0 \le \beta \le \alpha$ in $\check{Q}$.

This in particular shows that the restriction $Z^{\alpha,reg} \overset{\text{def}}{=} \mathbf{Z}^\alpha|\mathcal{H}^{\alpha,reg}_{C\times I}$ is reduced. The zastava space that we are interested in is its closure Z^α in the scheme $\mathbf{Z}^\alpha$. By definition, it is a reduced local space.

Remarks

(0) (Comparison of Terminology with [FM99]). The space Z over $\mathcal{H}_{C\times I}$ is projective so we can call it *projective zastava*. We also have open subspaces $Z\supseteq\mathcal{Z}\supseteq\overset{\circ}{\mathcal{Z}}$. What is called *open zastava* in [FM99] is $\overset{\circ}{\mathcal{Z}}$ which is obtained by replacing $\overline{S_0^{BD,+}} \cap \overline{S_\alpha^{BD,-}}$ with $S_0^{BD,+} \cap S_\alpha^{BD,-}$. For $C = \mathbb{A}^1$, this is the space of *based maps* into the flag variety $Map[(\mathbb{P}^1, \infty), (\mathcal{B}, \mathfrak{b}_-)]$. What is called *zastava* in [FM99] is the space $\mathcal{Z}$ for which one removes one closure and uses $S_0^{BD,+} \cap \overline{S_\alpha^{BD,-}}$. For $C = \mathbb{A}^1$, this is the space of *based quasimaps* $QMap[(\mathbb{P}^1, \infty), (\mathcal{B}, \mathfrak{b}_-)]$. We can call it *affine zastava* since it is affine.

(1) The diagonal points of $\mathcal{H}^\alpha_{C\times I}$ are of the form $\alpha p = (a_i p)_{i\in I}$ for a point $p \in C$ and $\alpha = \sum_{i\in I} a_i i \in \mathbb{N}[I]$. The unions of central fibers of zastava spaces $\cup_{\alpha\in\check{Q}_+} Z^\alpha_{\alpha p} \supseteq \cup_{\alpha\in\check{Q}_+} \mathcal{Z}^\alpha_{\alpha p}$ are by definitions the semi-infinite varieties $\overline{S_0}\supseteq S_0$ in $\underline{\mathcal{G}}(G)$.

Moreover, one can view $\underline{\mathcal{G}}(G)$ as the increasing union of $\overline{S_\beta} = z^{-\beta}\overline{S_0} \cong \overline{S_0}$, $\beta \in \check{Q}^+$. So the whole loop Grassmannian is a certain direct limit $\lim_{\to\, \check{Q}^+} \overline{S_0}$. As $\alpha \in \mathbb{N}[I]$ acts on $\overline{S_0}$ by z^α, we can rewrite this limit as $\underline{\mathcal{G}}(G) = \mathbb{Z}[I] \times_{\mathbb{N}[I]} \overline{S_0}$.

3.4.6 Reconstruction of Loop Grassmannians $\mathcal{G}(G)$ Associated to Groups

Here, we check that the Grassmannians $\mathcal{G}^P(I, \mathcal{Q})$ from Sect. 3.4.4 indeed generalize the loop Grassmannian $\underline{\mathcal{G}}(G)$ of semisimple groups.

Theorem *Let G be a simply connected simple group which is simply laced. Let I index the simple roots, and let $\mathcal{Q}'$ be the modification of the Cartan matrix by dividing by 2 on the diagonal.*[13] *Then $Z^\alpha(I, -\mathcal{Q}')$, $S(I, -\mathcal{Q}')$, $\mathcal{G}(I, -\mathcal{Q}')$ are naturally identified with the corresponding notions $Z^\alpha(G)$, $\overline{S_0}(G)$, $\mathcal{G}(G)$ for the group G.*

Proof We only need to prove equality $Z^\alpha(I, \mathcal{Q}) = Z^\alpha(G)$ since the other two spaces are obtained from zastavas in the same way (compare Sect. 3.4.4 and the Remark 3.4.5.1).

Let $\mathcal{T}^\alpha$ be the tautological bundle over $\mathcal{H}^\alpha_{C\times I}$ so that the fiber at a colored divisor $D = \sum_{i\in I} D_i i$ is the finite flat scheme $D = \sqcup_i D_i i \subseteq C \times I$. The interesting object is the relative Hilbert scheme $\mathcal{H}(\mathcal{T}^\alpha/\mathcal{H}^\alpha_{C\times I})$, its fiber at D is $\mathcal{H}(D) \overset{\text{def}}{=} \prod_{i\in I} \mathcal{H}_{D_i}$, where $\mathcal{H}(D_i)$ is the moduli of all subschemes of a finite flat scheme D_i. The point is

[13] The modification appears because we use the Hilbert scheme $\mathcal{H}_C$ rather than over powers of curves $\mathcal{C}_C$ (see the Remark 3.3.2(1)).

that according to Proposition 3.2.3, the fixed points subscheme $(Z^\alpha(G))^T$ is exactly $\mathcal{H}(D) \stackrel{\text{def}}{=} \prod_{i\in I} \mathcal{H}_{D_i}$.[14]

Now, we can apply the weak flatness Lemma 3.4.3. Choose $S\supseteq U$ to be $\mathcal{H}^\alpha_{C\times I}\supseteq\mathcal{H}^{\alpha,reg}_{C\times I}$. For Y/S, we choose the pullback of the Beilinson-Drinfeld Grassmannian $\mathcal{G}^{BD}(G)$ to $\mathcal{H}_{C\times I}$ (by the sum map). It carries a very ample line bundle $\mathcal{L}$ which is the pull back of $\mathcal{O}_\mathcal{G}(1)$ from $\mathcal{G}^{BD}(G)$.

For $F\subseteq Y$, take $(Z^\alpha(G))^T = \mathcal{H}(\mathcal{T}^\alpha/\mathcal{H}^\alpha_{C\times I})$. We have already argued that it is flat in the proof of Lemma 3.4.2. Finally, let X/U be $Z^{\alpha,reg}(G)$, the restriction of $Z^\alpha(G)$ to $\mathcal{H}^{\alpha,reg}_{C\times I}$.

We still need to check that the restriction map $\mathsf{L}_{X/U} \to \mathsf{L}_{F_U/U}$ is an isomorphism. Due to locality of $Z(G)$ and of the line bundle $\mathcal{L}$ on Y, we only need the claim at colored divisors D which are points ai.

Here, Z^i_{ai} is $\mathbb{P}^1$ with T-fixed points $\{L_0, L_{\check\alpha_i}\}$. So one just needs to check that the restriction of L to Z^i_{ai} is $\mathcal{O}_{\mathbb{P}^1}(1)$. Recall that $\mathsf{L} = \mathcal{O}_\mathcal{G}(1)$ is $\mathcal{O}_{\mathcal{G}(G)}(\mathcal{D}^G)$ for the divisor $\mathcal{D}^G$ in $\mathcal{G}(G)$ defined as the boundary of the open orbit $\mathcal{G}^0(G)$ of the negative congruence subgroup. So it suffices to see that $Z^i_{ai}\cap\mathcal{D}^G$ is a single point. This claim reduces to the SL_2 subgroup S_i of G corresponding to the simple root α_i – since $D^G\cap\mathcal{G}(S_i) = D^{S_i}$. In the SL_2 case, one easily checks explicitly in the lattice model for the loop Grassmannian (see the Example 3.2.3).

Now, Lemma 3.4.3 guarantees that for $Z^\alpha(G) = \overline{Z^{\alpha,reg}(G)} = \overline{X}$ and any $D \in \mathcal{H}^\alpha_{C\times I}$, the restriction map $\Gamma[Z^\alpha(G)_D, \mathcal{O}(1)] \to \Gamma[Z^\alpha(G)^T_D, \mathcal{O}(1)]$ is an isomorphism. This implies that $Z^\alpha(G)$ is a local projective space for the local vector bundle dual to $\mathcal{L}_{Z^\alpha(G)^T/\mathcal{H}_{C\times I}}$ (we use the notation from Lemma 3.4.3).

Now, $Z(G)^T$ has been identified with the Hilbert scheme for $\mathcal{T}/\mathcal{H}_{C\times I}$, and then the local vector bundles $\mathcal{L}_{Z^\alpha(G)^T/\mathcal{H}_{C\times I}}$ and $Ind(I, -\mathcal{Q}')$ on the two spaces are identified by Proposition 3.4.1. Therefore, the corresponding local projective spaces $Z(G)$ and $Z(I, -\mathcal{Q}')$ are the same.

Remark The "simply laced" restriction can be removed using the folding mechanism. Here, this means an action of a group γ on the data $(I, \mathcal{Q})$. Furthermore, if γ is allowed to act on a curve above $C\times I$, then one would include loop Grassmannians for twisted affine Lie algebras and the Galois action.

3.4.7 The "Homological" Aspect of $\mathcal{G}^P(I, \mathcal{Q})$

The standard interpretation of loop Grassmannians $\underline{\mathcal{G}}(G)$ is cohomological (Sect. 3.1.1). The construction Sect. 3.4 provides what one could view as a homological interpretation. First, for a torus $T = \mathbb{G}^I_m$, one builds $\underline{\mathcal{G}}(T)$ from the union of formal discs $d \times I$ in stages $d \times I \mapsto \mathcal{H}_{d\times I} \mapsto \underline{\mathcal{G}}(\mathbb{G}^I_m)$. Here, the

[14]While the quoted lemma is stated at a single point of C, we actually need a version of the lemma for the family $Z^\alpha(G) \to \mathcal{H}^\alpha_{C\times I}$. This is easy using the moduli description of semi-infinite stratifications from Sect. 3.4.5.

configuration space $\mathcal{H}_{d\times I}$ is the free commutative monoid generated by $d \times I$, and then one inverts the center of the disc to get $\underline{\mathcal{G}}(\mathbb{G}_m^I)$ as the free commutative group on $d \times I$ (remark 0 in Sect. 3.1.2). One repeats this procedure for a reductive group G with a Cartan $T = \mathbb{G}_m^I$ by adding a local line bundle L on $\mathcal{H}_{d\times I}$, to get $\underline{\mathcal{G}}(G)$ (or more generally $\mathcal{G}^P(I, \mathcal{Q})$). First, the *positive part* of the loop Grassmannian is the zastava space $Z^P(I, \mathcal{Q})$ built using the monoid $\mathcal{H}_{d\times I}$ in Sect. 3.4.2. Then $\mathcal{G}^P(I, \mathcal{Q})$ itself is obtained from $Z^P(I, \mathcal{Q})$ in Sect. 3.4.4 by inverting $\mathbb{N}[I]\subseteq\mathcal{H}_{d\times I}$. We mention that for a complete curve C, a reconstruction of $Bun_G(C)$ from C has been pursued long ago in [FS94].

4 Local Line Bundles from Quivers

We know that local line bundles L on configuration spaces $\mathcal{H}_{C\times I}$ correspond to quadratic forms $\mathcal{Q}$ (Sect. 3.4.4), and the forms $\mathcal{Q}$ with nonnegative integer coefficients clearly correspond to graphs. In this section, we construct local line bundles directly from graphs or quivers. The advantage is that such construction extends to the quantum setting (see [YZ16] and Sect. 4.5 below). In the quantum setting, the "commutative" configuration space $\mathcal{H}_{\mathbb{G}\times I}$ will be replaced with the "noncommutative", i.e., ordered, configuration space $\mathcal{C}_{\mathbb{G}\times I} \stackrel{\text{def}}{=} \sqcup_n (\mathbb{G} \times I)^n$. On the level of representations of quivers, the noncommutative configuration space corresponds to passing to complete flags in representations.

We start with the curve $\mathbb{G}$ which is the one-dimensional group corresponding to a cohomology theory A.[15] Then the Hilbert scheme $\mathcal{H}_{\mathbb{G}\times I}$ of points in $\mathbb{G} \times \mathrm{I}$ is obtained as the cohomological schematization $\mathfrak{A}(\mathrm{Rep}_Q)$ of the moduli $\mathcal{V}^I$ of I-graded finite-dimensional vector spaces.

In Sect. 4.1, we recall various categories of representations of quivers and their extension correspondences. The cotangent complexes for these correspondences are considered in Sect. 4.3.

The "classical" *local* and *biextension* line bundles $\mathsf{L}(Q, A)$ and $\mathcal{L}(Q, A)$ on $\mathcal{H}_{\mathbb{G}\times I}$ and $(\mathcal{H}_{\mathbb{G}\times I})^2$ are constructed as Thom line bundles of moduli of extensions of representations in Sect. 4.2. Here, $\mathsf{L}(Q, A)$ can be defined directly from the incidence quadratic form of the quiver Q.

In Sect. 4.4, we calculate Thom line bundles associated to the cotangent correspondence and the effect of dilations. Finally, in Sect. 4.5, we recall the construction of the quantum group $\mathcal{U}_{\mathcal{D}}(Q, A)$ from the cotangent correspondence, and this leads us to select a choice of quantization of the above "classical" line bundles from Sect. 4.2.

Remark This section is largely a retelling of the paper [YZ17]. That paper is primarily concerned with the construction of quantum affine groups in the language

[15] Here, $\mathbb{G}$ is defined over the ring of constants $A(\mathrm{pt})$ of the cohomology theory.

of preprojective algebras which is here viewed as the cotangent bundle of the moduli Rep_Q. This "symplectic" setting allows to "quantize" the notion of local line bundles and the construction of loop Grassmannian from local line bundles. The quantization comes from the action of the dilation torus $\mathcal{D}$ on representations (which is in turn defined by a choice of a Nakajima function $\mathbf{m}$ on the set of arrows of the double $\overline{Q}$ of the quiver Q).[16]

4.1 Quivers

Let Q be a quiver with finite sets I and H of vertices and arrows. For each arrow $h \in H$, we denote by h' (resp. h'') the tail (resp. head) vertex of h. The opposite quiver $Q^* = (I, H^*)$ has the same vertices, and the set of arrows H^* is endowed with a bijection $* : H \to H^*$ so that $h \mapsto h^*$ exchanges sources and targets. The double $\overline{Q}$ of the quiver Q has vertices I and arrows $H \sqcup H^*$.

Let $\mathcal{V}^{\mathrm{I}}$ be the moduli of finite-dimensional I-graded vector spaces $V = \oplus_{i \in I} V^i$. Let Rep_Q be the moduli of representations of Q. Its fiber at $V \in \mathcal{V}^{\mathrm{I}}$ is the vector space $\mathrm{Rep}_Q(V)$ of representations on V. This is the sum over $h \in H$ of $Rep_Q(V)_h = \mathrm{Hom}(V^{h'}, V^{h''})$. We usually denote $v = \dim(V) \in \mathbb{N}[I]$, and let G be $\mathrm{GL}(V)$ so that the connected component $\mathrm{Rep}_Q(v)$—given by representations of fixed dimension $v \in \mathbb{N}[I]$—is $G \backslash Rep_Q(V)$.

4.1.1 Dilation Torus $\mathcal{D}$

A choice of Nakajima's weight function $\mathbf{m} : H \coprod H^* \to \mathbb{Z}$ gives an action of $\mathbb{G}_m^2$ on

$$T^* \mathrm{Rep}_Q(V) = \mathrm{Rep}_{\overline{Q}}(V) = \mathrm{Rep}_Q(V) \oplus \mathrm{Rep}_{Q^*}(V).$$

Elements (t_1, t_2) act for each $h \in H$ on $\mathrm{Rep}_Q(V)_h$ by $t_1^{\mathbf{m}_h}$ and on $\mathrm{Rep}_{Q^*}(V)_{h^*}$ by $t_2^{\mathbf{m}_{h^*}}$. We also let $\mathbb{G}_m^2$ act on the Lie algebra $\mathfrak{g}$ of $\mathrm{GL}(V)$ by $t_1 t_2$.

We choose a subtorus $\mathcal{D}$ of $\mathbb{G}_m^2$ and require that the moment map for the $GL(V)$-action on $T^* \mathrm{Rep}_Q(V)$ is $\mathcal{D}$-equivariant. This means that on $\mathcal{D}$, we have $t_1^{\mathbf{m}(h)} t_2^{\mathbf{m}(h^*)} = t_1 t_2$ for any $h \in H$. In particular, the symplectic form on $T^* \mathrm{Rep}_Q(V)$ has weight $t_1 t_2$.

---

[16] While [YZ17] deals with the case of elliptic cohomology, some of its ideas appear in an earlier paper [YZ14] which was only concerned with affine groups $\mathbb{G}$. This allowed for a trivialization of Thom line bundles which accounts for a different presentation of functoriality of cohomology in that paper.

Example

1. Nakajima's construction of quantum affine algebra associated to Q uses $\mathcal{D} = \mathbb{G}_m$, the diagonal torus in $\mathbb{G}_m^2$ (see [Nak01, (2.7.1), (2.7.2)]). Here, the $\mathcal{D}$-weight of the symplectic form on $T^* \operatorname{Rep}_Q(V)$ and on $\mathfrak{g}$ is 2, and the condition on $\mathbf{m}$ is $\mathbf{m}(h) + \mathbf{m}(h^*) = 2$. If there are a arrows in Q from vertex i to j, we fix a numbering $h_1, \cdots, h_a$ of these arrows, and let

$$\mathbf{m}(h_p) := a + 2 - 2p, \ \ \mathbf{m}(h_p^*) := -a + 2p, \ \ \text{for } p = 1, \cdots, a.$$

2. In [SV13], the elliptic Hall algebra (the spherical double affine Hecke algebra of GL_∞) is obtained from the choice $\mathcal{D} = \mathbb{G}_m^2$ and $\mathbf{m} = 1$.

4.1.2 The Extension Correspondence for Quivers

The moduli $\mathcal{R} = Rep_Q$ is given by pairs of $V \in \mathcal{V}^{\mathrm{I}}$ and $a \in \operatorname{Rep}_Q(V)$. We denote the elements of $\mathcal{R}^m$ as sequences $(V_\bullet, a_\bullet)$ of pairs of $(V_i, a_i) \in \mathcal{R}$.

Let $\mathcal{F}^m$ be the moduli of m-step filtrations $F = (0 = F^0 \subseteq F^1 \subseteq \cdots \subseteq F^m = V)$ on objects V of $\mathcal{V}^{\mathrm{I}}$. Similarly, we consider the moduli of filtrations $\mathcal{F}^m\mathcal{R}$ of representations, the objects are triples of $V \in \mathcal{V}^{\mathrm{I}}$, representation a of Q on V and a compatible filtration $F \in \mathcal{F}^m\mathcal{R}(V)$. We denote the fiber of $\mathcal{F}^m$ at $V \in \mathcal{V}^{\mathrm{I}}$ by $\mathcal{F}^m(V)$ and the fiber of $\mathcal{F}^m Rep_Q$ at $F \in \mathcal{F}^m(V)$ by $\mathcal{F}^m\mathcal{R}(F) = Rep_Q(F)$.

The fiber $\mathcal{R}(V)$ of $\mathcal{R}$ at $V \in \mathcal{V}^{\mathrm{I}}$ is $\operatorname{Rep}_Q(V)$. Also, $\operatorname{Rep}_{Q^*}(V) = \mathcal{R}(V)^*$ and for $\overline{\mathcal{R}} = Rep_{\overline{Q}}$, we have $\overline{\mathcal{R}}(V) = T^*\mathcal{R}(V)$. By representations on a sequence $V_\bullet = (V_i)_{k=1}^m \in \mathcal{V}^m$, we mean a sequence of representations, say $\mathcal{R}(V_\bullet) = \oplus_{k=1}^m \mathcal{R}(V_k)$.

A decomposition $\mathbf{f}$ of $v \in \mathbb{N}[I]$ as $\mathbf{v}_1 + \cdots + \mathbf{v}_m$ gives the connected components $\mathcal{F}^{\mathbf{f}}$ and $\mathcal{F}^m\mathcal{R}(\mathbf{f})$, given by $\dim[Gr_F(V)] = \mathbf{f}$. The stabilizer P of a chosen $F \in \mathcal{F}^{\mathbf{f}}(V)$ is a parabolic in G and then $\mathcal{F}^{\mathbf{f}} \cong P\backslash G$.

Now, the m-step *extension correspondence* for $\mathcal{R}$ is

$$\mathcal{R}^m \xleftarrow{p} \mathcal{F}^m\mathcal{R} \xrightarrow{q} \mathcal{R}$$

where $p(V, a, F) = Gr_F(V, a)$ and $q(V, a, F) = (V, a)$. The obvious splitting $\oplus_1^m$ of p is given by sending $(V_\bullet, a_\bullet)$ to $(\oplus_1^m V_i, \oplus_1^m a_i, F)$ for $F_p = \oplus_{k=1}^p V_k$.

A filtration F on vector spaces A, B defines a filtration on $\operatorname{Hom}(A, B)$, where operator x is in F_d if $xF_pA \subseteq F_{p+d}B$ for all p. In particular, we get a filtration $F_d(A^*) = F_{-d-1}^\perp$, and the two filtrations on $\operatorname{Hom}(B, A) \cong Hom(A, B)^*$ coincide.

So a filtration $F \in \mathcal{F}^m(V)$ induces a filtration on $\operatorname{Rep}_Q(V) \subseteq \operatorname{End}(\oplus_{i \in I} V^i)$ with $x \in F_d \operatorname{Rep}(V)$ if $xF^pV^{h'} \subseteq F^{p+d}V^{h''}$. Then $F_0 \operatorname{Rep}_Q(V)$ is the space $\operatorname{Rep}_Q(F)$ of representations compatible with F and $Gr_0^F \operatorname{Rep}_Q(V) = \operatorname{Rep}(Gr_F V)$. Also, the two filtrations on $Rep_{\overline{Q}}(V) = T^* \operatorname{Rep}_Q(V)$ coincide.

4.2 Thom Line Bundles

4.2.1 Classical Thom Bundles for Quivers

As $\text{Rep}_Q(V)$ is quadratic in V, we define its bilinear version $\text{Rep}_Q(V_1, V_2) = \oplus_{h\in H} \text{Hom}(V_1^{h'}, V_2^{h''})$ for $V_i \in \mathcal{V}^{\text{I}}$. Let $v_i = \dim(V_i)$ and denote $L = GL(V_1) \times GL(V_2)$. Over $\mathfrak{A}_L = \mathbb{G}^{(v_1)} \times \mathbb{G}^{(v_2)}$, we define the line bundle

$$\mathcal{L}(Q, A)_{v_1,v_2} \stackrel{\text{def}}{=} \Theta_L(\text{Rep}_Q(V_1, V_2)).$$

Lemma

(a) For a quiver $Q = (I, H)$, the Thom line bundle $\mathsf{L}(Q, A) \stackrel{def}{=} \Theta[\text{Rep}_Q]$ is a local line bundle $\mathcal{O}(-\mathcal{Q})$ on $\mathfrak{A}(\text{Rep}_Q) = \mathcal{H}_{\mathbb{G}\times I}$, corresponding to the incidence quadratic form $\mathcal{Q}$ of the quiver.

(b) The line bundle $\mathcal{L}(Q, A)_{V_1,V_2}$ on $\mathbb{G}^{(v_1)} \times \mathbb{G}^{(v_2)}$ is bilinear in V_1, V_2 in the sense that for the addition map $\mathbb{S} : GL(U') \times GL(U'') \hookrightarrow GL(U'\oplus U'')$, one has

$$\mathcal{L}_{U',V} \boxtimes \mathcal{L}_{U'',V} \cong (\mathfrak{A}_{\mathbb{S}})^*(\mathcal{L}_{U'\oplus U'',V})$$

and the same for V.

Proof (a) For $V \in \mathcal{V}^{\text{I}}$ and $G = GL(V)$, as a G-module $\mathcal{R}(V) = \text{Rep}_Q(V)$ is $\oplus_H \mathcal{R}_h(V)$ for $\mathcal{R}_h(V) = \text{Hom}(V^{h'}, V^{h''})$. The corresponding connected component of Rep_Q is a vector bundle $G\backslash \text{Rep}_Q(V)$ over $\mathbb{B}(G)$. Then the ideal $\Theta_G(\mathcal{R}_h)$ in $\mathcal{O}_{\mathfrak{A}_G}$ is generated by the function $\mathfrak{l}_{ch(\mathcal{R}_h)}$ (defined in Sect. 2.2.1) corresponding to the character of $\mathcal{R}(h)$.

A system of coordinates x_s^i on each V^i, $i \in I$ gives a Cartan T in G such that a basis in $X^*(T)$ can be denoted by x_s^i. If $i \xrightarrow{h} j$, then the character of $\mathcal{R}_h$ is $ch(\mathcal{R}_h) = \sum_s \sum_t x_t^j (x_s^i)^{-1}$; hence, $\mathfrak{l}_{ch(\mathcal{R}_h)} = \prod_{s,t} \mathfrak{l}(x_t^j (x_s^i)^{-1})$, and the same divisor is given by $\prod_{s,t} \mathfrak{l}(x_t^j) - \mathfrak{l}(x_s^i)$ which is the equation of the (i, j)-diagonal in $\mathfrak{A}_T \cong \prod_{i\in I} \mathbb{G}^{\dim(V_i)}$.

(b) $\text{Hom}(U, V)$ is bilinear in U and V. We use the obvious observation that if V_i is a module for G_i for $1 \le i \le n$, then $\Theta_{\prod G_i}(\oplus V_i) \cong \boxtimes \Theta_{G_i}(V_i)$. By multiplicativity of Θ, this reduces to the claim that for a representation V of G, $\Theta_{G\times G'}(V\boxtimes\Bbbk) = \Theta_G(V)\boxtimes\mathcal{O}_{\mathfrak{A}_{G'}}$.

For this, we can assume that G, G' are reductive, and then they can be replaced by their Cartans T, T'. Then we can also assume that V_i are characters χ of T. But then $\text{Ker}(\chi\boxtimes\Bbbk) = \text{Ker}(\chi) \times T'$, and this implies the claim.

4.3 Cotangent Versions of the Extension Diagram

4.3.1 The (Co)tangent Functoriality

The *tangent complex* of a map of smooth spaces $f : \mathcal{X} \to \mathcal{Y}$ is $T(f) = [T\mathcal{X} \xrightarrow{df} f^*T\mathcal{Y}]_{-1,0}$ on $\mathcal{X}$, and the dual *cotangent complex* is $T^*(f) = [f^*T^*\mathcal{Y} \xrightarrow{d^*f} T^*\mathcal{X}]_{0,1}$. When f is an embedding, these are the (co)normal bundles $T(f) \cong N(f)$ and $T^*(f) = T^*_{\mathcal{X}}\mathcal{Y} = N(f)^*$. The *Thom line bundle of a map* f is $\Theta(f) \stackrel{\text{def}}{=} \Theta[T(f)] = \Theta(f^*T\mathcal{Y})\Theta(T\mathcal{X})^{-1}$. For the map f, the direct image of A-cohomology takes the form of $f_* : \Theta(f) \to A(\mathcal{Y})$ [GKV95].

The *cotangent functoriality* associates to $f : \mathcal{X} \to \mathcal{Y}$ the correspondence

$$T^*\mathcal{Y} \xleftarrow{\tilde{f}} f^*T^*\mathcal{Y} \xrightarrow{d^*f} T^*\mathcal{X}.$$

Therefore, any correspondence $A \xleftarrow{p} C \xrightarrow{q} B$ of smooth spaces gives two cotangent correspondences $T^*A \xleftarrow{\tilde{p}} p^*T^*A \xrightarrow{d^*p} T^*C \xleftarrow{d^*q} q^*T^*B \xrightarrow{\tilde{q}} T^*B$ that compose to the correspondence $p^*T^*A \times_{T^*C} q^*T^*B$.[17] Say, in the category of schemes this fibered product consists of all $c \in C$, $\alpha \in T^*_{p(c)}A$, $\beta \in T^*_{q(c)}B$ such that $d^*p\alpha = d^*q\beta$, so by passing to $(c, \alpha, -\beta)$, we identify it with $T^*_C(A \times B)$. Then the cotangent version of the original correspondence is $T^*A \xleftarrow{\dot{p}} T^*_C(A \times B) \xrightarrow{\dot{q}} T^*B$.

4.3.2 Stacks

If X is a smooth variety with an action of a group G, then $G\backslash X$ is a smooth stack whose tangent complex is $[\mathfrak{g} \to TX]_{-1,0}$ and the cotangent complex is $[T^*X \to \mathfrak{g}^*]_{0,1}$.

We will consider a map of smooth varieties $X_1 \xrightarrow{f} X_2$ and $G_1 \to G_2$ a compatible map of groups G_i acting on X_i. Then for $\mathcal{X}_i = G_i\backslash X_i$, one gets $F : \mathcal{X}_1 \to \mathcal{X}_2$. We will calculate its cotangent correspondence $T^*\mathcal{X}_2 \xleftarrow{\tilde{F}} F^*T^*\mathcal{X}_2 \xrightarrow{d^*F} T^*\mathcal{X}_1$. First, the Thom line bundles for stacky versions are the equivariant Thom bundles for f plus a change of equivariance factor $\Theta_{G_1}(\mathfrak{g}_1/\mathfrak{g}_2)$ defined as $\Theta_{G_1}([\mathfrak{g}_1 \to \mathfrak{g}_2]_{0,1})$.

Lemma

(a) $\Theta(F) = \Theta_{G_1}(f) \otimes \Theta_{G_1}(\mathfrak{g}_1/\mathfrak{g}_2)$.
(b) $\Theta(d^*F) = \Theta_{G_1}(d^*f) \otimes \Theta_{G_1}(\mathfrak{g}_1/\mathfrak{g}_2)$.
(c) $\Theta(\tilde{F}) = \Theta_{G_1}(f) \otimes \Theta_{G_1}(\mathfrak{g}_1/\mathfrak{g}_2)$.

[17] The fibered product has to be derived for the relevant base change to hold unless d^*p, d^*q are transversal.

(d) *The pullback map on cohomology* $A(T^*\mathcal{X}_2)\xrightarrow{(\widetilde{F})^*}A(F^*T^*\mathcal{X}_2)$ *is the same as* $A(\mathcal{X}_2)\xrightarrow{F^*}A(\mathcal{X}_1)$. *Also,* $(d^*F)^*$ *is identity on* $A(\mathcal{X}_1)$.

Proof The (co)tangent complexes of spaces $\mathcal{X}_i$ are calculated by formulas $T(G\backslash X) = G\backslash(T_G X)$ and $T^*(G\backslash X) = G\backslash(T^*_G X)$, where $T_G X = [\mathfrak{g}\times X \to TX]_{-1,0}$ and $T^*_G X = [T^*X \to \mathfrak{g}^*\times X]_{0,1}$.

The map F gives pullbacks $F^*(T\mathcal{X}_2) = \frac{X_1}{G_1}\times_{\frac{X_2}{G_2}}\frac{T_{G_2}X_2}{G_2} = \frac{f^*T_{G_2}X_2}{G_1}$ and $F^*(T^*\mathcal{X}_2) = \frac{f^*T^*_{G_2}X_2}{G_1}$.

(a) The tangent complex $T(F) = [T\mathcal{X}_1 \to F^*T\mathcal{X}_2]_{-1,0}$ of the map F comes from (the G_1-quotient of) the map of complexes $[\mathfrak{g}_1 \to TX_1] \to [\mathfrak{g}_2 \to TX_2]$ given by $TX_1\xrightarrow{df}f^*TX_2$ and $\mathfrak{g}_1 \to \mathfrak{g}_2$. When we view this as a bicomplex with horizontal and vertical degrees in $[-1,0]$, then $T(F)$ is its total complex $[\mathfrak{g}_1 \to TX_1\oplus\mathfrak{g}_2 \to f^*TX_2]_{-2,0}$, which is an extension of complexes $[TX_1\xrightarrow{df}f^*TX_2]_{-1,0} = T(f)$ and $[\mathfrak{g}_1\to\mathfrak{g}_2]_{-2,-1}$. So $\Theta(F)$ is as stated.

Now, the cotangent correspondence can be written as

$$\begin{array}{ccccc} T^*\mathcal{X}_2 & \xleftarrow{\widetilde{F}} & F^*(T^*\mathcal{X}_2) & \xrightarrow{d^*F} & T^*\mathcal{X}_1 \\ \big\downarrow{=} & & \big\downarrow{=} & & \big\downarrow{=} \\ \frac{T^*_{G_2}X_2}{G_2} & \longleftarrow & \frac{f^*T^*_{G_2}X_2}{G_1} & \longrightarrow & \frac{T^*_{G_1}X_1}{G_1}. \end{array}$$

(b) Write d^*F as $(d^*F)^o/G_1$ where $(d^*F)^o : f^*[T^*X_2 \to \mathfrak{g}_2^*] \to [T^*X_1 \to \mathfrak{g}_1^*]$ is a map of complexes viewed as a bicomplex with all horizontal and vertical degrees in $[-1,0]$. So $T((d^*F)^o)$ is the total complex $[f^*T^*X_2 \to T^*X_1\oplus\mathfrak{g}_2^* \to \mathfrak{g}_1^*]_{-2,0}$ which is an extension of complexes $[f^*T^*X_2\xrightarrow{d^*f}T^*X_1]_{-2,-1}$ and $[\mathfrak{g}_2^*\to\mathfrak{g}_1^*]_{-1,0}$. So $\Theta(d^*F) = \Theta_{G_1}((d^*F)^o) = \Theta_{G_1}(d^*f)\otimes\Theta_{G_1}([\mathfrak{g}_2^*\to\mathfrak{g}_1^*]_{-1,0}) = \Theta_{G_1}(d^*f)\otimes\Theta_{G_1}([\mathfrak{g}_1\to\mathfrak{g}_2]_{0,1})$, and then we use invariance of the Thom line bundle under duality.

(c) Denote the complex $T^*_{G_2}X_2 = [T^*X_2 \to \mathfrak{g}_2^*]_{0,1}$ by $\mathcal{V}$, and let $\eta : f^*\mathcal{V}\to\mathcal{V}$. Then the map $\widetilde{F}$ is given by η and the change of symmetry $G_1 \to G_2$. So part (a) says that $\Theta(\widetilde{F}) = \Theta_{G_1}(\eta)\otimes\Theta_{G_1}([\mathfrak{g}_1\to\mathfrak{g}_2]_{0,1})$. Let us denote $\pi : \mathcal{V}\to X_2$ and $\pi : f^*\mathcal{V}\to X_1$, and then $T(\eta) = [T(f^*\mathcal{V})\xrightarrow{d\eta}\eta^*T(\mathcal{V})]_{-1,0}$ is $\pi^*T(f)$. (One has $0\to\pi^*\mathcal{V}\to T\mathcal{V}\to\pi^*TX_1\to 0$ and $0\to\pi^*f^*\mathcal{V}\to Tf^*\mathcal{V}\to\pi^*TX_2\to 0$. Now the map of complexes is identity on subsheaves $\pi^*f^*\mathcal{V}\cong\eta^*\pi^*\mathcal{V}$ and what remains is $\pi^*TX_2\to\eta^*\pi^*TX_1 = \pi^*f^*TX_1$.)

So $\Theta_{G_1}(\eta) = \Theta_{G_1}(\pi^*T(f))$, and since π is contractible, this is $\Theta_{G_1}(f)$.

(d) After contracting complexes of vector bundles, the maps $\widetilde{F}$ and d^*F become, respectively, the map $X_2/G_2\xleftarrow{F}X_1/G_1$ and the identity on X_1/G_1.

4.4 The A-Cohomology of the Cotangent Correspondence for Extensions

We recall the construction of [YZ14] of a quantum group in the above setup. It originated from the study of affine quantum groups in [Nak01] and [SV13] and is closely related to [KS11].

4.4.1 Connected Components of the Cotangent Correspondence

Fixing $V \in \mathcal{V}^{\mathrm{I}}$ and $F \in \mathcal{F}^m(V)$, let $\mathrm{Gr}_F(V) = \oplus_{k=1}^m V_k$. We denote $G = \mathrm{GL}(V)$, $L = \prod_{k=1}^m \mathrm{GL}(V_k)$ the automorphism group of $Gr_F(V)$, and P is a parabolic subgroup of G with a Levi subgroup L. Let U be the unipotent radical of P. Denote the Lie algebras by $\mathfrak{p}$, $\mathfrak{l}$, $\mathfrak{u}$.

These choices fix the connected component of the correspondence $\mathcal{R}^m \xleftarrow{p} \mathcal{F}^m\mathcal{R} \xrightarrow{q} \mathcal{R} \to \mathcal{V}^{\mathrm{I}}$ given by

$$L\backslash\mathcal{R}(\mathrm{Gr}_F(V)) \xleftarrow{p} P\backslash\mathcal{R}(F) \xrightarrow{q} G\backslash\mathcal{R}(V) \ \to \ G\backslash\,\mathrm{pt}\,. \tag{1}$$

4.4.2 Line Bundles from the Cotangent Correspondence

The extension correspondence gives two cotangent correspondences

$$T^*\mathcal{R}^m \xleftarrow{\widetilde{p}} p^*T^*\mathcal{R}^m \xrightarrow{d^*p} T^*\mathcal{F}^m\mathcal{R} \xleftarrow{d^*q} q^*T^*\mathcal{R} \xrightarrow{\widetilde{q}} T^*\mathcal{R}. \tag{2}$$

These compose to a single correspondence as in Sect. 4.3.1 which is the *cotangent correspondence* of the extension correspondence. We will not consider it since we are calculating here its effect on cohomology and this is the composition of effects of the above two simpler correspondences.

Let $V \in \mathcal{V}^{\mathrm{I}}$ and $F \in \mathcal{F}^m(V)$. We write the fiber of the correspondence (1) over F as

$$\mathcal{R}(\mathrm{Gr}_F(V)) \xleftarrow{\overset{o}{p}} \mathcal{R}(F) \xrightarrow{\overset{o}{q}} \mathcal{R}(V).$$

The connected component of the diagram (2) determined by F takes the form

$$T^*(L\backslash\mathcal{R}^m(Gr\,F)) \xleftarrow{\widetilde{p}} p^*T^*(L\backslash\mathcal{R}^m(Gr\,F)) \xrightarrow{d^*p}$$

$$T^*(P\backslash\mathcal{F}^m\mathcal{R}(F)) \xleftarrow{d^*q} q^*T^*(G\backslash\mathcal{R}(V)) \xrightarrow{\widetilde{q}} T^*(G\backslash\mathcal{R}(V)). \tag{3}$$

Lemma *With notations as above (and the filtration on $\mathcal{R}(F)$ as in Sect. 4.1.2), we have*

$$\Theta(d^*p) \cong \Theta_L(\mathfrak{g}/\mathfrak{p})\otimes\Theta_L(F_{-1}\mathcal{R}(F)) \text{ and } \Theta(\widetilde{q}) \cong \Theta_L[\mathcal{R}(V)/\mathcal{R}(F)]\otimes\Theta_L(\mathfrak{g}/\mathfrak{p})^{-1}.$$

Proof According to the Lemma 4.3.2.b $\Theta(d^*p)$ is $\Theta_L(d^*\overset{o}{p})\otimes\Theta_L([\mathfrak{p}\to\mathfrak{l}]_{0,1})$. The second factor is $\Theta_L(\mathfrak{u})$, since $\mathfrak{u}\cong(\mathfrak{g}/\mathfrak{p})^*$ we can write it as $\Theta_L(\mathfrak{g}/\mathfrak{p})$. For the first factor, as $\overset{o}{p}:\mathcal{R}(F)\to\mathcal{R}(\mathrm{Gr}_F(V))$, we get $d^*\overset{o}{p}:\mathcal{R}(F)\times\mathcal{R}(\mathrm{Gr}_F(V))^*\to\mathcal{R}(F)\to\mathcal{R}(F)^*$ so up to a factor $\mathcal{R}(F)$, this is $\mathcal{R}(\mathrm{Gr}_F(V))^*\hookrightarrow\mathcal{R}(F)^*$ with the quotient $[F_{-1}\mathcal{R}(F)]^*$. So the first factor is $\Theta_L([F_{-1}\mathcal{R}(F)]^*)=\Theta_L(F_{-1}\mathcal{R}(F))$.

Again, by Lemma 4.3.2.c $\Theta(\widetilde{q})$ is $\Theta_P(\overset{o}{q})\otimes\Theta_P([\mathfrak{p}\to\mathfrak{g}]_{0,1})$ for the embedding $[\mathcal{R}(F)\xrightarrow{\overset{o}{q}}\mathcal{R}(V)]_{-1,0}$. So the first factor is $\Theta_L[\mathcal{R}(V)/\mathcal{R}(F)]$, and the second is $\Theta_L(\mathfrak{g}/\mathfrak{p})^{-1}$.

4.4.3 Dilations

Recall the action of the dilation torus $\mathcal{D}\subseteq\mathbb{G}_m^2$ from Sect. 4.4.3. The weight of the first $\mathbb{G}_m$-factor on $\mathcal{R}$ is prescribed by $\mathbf{m}$ while the second factor acts trivially. Then the $\mathcal{D}$-action on $T^*\mathcal{R}$ is uniquely determined by asking that the natural symplectic form on $T^*\mathcal{R}$ has weight t_1t_2. We denote the $\mathcal{D}$-character of weight t_1t_2 by ω, so that the $\mathcal{D}$-action on $T^*\mathcal{R}$ is twisted by ω. This gives rise to the following twisted version of (3),

$$T^*(L\backslash\mathcal{R}^m(Gr F))\otimes\omega\xleftarrow{\widetilde{p}}p^*T^*(L\backslash\mathcal{R}^m(Gr F))\otimes\omega\xrightarrow{d^*p}T^*(P\backslash\mathcal{F}^m\mathcal{R}(F))\otimes\omega \tag{4}$$

$$\xleftarrow{d^*q}q^*T^*(G\backslash\mathcal{R}(V))\otimes\omega\xrightarrow{\widetilde{q}}T^*(G\backslash\mathcal{R}(V))\otimes\omega.$$

The maps in the above diagram are equivariant with respect to $\mathcal{D}$.

Now, we analyze $\mathcal{D}$-action on the relative tangent complexes of F, d^*F, and $\widetilde{F}$. Lemma 4.3.2 applies to induced actions on cotangent bundles. When working $\mathcal{D}$-equivariantly, we need to add an ω-twist. This applies to the Lie algebra factors in Lemma 4.3.2 that come from the cotangent complexes. On the other hand, the Lie algebra factors that come from the change of symmetry are not affected as they only carry the adjoint action. To simplify notations, for any group H, we denote $H\times\mathcal{D}$ by $\widetilde{H}$. Then

$$\Theta_{\mathcal{D}}(d^*F)\cong\Theta_{\widetilde{G}_1}(d^*f)\otimes\Theta_{\widetilde{G}_1}((\mathfrak{g}_1/\mathfrak{g}_2)\otimes\omega) \text{ and } \Theta_{\mathcal{D}}(F)=\Theta_{\widetilde{G}_1}(f)\otimes\Theta_{\widetilde{G}_1}(\mathfrak{g}_1/\mathfrak{g}_2).$$

Therefore, $\Theta_{\mathcal{D}}(\widetilde{F})=\Theta_{\widetilde{G}_1}(f)\otimes\Theta_{\widetilde{G}_1}(\mathfrak{g}_1/\mathfrak{g}_2)$.

Lemma *With notations above:* $\Theta(d^*p) \cong \Theta_{\widetilde{L}}(\mathfrak{g}/\mathfrak{p}\otimes\omega)\otimes\Theta_{\widetilde{L}}(F_{-1}\mathcal{R}(F))$ *and* $\Theta(\widetilde{q}) \cong \Theta_{\widetilde{L}}[\mathcal{R}(V)/\mathcal{R}(F)]\otimes\Theta_{\widetilde{L}}(\mathfrak{g}/\mathfrak{p})^{-1}$.

Proof According to Lemma 4.3.2.b $\Theta(d^*p)$ is $\Theta_{\widetilde{L}}(d^*\overset{o}{p})\otimes\Theta_{\widetilde{L}}([\mathfrak{p}\to\mathfrak{l}]_{0,1}\otimes\omega)$. The second factor is $\Theta_{\widetilde{L}}(\mathfrak{u}\otimes\omega)$, since $\mathfrak{u}\cong(\mathfrak{g}/\mathfrak{p})^*$ we can write it as $\Theta_{\widetilde{L}}(\mathfrak{g}/\mathfrak{p}\otimes\omega)$.

Again, by Lemma 4.3.2.c, $\Theta(\widetilde{q})$ is $\Theta_{\widetilde{P}}(\overset{o}{q})\otimes\Theta_{\widetilde{P}}([\mathfrak{p}\to\mathfrak{g}]_{0,1})$ for the linear embedding $[\mathcal{R}(F)\overset{\overset{o}{q}}{\longrightarrow}\mathcal{R}(V)]_{-1,0}$. So the first factor is $\Theta_{\widetilde{L}}[\mathcal{R}(V)/\mathcal{R}(F)]$, and the second is $\Theta_{\widetilde{L}}(\mathfrak{g}/\mathfrak{p})^{-1}$, as it comes from the change of symmetries.

4.5 $\mathcal{D}$-Quantization of the Monoid $(\mathcal{H}_{\mathbb{G}\times I},+)$

Here, we recall the construction from [YZ17] of a deformation $(\mathcal{C}oh(\mathcal{H}_{\mathbb{G}\times I}),\star)$ of the convolution on the monoid $(\mathcal{H}_{\mathbb{G}\times I},+)$. The quantum group $U_{\mathcal{D}}(Q,A)$ and its positive part $U^+_{\mathcal{D}}(Q,A)$ were constructed in [YZ14, YZ16], as algebra objects in $(\mathcal{C}oh(\mathcal{H}_{\mathbb{G}\times I}),\star)$, and hence in particular as R-algebras.

4.5.1 Local and Biextension Line Bundles $\mathsf{L}_{\mathcal{D}}(Q,A)$ and $\mathcal{L}_{\mathcal{D}}(Q,A)$

These will be upgrades of $\mathsf{L}(Q,A)$ and $\mathcal{L}(Q,A)$ from Sect. 4.2.1. They will be constructed as special cases of line bundles associated to cotangent correspondences of extension moduli (3).

Case 1 The *biextension* line bundle $\mathcal{L}=\mathcal{L}_{\mathcal{D}}(Q,A)$ comes from $m=2$, i.e., the 2-step filtrations $\mathcal{F}^2(V)$ of V. For $\mathrm{Gr}_F(V)=V_1\oplus V_2$

$$\mathcal{L}_{V_1,V_2} := \Theta(d^*p)\otimes\Theta(\widetilde{q})$$

is a line bundle on $\mathfrak{A}_{\widetilde{L}}\cong\mathfrak{A}_{G(V_1)}\times\mathfrak{A}_{G(V_2)}\times\mathfrak{A}_{\mathcal{D}}$.

Case 2 Our quantum version $\mathsf{L}=\mathsf{L}_{\mathcal{D}}(Q,A)$ of the local line bundle $\mathsf{L}(Q,A)$ depends on a choice of a *type* of a complete flag $F\in\mathcal{F}^m(V)$ which is $\mathbf{f}=\dim(\mathrm{Gr}_F(V))\in(\mathbb{N}^I)^m$. Then

$$[\mathsf{L}_{\mathcal{D}}(Q,A)]_{V,\mathbf{f}} \overset{\text{def}}{=} \Theta(d^*p)\otimes\Theta(\widetilde{q})$$

is a line bundle on $\mathfrak{A}_{\widetilde{L}}=\mathbb{G}^{|V|}\times\mathfrak{A}_{\mathcal{D}}$, where $|V|=\sum_{i\in I}\dim(V^i)$ (here, the Levi subgroup L is a Cartan in $GL(V)$). It is called *the local line bundle.*

One easily sees that the restrictions of "quantum objects" $\mathsf{L}_{\mathcal{D}}(Q,A)$ and $\mathcal{L}_{\mathcal{D}}(Q,A)$ to $0\in\mathfrak{A}_{\mathcal{D}}$ are the classical Thom line bundles $\mathsf{L}(Q,A)$ and $\mathcal{L}(Q,A)$ from Sect. 4.2.1.

4.5.2 Convolutions and Biextensions

We recall the monoidal structure $\star$ on coherent sheaves on $\mathcal{H}_{\mathbb{G}\times I}\times\mathfrak{A}_{\mathcal{D}}$ (over the base scheme $\mathfrak{A}_{\mathcal{D}}$) from [YZ17].

For a smooth curve C, $\mathcal{H}_{C\times I}$ is a commutative monoid freely generated by C. The operation $\mathbb{S}:\mathcal{H}_{C\times I}\times\mathcal{H}_{C\times I}\to\mathcal{H}_{C\times I}$ is the addition of divisors ("symmetrization"). Since it is a finite map, it defines a convolution operation on the abelian category $\mathcal{C}oh(\mathcal{H}_{C\times I})$ of coherent sheaves by $\mathcal{F}*\mathcal{G}=\mathbb{S}_*(\mathcal{F}\boxtimes\mathcal{G})$.

The following definition can be found in [P, P126]. A line bundle $\mathcal{L}$ over $(\mathcal{H}_{C\times I})^2$ is a *biextension* (or *Poincare* line bundle) if we have the following isomorphism of line bundles

$$a_{x_1,x_2;y}:\mathcal{L}_{x_1+x_2,y}\to\mathcal{L}_{x_1,y}\otimes\mathcal{L}_{x_2,y},$$
$$a_{x;y_1,y_2}:\mathcal{L}_{x,y_1+y_2}\to\mathcal{L}_{x,y_1}\otimes\mathcal{L}_{x,y_2},$$

which satisfy the following cocycle conditions

(i) $a_{x_1+x_2,x_3;y}\circ(a_{x_1,x_2;y}\otimes\mathrm{id})=a_{x_1,x_2+x_3;y}\circ(\mathrm{id}\otimes a_{x_2,x_3;y})$,

(ii) $a_{x;y_1+y_2,y_3}\circ(a_{x;y_1,y_2}\otimes\mathrm{id})=a_{x;y_1,y_2+y_3}\circ(\mathrm{id}\otimes a_{x;y_2,y_3})$,

(iii) $(a_{x_1,x_2;y_1}\otimes a_{x_1,x_2;y_2})a_{x_1+x_2;y_1,y_2}=(a_{x_1;y_1,y_2}\otimes a_{x_2;y_1,y_2})a_{x_1,x_2;y_1+y_2}$.

This is equivalent to a central extension of the monoid $(\mathcal{H}_{C\times I},+)$ (or its group completion) by $\mathbb{G}_m$.

Now, $\mathcal{L}$ twists the convolution on $\mathcal{C}oh(\mathcal{H}_{C\times I})$ to another monoidal structure $\mathcal{F}\star\mathcal{G}\overset{\text{def}}{=}\mathbb{S}_*[(\mathcal{F}\boxtimes\mathcal{G})\otimes\mathcal{L}]$.

From now on, the curve C will be $\mathbb{G}=\mathfrak{A}_{\mathbb{G}_m}$.

Lemma *The line bundle $\mathcal{L}=\mathcal{L}_{\mathcal{D}}(Q,A)$ on $(\mathcal{H}_{C\times I})^2\times\mathfrak{A}_{\mathcal{D}}$ defined in Sect. 4.5.1.1 is an $\mathfrak{A}_{\mathcal{D}}$-family of biextension line bundles. This gives a "$\mathcal{D}$-twisted" convolution on $\mathcal{C}oh(\mathcal{H}_{\mathbb{G}\times I}\times\mathfrak{A}_{\mathcal{D}})$ by*

$$\mathcal{F}\star\mathcal{G}\overset{def}{=}\mathbb{S}_*[(\mathcal{F}\boxtimes_{\mathfrak{A}_{\mathcal{D}}}\mathcal{G})\otimes\mathcal{L}].$$

Proof We need to check that the quantum version of $\mathcal{L}$ is still a biextension. Notice that the quantum version has an extra factor $\Theta_{\mathcal{D}}(\mathfrak{g}/\mathfrak{p})$. However, since for $m=2$ the space $\mathfrak{g}/\mathfrak{p}$ is of the form $\mathrm{Hom}(V_1,V_2)$, the argument in the proof of Lemma 4.2.1.b applies again.

Proposition ([YZ17, Theorem A, Theorem 3.1])

(a) $(\mathcal{C}oh(\mathcal{H}_{\mathbb{G}\times I}\times\mathfrak{A}_{\mathcal{D}}),\star)$ is a monoidal category with a meromorphic braiding which is symmetric. The unit is the structure sheaf on $\mathcal{H}^0_{\mathbb{G}\times I}\times\mathfrak{A}_{\mathcal{D}}$.
(b) The structure sheaf on $\mathcal{H}_{\mathbb{G}\times I}\times\mathfrak{A}_{\mathcal{D}}$ is an algebra object in this category.

For any $\tau \in \mathfrak{A}_{\mathcal{D}}$, *we denote by* $\mathcal{L}_\tau$ *the restriction of* $\mathcal{L}$ *to* $\tau \in \mathfrak{A}_{\mathcal{D}}$, *and* $\mathcal{F} \star_\tau \mathcal{G} := \mathbb{S}_*[(\mathcal{F} \boxtimes_{\mathfrak{A}_{\mathcal{D}}} \mathcal{G}) \otimes \mathcal{L}_\tau]$.

Remark One way to motivate the $\mathcal{L}$-twisted convolution of coherent sheaves on $(\mathcal{H}_{\mathbb{G}\times I})^2 \times \mathfrak{A}_{\mathcal{D}}$ is to notice that when the cohomology theory A extends to constructible sheaves, then for a constructible $\mathcal{F}$ on a space X, the cohomology $A(\mathcal{F})$ is a coherent sheaf on $\mathfrak{A}(X)$. In this case, the A-cohomology functor intertwines the convolution of constructible sheaves on Rep_Q and the $\mathcal{L}$-twisted convolution of coherent sheaves on $\mathfrak{A}(\mathrm{Rep}_Q) = (\mathcal{H}_{\mathbb{G}\times I})^2 \times \mathfrak{A}_{\mathcal{D}}$. (This follows as in the proof of Lemma 4.4.2.)

4.5.3 Quantum Groups $U^+_{\mathcal{D}}(Q, A) \subset U_{\mathcal{D}}(Q, A)$

Now, we consider the setup of Sect. 4.4.2 with $V \in \mathcal{V}^{\mathrm{I}}$ and $F \in \mathcal{F}^m(V)$. Let $\mathbf{f} = \dim(\mathrm{Gr}_F(V)) \in (\mathbb{N}^I)^m$ be the type of the filtration F. Applying the cohomology theory A to the diagram (3), we have the following multiplication map associated to $\mathbf{f}$:

$$
\begin{aligned}
m_{\mathbf{f}} := (\tilde{q}_*) \circ (d^* q^*) \circ (d^* p_*) \circ (\tilde{p}^*) : \mathbb{S}_*(\Theta(\tilde{q}) \otimes \Theta(d^* p)) \to \\
A_{\mathcal{D}}(T^*(G \backslash \mathcal{R}(V))) \cong \mathcal{O}_{\mathfrak{A}_G \times \mathfrak{A}_{\mathcal{D}}}, \qquad (5)
\end{aligned}
$$

where $\mathbb{S} : \mathfrak{A}_L \to \mathfrak{A}_G$ is the symmetrization map.

Let $Sph(V)$ be the set of types v of filtrations in $\mathcal{F}^m(V)$ consisting of complete flags (so $m = |v| \overset{\mathrm{def}}{=} \sum_{i\in I} v^i$). We define $U^+_{\mathcal{D}}(Q, A)$ so that on the connected component $\mathfrak{A}_G \times \mathfrak{A}_{\mathcal{D}}$,

$$
(U^+_{\mathcal{D}}(Q, A))_V \overset{\mathrm{def}}{=} \sum_{\mathbf{f}\in Sph(V)} \mathrm{Image}(m_{\mathbf{f}}) \subseteq A_{\widetilde{G}} = \mathcal{O}_{\mathfrak{A}_G \times \mathfrak{A}_{\mathcal{D}}}.
$$

Lemma

1. *The coherent sheaf* $U^+_{\mathcal{D}}(Q, A)$ *is an ideal sheaf on* $\mathfrak{A}_G \times \mathfrak{A}_{\mathcal{D}}$.
2. $U^+_{\mathcal{D}}(Q, A)$ *is an* $\mathfrak{A}_{\mathcal{D}}$*-family of algebras in the monoidal categories* $(Coh(\mathcal{H}_{\mathbb{G}\times I}), \star_\tau)$.

Proof As each $m_{\mathbf{f}}$, for $\mathbf{f} \in Sph(V)$, is a morphism of coherent sheaves, the image $\mathrm{Image}(m_{\mathbf{f}})$ is a coherent subsheaf in $\mathcal{O}_{\mathfrak{A}_G \times \mathfrak{A}_{\mathcal{D}}}$. Since $Sph(V)$ is a finite set, $(U^+_{\mathcal{D}}(Q, A))_V$ is a sum of finitely many coherent subsheaves, so it is itself a coherent subsheaf of $\mathcal{O}_{\mathfrak{A}_G \times \mathfrak{A}_{\mathcal{D}}}$. A coherent subsheaf of the structure sheaf is a sheaf of ideals, hence so is $(U^+_{\mathcal{D}}(Q, A))_V$.

For (2), the algebra structure on $(U^+_{\mathcal{D}}(Q, A))_V$ is defined using $m_{\mathbf{f}}$, where F is the two-step filtrations in Sect. 4.5.1 **Case 1**.

The sheaf $U^+_{\mathcal{D}}(Q, A)$ on $\mathfrak{A}_G \times \mathfrak{A}_{\mathcal{D}}$ is denoted by $\mathcal{P}^{sph}$ in [YZ14], since it is the spherical subalgebra of the cohomological Hall algebra of preprojective algebra.

The affine quantum group $U_{\mathcal{D}}(Q, A)$ associated to the quiver Q and the cohomology theory A is defined in [YZ16] as the Drinfeld double of $U^+_{\mathcal{D}}(Q, A)$. The quantization parameters of $U_{\mathcal{D}}(Q, A)$ are given by $\mathfrak{A}_{\mathcal{D}}$. This Drinfeld double was constructed in [YZ16] using a comultiplication and a bialgebra pairing on an extended version of $U^+_{\mathcal{D}}(Q, A)$. $U^+_{\mathcal{D}}(Q, A)$ itself also has a coproduct but in the meromorphic braided tensor category $(\mathcal{C}oh(\mathcal{H}_{\mathbb{G}\times I}), \star)$ [YZ17]. The affine quantum group $U_{\mathcal{D}}(Q, A)$ acts on the corresponding A-homology of the Nakajima quiver varieties (see [YZ14, YZ16]), generalizing a construction of Nakajima [Nak01].

4.5.4 Local Line Bundle from Zastava and $U^+_{\mathcal{D}}(Q, A)$

In this section, we associate to a quiver Q a different integral form $\underline{U}^+_{\mathcal{D}}(Q, A)$ of the quantum group $U^+_{\mathcal{D}}(Q, A)$. There is an algebra homomorphism $U^+_{\mathcal{D}}(Q, A) \to \underline{U}^+_{\mathcal{D}}(Q, A)$, which becomes an isomorphism after a certain localization to be explained below.

Recall from Sect. 3.4.1 that for a semisimple simply laced group of simply connected type, the restriction from the loop Grassmannian gives a local line bundle on $\mathcal{H}_{C\times I}$ related to the Cartan quadratic form. On the other hand, a quiver Q produces the local line bundle $\mathsf{L}(Q, A)$ corresponding to the incidence quadratic form of Q (Sect. 4.2.1). So when Q is the corresponding Dynkin graph, the quadratic forms differ on the diagonal. Then the new quantum group $\underline{U}^+_{\mathcal{D}}(Q, A)$ will be related to the Cartan quadratic form.

To define $\underline{U}^+_{\mathcal{D}}(Q, A)$ as an algebra object in a monoidal category, we modify the local and biextension line bundles from Sect. 4.5.1. We follow the notations from Sect. 4.4.3. Let $V \in \mathcal{V}^{\mathrm{I}}$ and $F \in \mathcal{F}^m(V)$. Let $\mathbf{f} = \dim(\mathrm{Gr}_F(V)) \in (\mathbb{N}^I)^m$ be the type of the filtration F. Let G be the automorphism group of V and P be the parabolic subgroup preserving F and L the automorphism group of $\mathrm{Gr}_F(V)$. Consider the $\tilde{L}$-representation $(\mathfrak{u} \oplus \mathfrak{g}/\mathfrak{p}) \otimes \omega$.

Now assume that $\mathbf{f} \in Sph(V)$, i.e., it is a type of a complete flag. By the invariance under the symmetric group, $\Theta_{\tilde{L}}((\mathfrak{u} \oplus \mathfrak{g}/\mathfrak{p}) \otimes \omega)$ (as a line bundle on $\mathfrak{A}_L \times \mathfrak{A}_{\mathcal{D}}$), is obtained from pullback of a line bundle on $\mathfrak{A}_G \times \mathfrak{A}_{\mathcal{D}}$, which by an abuse of notation is denoted by $\Theta_{\tilde{G}}((\mathfrak{u} \oplus \mathfrak{g}/\mathfrak{p}) \otimes \omega)$. Twisting by this line bundle, we get from (5) the following map:

$$m_{\mathbf{f}} := (\tilde{q}_*) \circ (d^* q^*) \circ (d^* p_*) \circ (\tilde{p}^*) \tag{6}$$

with

$$m_{\mathbf{f}} : \mathbb{S}_*(\Theta(\tilde{q}) \otimes \Theta(d^* p) \otimes \Theta_{\tilde{L}}((\mathfrak{u} \oplus \mathfrak{g}/\mathfrak{p}) \otimes \omega))^{-1} \to \Theta_{\tilde{G}}((\mathfrak{u} \oplus \mathfrak{g}/\mathfrak{p}) \otimes \omega)^{-1}.$$

We define $(\underline{\mathsf{L}}_{\mathcal{D}}(Q, A))_{V,\mathbf{f}}$ to be $\Theta(\tilde{q}) \otimes \Theta(d^* p) \otimes \Theta_{\tilde{L}}((\mathfrak{u} \oplus \mathfrak{g}/\mathfrak{p}) \otimes \omega)^{-1}$. We define $\underline{U}^+_{\mathcal{D}}(Q, A)$ so that on the connected component of $\mathfrak{A}_G \times \mathfrak{A}_{\mathcal{D}}$ containing V,

$$(\underline{U}^{+}_{\mathcal{D}}(Q,A))_V \stackrel{\text{def}}{=} \sum_{\mathfrak{f}\in Sph(V)} \text{Image}(m_{\mathfrak{f}}) \subseteq \Theta_{\tilde{G}}((\mathfrak{u}\oplus\mathfrak{g}/\mathfrak{p})\otimes\omega)^{-1}.$$

Similarly, we have the modified biextension line bundle $\underline{\mathcal{L}}_{\mathcal{D}}(Q,A)$ coming from the 2-step filtrations $\mathcal{F}^2(V)$ of V. For $\text{Gr}_F(V) = V_1 \oplus V_2$

$$\underline{\mathcal{L}_{V_1,V_2}} := \Theta_{\tilde{L}}(\mathfrak{g}/\mathfrak{p}\otimes\omega)^{-1} \otimes \Theta_{\tilde{L}}(F_{-1}\mathcal{R}(F)) \otimes \Theta(\tilde{q})$$

is a line bundle on $\mathfrak{A}_{\tilde{L}} \cong \mathfrak{A}_{G(V_1)} \times \mathfrak{A}_{G(V_2)} \times \mathfrak{A}_{\mathcal{D}}$.

Restrictions of these "quantum objects" $\underline{\mathsf{L}}_{\mathcal{D}}(Q,A)$ and $\underline{\mathcal{L}}_{\mathcal{D}}(Q,A)$ to $0 \in \mathfrak{A}_{\mathcal{D}}$ gives line bundles on $\mathcal{H}_{C\times I}$ and $\mathcal{H}_{C\times I}{}^2$, denoted by $\underline{\mathsf{L}}(Q,A)$ and $\underline{\mathcal{L}}(Q,A)$, respectively.

Similar to Sect. 4.5.3, $\underline{\mathcal{L}}_{\mathcal{D}}(Q,A)$ defines a family of monoidal structure on $\mathcal{H}_{\mathbb{G}\times I}$, denoted by $(\mathcal{C}oh(\mathcal{H}_{\mathbb{G}\times I}), \underline{\star}_\tau)$.

Proposition *Let Q be a simply laced Dynkin quiver, with Cartan quadratic form $\underline{Q}$ and the simply connected group G.*

1. *The local line bundle $\underline{\mathsf{L}}(Q,A)$ on $\mathcal{H}_{C\times I}$ is isomorphic to $AJ^*\mathcal{O}_{\mathcal{G}(G)}(1)$.*
2. *$\underline{U}^{+}_{\mathcal{D}}(Q,A)$ is an $\mathfrak{A}_{\mathcal{D}}$-family of algebras in the monoidal categories $(\mathcal{C}oh(\mathcal{H}_{\mathbb{G}\times I}), \underline{\star}_\tau)$.*

Proof The proof of (1) is similar to that of Lemma 4.2.1. By Lemma 4.2.1, the local line bundle $\mathsf{L}(Q,A)$ is associated to the incidence quadratic form $\mathcal{Q}$ of Q. On the component associated to $\mathbf{f} \in Sph(V)$, tensoring $\mathsf{L}(Q,A)$ with $\Theta_G(\mathfrak{u}\oplus\mathfrak{g}/\mathfrak{p})$, we get $\underline{\mathsf{L}}(Q,A)$, which is associated to the Cartan quadratic form $\underline{Q}$. On the other hand, when G is the simply connected group whose Dynkin diagram is Q, then $AJ^*\mathcal{O}_{\mathcal{G}(G)}(1)$ is associated to $\underline{Q}$ by Proposition 3.4.1.

Proof of (2) is similar to Sect. 4.5.3. The algebra structure on $(\underline{U}^{+}_{\mathcal{D}}(Q,A))_V$ is defined using $m_{\mathfrak{f}}$, where taking F to be a 2-step filtration gives $\underline{\mathcal{L}}_{\mathcal{D}}(Q,A)$.

Remarks

(1) There is an algebra homomorphism $U^{+}_{\mathcal{D}}(Q,A) \to \underline{U}^{+}_{\mathcal{D}}(Q,A)$. Topologically, for $V \in \mathcal{V}^{\mathrm{I}}$, the map $U^{+}_{\mathcal{D}}(Q,A))_V \to (\underline{U}^{+}_{\mathcal{D}}(Q,A))_V$ is induced by

$$z^* z_* : \Theta_{\tilde{L}}((\mathfrak{u}\oplus\mathfrak{g}/\mathfrak{p})\otimes\omega) \to A_{\mathcal{D}}(T^*(L\,\mathcal{R}^m(\text{Gr}\,F)))$$

where z is the zero section of the vector bundle on $T^*(L\backslash\mathcal{R}^m(\text{Gr}\,F))$ which is the pullback of $(\mathfrak{u}\oplus\mathfrak{g}/\mathfrak{p})\otimes\omega$ form $\tilde{L}\backslash$ pt. In particular, the map becomes an isomorphism after inverting the Euler class of $(\mathfrak{u}\oplus\mathfrak{g}/\mathfrak{p})\otimes\omega$.

(2) The shuffle formula for $\underline{U}^+_{\mathcal{D}}(Q,A)$ is similar to that of $U^+_{\mathcal{D}}(Q,A)$ given in [YZ16, § 1.2].[18]
(3) The integral form $\underline{U}^+_{\mathcal{D}}(Q,A)$ has originally appeared in [BFN19, Appendix B]. The isomorphism can be seen directly by comparing the shuffle formula for $\underline{U}^+_{\mathcal{D}}(Q,A)$ with the formula [FT18, (3.6.3)].

5 Loop Grassmannians $\mathcal{G}^P_{\mathcal{D}}(Q,A)$ and Quantum Locality

In the preceding Sect. 4, we have attached to a quiver $Q = (I, H)$ and a cohomology theory A a local line bundle $\mathsf{L}(Q,A)$ on the colored configuration space $\mathcal{H}_{\mathbb{G}\times I}$ of the curve $\mathbb{G}$ given by A (Lemma 4.2.1.a). Modified as in Sect. 4.5.4, in Sect. 3.4 the local line bundle $\underline{\mathsf{L}}(Q,A)$ can be used to produce a "loop Grassmannian" $\mathcal{G}(Q,A)$ over $\mathcal{H}_{\mathbb{G}\times I}$.

The local line bundle $\mathsf{L}(Q,A)$ is closely related to the biextension line bundle $\mathcal{L}(Q,A)$ from Lemma 4.2.1.b. In Sect. 4, we have also recalled the construction of the affine quantum groups $\underline{U}^+_{\mathcal{D}}(Q,A)$ and used this to select the "correct" quantizations $\underline{\mathsf{L}}_{\mathcal{D}}(Q,A)$ and $\underline{\mathcal{L}}_{\mathcal{D}}(Q,A)$ of the above line bundles, on the basis of relation to this quantum group (Sect. 4.5.1).

While pieces $\underline{\mathsf{L}}(Q,A)_\alpha$ of the classical local line bundle depend on $\alpha \in \mathbb{N}[I]$ parameterizing connected components of $\mathcal{H}_{\mathbb{G}\times I}$, the pieces $\underline{\mathsf{L}}_{\mathcal{D}}(Q,A)_{\mathbf{i}}$ of the quantum version depends on a choice of $\mathbf{i} \in I^{\mathbb{N}}$ (Sect. 4.5.1). This really means that we are dealing with the noncommutative (ordered) configuration spaces $\mathcal{C} = \mathcal{C}_{\mathbb{G}\times I} = \sqcup\, (\mathbb{G}\times I)^n$ so that each $\alpha \in \mathbb{N}[I]$ is refined to all $\mathbf{i} = (i_1, \dots, i_n) \in I^n$ with $\sum i_p = \alpha$. The connected components given by all refinements $\mathbf{i}$ of the same α are related by the meromorphic braiding from [YZ17]. So the information carried by all refinements $\mathbf{i}$ of α is (only) generically equivalent.

Altogether, $\mathcal{G}_{\mathcal{D}}(Q,A)$ can still be constructed by the same prescription as in the case of $\mathcal{G}(Q,A)$. However, the local line bundle $\underline{\mathsf{L}}_{\mathcal{D}}(Q,A)$ now lives on the larger ("noncommutative") configuration space $\mathcal{C}_{\mathbb{G}\times I}$. The zastava space $Z_{\mathcal{D}}(Q,A)$ over $\mathcal{C} = \mathcal{C}_{\mathbb{G}\times I}$ is first defined generically in $\mathcal{C}$ where fibers are products of projective lines. Then the singularities of the locality structure prescribe how fibers degenerate. Finally, passing from the zastava space to loop Grassmannian is given by the procedure of extending the free monoid on I to the free group on I.

Altogether, the key difference in the quantum case is seen in the configuration space. It has more connected components (but they are related by braiding), and the singularities of locality structure (hence also the notion of locality) are now the diagonals shifted by the quantum parameter.

[18] Placing the numerator of the factor fac_1 in Equation [YZ14, (2)] on the denominator to get the corresponding factor in the shuffle formula for $\underline{U}^+_{\mathcal{D}}(Q,A)$. The homomorphism from (1) is on the level of shuffle algebras the multiplication by the Euler class of $(\mathfrak{u}\oplus\mathfrak{g}/\mathfrak{p})\otimes\omega$.

5.1 The "Classical" Loop Grassmannians $\mathcal{G}^P(Q, A)$

The choice of A influences the space $\mathcal{G}^P(Q, A)$ only through the curve $\mathbb{G}$. Whenever $\mathbb{G}$ is a formal group, then the orientation $\mathfrak{l}$ of A identifies $\mathbb{G}$ with the coordinatized formal disc d.

However, since the loop Grassmannian $\underline{\mathcal{G}}(\mathbb{G}_m)$ is the free commutative group indscheme generated by d, the group law on d given by A induces a commutative ring structure on the loop Grassmannian $\underline{\mathcal{G}}(\mathbb{G}_m)$. This is the group algebra of the group $\mathbb{G}$ taken in algebraic geometry.

Remark The universal Witt ring has the same nature, and it is the homology $\mathbb{H}_*(\mathbb{A}^1, 0)$ of the multiplicative monoid $(\mathbb{A}^1, 0, ; \cdot)$ in pointed spaces. Observations of this nature have already been made in [BZ95, Str00, No09].

5.2 Quantization Shifts Diagonals

Any Thom line bundle is the ideal sheaf of the corresponding Thom divisor. While the Thom divisor corresponding to $\mathsf{L}(Q, A)$ is a combination of diagonals of $\mathcal{H} = \mathcal{H}_{d\times I}$, the quantization shifts these diagonals in the configuration space $\mathcal{C} = \mathcal{C}_{d\times I}$.

We first examine how an added action of a torus $\mathcal{D}$ affects the Thom divisor in general (Sect. 5.2.1), and then we specialize this to the local line bundle $\underline{\mathsf{L}}_{\mathcal{D}}(Q, A)$ in Sect. 5.2.2.

5.2.1 Deformation of a Thom Divisor from an Additional Torus $\mathcal{D}$

For a representation E of a product $\widetilde{G} = G\times\mathcal{D}$, we can view the line bundle $\Theta_{\widetilde{G}}(E)$ on $\mathfrak{A}_G\times\mathfrak{A}_{\mathcal{D}}$ as a family of line bundles $\Theta_{\widetilde{G}}(E)_\tau$ (for $\tau\in\mathfrak{A}_{\mathcal{D}}$), on $\mathfrak{A}_G \overset{\tau}{\hookrightarrow} \mathfrak{A}_G\times\mathfrak{A}_{\mathcal{D}}$. If E contains no trivial characters of a Cartan T, we will see that this deformation lifts to divisors.

First, consider the case when G is a torus T and $E = \chi\boxtimes\zeta^{-1}$ for characters χ, ζ of $T, \mathcal{D}$ (so $\chi\neq 0$). Then for any $\tau\in\mathfrak{A}_{\mathcal{D}}$, the restriction $\Theta_{\widetilde{T}}(E)_\tau$ to $\mathfrak{A}_T$ is the ideal sheaf of the divisor

$$\mathrm{Ker}(\mathfrak{A}_{\chi\boxtimes\zeta^{-1}})\cap[\mathfrak{A}_T\times\tau] \;=\; {\mathfrak{A}_\chi}^{-1}(\mathfrak{A}_\zeta(\tau))\subset\mathfrak{A}_T.$$

Here, $\chi : T\to\mathbb{G}_m$ induces the homomorphism $\mathfrak{A}_\chi : \mathfrak{A}_T\to\mathfrak{A}_{\mathbb{G}_m} = \mathbb{G}$ as in Sect. 2.2.1, and ${\mathfrak{A}_\chi}^{-1}(\mathfrak{A}_\zeta(\tau))$ is a divisor in $\mathfrak{A}_T$. For $\tau = 0$, this is the divisor $\mathrm{Ker}(\mathfrak{A}_\chi)$ whose ideal sheaf is $\Theta_T(\chi)$, and in general, ${\mathfrak{A}_\chi}^{-1}(\mathfrak{A}_\zeta(\tau))$ is its torsor which we think of as a shift of $\mathrm{Ker}(\mathfrak{A}_\chi) = {\mathfrak{A}_\chi}^{-1}(0)$ by $\mathfrak{A}_\zeta(\tau)\in\mathbb{G}$.

Now, for any reductive group G with a Cartan T and Weyl group W, we decompose E according to $\mathcal{D}$-action as $E = \oplus_{\zeta \in X^*(\mathcal{D})} (E_\zeta \boxtimes \zeta^{-1})$, for some G-modules E_ζ. Then $\Theta_G(E_\zeta)$ is the ideal sheaf of some divisor, denoted by $D(E_\zeta)$, in $\mathfrak{A}_G = \mathfrak{A}_T//W$. As T-representations, we have the decomposition $E_\zeta|_T = \oplus_{\chi \in X^*(T)} [E_\zeta : \chi]\chi$. Therefore, the divisor $D(E_\zeta)$ is a sum over $\chi \in X^*(T)$ of divisors $[E_\zeta : \chi]\cdot \mathrm{Ker}(\mathfrak{A}_\chi)$. Now, for any $\tau \in \mathfrak{A}_{\mathcal{D}}$, $\Theta_{\widetilde{G}}(E)_\tau$ is the ideal sheaf of the shifted divisors $D(E_\zeta) + \mathfrak{A}_\zeta(\tau)$ of $D(E_\zeta)$.

5.2.2 Quantum Diagonals

In our quiver setting, each $h \in H \sqcup H^*$ defines (via the Nakajima function $\mathbf{m}$) a character $\mu_h \in X^*(\mathcal{D})$, by which $\mathcal{D}$ acts on the component $Rep_{\overline{Q}}(V)_h$ of $Rep_{\overline{Q}}(V)$.

For ${}_1V$, ${}_2V$ in $\mathcal{V}^{\mathrm{I}}$ for each $i \in I$ choose coordinates ${}_sx^i_p$ on ${}_sV^i$ hence a decomposition of ${}_sV^i$ into lines ${}_sV^i_p$. This gives Cartans $T_s \subseteq G_s = GL({}_sV)$ with a basis ${}_sx^i_p$ of $X^*(T_s)$. Then on the line $\mathrm{Hom}({}_1V^i_p, {}_2V^j_q)$, the torus $\widetilde{T} \overset{\mathrm{def}}{=} T_1 \times T_2 \times \mathcal{D}$ acts by ${}_2x^j_q \cdot ({}_1x^i_p)^{-1} \cdot \mu_h$, so its Thom divisor is given by vanishing of $\mathfrak{A}_{{}_2x^j_q} + \mathfrak{A}_{\mu_h} - \mathfrak{A}_{{}_1x^i_p}$ in $\mathfrak{A}_{T_1 \times T_2 \times \mathcal{D}}$. Therefore, the Thom divisor of the $T_1 \times T_2 \times \mathcal{D}$-module $\mathrm{Rep}_{\overline{Q}}({}_1V, {}_2V)_h$ is the shifted diagonal

$$\Delta^{v_1,v_2}_h(\tau) \overset{\mathrm{def}}{=} \Delta^{v_1,v_2}_{h',h''} + (0, \tau_h) \subset \mathfrak{A}_{T_1} \times \mathfrak{A}_{T_2}.$$

Here, $\tau_h = \mathfrak{A}_{\mu_h}(t)$ depends on h, and $\Delta^{v_1,v_2}_{h',h''} \subset \mathfrak{A}_{T_1} \times \mathfrak{A}_{T_2}$ is the diagonal divisor defined by vanishing of $\prod_{p,q}(\mathfrak{A}_{{}_2x^j_q} - \mathfrak{A}_{{}_1x^i_p})$, and the shift $\Delta^{v_1,v_2}_{h',h''} + (0, \tau_h)$ means that for ${}_2V^j$, we use the embedding of $\mathbb{G} = \mathfrak{A}_{\mathbb{G}_m}$ into $\mathfrak{A}_{GL({}_2V^j)}$ via $\mathbb{G}_m = Z(GL({}_2V^j))$ and the corresponding addition action of $\mathbb{G}$ on $\mathfrak{A}_{GL({}_2V^j)}$.

Consider the diagonal Δ^{v_1,v_2}_i of $\mathbb{G}^{v_1^i} \times \mathbb{G}^{v_2^i}$ given by the vanishing of $\prod_{p,q}(\mathfrak{A}_{{}_1x^i_p} - \mathfrak{A}_{{}_2x^i_q})$. Let $\Delta^{v_1,v_2}_i(\tau) \overset{\mathrm{def}}{=} \Delta^{v_1,v_2}_i + (\tau, 0)$, where $\tau = \mathfrak{A}_\omega(t)$. The character $\omega \in X^*(\mathcal{D})$ is as before. The shift $\Delta^{v_1,v_2}_{h',h''} + (\tau, 0)$ means that for ${}_1V^j$, we use the embedding of $\mathbb{G} = \mathfrak{A}_{\mathbb{G}_m}$ into $\mathfrak{A}_{GL({}_1V^i)}$ via $\mathbb{G}_m = Z(GL({}_1V^i))$ and the corresponding addition action of $\mathbb{G}$ on $\mathfrak{A}_{GL({}_1V^i)}$, given by the vanishing of $\prod_{p,q}(\mathfrak{A}_{{}_1x^i_p} - \mathfrak{A}_{{}_2x^i_q} + \mathfrak{A}_\omega)$.

We will say that for $\tau \in \mathfrak{A}_{\mathcal{D}}$, and $D_s = (D^i_s)_{i \in I} \in \mathbb{G}^{|v_s|}$ for $s = 1, 2$; the pair (D_1, D_2) is $(\mathbf{m}, \tau)$-*disjoint* if (D_1, D_2, τ) and (D_2, D_1, τ) do not lie in any of the shifted diagonals $\Delta^{v_1,v_2}_h(\tau)$, Δ^{v_1,v_2}_i, $\Delta^{v_1,v_2}_i(\tau)$. Equivalently, for any $i \in I$, the divisors $D^i_1 \pm \tau$ and D^i_2 are disjoint, and D^i_1 and D^i_2 are disjoint; for each $h : h' \to h''$ in $\overline{H}$, $D^{h''}_2 \pm \tau_h$ and $D^{h'}_1$ are disjoint.

5.3 Quantum Locality

We will now consider locality in the setting of the (noncommutative) monoid $\mathcal{C}_{\mathbb{G}\times I}$ freely generated by $\mathbb{G} \times I$.

Let $\mathcal{C}_I$ be the free monoid on I, so elements are ordered sequences $\gamma = i_1 i_2 \cdots i_N$ of elements in I. The product of $\gamma = i_1 i_2 \cdots i_N, \gamma' = j_1 j_2 \cdots j_{N'}$, is the concatenation $\gamma + \gamma' = i_1 i_2 \cdots i_N j_1 j_2 \cdots j_{N'}$.

Let $\mathcal{C}_{\mathbb{G}\times I} = \sqcup\, (\mathbb{G} \times I)^n = \sqcup_{\gamma \in \mathcal{C}_I} \mathbb{G}^\gamma$ be the indscheme monoid freely generated by $\mathbb{G} \times I$, with connected components labeled by $\mathcal{C}_I$. The natural projection, from the free monoid to the free commutative monoid is denoted $\varpi : \mathcal{C}_{\mathbb{G}\times I} \to \mathcal{H}_{\mathbb{G}\times I}$.

We will use the notation $\mathcal{L} = \mathcal{L}_\mathcal{D}(Q, A)$ both for the biextension line bundle defined on $\mathcal{H}^2_{\mathbb{G}\times I} \times \mathfrak{A}_\mathcal{D}$ in Sect. 4.5.1 and also for its pullback to $\mathcal{C}^2_{\mathbb{G}\times I} \times \mathfrak{A}_\mathcal{D}$. For $\gamma', \gamma'' \in \mathcal{C}_I$, we denote by $\mathcal{L}_{\gamma',\gamma''}$ its restriction to the component $\mathbb{G}^{\gamma'} \times \mathbb{G}^{\gamma''} \times \mathfrak{A}_\mathcal{D}$,

5.3.1 m-Locality

An **m**-*locality structure* on a vector bundle K on $\mathcal{C}_{\mathbb{G}\times I} \times \mathfrak{A}_\mathcal{D}$ is a consistent system of isomorphisms

$$(K_{\gamma_1,\tau} \boxtimes K_{\gamma_2,\tau}) \otimes \mathcal{L}_{\gamma_1,\gamma_2} \cong K_{\gamma_1+\gamma_2,\tau} \quad \text{for } \tau \in \mathfrak{A}_\mathcal{D}.$$

Any **m**-locality structure on K implies an algebra structure on K in the monoidal category $\big(Coh(\mathcal{C}_{\mathbb{G}\times I} \times \mathfrak{A}_\mathcal{D}), \star\big)$ (by the biextension property of $\mathcal{L}$). In this way, an **m**-locality structure on K is the same as a structure of a $\star$-algebra, whose multiplications are isomorphisms.[19]

Example The line bundle $\mathsf{L} = \mathsf{L}_\mathcal{D}(Q, A)$ constructed component-wise in Sect. 4.5.1 **Case 2**, is a line bundle over $\mathcal{C}_{\mathbb{G}\times I} \times \mathfrak{A}_\mathcal{D}$, and it has a natural **m**-locality structure. So is the modification $\underline{\mathsf{L}}_\mathcal{D}(Q, A)$ as in Sect. 4.5.4. We will write the proof only generically:

Lemma *The line bundle* L *on* $\mathcal{C}_{\mathbb{G}\times I} \times \mathfrak{A}_\mathcal{D}$ *defined in Sect. 4.5.1* ***Case 2*** *has the property that* $(D_1, D_2) \in (\mathcal{C}_{\mathbb{G}\times I})^2$ *is* $(\mathbf{m}, \tau)$-disjoint *for* $\tau \in \mathfrak{A}_\mathcal{D}$, *and then there is a canonical identification of fibers*

$$\mathsf{L}_{D_1+D_2,\tau} \cong \mathsf{L}_{D_1,\tau} \otimes \mathsf{L}_{D_2,\tau}.$$

Proof Let $V = V_1 \oplus V_2$ in $\mathcal{V}^\mathrm{I}$ and $G_i = GL(V_i)$ and $G = GL(V)$. Choose a Cartan T_i in G_i. Then $\mathrm{Rep}_{\overline{Q}}(V) \cong \mathrm{Rep}_{\overline{Q}}(V_1) \oplus \mathrm{Rep}_{\overline{Q}}(V_2) \oplus \mathrm{Rep}_{\overline{Q}}(V_1, V_2) \oplus \mathrm{Rep}_{\overline{Q}}(V_2, V_1)$

[19] Notice that this is stronger than the standard definition of locality which only requires such isomorphism over the regular part of the configuration space where $\mathcal{L}$ happens to trivialize by Sect. 5.2.

gives

$$\Theta_{\widetilde{G}}[\mathrm{Rep}_{\overline{Q}}(V)]\otimes\Theta_{\widetilde{G}}[\mathrm{Rep}_{\overline{Q}}(V_1)]^{-1}\otimes\Theta_{\widetilde{G}}[\mathrm{Rep}_{\overline{Q}}(V_2)]^{-1}$$
$$\cong \Theta_{\widetilde{G}}[\mathrm{Rep}_{\overline{Q}}(V_1, V_2)]\otimes\Theta_{\widetilde{G}}[\mathrm{Rep}_{\overline{Q}}(V_2, V_1)].$$

Now the disjointness condition implies that the last two factors have canonical trivializations at (D_1, D_2, τ). A similar statement holds for $\Theta_{\widetilde{L}}(\mathfrak{g}/\mathfrak{p} \otimes \omega)$, and $\Theta_{\widetilde{L}}(\mathfrak{g}/\mathfrak{p})^{-1}$, where $\mathfrak{g}/\mathfrak{p} = \oplus_{i=1}^{2}\mathfrak{g}_i/\mathfrak{p}_i$.

Therefore, $\mathcal{L}_{D_1,D_2} = \Theta(d^*p) \otimes \Theta(\widetilde{q})$ has a canonical trivialization when (D_1, D_2) is $(\mathbf{m}, \tau)$-disjoint. The claim now follows from the identification $(\mathsf{L}_{D_1,\tau} \boxtimes \mathsf{L}_{D_2,\tau}) \otimes \mathcal{L}_{D_1,D_2} \cong \mathsf{L}_{D_1+D_2,\tau}$.

Remark The quantum local line bundle L is in a sense a localization of the quantum group $U^+_{\mathcal{D}}(Q, A)$ to the noncommutative configuration space $\mathcal{C}_{\mathbb{G}\times I}$. By its definition, the α-weight space $U^+_{\mathcal{D}}(Q, A)(\alpha)$ is a sum of contributions from all refinements $\gamma \in \mathcal{C}_I$ of a given $\alpha \in \mathbb{N}[I]$.[20]

In the classical case $\mathcal{D} = 1$, for all γ above, $\alpha\mathsf{L}_\gamma$ are the same, so the sum $U^+_{\mathcal{D}}(Q, A)$ is the line bundle L. However, upon quantization, there is a genuine dependence on γ, and one has to take the sum of all contributions in order to construct a subalgebra.

Example In the case when I is a point (the "sl_2-case"), then $\mathcal{C}_I = \mathbb{N}[I] = \mathbb{N}$; hence, $\mathcal{C}_{\mathbb{G}\times I}$ is the system $\sqcup_{n\in\mathbb{N}}\mathbb{G}^n$ of Cartesian powers of $\mathbb{G}$. Then $\underline{\mathsf{L}}_n = \varpi_n^*(\underline{U}^+_{\mathcal{D}}(Q, A)_n)$.

5.3.2 Some Expectations

The above construction of loop Grassmannians is of "existential" nature, with hidden difficulties of explicit computations. We hope to ameliorate this difficulty by some equivalent descriptions. Our construction is based on "abelianization" (as we construct sections of $\mathcal{O}(1)$ on the loop Grassmannian from the same objects for a Cartan subgroup) and on locality (as we interpret equations of the projective embedding of the Grassmannian as locality conditions).

We would like to describe these equations in more standard terms by constructing a central extension of the quantum group $U_{\mathcal{D}}(Q, A)$ and its action on sections of $\mathcal{O}(1)$. Here, the central extension should appear as one extends the "quantum local" line bundle $\underline{\mathsf{L}}_{\mathcal{D}}(Q, A)$ from the analogue $\mathcal{C}_{\mathbb{G}\times I}$ of $\mathcal{H}_{\mathbb{G}\times I}$ to an analogue of $\mathcal{G}(T)$.

One could also try to construct the graded algebra of section of line bundles $\mathcal{O}(m)$ by choosing the poset P in $\mathcal{G}^P_{\mathcal{D}}(Q, A)$ to be $1 < \cdots < m$.

---

[20] One formal way to say it is that $U^+_{\mathcal{D}}(Q, A)_\alpha$ is the smallest subsheaf on $\mathbb{G}^{(\alpha)}$ such that its pullback to each refinement $\mathcal{C}_\gamma$ contains L_γ.

Appendix 1: Loop Grassmannians with a Condition

We recall a general technique providing modular description of some parts of loop Grassmannian. This allows us to finish (in Appendix "Proof of the Proposition 3.2.3") the proof of Proposition 3.2.3 on T-fixed points in closures of semi-infinite orbits.

Moduli of Finitely Supported Maps

Here, we recall some elements of Drinfeld's notion of loop Grassmannians with a geometric ("asymptotic") condition. This material will be covered in more details elsewhere. We will fix a smooth curve C.

We are interested in various moduli of G-torsors over a curve C that are local spaces over C. As observed by Beilinson and Drinfeld, the relevant spaces Y are usually of the form $\mathcal{M}_{\mathcal{Y}}(C)$, the moduli of finitely supported, i.e., *generically trivialized* maps into some pointed stack $(\mathcal{Y}, \mathrm{pt})$ built from G. (We usually omit the point pt from notation.)

The Subfunctor $\underline{\mathcal{G}}(G, Y) \subseteq \underline{\mathcal{G}}(G)$ Given by "Condition Y"

Let C be a smooth connected curve with the generic point η_C. Let G be an algebraic group and (Y, y) a pointed scheme with a G-action on Y. This gives a pointed stack $(\mathcal{Y}, *)$ with $\mathcal{Y} = G\backslash Y$. Consider the moduli of maps of pairs $Map[(C, \eta_C), (\mathcal{Y}, *)]$. Denote by $\underline{\mathcal{G}}(G, Y)$ the space over $\mathcal{H}_C$ with the fiber at $D \in \mathcal{H}_C$ given by the maps $f \in Map[(C, \eta_C), (\mathcal{Y}, *)]$ that are defined off D. This is a *factorization space* (Sect. 3.3).

If the orbit Gy is open in Y and its boundary $\partial(Gy)$ is a union of divisors Y_i, $i \in I$, to any $f \in Map[(C, \eta_C), (\mathcal{Y}, *)]$ one can associate an I-colored finite subscheme $f^{-1}(\partial *) \stackrel{\text{def}}{=} (f^{-1} G\backslash Y_i)_{i \in I}$. Then we define $\underline{\mathcal{G}}(G, Y; I)$ to be $Map[(C, \eta_C), (\mathcal{Y}, *)]$ considered as a space over $\mathcal{H}_{C\times I}$. This is an I-colored local space (Sect. 3.3).

Examples

(a) When Y is a point, $\underline{\mathcal{G}}(G, \mathrm{pt})$ is the loop Grassmannian $\underline{\mathcal{G}}(G)$.
(b) When $G = \mathbb{G}_m$ and $(Y, y) = (\mathbb{A}^1, 1)$, I is a point and $\underline{\mathcal{G}}(\mathbb{G}_m, \mathbb{A}^1, I) = Map[(C, \eta_C), (\mathbb{G}_m\backslash\mathbb{A}^1, *)]$ is the space of effective divisors on C, i.e., the Hilbert scheme $\mathcal{H}_C$.

Lemma *Let the scheme Y be separated.*

(a) $\underline{\mathcal{G}}(G, Y)$ *is a subfunctor of* $\underline{\mathcal{G}}(G)$*. If* Y *is also affine, then* $\underline{\mathcal{G}}(G, Y)$ *is closed in* $\underline{\mathcal{G}}(G)$*.*

(b) For a subgroup $K \subseteq G$*, the intersection with* $\underline{\mathcal{G}}(K) \subseteq \underline{\mathcal{G}}(G)$ *reduces the condition* Y *to the condition* $\overline{Ky} \subseteq Y$*:*

$$\underline{\mathcal{G}}(G, Y) \cap \underline{\mathcal{G}}(K) = \underline{\mathcal{G}}(K, \overline{Ky}).$$

The Closure of S_0

It is well known that G/N is quasi-affine, i.e., it is an open part of its affinization $(G/N)^{\mathrm{aff}}$. We will consider it with the base point $y = eN$.

Proposition *Let* G *be of simply connected type.*

(a) The scheme of T*-fixed points* $\underline{\mathcal{G}}(G, (G/N)^{aff})^T$ *is* $\mathcal{H}_{d\times I}$*.*

(b) The closure $\overline{S_0}$ *is the reduced part* $\underline{\mathcal{G}}[G, (G/N)^{aff}]_{red}$ *of the loop Grassmannian with the condition* $(G/N)^{aff}$*.*

Proof

(a) The fixed points $\underline{\mathcal{G}}(G)^T$ are known to be $\underline{\mathcal{G}}(T)$ so $\underline{\mathcal{G}}(G, (G/N)^{\mathrm{aff}})^T = \underline{\mathcal{G}}(G, (G/N)^{\mathrm{aff}}) \cap \underline{\mathcal{G}}(T)$. This has been identified in Lemma 5.3.2.b with $\underline{\mathcal{G}}(T, \overline{Ty})$ where y is the base point eN of $(G/N)^{\mathrm{aff}}$. So $Ty = B/N$. When G is simply connected, $\prod_{i\in I} \check{\alpha}_i : \mathbb{G}_m^I \xrightarrow[\cong]{} T \cong B/N$. This extends to an identification of the closure of B/N in $(G/N)^{\mathrm{aff}}$ (a T-variety) with $(\mathbb{A}^1)^I$ (a $\mathbb{G}_m^I$-variety). Now, $\underline{\mathcal{G}}(T, \overline{Ty}) \cong \underline{\mathcal{G}}(\mathbb{G}_m, \mathbb{A}^1)^I$ is identified with $\mathcal{H}_{d\times I}$ in the example 2 in Sect. 5.3.2.

(b) Since $(G/N)^{\mathrm{aff}}$ is affine, $\underline{\mathcal{G}}(G, (G/N)^{\mathrm{aff}})$ is closed in $\underline{\mathcal{G}}(G)$ (Lemma 5.3.2.a). Since $\underline{\mathcal{G}}(G, (G/N)^{\mathrm{aff}})$ contains $\underline{\mathcal{G}}(G, G/N) = \underline{\mathcal{G}}(N) = S_0$, its reduced part contains $\overline{S_0}$. Since the stabilizer of the base point of $(G/N)^{\mathrm{aff}}$) is N, $\underline{\mathcal{G}}(G, (G/N)^{\mathrm{aff}}) \subseteq \underline{\mathcal{G}}(G)$ is $N_{\mathcal{K}}$-invariant. Then the reduced part $\underline{\mathcal{G}}(G, (G/N)^{\mathrm{aff}})_{red}$ has a stratification by $N_{\mathcal{K}}$-orbits S_λ for $\lambda \in X_*(T)$ such that L_λ lies in $\underline{\mathcal{G}}(G, (G/N)^{\mathrm{aff}})_{red}$.

We have, according to Part (a), that $\underline{\mathcal{G}}(G, (G/N)^{\mathrm{aff}})^T$ is $\mathcal{H}_{d\times I}$. So $\underline{\mathcal{G}}(G, (G/N)^{\mathrm{aff}})_{red} = \overline{S_0}$.

Proof of the Proposition 3.2.3

(a) According to Proposition 5.3.2.a, we have $\overline{S_0} = \underline{\mathcal{G}}(G, (G/N)^{\mathrm{aff}})_{red}$. Therefore, $\overline{S_0}^T \subseteq \underline{\mathcal{G}}(G, (G/N)^{\mathrm{aff}})^T$ which is $\mathcal{H}_{d\times I}$ by Proposition 5.3.2.b.

To see that $\mathcal{H}_{d\times I} \subseteq \underline{\mathcal{G}}(G)$ lies in $\overline{S_0}$, we denote by $G_i \subseteq G$ the connected three-dimensional subgroup corresponding to $i \in I$. Then $\overline{S_0}$ contains the corresponding object $\overline{S_0}(G_i)$ for G_i, and since we have already checked the proposition for SL_2, this is $\mathrm{AJ}^{G_i}(\mathcal{H}_d)$, i.e., $\mathrm{AJ}^G(\mathcal{H}_{d\times i})$.

It remains to prove that $\overline{S_0}^T \subseteq \underline{\mathcal{G}}(T)$ is closed under the product in $\underline{\mathcal{G}}(T)$ (because $\mathrm{AJ}(\mathcal{H}_{d\times I})$ is the product of all $\mathrm{AJ}(\mathcal{H}_{d\times i})$). However, the product in $\underline{\mathcal{G}}(T)$ can be realized using fusion in $\mathcal{G}(T)$.[21] So it suffices to notice that $\overline{S_0}$ is the fiber at a point $a = 0$ in a curve $C = \mathbb{A}^1$ of a factorization space $\overline{\mathcal{G}(G, G/N)}$ which is defined as the closure of the factorization subspace $\mathcal{G}(N) \cong \mathcal{G}(G, G/N) \subseteq \mathcal{G}(G)$.

(b–c) The part (a) of Proposition 3.2.3 gives a factorization of $\overline{S_0}^T$ as a product $\mathcal{H}_{d\times I} \cong \prod_{i\in I} (\mathcal{H}_d)^I$ over contributions from all $i \in I$. One therefore also has such factorization for $\overline{S_\alpha^-}^T$ and obviously for the connected components $\mathcal{G}(T)_\beta$. This reduces Parts (b) and (c) of the proposition to the SL_2 case. This case has already been checked by explicit calculation following Proposition 3.2.3. □

Appendix 2: Calculation of Thom Line Bundles from [YZ17]

In [YZ17], one uses a different convolution diagram. The only essential difference is the map ι described below. We check that it gives the same Thom line bundle as the calculation in Sect. 4.4.2 which used the dg cotangent correspondence. We will recall without character formulas how computations of Thom line bundles were made in [YZ17]. For calculational reasons, one uses an extra variety $X = G \times_P Y$ for $Y = \mathrm{Rep}_{\underline{Q}}(\mathrm{Gr}_F(V))$, and then a nonlinear map ι accounts for the difference between ambiental embeddings $T^*_{\underline{G}} X \subseteq T^*X$ and $T^*_{\underline{L}} Y \subseteq T^*Y$.

The notations are as in Sect. 4.4.2. Denote the elements of $Y = \mathrm{Rep}(V_\bullet) = \oplus_{k=1}^m \mathrm{Rep}_Q V_k$ and $Y^* = \mathrm{Rep}_{Q^*}(V_\bullet)$ by y and y^*. The moment map $\mu : T^*Y \to \mathfrak{l}^* \cong \mathfrak{l}$ is given by the projection of the commutator to $\mathfrak{l}$

$$\mu(y, y^*) = [y, y^*]_{\mathfrak{l}} \stackrel{\mathrm{def}}{=} \Big(\sum_{\{h\in H,\ h'=i\}} y_h y_h^* - \sum_{\{h\in H: h''=i\}} y_h^* y_h \Big)_{i\in I}.$$

[21] For $C = \mathbb{A}^1$, we have a canonical trivialization of $\mathcal{G}(G) \to \mathcal{H}_C$ over $C = \mathcal{H}^1_C$, as $\underline{\mathcal{G}}(G)$. Now, consider the pullback $\mathcal{G}_{C^2}(G) \stackrel{\mathrm{def}}{=} C^2 \times_{C^{[2]}} \mathcal{G}_{\mathcal{H}_{C\times I}}(G)$ of the restriction of $\mathcal{G}_{\mathcal{H}_{C\times I}}(G)$ to $\mathcal{H}^2 = C^2$. The locality identifies it over $C^2 - \Delta_C$ with the constant bundle $\underline{\mathcal{G}}(G)^2$. By *fusion of* $u, v \in \underline{\mathcal{G}}(G)$, we mean the limit (when it exists) over the diagonal of the constant section (u, v) which is defined off the diagonal.

The story in [YZ17] is told in terms of singular subvarieties $\mu^{-1}(0) \subseteq T^*Y$ (for a group L acting on a smooth variety Y), and the functoriality of cohomology is constructed in terms of ambiental smooth varieties T^*Y. The difference here is that we derive the cotangent correspondence mechanically from the original correspondence. For instance, this makes the associativity of multiplication follow manifestly from associativity of the extension correspondence.

Let $W = G \times_P \mathcal{R}(F)$ with projection to $X' = \mathcal{R}(V)$. Let $Z := T^*_W(X \times X')$. We have the following correspondence in [YZ14, Section 5.2]

$$G \times_P T^*Y \overset{\iota}{\hookrightarrow} T^*X \overset{\phi}{\longleftarrow} Z \overset{\psi}{\longrightarrow} T^*X' \tag{7}$$

The maps are the natural ones, which we further describe below.

Let U be the unipotent radical of P. Denote the Lie algebras by $\mathfrak{p}, \mathfrak{l}, \mathfrak{u}$. Denote the natural projections by $\pi : P \to L$, $\pi : \mathfrak{p} \to \mathfrak{l}$, and $\pi' : \mathfrak{p} \to \mathfrak{u}$.

For any associated G-bundle $\mathcal{E} = G \times_P E$, we denote the fiber at the origin by $\mathcal{E}_0 = E$. Then $T^*X \cong G \times_P (T^*X)_0$ and the L-variety $(T^*X)_0$ is (by Yang and Zhao [YZ14, Lemma 5.1 (a)])

$$(T^*X)_0 \overset{\text{def}}{=} \{(c, y, y^*) \mid c \in \mathfrak{p}, (y, y^*) \in T^*Y, \text{ such that } \mu(y, y^*) = \pi(c)\}. \tag{8}$$

Lemma

(a) We have an isomorphism of L-varieties $\mathfrak{u} \times T^*Y \cong (T^*X)_0$ over G/P by $(u, y, y^*) \mapsto (u + \mu(y, y^*), y, y^*)$.

(b) This makes T^*X into a G-equivariant vector bundle over G/P the sum of $T^*(G/P)$ and $G \times_P T^*Y$.

Proof In (a), the inverse map is $(c, y, y^*) \mapsto (\pi'(c), y, y^*)$. In (b), we use $T^*(G/P) \cong G \times_P \mathfrak{u}$.

The map $G \times T^*Y \to T^*X$ defined as $(g, y, y^*) \mapsto (g, \mu(y, y^*), y, y^*)$ induces a well-defined map $\iota : G \times_P T^*Y \to T^*X$. By Yang and Zhao [YZ14, Lemma 5.1], we have the isomorphism

$$Z := T^*_W(X \times X') \cong G \times_P \operatorname{Rep}_{\overline{Q}}(F)$$

with $\psi(g, x, x^*) \mapsto {}^g(x, x^*)$ for $g \in G$ and $(x, x^*) \in \operatorname{Rep}_{\overline{Q}}(V)$. So the map ψ is a composition of the inclusion ψ' of vector bundles over G/P and the conjugation action ψ'' (which acts by the same formula as ψ) and the diagram is

$$G \times_P T^*Y \overset{\iota}{\hookrightarrow} T^*[G \times_P \operatorname{Rep}(\operatorname{Gr}_F(V))] \overset{\phi}{\longleftarrow} Z \overset{\psi'}{\subseteq} G \times_P \operatorname{Rep}_{\overline{Q}}(V) \overset{\psi''}{\longrightarrow} \operatorname{Rep}_{\overline{Q}}(V).$$

Lemma *The Thom line bundles $\Theta_{\widetilde{G}}(\psi')$, $\Theta_{\widetilde{G}}(\psi'')$ and $\Theta_{\widetilde{G}}(\iota)$ are, respectively, the line bundles*

$$\Theta_{\widetilde{L}}[(F_\infty/F_0)\,\mathrm{Rep}_{\overline{Q}}(V)],\quad \Theta_{\widetilde{L}}(\mathfrak{g}/\mathfrak{p})^{-1}\quad \textit{and}\quad \Theta_{\widetilde{L}}(\mathfrak{p}^\perp)=\Theta_{\widetilde{L}}(\mathfrak{g}/\mathfrak{p}\otimes\omega),$$

In particular, $\Theta(d^*p)\otimes\Theta(\widetilde{q})\cong\Theta_L(\iota)\otimes\Theta_L(\psi)$.

Proof If $\mathcal{S}$ is one of the first four spaces in the diagram, then $\mathfrak{A}_{\widetilde{G}}(\mathcal{S})=\mathfrak{A}_{\widetilde{L}}$ since $\mathcal{S}=G\times_P\mathcal{S}_0$ for the fiber $\mathcal{S}_0$ which is an affine space. In particular, for a map $\eta\in\{\iota,\psi',\psi''\}$, the line bundle $\Theta_{\widetilde{G}}(\eta)$ on $\mathfrak{A}_{\widetilde{L}}$ is $\Theta_{\widetilde{L}}(T(\eta)_0)$.

(1) Vector bundle $T(\iota)$ is the normal bundle $N(\iota)$. According to the Lemma 5.3.2, it is isomorphic to $G\times_P -$ of the $\widetilde{L}$-module $(\mathfrak{g}/\mathfrak{p})^*\otimes\omega=\mathfrak{p}^\perp\otimes\omega$.
(2) Similarly, $T(\psi')$ is the normal bundle $N(\psi')$, and the fiber $T(\psi')_0$ is $(F/F_0)\,\mathrm{Rep}_Q(V)$.
(3) The equality $\Theta_{\widetilde{G}}(\psi'')=\Theta_{\widetilde{P}}[\mathfrak{g}/\mathfrak{p}]^{-1}$ is clear.

Corollary

(a) $\Theta_L[F_\infty/F_0(\mathrm{Rep}_{\overline{Q}}(V)]=\Theta_L[\mathrm{Rep}_Q(V)-Gr_0^F(\mathrm{Rep}_Q(V)]$.
(b) Consider the case when Q has no loop edges, and the filtration type $\mathbf{v}$ is a flag, i.e., $\mathbf{v}_k\in I$ for all k. Then $Gr_0\,\mathrm{Rep}(V)=0$ and $\Theta_L(\psi')=\Theta_G(Rep_Q V)$.

Proof

(a) A filtration on V induces a family of filtrations, compatible with the decomposition $\mathrm{Rep}_{\overline{Q}}(V)=\mathrm{Rep}_Q(V)\oplus\mathrm{Rep}_{Q^*}(V)$ and with the L-equivariant identification $\mathrm{Rep}_{Q^*}(V)\cong[\mathrm{Rep}_Q(V)]^*$. Therefore, the claim follows from

$$\frac{F_\infty}{F_0}[(\mathrm{Rep}_Q(V)^*]=[\frac{F_{-1}}{F_{-\infty}}(\mathrm{Rep}_Q(V)]^*$$

and the invariance of Thom line bundles under duality of vector bundles.
(b) follows since $Gr_0^F\,\mathrm{Rep}_Q(V)=0$ under the assumption on Q. The reason is that $Gr_0^F\,\mathrm{Rep}_Q(V)=\oplus\,\mathrm{Rep}_Q(Gr_p V)$, and all $Gr_p(V)\in\mathcal{V}^{\mathrm{I}}$ are one-dimensional.

Acknowledgments I.M. thanks Zhijie Dong for long-term discussions on the material that entered this work. We thank Misha Finkelberg for pointing out errors in earlier versions. His advice and his insistence have led to a much better paper. A part of the writing was done at the conference at IST (Vienna) attended by all coauthors. We therefore thank the organizers of the conference and the support of ERC Advanced Grant Arithmetic and Physics of Higgs moduli spaces No. 320593. The work of I.M. was partially supported by NSF grants. The work of Y.Y. was partially supported by the Australian Research Council (ARC) via the award DE190101231. The work of G.Z. was partially supported by ARC via the award DE190101222.

References

[BK10] Pierre Baumann, Joel Kamnitzer, *Preprojective algebras and MV polytopes*, Represent. Theory 16 (2012), 152–188. arXiv:1009.2469.

[BZ95] David Ben-Zvi, *An introduction to formal algebra*, Master's thesis, Harvard University, January 1995.

[BD04] Alexander Beilinson, and Vladimir Drinfeld. *Chiral Algebras*. Colloquium Publications 51, 2004.

[BFN19] Alexander Braverman, Michael Finkelberg, Hiraku Nakajima, *Coulomb branches of $3d\mathcal{N} = 4$ quiver gauge theories and slices in the affine Grassmannian* (with appendices by Alexander Braverman, Michael Finkelberg, Joel Kamnitzer, Ryosuke Kodera, Hiraku Nakajima, Ben Webster, and Alex Weekes), Advances in Theoretical and Mathematical Physics 23 (2019), no. 1, 75–166

[CC81] Contou-Carrère, Carlos E., *Corps de classes local geometrique relatif.* C. R. Acad. Sci. Paris Sér. I Math. 292 (1981), no. 9, 481–484.

[CZZ14] B. Calmès, K. Zainoulline, and C. Zhong, *Equivariant oriented cohomology of flag varieties*, Documenta Math. Extra Volume: Alexander S. Merkurjev's Sixtieth Birthday (2015), 113–144. arxiv 1409.7111

[D18] Z. Dong, *A relation between Mirković-Vilonen cycles and modules over preprojective algebra of Dynkin quiver of type ADE*, arxiv 1802.01792.

[FT18] M. Finkelberg and A. Tsymbaliuk, *Shifted quantum affine algebras: integral forms in type A* (with appendices by Alexander Tsymbaliuk and Alex Weekes) Arnold Mathematical Journal (2019) 5:197–283, arxiv1811.12137.

[FM99] M.Finkelberg, I. Mirković, *Semi-infinite flags I, the case of a global curve* $\mathbb{P}^1$, Differential Topology, Infinite-Dimensional Lie Algebras, and Applications: D. B. Fuchs' 60th Anniversary Collection, Editors A. Astashkevich, S. Tabachnikov, AMS series Adv. in the Math. Sci., Vol. 194 (1999).

[FS94] B. L. Feigin, A. V. Stoyanovsky, *A Realization of the Modular Functor in the Space of Differentials and the Geometric Approximation of the Moduli Space of G-Bundles*, Funktsional. Anal. i Prilozhen., 1994, Volume 28, Issue 4, 42–65

[FK] I.B. Frenkel, V.G. Kac, *Basic representations of affine Lie algebras and dual resonance models*. Invent Math 62, 23–66 (1980). Invent. Math. 62.

[GKV95] V. Ginzburg, M. Kapranov, and E. Vasserot, *Elliptic algebras and equivariant elliptic cohomology*, Preprint, (1995). arxiv9505012

[KS11] M. Kontsevich, Y. Soibelman, *Cohomological Hall algebra, exponential Hodge structures and motivic Donaldson-Thomas invariants*, Commun. Number Theory Phys. **5** (2011), no. 2, 231–352. MR 2851153

[Lev15] M. Levine, *Motivic Landweber exact theories and their effective covers*. Homology Homotopy Appl. 17 (2015), no. 1, 377–400.

[LM07] M. Levine, F. Morel, *Algebraic cobordism theory*, Springer, Berlin, 2007. MR 2286826

[M14] I. Mirković, *Loop Grassmannians in the framework of local spaces over a curve*. Recent advances in representation theory, quantum groups, algebraic geometry, and related topics, 215–226, Contemp. Math., 623, Amer. Math. Soc., Providence, RI, 2014. MR 3288629

[M17] I. Mirković, *Some extensions of the notion of loop Grassmannians*. Rad Hrvat. Akad. Znan. Umjet. Mat. Znan., the Mardešić issue. No. 532, 53–74 (2017). [At people.math.umass.edu/~mirkovic]

[MV07] Mirković, I.; Vilonen, K., *Geometric Langlands duality and representations of algebraic groups over commutative rings*. Ann. of Math. (2) 166 (2007), no. 1, 95–143.

[Nak01] H. Nakajima, *Quiver varieties and finite dimensional representations of quantum affine algebras*, J. Amer. Math. Soc. **14** (2001), no. 1, 145–238, MR1808477

[No09] Justin Noel, *Generalized Witt Schemes in Algebraic Topology*. http://www.nullplug.org/publications/generalized-witt-schemes.pdf

[P] A. Polishchuk, *Abelian varieties, theta functions and the Fourier transform*. Cambridge Tracts in Mathematics, **153**. Cambridge University Press, Cambridge, 2003. xvi+292 pp.

[Se] Segal, G. *Unitary representations of some infinite dimensional groups*. Commun. Math. Phys. 80, 301–342 (1981)

[Str00] N.P. Strickland, *Formal schemes and formal groups*, Homotopy invariant algebraic structures (Baltimore, MD, 1998), 263–352, Contemp. Math., 239, Amer. Math. Soc., Providence, RI, 1999.

[SV13] O. Schiffmann, E. Vasserot, *The elliptic Hall algebra and the K-theory of the Hilbert scheme of* $\mathbb{A}^2$. Duke Math. J. **162** (2013), no. 2, 279–366. MR3018956

[TZ19] James Tao, Yifei Zhao, *Extensions by* K_2 *and factorization line bundles*, arXiv:1901.08760.

[YZ14] Y. Yang, G. Zhao, *The cohomological Hall algebra for a preprojective algebra*, Proc. Lond. Math. Soc. Volume 116, Issue 5 (2018), Pages 1029–1074. arxiv1407.7994.

[YZ16] Y. Yang and G. Zhao, *Cohomological Hall algebras and affine quantum groups*, Selecta Math., Vol. 24, Issue 2 (2018), pp. 1093–1119. arxiv1604.01865

[YZ17] Y. Yang, G. Zhao, *Quiver varieties and elliptic quantum groups*, preprint. arxiv1708.01418.

[ZZ14] G. Zhao and C. Zhong, *Geometric representations of the formal affine Hecke algebra*, Adv. Math. 317 (2017), 50–90. arxiv1406.1283

[ZZ15] G. Zhao and C. Zhong, *Elliptic affine Hecke algebras and their representations*, preprint, 2015. arxiv1507.01245

[Zhu07] Xinwen Zhu. *Affine Demazure modules and* T*-fixed point subschemes in the affine Grassmannian.* Adv. Math. 221 (2009), no. 2, 570–600. arXiv:0710.5247.

[Zhu16] Xinwen Zhu, *An introduction to affine Grassmannians and the geometric Satake equivalence*. IAS/Park City Mathematics Series (2016). arXiv:1603.05593.

Symplectic Resolutions for Multiplicative Quiver Varieties and Character Varieties for Punctured Surfaces

Travis Schedler and Andrea Tirelli

To Sasha Beilinson and Victor Ginzburg on the occasion of their 60th birthdays, with admiration

Contents

T. Schedler (✉) · A. Tirelli
Department of Mathematics, Imperial College, London, United Kingdom
e-mail: t.schedler@imperial.ac.uk; a.tirelli15@imperial.ac.uk

V. Baranovsky et al. (eds.), *Representation Theory and Algebraic Geometry*, Trends in Mathematics, https://doi.org/10.1007/978-3-030-82007-7_9

1 Introduction

1.1 Motivation

This paper is devoted to the study of the symplectic algebraic geometry of coarse moduli spaces of (semistable) representations of multiplicative preprojective algebras. These can be thought of as multiplicative analogues of Nakajima quiver varieties [Nak94], which includes character varieties of (open) Riemann surfaces. In particular, our attention is focused on tackling two main problems: The first is to understand whether these multiplicative quiver varieties are symplectic singularities, as defined by Beauville in [Bea00];[1] the second is to classify all the possible cases in which they admit symplectic resolutions.

Multiplicative preprojective algebras were first defined by Crawley-Boevey and Shaw in [CBS06], with the aim of better understanding Katz's middle convolution operation for rigid local systems, [Kat96]. Another important application contained in the seminal paper [CBS06] is the solution of (one direction of) the multiplicative Deligne-Simpson problem in terms of the root data of a certain star-shaped quiver. Moduli spaces of representations, in the sense of King [Kin94], of these algebras give rise to the so-called multiplicative quiver varieties. Ordinary (Nakajima) quiver varieties have appeared in numerous places in representation theory, algebraic geometry, and mathematical physics; their homology theories are closely related to the representation theory of Kac–Moody Lie algebras [Nak94], and their quantum cohomology is closely related to quantum R-matrices and the Casimir connection [MO19]. A number of authors have studied multiplicative quiver varieties since their definition: Among others, Jordan [Jor14] considered quantizations of such varieties from a representation theoretic point of view by

[1] Since submission of this article, in [KS], the first author and D. Kaplan resolve this in the affirmative for all multiplicative quiver varieties. In particular, they are all normal.

constructing flat q-deformations of the algebra of differential operators on certain affine spaces; a more geometric approach was used by Yamakawa in [Yam08], where a symplectic structure on these moduli spaces was defined and studied. Some of these results will be recalled in the next sections of the present paper. More recently, (derived) multiplicative preprojective algebras appeared in the study of wrapped Fukaya categories of certain Weinstein 4-manifolds constructed by plumbing cotangent bundles of Riemann surfaces: See [EL19]. In the recent work of Chalykh and Fairon [CF17], multiplicative quiver varieties are used to construct a new integrable system generalizing the Ruijsenaars–Schneider system, which plays a central role in supersymmetric gauge theory and cyclotomic DAHAs. Moreover, in work of McBreen–Webster [MW18] and McBreen–Gammage–Webster [MGW19], related to [BK16, §7], mirror symmetry is studied for multiplicative hypertoric varieties, which include multiplicative quiver varieties when the dimensions are one. Finally, mixed Hodge polynomials of character varieties and quiver varieties were studied in the groundbreaking work of Hausel, Lettelier, and Rodriguez-Villegas using arithmetic methods [HLRV11, HLRV13a, HLRV13b]; they suggested that similar methods should apply to general multiplicative quiver varieties. Given their appearance in so many different contexts, it seems natural to perform a careful analysis of multiplicative quiver varieties from the point of view of symplectic algebraic geometry.

The subject of symplectic resolutions and the more general symplectic singularities (the latter dating to Beauville [Bea00]) has recently gained importance in many areas of mathematics and physics. Their quantizations subsume many of the important examples of algebras appearing in representation theory (Cherednik and symplectic reflection algebras, D-modules on flag varieties and representations of Lie algebras, quantized hypertoric and quiver algebras, etc.). There is a growing theory of symplectic duality, or three-dimensional physical mirror symmetry ([BFN18, Nak16, CHZ14, BLPW16] and many others), between pairs of these varieties. Pioneering work of Braverman, Maulik, and Okounkov [BMO11] (continued in the aforementioned [MO19] and in many other places) shows that their quantum cohomology is also deeply tied to connections arising in representation theory, related to derived autoequivalences of duals in the sense of homological mirror symmetry (two-dimensional field theories). Since, as mentioned before, quiver varieties play such an important role here, it is expected that multiplicative quiver varieties will as well. Moreover, the varieties in question are instances of moduli spaces parametrizing geometric objects. The study of such spaces and their singularities is, in general, important in algebraic geometry.

For all of these reasons, it is natural to ask when multiplicative quiver varieties have symplectic singularities and admit symplectic resolutions. We largely answer these questions, leaving a couple cases (the so-called "(2,2)"-cases related to O'Grady's examples [O'G99], and the so-called isotropic cases, which are multiplicative analogues of symmetric powers of du Val singularities), that appear to require local structure theory. Our methods generalize those of [BS21], which we largely follow. They build on Crawley-Boevey and Shaw's pioneering work on multiplicative quiver varieties (and extensions by Yamakawa [Yam08]) and apply (as in [BS21]) Drezet's factoriality criteria [Dre91] and Flenner's theorem [Fle88] on extendability of differential forms beyond codimension four.

1.2 Summary of Results on Character Varieties

Since they are the easiest to state and perhaps of the broadest interest, we first explain the results on character varieties that follow from our considerations on multiplicative quiver varieties. Fix a connected compact Riemann surface X of genus $g \geq 0$, let $S = \{p_1, \dots, p_k\} \subset X$ be a subset of $k \geq 0$ points, and fix a tuple $\mathcal{C} = (\mathcal{C}_1, \dots, \mathcal{C}_k)$ of conjugacy classes $\mathcal{C}_i \subset GL_n(\mathbb{C}), i = 1, \dots, k$. Let $X^\circ := X \setminus \{p_1, \dots, p_k\}$ be the corresponding punctured surface, and let γ_i be the homotopy class in $\pi_1(X^\circ)$ of some choice of loop around the puncture p_i (having the same free homotopy class as a small counterclockwise loop around p_i). We define the character variety of X° with monodromies in $\overline{\mathcal{C}_i}$ as follows:

$$\mathcal{X}(g, k, \overline{\mathcal{C}}) := \{\chi : \pi_1(X^\circ) \to GL_n \mid \chi(\gamma_i) \in \overline{\mathcal{C}_i}\} /\!/ GL_n. \tag{1.2.1}$$

As recalled in Sect. 3 below, this has the structure of an affine algebraic variety. Note that X (or X°) does not appear in the notation on the left-hand side, since the result does not depend on the choice of X up to isomorphism (only the identification of $\pi_1(X^\circ)$ is relevant).

Observe that in order for this character variety to be non-empty, we must have $\prod_{i=1}^k \det(\mathcal{C}_i) = 1$, where we let $\det(\mathcal{C}_i)$ be defined as the determinant of any element of $\mathcal{C}_i$. Let us assume this from now on. Given $m \geq 1$, we let $m \cdot \mathcal{C} = (\mathcal{C}_1^{\oplus m}, \dots, \mathcal{C}_k^{\oplus m})$. We call $\mathcal{C}$ *q-divisible* if $\mathcal{C} = m \cdot \mathcal{C}'$ for $m \geq 2$ and $\prod_{i=1}^k \det(\mathcal{C}_i') = 1$. Call it q-indivisible if it is not q-divisible. Below, q-indivisibility will be the most important criterion for the existence of symplectic resolutions for $\mathcal{X}(g, k, \overline{\mathcal{C}})$.

For each $\mathcal{C}_i$, let the minimal polynomial of any $A \in \mathcal{C}_i$ be $(x - \xi_{i,1}) \cdots (x - \xi_{i,w_i})$, ordered so that the sequence $\alpha_{i,j} := \operatorname{rank}(A - \xi_{i,1}) \cdots (A - \xi_{i,j})$ has the property that $\alpha_{i,j} - \alpha_{i,j+1}$ is nonincreasing in j (for $0 \leq j \leq w_i - 1$, setting $\alpha_{i,0} = n$). This is possible since the nonincreasing property obviously holds when all the $\xi_{i,j}$ are equal. The following quantities will have importance for us:

$$\ell := \sum_i \alpha_{i,1}, \quad p(\alpha) := 1 + n^2(g-1) + n\ell + \sum_{i=1}^k \sum_{j=1}^{w_i - 1} \alpha_{i,j}\alpha_{i,j+1} - \sum_{i=1}^k \sum_{j=1}^{w_i} \alpha_{i,j}^2. \tag{1.2.2}$$

The quantity $2p(\alpha)$ is the "expected dimension" of the character variety, which is its actual dimension in many cases, as explained below.

Our main results on character varieties can be summarized as follows. We divide separately into the genus 0 and the positive genus cases.

Recall here that a *symplectic singularity* is a normal variety X equipped with a symplectic structure ω_{reg} on the smooth locus X_{reg} such that for any (or equivalently every) resolution of singularities $\rho : \widetilde{X} \to X$, $\rho^*\omega_{\text{reg}}$ extends to a regular two-form $\widetilde{\omega} \in \Omega^2(\widetilde{X})$. The map ρ is furthermore a symplectic resolution if $\widetilde{\omega}$ is nondegenerate.

Theorem 1.1 *Let $g = 0$ and fix n and conjugacy classes $C_1, \ldots, C_k \subseteq GL_n(\mathbb{C})$ as above.*

- *If $\ell < 2n$, then one of the following exclusive possibilities occur and can be computed by an explicit algorithm:*
  - *$\mathcal{X}(0, k, \overline{C})$ is empty.*
  - *$\mathcal{X}(0, k, \overline{C})$ is a point.*
  - *There is a canonical datum $(n', k', C_1', \ldots, C_{k'}', \iota)$ of $n' < n$, $k' \le k$, and conjugacy classes $C_1', \ldots, C_{k'}' \subseteq GL_{n'}(\mathbb{C})$ such that ℓ' (defined as above) satisfies $\ell' \ge 2n'$, and an isomorphism $\iota : \mathcal{X}(0, k', \overline{C}') \to \mathcal{X}(0, k, \overline{C})$.*

  Suppose, therefore, that $\ell \ge 2n$.
- *If C is q-indivisible, then $\mathcal{X}(0, k, \overline{C})$ admits a projective symplectic resolution (via geometric invariant theory). Therefore, its normalization is a symplectic singularity.*[2] *Moreover,* $\dim \mathcal{X}(0, k, \overline{C}) = 2p(\alpha)$.
- *Suppose that C is q-divisible. Then, unless one of the conditions listed after Corollary 1.12 is satisfied (for $k \le 5$), the normalization of $\mathcal{X}(0, k, \overline{C})$ is a symplectic singularity which does not admit a symplectic resolution (in fact, it contains a singular terminal factorial open subset).*

As mentioned in the theorem, the technique used to show nonexistence of symplectic resolutions is by identifying an open singular factorial terminal subset. It is well known that singular factorial terminal varieties cannot admit crepant resolutions, and hence not symplectic resolutions. Indeed, by Van der Waerden purity, any resolution of a singular factorial variety has exceptional locus which is a divisor. By definition, any crepant resolution of a terminal variety has exceptional locus of codimension at least two. Put together, there is no crepant resolution of a singular factorial terminal variety.

Remark 1.2 Note that, when $k \le 2$ in genus zero, the character variety is always a point (or empty).

Theorem 1.3 *Suppose that $g \ge 1$. Then the following holds:*

- *If C is q-indivisible, then $\mathcal{X}(g, k, \overline{C})$ admits a projective symplectic resolution (via geometric invariant theory). Therefore, its normalization is a symplectic singularity. Moreover, it has dimension $2p(\alpha)$.*

[2] As mentioned in Sect. 1.1, it is now known that these and the other varieties in this article are normal.

- *If C is q-divisible, then unless one of the following conditions is satisfied, the normalization of $\mathcal{X}(g, k, \overline{C})$ is a symplectic singularity which does not admit a symplectic resolution (in fact, it contains a singular terminal factorial open subset):*

  (a) $g = 2, k = 0$, *and* $n = 2$;
  (b) $g = 1, k = 0$;
  (c) $g = 1, k = 1, w_1 = 2$, *and* $\alpha_{1,1} = p$, *with* p *prime.*

  Moreover, in all cases except case (b), $\dim \mathcal{X}(g, k, \overline{C}) = 2p(\alpha)$.

The proofs of these theorems is given in Sect. 6.7; they are consequences of our main results on multiplicative quiver varieties (particularly Corollary 6.28).

Remark 1.4 Actually, the results above (slightly modified) should also apply to twisted character varieties, where we replace $\pi_1(X^\circ)$ by a finite central extension, corresponding to setting the relation $\prod_{i=1}^{g}[A_i, B_i]\prod_{j=1}^{k} M_j$ to be a root of unity times the identity matrix. To prove, such a statement would require a straightforward generalization of [CB13] and of Sect. 3 below. With this in hand, these results would follow from Corollary 6.28 just as before. For some more details, see Sect. 1.5 of the introduction, where we describe roughly how to translate this corollary into the setting of twisted character varieties.

1.3 Multiplicative Quiver Varieties with Special Dimension Vectors

Recall that a quiver Q is a directed graph. We let Q_0 denote the set of vertices and Q_1 the set of arrows (=edges). Given Q together with a tuple of nonzero complex numbers $q \in (\mathbb{C}^\times)^{Q_0}$, one can define the multiplicative preprojective algebra $\Lambda^q(Q)$, over the semisimple ring $\mathbb{C}^{Q_0}$ (see Sect. 2.2 below). To a representation, we associate a dimension vector in $\mathbb{N}^{Q_0}$. Given furthermore a stability parameter $\theta \in \mathbb{Z}^{Q_0}$, one can define a variety, denoted $\mathcal{M}_{q,\theta}(Q, \alpha)$, which is a coarse moduli space of θ-semistable representations of $\Lambda^q(Q)$ of dimension vector α. It is natural to ask what the dimension vectors of θ-stable representations are. Toward this end, one considers a combinatorially defined subset $\Sigma_{q,\theta} \subseteq \mathbb{N}^{Q_0}$ of the set of all possible dimension vectors (defined in Sect. 2.8 below). It has the property that for $\alpha \in \Sigma_{q,\theta}$, the θ-stable locus is dense in $\mathcal{M}_{q,\theta}(Q, \alpha)$ (and it is always open). However, it is unknown in general if $\mathcal{M}_{q,\theta}(Q, \alpha)$ is non-empty. It is expected, but not known, that these conversely describe all dimension vectors of stable representations, that is:

$$\text{If there is a } \theta\text{-stable representation of } \Lambda^q(Q) \text{ of dimension } \alpha \in \mathbb{N}^{Q_0}, \text{ then } \alpha \in \Sigma_{q,\theta}. \tag{*}$$

In the case $\theta = 0$, Crawley-Boevey kindly pointed out a work in progress with Hubery toward a proof of (*). We prove a weakened version of (*) below

(Corollary 6.18), replacing $\Sigma_{q,\theta}$ by a larger set. Note that if (*) holds and furthermore $\mathcal{M}_{q,\theta}(Q,\alpha) \neq \emptyset$ for all $\alpha \in \Sigma_{q,\theta}$, then put together, we would obtain a characterization of the set $\Sigma_{q,\theta}$: In this case, $\alpha \in \Sigma_{q,\theta}$ if and only if there exists a θ-stable representation of dimension α. However, this is, again, unknown.

To define $\mathcal{M}_{q,\theta}(Q,\alpha)$, we require $\alpha \cdot \theta = 0$, and for it to be non-empty, we require that $q^\alpha := \prod_{i\in Q_0} q_i^{\alpha_i} = 1$. Let $N_{q,\theta} := \{\alpha \in \mathbb{N}^{Q_0} \mid q^\alpha = 1, \alpha \cdot \theta = 0\}$. We call a vector $\alpha \in N_{q,\theta}$ *q-indivisible* if $\frac{1}{m}\alpha \notin N_{q,\theta}$ for any $m \geq 2$. Equivalently, writing $\alpha = m\beta$ for $m = \gcd(\alpha_i)$, we have that q^β is a primitive m-th root of unity. Note that if $\alpha \in N_{q,\theta}$ is indivisible, it is clearly q-indivisible, although the converse does not hold in general. (Unlike in the case of character varieties, here the q in "q-(in)divisible" refers to an actual parameter; see Remark 1.10 for an explanation how the two notions nonetheless coincide.)

We denote by p the following function:

$$p : \mathbb{N}^{Q_0} \to \mathbb{Z}, \quad p(\alpha) = 1 - \frac{1}{2}(\alpha, \alpha) \geq 0,$$

where $(-,-)$ denotes the Cartan–Tits form associated to the quiver Q (see Sect. 2.1 for more details). Geometrically, $2p(\alpha)$ gives the "expected dimension" of $\mathcal{M}_{q,\theta}(Q,\alpha)$ (which is the actual dimension if $\alpha \in \Sigma_{q,\theta}$ and $\mathcal{M}_{q,\theta}(Q,\alpha) \neq \emptyset$; see Remark 2.20 below). If $p(\alpha) = 1$, that is, $(\alpha,\alpha) = 0$, then α is called *isotropic*. Otherwise, it is called *anisotropic*.

One of the main results of this paper, proved in Sect. 4, is the following:

Theorem 1.5 *Let $\alpha \in \Sigma_{q,\theta}$ and assume that $\alpha \neq 2\beta$ for $\beta \in N_{q,\theta}$ and $p(\beta) = 2$. Then, assuming it is non-empty, $\mathcal{M}_{q,\theta}(Q,\alpha)$ satisfies the following:*

- *Its normalization is a symplectic singularity.*
- *If α is q-indivisible, then for suitable generic θ', it admits a symplectic resolution of the form $\mathcal{M}_{q,\theta'}(Q,\alpha) \to \mathcal{M}_{q,\theta}(Q,\alpha)$.*
- *If $\alpha = m\beta$ for $\beta \in \Sigma_{q,\theta}$ and $m \geq 2$, and $\mathcal{M}_{q,\theta}(Q,\beta) \neq \emptyset$, then $\mathcal{M}_{q,\theta}(Q,\alpha)$ does not admit a symplectic resolution. Moreover, for suitable generic θ', $\mathcal{M}_{q,\theta'}(Q,\alpha)$ is a singular factorial terminalization. In fact, $\mathcal{M}_{q,\theta}(Q,\alpha)$ itself contains a singular, factorial, terminal open subset.*

Implicit in Theorem 1.5 is the fact (see Lemma 2.12 and Corollary 2.21 below) that for all $\alpha \in \Sigma_{q,\theta}$, $\mathcal{M}_{q,\theta'}(Q,\alpha) \to \mathcal{M}_{q,\theta}(Q,\alpha)$ is a projective birational Poisson morphism for suitable θ'. This implies, by definition, that it is a symplectic resolution if the source is smooth symplectic. In the last part of the theorem, by singular factorial terminalization, we mean a projective birational Poisson morphism with source a singular factorial terminal variety.

In the case of generic θ, the theorem can be simplified as follows, avoiding the need to check if a vector is in $\Sigma_{q,\theta}$. First, note that $\Sigma_{q,\theta}$, by definition, is a subset of the set of *roots* for the quiver (which in turn equals the set of roots of the associated Kac–Moody Lie algebra in suitable cases). The real roots are those vectors obtained from elementary vectors $e_i, i \in Q_0$ by simple reflections $\alpha \mapsto \alpha - (\alpha, e_i)e_i$;

the imaginary roots are those obtained by such reflections from nonnegative (or nonpositive) vectors with connected support and nonpositive Cartan pairing with all e_i.

Corollary 1.6 *Fix an imaginary root α for Q. Let q be such that $q^\alpha = 1$. Let θ be generic (inside the hyperplane $\{\theta \cdot \alpha = 0\}$). Then:*

(i) We have $\alpha \in \Sigma_{q,\theta}$ if and only if α is q-indivisible or anisotropic. If α is q-indivisible, $\mathcal{M}_{q,\theta}(Q, \alpha)$ is smooth symplectic.
(ii) Assume α is q-divisible and anisotropic. Moreover, assume that $\alpha \neq 2\beta$ for $p(\beta) = 2$ and $q^\beta = 1$. Then $\mathcal{M}_{q,\theta}(Q, \alpha)$ is a (normal) symplectic singularity.
(iii) Under the assumptions of (ii), we have the following:

- *If there exists a θ-stable representation of $\Lambda^q(Q)$ of dimension $\frac{1}{m}\alpha$, for some $m \geq 2$, then $\mathcal{M}_{q,\theta}(Q, \alpha)$ is singular, factorial, and terminal and hence does not admit a symplectic resolution.*
- *If, on the other hand, there are no θ-stable representations of $\Lambda^q(Q)$ of dimension $r\alpha$ for all rational $r < 1$, then $\mathcal{M}_{q,\theta}(Q, \alpha)$ is smooth.*

Note that in the general case with $\alpha \in \Sigma_{q,\theta}$, the above corollary always describes the source of projective birational Poisson morphisms obtained by suitably varying θ.

Remark 1.7 Note that every $r\alpha, r \in \mathbb{Q}_{<1}$ appearing in the theorem is also in $\Sigma_{q,\theta}$, by part (i). Thus, if there exists a θ-stable representation of every dimension in $\Sigma_{q,\theta}$, then in part (iii), we are necessarily in the first case. This condition holds in the additive case (with $\lambda \in \mathbb{R}^{Q_0}$), by Bellamy and Schedler [BS21], but we don't have any other evidence that this holds here. Also, note that the two cases are not exhaustive, so it could happen that there are some stable representations of dimension $r\alpha$ but not when $r = \frac{1}{m}$. In this (unexpected) situation, it would require more detailed analysis to determine whether a symplectic resolution exists.

Remark 1.8 In the case left out of the theorem, where $\alpha = 2\beta$ for some $\beta \in N_{q,\theta}$ satisfying $p(\beta) = 2$ (we call this the "(2, 2)-case"), we conjecture, as in the special case of character varieties of rank two local systems on genus two surfaces handled in [LS06, BS], that $\mathcal{M}_{q,\theta}(Q, \alpha)$ has a symplectic resolution obtained by blowing up the singular locus of $\mathcal{M}_{q,\theta'}(Q, \alpha)$, for suitably generic θ'. However, in order to prove this, it is necessary to understand the étale local structure of $\mathcal{M}_{q,\theta}(Q, \alpha)$ while at the moment, all of our techniques are global in nature. In Sect. 7, we discuss an approach to understand the local structure of the multiplicative quiver varieties, based on the conjectural 2-Calabi–Yau property of the multiplicative preprojective algebra for non-Dynkin quivers.

Remark 1.9 Note that, as part of Corollary 1.6, when θ is generic (and $\alpha \in \Sigma_{q,\theta}$), we prove normality of the variety $\mathcal{M}_{q,\theta}(Q, \alpha)$; see Proposition 4.10. Moreover, we conjecture normality for all θ (as well as for the (2, 2)-case). Such a result requires a local understanding of the varieties $\mathcal{M}_{q,\theta}(Q, \alpha)$, which would again follow from the conjectural 2-Calabi–Yau property for $\Lambda^q(Q)$ when Q is not Dynkin; see Sect. 7.

1.4 Character Varieties as (Open Subsets of) Multiplicative Quiver Varieties

In Sect. 3, extending results of [CBS06] and [Yam08], we explain how character varieties identify as natural open subsets of the multiplicative quiver varieties for crab-shaped quivers (Theorem 3.6), also known as "comet-shaped" in [HLRV11]. Namely, the character variety identifies as an open subset of a multiplicative quiver variety for the crab-shaped quiver described in Sect. 1.2, with appropriate parameter $q \in (\mathbb{C}^\times)^{Q_0}$. The open subset is defined by requiring the loops in the original (undoubled) quiver to act invertibly. In particular, in the genus zero case, the character variety equals the multiplicative character variety.

Remark 1.10 By the above correspondence, a collection of conjugacy classes $\mathcal{C} \subseteq GL_n(\mathbb{C}^\times)$ is q-divisible in the sense of Sect. 1.2 (where q is not yet a parameter) if and only if, for the associated quiver Q, dimension vector $\alpha \in \mathbb{N}^{Q_0}$, and parameter $q \in (\mathbb{C}^\times)^{Q_0}$, the vector α is q-divisible in the sense of Sect. 1.3. We hope that this abuse of notation aids understanding.

It is then an interesting question which character varieties exhibit the different properties discussed above, in particular, which ones are the "(2,2)"-cases where an "O'Grady"-type resolution is expected (see Remark 1.8)? Their classification is achieved in Theorems 5.1 and 5.3: All of these are in the genus zero case (i.e., they are star-shaped quivers), with three to five punctures and particular monodromy conditions, as classified in Theorem 5.1, except for two cases in Theorem 5.3. The latter cases correspond to once-punctured tori and to closed genus two surfaces (with even rank and rank two local systems, respectively, the former having particular monodromy about the puncture).

1.5 General Dimension Vectors

Although it is difficult to study directly quiver varieties of dimensions $\alpha \notin \Sigma_{q,\theta}$, in the additive setting, this issue is alleviated by Crawley-Boevey's canonical decomposition, expressing an arbitrary variety as a product of varieties for dimension vectors in $\Sigma_{q,\theta}$ [CB02, Theorem 1.1] (extended to $\theta \neq 0$ in [BS21, Proposition 2.1]). In Theorem 6.17 below, we provide a version of this decomposition in the multiplicative setting using reflection functors, following the proof of [CB02], which is weaker in the sense that the dimension vectors of the factors need not be in $\Sigma_{q,\theta}$, and hence the factors could further decompose (although it is not known in general if they do). One of the reasons why we must give the weaker statement is the unavailability of (*); see Sect. 7.2 for more details. Along with this, we prove a more general sufficient criterion for varying θ to produce a symplectic resolution (Theorem 6.23), that does not require dimension vectors to be in $\Sigma_{q,\theta}$. Using these results, in Theorem 6.27, we are able to extend Theorem 1.5 to general dimension vectors. The content of Theorems 6.17 and 6.27 can be summarized in the following. Here, $\widetilde{\Sigma}_{q,\theta}$ is a larger set than $\Sigma_{q,\theta}$, consisting of roots for which a certain multiplicative moment map is flat; $\Sigma^{\mathrm{iso}}_{q,\theta} \subseteq \Sigma_{q,\theta}$ is the subset of isotropic roots. See Sects. 2 and 6 for details on these definitions. For any subset $X \subseteq \mathbb{N}^{Q_0}$, let $\mathbb{N}_{\geq 2} \cdot X := \{m\alpha \mid m \geq 2, \alpha \in X\}$.

Theorem 1.11 *Assume that $\mathcal{M}_{q,\theta}(Q,\alpha)$ is non-empty.*

(i) There is a decomposition $\alpha = \beta^{(1)} + \cdots + \beta^{(k)}$ with $\beta^{(i)} \in \widetilde{\Sigma}_{q,\theta} \cup \mathbb{N}_{\geq 2} \cdot \Sigma^{iso}_{q,\theta}$, such that the direct sum map produces an isomorphism (of reduced varieties):

$$\prod_{i=1}^{k} \mathcal{M}_{q,\theta}(Q,\beta^{(i)}) \xrightarrow{\sim} \mathcal{M}_{q,\theta}(Q,\alpha).$$

(ii) Assume that this decomposition has neither elements $\beta^{(i)} \in \mathbb{N}_{\geq 2} \cdot \Sigma^{iso}_{q,\theta}(Q,\alpha)$ nor $\beta^{(i)} = 2\alpha$ for $\alpha \in N_{q,\theta}$ and $p(\alpha) = 2$. Then:

- *The normalization of $\mathcal{M}_{q,\theta}(Q,\alpha)$ is a symplectic singularity.*
- *Each factor $\mathcal{M}_{q,\theta}(Q,\beta^{(i)})$ with $\beta^{(i)} \notin \Sigma_{q,\theta}$ admits a symplectic resolution.*
- *If for any factor $\beta^{(i)}$ there exists a θ-stable representation of dimension $\gamma^{(i)} = \frac{1}{m}\beta^{(i)}$ with $m \geq 2$, then $\mathcal{M}_{q,\theta}(Q,\alpha)$ does not admit a symplectic resolution. In fact, it has an open, singular, terminal, factorial subset.*

Putting everything together, in Corollary 6.28, we are able to give a classification of crab-shaped settings whose multiplicative quiver varieties admit symplectic resolutions. By Theorem 3.6, we also deduce the corresponding statement for character varieties (Theorems 1.1 and 1.3), which are open subsets of these varieties, for $\theta = 0$ and for certain values of the parameter q. To state the result, first recall that the Jordan quiver is the quiver with one vertex and one arrow (a loop). The *fundamental region* $\mathcal{F}(Q)$ consists of those nonzero vectors $\alpha \in \mathbb{N}^{Q_0}$ with connected support and with $(\alpha, e_i) \leq 0$ for all i. As we explain below, by applying certain reflection functors, we can reduce to this case. We give a simplified version of the statement of Corollary 6.28 below; see the full statement for precise details.

Corollary 1.12 *Let Q be a crab-shaped quiver and $\alpha \in N_{q,\theta}$ a vector in the fundamental region with $\alpha_i > 0$ for all $i \in Q_0$. Further assume that (Q,α) is* **not** *one of the following cases:*

(a) $\beta := \frac{1}{2}\alpha$ is integral, $q^\beta = 1$, and (Q,β) is one of the quivers in Theorems 5.1 and 5.3;

(b) Q is affine Dynkin of type $\tilde{A}_0$ (i.e., the Jordan quiver with one vertex and one arrow), $\tilde{D}_4$ or $\tilde{E}_6, \tilde{E}_7, \tilde{E}_8$) and α is a q-divisible multiple of the indivisible imaginary root δ of Q.

Then:

- *The normalization of $\mathcal{M}_{q,\theta}(Q,\alpha)$ is a symplectic singularity.*
- *If α is q-indivisible, $\mathcal{M}_{q,\theta}(Q,\alpha)$ admits a symplectic resolution;*

- *If α is q-divisible, and α is not: (c) A prime multiple of one of the quivers listed in Theorem 6.16.(b2) below (a framed affine Dynkin quiver with dimension vector $(1, m\delta)$ with $m\delta$ q-divisible) with $\theta \cdot \delta = 0$, then $\mathcal{M}_{q,\theta}(Q, \alpha)$ does not admit a symplectic resolution (it contains an open singular factorial terminal subset).*

Thus, after reducing to the fundamental region, a symplectic resolution exists if and only if the dimension vector is q-indivisible, unless we are in one of the following three open cases:

(a) Twice one of the dimension vectors appearing in Theorems 5.1 and 5.3 below, which correspond to one of certain (twisted) character varieties of a sphere with 3–5 punctures (with rank at most 24), of a once-punctured torus (of rank 4) or a closed genus two surface (of rank 2)
(b) a q-divisible imaginary root on an affine Dynkin quiver (of type $\tilde{A}_0$, $\tilde{D}_4$, $\tilde{E}_6$, $\tilde{E}_7$, or $\tilde{E}_8$), which corresponds to either a (twisted) character variety of a closed torus (type A), or a (twisted) character variety of a sphere with 3 or 4 punctures (type E or D, respectively) with particular rank and monodromy conditions
(c) A prime multiple of the vector $(1, \ell\delta)$ on a framed affine Dynkin quiver (again of type $\tilde{A}_0$, $\tilde{D}_4$, $\tilde{E}_6$, $\tilde{E}_7$, or $\tilde{E}_8$) with $\theta \cdot \delta = 0$, which corresponds again to a certain character variety of a once-punctured torus or a sphere with 4 or 5 punctures.

Here, when we say "correspond," we mean precisely that, for $\theta = 0$, the multiplicative quiver varieties equal the given (twisted) character varieties, whereas for the genus ≥ 1 case, the latter is the open subset of the former where the transformations corresponding to loops in the undoubled quiver are invertible. Setting $\theta \neq 0$ gives a partial resolution (which may be an actual one, as in case of θ generic and α q-indivisible).

Remark 1.13 For the ordinary (untwisted) character varieties of closed genus one or two surfaces appearing in the lists for cases (b) and (a) above, a symplectic resolution exists; see, for example, [BS]. The proof makes use of Poincaré–Verdier duality for closed surfaces; perhaps, a suitable generalization of this for orbifolds would allow us to extend those results to the orbifold case. If so, we could remove $\tilde{A}_0$ from case (b) and the quiver with one vertex and two loops from (a).

1.6 Outline of the Paper

The outline of the paper is as follows: In Sect. 2, we recall some basic facts about quivers and root systems and establish the notation that shall be used throughout the paper. We then recall the definition of multiplicative preprojective algebras and outline some of their algebraic properties. These are needed in the construction, via Geometric Invariant Theory (GIT), of their moduli spaces of semistable representations, following [Kin94]. In Definition 2.18, we introduce the fundamental combinatorially defined subset $\Sigma_{q,\theta}$ of roots appearing in our main results (which is expected to contain, if not equal, the dimension vectors of θ-stable representations of the multiplicative preprojective algebra). We extend some properties of multiplicative quiver varieties with dimension vector in $\Sigma_{q,\theta}$ that were originally formulated and proved in [CBS06] in the case of a trivial stability condition to the general case.

In Sect. 3, we prove that for quivers of special type, namely, those which are crab-shaped (see Fig. 1), there is an isomorphism between (an open subset of) the corresponding multiplicative quiver variety and a character variety arising from considering representations of the fundamental group of a punctured Riemann surface where the monodromies of loops around the punctures are assumed to lie in the closure of certain conjugacy classes. In order to build such a correspondence, we exploit [CBS06, Lemma 8.2 and Theorem 1.1]. Note that an instance of the correspondence between multiplicative quiver varieties and local systems on punctured surfaces already appeared in [Yam08], where a proof is given for the case of the punctured projective line. Our result applies to all genera. Thanks to this correspondence, and to the results proved in Sect. 4, we are able to extend the work of Bellamy and the first author from closed Riemann surfaces to open ones. Another interesting aspect of this correspondence is that it could be conjecturally combined with the non-abelian Hodge theorem to extend the main results of [Tir19], proved by the second author, in the context of moduli spaces of parabolic Higgs bundles. More details on this topic are provided in Sect. 7, where possible future research directions of the present work are discussed.

Section 4 contains the proof of Theorem 1.5. A careful study of the singularities of multiplicative quiver varieties is carried out. First, we show that the smooth locus is precisely the θ-stable locus. The remaining part of the section is devoted to the study of the nature of the singular locus. To this end, we use techniques from the work [BS21] of Bellamy and the first author to prove that, under suitable hypotheses, the singularities are symplectic. We also prove that, under certain conditions, the moduli space $\mathcal{M}_{q,\theta}(Q,\alpha)$ contains an open subset which is singular, factorial, and terminal. As a consequence, $\mathcal{M}_{q,\theta}(Q,\alpha)$ does not admit a symplectic resolution. Moreover, for generic θ, we see that the open subset is the entire variety. The only case left out by Theorem 1.5, when $\alpha = 2\beta$ for $\beta \in N_{q,\theta}$ and $p(\beta) = 2$, is more subtle than the others. The corresponding result in the context of ordinary quiver varieties, treated in [BS21], is based on the study of the local structure of such varieties. In our case, such a tool is still not available but will hopefully be the object of future research.

In Sect. 5, namely, in Theorems 5.1 and 5.3, we combinatorially classify all the pairs (Q,α), formed by a crab-shaped quiver and a corresponding dimension vector in the fundamental region such that $(p(\gcd(\alpha)^{-1}\alpha),\ \gcd(\alpha)) = (2,2)$. This is relevant as these are the cases expected, for generic θ, to admit "O'Grady"-type resolutions (i.e., by blowing up the singular locus). This is also important since by Theorem 1.11, it allows us to recognize whether a dimension vector in the fundamental region is expected to admit a symplectic resolution or not.

In Sect. 6, we face the problem of existence of symplectic resolutions of multiplicative quiver varieties for general dimension vectors. In order to do so, we follow the approach of Bellamy and the first author. In particular, we prove that a multiplicative quiver variety has a canonical decomposition into natural factors; see Theorem 6.17. This can be viewed as a multiplicative analogue of Crawley-Boevey's decomposition [CB02], and we follow his proof, obtaining some more factors due to the unavailability of (*) and some local structure results. Our result makes it possible to solve the problem by understanding it only at the level of such indecomposable factors, which are multiplicative quiver varieties with particular dimension vectors. We then extend the GIT construction of symplectic resolutions by varying θ to dimension vectors not in Σ (Theorem 6.23); this includes multiplicative analogues of framed quiver varieties such as Hilbert schemes of $\mathbb{C}^2$ and of hyperkähler almost locally Euclidean spaces. In Theorem 6.27, we make use of our canonical decomposition and, modulo some cases for which the question remains still open, we classify all multiplicative quiver varieties with arbitrary dimension vector that admit a symplectic resolution. As an application of this result, by restricting to crab-shaped quivers, we give an explicit classification of the character varieties of punctured surfaces admitting symplectic resolutions (Corollary 6.28), combining Theorem 6.27 and the results of Sect. 3.

Last, Sect. 7 contains some open questions which naturally arise from the study carried out in the present paper. One open question which would be interesting to tackle regards the possibility to extend the main results of [Tir19] starting from the correspondence outlined in Sect. 3: In Sect. 7, we provide some details on this topic by describing which moduli space one would need to consider, via the non-abelian Hodge theorem in the non-compact case [Sim90], and we conjecture a generalization of the Isosingularity theorem for such moduli spaces. To conclude, we outline a general setting and of pose a number of questions which should generalize the work of the present and many other papers, for example, [AS18, BS21, BS, KL07, Tir19]: It seems that many of the techniques exploited in the mentioned works are particular instances of theorems which conjecturally hold in the context of moduli spaces of semistable objects in 2-Calabi–Yau categories, under suitable hypotheses. This assertion is motivated also by the work of Bocklandt, Galluzzi, and Vaccarino, [BGV16], who studied moduli spaces of representations of 2-Calabi–Yau algebras and proved that such varieties locally look like representations of (ordinary) preprojective algebras. This seems to be a singular, local, underived version of the phenomenon that representation varieties of Calabi–Yau algebras are (shifted) symplectic (as announced by Brav and Dyckerhoff; see a similar result in [Yeu]).

2 Multiplicative Quiver Varieties

In this section, we give the definition of multiplicative quiver varieties following [CBS06] and recall some basic properties of such moduli spaces which will be useful in the arguments of the proof of our main theorems. In addition to these known results, we prove a new one, concerning the normality of the aforementioned varieties.

Throughout the paper, we work over the field $\mathbb{C}$ of complex numbers.

2.1 Preliminaries on Quivers and Root Systems

We recall the basic definitions and fix the notations from the theory of quiver representations. Let Q be a finite quiver (= directed graph with finitely many vertices and edges). We let Q_0 and Q_1 denote the set of vertices and the set of arrows (= edges) of Q, respectively. Moreover, for an arrow $a \in Q_1$, let $h(a)$ and $t(a)$ denote the head and the tail of a, respectively. For a dimension vector $\alpha \in \mathbb{N}^{Q_0}$, we will denote by $\operatorname{Rep}(Q, \alpha)$ the space of representations of Q of dimension α, which is naturally acted upon by the group $\operatorname{GL}(\alpha) := \prod_{i \in Q_0} \operatorname{GL}(\alpha_i)$.

The coordinate vector at vertex i is denoted e_i. The set $\mathbb{N}^{Q_0}$ of dimension vectors is partially ordered by $\alpha \geq \beta$ if $\alpha_i \geq \beta_i$ for all i, and we say that $\alpha > \beta$ if $\alpha \geq \beta$ with $\alpha \neq \beta$. The *support* of a vector α is the set of $i \in Q_0$ with $\alpha_i \neq 0$; α is called *sincere* if its support is all of Q_0. The Euler (or Ringel) form on $\mathbb{Z}^{Q_0}$ is defined by

$$\langle \alpha, \beta \rangle = \sum_{i \in Q_0} \alpha_i \beta_i - \sum_{a \in Q_1} \alpha_{t(a)} \beta_{h(a)}.$$

Let $(\alpha, \beta) = \langle \alpha, \beta \rangle + \langle \beta, \alpha \rangle$ denote the corresponding Cartan (or Tits) form and set $p(\alpha) = 1 - \langle \alpha, \alpha \rangle$. The fundamental region $\mathcal{F}(Q)$ is the set of nonzero $\alpha \in \mathbb{N}^{Q_0}$ with connected support and with $(\alpha, e_i) \leq 0$ for all i. For $q \in (\mathbb{C}^\times)^{Q_0}$ and $\alpha \in \mathbb{N}^{Q_0}$, let $q^\alpha := \prod_{i \in Q_0} q_i^{\alpha_i}$.

If i is a loop-free vertex, so $p(e_i) = 0$, there is a reflection $s_i : \mathbb{Z}^{Q_0} \to \mathbb{Z}^{Q_0}$ defined by $s_i(\alpha) = \alpha - (\alpha, e_i)e_i$. The real roots (respectively, imaginary roots) are the elements of $\mathbb{Z}^{Q_0}$ which can be obtained from the coordinate vector at a loop-free vertex (respectively, $\pm$ an element of the fundamental region) by applying some sequence of reflections at loop-free vertices. Let R^+ denote the set of positive roots. Recall that a root β is *isotropic* imaginary if $p(\beta) = 1$ and *anisotropic* imaginary if $p(\beta) > 1$. We say that a dimension vector α is *indivisible* if the greatest common divisor of the α_i is one.

2.2 Multiplicative Preprojective Algebras

We now define the quiver algebras whose moduli of representations are the varieties of interest in the present paper. To this purpose, let Q be a finite quiver, fixed once and for all in this section. First, recall that for a vector $\lambda \in \mathbb{C}^{Q_0}$, the *deformed preprojective algebra* $\Pi^\lambda(Q)$ is the quotient of the path algebra $\mathbb{C}\overline{Q}$ of the doubled quiver $\overline{Q}$ by the relation

$$\sum_{x\in Q_1}[x,x^*]=\sum_{i\in Q_0}\lambda_i e_i,$$

where x^* denotes the dual loop to x in $\overline{Q}_1$; it is well known that Nakajima quiver varieties can be interpreted as moduli spaces of (θ-semistable) representations of such algebras. As one might expect, the defining relation for multiplicative preprojective algebras is a multiplicative analogue of the above equation: Choose $q \in (\mathbb{C}^\times)^{Q_0}$ and define $A(Q)$ to be the universal localization of the path algebra $\mathbb{C}\overline{Q}$ such that $1+xx^*$ and $1+x^*x$ are invertible, for $x \in \overline{Q}_1$. Then, following [CBS06, Definition 1.2], the *multiplicative preprojective algebra* $\Lambda^q(Q)$ is defined as the quotient of $A(Q)$ by the relation

$$\prod_{x\in\overline{Q}_1}^{<}(1+xx^*)^{\varepsilon(x)}=\sum_{i\in Q_0}q_i e_i,$$

where $\varepsilon(x)$ equals 1 if $x \in Q_1$ and -1 otherwise and the product is ordered by an arbitrary choice of ordering "$<$" on $\overline{Q}_1$. It is known, by [CBS06, Theorem 1.4] that up to isomorphism, $\Lambda^q(Q)$ does not depend on the orientation of the quiver or the chosen ordering on $\overline{Q}_1$. When the quiver Q is clear from the context, we will use the shortened notation Λ^q in place of $\Lambda^q(Q)$.

Analogously to the additive case mentioned above, representations of $\Lambda^q(Q)$ are representations of the underlying quiver $\overline{Q}$, $\{(V_i)_{i\in\overline{Q}_0},(\phi_a)_{a\in\overline{Q}_1}\}$, satisfying the additional relations:

$$\mathrm{Id}_{V_{h(a)}}+\phi_a\phi_a^* \text{ is an invertible endomorphism of } V_{h(a)} \text{ for all } a\in\overline{Q}_1$$

$$\prod_{a\in\overline{Q}_1,h(a)=i}(\mathrm{Id}_{V_{h(a)}}+\phi_a\phi_a^*)^{\varepsilon(a)}=q_i\mathrm{Id}_{V_i} \text{ for all } i\in Q_0,$$

where for an edge, $a \in \overline{Q}_1$, ϕ_a^* denotes the linear map ϕ_{a^*}, a^* being the dual edge of a.

For a positive vector $\alpha \in \mathbb{N}^{Q_0}$, we denote by $\mathrm{Rep}(\Lambda^q,\alpha)$ the set of representations of Λ^q with $V_i=\mathbb{C}^{\alpha_i}$ for all i. This can be given an obvious affine scheme structure via the subset of matrices satisfying the obvious polynomial equations. We will work below with the reduced subvariety of this affine scheme.

Remark 2.1 By taking determinants of the defining relation for the multiplicative preprojective algebra, one can easily see that if Λ^q has a representation of dimension vector α, then $q^\alpha=1$, which, thus, is a necessary condition to be satisfied in order to have a non-empty moduli space.

The following results, which will be used in the next sections, are proved in [CBS06]. It is worth pointing out that even though we work over $\mathbb{C}$, these statements hold true over an arbitrary field $\mathbb{K}$.

Proposition 2.2 *If X and Y are finite-dimensional representations of Λ^q, then*

$$\dim \operatorname{Ext}^1_{\Lambda^q}(X, Y) = \dim \operatorname{Hom}_{\Lambda^q}(X, Y) + \operatorname{Hom}_{\Lambda^q}(Y, X) - (\underline{\dim} X, \underline{\dim} Y)$$

The following result concerns the geometry of the space $\operatorname{Rep}(\Lambda^q, \alpha)$ of representations of the algebra Λ^q, when a dimension vector $\alpha \in \mathbb{N}^{Q_0}$ is fixed. Define g_α as $g_\alpha := -1 + \sum_{i \in Q_0} \alpha_i^2$.

Proposition 2.3 $\operatorname{Rep}(\Lambda^q, \alpha)$ *is an affine variety, and every irreducible component has dimension at least $g_\alpha + 2p(\alpha)$. The subset $T \subset \operatorname{Rep}(\Lambda^q, \alpha)$ of representations X with trivial endomorphism algebra,* $\operatorname{End}(X) = \mathbb{C}$, *is open and, if non-empty, smooth of dimension $g_\alpha + 2p(\alpha)$.*

2.3 Reflection Functors for $\Lambda^q(Q)$

As in the additive case, one can define reflection functors for the multiplicative preprojective algebra $\Lambda^q(Q)$: Let v a loop-free vertex in Q and define

$$u_v : (\mathbb{C}^\times)^{Q_0} \to (\mathbb{C}^\times)^{Q_0}, \quad u_v(q)_w = q_v^{-(e_v, e_w)} q_w.$$

It is easy to see that the map u_v satisfies the following identity:

$$(u_v(q))^\alpha = q^{s_v(\alpha)},$$

where s_v is the reflection map defined in Sect. 2.1. The main result concerning such maps is analogous to the properties of reflections functors for $\Pi^\lambda(Q)$.

Proposition 2.4 *[CBS06, Theorem 1.7] If v is a loop-free vertex and $q_v \neq 1$, then there is an equivalence of categories F_q from the category of representations of Λ^q to the category of representations of $\Lambda^{u_v(q)}$, acting on dimension vectors through the reflection s_v. The inverse equivalence is given by $F_{u_v(q)}$.*

We will need also reflections on θ. Define

$$r_v : \mathbb{Z}^{Q_0} \to \mathbb{Z}^{Q_0}, \quad r_v(\theta)_w = \theta_w - (e_v, e_w)\theta_v.$$

Definition 2.5 The map $(q, \theta, \alpha) \mapsto (u_v(q), r_v(\theta), s_v(\alpha))$, is called a reflection. If $\theta_v \neq 0$ or $q_v \neq 1$, it is called an *admissible reflection.*

We will explain below isomorphisms of multiplicative quiver varieties, due to Yamakawa, which are closely related to the above equivalence.

2.4 Moduli of Representations of $\Lambda^q(Q)$

We shall now outline the construction of the varieties of interest for the present work. As mentioned above, the general definition involves a stability condition $\theta \in \mathbb{Z}^{Q_0}$, which we fix for the rest of this section.

The seminal work of King [Kin94] allows one to define the notion of θ-semistability for modules over Λ^q:

Definition 2.6 Let M be a finite-dimensional representation of Λ^q such that $\underline{\dim} M \cdot \theta = 0$. The module M is said to be *θ-semistable* if for any submodule $N \subset M$

$$\theta \cdot \underline{\dim}\, N \leq 0.$$

The module M is said to be *θ-stable* if the strict inequality holds. Finally, M is said to be *θ-polystable* if it is a direct sum of θ-stable representations. Given a set (or scheme) X of representations, let $X^{\theta-s}$ and $X^{\theta-ss}$ denote the θ-stable and θ-semistable loci, respectively. We will use the notation $\operatorname{Rep}^{\theta-s}(Q,\alpha) := \operatorname{Rep}(Q,\alpha)^{\theta-s}$ and similarly for $\theta-ss$.

Remark 2.7 By [Kin94, Proposition 3.1], one has that the above definition of stability coincides with the usual one coming from GIT: Indeed, consider the character

$$\chi_\theta : \mathrm{GL}(\alpha) \to \mathbb{C}^\times, \qquad (g_i)_{i\in Q_0} \mapsto \prod_{i\in Q_0} (\det g_i)^{-\theta_i}.$$

It defines a linearization on the trivial line bundle $\operatorname{Rep}(Q,\alpha) \times \mathbb{C}$ of the action of $\mathrm{GL}(\alpha)$ on $\operatorname{Rep}(Q,\alpha)$; thus, one can define the notion of χ_θ-(semi)stability à la Mumford, [MFK02]. The aforementioned result of King proves that M is θ-(semi)stable if and only if it is χ_θ-(semi)stable.

Using the notion above, one can construct the moduli space of (semistable) representations of Λ^q of dimension α as follows (see [Yam08, §2], for the details): define

$$\operatorname{Rep}^\circ(\overline{Q},\alpha) = \{\phi \in \operatorname{Rep}(\overline{Q},\alpha) \mid \det(1+\phi_a\phi_a^*) \neq 0,\ a \in \overline{Q}_1\}.$$

Here and in the following, for $\phi \in \operatorname{Rep}(\overline{Q},\alpha)$, we let $\phi_a, a \in \overline{Q}_1$ denote the component linear maps. One can then consider the map

$$\Phi : \operatorname{Rep}^\circ(\overline{Q},\alpha) \longrightarrow \mathrm{GL}(\alpha),$$

defined by the formula

$$\Phi(\phi) = \prod_{a \in \overline{Q}_1}^{<} (1 + \phi_a \phi_a^*)^{\varepsilon(j)}.$$

Let us identify $\mathbb{C}^\times$ also with the scalar matrices in $\mathrm{GL}(\alpha_i)$, and hence $(\mathbb{C}^\times)^{Q_0}$ also with a subset of $\mathrm{GL}(\alpha)$. Fixing $q \in (\mathbb{C}^\times)^{Q_0}$, one has that $\mathrm{Rep}(\Lambda^q(Q), \alpha)$ is the set-theoretic preimage $\Phi^{-1}(q)$. Thus, one can give the following

Definition 2.8 The *multiplicative quiver variety* $\mathcal{M}_{q,\theta}(Q, \alpha)$ is the GIT quotient

$$\mathcal{M}_{q,\theta}(Q, \alpha) := (\mathrm{Rep}^{\theta-ss}(\overline{Q}, \alpha) \cap \Phi^{-1}(q)) /\!\!/ \mathrm{GL}(\alpha).$$

Remark 2.9 The reason for the terminology in the previous definition is apparent: The equations defining the multiplicative preprojective relation are modifications of the ones used to define the usual *deformed preprojective algebras*, whose moduli of (semistable) representations are Nakajima quiver varieties.

It is worth recalling a fundamental result of King, which gives a moduli-theoretic interpretation—in the sense of (representable) moduli functors—to $\mathcal{M}_{q,\theta}(Q, \alpha)$.

Theorem 2.10 ([Kin94, Propositions 3.1 and 3.2]) *Assume $\theta \in \mathbb{Z}^{Q_0}$. Then, $\mathcal{M}_{q,\theta}(Q, \alpha)$ is a coarse moduli space for families of θ-semistable representations up to S-equivalence.*

Here, two θ-semistable representations are S-equivalent if and only if they have the same composition factors into θ-stable representations (i.e., they have filtrations whose subquotients are isomorphic θ-stable representations). This means that every point in $\mathcal{M}_{q,\theta}(Q, \alpha)$ has a unique representative which is θ-polystable, up to isomorphism.

Precisely as in [BS21, Lemma 2.4], we have the following instance of the well-known principle of GIT:

Definition 2.11 We say that $\theta' \geq \theta$ if every θ'-semistable representation of Λ^q is also θ-semistable.

Note that $\theta' \geq \theta$ is implied if the purely combinatorial condition holds that $\theta \cdot \beta > 0$ implies $\theta' \cdot \beta > 0$ for all $\beta < \alpha$.

Lemma 2.12 ([BS21, Lemma 2.4]) *Let $\alpha \in N_{q,\theta}$ be such that $\mathcal{M}_{q,\theta}(Q, \alpha) \neq \emptyset$. Take $\theta' \geq \theta$. Then we have a projective Poisson morphism $\mathcal{M}_{q,\theta'}(Q, \alpha) \to \mathcal{M}_{q,\theta}(Q, \alpha)$ induced by the inclusion $\Phi^{-1}(q)^{\theta'-ss} \subseteq \Phi^{-1}(q)^{\theta-ss}$.*

We caution that this morphism need not be surjective (and indeed the source could be empty when the target is not). However, in many cases, as we will see, it produces a symplectic resolution.

2.5 Reflection Isomorphisms

There is a multiplicative analogue of the Lusztig–Maffei–Nakajima reflection isomorphisms of quiver varieties (see in particular [Maf02, Theorem 26]), due to Yamakawa, which makes use of the reflection functors F_q. Let us extend the definition of $\mathcal{M}_{q,\theta}(Q,\alpha)$ to $\alpha \in \mathbb{Z}^{Q_0}$ by setting it to be empty in the case that $\alpha_i < 0$ for some i.

Theorem 2.13 ([Yam08, Theorem 5.1]) *An admissible reflection* $(q,\theta,\alpha) \mapsto (u_v(q), r_v(\theta), s_v(\alpha))$ *induces an isomorphism of multiplicative quiver varieties,* $\mathcal{M}_{q,\theta}(Q,\alpha) \cong \mathcal{M}_{u_v(q),r_v(\theta)}(Q, s_v(\alpha))$.

2.6 Poisson Structure on $\mathcal{M}_{q,\theta}(Q,\alpha)$

In order to construct a Poisson structure on $\mathcal{M}_{q,\theta}(Q,\alpha)$, we shall use the theory of quasi-Hamiltonian reductions, first developed in [AMM98] for the case of real manifolds, and then treated by Boalch, [Boa07], and Van den Bergh [Van08a, Van08b] in the holomorphic and algebraic settings. To this end, note that the map Φ defined above is a group valued moment map for the quasi-Hamiltonian action of $\mathrm{GL}(\alpha)$ on $\mathrm{Rep}^{\circ}(\overline{Q},\alpha)$. Thus, the variety $\mathcal{M}_{q,\theta}(Q,\alpha)$ can be considered as the quasi-Hamiltonian reduction of $\mathrm{Rep}^{\circ}(\overline{Q},\alpha)$ modulo the action of $\mathrm{GL}(\alpha)$. From the properties of such a reduction, we obtain that $\mathcal{M}_{q,\theta}(Q,\alpha)$ is a Poisson variety. Moreover, defining

$$\mathcal{M}^{s}_{q,\theta}(Q,\alpha) := (\mathrm{Rep}^{\theta-s}(\overline{Q},\alpha) \cap \Phi^{-1}(q))/\,\mathrm{GL}(\alpha),$$

where $\mathrm{Rep}^{\theta-s}(\overline{Q},\alpha) \subset \mathrm{Rep}^{\theta-ss}(\overline{Q},\alpha)$ denotes the θ-stable locus, one has the following result, which will be crucial in proving that $\mathcal{M}_{q,\theta}(Q,\alpha)$ is a symplectic singularity. Note that in the above definition, the quotient is the usual orbit space, if we replace $\mathrm{GL}(\alpha)$ by $\mathrm{PGL}(\alpha) = \mathrm{GL}(\alpha)/\mathbb{C}^{\times}$, as a point in the stable locus has trivial stabilizer group under $\mathrm{PGL}(\alpha)$.

Proposition 2.14 ([Yam08, Theorem 3.4]) $\mathcal{M}^{s}_{q,\theta}(Q,\alpha)$*, if non-empty, is an equidimensional algebraic symplectic manifold and its dimension is* $2p(\alpha)$.

2.7 Stratification by Representation Type

An important result proved in [CBS06, §7] concerns a natural stratification of the affine variety $\mathcal{M}_{q,0}(Q,\alpha)$ which parametrizes semisimple representations of the algebra Λ^q. This stratification and its generalization, proved below, to the case of θ-semistable representations are important in order to understand the singular locus of $\mathcal{M}_{q,\theta}(Q,\alpha)$.

Consider $M \in \mathrm{Rep}^{\theta-ss}(\Lambda^q, \alpha)$. Replace it by the unique θ-polystable representation which is S-equivalent to it (see the discussion after Theorem 2.10). M is then said to be of *representation type* $\tau = (k_1, \beta^{(1)}; \ldots; k_r, \beta^{(r)})$ if it can be decomposed into the direct sum $M \cong M_1^{k_1} \oplus \cdots \oplus M_r^{k_r}$, where M_i is a θ-stable representation of Λ^q of dimension vector $\beta^{(i)}$, $i = 1, \ldots, r$, and $M_i \ncong M_j$ for $i \neq j$.

Proposition 2.15 *If τ is a representation type for Λ^q, then the set $C^{\tau}_{q,\theta}(Q, \alpha)$ of θ-semistable representations of type τ is a locally closed subset of $\mathcal{M}_{q,\theta}(Q, \alpha)$, which, if non-empty, has dimension $\sum_{i=1}^{r} 2p(\beta^{(i)})$. $\mathcal{M}_{q,\theta}(Q, \alpha)$ is the disjoint union of the strata $C^{\tau}_{q,\theta}(Q, \alpha)$, where τ runs over the set of representation types that can occur for Λ^q.*

Proof First, note that the case when $\theta = 0$ is treated in [CBS06] and proved in Lemma 7.1 therein. For the case when $\theta \neq 0$, we use the same arguments. Indeed, the fact that $\mathcal{M}_{q,\theta}(Q, \alpha)$ is a disjoint union of subsets of a fixed representation type is immediate from the fact that the decomposition of a θ-polystable module into θ-stable modules is unique. This, in turn, holds because for θ-stable modules M and N, we have $\dim \mathrm{Hom}(M, N) \leq 1$, with equality if and only if M and N are isomorphic. Moreover, to prove that each $C^{\tau}_{q,\theta}(Q, \alpha)$ is locally closed and of the dimension prescribed by the lemma, one can adapt the proof [CB01, Theorem 1.3]: Indeed, those arguments can be repeated in this case as well, replacing $\mathrm{Rep}(\overline{Q}, \alpha)$ with $\mathrm{Rep}^{\theta-ss}(\overline{Q}, \alpha)$, $\mu_{\alpha}^{-1}(\lambda)$ with $\Phi^{-1}(q)$, the word " (semi)simple" with "θ-(semi)stable" in the proof, and noting that everything goes through in the same way because $\mathrm{Rep}^{\theta-ss}(\overline{Q}, \alpha)$ is open in $\mathrm{Rep}(\overline{Q}, \alpha)$. The only difference is that in this case, we do not claim irreducibility, since $\mathrm{Rep}^{\theta-ss}(\overline{Q}, \beta)$ is not known to be irreducible. □

We will need also the following property of $\mathrm{Rep}^{\theta-ss}(\Lambda^q(Q), \alpha)$:

Lemma 2.16 *Every irreducible component of* $\mathrm{Rep}^{\theta-ss}(\Lambda^q(Q), \alpha)$ *has dimension at least $g_\alpha + 2p(\alpha)$, and the set of θ-stable representations form an open subset of* $\mathrm{Rep}(\Lambda^q(Q), \alpha)$ *which, if non-empty, is smooth of dimension $g_\alpha + 2p(\alpha)$.*

Proof For the first part, Lemma 6.2 in [CB03] proves the statement in the case when $\theta = 0$, of which the above result is a consequence since $\mathrm{Rep}^{\theta-ss}(\Lambda^q(Q), \alpha)$ is an open subset of $\mathrm{Rep}(\Lambda^q(Q), \alpha)$: Indeed, every irreducible component of the former variety is contained in only one irreducible component of the latter and, hence, the dimension estimate holds. For the second part, one just needs to note that if X is a θ-stable representation, then $\mathrm{End}(X) = \mathbb{C}$ and, hence, by Crawley-Boevey and Shaw [CBS06, Theorem 1.10] defines a smooth point of $\mathrm{Rep}(\Lambda^q(Q), \alpha)$, which implies that it is a smooth point of $\mathrm{Rep}^{\theta-ss}(\Lambda^q(Q), \alpha)$. □

For the proof of the following proposition, apply the strategy carried out in [CB03, §6, 7] and [CBS06, §7]: The only change is that in the definition of representation of *top-type*, one has to replace the word "simple" with the word "θ-stable" and use Proposition 2.15 instead of [CBS06, Lemma 7.1] and Lemma 2.16 instead of [CBS06, Theorem 1.1].

Proposition 2.17 *The inverse image in* $\operatorname{Rep}^{\theta-ss}(\Lambda^q(Q),\alpha)$ *of the stratum of representations of type* $\tau = (k_1, \beta^{(1)}; , \ldots ; k_r, \beta^{(r)})$ *has dimension at most* $g_\alpha + p(\alpha) + \sum_{l=1}^{r} p(\beta^{(l)})$.

2.8 The Set $\Sigma_{q,\theta}$

As mentioned in the introduction, the dimension vectors of stable representations are closely related to the following combinatorially defined set, which is the multiplicative analogue of the set Σ_λ introduced by Crawley-Boevey in [CB02] and extensively used in [BS21]:

Definition 2.18 Fix $q \in (\mathbb{C}^\times)^{Q_0}$ and $\theta \in \mathbb{Z}^{Q_0}$ and set $N_{q,\theta} := \{\alpha \in \mathbb{N}^{Q_0} \mid q^\alpha = 1, \alpha \cdot \theta = 0\}$. Define $R^+_{q,\theta} := R^+ \cap N_{q,\theta}$. Then,

$$\Sigma_{q,\theta} := \left\{ \alpha \in R^+_{q,\theta} \;\middle|\; p(\alpha) > \sum_{i=1}^{r} p\left(\beta^{(i)}\right) \text{ for any decomposition} \right.$$

$$\left. \alpha = \beta^{(1)} + \cdots + \beta^{(r)} \text{ with } r \geq 2,\ \beta^{(i)} \in R^+_{q,\theta} \right\}.$$

When $\theta = 0$, we shall use the shortened notation Σ_q in place of $\Sigma_{q,0}$.

The following is an extension of [CBS06, Theorem 1.11] to the case $\theta \neq 0$.

Proposition 2.19 *Let* $\alpha \in \Sigma_{q,\theta}$. *Then, if non-empty,* $\operatorname{Rep}^{\theta-ss}(\Lambda^q(Q),\alpha)$ *is a complete intersection in* $\operatorname{Rep}^{\theta-ss}(Q,\alpha)$, *equidimensional of dimension* $g_\alpha + 2p(\alpha)$. *The locus of* θ*-stable representations* $\operatorname{Rep}^{\theta-s}(\Lambda^q(Q),\alpha)$ *is dense inside* $\operatorname{Rep}^{\theta-ss}(\Lambda^q(Q),\alpha)$.

Proof This is a direct consequence of Lemma 2.16 and Proposition 2.17, by the definition of $\Sigma_{q,\theta}$. □

Remark 2.20 Note that a consequence of the above proposition is that if $\pi : \operatorname{Rep}^{\theta-ss}(\Lambda^q(Q),\alpha) \to \mathcal{M}_{q,\theta}(Q,\alpha)$ is the projection map, the image $\mathcal{M}^s_{q,\theta}(Q,\alpha)$ of the stable locus is dense in the moduli space $\mathcal{M}_{q,\theta}(Q,\alpha)$. As a corollary of this and Proposition 2.14 (or Proposition 2.15), one has that every component of $\mathcal{M}_{q,\theta}(Q,\alpha)$ has dimension $2p(\alpha)$.

A useful corollary of the proposition is the following criterion for birationality of the maps $\mathcal{M}_{q,\theta'}(Q,\beta) \to \mathcal{M}_{q,\theta}(Q,\beta)$. Together with Lemma 2.12, this explains that these maps will be resolutions of singularities when the source is smooth.

Corollary 2.21 *Let* $\alpha \in \Sigma_{q,\theta}$ *be such that* $\mathcal{M}_{q,\theta}(Q,\alpha) \neq \emptyset$. *Take* $\theta' \geq \theta$ *such that every* θ*-stable representation is* θ'*-stable. Then the morphism* $\mathcal{M}_{q,\theta'}(Q,\beta) \to \mathcal{M}_{q,\theta}(Q,\beta)$ *is birational.*

Remark 2.22 Note that $\theta' \geq \theta$ is guaranteed if whenever $\beta < \alpha$, then $\theta \cdot \beta > 0$ implies $\theta' \cdot \beta > 0$. Similarly, the assumption that every θ-stable representation is θ'-stable is implied if for $\beta < \alpha$, then $\theta \cdot \beta < 0$ implies $\theta' \cdot \beta < 0$. To find θ' satisfying these conditions, first note that they will be satisfied for rational stability conditions $\theta' \in \mathbb{Q}^{Q_0}$ sufficiently close to θ. But they hold for a rational vector if and only if they hold for an integral multiple.

Proof By Definition 2.11, $\mathrm{Rep}^{\theta'-ss}(\Lambda^q, \alpha)$ is a subset of $\mathrm{Rep}^{\theta-ss}(\Lambda^q, \alpha)$, and it is open. By assumption, the locus $\mathrm{Rep}^{\theta-s}(\Lambda^q, \alpha)$ is open in $\mathrm{Rep}^{\theta'-ss}(\Lambda^q, \alpha)$. It is also dense, since it is dense in $\mathrm{Rep}^{\theta-ss}$. Therefore, the locus $\mathcal{M}^{\theta-s}_{q,\theta'}(Q, \beta)$ is open and dense in $\mathcal{M}_{q,\theta'}(Q, \beta)$. As the stable $\mathrm{GL}(\alpha)$-orbits are closed, $\mathcal{M}^{\theta-s}_{q,\theta'}(Q, \beta)$ maps isomorphically to $\mathcal{M}^{s}_{q,\theta}(Q, \beta)$. As the latter is dense in $\mathcal{M}_{q,\theta}(Q, \beta)$, we conclude the desired birationality. □

Using the above results, one can derive an important geometric property of the moduli space $\mathcal{M}_{q,\theta}(Q, \alpha)$. For reasons which are clear in the proof of the proposition, we assume that a certain codimension estimate holds. As usual, let $\pi : \mathrm{Rep}^{\theta-ss}(\Lambda^q, \alpha) \to \mathcal{M}_{q,\theta}(Q, \alpha)$ denote the quotient map.

Lemma 2.23 *Assume $\alpha \in \Sigma_{q,\theta}$ and let τ be a stratum. The following inequality holds true:*

$$\mathrm{codim}_{\mathrm{Rep}^{\theta-ss}(\Lambda^q,\alpha)}(\pi^{-1}(C^{\tau}_{q,\theta}(Q, \alpha))) \geq \frac{1}{2}\mathrm{codim}_{\mathcal{M}_{q,\theta}(Q,\alpha)}(C^{\tau}_{q,\theta}(Q, \alpha)).$$

Proof By Proposition 2.19, one has that

$$\mathrm{codim}(\pi^{-1}(C^{\tau}_{q,\theta}(Q, \alpha))) = g_{\alpha} + 2p(\alpha) - \dim \pi^{-1}(C^{\tau}_{q,\theta}(Q, \alpha)).$$

Moreover, from Proposition 2.17, it follows that

$$\begin{aligned} g_{\alpha} + 2p(\alpha) - \dim \pi^{-1}(C^{\tau}_{q,\theta}(Q, \alpha)) &\geq g_{\alpha} + 2p(\alpha) - g_{\alpha} - p(\alpha) - \textstyle\sum_{l=1}^{r} p(\beta^{(l)}) \\ &= p(\alpha) - \textstyle\sum_{l=1}^{r} p(\beta^{(l)}). \end{aligned}$$

On the other hand, by Proposition 2.15, one has that

$$p(\alpha) - \sum_{l=1}^{r} p(\beta^{(l)}) = \frac{1}{2}\left(\dim \mathcal{M}_{q,\theta}(Q, \alpha) - \dim C^{\tau}_{q,\theta}(Q, \alpha)\right),$$

which, combined with the above inequality, leads to the desired statement. □

By taking the minimum of these codimensions, we immediately conclude:

Corollary 2.24 *Let Z denote the complement inside $\mathcal{M}_{q,\theta}(Q, \alpha)$ of the set of θ-stable representations $\mathcal{M}^{s}_{q,\theta}(Q, \alpha)$, that is, Z is the union of all the non-open strata of $\mathcal{M}_{q,\theta}(Q, \alpha)$. Then, the following inequality holds:*

$$\operatorname{codim} \pi^{-1}(Z) \geq \frac{1}{2} \min_{\tau \neq (1,\alpha)} \operatorname{codim} C^{\tau}_{q,\theta}(Q, \alpha).$$

Proposition 2.25 *Consider $\alpha \in \Sigma_{q,\theta}$ and assume that all strata in the non-empty multiplicative quiver variety $\mathcal{M}_{q,\theta}(Q, \alpha)$ have codimension at least 4, that is, assume that*

$$\min_{\tau \neq (1,\alpha)} \left(\dim \mathcal{M}_{q,\theta}(Q, \alpha) - \dim C^{\tau}_{q,\theta}(Q, \alpha)) \right) \geq 4.$$

Then the variety $\mathcal{M}_{q,\theta}(Q, \alpha)$ is normal.

Proof The arguments to prove the above statement are analogous to the ones used in [BS]. In particular, we shall use a criterion proved by Crawley-Boevey, [CB03, Corollary 7.2]. We first deal with the case when $\theta = 0$ and then explain how to adapt the arguments for general θ. When $\theta = 0$, $\mathcal{M}_{q,0}(Q, \alpha)$ is the categorical quotient $\operatorname{Rep}(\Lambda^q, \alpha) /\!\!/ \operatorname{GL}(\alpha)$ of an affine variety modulo a reductive group. Thus, we only need to show that $\operatorname{Rep}(\Lambda^q, \alpha)$ satisfies Serre's condition (S_2) and that certain codimension estimates hold true. The first condition is ensured by the fact that by Crawley-Boevey and Shaw [CBS06, Theorem 1.11] (the case $\theta = 0$ of Proposition 2.19), $\operatorname{Rep}(\Lambda^q, \alpha)$ is a complete intersection and, hence, Cohen-Macaulay, which indeed implies condition (S_2). Now, denote by S the open subset $S \subset \mathcal{M}_{q,0}(Q, \alpha)$ of simple representations, which is non-empty by our assumption. S is contained in the smooth locus and hence is normal. Moreover, let Z denote its complement in $\mathcal{M}_{q,0}(Q, \alpha)$ and denote with $\pi : \operatorname{Rep}(\Lambda^q, \alpha) \to \mathcal{M}_{q,0}(Q, \alpha)$ the quotient map; then by Corollary 2.24, one has

$$\dim \operatorname{Rep}(\Lambda^q, \alpha) - \dim \pi^{-1}(Z) \geq \frac{1}{2} \min_{\tau \neq (1,\alpha)} (\dim \mathcal{M}_{q,0}(Q, \alpha) - \dim C^{\tau}_{q,0}(Q, \alpha)),$$

and the right-hand side is greater or equal than two by assumption. Thus, all the hypotheses of [CB03, Corollary 7.2] are satisfied, and we can conclude that $\mathcal{M}_{q,0}(Q, \alpha)$ is normal. For the case when $\theta \neq 0$, keeping in mind that normality is a local property, we fix a point $x \in \mathcal{M}_{q,\theta}(Q, \alpha)$ and aim at proving normality at x. This is achieved by choosing an open neighborhood V of x such that the restriction to $\pi^{-1}(V)$ of the projection morphism $\pi^{-1}(V) \to V$ is an affine quotient (note that this can be done thanks to the properties of the GIT construction). One can now repeat the same arguments as for the $\theta = 0$ case, noting that by Proposition 2.15, the estimates above hold true also in this more general setting: Being Cohen-Macaulay is a local statement and thus the previous part of the proof ensures that $\pi^{-1}(V)$, which is open in $\operatorname{Rep}(\Lambda^q, \alpha)$, satisfies this property. Moreover, defining S_θ to be the subset of V of θ-stable representations, then one may proceed as in the first part of the proof to obtain the desired conclusion. □

Remark 2.26 In the next sections, we will examine some cases in which the technical assumption in the previous result is satisfied, thus giving explicit examples of when $\mathcal{M}_{q,\theta}(Q, \alpha)$ is normal.

Finally, for the sequel, we will have to consider the following analogue of divisibility:

Definition 2.27 A dimension vector $\alpha \in N_{q,\theta}$ is said to be q-indivisible if $\frac{1}{m}\alpha \notin N_{q,\theta}$ for all $m \geq 2$. Equivalently, for $\alpha = m\beta$ and β indivisible, then q^{β} is a primitive m-th root of unity.

3 Punctured Character Varieties as Multiplicative Quiver Varieties

In this section, we explain how it is possible to realize certain character varieties as particular examples of multiplicative quiver varieties by considering quivers of special type, the so-called crab-shaped quivers. Such character varieties parametrize representations of the fundamental group of a compact Riemann surface with a finite number of punctures, where the monodromies at closed loops around such punctures are fixed to lie in (the closure of) certain conjugacy classes. We use the language of quiver Riemann surfaces introduced by Crawley-Boevey in [CB13]. Moreover, in what follows, we shall adopt the term *punctured character variety* to refer to the character variety of a Riemann surface with punctures.

Fix a connected compact Riemann surface X of genus $g \geq 0$, let $S = \{p_1, \ldots, p_k\} \subset X$ be the set of punctures, and fix a tuple $\mathcal{C} = (\mathcal{C}_1, \ldots, \mathcal{C}_k)$ of conjugacy classes $\mathcal{C}_i \subset GL_n(\mathbb{C})$, $i = 1, \ldots, k$. Recall that the fundamental group $\pi_1(X \setminus S)$ of the punctured surface $X \setminus S$ admits the following presentation:

$$\pi_1(X\backslash S) = \langle a_1, \ldots, a_g, b_1, \ldots, b_g, c_1, \ldots, c_k \mid [a_1, b_1] \cdots [a_g, b_g] c_1 \cdots c_k = 1 \rangle,$$

where $[a, b] = aba^{-1}b^{-1}$ denotes the commutator. Note that the generators $c_1, \ldots, c_k$ represent homotopy classes of closed loops around the punctures, in the same free homotopy classes as small counterclockwise loops around the punctures. Thus, a representation of $\pi_1(X \setminus S)$ whose monodromies about the punctures are in the conjugacy classes $\mathcal{C}_i$ is given by a tuple of matrices $(A_1, \ldots, A_g, B_1, \ldots, B_g, C_1, \ldots, C_k) \in GL_n(\mathbb{C})^{2g} \times \mathcal{C}_1 \times \cdots \times \mathcal{C}_k$, satisfying the relation

$$\prod_{i=1}^{g} [A_i, B_i] \prod_{j=1}^{k} C_j = I.$$

Given the above, from the fact that isomorphic representations correspond to conjugate matrices, one has that the character variety $\mathcal{X}(g, k, \overline{\mathcal{C}})$ associated to

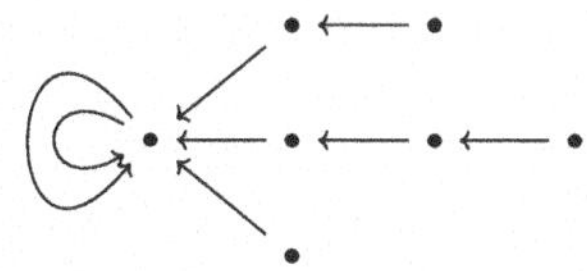

Fig. 1 A crab-shaped quiver with 2 loops and 3 legs, of length 2, 3, and 1, respectively

the pair (X, S) and monodromies lying in the conjugacy classes fixed above is isomorphic to the affine quotient

$$\mathcal{X}(g,k,\overline{\mathcal{C}}) := \{(A_1,\dots,A_g,B_1,\dots,B_g,C_1,\dots,C_k) \in GL_n(\mathbb{C})^{2g} \times \overline{\mathcal{C}}_1 \times \cdots \times \overline{\mathcal{C}}_k \mid \prod_{i=1}^{g}[A_i,B_i]\prod_{j=1}^{k}C_j = I\}/\!\!/ GL_n(\mathbb{C}).$$

Remark 3.1 Note that the closures $\overline{\mathcal{C}}_i$ are affine varieties, and hence, the quotient is indeed that of an affine variety by an algebraic group.

We shall now explain how to realize the variety $\mathcal{X}(g,k,\overline{\mathcal{C}})$ as an open subset of a multiplicative quiver variety, using an equivalence of categories proved in [CB13]. As mentioned above, such a correspondence holds when one considers the so-called *crab-shaped* quivers (called "comet-shaped" in [HLRV11]), that is, quivers such that there exists a vertex v satisfying the following condition: The set of arrows is formed by loops at v and a finite number of *legs* ending at v. See Fig. 1. A *star-shaped* quiver is a crab-shaped quiver with no loops.

For the remainder of this section, the following notation will be used: g, for the number of loops around the central vertex; k for the number of legs, and l_i, for $i = 1,\dots,k$, for the length of the i-th leg. As we shall see, g contains the information regarding the genus of the surface, while the integers k and l_i encode information about the (prescribed) conjugacy classes of the monodromies of the loops around the punctures.

Definition 3.2 [CB13, §2] A *Riemann surface quiver* Γ is a quiver whose set of vertices has the structure of a Riemann surface X with finitely many connected components. Γ is said to be *compact* if X is compact. A point $p \in X$ is called *marked* if it is a head or a tail of an arrow of Γ.

Definition 3.3 Given a Riemann surface quiver Γ, the *component quiver* $[\Gamma]$ of Γ, is the quiver whose set of vertices is the set of connected components of Γ and arrows given by $[a] : [p] \to [q]$ for any arrow $a : p \to q$, where p and q are points of X and $[p]$ denotes the connected component of X containing p.

Remark 3.4 Although, by definition, there are in general infinitely many vertices, we will consider (Riemann surface) quivers with finitely many arrows.

Following closely [CB13, §5, §8], starting from a Riemann surface quiver Γ, it is possible to define two categories of representations, $\mathrm{Rep}_\sigma(\pi(\Gamma))$ and $\mathrm{Rep}\,\Lambda^q([\Gamma])$, whose equivalence is the key point to proving the correspondence between multiplicative quiver varieties and punctured character varieties. Fix a quiver Riemann surface Γ, and let $\{X_i\}_{i\in I}$ the set of connected components of the underlying Riemann surface X. For each $i \in I$, let D_i be the set of marked points of Γ contained in X_i. Moreover, let $D = \cup_i D_i$: fix $\sigma \in (\mathbb{C}^\times)^D$, $b_i \in X_i \setminus D_i$, and for each $p \in D_i$, fix a loop $l_p \in \pi_1(X_i \setminus D_i, b_i)$ around p.

$\mathrm{Rep}_\sigma \pi(\Gamma)$ is defined to be the category whose objects are given by collections $(V_i, \rho_i, \rho_a, \rho_a^*)$ consisting of representations $\rho_i : \pi_1(X_i \setminus D_i, b_i) \to \mathrm{GL}(V_i)$, for $i \in I$ and linear maps $\rho_a : V_i \to V_j$ and $\rho_a^* : V_j \to V_i$ for each arrow $a : p \to q$ in Γ, where $X_i = [p]$ and $X_j = [q]$, satisfying

$$\sigma_p^{-1}\rho_i(\ell_p)^{-1} = 1_{V_i} + \rho_a^*\rho_a \quad \text{and} \quad \sigma_q\rho_j(\ell_q) = 1_{V_j} + \rho_a\rho_a^*$$

and whose morphisms are the natural ones.

Consider the component quiver $[\Gamma]$, and define Q to be the quiver obtained from $[\Gamma]$ by adjoining g_i loops at each vertex i, where g_i is the genus of X_i. Moreover, define $q \in (\mathbb{C}^\times)^I$ by $q_i = \prod_{p\in D_i} \sigma_p$. We define $\mathrm{Rep}\,\Lambda^q([\Gamma])'$ to be the category of representations of the multiplicative preprojective algebra $\Lambda^q(Q)$ in which the linear maps representing the added loops in Q (but not their reverse loops in $\overline{Q}$) are invertible.

Lemma 3.5 ([CB13, Proposition 2]) *There is an equivalence of categories*

$$\mathrm{Rep}_\sigma \pi(\Gamma) \simeq \mathrm{Rep}\,\Lambda^q([\Gamma])'.$$

This induces a $\mathrm{GL}(\alpha)$*-equivariant isomorphism of affine algebraic varieties,*

$$\mathrm{Rep}_\sigma(\pi(\Gamma), \alpha) \xrightarrow{\sim} \mathrm{Rep}(\Lambda^q([\Gamma])', \alpha),$$

defined as the collections of representations with $V_i = \mathbb{C}^{\alpha_i}$ *for all* i.

Proof The first statement is precisely [CB13, Proposition 2]. For the second, both $\mathrm{Rep}_\sigma(\pi(\Gamma), \alpha)$ and $\mathrm{Rep}(\Lambda^q([\Gamma])', \alpha)$ are acted upon by the group $\mathrm{GL}(\alpha)$, and the above equivalence of categories implies that there is a $\mathrm{GL}(\alpha)$-equivariant bijection as desired. Moreover, $\mathrm{Rep}_\sigma(\pi(\Gamma), \alpha)$ and $\mathrm{Rep}(\Lambda^q([\Gamma]', \alpha))$ are easily seen to be affine algebraic varieties, defined as tuples of matrices satisfying certain polynomial relations, with certain polynomials inverted. To see that the above map is a $\mathrm{GL}(\alpha)$-equivariant algebra isomorphism, observe that the proof of [CB13, Proposition 2] uses explicit invertible polynomial formulae. □

In order to explain how the above equivalence of categories implies the correspondence between character varieties and preprojective algebras, we shall explain how it is possible to encode the datum of a number of conjugacy classes into a star-shaped quiver. We follow [CBS06, §8] and [CB04, §2]: Fix k conjugacy classes $C_1, \ldots, C_k$ in $GL_n(\mathbb{C})$, for $k \geq 1$. We can encode the datum of such conjugacy classes in a combinatorial object as follows: Take $A_i \in C_i$ and let $w_i \geq 1$ be the degree of its minimal polynomial, for $i = 1, \ldots, k$; choose elements $\xi_{ij} \in \mathbb{C}^\times$, $1 \le i \le k$, $1 \le j \le w_i$, such that

$$(A_i - \xi_{i1} I) \cdot \cdots \cdot (A_i - \xi_{i w_i} I) = 0.$$

The closure of the conjugacy class $\mathcal{C}_i$ is then determined by the ranks of the partial products

$$\alpha_{ij} = \operatorname{rank}(A_i - \xi_{i1} I) \cdot \cdots \cdot (A_i - \xi_{ij} I),$$

for $A_i \in \mathcal{C}_i$ and $1 \le j \le w_i - 1$. In addition, if we set $\alpha_0 = n$, we get a dimension vector α for the following quiver Q_w

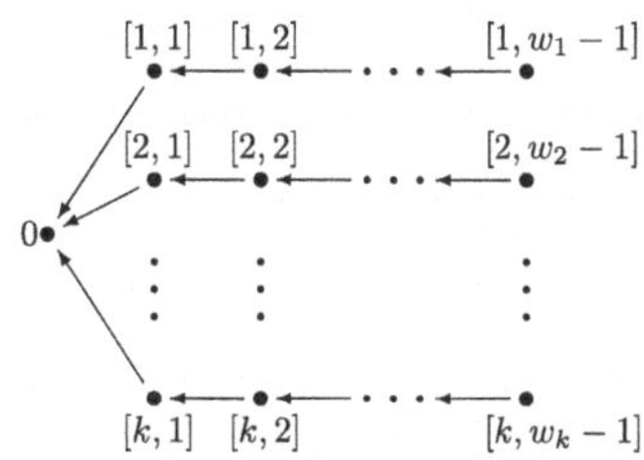

Now, for every $i \in \{1, \ldots, k\}$, let t_i be a nonnegative integer, and define a Riemann surface quiver Γ as follows: Its underlying Riemann surface X is given by the disjoint union

$$X = X_0 \sqcup \bigsqcup_{i \in \{1,\ldots,k\}, j \in \{1,\ldots,t_i\}} \mathbb{P}^1_{i,j},$$

where X_0 is an arbitrary closed Riemann surface of genus g (the choice does not matter), and $\mathbb{P}^1_{i,j}$ is simply a copy of $\mathbb{P}^1$ for the index (i, j). For each pair of indices (i, j), fix a point $p_{i,j} \in \mathbb{P}^1_{i,j}$, and for $i = 1, \ldots, k$, fix distinct points $p_i \in X_0$. Define D as before to be

$$D = \{p_{i,j}\} \cup \{p_l\}.$$

The arrows of Γ are listed as follows (Fig. 2):

- $a_{i,0} : p_{i,1} \to p_i$, for $i = 1, \ldots, k$;
- $a_{i,j} : p_{i,j+1} \to p_{i,j}$, for $i = 1, \ldots, k$, $j = 1, \ldots, t_i - 1$.

Unfolding the definition for the objects of the category $\operatorname{Rep}_\sigma \pi(\Gamma)$, one has that for such a Riemann surface quiver Γ, these are representations

$$\rho : \pi_1(X_0 \setminus \{p_1, \ldots, p_k\}) \to \mathrm{GL}(V),$$

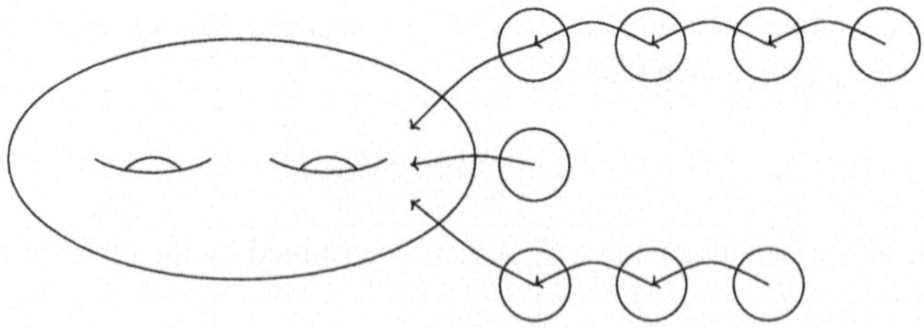

Fig. 2 An example of a Riemann surface quiver associated with a tuple of conjugacy classes

and linear maps

$$\rho_{i,0} : V_{i,1} \to V, \quad \rho_{i,j} : V_{i,j+1} \to V_{i,j}$$

and

$$\rho^*_{i,0} : V \to V_{i,1}, \quad \rho^*_{i,j} : V_{i,j} \to V_{i,j+1},$$

for $i = 1 \ldots, k$ and $j = 1, \ldots, t_i - 1$, such that if $l_i \in \pi_1(X_0 \setminus \{p_1, \ldots, p_k\})$ is the loop around p_i, $i = 1, \ldots, k$, the linear automorphism $\rho(l_i)$ satisfies the condition

$$\sigma_i \rho(l_i) = 1_V + \rho_{i,0}\rho^*_{i,0}$$

and the linear maps $\rho_{i,j}$ and $\rho^*_{i,j}$ satisfy the equations:

$$\begin{aligned} \sigma^{-1}_{i,j+1} 1_{V_{i,j+1}} &= 1_{V_{i,j+1}} + \rho^*_{i,j}\rho_{i,j}, \\ \sigma_{i,l} 1_{V_{i,l}} &= 1_{V_{i,l}} + \rho_{i,l}\rho^*_{i,l}, \end{aligned}$$

for $i = 1, \ldots, k$, $j = 1, \ldots, t_i - 1$, and $l = 1, \ldots, t_i$, which, setting $j = l - 1$ and summing the equations involving operators on the same space $V_{i,l}$, can be rewritten as

$$\begin{aligned} \rho(l_i) &= \sigma_i^{-1} 1_V + \sigma_i^{-1} \rho_{i,0}\rho^*_{i,0}, \\ \rho^*_{i,j-1}\rho_{i,j-1} - \rho_{i,j}\rho^*_{i,j} &= (\sigma^{-1}_{i,j} - \sigma_{i,j}) 1_{V_{i,j}}, \\ \sigma_{i,t_i} 1_{V_{i,t_i}} &= 1_{V_{i,t_i}} + \rho_{i,t_i}\rho^*_{i,t_i}, \end{aligned}$$

for $i = 1, \ldots, k$ and $j = 1, \ldots, t_i - 1$. Now, we specialize to the case $t_i = w_i - 1$ and assume that $\dim V_{i,j} = \alpha_{i,j}$ and $\dim V = n$, where w_i and $\alpha_{i,j}$ are defined as before. Through some simple algebraic computations, it is possible to see that given $\xi_{i,j}$ as before, it is possible to find corresponding $\sigma_{i,j}$, defined as

$$\sigma_0 = \frac{1}{\prod_{i=1}^k \xi_{i,1}}, \quad \sigma_{i,j} = \frac{\xi_{i,j}}{\xi_{i,j+1}},$$

such that the above sets of equations can be rewritten in terms of linear operators $\phi_{i,j}$ and $\psi_{i,j}$, for $i = 1, \ldots, k$ and $j = i, \ldots, w_i - 1$,

$$V \underset{\psi_{i1}}{\overset{\phi_{i1}}{\rightleftarrows}} V_{i1} \underset{\psi_{i2}}{\overset{\phi_{i2}}{\rightleftarrows}} V_{i2} \underset{\psi_{i3}}{\overset{\phi_{i3}}{\rightleftarrows}} \ldots \underset{\psi_{i,w_i-1}}{\overset{\phi_{i,w_i-1}}{\rightleftarrows}} V_{i,w_i-1}$$

satisfying

$$\begin{aligned} \rho(l_i) - \psi_{i1}\phi_{i1} &= \xi_{i1} 1_V \\ \phi_{ij}\psi_{ij} - \psi_{i,j+1}\phi_{i,j+1} &= (\xi_{i,j+1} - \xi_{ij}) 1_{V_{ij}} \qquad (1 \leq j < w_i - 1) \\ \phi_{i,w_i-1}\psi_{i,w_i-1} &= (\xi_{i,w_i} - \xi_{i,w_i-1}) 1_{V_{i,w_i-1}}, \end{aligned}$$

which, by Crawley-Boevey [CB04, Theorem 2.1], implies that $\rho(l_i)$ lies in the closure of the conjugacy class $\mathcal{C}_i$, $i = 1, \ldots, k$. In fact, this theorem says that this is a necessary and sufficient condition; thus, given a representation

$$\rho : \pi_1(X_0 \setminus \{p_1, \ldots, p_k\}) \to \mathrm{GL}(V),$$

where $\dim V = n$ and $\rho(l_i) \in \overline{\mathcal{C}_i}$, for prescribed conjugacy classes $\mathcal{C}_1, \ldots, \mathcal{C}_k$ in $GL_n(\mathbb{C})$, we can find linear maps $\rho_{i,j}$ and $\rho^*_{i,j}$ and vector spaces $V_{i,j}$ of dimension $\alpha_{i,j}$ as above, such that the tuple $(V, \rho, V_{i,j}.\rho_{i,j}, \rho^*_{i,j})$ is an object of the category $\mathrm{Rep}_\sigma \pi(\Gamma)$. Then, combining this with Lemma 3.5, one has the following result.

Theorem 3.6 *There is an isomorphism between the character variety $\mathcal{X}(g, k, \overline{\mathcal{C}})$ and the affine quotient $\widetilde{\mathcal{M}}_{q,0}([\Gamma], \alpha) := \mathrm{Rep}\ \Lambda^q([\Gamma], \alpha)' /\!/ \mathrm{GL}(\alpha)$.*

Remark 3.7 From its definition, one can see that $\mathrm{Rep}\ \Lambda^q([\Gamma], \alpha)'$ is an open affine $\mathrm{GL}(\alpha)$-invariant subset of $\mathrm{Rep}(\Lambda^q[\Gamma], \alpha)$ which is obtained by inverting certain $\mathrm{GL}(\alpha)$-invariant functions (the determinants of the linear transformations corresponding to loops of the undoubled quiver at the node). Since the quotient in the affine case, with G reductive, is obtained by passing to G-invariant functions, that is, $\mathrm{Spec}\ B /\!/ G = \mathrm{Spec}\ B^G$, we deduce that the affine quotient $\widetilde{\mathcal{M}}_{q,0}([\Gamma], \alpha)$ can be identified with an open subset of the multiplicative quiver variety $\mathcal{M}_{q,0}([\Gamma], \alpha)$. This is important because, as outlined in the following section, in order to show the nonexistence of symplectic resolutions, we prove that certain such varieties contain an open subset which is factorial and terminal.

Remark 3.8 We note that, in the star-shaped case, this result follows from [CBS06, Section 8]. Moreover, in the general case, Yamakawa proves a similar result to the one obtained in this section in the language of local systems on punctured surfaces; see [Yam08, Theorem 4.14] for more details.

4 Singularities of Multiplicative Quiver Varieties

Throughout this section, which is devoted to the study of the singularities $\mathcal{M}_{q,\theta}(Q,\alpha)$ and to the proof of Theorem 1.5, we use the notation introduced in Sect. 2.

In order to carry out this analysis, in Sect. 4.1, we describe the singular locus of the varieties in question. As one might expect, for $\alpha \in \Sigma_{q,\theta}$, this is given by the locus of strictly semistable representations. This follows because these varieties are Poisson, the stable locus is symplectic and smooth, and its complement has codimension at least two. Since a generically nondegenerate Poisson structure on a smooth variety can only degenerate along a divisor (the vanishing locus of the Pfaffian of the Poisson bivector), we conclude that the entire smooth locus is nondegenerate. Since the strictly semistable locus is degenerate, it must therefore be singular. Moreover, in the case where α is q-indivisible, a symplectic resolution can be obtained by varying θ, by Lemma 2.12, and Corollary 2.21 (and Remark 2.22). These arguments, spelled out below, prove the second statement of Theorem 1.5.

In Sect. 4.4, we complete the proof of Theorem 1.5 by considering strata of representation type $\nu\beta$, where $\alpha = n\beta$ and ν is a partition of n. We compute their codimension. As a consequence, taking β to be q-indivisible, for suitable $\theta' \geq \theta$, $\mathcal{M}_{q,\theta'}(Q,\alpha)$ has singularities in codimension ≥ 4. Hence, by Flenner's theorem [Fle88], its normalization is a symplectic singularity, which proves the first statement of Theorem 1.5. Finally, we show, using Drezet's criterion of factoriality, that the singularities along most strata $\nu\beta$ are factorial and terminal. This proves the final statement of Theorem 1.5. Note that Sect. 4.4 closely follows [BS21], where the analogous strata are considered for ordinary quiver varieties.

4.1 Singular Locus of $\mathcal{M}_{q,\theta}(Q,\alpha)$ for $\alpha \in \Sigma_{q,\theta}$

Before proving the main statement, we need a well-known result which is valid for any variety endowed with a Poisson structure.

Lemma 4.1 *Let X be a smooth variety and $\pi \in \wedge^2 TX$ a generically nondegenerate Poisson bivector. Let D the degenerate locus of π. Then, if non-empty, D is a divisor.*

Proof By generic nondegeneracy, $\dim X$ has to be even; therefore, $\dim X = 2d$. Define the top polyvector field $\gamma = \wedge^d \pi$. Then, D coincides with the zero locus of γ. On the other hand, γ is a section of a line bundle, and therefore, its zero locus is a divisor (if non-empty). □

This implies the following criterion for the singular locus of a Poisson variety:

Corollary 4.2 *Let X be a Poisson variety which is smooth and symplectic in the complement of a closed Poisson subvariety $Z \subseteq X$ which has codimension at least two everywhere. Then Z equals the set-theoretic singular locus of X.*

Proof Suppose for a contradiction that X is smooth at a point $z \in Z$. Since Z is a closed Poisson subvariety, the Poisson structure of X is degenerate at z. It follows from Lemma 4.1 that the degeneracy locus of X has codimension 1 at z. However, this locus is contained in Z, which has codimension at least two at z. This is a contradiction. □

Proposition 4.3 *Let $\alpha \in \Sigma_{q,\theta}$. The smooth locus of $\mathcal{M}_{q,\theta}(Q, \alpha)$ is $\mathcal{M}^s_{q,\theta}(Q, \alpha)$.*

Proof By Proposition 2.14, $\mathcal{M}^s_{q,\theta}(Q, \alpha)$ is smooth and symplectic. Let Z be the complement. It is the union of all the non-open strata of $\mathcal{M}_{q,\theta}(Q, \alpha)$. There are finitely many, and these all have purely even dimension; hence, Z has codimension at least two everywhere (as $\mathcal{M}^s_{q,\theta}$ is dense and it has purely even dimension, $2p(\alpha)$). Furthermore, we claim that Z is a Poisson subvariety, that is, all Hamiltonian vector fields are tangent to it. Indeed, Hamiltonian vector fields descend from $\mathrm{GL}(\alpha)$-invariant Hamiltonian vector fields on representation varieties. These integrate to formal automorphisms which commute with the G-action, which hence preserve the stratification by conjugacy classes of stabilizer. Therefore, the hypotheses of Corollary 4.2 are satisfied, and the statement follows. □

Remark 4.4 It is reasonable to ask if a stronger statement is true, which makes sense for general α: Are the connected components of the representation-type strata the symplectic leaves? Equivalently, do the Hamiltonian vector fields span the tangent spaces to the representation type strata? If so, then (a) $\mathcal{M}_{q,\theta}(Q, \alpha)$ has finitely many symplectic leaves, and (b) the representation type strata are all smooth. The converse statement also holds: If a stratum is smooth and it is a union of finitely many symplectic leaves, its Poisson structure must be nondegenerate outside a locus of codimension at least two. So Lemma 4.1 implies that it is actually nondegenerate.

Let us comment briefly on conditions (a) and (b). First, if $\mathcal{M}_{q,\theta}(Q, \alpha)$ is a symplectic singularity, it has finitely many symplectic leaves, by Kaledin [Kal06, Theorem 2.5]. Next, for a representation type $\tau = (k_1, \beta^{(1)}; \ldots; k_r, \beta^{(r)})$, the direct sum map produces a surjection $(\mathcal{M}^s_{q,\theta}(Q, \beta^{(1)}) \times \cdots \times \mathcal{M}^s_{q,\theta}(Q, \beta^{(r)}))^{\mathrm{dist}} \to C^\tau_{q,\theta}(Q, \alpha)$ with smooth source, where the dist refers to the open subset where the elements of the i-th and j-th factors are unequal for all distinct i and j. It seems reasonable to expect this to be a covering, in which case the stratum is smooth.

4.2 Generalities on Symplectic Singularities

In order to prove the first statement of Theorem 1.5, we need a criterion for the normalization of a variety to have symplectic singularities. This is an extension of [BS21, Lemma 6.12], using [Kal09, Theorem 1.5]:

Proposition 4.5 *Let X be a Poisson variety and assume that $\pi : Y \to X$ is a proper birational Poisson morphism from a variety Y with symplectic singularities. Then the normalization X' of X has symplectic singularities. Moreover, the induced map $\pi : Y \to X'$ is Poisson.*

Proof In [BS21, Lemma 6.12], the result is proved under the assumption that X is in fact normal. To conclude the lemma from this result, we may apply [Kal09, Corollary 1.4, Theorem 1.5]. By these results (and their proofs), given a Poisson variety X, the normalization X' has a unique Poisson structure such that the normalization map $\nu : X' \to X$ is a Poisson morphism. The map π factors through ν, and the induced map $\pi' : Y \to X'$ must be Poisson, since the Poisson bracket on $\mathcal{O}_{X'}$ is the unique extension of the Poisson bracket on $\mathcal{O}_X$ to a biderivation $\mathcal{O}_{X'} \times \mathcal{O}_{X'} \to \mathcal{O}_Y$. Then the fact that X' has symplectic singularities follows from [BS21, Lemma 6.12]. □

Remark 4.6 For convenience, we will apply this result even in the case where π is a symplectic resolution. However, in this case, the statement follows from definitions, without really requiring the results of [BS21, Kal09], as follows. The map $\pi : Y \to X$ factors through $\pi' : Y \to X'$, which induces on X' a unique Poisson structure such that π' is Poisson; as Y is nondegenerate and its symplectic form is pulled back from X', X' must also be nondegenerate on the smooth locus. By definition, X' is then a symplectic singularity. Since π is dominant, the Poisson structure on X is uniquely determined from the one on Y and must be the one obtained from X' via the inclusion $\mathcal{O}_X \to \mathcal{O}_{X'}$. This proves the last statement.

Remark 4.7 Actually, in the above proposition, the biconditional holds: X has symplectic singularities if and only if Y does. Moreover, one can generalize to the case where X is a non-reduced Poisson scheme: In this case, the map π factors through the reduced subvariety X^{red}, which is canonically Poisson by Kaledin [Kal09, Corollary 1.4].

4.3 The q-Indivisible Case

We now prove the second statement of Theorem 1.5. Suppose that $\beta \in \Sigma_{q,\theta}$ is q-indivisible.

First, suppose that β is real. In this case, by Crawley-Boevey and Shaw [CBS06, Theorem 2.1], $\Lambda^q(Q)$ admits a simple rigid representation X, and any other representation Y of the same dimension must be isomorphic to X, which means that the variety $\mathcal{M}_{q,\theta}(Q, \beta)$ is a point. So there is nothing to prove.

Next, suppose that β is imaginary. In this case, one may proceed as follows: By choosing a generic stability parameter $\theta' \geq \theta$, there is a projective symplectic resolution

$$\pi : \mathcal{M}_{q,\theta'}(Q, \beta) \longrightarrow \mathcal{M}_{q,\theta}(Q, \beta).$$

Indeed, by Yamakawa [Yam08, Proposition 3.5], for θ' generic, the stable locus $\mathcal{M}^s_{q,\theta'}(Q,\beta)$, which is smooth, coincides with the semistable locus. Hence, we can find $\theta' \geq \theta$ such that $\mathcal{M}_{q,\theta'}$ is smooth and symplectic. Moreover, the fact that the morphism π exists and is projective and Poisson follows, in the $\theta = 0$ case, from the very definitions of affine and GIT quotient and, for general θ, from Lemma 2.12. Finally, birationality of π is ensured by Corollary 2.21 (and Remark 2.22). Thus, we can conclude that $\mathcal{M}_{q,\theta}(Q,\beta)$ admits a symplectic resolution, given by the morphism π. By Proposition 4.5 (or Remark 4.6), this implies that the normalization of $\mathcal{M}_{q,\theta}(Q,\beta)$ has symplectic singularities.

4.4 The q-Divisible Case

In this subsection, we prove the first and third statements of Theorem 1.5. We may assume that α is q-divisible: This is automatic in the third part, whereas in the first part, the result follows from the second part (proved in the preceding subsection) in the q-indivisible case. This means that α is anisotropic, by the following result:

Lemma 4.8 *Let $\alpha \in N_{q,\theta}$ be q-divisible. Then $\alpha \in \Sigma_{q,\theta}$ only if α is anisotropic. Conversely, if $\alpha = m\beta$ and $\beta \in \Sigma_{q,\theta}$ is anisotropic, then $\alpha \in \Sigma_{q,\theta}$.*

Proof This is a generalization of [CB02, Proposition 1.2] (in view of Remarks 6.7 and 6.8), with the same proof. For details, see Corollary 6.21 below (whose proof is independent of any of the results of this section). □

Recall that a *weighted partition* of n is a sequence $\nu = (l_1, \nu_1; \ldots; l_k, \nu_k)$ such that $\nu_1 \geq \cdots \geq \nu_k$ and $\sum_{i=1}^k l_i \nu_i = n$. If ν is a partition of n, we shall denote by $\nu\beta$ the representation type $(l_1, \nu_1\beta; \ldots, l_k, \nu_k\beta)$.

Lemma 4.9

(1) The set $\Sigma_{q,\theta}$ contains $\{m\beta \mid m \geq 1\}$.
(2) $\dim C^{\nu\beta}_{q,\theta}(Q,n\beta) = 2\left(k + (p(\beta)-1)\sum_{i=1}^k \nu_i^2\right)$.
(3) For $(p(\beta),n) \neq (2,2)$, $\dim \mathcal{M}_{q,\theta}(Q,n\beta) - \dim C^{\nu\beta}_{q,\theta}(Q,n\beta) \geq 4$ for all $\nu \neq (1,n)$.
(4) For $(p(\beta),n) \neq (2,2)$ and $\nu \neq (1,n)$, one has $\dim \mathcal{M}_{q,\theta}(Q,n\beta) - \dim C^{\nu\beta}_{q,\theta}(Q,n\beta) \geq 8$ unless one of the following holds: (i) $(p(\beta),n) = (2,3)$ and $\nu = (1,2;1,1)$; (ii) $(p(\beta),n) = (3,2)$ and $\nu = (1,1;1,1)$.

Proof The arguments are completely analogous to those of [BS21, Lemma 6.1], except here that we use the dimension estimates given by Proposition 2.15. The first statement is a consequence of Lemma 4.8. □

Note that the above result has the following interesting consequence.

Proposition 4.10 *Assume that all θ-stable representations of dimension $\gamma < n\beta$ have $\gamma = m\beta$ for some m. Moreover, assume that $(p(\beta), n) \neq (2, 2)$. Then, $\mathcal{M}_{q,\theta}(Q, n\beta)$ is normal.*

For example, the first condition holds if β is q-indivisible and θ is generic.

Proof This is an immediate consequence of Proposition 2.25 and point (3) of Lemma 4.9, given that, by assumption on θ, all strata except for the open one have codimension greater than 4. □

Now, let $\alpha \in \Sigma_{q,\theta}$ be q-divisible. Write $\alpha = n\beta$ for β q-indivisible and $n \geq 2$. For generic $\theta' \geq \theta$, the only strata of $\mathcal{M}_{q,\theta'}(Q, \alpha)$ are those of the form $\nu\beta$, which appear in Lemma 4.9. If $(p(\beta), n) \neq (2, 2)$, then, taking into account Remark 2.20, all non-open strata have codimension at least four by Lemma 4.9.(3). Therefore, $\mathcal{M}_{q,\theta'}(Q, \alpha)$ is a symplectic singularity by Flenner's Theorem [Fle88]. Now, the map $\mathcal{M}_{q,\theta'}(Q, \alpha) \to \mathcal{M}_{q,\theta}(Q, \alpha)$ is birational, projective, and Poisson by Corollary 2.21 (and Remark 2.22), and Lemma 2.12. Therefore, the normalization of $\mathcal{M}_{q,\theta}(Q, \alpha)$ is itself a symplectic singularity by Proposition 4.5. This proves the first statement of Theorem 1.5.

It remains to prove the final statement of Theorem 1.5. For this purpose, assume that $\alpha = n\beta$ for $n \geq 2$ and that $\beta \in N_{q,\theta}$ (not necessarily q-indivisible or in $\Sigma_{q,\theta}$), such that there exists a θ-stable representation of dimension β. Let U be the union of all the strata indexed by $\nu\beta$ for ν a weighted partitions of n,

$$U := \bigcup_{\nu} C^{\nu\beta}_{q,\theta}(Q, \alpha).$$

As well as for the previous lemma, to prove the following result, one can repeat *verbatim* the arguments in [BS21, Lemma 6.2].

Lemma 4.11 *The subset U is open in $\mathcal{M}_{q,\theta}(Q, \alpha)$. If θ is generic and β is q-indivisible, this subset is the entire variety.*

In order to prove that U is factorial, we shall follow the approach of [BS21], which was itself inspired by results of Drezet [Dre91] on factoriality of points in moduli spaces of semistable sheaves on rational surfaces. Assuming the notation above, with $\pi : \operatorname{Rep}^{\theta-ss}(\Lambda^q, \alpha) \to \mathcal{M}_{q,\theta}(Q, \alpha)$ denoting the quotient map, define $V := \pi^{-1}(U)$. We aim at proving that V is a local complete intersection and that it is factorial and normal. We shall then descend the factoriality property to the subvariety U.

Proposition 4.12 *V is a local complete intersection, factorial, and normal.*

The proof of this proposition follows closely the arguments used in [BS21, Proposition 6.5].

Proof of Proposition 4.12 Since V is open inside $\operatorname{Rep}^{\theta-ss}(\Lambda^q, \alpha)$, Proposition 2.19 implies that it is a local complete intersection. To prove normality and factoriality, recall that a local complete intersection satisfies Serre's S_2 property, so Serre's criterion implies that it is normal if it is smooth outside a locus of codimension at least 2. Moreover, by a result of Grothendieck [KLS06, Theorem 3.12], a local complete intersection which is smooth outside a locus of codimension at least 4 is factorial. Put together, to show that V is normal and factorial, it suffices to show that it is smooth outside of a locus of codimension at least 4. For this, one can repeat verbatim the arguments used in [BS21, Proposition 6.5], replacing Corollary 6.4 and Lemma 6.1 with Lemmas 2.23 and 4.9, respectively. □

In order to descend factoriality from V to U we use Drezet's method. In particular, [BS21, Theorem 6.7] holds true in this context as well with no change in the proof of the result, as we have already made sure that all of the tools used there are still applicable here, Proposition 4.12 being the most important one. Thus, the corresponding statement of [BS21, Corollary 6.9] is the following:

Theorem 4.13 *U is a factorial variety.*

We omit the proof, as it is exactly the same as in [BS21, Corollary 6.9].

Using the previous theorem and the estimates on the codimension of the singular locus, one can conclude that $\mathcal{M}_{q,\theta}(Q,\alpha)$ does not admit a symplectic resolution. We state this formally below, where we also recall our running hypotheses for the reader's convenience.

Theorem 4.14 *Let $\alpha = n\beta \in \Sigma_{q,\theta}$ be anisotropic imaginary, for $n \geq 2$, such that there exists a θ-stable representation of Λ^q of dimension β, and $(p(\beta), n) \neq (2, 2)$. Then $\mathcal{M}_{q,\theta}(Q,\alpha)$ has an open subset which is factorial, terminal, and singular. Hence, it does not admit a symplectic resolution. Moreover, if θ is generic and α is q-indivisible, then this open subset is the entire variety.*

Proof The subset U is singular, since it contains the non-open stratum (n, β). It is factorial by Theorem 4.13. Under the assumptions, the singular strata in U all have codimension at least four, hence also the singular locus. Thus, U is terminal by Namikawa [Nam], since it has symplectic singularities and the singular locus has codimension at least four. □

This completes the proof of the third and final statement of Theorem 1.5.

4.5 *Proof of Corollary 1.6*

Write $\alpha = m\beta$ for β q-indivisible. Note that for θ generic, the only possible decompositions of α are into multiples of β. If α is q-indivisible, it therefore follows trivially that $\alpha \in \Sigma_{q,\theta}$. Since the only stratum in $\mathcal{M}_{q,\theta}(Q,\alpha)$ is the open one of stable representations, it also follows from Proposition 2.14 that $\mathcal{M}_{q,\theta}$ is smooth symplectic. Suppose that α is q-divisible. It then follows from Lemma 4.8 that α is in $\Sigma_{q,\theta}$ if and only if it is anisotropic. This completes the proof of part (i).

Part (ii) follows from Proposition 4.10 and Theorem 1.5. The first statement of part (iii) follows from Theorem 1.5. Finally, the last statement follows from

Proposition 2.14 because in this case, there is only one stratum in $\mathcal{M}_{q,\theta}(Q,\alpha)$, consisting of θ-stable representations.

4.6 The Anisotropic Imaginary $(p(\alpha), n) = (2, 2)$ Case

The only case left out in this analysis is that of $2\alpha \in \Sigma_{q,\theta}$ for $\alpha \in N_{q,\theta}$ satisfying $p(\alpha) = 2$. The analogous question of existence of a symplectic resolution in the setting of Nakajima quiver varieties is settled in [BS21, Theorem 1.6], where it is shown that for generic θ, blowing up the ideal sheaf defining the singular locus gives a symplectic resolution of singularities. This is achieved by showing that, étale locally, the variety is isomorphic to the product of $\mathbb{C}^4$ with the closure of the six-dimensional nilpotent orbit closure in $\mathrm{Sp}(\mathbb{C}^4)$: see [BS21, Theorem 5.1] and the references therein. Given this, one might conjecture that an analogous result holds for multiplicative quiver varieties, and such a result should be proved by studying the étale local structure of the variety. In fact, by Artin's approximation theorem [Art69], it would be sufficient to give a description of the formal neighborhood of a point. This will be discussed in a future work. For more details, see Sect. 7.

5 Combinatorics of Multiplicative Quiver Varieties

In this section, we study some combinatorial problems which are related to the geometry of multiplicative quiver varieties. Indeed, an interesting problem is to classify all the possible "(2, 2)-cases": These are the main q-divisible cases for which we conjecture that there exists a symplectic resolution. In the next subsection, we carry out these computations in the case of crab-shaped quivers (which we defined in Sect. 3). We shall see how most of the (2, 2)-cases occur in the case of star-shaped quivers, that is, where there are no loops, so that the corresponding surface has genus zero. It is also important to point out that the classification below yields an explicit classification of the crab-shaped quivers for which symplectic resolutions exist, or are conjectured to exist: See Corollary 6.28 in the next section for details on this.

5.1 (2,2) Cases for Crab-Shaped Quivers

The analysis is based on some standard numerical arguments and the constraints on the dimension vector α for it to satisfy the conditions of Sect. 3, that is, it has to represent the multiplicities of the eigenvalues in the prescribed conjugacy class.

Theorem 5.1 *There are exactly 13 pairs (Q,α), where Q is a star-shaped quiver as in Sect. 3 and $\alpha \in \mathcal{F}(Q)$ is in the fundamental region, such that $p(\alpha) = 2$. Such* *pairs are depicted as follows, where a vertex is substituted by the corresponding entry of the dimension vector:*

$$\begin{array}{ccccc} 1 & & 1 & & 1\\ & \nwarrow & \uparrow & \nearrow & \\ 1 & \longleftarrow & 2 & \longrightarrow & 1\end{array} \tag{5.1.1}$$

$$\begin{array}{ccccccc} 1 & & & & 2 & \longrightarrow & 1\\ & \nwarrow & & \nearrow & & & \\ 1 & \longleftarrow & 3 & \longrightarrow & 2 & \longrightarrow & 1\end{array} \tag{5.1.2}$$

$$\begin{array}{ccccccccc} 2 & & & & 2 & & & & \\ & \nwarrow & & \nearrow & & & & & \\ 1 & \longleftarrow & 4 & \longrightarrow & 3 & \longrightarrow & 2 & \longrightarrow & 1\end{array} \tag{5.1.3}$$

$$\begin{array}{ccccccccccc} & & & & & & 3 & \longrightarrow & 2 & \longrightarrow & 1\\ & & & & & \nearrow & & & & & \\ 1 & \longleftarrow & 2 & \longleftarrow & 4 & \longrightarrow & 3 & \longrightarrow & 3 & \longrightarrow & 1\end{array} \tag{5.1.4}$$

$$\begin{array}{ccccccc} 2 & & & & 2 & & \\ & \nwarrow & & \nearrow & & & \\ 2 & \longleftarrow & 4 & \longrightarrow & 2 & \longrightarrow & 1\end{array} \tag{5.1.5}$$

$$\begin{array}{ccccccccccc} & & & & 4 & \longrightarrow & 3 & \longrightarrow & 2 & \longrightarrow & 1\\ & & & \nearrow & & & & & & & \\ 2 & \longleftarrow & 5 & \longrightarrow & 4 & \longrightarrow & 3 & \longrightarrow & 2 & \longrightarrow & 1\end{array} \tag{5.1.6}$$

$$\begin{array}{ccccccccccccc} & & & & 1 & & & & & & & & \\ & & & & \uparrow & & & & & & & & \\ & & & & 3 & & & & & & & & \\ & & & & \uparrow & & & & & & & & \\ 1 & \longleftarrow & 3 & \longleftarrow & 5 & \longrightarrow & 4 & \longrightarrow & 3 & \longrightarrow & 2 & \longrightarrow & 1\end{array} \tag{5.1.7}$$

$$\begin{array}{ccccccccccccccccc} & & & & & & & & & & 3 & & & & & & \\ & & & & & & & & & & \uparrow & & & & & & \\ 1 & \longleftarrow & 2 & \longleftarrow & 3 & \longleftarrow & 4 & \longleftarrow & 5 & \longleftarrow & 6 & \longrightarrow & 4 & \longrightarrow & 2 & \longrightarrow & 1\end{array} \tag{5.1.8}$$

$$\begin{array}{ccccccccccc} & & & & 2 & & & & & & \\ & & & & \uparrow & & & & & & \\ & & & & 4 & & & & & & \\ & & & & \uparrow & & & & & & \\ 2 & \longleftarrow & 4 & \longleftarrow & 6 & \longrightarrow & 4 & \longrightarrow & 2 & \longrightarrow & 1\end{array} \tag{5.1.9}$$

$$\begin{array}{ccccccccccccccccccc} & & & & 4 & & & & & & & & & & & & & & \\ & & & & \uparrow & & & & & & & & & & & & & & \\ 2 & \longleftarrow & 5 & \longleftarrow & 8 & \longrightarrow & 7 & \longrightarrow & 6 & \longrightarrow & 5 & \longrightarrow & 4 & \longrightarrow & 3 & \longrightarrow & 2 & \longrightarrow & 1\end{array} \tag{5.1.10}$$

$$\begin{array}{ccccccccccccccc} & & & & & & & & 4 & & & & & & \\ & & & & & & & & \uparrow & & & & & & \\ 1 & \longleftarrow & 2 & \longleftarrow & 4 & \longleftarrow & 6 & \longleftarrow & 8 & \longrightarrow & 6 & \longrightarrow & 4 & \longrightarrow & 2\end{array} \tag{5.1.11}$$

$$\begin{array}{c} \qquad\qquad 5 \\ \qquad\qquad \uparrow \\ 1 \longleftarrow 4 \longleftarrow 7 \longleftarrow 10 \longrightarrow 8 \longrightarrow 6 \longrightarrow 4 \longrightarrow 2 \end{array} \tag{5.1.12}$$

$$\begin{array}{c} 6 \\ \uparrow \\ 4 \longleftarrow 8 \longleftarrow 12 \longrightarrow 10 \longrightarrow 8 \longrightarrow 6 \longrightarrow 4 \longrightarrow 2 \longrightarrow 1 \end{array} \tag{5.1.13}$$

Remark 5.2 It is important to highlight that quivers (5.1.5), (5.1.9), (5.1.11), (5.1.13) are the framed affine Dynkin quivers $\tilde{D}_4$, $\tilde{E}_6$, $\tilde{E}_7$, $\tilde{E}_8$, respectively, with dimension vector given by $(2\delta, 1)$, where δ is the minimal isotropic imaginary root of the corresponding quiver. See Remark 5.5 for the significance of this.

Proof of Theorem 5.1 Note that $p(\alpha) = 2$ if and only if $\langle \alpha, \alpha \rangle = -1$. Let us calculate the value of $\langle \alpha, \alpha \rangle$ explicitly, for α a general dimension vector. The general star-shaped quiver has g loops and k legs, each of which has l_i arrows, $i = 1, \ldots, k$. We have

$$\langle \alpha, \alpha \rangle = (1-g)n^2 + \sum_{i,j} \alpha_{i,j}^2 - n \sum_{i=1}^{k} \alpha_{i,1} - \sum_{i=1}^{k} \sum_{j=1}^{l_i - 1} \alpha_{i,j}\alpha_{i,j+1}.$$

Assume now that $\alpha \in \mathcal{F}(Q)$ and that $\langle \alpha, \alpha \rangle = -1$; then given that $\langle \alpha, \alpha \rangle = \sum_{i \in Q_0} \alpha_i \langle \alpha, e_i \rangle = \sum_{i \in Q_0} \alpha_i \langle e_i, \alpha \rangle$, this implies that there can only be two possibilities:

(a) There exists a unique vertex $i \in Q_0$ such that either $\alpha_i = 1$ and $(\alpha, e_i) = -2$, or $\alpha_i = 2$ and $(\alpha, e_i) = -1$, with $(\alpha, e_j) = 0$ for $j \neq i$; this implies that, denoted by $\mathrm{Adj}(i)$, the set of vertices which are adjacent to i, one has $\sum_{j \in \mathrm{Adj}(i)} \alpha_j = 5$ for $\alpha_i = 2$ and $\sum_{j \in \mathrm{Adj}(i)} \alpha_j = 4$ for $\alpha_i = 1$.
(b) There are two distinct vertices i and i' such that $(\alpha, e_i) = (\alpha, e_{i'}) = -1$ and $\alpha_i = \alpha_{i'} = 1$, with $(\alpha, e_j) = 0$ for $j \neq i, i'$. In this case, one has $\sum_{j \in \mathrm{Adj}(k)} \alpha_j = 3$ for $k = i, i'$.

In this case, if i or i' is the central vertex, then the only possibility is given by the quiver (1) in the statement of the theorem. Otherwise, if v is the central vertex, then $(\alpha, e_v) = 0$, which implies that $\sum_{j \in \mathrm{Adj}(v)} \alpha_j = 2n$, where $\alpha_v = n$: Indeed,

$$0 = (\alpha, e_v) = \langle \alpha, e_v \rangle + \langle e_v, \alpha \rangle = n - \sum_{k \to v} \alpha_k + n - \sum_{v \to l} \alpha_l = 2n - \sum_{j \in \mathrm{Adj}(v)} \alpha_j.$$

Now, fix a branch along which none of the special vertices i and i' appear, let l be its length, and let $\beta_0 = n, \beta_1, \ldots, \beta_l$ be the components of the vector α along the branch.

Then using that $(\alpha, e_j) = 0$ for $j \neq i, i'$, we get the recursive formula

$$2\beta_j = \beta_{j-1} + \beta_{j+1},$$

for $j = 1, \dots, l-1$, and also $\beta_{l-1} = 2\beta_l$, which implies that

$$\beta_j = (l+1-j)\beta_l.$$

Therefore, the branch has the form

$$n \longrightarrow n-c \longrightarrow n-2c \longrightarrow \dots \longrightarrow c,$$

where c is a positive integer such that $c|n$. Moreover, in order for condition (a) to be satisfied, there has to be one branch ending with one of the following

$$5 \longrightarrow 2, \quad 4 \longrightarrow 2 \longrightarrow 1, \quad 4 \longrightarrow 1,$$

and, thus, having the form

$$\begin{aligned} &n-3 \longrightarrow \dots, \longrightarrow 5 \longrightarrow 2, \\ &\qquad n-2 \longrightarrow \dots, \longrightarrow 4 \longrightarrow 2 \longrightarrow 1, \\ &\qquad\qquad n-3 \longrightarrow \dots, \longrightarrow 4 \longrightarrow 1 \end{aligned} \tag{§}$$

respectively; for condition (b), there have to be two branches ending as

$$3 \longrightarrow 1,$$

having the form

$$n-2 \longrightarrow \dots \longrightarrow 3 \longrightarrow 1. \tag{§§}$$

Therefore, we are left to consider a star-shaped quiver where all but one or two branches are as follows:

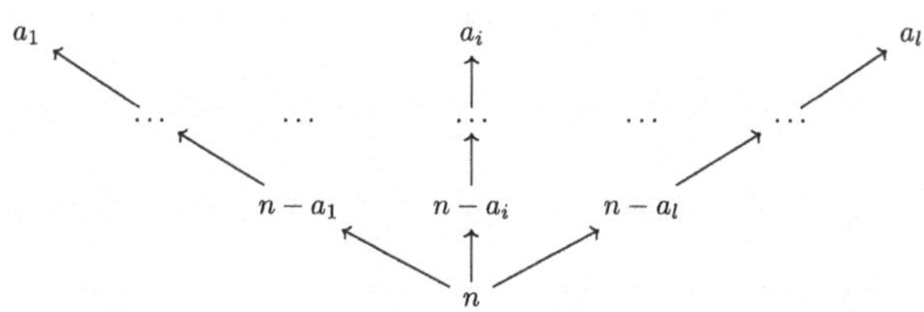

Moreover, if the quiver satisfies condition a), then $l = k - 1$ and there is an additional branch having one of the forms in (§); on the other hand, if the quiver is as in case b), then $l = k - 2$ and there are two additional legs of the form described by (§§).

We shall now use some numerical arguments to prove that, among all such possibilities, only the ones listed in the statement of the theorem can actually occur.

First, let us spell out how the equality $0 = (\alpha, e_v)$ can be rephrased: One has that

$$0 = (\alpha, e_v) \iff 2n = \sum_{i=1}^{k}(n - a_i) \iff \sum_{i=1}^{k} a_i = (k-2)n.$$

Therefore, one has the following possibilities:

(a) In the cases of a branch ending with $5 \longrightarrow 2$ or $4 \to 1$, the equality $0 = (\alpha, e_v)$ reads as

$$\frac{3}{n} + \sum_{i=1}^{k-1} \frac{1}{n/a_i} = k - 2,$$

where $n \equiv 2 \pmod 3$ and $n \equiv 1 \pmod 3$, respectively, and $n > a_i \geq 2$, $a_i | n$ for every i; these shall be mentioned in the following as cases a.1 and a.2. On the other hand, for a branch ending with $4 \longrightarrow 2 \longrightarrow 1$, we have

$$\frac{2}{n} + \sum_{i=1}^{k-1} \frac{1}{n/a_i} = k - 2, \tag{5.1.14}$$

where n has to be even and $a_i | n$; this is renamed as case a.3.

(b) There are two branches $3 \longrightarrow 1$ and $0 = (\alpha, e_v)$ is equivalent to

$$\frac{4}{n} + \sum_{i=1}^{k-2} \frac{1}{n/a_i} = k - 2,$$

and $a_i | n$ for every i and n has to be odd, and $a_i < n$, for every i.

In cases a.1 and a.2, one has that $n \geq 4$ which forces $k \leq 4$: Indeed, one has that for $n \geq 4$ $n/a_i \geq 2$ and, therefore,

$$\frac{3}{n} + \sum_{i=1}^{k-1} \frac{1}{n/a_i} \leq \frac{3}{4} + \frac{k-1}{2} < \frac{k+1}{2},$$

which implies that

$$k - 2 < \frac{k+1}{2},$$

and thus $k \leq 4$. For $k = 4$ and $n = 4$, it is easily checked that quiver (5.1.3) in the statement of the result is the only possibility. If $k = 3$, then the following inequality holds:

$$\left(\frac{1}{2}+\frac{1}{4}\right)n \geq \left(1-\frac{3}{n}\right)n = n-3,$$

which forces $n \leq 12$; one can check that case (a) cannot be realized for $n = 4, 7, 11$ and that the cases $n = 5, 8, 10$ give quivers (5.1.6), (5.1.10), and (5.1.12), respectively. Next, for case a.3, one has: n even and $k \leq 4$: Indeed, since $n \geq 4$, from Eq. (5.1.14), one has that

$$\frac{2}{4}+\frac{k-1}{2} \geq k-2,$$

which implies that

$$\frac{k}{2} \leq 2.$$

If $k = 4$ and $n = 4$, then one gets quiver (5.1.5). If $k = 3$, then $n \leq 12$: Indeed, from Eq. (5.1.14), we have

$$\frac{2}{n}+\frac{1}{2}+\frac{1}{3} \geq 1$$

This leads to quiver (5.1.4) for $n = 4$, quivers (5.1.8) and (5.1.9) for $n = 6$, quiver (5.1.11) for $n = 8$, and quiver (5.1.13) for $n = 12$.

We turn now to case (b): n is odd and ≥ 3; this implies that $k \leq 4$: Indeed, in the same way as the previous case, Eq. (5.1.14) gives

$$\frac{4}{3}+\frac{k-2}{3} \geq k-2 \implies \frac{8}{3} \geq \frac{2k}{3}.$$

Therefore, setting $k = 4$ forces $n = 3$, which leads to quiver (5.1.2). When $k = 3$ one has that $n \leq 6$, which implies that $n = 3$ or $n = 5$. One checks that $n = 3$ is impossible, whereas $n = 5$ gives quiver (5.1.7). Since we have dealt with all the possible cases, the proof is complete. □

Theorem 5.3 *Assume that $g \geq 1$. Then, the only pairs (Q, α), where Q is a crab-shaped quiver and $\alpha \in \mathcal{F}(Q)$ is such that $\langle \alpha, \alpha \rangle = -1$ are the following:*

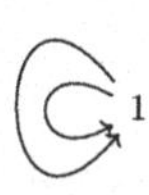

(5.1.15)

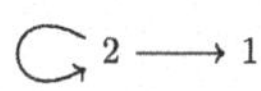

(5.1.16)

Remark 5.4 Parallel to Remark 5.2, in the second case above, the quiver and dimension vector are also of the form $(2\delta, 1)$ where $\delta = (1)$ is the primitive imaginary root of affine type $\widetilde{A}_0$ (the Jordan quiver with one vertex and one arrow).

Proof of Theorem 5.3 As in the arguments of the previous theorem, we see that if v is the central vertex and $\alpha \in \mathcal{F}$, then there are three possibilities for the value of (α, e_v), that is, (α, e_v) can be either 0, -1, or -2. In general, one has that

$$(\alpha, e_v) = 2(1 - g)n - \sum_{i=1}^{k} \alpha_{i,1}.$$

If $k = 0$, then by the argument leading to cases (a) and (b) in the proof of Theorem 5.1, one must have $n = 1$ and $g = 2$, which gives the first quiver of the statement of the result. Thus, one is left to show that for $k \geq 1$, there are no crab-shaped quivers satisfying the mentioned conditions other than the second quiver in the statement of the theorem. If $k \geq 1$, $n > 1$, which implies that either $(\alpha, e_v) = 0$ or -1. In the first case, we get that

$$2(1 - g)n = \sum_{i=1}^{k} \alpha_{i,1},$$

but $g \geq 1$, which gives $\sum_{i=1}^{k} \alpha_{i,1} \leq 0$, a contradiction. If $(\alpha, e_v) = -1$, then one must have $n = 2$ and $\alpha_{i,1} = 1$ for $i = 1, \ldots, k$ and $g = 1$. This implies that

$$k = 5 - 4g,$$

which forces $k = 1$. Therefore, we get the quiver

$$\circlearrowleft 2 \longrightarrow 1.$$

Since we dealt with all the possible cases, the proof is complete. □

Remark 5.5 Note that to get a list of all the (2, 2)-cases, one has to take each of the pairs (Q, α) drawn above and consider the pair $(Q, 2\alpha)$. Moreover, given $q \in (\mathbb{C}^\times)^{Q_0}$ and $\theta \in \mathbb{Z}^{Q_0}$, it follows from Theorem 6.16 below that the given α are in $\Sigma_{q,\theta}$ (since they are already in $\mathcal{F}(Q)$) if and only if the following are satisfied: (a) They are in $N_{q,\theta}$, that is, $q^\alpha = 1$ and $\theta \cdot \alpha = 0$, and (b) in the $(2\delta, 1)$ cases (mentioned in Remarks 5.2 and 5.4), $\delta \notin N_{q,\theta}$, that is, $q^\delta \neq 1$ or $\theta \cdot \delta \neq 0$.

6 General Dimension Vectors and Decomposition

One fundamental tool in the classification theorem [BS21, Theorem 1.4] is the canonical decomposition of a dimension vector of a quiver variety into summands which lie in $\Sigma_{\lambda,\theta}$, which is the additive version of the set $\Sigma_{q,\theta}$ defined in this paper (one just needs to replace the condition $q^\alpha = 1$ with $\lambda \cdot \alpha = 0$). This appears in

Crawley-Boevey's canonical decomposition in the additive case (extended to the case $\theta \neq 0$ in [BS21]). Combinatorially, it says:

Lemma 6.1 *[CB02, Theorem 1.1], [BS21, Proposition 2.1] Let $\alpha \in \mathbb{N}R^+_{\lambda,\theta}$. Then α admits a unique decomposition $\alpha = n_1\sigma^{(1)} + \cdots + n_k\sigma^{(k)}$ as a sum of elements $\sigma^{(i)} \in \Sigma_{\lambda,\theta}$ such that any other decomposition of α as a sum of elements from $\Sigma_{\lambda,\theta}$ is a refinement of this decomposition.*

Geometrically, the statement (together with the consequence for symplectic resolutions) is:

Theorem 6.2 *[CB02, Theorem 1.1], [BS21, Theorem 1.4] The symplectic variety $\mathcal{M}_{\lambda,\theta}(Q,\alpha) = \mu^{-1}(\lambda)^{\theta-ss} /\!/ \operatorname{GL}(\alpha)$ is isomorphic to the product*

$$\mathcal{M}_{\lambda,\theta}(Q,\alpha) \cong \prod_{i=1}^{k} S^{n_i} \mathcal{M}_{\lambda,\theta}(Q,\sigma^{(i)}).$$

Moreover, it admits a symplectic resolution if and only if each $\mathcal{M}_{\lambda,\theta}(Q,\sigma^{(i)})$ admits a symplectic resolution.

For multiplicative quiver varieties, the combinatorial statement still holds, but it is not clear that such a geometric decomposition holds. We instead prove a weaker statement, which gives a decomposition into factors which might not be minimal but still has all of the needed properties. Moreover, the resulting classification of symplectic resolutions is the same statement as if the canonical decomposition as above held. As a result, we are able to generalize Theorem 1.5 to the case of general dimension vectors (Theorem 6.27) and give its specialization to the crab-shaped case (Corollary 6.28). To complete the proof, we need to establish that $\mathcal{M}_{q,\theta'}(Q,\alpha) \to \mathcal{M}_{q,\theta}(Q,\alpha)$ is a symplectic resolution for many α not in $\Sigma_{q,\theta}$ (Theorem 6.23), in order to handle such factors appearing in the decomposition. In the additive case, such resolutions include the Hilbert schemes of points in $\mathbb{C}^2$ and in hyperkähler ALE spaces (i.e., minimal resolutions of du Val singularities).

6.1 Flat Roots

In order to write a product decomposition in the multiplicative setting, the dimension vectors for the factors need to be more general than those in $\Sigma_{q,\theta}$. The dimension vectors turn out to include "flat roots," which are those for which the moment map is flat (this is true for roots in $\Sigma_{q,\theta}$). This condition is also very important in order to have a geometric understanding of the varieties.

Definition 6.3 A vector $\alpha \in N_{q,\theta}$ is called *flat* if, for every decomposition $\alpha = \alpha^{(1)} + \cdots + \alpha^{(m)}$ with $\alpha^{(i)} \in R^+_{q,\theta}$, we have $p(\alpha) \geq p(\alpha^{(1)}) + \cdots + p(\alpha^{(m)})$. Let $\widetilde{\Sigma}_{q,\theta}$ be the set of flat roots.

Remark 6.4 As in [CB01, Theorem 1.1], we could alternatively have made the definition only requiring $\alpha^{(i)} \in N_{q,\theta}$. Indeed, it follows from the proof of the decomposition theorem (Theorem 6.17) below that if $\alpha \in N_{q,\theta}$, then there is a decomposition $\alpha = \beta^{(1)} + \cdots + \beta^{(k)}$ with each $\beta^{(i)}$ either in $\Sigma_{q,\theta}$ or of the form $\beta^{(i)} = m\gamma, \gamma \in \Sigma^{\text{iso}}_{q,\theta}$, satisfying $p(\alpha) \leq p(\beta^{(1)}) + \cdots + p(\beta^{(k)})$. Hence, if we know the inequality when the $\alpha^{(i)} \in R^+_{q,\theta}$, we also know it when the $\alpha^{(i)} \in N_{q,\theta}$.

The definition has the following interpretation. Let $\mathrm{SL}(\alpha) := \{g \in \mathrm{GL}(\alpha) \mid \prod_{i \in Q_0} \det(g_i) = 1\} \subset \mathrm{GL}(\alpha)$. Note that Φ_α factors through the inclusion $\mathrm{SL}(\alpha) \to \mathrm{GL}(\alpha)$; let $\overline{\Phi_\alpha} : \mathrm{Rep}^\circ(\overline{Q}, \alpha)^{\theta-ss} \to \mathrm{SL}(\alpha)$ be the factored map.

Proposition 6.5 *If α is a flat root, then $\overline{\Phi}_\alpha$ is flat over a neighborhood of q. In particular, $\Phi_\alpha^{-1}(q)^{\theta-ss}$ is a complete intersection and is equidimensional of dimension $g_\alpha + 2p(\alpha)$.*

Proof The second statement follows from the argument of Proposition 2.19 (following [CB01, Theorem 1.11]): All arguments go through with the strict inequality replaced by the non-strict one, except that no statement can be deduced about the stable (or simple) representations forming a dense subset. For the first statement, concerning flatness, note that $\dim \overline{\Phi}_\alpha^{-1}(q)^{\theta-ss} = \dim \mathrm{Rep}^\circ(\overline{Q}, \alpha)^{\theta-ss} - \dim \mathrm{SL}(\alpha)$. Then the statement follows from the following general considerations. Suppose we are given a morphism of varieties $f : X \to Y$, with X equidimensional. By upper semicontinuity of the fiber dimension, the minimum fiber dimension is $\dim X - \dim Y$, and the locus in Y where the fibers have this minimal dimension is open. Next, it is a standard fact that a morphism from a Cohen–Macaulay variety X to a smooth variety Y is flat if and only if for every $x \in X$, with $y = f(x) \in Y$, one has the equality $\dim_x X = \dim_y Y + \dim f^{-1}(y)$. It follows that f is flat over the open locus where the fibers have minimum dimension. Now, back to the situation at hand, by the second statement of the proposition, the minimum dimension is attained over $q \in \mathrm{SL}(\alpha)$. As the domain and codomain are equidimensional and smooth, the aforementioned open locus is a neighborhood of q over which $\overline{\Phi}_\alpha$ is flat. □

Putting this together with Proposition 2.17, we conclude the following analogue of the last statement of Proposition 2.19:

Corollary 6.6 *For α a flat root, a dense subset of $\Phi_\alpha^{-1}(q)^{\theta-ss}$ is given by the union of preimages of strata of types $(1, \beta^{(1)}; \ldots; 1, \beta^{(r)})$ with $p(\alpha) = \sum_{i=1}^r p(\beta^{(i)})$.*

Observe that $\Sigma_{q,\theta} \subseteq \widetilde{\Sigma}_{q,\theta}$. The opposite inclusion does not hold: For instance, with $(q, \theta) = (1, 0)$, one can take the quiver with two vertices and two arrows, one a loop at the first vertex, and the other an arrow to the second vertex. Then the dimension vector $(m, 1)$ is flat for all m but only in $\Sigma_{q,\theta}$ for $m = 1$.

Remark 6.7 Note that, for every pair q, θ, there always exists q' such that $N_{q',0} = N_{q,\theta}$, and hence $\Sigma_{q,\theta} = \Sigma_{q',0}$ and $\widetilde{\Sigma}_{q,\theta} = \widetilde{\Sigma}_{q',0}$. Indeed, let $z \in \mathbb{C}^\times$ be a multiplicatively independent element from the q_i (i.e., $\langle z, q_i \rangle / \langle q_i \rangle$ is infinite cyclic, where $\langle - \rangle$ denotes the multiplicative group generated by the given elements). Set $q'_i := q_i z^{\theta_i}$. Then q' has the desired properties.

Remark 6.8 Similarly, given any parameters in the additive case (for ordinary quiver varieties), $\lambda \in \mathbb{C}^{Q_0}, \theta \in \mathbb{Z}^{Q_0}$, we also can construct $q' \in (\mathbb{C}^\times)^{Q_0}$ such that the sets N and Σ correspond. More precisely, letting $N^a_{\lambda,\theta}, \Sigma^a_{\lambda,\theta}, \widetilde{\Sigma^a}_{\lambda,\theta}$ denote the sets defined for the additive case, this means that $N^a_{\lambda,\theta} = N_{q',0}$, $\Sigma^a_{\lambda,\theta} = \Sigma_{q',0}$, and $\widetilde{\Sigma^a}_{\lambda,\theta} = \widetilde{\Sigma}_{q',0}$.

We recall, following [CB01] and [Su06], how to classify flat roots in terms of the fundamental region.

Definition 6.9 We say that the transformation $\alpha \mapsto s_v(\alpha)$ is a (-1)-*reflection* if $s_v(\alpha) = \alpha - e_v$.

We point out a useful geometric consequence of this definition:

Proposition 6.10 *Suppose that* $\alpha \mapsto s_v(\alpha)$ *is a* (-1)-*reflection and that* $q_v = 1$ *and* $\theta_v = 0$. *Then there is a reflection isomorphism* $\mathcal{M}_{q,\theta}(s_v(\alpha)) \xrightarrow{\sim} \mathcal{M}_{q,\theta}(\alpha)$.

Proof There is an obvious map $\mathcal{M}_{q,\theta}(\alpha - e_v) \to \mathcal{M}_{q,\theta}(\alpha)$, given by $\rho \mapsto \rho \oplus \mathbb{C}_v$, where $\mathbb{C}_v$ is the trivial representation (all arrows act as zero). We claim that it is an isomorphism. In the decomposition of any θ-polystable representation of dimension α into stable representations, at least one factor must have dimension vector which has positive pairing with e_v. By Crawley-Boevey and Shaw [CBS06, Lemma 5.1] (in the case $\theta = 0$, which extends to the general case by replacing simple representations by θ-stable ones), this summand must be $\mathbb{C}_v$ itself. Therefore, the obvious map is an isomorphism. □

Definition 6.11 Given $\alpha \in R^+_{q,\theta}$, call a sequence $v_1, \ldots, v_m \in Q_0$ a reflecting sequence if, setting

$$(q^{(i)}, \theta^{(i)}, \alpha^{(i)}) := (u_{v_i} \cdots u_{v_1}(q), r_{v_i} \cdots r_{v_1}(\theta), s_{v_i} \cdots s_{v_1}(q)),$$

we have (a) $\alpha^{(m)} \in \mathcal{F}(Q) \cup \{e_v \mid v \in Q_0\}$ and (b) $\alpha^{(i)}_{v_i} < \alpha^{(i-1)}_{v_i}$ for all i.

Lemma 6.12 *A reflecting sequence always exists.*

Proof By definition, α is a root if and only if there exists a sequence of reflections at loop-free vertices taking α to either the fundamental region or to an elementary root e_v (it is imaginary in the former case and real in the latter case). Now, given $\alpha \in \mathbb{N}^{Q_0}$, let $N_\alpha := |\{\beta \in R^+ \text{ real} \mid (\alpha, \beta) > 0\}|$. Then each reflection satisfying (b) decreases N_α by one, and a nontrivial reflection not satisfying (b) increases N_α by one. Now, assume that $\alpha \in R^+_{q,\theta}$. Then $N_\alpha < \infty$. Since $s_v(R^+ \setminus \{e_v\}) \subseteq R^+$, an arbitrary sequence of reflections satisfying (b) will remain in $\mathbb{N}^{Q_0}$. Thus, if α is real, an arbitrary $N_\alpha - 1$ reflections satisfying (b) will send α to e_v for some $v \in Q_0$, and if α is imaginary, then an arbitrary N_α reflections satisfying (b) will take α to $\mathcal{F}(Q)$. □

As in [Su06, Theorem 1.2], we have the following.

Theorem 6.13 *Let $\alpha \in R^+_{q,\theta}$. Pick any sequence of vertices $v_1, \dots, v_m \in Q_0$ such that, for $(q^{(i)}, \theta^{(i)}, \alpha^{(i)}) := (u_{v_i} \cdots u_{v_1}(q), r_{v_i} \cdots r_{v_1}(\theta), s_{v_i} \cdots s_{v_1}(q))$, we have $\alpha^{(m)} \in \mathcal{F}(Q)$ and $\alpha^{(i)}_{v_i} < \alpha^{(i-1)}_{v_i}$. Then α is flat if and only if (a) $\alpha^{(m)} \in \mathcal{F}(Q)$ is flat, and (b) for every i, either (b1) $(q^{(i)}, \theta^{(i)}, \alpha^{(i)})$ is an admissible reflection (Definition 2.5) of $(q^{(i-1)}, \theta^{(i-1)}, \alpha^{(i-1)})$ (i.e., $q^{(i-1)}_{v_i} \neq 1$ or $\theta^{(i-1)}_{v_i} \neq 0$), or (b2) $\alpha^{(i)}$ is a (-1)-reflection of $\alpha^{(i-1)}$.*

Analogously to [Su06], we will show below that $\alpha^{(m)} \in \mathcal{F}(Q)$ is flat if and only if it is not of the form $m\ell\delta$ for δ the minimal imaginary root of an affine Dynkin subquiver, $m \geq 2$, and $\ell \geq 1$ is such that q^δ is a primitive ℓ-th root of unity.

The theorem actually gives an algorithm to determine if a root is flat, by playing a variant of the numbers game [Moz90] (with a cutoff in the inadmissible case as in [GS11]).

Proof of Theorem 6.13 Under an admissible reflection, the condition of being flat does not change since $s_{v_i} : R^+_{q^{(i-1)},\theta^{(i-1)}} \to R^+_{q^{(i)},\theta^{(i)}}$ is a bijection (as $e_{v_i} \notin R^+_{q^{(i-1)},\theta^{(i-1)}}$), cf. [CB01, Lemma 5.2]. Here, we let $q^{(0)} := q$ and $\theta^{(0)} := \theta$. If we apply a (-1)-reflection, we claim that the condition of being flat does not change. We only have to show that if $\alpha^{(i)} \in \widetilde{\Sigma}_{q^{(i)},\theta^{(i)}}$, then also $\alpha^{(i-1)} \in \widetilde{\Sigma}_{q^{(i-1)},\theta^{(i-1)}}$, since the converse follows immediately from the definition of flat. Suppose on the contrary that $\alpha^{(i-1)} = \beta^{(1)} + \cdots + \beta^{(k)}$ is a decomposition with $\beta^{(j)} \in R^+_{q^{(i-1)},\theta^{(i)}}$ and $p(\alpha^{(i-1)}) < p(\beta^{(1)}) + \cdots + p(\beta^{(k)})$. Since $(\alpha^{(i-1)}, e_{v_i}) = 1$, there must exist j with $(\beta^{(j)}, e_{v_i}) \geq 1$ and hence also $\beta^{(j)}_{v_i} \geq 1$. Set $\gamma^{(\ell)} := \beta^{(\ell)} - \delta_{\ell j} e_{v_i}$ for all ℓ. Note that $p(\gamma^{(j)}) \geq p(\beta^{(j)})$, with equality if and only if $(\beta^{(j)}, e_{v_i}) = 1$. Then

$$p(\alpha^{(i)}) = p(\alpha^{(i-1)}) < p(\beta^{(1)}) + \cdots + p(\beta^{(k)}) \leq p(\gamma^{(1)}) + \cdots + p(\gamma^{(k)})$$

so that $\alpha^{(i)} \notin \widetilde{\Sigma}_{q^{(i)},\theta^{(i)}}$. We have proved the contrapositive.

It remains to show that if $(\alpha^{(i-1)}, e_{v_i}) > 1$ and $(q_{v_i}, \theta_{v_i}) = (1, 0)$, then $\alpha^{(i-1)} \notin \widetilde{\Sigma}_{q^{(i-1)},\theta^{(i-1)}}$ In this case, there can be no loops at v_i so that e_{v_i} is a real root. Moreover, $\alpha^{(i-1)}_{v_i} \geq 1$. Then, $p(\alpha^{(i-1)} - e_{v_i}) = p(\alpha^{(i-1)} - e_{v_i}) + p(e_{v_i}) > p(\alpha^{(i-1)})$. So $\alpha^{(i-1)}$ is not flat. □

Remark 6.14 The same theorem as above applies in the additive case, to characterize the analogous set $\widetilde{\Sigma}_{\lambda,\theta}$ of flat roots. Also, note that when $\theta = 0$, the above proof simplifies the proof of [Su06, Theorem 1.2], since it does not require the classification [CB01, Theorem 8.1] of roots in $\mathcal{F}(Q) \setminus \Sigma_{\lambda,0}$.

Remark 6.15 Thanks to [GS11, Theorem 3.1], the condition that any (or every) reflecting sequence consists only of admissible and (-1)-reflections is equivalent to the condition that for every real root $\beta \in R^+_{q,\theta}$, we have $(\alpha, \beta) \leq 1$.

6.2 Fundamental and Flat Roots Not in $\Sigma_{q,\theta}$

To complete the characterization, we need to determine the set $\mathcal{F}(Q) \setminus \widetilde{\Sigma}_{q,\theta}$. This follows from [CB01, Theorem 8.1], which computes $\mathcal{F}(Q) \setminus \Sigma_\lambda$ (in the additive case, but which extends to the present setting). We state a sharper and more general version:

Theorem 6.16 *A root in $R^+_{q,\theta}$ is not in $\Sigma_{q,\theta}$ if and only if, applying a reflecting sequence as in Theorem 6.13, either one of the reflections is inadmissible, or the resulting element of $\mathcal{F}(Q)$ is one of the following:*

(a) $m\ell\delta$ with δ the indivisible imaginary root for an affine Dynkin subquiver, $m \geq 2$, and ℓ is such that q^δ is a primitive ℓ-th root of unity; or

(b) The support of α is $J \sqcup K$ for $J, K \subseteq Q_0$ disjoint subsets with exactly one arrow in $\overline{Q}_1$ from a vertex $j \in J$ to a vertex $k \in K$, $\alpha_k = 1$, and either:

(b1) $\alpha_j = 1$, and $\alpha|_J \in R^+_{q,\theta}$, or

(b2) $Q|_J$ is affine Dynkin and $\alpha|_J = m\delta$ for some $m \geq 2$, with $\delta \in R^+_{q,\theta}$ indivisible and j an extending vertex of $Q|_J$.

Moreover, the root is not in $\widetilde{\Sigma}_{q,\theta}$ if and only if one of the reflections is neither admissible nor a (-1)-reflection, or the resulting element of $\mathcal{F}(Q)$ is in case (a).

Proof First, it is clear that if an inadmissible reflection is applied in the sequence, $p(\alpha) \leq p(s_i\alpha) + p(\alpha - s_i\alpha)$ shows that α is not in $\Sigma_{q,\theta}$. So we can assume all reflections are admissible. By Crawley-Boevey [CB01, Theorem 8.1] in the additive case (with $\theta = 0$), there is a sequence of admissible reflections resulting in one of the given cases (which is not assumed to be in $\mathcal{F}(Q)$). The proof in *loc. cit.* extends verbatim to our case, replacing N_λ by $N_{q,\theta}$. Although the additive statement has $\ell = 1$ (or in characteristic p has $\ell = p$), for us, we note that if q^δ is a primitive ℓ-th root of unity, that is, $\ell\delta$ is q-indivisible, then $\ell\delta \in \Sigma_{q,\theta}$, since every element $\beta \in N_{q,\theta}$ with $\beta < \ell\delta$ is real. Note that in [CB01, Theorem 8.1], the condition in (b2) that $\delta \in R^+_\lambda$ is not stated (since its goal is to produce non-exhaustive necessary conditions for $\alpha \in \Sigma$), but it follows from the proof that it is also a necessary condition for $\alpha \notin \Sigma$.

We claim that in fact, we can take the result to be in $\mathcal{F}(Q)$. Note that, in [CB01], it is not required in conditions (a), (b) that α be in $\mathcal{F}(Q)$. However, we already know that in order to have $\alpha \in \Sigma_{q,\theta}$, there must be an admissible reflection sequence taking α to $\mathcal{F}(Q)$. Thus, we may assume that $\alpha \in \mathcal{F}(Q)$ and moreover that it is sincere. Then, applying [CB01, Theorem 8.1], there is a further sequence of admissible reflections taking α to one of the forms above. After this, we can apply an admissible reflection sequence to get back to an element of $\mathcal{F}(Q)$, necessarily α again. It is clear that doing so will not change the form as above, since α was assumed to be sincere, so the reflections cannot shrink the support of α; note that in cases (a) and (b2), this means that no reflections will be applied to the coefficients of the multiple of δ.

For the converse, it remains to verify that cases (a) and (b1), (b2) are not in $\Sigma_{q,\theta}$. In case (a), $p(m\delta) = 1 < mp(\delta) = m$. In case (b1), $p(\alpha) = p(\alpha|_J) + p(\alpha|_K)$. Finally, in case (b1), $p(\alpha) = p(\delta) + p(\alpha - \delta)$.

The statement about flat roots follows from Theorem 6.13 together with the observation that case (a) is not flat (as $p(m\ell\delta) < mp(\ell\delta)$), whereas cases (b1) and (b2) are flat: This follows from [Su06, Theorem 1.1], where it is shown that (b1) and (b2) are already flat as elements of $\Sigma_{1,0}$ (which is stronger). □

6.3 Canonical Decompositions

Let $\Sigma^{\text{iso}}_{q,\theta} \subseteq \Sigma_{q,\theta}$ be the subset of isotropic imaginary roots. We use the notation $\mathbb{N}_{\geq 2} \cdot \Sigma^{\text{iso}}_{q,\theta} := \{m\alpha \mid m \geq 2, \alpha \in \Sigma^{\text{iso}}_{q,\theta}\}$.

Theorem 6.17

(i) Given $\alpha \in N_{q,\theta}$, there exists a unique decomposition $\alpha = \alpha^{(1)} + \cdots + \alpha^{(m)}$ with $\alpha^{(i)} \in \Sigma_{q,\theta}$ such that any other such decomposition is a refinement of this one.

(ii) There is also a unique decomposition $\alpha = \beta^{(1)} + \cdots + \beta^{(k)}$ with $\beta^{(i)} \in \widetilde{\Sigma}_{q,\theta} \cup \mathbb{N}_{\geq 2} \cdot \Sigma^{iso}_{q,\theta}$, satisfying the properties:

(a) Every element $\beta^{(i)}$ is of one of the following three types:

(1) $\beta^{(i)} \in \Sigma_{q,\theta}$

(2) $\beta^{(i)} \in \mathbb{N}_{\geq 2} \cdot \Sigma^{iso}_{q,\theta}$

(3) $\beta^{(i)} \in \widetilde{\Sigma}_{q,\theta} \setminus \Sigma_{q,\theta}$; moreover, there is an admissible reflection sequence taking $\beta^{(i)}$ to an element of the fundamental region having only decompositions of the form (b2) in Theorem 6.16.

(b) Any other decomposition into $\widetilde{\Sigma}_{q,\theta} \cup \mathbb{N}_{\geq 2} \cdot \Sigma^{iso}_{q,\theta}$ satisfying (a) is a refinement of this one.

(iii) The decomposition in (i) is a refinement of the one in (ii), obtained uniquely, after applying admissible reflections, by performing decompositions $\alpha = \alpha|_J + \alpha|_K$ of type (b2) in Theorem 6.16.

(iv) The direct sum map produces a Poisson isomorphism (of reduced varieties), using the decomposition in (ii),

$$\prod_{i=1}^{k} \mathcal{M}_{q,\theta}(Q, \beta^{(i)}) \xrightarrow{\sim} \mathcal{M}_{q,\theta}(Q, \alpha).$$

Notice the following immediate consequence (of the decomposition in (ii)), giving a weakened version of (*):

Corollary 6.18 *If α is the dimension of a θ-stable representation of Λ^q, then one of the following three cases must hold: (1) $\alpha \in \Sigma_{q,\theta}$; (2) $\alpha \in \mathbb{N}_{\geq 2}\Sigma_{q,\theta}^{iso}$; (3) α is obtained by admissible reflections from an element in the fundamental region having only type (b2) decompositions in Theorem 6.16.*

Remark 6.19 In the additive situation, the fact that cases (2) and (3) in the preceding corollary cannot occur was very recently given a simpler, unified proof in [CBH19].

Note that in the real case $\alpha \in \Sigma_{q,\theta}$, it is obvious from the properties of admissible reflections that α is the dimension of a θ-stable representation (see [CBS06, Theorem 1.9] for a stronger statement). In the imaginary case, following (*), we only expect a stable representation if $\alpha \in \Sigma_{q,\theta}$, but it is not at all clear how to prove its existence.

In the proof of the theorem, we will produce also an algorithm for constructing the $\beta^{(i)}$, by a sequence of reflections and subtracting roots e_i, with the end result an element in $\mathcal{F}(Q)$ whose connected components give the imaginary $\beta^{(i)}$. For the real roots, we obtain the unique decomposition into real roots in $\Sigma_{q,\theta}$ (as a real root which is the sum of multiple real roots cannot be in $\Sigma_{q,\theta}$).

Remark 6.20 Observe that essentially the same proof as that provided below of Theorem 6.17 was given in [CB02] in the context of Nakajima quiver varieties, and indeed, the result above holds in that setting. However, due to the simplifying properties of that case (such as expectation (*) holding, and q-divisibility coinciding with ordinary divisibility), Crawley-Boevey is able to show that the product decomposition in (iv) always refines to one using the decomposition of (i). Hence, the statement given in [CB02] is substantially simpler, eliminating parts (ii) and (iii).

Proof of Theorem 6.17 We first obtain the existence of the desired decompositions in (i) and (ii) satisfying (iii) and (iv). We prove this by induction on the sum of the entries of α. If there is a vertex $v \in Q_0$ at which $(\alpha, e_v) > 0$ and either $q_v \neq 1$ or $\theta_v \neq 0$, then we can apply an admissible reflection. Since admissible reflections preserve the set of flat roots, the statements follow from Theorem 2.13 and the induction hypothesis. So suppose that there is no such vertex. Instead, suppose that $v \in Q_0$ is such that $(\alpha, e_v) > 0$ but $q_v = 1, \theta_v = 0$. Then every decomposition of α into elements of $\widetilde{\Sigma}_{q,\theta}$ must have an element having positive Cartan pairing with e_v. This cannot happen by definition for the imaginary roots. Since e_v is the only real root in $\Sigma_{q,\theta}$ with positive pairing with e_v, this implies that e_v must appear as a summand of every decomposition of types (i) and (ii). Similarly, the argument of the proof of Proposition 6.10 shows that in this case, the direct sum map yields an isomorphism $\mathcal{M}_{q,\theta}(Q, \alpha - e_v) \times \mathcal{M}_{q,\theta}(Q, e_v) \xrightarrow{\sim} \mathcal{M}_{q,\theta}(Q, \alpha)$. We can apply the induction hypothesis to $\alpha - e_v$.

This reduces the theorem to the case that $(\alpha, e_i) \leq 0$ for all i. In this case, we can decompose α into its connected components. By Theorem 6.16, all of these are flat roots, except for elements of the form $m\ell\delta$ with $\ell\delta \in \Sigma_{q,\theta}^{\text{iso}}$ and $m \geq 2$.

It remains to show that given the situation (b1) of Theorem 6.16, we get a decomposition of our moduli space. This follows from the arguments of [CB01, §10, II]. We briefly repeat them for the reader's convenience, adapting them to our situation (see *loc. cit.* for details). Restrict the quiver to the vertices $\{j\} \cup K$ and the arrows incident only to these vertices. Let a, a^* be the pair of reverse arrows between j and k; without loss of generality suppose $a : j \to k$. Adding the relations at all the vertices of K, and using that $\alpha|_K \in N_{q,\theta}$ (since α and $\alpha|_J$ are), we obtain that a^*a is equal to a sum of commutators. It thus has trace one, and since $\alpha_j = 1$, it is zero. Therefore, either a^* or a acts by zero. In the former case, we obtain a quotient representation with dimension vector $\alpha|_K$; in the latter, we obtain a subrepresentation with this dimension. These representations are θ-semistable since $\theta \cdot \alpha|_K = 0$, and the original representation was θ-semistable. So the polystabilization of our representation decomposes into a direct sum of representations with dimension vectors $\alpha|_J$ and $\alpha|_K$. Therefore, the direct sum map $\mathcal{M}_{q,\theta}(Q, \alpha|_J) \times \mathcal{M}_{q,\theta}(Q, \alpha|_K) \to \mathcal{M}_{q,\theta}(Q, \alpha)$ is an isomorphism.

This yields the desired decomposition in (ii), as well as the isomorphism in (iv), and the decomposition in (iii).

Let us prove that the decompositions are unique. For (i), this is done in [CB02, Theorem 1.1]; the proof carries over verbatim, replacing Σ_λ by $\Sigma_{q,\theta}$ and adapting all notions. Let us consider (ii), whose proof is similar. We can obviously assume that α is sincere. We first claim that the uniqueness statement is unaffected by applying the aforementioned reduction to the fundamental region. It is obvious that applying admissible reflections does not change the statement. So we only have to show that if $(\alpha, e_v) > 0$, then every decomposition of type (ii) includes e_v. In this case, any decomposition of the form (ii) must have $(\beta^{(i)}, e_v) > 0$ for some v. If $\beta^{(i)} \neq e_v$, then we must have $\beta^{(i)} \notin \mathbb{N} \cdot \Sigma_{q,\theta}$. Thus, $\beta^{(i)} \in \widetilde{\Sigma}_{q,\theta}$, and by assumption (3), it must be related by admissible reflections to an element of the fundamental region. Therefore, it does not have a positive pairing with any real roots. This is a contradiction. We therefore obtain that $\beta^{(i)} = e_v$ for some i. Thus, the uniqueness statement for α is equivalent to that for $\alpha - e_v$, as desired.

This reduces us to the case that α is in the fundamental region. We clearly can get a decomposition in a unique way by iteratively replacing α by the sum $\alpha|_J + \alpha|_K$ as in (b1) of Theorem 6.16. We only have to show that every decomposition of the form (ii) refines such a decomposition of type (b1). For a contradiction, suppose α is in the fundamental region, and we have a decomposition of type (b1) with sets J and K, but also a decomposition of type (ii) with some $\beta^{(i)}$ not supported entirely on J or K. After applying admissible reflections to $\beta^{(i)}$, this property continues to hold, since we cannot perform admissible reflections at the vertices j and k. So we can assume $\beta^{(i)}$ is itself in the fundamental region. This contradicts our assumptions on $\beta^{(i)}$. □

As a consequence, we obtain the following description of divisibility criteria for elements of $\widetilde{\Sigma}_{q,\theta}$ and $\Sigma_{q,\theta}$, analogous to [BS21, Theorem 2.2]:

Corollary 6.21 *Let $\alpha = m\beta$ for $\beta \in R^+_{q,\theta}$ q-indivisible and imaginary, and $m \geq 2$. Then, $\alpha \in \widetilde{\Sigma}_{q,\theta}$ if and only if $\beta \in \widetilde{\Sigma}_{q,\theta}$, β is anisotropic, and a reflecting sequence taking β to the fundamental region involves only admissible reflections. In this case, also $\alpha \in \Sigma_{q,\theta}$.*

In particular, for $\gamma \in \widetilde{\Sigma}_{q,\theta}$, every rational multiple $r\gamma \in N_{q,\theta}$ for $r \in Q_{\leq 1}$ is also in $\widetilde{\Sigma}_{q,\theta}$.

Proof This follows from the classification of flat roots in Theorems 6.13, 6.16, and 6.17. Observe simply that if a reflection sequence for β involves an inadmissible (-1)-reflection, then the same sequence for α involves an inadmissible (-2)-reflection (which is not allowed). On the other hand, if only admissible reflections are allowed, then $m\beta$ will also be flat unless β is isotropic. In the anisotropic case, $m\beta \in \Sigma_{q,\theta}$, since the decompositions of type (b1) and (b2) cannot occur for a divisible vector. The last statement then follows by considering γ and $r\gamma$ as multiples of a common vector. □

Remark 6.22 Although we are working with reduced varieties throughout the paper, we emphasized this in Theorem 6.17(iv) because it is not completely clear that the reflection isomorphisms in [Yam08, Theorem 5.1] are defined scheme-theoretically. On the other hand, this is the only obstacle here. That is, if these isomorphisms are defined scheme-theoretically, then the proof would appear to extend to this case, that is, to not-necessarily-reduced multiplicative quiver schemes.

6.4 Symplectic Resolutions for q-Indivisible Flat Roots

Theorem 6.23 *Suppose that $\alpha \in \widetilde{\Sigma}_{q,\theta}$ is q-indivisible and $\mathcal{M}_{q,\theta}(Q,\alpha)$ is non-empty. Then for suitable $\theta' \geq \theta$, $\mathcal{M}_{q,\theta'}(Q,\alpha) \to \mathcal{M}_{q,\theta}(Q,\alpha)$ is a symplectic resolution.*

Remark 6.24 Observe that the theorem also holds in the additive setting, where the result is also interesting. Indeed, it explains and generalizes the technique of framing used to construct resolutions such as Hilbert schemes of $\mathbb{C}^2$ or of hyperkähler ALE spaces. In the former case, the quiver is again the framed Jordan quiver (with two vertices and two arrows, a loop at the first vertex, and an arrow from the second to the first vertex). The dimension vector is $\alpha = (m, 1)$. The theorem recovers the well-known statement that taking $\theta \neq 0$ gives a symplectic resolution of the singularity $\mathrm{Sym}^m \mathbb{C}^2$ in the additive case; this identifies with $\mathrm{Hilb}^m \mathbb{C}^2$. In the multiplicative case for the same quiver, by Theorem 3.6 and Remark 3.7, after localization, we obtain a resolution of the character variety of the once-punctured torus in the multiplicative case for rank m local systems with unipotent monodromy A satisfying $\mathrm{rk}(A - I) \leq 1$.

Proof of Theorem 6.23 In view of Lemma 2.12, we only have to show that we can find $\theta' \geq \theta$ such that $\mathcal{M}_{q,\theta'}(Q,\alpha)$ is smooth and $\mathcal{M}_{q,\theta'}(Q,\alpha) \to \mathcal{M}_{q,\theta}(Q,\alpha)$ is birational. We will make use of the combinatorial analysis of [CB01, Section 8]. Note that if $\alpha \in \Sigma_{q,\theta}$, then the result follows from the discussion in Sect. 4.1 so we can assume that this is not the case.

Let us take θ' generic such that the conditions of Remark 2.22 are satisfied (i.e., θ' is an integral multiple of a generic rational stability condition in a small neighborhood of θ). In particular, this means that $\theta' \geq \theta$, every θ-stable representation is θ'-stable, and $\theta' \cdot \beta \neq 0$ for any $\beta < \alpha$ with $\beta \in N_{q,\theta}$.

By Corollary 6.6, for each connected component of $\mathcal{M}_{q,\theta}(Q,\alpha)$, there is a dense stratum of the form $(1, \beta^{(1)}; \ldots; 1, \beta^{(r)})$ with $p(\alpha) = \sum_{i=1}^{r} p(\beta^{(i)})$. We need to show that each representation ρ in such a stratum is in the boundary of a unique $\mathrm{GL}(\alpha)$-orbit in $\mathrm{Rep}^{\theta'-s}(\Lambda^q(Q), \alpha)$. Equivalently, we must show that there is a unique θ'-stable representation ρ' up to isomorphism such that ρ is the θ-polystabilization of ρ'.

We will prove the statement by induction on α, with respect to the partial ordering $\leq$. First, applying admissible and (-1)-reflections, we reduce to the case that α is in the fundamental region. Indeed, it is clear from Theorem 2.13 and Proposition 6.10 that applying these reflections causes no harm. Note that each (-1)-reflection will modify stratum types by removing a real root from the type; once we are in the fundamental region, no real roots will appear.

We first show uniqueness. If ρ' is as above, then suppose that there is an exact sequence of θ-semistable representations of the form $0 \to \psi \to \rho' \to \phi \to 0$. By our assumptions, the dimension vectors of ψ and ϕ are sums of complementary subsets of the $\beta^{(i)}$. It follows from the proof of Theorem 6.16 (using [CB02, Section 8]) that there is a corresponding decomposition of α as in Theorem 6.16, of type (b1) or (b2), with the following property: In type (b1), $\alpha^{(1)} := \alpha|_J$ and $\alpha^{(2)} := \alpha|_K$ are the dimension vectors of ψ and ϕ, in either order, or in type (b2), $\alpha^{(1)} := \delta$ and $\alpha^{(2)} := \alpha - \delta$ are these dimension vectors, again in some order. Note that the ordering of the $\alpha^{(i)}$ is fixed by the conditions that $\dim \psi \cdot \theta < 0$ and $\dim \phi \cdot \theta > 0$. Next, since $(\alpha^{(1)}, \alpha^{(2)}) = -1$, it follows from Proposition 2.2 that $\dim \mathrm{Ext}^1(\psi, \phi) = \dim \mathrm{Ext}^1(\phi, \psi) = 1$. So the extension ρ' is uniquely determined, up to isomorphism, from ψ and ϕ.

We claim that ψ and ϕ are uniquely determined from their dimension vectors up to isomorphism. We give the argument for ψ; the one for ϕ is symmetric. Let $C \subseteq \{1, \ldots, r\}$ be a subset of indices such that $\dim \psi = \sum_{i \in C} \beta^{(i)}$. (This set is unique except in case (b2) with $\dim \psi = \alpha^{(2)} = \alpha - \delta$.) There exists a unique $i \in C$ such that $(\beta^{(i)}, \dim \phi) = -1$ (since $(\alpha^{(1)}, \alpha^{(2)}) = -1$). Now, define $\theta'' := \theta'|_{\mathrm{supp} \dim \psi} - (\theta' \cdot \dim \psi) e_v$, where v is the unique vertex in $\mathrm{supp} \dim \psi$ which has nonzero Cartan pairing with $\mathrm{supp} \dim \phi$ (so $(\dim \psi)_v = 1$). By construction, $\theta'' \cdot \dim \psi = 0 = \theta'' \cdot \dim \phi$. Moreover, $\theta'' \cdot \beta^{(j)} = \theta' \cdot \beta^{(j)}$ for $j \in C \setminus \{i\}$. We claim that ψ is θ''-stable. By definition of θ'-stability, every nonzero submodule η of ψ satisfies $\theta' \cdot \dim \eta < 0$. Now, if $\beta^{(i)} \not\leq \dim \eta$, then $\theta'' \cdot \dim \eta = \theta' \cdot \dim \eta < 0$. On the other hand, if η is a proper submodule of ψ with $\beta^{(i)} \leq \dim \eta$, then ψ/η is a nonzero quotient module with $\beta^{(i)} \not\leq \dim(\psi/\eta)$. Then $(\dim(\psi/\eta), \dim \phi) = 0$. By Proposition 2.2, $\mathrm{Ext}^1(\phi, \psi/\eta) = 0$. Therefore, we have an exact sequence $0 \to \eta \to \rho' \to (\psi/\eta) \oplus \phi \to 0$. As a consequence, ψ/η itself is a quotient module of ρ'. It follows that $\theta' \cdot \dim(\psi/\eta) > 0$. Therefore, $\theta'' \cdot \dim(\psi/\eta) = \theta' \cdot (\psi/\eta) > 0$. Therefore, $\theta'' \cdot \eta < 0$. We conclude that ψ is θ''-stable, as desired. By induction on α, ψ is then uniquely determined up to isomorphism.

This completes the proof of uniqueness. We move on to existence, which is similar. Begin with a decomposition given by Theorem 6.16 of type (b1) or (b2). Let us keep the notation $\alpha^{(1)}, \alpha^{(2)}$ defined above. The same construction as above yields modifications $\theta^{(1)}, \theta^{(2)}$ of θ' such that $\theta^{(i)} \cdot \alpha^{(i)} = 0$. By induction, we can take $\theta^{(i)}$-stable representations ϕ, ψ of dimension vectors $\alpha^{(i)}$. Then since $(\alpha^{(1)}, \alpha^{(2)}) = -1$, $\dim \mathrm{Ext}^1(\phi, \psi) = \dim \mathrm{Ext}^1(\psi, \phi) = 1$. Assume that $\theta' \cdot \phi < 0$; otherwise, swap ϕ and ψ. Then form a nontrivial extension $0 \to \phi \to \rho' \to \psi \to 0$. The same computation as above guarantees that ρ' is θ'-stable. □

Corollary 6.25 *In the situation of the proposition, the normalization of $\mathcal{M}_{q,\theta}(Q, \alpha)$ is a symplectic singularity.*

Proof This follows since we have constructed a symplectic resolution (see Proposition 4.5 or Remark 4.6). □

Remark 6.26 Note that the main step of the proof is to show that $\mathcal{M}_{q,\theta'}(Q, \alpha) \to \mathcal{M}_{q,\theta}(Q, \alpha)$ is birational for suitable $\theta' \geq \theta$. For this, we did not need the hypothesis that α is q-indivisible. On the other hand, by Theorems 6.13 and 6.16, when $\alpha \in \widetilde{\Sigma_{q,\theta}} \setminus \Sigma_{q,\theta}$, α is actually indivisible (not merely q-indivisible). For $\alpha \in \Sigma_{q,\theta}$, the birationality statement is Corollary 2.21, which is easy. (Moreover, the full statement of Theorem 6.23 was established for $\alpha \in \Sigma_{q,\theta}$ in Sect. 4.1.) So it does not really add anything to state the birationality property without the q-indivisibility hypothesis.

6.5 Symplectic Resolutions for General α

Theorem 6.27 *Assume that $\mathcal{M}_{q,\theta}(Q, \alpha)$ is non-empty and that the decomposition of Theorem 6.17(ii) has no elements $\beta^{(i)}$ of the forms (a) $\beta^{(i)} = 2\gamma$ for $\gamma \in N_{q,\theta}$ and $p(\gamma) = 2$, or (b) $\beta^{(i)} = m\gamma$ for $m \geq 2$ and $\gamma \in \Sigma^{iso}_{q,\theta}$. Then:*

- *The normalization of $\mathcal{M}_{q,\theta}(Q, \alpha)$ is a symplectic singularity.*
- *Each factor $\mathcal{M}_{q,\theta}(Q, \beta^{(i)})$ with $\beta^{(i)} \notin \Sigma_{q,\theta}$ admits a symplectic resolution.*
- *If for any factor $\beta^{(i)}$ there exists a θ-stable representation of dimension $\gamma^{(i)} = \frac{1}{m}\beta^{(i)}$ with $m \geq 2$, then $\mathcal{M}_{q,\theta}(Q, \alpha)$ does not admit a symplectic resolution. In fact, it has an open, singular, terminal, factorial subset.*

Proof The first statement follows if we show that the normalization of each factor $\mathcal{M}_{q,\theta}(Q, \beta^{(i)})$ is a symplectic singularity. For the factors such that $\beta^{(i)}$ is in $\Sigma_{q,\theta}$, this is a consequence of Theorem 1.5. For $\beta^{(i)} \notin \Sigma_{q,\theta}$, after applying admissible reflections, it follows from Theorem 6.16 that it is indivisible. Hence, $\beta^{(i)}$ is itself indivisible. By our assumptions, $\beta^{(i)}$ is flat. The result then follows from Theorem 6.23. This also proves the second statement.

We proceed to the third statement. Under the hypotheses, since we have excluded the isotropic and (2, 2)-cases, an open subset of $\mathcal{M}_{q,\theta}(Q, \beta^{(j)})$ is factorial terminal singular by Theorem 1.5 (see Theorem 4.14). Hence so is an open subset of $\mathcal{M}_{q,\theta}(Q,\alpha)$, which therefore does not admit a symplectic resolution. □

6.6 Classifications of Symplectic Resolutions of Punctured Character Varieties

Here, we combine the results of this section and Theorem 5.1 to get a classification of all the character varieties of punctured surfaces which admit a symplectic resolution, modulo the conjectural results of the $(2,2)$-cases. As explained in Sect. 3, in order to get such a result, it suffices to consider multiplicative quiver varieties of crab-shaped quivers, where the parameter q and the dimension vector α are chosen in an appropriate way; see Theorem 3.6.

Let Q be a crab-shaped quiver, $q \in (\mathbb{C}^\times)^{Q_0}$ and $\alpha \in N_{q,\theta}$, and consider the corresponding quiver variety $\mathcal{M}_{q,\theta}(Q,\alpha)$. Then, if $\alpha \notin \mathcal{F}(Q)$, we can apply the algorithm of Theorem 6.17 and obtain a decomposition where the dimension vectors of the factors are in the fundamental region (such dimension vectors are the connected components of the reflection of α). Moreover, note that in the crab-shaped case, all the vector components not containing the central vertex are Dynkin quivers of type A: Therefore, the associated multiplicative quiver variety is just a point. This implies that we can assume, without loss of generality, that α be sincere ($\alpha_i > 0$ for all i) and in the fundamental region. After having performed this reduction, we can prove the following.

Corollary 6.28 *Let Q be a crab-shaped quiver and $\alpha \in N_{q,\theta}$ a sincere vector in the fundamental region. Further assume that (Q,α) is **not** one of the following cases:*

(a) $\beta = \frac{1}{2}\alpha \in N_{q,\theta}$ and (Q,β) is one of the quivers in Theorems 5.1 and 5.3.
(b) Q is affine Dynkin (of type $\tilde{A}_0$ (i.e., the Jordan quiver with one vertex and one arrow), $\tilde{D}_4$ or $\tilde{E}_6, \tilde{E}_7, \tilde{E}_8$) and α is a q-divisible multiple of the indivisible imaginary root δ of Q.

Then:

- *The normalization of $\mathcal{M}_{q,\theta}(Q,\alpha)$ is a symplectic singularity.*
- *If α is q-indivisible, $\mathcal{M}_{q,\theta}(Q,\alpha)$ admits a symplectic resolution.*
- *If $\alpha = m\beta$ for $m \geq 2$ and there exists a θ-stable representation of $\Lambda^q(Q)$ of dimension β, then $\mathcal{M}_{q,\theta}(Q,\alpha)$ does not admit a symplectic resolution (it contains an open singular factorial terminal subset).*
- *In the case that α is q-divisible, the condition of the preceding part is always satisfied, except possibly in the case: (c) $Q = Q^e \cup \{*\}$ is an affine Dynkin quiver Q^e of type $\tilde{A}_0, \tilde{D}_4, \tilde{E}_6, \tilde{E}_7$, or $\tilde{E}_8$ together with an additional vertex $\{*\}$ and an additional arrow from this vertex to one with dimension vector 1 in δ, and α has the form $(p, p\ell\delta)$ for $p \geq 2$ a prime, $\ell\delta$ q-indivisible. Here, δ denotes the indivisible imaginary root of Q^e.*

Remark 6.29 In the final part of the corollary, expectation (*) from the introduction predicts that the exception indeed fails to satisfy the conditions of the preceding part. Nonetheless, we believe that, also in this case, there should not exist a projective symplectic resolution; this would be implied by Conjecture 7.15 in the appendix together with the consequence stated after it, thanks to [BS21, Theorem 1.5].

Proof of Corollary 6.28 By Theorem 6.17, we know that α is flat unless it is a positive integral multiple of an isotropic root, excluded in case (b). Then the first and third statements follow from Theorems 1.5 and the second from 6.23. For the fourth statement, we apply Theorem 6.16. to see that, in the crab-shaped case, the dimension vector can only be in the fundamental region but not in $\Sigma_{q,\theta}$ if the quiver is a framed affine Dynkin quivers of types $\tilde{A}_0$, $\tilde{D}_4$, $\tilde{E}_6$, $\tilde{E}_7$, and $\tilde{E}_8$, and the dimension vector is $(1, \ell\delta)$. Thus, only prime multiples of this vector can be q-divisible but have no factor in $\Sigma_{q,\theta}$.

It then remains only to show that $\mathcal{M}_{q,\theta}(Q, \beta) \neq \emptyset$ for $\beta \in \Sigma_{q,\theta}$, not of the form $(1, \ell\delta)$ for a framed affine Dynkin quiver. Let us first assume that $\theta = 0$. In the star-shaped case, the result follows from [CBS06, Theorem 1.1]. In the crab-shaped case with $g > 0$ loops at the central vertex, with $\theta = 0$, it suffices by Theorem 3.6 and Remark 3.7 to show that for all conjugacy classes $\mathcal{C}_1, \ldots, \mathcal{C}_m \subset \mathrm{GL}(n, \mathbb{C})$ with product of determinants equal to one, there exists a solution to the equations $[A_1, B_1] \cdots [A_g, B_g] = C_1 \cdots C_m$ for $C_i \in \mathcal{C}_i$. This follows because there is a solution to the equation $[X_1, Y_1] = C$ for arbitrary $C \in \mathrm{SL}(n, \mathbb{C})$ (by Thompson [Tho61, Theorems 1, 2]).

Now, assume $\theta \neq 0$. In most cases (excepting the case of one loop and one branch), one can extend [CBS06, Theorem 1.1] to this case using [Yam08, §4.3, 4.4]; however, we may give a more direct argument. Since we are not in the situation of a framed affine Dynkin quiver with dimension vector $(1, \ell\delta)$, note that $\alpha \in \Sigma_{q,0}$ as well, so from the $\theta = 0$ case and [CBS06, Theorem 1.11], we know that there exists a simple representation of Λ^q of dimension α. This is automatically θ-stable, so we obtain the desired non-emptiness statement. □

Remark 6.30 The assumptions made in the above theorem relate to the fact that in cases (*a*) and (*b*), it is still unknown whether a symplectic resolution exists, as this problem seems to be solvable only through a deep understanding of the local structure of the variety. Nonetheless in case (a), we expect such symplectic resolutions to exist and to be constructible by using analogous techniques to the ones used by Bellamy and the first author in [BS21, Theorem 1.6] (see Remark 1.8).

6.7 *Proof of Theorems 1.1 and 1.3*

Theorems 1.1 and 1.3 follow from Corollary 6.28, together with Theorem 3.6, as follows.

First, we claim that q-divisibility for the collection of conjugacy classes $\mathcal{C}$ coincides with the same-named property for the dimension vector α of the corresponding crab-shaped quiver. To see this, first note that $m \cdot C$ indeed corresponds to $m \cdot \alpha$. So we only have to show that the condition that $\prod_i \det C_i = 1$ is equivalent to $q^\alpha = 1$. This is true by construction.

Next, we claim that, for $g = 0$, the condition $\ell \geq 2n$ of Theorem 1.1 is equivalent to the condition that $\alpha \in \mathcal{F}(Q)$, whereas for $g > 0$, we have $\alpha \in \mathcal{F}(Q)$ unconditionally. By the chosen ordering of the $\xi_{i,j}$, we have $(\alpha, e_i) \leq 0$ for all $i \in Q_0$ except possibly the node. There, the condition $\ell \geq 2n$ is equivalent to $(\alpha, e_i) \leq 0$. On the other hand, when $g \geq 1$, then as there is a loop at the node, it is automatic that $(\alpha, e_i) \leq 0$ for $i \in Q_0$ the node, and hence in this case $\alpha \in \mathcal{F}(Q)$ automatically.

We now claim that the dimension of $\mathcal{X}(g, k, C)$ equals $2p(\alpha)$ when the quiver is not Dynkin or affine Dynkin and moreover $\ell \geq 2n$ or $g \geq 1$. This follows from Theorems 6.16 and 3.6 (see also Remark 3.7), provided that α is not both q-divisible and isotropic. However, the latter conditions, for $\alpha \in \mathcal{F}(Q)$, are equivalent to saying that the graph Γ_C is affine Dynkin and α is q-divisible.

With the preceding claims established, we proceed to the proof of the theorems. Note that applying reflections as earlier in this section preserves the property that a dimension vector is one corresponding to a character variety (by Theorem 3.6). So we can always reduce to case that $\alpha \in \mathcal{F}(Q)$, unless we end up with something with Dynkin support (hence, $\mathcal{X}(g, k, C)$ is a point) or something where α becomes negative (hence, $\mathcal{X}(g, k, C)$ is empty). This proves the first part of Theorem 1.1. The remaining assertions of the theorems follow from the above claims, which allow us to translate Corollary 6.28 into the given results via Theorem 3.6.

7 Open Questions and Future Directions

In this section, we pose some questions concerning the cases which are left out from the analysis carried out in the previous sections. This includes the question of to what extent the decomposition in Theorem 6.17 can be refined, as in the additive case in [CB02].

One interesting direction of research which naturally arises from the results proved in this paper and the work [Tir19] of the second author is the study of analogous problems in the context of the Higgs bundle moduli spaces, which appear in the picture via the non-abelian Hodge correspondence.

We also say a few words on how one might hope to study the local structure of formal moduli spaces of polystable objects in a 2-Calabi–Yau category and prove that, under suitable conditions, formal neighborhoods of such moduli spaces are quiver varieties associated to a quiver which arises from the deformation theory of the objects parametrized by the moduli spaces. This is relevant to the present context as it would make it possible to give an alternative and more insightful proof of Proposition 4.3, as explained at the beginning of Sect. 4.

Before getting into these issues, we begin by discussing some cases where the multiplicative quiver varieties are known to be non-empty.

7.1 Non-emptiness of Multiplicative Quiver Varieties

As explained in the previous sections, one of the subtleties in the study of multiplicative quiver varieties is the fact that it is not known in general when they are non-empty (nor how many connected components they have). On the other hand, there are special cases in which non-emptiness can be shown. For example, when $q = 1$, then, for any quiver Q and any vector $\alpha \in \mathbb{N}^{Q_0}$, the zero representation is a suitable element of $\text{Rep}(\Lambda^q, \alpha)$, since the invertibility condition is automatically satisfied as well as the multiplicative preprojective relation. Thus, $\mathcal{M}_{1,0}(Q, \alpha) \neq \emptyset$. More generally, for every real root $\beta \in R^+_{q,\theta}$, then by applying reflection sequences as in Sect. 6.1 (see also the discussion after Corollary 6.18), we conclude that $\mathcal{M}_{q,\theta}(Q, \beta) \neq \emptyset$. As a result, if α can be expressed as a sum of real roots in $R^+_{q,\theta}$ (not necessarily coordinate vectors), then $\mathcal{M}_{q,\theta}(Q, \alpha) \neq \emptyset$.

Another important and less trivial case in which we are guaranteed that $\mathcal{M}_{q,\theta}(Q, \alpha)$ is non-empty is when Q is crab-shaped (for arbitrary $\alpha \in N_{q,\theta}$): This follows from the arguments of the proof of Corollary 6.28 (relying on [CB13] and [Tho61]). We remark that, by the arguments of [Yam08, §4.3, 4.4], relying on the correspondence between character varieties of punctured surfaces and moduli of parabolic bundles and [Ina13, §5], it follows that these varieties are in fact irreducible except possibly in certain cases of a crab-shaped quiver with a single loop (when, after reducing to the fundamental region, the support includes exactly one branch).

We can also ask when the stable locus $\mathcal{M}^s_{q,\theta}(Q, \alpha) \neq \emptyset$. Note that an answer to this question for all α also answers the question of non-emptiness of the entire locus, since every point in $\mathcal{M}_{q,\theta}(Q, \alpha)$ is represented by a polystable representation. More explicitly, $\mathcal{M}_{q,\theta}(Q, \alpha) \neq \emptyset$ if and only if α can be represented as a sum of roots $\alpha^{(i)}$ for which $\mathcal{M}^s_{q,\theta}(Q, \alpha^{(i)}) \neq \emptyset$. Note that when $\alpha \in \Sigma_{q,\theta}$, then non-emptiness of $\mathcal{M}_{q,\theta}(Q, \alpha)$ is equivalent to that of $\mathcal{M}^s_{q,\theta}(Q, \alpha)$, by Proposition 2.19. Our expectation (*) says that $\mathcal{M}^s_{q,\theta}(Q, \alpha) \neq \emptyset$ implies $\alpha \in \Sigma_{q,\theta}$.

7.2 Refined Decompositions for Multiplicative Quiver Varieties

Recall Crawley-Boevey's canonical decomposition in the additive case (Theorem 6.2, Lemma 6.1). It is useful to ask to what extent such a decomposition holds in the multiplicative setting, refining the one of Theorem 6.17. Let $\beta^{(i)}, \alpha^{(i,j)}$ be as in Theorem 6.17, and group together the $\alpha^{(i,j)}$ that are equal, yielding distinct $\gamma^{(i,j)}$ each occurring $r_{i,j} \geq 1$ times. Note that when $\gamma^{(i,j)}$ is anisotropic, then $r_{i,j} = 1$, since $r_{i,j}\gamma^{(i,j)} \in \Sigma_{q,\theta}$, by the uniqueness of the decomposition in Theorem 6.17.(i).

Conjecture 7.1 *We have a decomposition as follows:*

$$\mathcal{M}_{q,\theta}(Q, \alpha) \cong \prod_{i,j} S^{r_{i,j}} \mathcal{M}_{q,\theta}(Q, \gamma^{(i,j)}). \tag{7.2.1}$$

The following proposition partly resolves the conjecture modulo expectation (*).

Proposition 7.2 *If (*) holds, then the decomposition of Theorem 6.17.(iv) refines to one of the form*

$$\mathcal{M}_{q,\theta}(Q,\alpha) \cong \prod_{i,j} \mathcal{M}_{q,\theta}(Q, r_{i,j}\gamma^{(i,j)}). \tag{7.2.2}$$

Moreover, in this case, the direct sum map

$$S^{r_{i,j}}\mathcal{M}_{q,\theta}(Q,\gamma^{(i,j)}) \to \mathcal{M}_{q,\theta}(Q, r_{i,j}\gamma^{(i,j)}) \tag{7.2.3}$$

is surjective.

Proof It suffices to decompose each of the $\mathcal{M}_{q,\theta}(Q,\beta^{(i)})$. By Proposition 6.16, the first statement follows by the arguments of [CB02, Section 5] verbatim, replacing simple representations by θ-stable ones. For the second statement, if $r_{i,j} > 1$, then $\gamma^{(i,j)}$ is isotropic. Then, the canonical decomposition of $r_{i,j}\gamma^{(i,j)}$ appearing in Theorem 6.17.(i) is just as a sum of $r_{i,j}$ copies of $\gamma^{(i,j)}$. Thus, the statement follows from Theorem 6.17.(i) and expectation (*), since every representation in $\mathcal{M}_{q,\theta}(Q, r_{i,j}\gamma^{(i,j)})$ is represented by a polystable one. □

Therefore, modulo (*), Conjecture 7.1 reduces to the following statement: *The natural map* (7.2.3) *an isomorphism.* This should have a positive answer if Conjecture 7.15 holds, since as explained in Sect. 7.5, the multiplicative quiver varieties would formally locally be additive quiver varieties, compatibly with the direct sum map; then the statement reduces to Theorem 6.2.

Example 7.3 Suppose that $\beta^{(i)}$ is the following dimension vector supported on a framed type $\widetilde{E}_6$ quiver:

$$\begin{array}{ccccccccccc} & & & & n & & & & & & \\ & & & & \uparrow & & & & & & \\ & & & & 2n & & & & & & \\ & & & & \uparrow & & & & & & \\ n & \longleftarrow & 2n & \longleftarrow & 3n & \longrightarrow & 2n & \longrightarrow & n & \longrightarrow & 1 \end{array} \tag{7.2.4}$$

Then, by the star-shaped case of Theorem 3.6 (proved in [CBS06, Section 8]), the variety $\mathcal{M}_{1,0}(Q,\beta^{(i)})$ is isomorphic to the character variety of rank $3n$ local systems on the three-punctured sphere $\Sigma_{0,3} = \mathbb{P}^1 \setminus \{0, 1, \infty\}$ with unipotent monodromies: About the first two punctures, there should be n Jordan blocks of size three (or some refinement), and about the third puncture, there should be $n-2$ Jordan blocks of size three, one Jordan block of size four, and one of size two (or some refinement). On the other hand, $\mathcal{M}_{1,0}(Q,\delta)$ is the character variety of rank 3 local systems on $\Sigma_{0,3}$ with arbitrary unipotent monodromies. Conjecture 7.1 then asks whether the first variety is isomorphic to the n-th symmetric power of the second; it does not seem so obvious that this should be the case.

One of the difficulties in trying to adapt the proof of the analogous statement to Conjecture 7.1 in the additive case [CB02, Section 3] is that, in the multiplicative case, it is no longer guaranteed that one of the components of $\gamma^{(i,j)}$ equals one (since $\gamma^{(i,j)}$ need only be q-indivisible, not indivisible). It seems it may be a better approach to prove Conjecture 7.15, as stated above.

Note finally that the proof of Theorem 6.2 [BS21, Theorem 1.4], in the additive case, relied on hyperkähler twistings, for which one needs to assume that the parameter λ is real. In fact, some of the issues we face (such as expectation (*)) are not yet resolved, to our knowledge, in the general additive case where both $\lambda \notin \mathbb{R}$ and $\theta \neq 0$.

7.3 Symplectic Resolutions and Singularities

In view of our results and the flexibility of symplectic singularities, as well as the relationships between multiplicative and additive quiver varieties, we propose the following:

Conjecture 7.4 *Every multiplicative quiver variety is a symplectic singularity.*

Note that a product of Poisson varieties is a symplectic singularity if and only if each of the factors is (because normality and being symplectic on the smooth locus have this property, and in the definition of symplectic singularity it is equivalent to check the extension property for one or all resolutions of singularities). Therefore, the conjecture reduces to the case of factors appearing in Theorem 6.17 and, if (*) holds, to the case $\alpha \in \Sigma_{q,\theta} \cup \mathbb{N}_{\geq 2}\Sigma^{\text{iso}}_{q,\theta}$ by Proposition 7.2. If Conjecture 7.1 holds, then we can furthermore reduce to the case $\alpha \in \Sigma_{q,\theta}$. In that case, by Theorem 1.3, the only issue is normality, which would be implied if the variety is formally locally an additive one, as predicted by Conjecture 7.15 and the discussion thereafter.

Next, we ask to what extent the property of having a symplectic resolution is equivalent to the same property for the factors.

Question 7.5

(i) Is it true that $\mathcal{M}_{q,\theta}(Q,\alpha)$ admits a symplectic resolution if and only if each of the factors $\mathcal{M}_{q,\theta}(Q,\beta^{(i)})$ does?

(ii) Suppose that (*) holds. Is it true that $\mathcal{M}_{q,\theta}(Q,\alpha)$ admits a symplectic resolution if and only if each of the factors $\mathcal{M}_{q,\theta}(Q, r_{i,j}\gamma^{(i,j)})$, appearing in Proposition 7.2, does?

Note that when $r_{i,j} > 1$, then $\gamma^{(i,j)} \in \Sigma^{\text{iso}}_{q,\theta}$, and hence is q-indivisible. Therefore, in this case, $\mathcal{M}_{q,\theta}(Q,\gamma^{(i,j)})$ has a symplectic resolution by varying θ, by Lemma 2.12 and Corollary 2.21. Since it is a surface, so does its $r_{i,j}$-th symmetric power, by the corresponding Hilbert scheme. Therefore, if Conjecture 7.1 holds, we can ignore these factors in (ii) above and only consider the ones with $r_{i,j} = 1$. Also notice that all q-indivisible factors, including those with $\gamma^{(i,j)} \notin \Sigma_{q,\theta}$, admit symplectic resolutions. Also, if any factor $\gamma^{(i,j)}$ appears which is a ≥ 2 multiple of the dimension vector of a θ-stable representation, there can be no symplectic resolution of $\mathcal{M}_{q,\theta}(Q,\alpha)$, nor of $\mathcal{M}_{q,\theta}(Q,\gamma^{(i,j)})$. So again, for the question (ii), it is enough to consider only the factors $\gamma^{(i,j)}$ which are anisotropic, q-divisible, and not a ≥ 2 multiple of the dimension vector of a θ-stable representation.

7.4 Moduli of Parabolic Higgs Bundles and the Isosingularity Theorem

We restrict the attention to the case of crab-shaped quivers, the corresponding multiplicative quiver varieties of which, as explained in Sect. 3, lead to the study of character varieties of (possibly non-compact) Riemann surfaces.

Character varieties of compact Riemann surfaces are important, among many reasons, as they appear as the *Betti side* of the non-abelian Hodge correspondence, which is a series of results that establish isomorphisms between apparently unrelated moduli spaces; see [Sim94a]. Such a correspondence holds also in the case of non-compact curves, thanks to the work of Simpson; see [Sim90].

For the compact case, in [Tir19], the second author exploited a fundamental result of Simpson, called the Isosingularity theorem, [Sim94b], to show how the statements proved in [BS] could be translated to the *Dolbeault side* of the non-abelian Hodge correspondence, that is, to the moduli spaces of semistable Higgs bundles of degree 0.

In the light of the results of this paper and the non-abelian Hodge correspondence in the non-compact setting, it is a natural question to ask whether an analogue of the main theorems of [Tir19] holds for the Dolbeault moduli spaces defined on complex curves with punctures, which turn out to be the moduli spaces of *parabolic* Higgs bundles. Before stating some conjectural results, we recall the relevant definitions. Motivated by the work of Simpson, [Sim90], we recall filtered local systems, following [Yam08, §4], which gives a slightly different but nonetheless equivalent definition from the one given in Simpson, [Sim90].

Definition 7.6 [Yam08, Definition 4.5] Let X be a compact Riemann surface and $D \subset X$ be a finite subset. Let L be a local system on $X \setminus D$. For a collection of nonnegative integers $l = (l_p)_{p \in D}$, a *filtered structure* on L of *filtration type* l is a collection $(U_p, F_p)_{p \in D}$, where for each $p \in D$:

(i) U_p is a neighborhood of p in X (we set $U_p^* := U_p \setminus \{p\}$); and
(ii) F_p is a filtration

$$L|_{U_p^*} = F_p^0(L) \supset F_p^1(L) \supset \cdots \supset F_p^{l_p}(L) \supset F_p^{l_p+1}(L) = 0$$

by local subsystems of $L|_{U_p^*}$.

Two filtered structures $(U_p, F_p)_{p\in D}$, $(U'_p, F'_p)_{p\in D}$ of the same filtration type are *equivalent* if for each $p \in D$, there exists a neighborhood $V_p \subset U_p \cap U'_p$ of p such that F_p and F'_p coincide on V_p^*. A local system L together with an equivalence class of filtered structures $F = [(U_p, F_p)_{p\in D}]$ is called a *filtered local system* on (X, D) of filtration type l.

From [Yam08] one has also the following definition of (semi)stability.

Definition 7.7 [Yam08, Definition 4.6] Let (L, F) be a filtered local system on (X, D) of filtration type l. Let $\beta = (\beta_p^j \mid p \in D,\ j = 0, \ldots, l_p)$ be a collection of rational numbers satisfying $\beta_p^i < \beta_p^j$ for any p and $i < j$—such a collection is called a *weight*. The pair (L, F) is said to be β-*semistable* if for any nonzero proper local subsystem $M \subset L$, the following inequality holds:

$$\sum_{p\in D}\sum_j \beta_p^j \frac{\operatorname{rank}\left(M \cap F_p^j(L)\right)/\left(M \cap F_p^{j+1}(L)\right)}{\operatorname{rank} M} \le \sum_{p\in D}\sum_j \beta_p^j \frac{\operatorname{rank}\left(F_p^j(L)/F_p^{j+1}(L)\right)}{\operatorname{rank} L}.$$

(L, F) is β-*stable* if the strict inequality always holds.

Yamakawa established a correspondence between semistable filtered local systems and multiplicative quiver varieties of star-shaped quivers. This is, as mentioned, a particular case of the correspondence outlined in Sect. 3. To shed more light on this correspondence, we spell out the correspondence between the parameters q, α, θ defining a multiplicative quiver variety $\mathcal{M}_{q,\theta}(Q, \alpha)$ of a star-shaped quiver and the weights β of a filtered local system (L, F): Start with a crab-shaped quiver Q with vertex set $Q_0 = \{0, (i, j)_{i\in\{1,\ldots,n\}, j\in\{1,\ldots,l_i\}}\}$—that is, Q has n legs, each of which has length l_i, for $i = 1, \ldots, n$. Moreover, let α be a dimension vector, $\alpha \in \mathbb{N}^{Q_0}$, $\xi \in (\mathbb{C}^\times)^{Q_0}$ a collection of nonzero complex numbers, and $\beta \in \mathbb{Q}^{Q_0}$ a collection of rational numbers. Then, on one side, one can consider filtered local systems (L, F) on $(\mathbb{P}^1, \{p_1, \ldots, p_n\})$, where $p_i, i = 1, \ldots, n$ are pairwise distinct points in $\mathbb{P}^1$, with stability parameter β and such that:

(1) $\operatorname{rank}(L) = \alpha_0$,
(2) $\dim F_{p_i}^j(L) = \alpha_{i,j}$,
(3) the local monodromy of $F_{p_i}^j(L)/F_{p_i}^{j+1}(L)$ around p_i is given by the scalar multiplication by $\xi_{p_i}^j$ for all i, j.

On the other side, one can consider the multiplicative quiver variety $\mathcal{M}_{q,\theta}(Q, \alpha)$, where Q and α are as above and q and θ are given as

$$q_0 := \prod_i (\xi_{p_i}^0)^{-1}, \quad q_{i,j} = \xi_{p_i}^{j-1}/\xi_{p_i}^j$$

$$\theta_0 := \frac{\sum_{i,j} \theta_{i,j}\alpha_{i,j}}{\alpha_0}, \quad \theta_{i,j} = \beta_{p_i}^j - \beta_{p_i}^{j-1}.$$

The other main concept in the non-abelian Hodge correspondence on non-compact curves is that of parabolic Higgs bundle, which we recall below (note that in [Sim90], the term *filtered Higgs bundle* is used instead).

Definition 7.8 Let X and D be a compact Riemann surface and a reduced divisor on X, respectively. Let $E \to X$ be a holomorphic vector bundle on X. A *parabolic structure* on E is the datum of weighted flags $(E_{i,p}, \alpha_{i,p})_{p \in D}$

$$E_p = E_{1,p} \supseteq E_{2,p} \supseteq \cdots \supseteq E_{l+1,p} = 0,$$

$$0 \leq \alpha_{1,p} < \cdots < \alpha_{l,p} < 1,$$

for each $p \in D$. A *morphism of parabolic vector bundles* is a morphism of holomorphic vector bundles which preserves the parabolic structure at every point $p \in D$.

Definition 7.9 Given a parabolic bundle $(E, (E_{i,p}, \alpha_{i,p})_{p \in D})$, its *parabolic degree* is defined to be

$$\text{pardeg}(E) = \deg(E) + \sum_{p \in D} \sum_i m_i(p) \alpha_{i,p},$$

where $m_i(p) = \dim E_{i,p} - \dim E_{i,p+1}$ is called the *multiplicity* of $\alpha_{i,p}$.

Remark 7.10 Given the notion of parabolic degree, stability and semistability of a parabolic bundle are defined as in the case of vector bundles, using the parabolic degree in place of the ordinary degree; see [LM10, §2] for more details and an outline on some geometric properties of the corresponding moduli spaces.

Definition 7.11 A *parabolic Higgs bundle* on (X, D) is a parabolic bundle $(E, (E_{i,p}, \alpha_{i,p})_{p \in D})$ together with a meromorphic map $\Phi : E \to E \otimes K_X$ with poles of order at most 1 at the points $p \in D$. The residue of Φ at marked points is assumed to preserve the corresponding filtration.

From the Riemann-Hilbert correspondence, we know that representations of the fundamental group of a punctured surface with fixed monodromies correspond bijectively to filtered local systems. Moreover, in [Sim90], the following theorem is proved.

Theorem 7.12 *[Sim90, Theorem, p. 718] There is a one-to-one correspondence between (stable) filtered local systems and (stable) parabolic Higgs bundles of degree zero.*

Even though we will not go into the details of this correspondence, we shall at least explain how it works at the level of parameters α and β. To this purpose, let X be a compact Riemann surface and $D \subset X$ a finite subset of distinct points of X. Fixing $p \in D$, consider the sets

$$\{(\lambda, \alpha_p^j) \in \mathbb{C} \times [0, 1) \mid \text{the action of } \mathrm{Res}_p\Phi \text{ on } E_{j,p}/E_{j-1,p} \text{ has an eigenvalue } \lambda\},$$

$$\left\{ (\xi, \beta_p^k) \in \mathbb{C}^\times \times \mathbb{R} \;\middle|\; \begin{array}{l} \text{the monodromy of } F_p^k(L)/F_p^{k+1}(L) \\ \text{along a simple loop around } p \text{ has an eigenvalue } \xi \end{array} \right\}.$$

Then the correspondence between these two sets is explicitly given by $(\lambda, \alpha) \mapsto (\xi, \beta)$, where

$$\beta := \alpha - \Re\lambda, \qquad \xi := \exp(-2\pi\sqrt{-1}\lambda).$$

Another fundamental result of Simpson that is crucially used in [Tir19] is the Isosingularity theorem, which, roughly speaking, states that the moduli spaces of the non-abelian Hodge theorem in the compact case, that is, with no punctures, are étale isomorphic at corresponding points. It is still not known whether the same result holds in the non-compact case.

Conjecture 7.13 *The Isosingularity theorem holds between the moduli space of semistable filtered local systems for fixed parameters and the moduli space of semistable parabolic Higgs bundles of degree zero with corresponding parameters.*

From the above conjectural result, in combination with the results proved in Sect. 4 and Conjecture 7.15 below, one should be able to deduce the following:

Conjecture 7.14 *The moduli space of semistable parabolic Higgs bundles of degree zero with fixed parameters is a symplectic singularity, admitting a symplectic resolution if and only if the corresponding character variety admits a symplectic resolution.*

A possible strategy to prove the results listed above would be to study the local structure of moduli spaces of (semistable) objects in Calabi–Yau categories. More details are provided in the next subsection.

7.5 Moduli Spaces in 2-Calabi–Yau Categories

As mentioned above, the problem of studying the singularities of a variety can be carried out by analyzing the local structure of the variety itself around a point.

This method is powerful in certain cases, for example, when one is able to prove some locally étale isomorphism between the variety of interest and another variety whose singularities are well known: For example, this is carried out by Kaledin and Lehn in [KL07] and later by Arbarello and Saccà in [AS18], where they prove that given a strictly semistable bundle in the moduli space of semistable sheaves on a K3 surface with a fixed nongeneric polarization, there exists an étale neighborhood around that point that is isomorphic to an affine quiver variety, which depends on the point itself. Similar computations, which find their inspiration from [KL07], were also performed by the Bellamy and the first author in [BS21, BS] in the context of quiver and character varieties.

The fact that such a technique can be used and gives the same results in so many apparently different situations suggests that these are indeed particular cases of a series of theorems which should apply in much greater generality, namely, in the context of 2-Calabi–Yau categories. In the work [BGV16], the authors carry out a detailed study of the deformation theory of representation spaces of 2-Calabi–Yau algebras, and they show that among all semisimple representations, the ones that correspond to smooth points are precisely the simple ones. It is known that when Q is a Dynkin quiver and $q = 1$, there is an isomorphism

$$\Lambda^1(Q) \cong \Pi^0(Q),$$

between the multiplicative preprojective algebra and the additive preprojective algebra, as shown in [CB13, Corollary 1]. Moreover, given that the additive preprojective algebras of Dynkin quivers with at least one arrow have infinite homological dimension, the above isomorphism suggests that, in general, multiplicative preprojective algebras are not 2-Calabi–Yau. On the other hand, a conjectural statement can be made for the case of non-Dynkin quivers.

Conjecture 7.15 *Let Q be a connected non-Dynkin quiver and $q \in (\mathbb{C}^\times)^{Q_0}$. Then $\Lambda^q(Q)$ is a 2-Calabi–Yau algebra.*

Assuming the above conjecture, then [BGV16, Theorem 6.3, 6.6] implies the following for $\theta = 0$: Given a dimension vector $\alpha \in \Sigma_{q,\theta}$ and a point $x \in \mathcal{M}_{q,\theta}(Q, \alpha)$, then formally locally around x, we have an isomorphism

$$\widehat{\mathcal{M}_{q,\theta}(Q,\alpha)_x} \cong \widehat{\mathcal{M}_{0,0}(Q',\alpha')_0}$$

for some appropriate quiver Q' and dimension vector α'. If we can generalize this to arbitrary θ and prove the conjecture, there would be many interesting consequences. First, it would make it possible to handle the (2, 2)-case, where one may construct a symplectic resolution by first performing GIT (replacing θ by suitably generic θ') and then performing a blowup of the singular locus. Moreover, this result would imply normality for $\mathcal{M}_{q,\theta}(Q, \alpha)$, without any assumption on α and θ.

Secondly, it would be interesting to extend such arguments and results to coarse moduli spaces of objects in 2-Calabi–Yau categories. This would indeed make it possible to reduce the study of singularities of a numerous class of moduli spaces to answering the following question: *does the moduli space parametrize objects of a 2-Calabi–Yau category?*

Knowing more about the singularities of coarse moduli spaces of objects in Calabi–Yau categories might allow one to prove a generalization of the Isosingularity theorem in the non-compact case, provided that a suitable version of the 2-Calabi–Yau condition holds for the category of parabolic Higgs bundles.

7.6 Character Varieties and Higgs Bundles for Arbitrary Groups

Another interesting problem would be to analyze whether the results of the present paper can extend to character varieties of (punctured) Riemann surfaces with representations in arbitrary groups and, via the non-abelian Hodge correspondence, to the moduli spaces of (parabolic) Higgs principal bundles. Since the results on punctured character varieties contained in the present paper are deduced based on the correspondence described in Sect. 3, which heavily relies on the fact that one considers representations and conjugacy classes inside $\mathrm{GL}(n, \mathbb{C})$, it seems unlikely that the techniques used here could be applied in the more general setting of G-representations. On the other hand, a possible approach could be the one outlined in the previous subsection. To this end, the question to be answered would be the following: *given an algebraic group G, are there cases other than $G = GL_n$, where the category of morphisms $\rho : \pi_1(X) \to G$ is 2-Calabi–Yau?*

Acknowledgments This work is part of the second author's PhD thesis. We thank Gwyn Bellamy, William Crawley-Boevey, Ben Davison, Emilio Franco, and Marina Logares for many useful conversations and enlightening comments on the topic of this paper and Davison and Johannes Nicaise for their careful reading. The second author would like to thank Jacopo Stoppa and the Department of Mathematics at SISSA, Trieste, for providing an excellent and stimulating working environment. The work of the second author was supported by the Engineering and Physical Sciences Research Council [EP/L015234/1]. The EPSRC Centre for Doctoral Training in Geometry and Number Theory (The London School of Geometry and Number Theory), University College London, and Imperial College London. The first author would like to thank the Hausdorff Institute for Mathematics and the Max Planck Institute for Mathematics in Bonn, where some of this research was carried out. We would also like to thank the anonymous referee for his/her useful suggestions.

References

[AMM98] A. Alekseev, A. Malkin, and E. Meinrenken, *Lie group valued moment maps*, J. Differential Geom. **48** (1998), no. 3, 445–495.

[AS18] E. Arbarello and G. Saccà, *Singularities of moduli spaces of sheaves on K3 surfaces and Nakajima quiver varieties*, Adv. Math. **329** (2018), 649–703.

[Art69] M. F. Artin, *Algebraic approximation of structures over complete local rings*, Publications Mathématiques de l'IHÉS **36** (1969), 23–58.

[Bea00] A. Beauville, *Symplectic singularities*, Inventiones mathematicae **139** (2000), no. 3, 541–549.

[BS] G. Bellamy and T. Schedler, *Symplectic resolutions of character varieties*, arXiv:1909.12545, accepted to Geom. Top.

[BS21] Gwyn Bellamy and Travis Schedler, *Symplectic resolutions of quiver varieties*, Selecta Math. (N.S.) **27** (2021), no. 3, Paper No. 36, 50.

[BK16] R. Bezrukavnikov and M. Kapranov, *Microlocal sheaves and quiver varieties*, Ann. Fac. Sci. Toulouse Math. (6) **25** (2016), no. 2–3, 473–516.

[Boa07] P. Boalch, *Quasi-Hamiltonian geometry of meromorphic connections*, Duke Math. J. **139** (2007), no. 2, 369–405.

[BGV16] R. Bocklandt, F. Galluzzi, and F. Vaccarino, *The Nori-Hilbert scheme is not smooth for 2-Calabi-Yau algebras*, J. Noncommut. Geom. **10** (2016), no. 2, 745–774.

[BLPW16] T. Braden, A. Licata, N. Proudfoot, and B. Webster, *Quantizations of conical symplectic resolutions II: category $\mathcal{O}$ and symplectic duality*, Astèrique **384** (2016), 75, 179.

[BFN18] Alexander Braverman, Michael Finkelberg, and Hiraku Nakajima, *Towards a mathematical definition of Coulomb branches of 3-dimensional $\mathcal{N} = 4$ gauge theories, II*, Adv. Theor. Math. Phys. **22** (2018), no. 5, 1071–1147.

[BMO11] A. Braverman, D. Maulik, and A. Okounkov, *Quantum cohomology of the Springer resolution*, Adv. Math. **227** (2011), no. 1, 421–458.

[CF17] O. Chalykh and M. Fairon, *Multiplicative quiver varieties and generalised Ruijsenaars-Schneider models*, Journal of Geometry and Physics **121** (2017), 413–437.

[CB01] W. Crawley-Boevey, *Geometry of the moment map for representations of quivers* Compositio Math. **126** (2001), no. 3, 257–293.

[CB02] ——, *Decomposition of Marsden-Weinstein reductions for representations of quivers*, Compositio Math. **130** (2002), no. 2, 225–239.

[CB03] ——, *Normality of Marsden-Weinstein reductions for representations of quivers*, Math. Ann. **325** (2003), no. 1, 55–79.

[CB04] ——, *Indecomposable parabolic bundles and the existence of matrices in prescribed conjugacy class closures with product equal to the identity*, Publ. Math. Inst. Hautes Études Sci. (2004), no. 100, 171–207.

[CB13] ——, *Monodromy for systems of vector bundles and multiplicative preprojective algebras*, Bull. Lond. Math. Soc. **45** (2013), no. 2, 309–317.

[CBH19] W. Crawley-Boevey and A. Hubery, *A New Approach to Simple Modules for Preprojective Algebras*, Algebras and Representation Theory (2019).

[CBS06] W. Crawley-Boevey and P. Shaw, *Multiplicative preprojective algebras, middle convolution and the Deligne-Simpson problem*, Adv. Math. **201** (2006), no. 1, 180–208.

[CHZ14] S. Cremonesi, A. Hanany, and A. Zaffaroni, *Monopole operators and hilbert series of coulomb branches of 3d $\mathcal{N} = 4$ gauge theories*, J. High Energy Phys. **005** (2014).

[Dre91] J.-M. Drezet, *Points non factoriels des variétés de modules de faisceaux semi-stables sur une surface rationnelle*, J. Reine Angew. Math. **413** (1991), 99–126.

[EL19] Tolga Etgü and Yankı Lekili, *Fukaya categories of plumbings and multiplicative preprojective algebras*, Quantum Topol. **10** (2019), no. 4, 777–813.

[Fle88] H. Flenner, *Extendability of differential forms on nonisolated singularities*, Invent. Math. **94** (1988), no. 2, 317–326.

[GS11] Q. R. Gashi and T. Schedler, *On dominance and minuscule Weyl group elements*, J. Algebraic Combin. **33** (2011), no. 3, 383–399.

[HLRV11] T. Hausel, E. Letellier, and F. Rodriguez-Villegas, *Arithmetic harmonic analysis on character and quiver varieties*, Duke Math. J. **160** (2011), no. 2, 323–400.

[HLRV13a] ——, *Arithmetic harmonic analysis on character and quiver varieties II*, Adv. Math. **234** (2013), 85–128.

[HLRV13b] T. Hausel, E. Letellier, and F. Rodriguez-Villegas, *Positivity for Kac polynomials and DT-invariants of quivers*, Ann. of Math. (2) **177** (2013), no. 3, 1147–1168.

[Ina13] M.-A. Inaba, *Moduli of parabolic connections on curves and the Riemann-Hilbert correspondence*, J. Algebraic Geom. **22** (2013), no. 3, 407–480.

[Jor14] D. Jordan, *Quantized multiplicative quiver varieties*, Adv. Math. **250** (2014), 420–466.

[Kal06] D. Kaledin, *Symplectic singularities from the Poisson point of view*, J. Reine Angew. Math. **600** (2006), 135–156.

[KL07] D. Kaledin and M. Lehn, *Local structure of hyperkähler singularities in O'Grady's examples*, Mosc. Math. J. **7** (2007), no. 4, 653–672, 766–767.

[KLS06] D. Kaledin, M. Lehn, and C. Sorger, *Singular symplectic moduli spaces*, Invent. Math. **164** (2006), no. 3, 591–614.

[Kal09] D. B. Kaledin, *Normalization of a Poisson algebra is Poisson*, Tr. Mat. Inst. Steklova **264** (2009), no. Mnogomernaya Algebraicheskaya Geometriya, 77–80.

[Kat96] N. M. Katz, Rigid local systems, Annals of Mathematics Studies, vol. 139, Princeton University Press, Princeton, NJ, 1996.

[Kin94] A. D. King, *Moduli of representations of finite-dimensional algebras*, Quart. J. Math. Oxford Ser. (2) **45** (1994), no. 180, 515–530.

[KS] Daniel Kaplan and Travis Schedler, *The 2-Calabi-Yau property for multiplicative preprojective algebras*, arXiv:1905.12025.

[LS06] M. Lehn and C. Sorger, *La singularité de O'Grady*, J. Algebraic Geom. **15** (2006), no. 4, 753–770.

[LM10] M. Logares and J. Martens, *Moduli of parabolic Higgs bundles and Atiyah algebroids*, J. Reine Angew. Math. **649** (2010), 89–116.

[Maf02] A. Maffei, *A remark on quiver varieties and Weyl groups*, Ann. Sc. Norm. Super. Pisa Cl. Sci. (5) **1** (2002), no. 3, 649–686.

[MO19] D. Maulik and A. Okounkov, *Quantum groups and quantum cohomology*, Astérisque (2019), no. 408, ix+209.

[MGW19] M. McBreen, B. Gammage, and B. Webster, *Homological Mirror Symmetry for Hypertoric Varieties II*, arXiv:1903.07928, March 2019.

[MW18] M. McBreen and B. Webster, *Homological Mirror Symmetry for Hypertoric Varieties I*, arXiv:1804.10646, April 2018.

[Moz90] S. Mozes, *Reflection processes on graphs and Weyl groups*, J. Combin. Theory Ser. A **53** (1990), no. 1, 128–142.

[MFK02] D. Mumford, J. Fogarty, and F. Kirwan, *Geometric Invariant Theory*, Ergebnisse der Mathematik und ihrer Grenzgebiete. 2. Folge, Springer Berlin Heidelberg, 2002.

[Nak16] H. Nakajima, *Towards a mathematical definition of Coulomb branches of 3-dimensional $\mathcal{N} = 4$ gauge theories*, I, Adv. Theor. Math. Phys. **20** (2016), no. 3, 595–669.

[Nak94] H. Nakajima, *Instantons on ALE spaces, quiver varieties, and Kac-Moody algebras*, Duke Math. J. **76** (1994), no. 2, 365–416.

[Nam] Y. Namikawa, *A note on symplectic singularities*, arXiv:math/0101028.

[O'G99] K. G. O'Grady, *Desingularized moduli spaces of sheaves on a K3*, J. Reine Angew. Math. **512** (1999), 49–117.

[Sim90] C. T. Simpson, *Harmonic bundles on noncompact curves*, J. Amer. Math. Soc. **3** (1990), no. 3, 713–770.

[Sim94a] C. T. Simpson, *Moduli of representations of the fundamental group of a smooth projective variety I*, Publications Mathématiques de l'IHÉS **79** (1994), 47–129 (eng).

[Sim94b] ——, *Moduli of representations of the fundamental group of a smooth projective variety II*, Publications Mathématiques de l'IHÉS **80** (1994), 5–79 (eng).

[Su06] X. Su, *Flatness for the moment map for representations of quivers*, J. Algebra **298** (2006), no. 1, 105–119.

[Tho61] R. C. Thompson, *Commutators in the special and general linear groups*, Trans. Amer. Math. Soc. **101** (1961), 16–33.

[Tir19] A. Tirelli, *Symplectic resolutions for Higgs moduli spaces*, Proc. Amer. Math. Soc. **147** (2019), no. 4, 1399–1412.

[Van08a] M. Van den Bergh, *Double Poisson algebras*, Trans. Amer. Math. Soc. **360** (2008), no. 11, 5711–5769.

[Van08b] ——, *Non-commutative quasi-Hamiltonian spaces*, Poisson geometry in mathematics and physics, Contemp. Math., vol. 450, Amer. Math. Soc., Providence, RI, 2008, pp. 273–299.

[Yam08] D. Yamakawa, *Geometry of multiplicative preprojective algebra*, Int. Math. Res. Pap. IMRP (2008), 1–77.

[Yeu] W.-K. Yeung, *Weak Calabi-Yau structures and moduli of representations*, arXiv:1802.05398.

The manufacturer's authorised representative in the EU is Springer Nature Customer Service Centre GmbH, Europaplatz 3, 69115 Heidelberg, Germany. If you have any concerns regarding our products, please contact ProductSafety@springernature.com

Printed and bound by CPI Group (UK) Ltd, Croydon, CR0 4YY
31/07/2026
02182966-0003